3rd International Symposium on Current Progress in Functional Materials

IOP Conference Series: Materials Science and Engineering
Volume 763

Depok, Indonesia
8 - 9 August 2018

Editors:

Ivandini T. Anggraningrum
Djoko Triyono
Young-ill Lee
Yan Mulyana
A. Denny Handoko

ISBN: 978-1-7138-4704-5
ISSN: 1757-8981

Printed from e-media with permission by:

Curran Associates, Inc.
57 Morehouse Lane
Red Hook, NY 12571

Some format issues inherent in the e-media version may also appear in this print version.

Institute of Physics
Dirac House, Temple Back
Bristol BS1 6BE UK

Phone: 44 1 17 929 7481
Fax: 44 1 17 920 0979

techtracking@iop.org

Additional copies of this publication are available from:

Curran Associates, Inc.
57 Morehouse Lane
Red Hook, NY 12571 USA
Phone: 845-758-0400
Fax: 845-758-2633
Email: curran@proceedings.com
Web: www.proceedings.com

TABLE OF CONTENTS

IOP Conf. Series: Materials Science and Engineering **763** (2020) 011001 doi:10.1088/1757-899X/763/1/011001

Preface from the editors

We are very pleased that the third **International Symposium on Current Progress in Functional Materials 2018 (ISCPFM 2018)** has been successfully organized. This event is an annual program of Faculty of Mathematics and Natural Sciences, Universitas Indonesia, intended to provide the academics, scholars as well as scientists and students a solid forum platform to discuss their research activities and to exchange new ideas in functional material field.

The major conference theme was *Functional Materials for Technologies Development*. The global changes all over the world are resulting in the actual requirement for bright ideas sharing and encouraging high awareness of this relevant research field. A tremendous growth of functional materials is potential in creating improvement in various fields of technology. The responses to the call-for-papers had been overwhelming. More than 100 papers have been presented. The numbers of delegates were also gratifying, showing the conference has attracted a high level attention. We highly appreciate their great contribution for this event. This proceedings' volume facilitates the readers to engage with the selected manuscripts that were presented during the 3rd ISCPFM event. There are 76 papers published here covered a more wide range of topics and issues, including low-dimensional, nano and 2D materials, metals, metal oxides, polymers, ceramics and composite materials, biomaterials as well as computation and simulation of materials.

Finally, it is appropriate to extend honour to our fellow colleagues in the advisory and scientific committee for their effort in securing a substantial input of paper and encouraging participation from all parties. We are also indebted to the sponsor Universitas Indonesia to support this 3rd Symposium and also to symposium committee members. Last but not least, we thank all the participants and authors.

Depok, 7 November 2019

Editors,
Ivandini T Anggraningrum
Djoko Triyono
Young-ill Lee
Yan Mulyana
A Denny Handoko

IOP Conf. Series: Materials Science and Engineering **763** (2020) 011003 doi:10.1088/1757-899X/763/1/011003

Organizing Committee

Advisor	:	Prof. Dr. Ir. Muhammad Anis, M.Met
		Prof. Dr.rer.nat. Rosari Saleh
Steering Committee	:	Dr.rer.nat. Abdul Haris
		Dr. tech. Djoko Triyono
		Dr. Rokhmatuloh
Conference Chair	:	Prof. Dr. Ivandini T Anggraningrum
Program Chair	:	Dr. Ratna Yuniati
Committee	:	Lusitra Munisa, M.Si
		Yustin S Dewi, S.Si
		Adibah, S.Si
		Isnaini Rahmawati, S.Si.
		Ardiansyah Taufik, S.Si.
		Lisna Arumsari, A.Md
		M. Hidayat, S.Sos.
		Fariz Hafizi Roihan, S. Sn.

Conference Proceedings Committee

Dr. Djoko Triyono
Prof. Dr. Ivandini T Anggraningrum
Lusitra Munisa, M.Si.
Adibah, S.Si.
Yustin S Dewi, S.Si.
Lisna Arumsari, A.Md.

ISCPFM 2018

The International Symposium on Current Progress in Functional Materials 2018 (ISCPFM 2018)

The Sponsor

UNIVERSITAS INDONESIA

The conference is held by

FMIPA

Faculty of Mathematics and Natural Sciences
Universitas Indonesia
2018

Name	Email
Abdul Hadi	abdul.hadi41@sci.ui.ac.id
Adira Nofeadri Ryofi	adira.nofeadri51@sci.ui.ac.id
Adistya Maulidya	adistyamaulidya43@gmail.com
Aditya Wahyu Anugrah	aditdreams@gmail.com
Adrian Ferrariski	adrianferrariski@gmail.com
Agustino Zulys	zulys@ui.ac.id
Ahmad Aly Irfan	ahmad.aly@sci.ui.ac.id
Aminah Umar	aminah.umar@sci.ui.ac.id
Antonius Herry Cahyana	herrykim@ui.ac.id
Anugrah Cahyo Widodo	anugrahnunu96.an@gmail.com
Aprilia Inggri Astuti	apriliainggriastuti@gmail.com
Arbhyando Tri Putrananda	arbhyando@gmail.com
Ardiansyah Taufik	ardiansyahtaufik70@gmail.com
Arya Maulana Ichsan	aryamaulana1995@gmail.com
Asep Saefumillah	asep.saefumillah@sci.ui.ac.id
Asman Kumik	asmankumik@gmail.com
Aulia Dyah Hutami K	aulia.dyah41@gmail.com
Azhar Yogaswara Sulistyawan	azharygswr@gmail.com
Baiq Hijratul Hasanah	hijr.nana@gmail.com
Bayu Ardiansah	bayu.ardiansah@sci.ui.ac.id
Bella Andreana	bella.andreana@ui.ac.id
Benni F Ramadhoni	benni.ramadhoni@gmail.com
Budiawan	dr.budiawan@gmail.com
Carina Aulia Wijayanti	carinaulia@gmail.com
Charnela Mardani	charnela.mardani@gmail.com
Cheyenne Sekarharum	cheyenne_abdulkarim@outlook.com
Cuk Imawan	cuk.imawan@sci.ui.ac.id
Dede Djuhana	dede.djuhana@sci.ui.ac.id
Dessy Yulia	dyulia09@gmail.com
Desy Maulina	desy.maulina1992@gmail.com
Dhawud Sabilur Razaq	cakalang.99@gmail.com
Dicky Rezky Munazat	dicky.rezky@sci.ui.ac.id
Dimas Nugraha	dmasngr@gmail.com
Dita Arifa Nurani	d.arifa@sci.ui.ac.id
Dita Arifa Nurani	d.arifa@sci.ui.ac.id
Djati Handoko	djati.handoko@ui.ac.id
Djoko Triyono	djoko.triyono@sci.ui.ac.id
Dyah Utami Cahyaning Rahayu	dyahutamicr@sci.ui.ac.id
Dyah Utami Cahyaning Rahayu	dyahutamicr@sci.ui.ac.id
Efta Yudiarsah	e.yudiarsah@sci.ui.ac.id
Elise Natalia Manurung	elisenataliamanurung@gmail.com
Endang Saepudin	endang.saefudin@sci.ui.ac.id
Engga Saputra Hasan Basri	enggasaputra88@gmail.com
Erlina Yustanti	erlina.yustanti68@gmail.com
Esri Nawati Sinurat	esry.nawati@yahoo.com
Farid Mujiantono	fmujiantono@gmail.com
Fathiya Rahmani	fathiya.rahmani@gmail.com
Feby Zilvania	feby.zilvania@sci.ui.ac.id
Firsty Aprielta Aribah	firsty.aprielta@ui.ac.id
Frandi Yuanda	frandy.y@gmail.com
Gita Andani	gita.andani@ui.ac.id
Gita Rahmaningsih	grahmaningsih@gmail.com
Haickel Franklyn	haickel.franklyn@ui.ac.id
Hanifah	hanifah.irham@gmail.com
Helmiyati	helmiyt@gmail.com
Hsiang-Lin Liu	hliu@ntnu.edu.tw
Ilham faturachman	ilhamfaturachmn@gmail.com
Iman Abdullah	iman.abdullah@sci.ui.ac.id
Indah Lestari	indahshawty@gmail.com
Indrianita Lionardi	amazingrace.indri@gmail.com
Irfan Habib Ramadhan	irfan.habib@sci.ui.ac.id
Ivandini T Anggraningrum	ivandini.tri@sci.ui.ac.id
Jarnuzi Gunlazuardi	jarnuzi@ui.ac.id
Jauharuddin Luthfi Al Jabbar	luthfi.chem@gmail.com
Khoirul Umam	umam.94@gmail.com
Lenny Apriyanti	lennyapriyanti@yahoo.com
Lia Trisnawati	lia.trisnawati@ui.ac.id
Listya Eka Anggraini	anggrainiekalistya@gmail.com
Maisarah	sarahjufri24@gmail.com
Margaretha Indra Pratiwi	mindrapratiwi@yahoo.com
Margarita Claudya Maida	mclaudyamaida@gmail.com
Mariska Winda Asrini	ichamariska41@gmail.com
Masripah Triharto	masrifaptbn@gmail.com
Millati Hanifah Saprudin	millatihanifah96@gmail.com
Miyah Sofani	miyahsofani@gmail.com

IOP Conf. Series: Materials Science and Engineering **763** (2020) 011005 doi:10.1088/1757-899X/763/1/011005

Monica Putri Monic	monicaonic16@gmail.com
Muhammad Faiz Rizqullah Hasian Rambey	muhammad.faiz51@sci.ui.ac.id
Muhammad Fariz Mulyanto	m.fariz@sci.ui.ac.id
Muhammad Irsyad Amzy	muhammad.irsyad43@ui.ac.id
Muhammad Sujak MK	muhammad.sujak@sci.ui.ac.id
Muhammad Yose Rizal	muhammadyoserzl@gmail.com
Munawar Khalil	mkhalil@sci.ui.ac.id
Muqtaf Najich Abdillah	muqtafabdillah@gmail.com
Mutiara Ayu Annisa	mutiara.ayu41@sci.ui.ac.id
Nandira Siti Aisyah	nandirasa@gmail.com
Nesy Fitria	nesyfitria@gmail.com
Nia Nurmalasari	leeniami@gmail.com
Nurhidayati Sari	nurhidayatifs@gmail.com
Nursita Rihhadatul Aisy	aisy.nursita2501@gmail.com
Prastika K. Jiwanti	prastika@chem.keio.ac.jp
Puteri Aprilia Sitadevi	puteriaprilia96@gmail.com
Puti Rahmi Arief,	putirahmi@yahoo.com
Qoimatul Mustaghfiroh	qoimatul.mustaghfiroh@sci.ui.ac.id
Raden Iqrafia Ashna	iqrafia@gmail.com
Raka Firdaus	drraka4@gmail.com
Randi Subekti	randiisubekti@yahoo.com
Rani Puspitasari	raniaritonanh.ckh@gmail.com
Resi Annisaa Rubianti	resi.annisaa@gmail.com
Rian Tri Laksono Suropati	rian.tri@ui.ac.id
Rinaldo Sitorus	rinaldositorus91@outlook.com
Rizky Wira Yudanto	rizky.wira@ui.ac.id
Rizqa Dhafiningtia	devindolphin23@gmail.com
Russel Howe	r.howe@abdn.ac.uk
Saskia Andiane Hidayat	andianesaskia@gmail.com
Senia Budiana	seniabudiana@gmail.com
Shanker Kalakotla	shankerkalakotla@gmail.com
Siswanto Adi Wijanarko	siswanto.wijanarko@gmail.com
Siswati Setiasih	siswati.setiasih@sci.ui.ac.id
Siti Annissa Andrani	sitiannissa46@gmail.com
Sitti Fachramiya Fhaza	sitti.fachramiya@ui.ac.id
Sulaiman Hawibowo	bowo992010@gmail.com
Suwardiyanto	antok.fmipa@unej.ac.id
Syarfina Andini	syarfinaa@gmail.com
Tae-Hyun Yoon	taeyoon@hanyang.ac.kr
Tasya Kamilia Zahrah	tasyakamilia136@gmail.com
Tia Agustiany	tia.agustiany@sci.ui.ac.id
Trisnaningsih Affendi	trisnaningsihaffendi741@gmail.com
Ulfah Hartina	ulfahhartina12@gmail.com
Ummaira Fadhilah	ummairafadhilah@gmail.com
Umu Hanifah	umuhanifah95@gmail.com
Wanti Megawati	wanti.megawati@ui.ac.id
Widya Amandha Hidayani	widyaamandha18@gmail.com
Wulan Chairunisa	wulanchairunisa696@gmail.com
Yan Mulyana	ymulyana@tmu.ac.jp
Yesi Pratiwi	jazalea15@gmail.com
Yoki Yulizar	yokiy@sci.ui.ac.id
Yu-Chiang Chao	ycchao@ntnu.edu.tw
Yulia Dwi Susanti	dwyulia24@gmail.com
Yuni K. Krisnandi	yuni.krisnandi@sci.ui.ac.id
Yuris Diksy	yurizyui@gmail.com
Yusup Supriyadi	yusupsupriyadi92@gmail.com

IOP Conf. Series: Materials Science and Engineering **763** (2020) 011006 doi:10.1088/1757-899X/763/1/011006

Peer review statement

All papers published in this volume of *IOP Conference Series: Materials Science and Engineering* have been peer reviewed through processes administered by the proceedings Editors. Reviews were conducted by expert referees to the professional and scientific standards expected of a proceedings journal published by IOP Publishing.

Modification of boron-doped diamond with gold-palladium nanoparticles for CO_2 electroreduction

A Kumik, T A Ivandini and R Wibowo

Department of Chemistry, Faculty of Mathematics and Natural Sciences, Universitas Indonesia, Kampus UI Depok, Depok 16424, Indonesia

Corresponding author's e-mail: ivandini.tri@sci.ui.ac.id

Abstract. Modification of boron-doped diamond (BDD) with gold-palladium nanoparticles (AuPdNPs) was performed for an application in electrochemical reduction of CO_2. Preparation of the colloidal AuPdNPs involved two steps, including the synthesis of gold nanoparticles (AuNPs) and the growth of palladium on the surface of AuNPs. BDD, which had been terminated with allylamine under UV light irradiation for 6 h, then, was immersed in the colloidal AuPdNPs. Characterization of the nanoparticles by using Tunneling Electron Microscopy (TEM) showed flowers-like nanoparticles with an average size of 40 nm, while characterization of the modified BDD showed the Au and Pd surface coverage of 4.4 % and 6.0 %, respectively, on BDD surface. The electrochemical reduction of CO_2 at the modified BDD was performed in a two-compartments cell with 0.1 M NaCl solution in the cathode chamber and 0.1 M Na_2SO_4 solution in the anode chamber. By applying the potential -0.9 V (vs. Ag/AgCl) in the system for 60 min, it showed the faradaic efficiency of the system at approximately 84.65 % with acetic acid as one of the products.

Keywords: CO_2 electroreduction, gold-palladium, boron-doped diamond, surface modification.

1. Introduction

The increase of CO_2 concentration in atmosphere attracts the attention of scientists to convert CO_2 into value added products like CH_4, HCOOH, CH_3COOH and CH_3OH [1]. Boron-doped diamond (BDD) is one of the promising electrodes for CO_2 conversion, due to its wide range of working potential as well as chemically and mechanically high inertness [2]. Research on the electrochemical conversion of CO_2 was firstly reported by Hori *et al.* [3] who reported the conversion of CO_2 into formic acid using some types of working electrodes, such as lead, indium, cadmium and tin. Meanwhile, BDD had been applied for electroreduction of CO_2 to produce formaldehyde and formic acid with high Faradaic efficiency [4]. Furthermore, modification of BDD with some metals for electrochemical reduction of CO_2 had also been reported with some promising results [5,6].

On the other hand, metal nanoparticles are interesting due to their unique properties in optical, catalytic, electronic and magnetic characteristics. Among various nanostructures, flower is one of an attractive shape because it has an anisotropic structure, which not only provides optical capability but also serves as a catalyst where the activity and selectivity depend on its morphology [7-8]. Moreover, nanoparticles of bimetallic metals, such as AuPd, have been reported to provide a higher catalytic activity than both of its mono metals, i.e. Au and Pd nanoparticles [9]. AuPd nanoparticles were also reported as superior catalyst in trichloroethene hydrodechlorination [10] as well as in the electrolytic study of CO_2 to produce hydrocarbons with C2 to C5 [11,12].

In this study, gold-palladium core-shell nanoparticles were prepared to modify the BDD electrode in order to study the catalytic effect of composite AuPd-modified BDD in electrochemical conversion of CO_2.

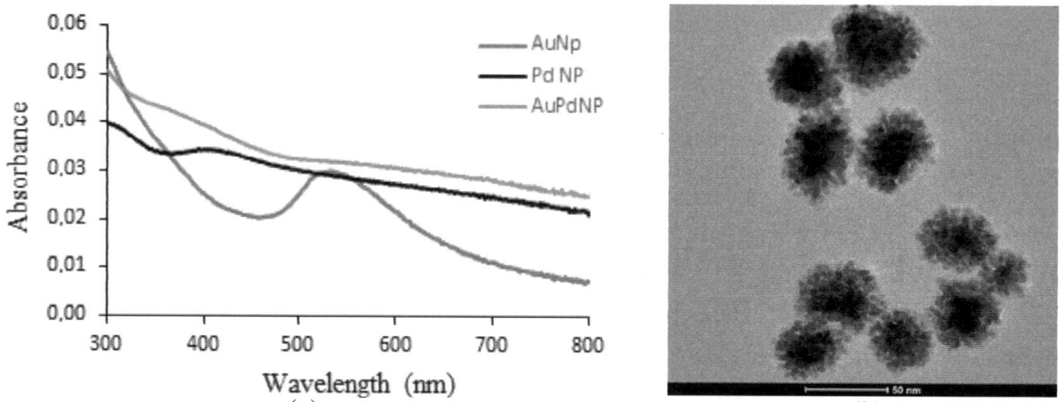

(a) (b)

Figure 1. (a) UV-vis spectra of Au NPs (red line), Pd NPs (black line), and AuPdNPs (green line) and (b) TEM image of AuPdNPs

2. Experimental section

2.1. Materials and instruments

BDD with a B:C ratio of 1:1000 was prepared by using microwave plasma-assisted chemical vapor deposition [13]. Both gold and palladium precursors ($HAuCl_4$. $4H_2O$ and $PdCl_2$) were purchased from WAKO chemicals, Japan, while sodium citrate, ascorbic acid, sodium chloride, allylamine and other chemicals were purchased from Sigma-Aldrich.

Electrochemical parameters were conducted using a three-electrodes cell by using a potentiostat (EDAQ E-Corder 410). A platinum wire (Nilaco, Japan) was used as the counter electrode, while an Ag/AgCl system (BAS Inc., Japan) as the reference electrode. Some instruments were used to characterize the nanoparticles and the BDD surface, including UV-Vis spectrophotometry (Thermo Scientific Multisam Go), Transmission Electron Microscopy (TEM) (Microscope Tecnai 200 kV D2360 SuperTwin), and Scanning Electron Microscopy (SEM) with Energy Dispersive X-ray (EDX) (JEOL/EO JCM-6000Plus, Japan).

2.2. Preparation of AuPd-modified BDD electrode

AuPd nanoparticles (AuPdNPs) was prepared through the gold seed mediated method as previously reported with some modification [14]. Briefly, 30 mL of colloidal AuNP was added into 15 mL of 1.0 mM H_2PdCl_4 and ultrapure water of 5 mL. The mixture was cooled at 4 °C in ice bath, then 6 mL of 10 mM ascorbic acid was added and stirred for around 15 min to obtain purple colloid [15]. Prior to modification, the BDD surface was changed into N-terminated BDD (N-BDD) by immersing in allylamine under UV light irradiation (λ = 254 nm) [16]. The N-BDD was then immersed in the prepared colloidal AuPdNPs for 12 h to attach the nanoparticles on the surface of BDD.

2.3. Electroreduction of CO_2

The electrochemical experiments were carried out in a cell with two compartments with the AuPdNPs-modified BDD was used as the working electrode. Chronoamperometric method was applied. The electrolytes in the anodic chamber (0.1 M Na_2SO_4) and in the cathodic chamber (0.1 M NaCl) were saturated with N_2 for 30 min. Then, the one in the cathodic chamber was bubbled with CO_2 for 60 min. The applied potential of -0.9 V (V vs. Ag/AgCl) was conducted. The products were then analyzed using HPLC and GC.

3. Results and Discussion

3.1. Synthesis of AuPdNPs

AuPdNPs were synthesized using seeds mediated growth method, in which gold nanoparticles (AuNPs) were used as the core for palladium growth. Solution of H_2PdCl_4 was used as the precursor of Pd. When $NaBH_4$ solution was added, $PdCl_4$ ions were reduced to form PdNPs. Figure 1a shows UV-

IOP Conf. Series: Materials Science and Engineering **763** (2020) 012001 doi:10.1088/1757-899X/763/1/012001

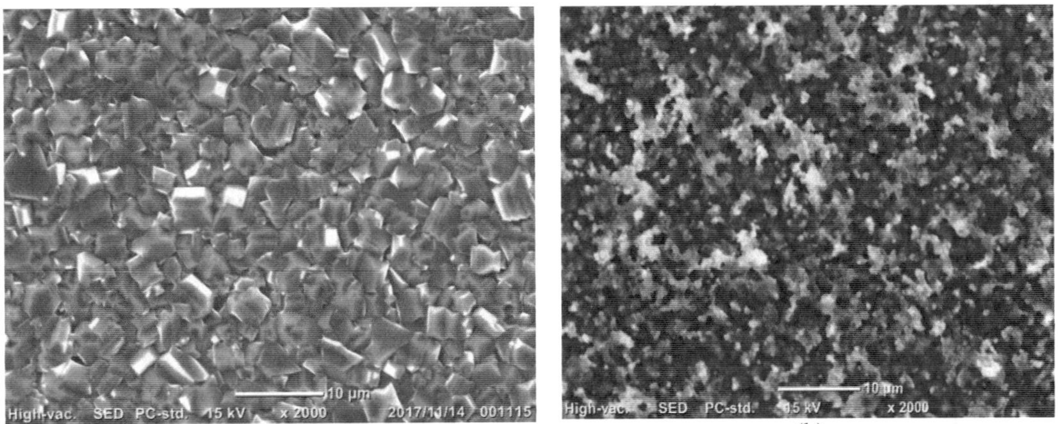

(a) (b)

Figure 2. SEM images (with 2000 times magnification) (a) BDD and (b) AuPd-BDD

Vis spectra of the colloidal of AuNPs, palladium nanoparticles (PdNPs), and AuPdNPs. The figure shows that AuNPs has a typical absorption in the visible region of 552 nm, reflecting the surface plasmon resonance of AuNPs [15]. This peak was not found in the spectrum of PdNPs nor at AuPdNPs. Therefore, the formation of AuPdNPs was predicted due to the covering of PdNPs at the AuNPs core. The results in figure 1b showed flowers-like AuPdNPs, confirming that the synthesized AuPdNPs was core-shell nanoparticles, with Au as the core and Pd as the shell.

Meanwhile, the TEM characterization of AuPdNPs was applied to determine the size and shape of the particles. Figure 1b shows the enriched palladium clusters attached onto the seeds of gold surface, forming a core-shell AuPd nanoparticle in flower-like shape. The average particle size of AuPdNPs was around 40 nm.

3.2. Characterization of AuPd-modified BDD by using SEM-EDS

SEM-EDS characterization of AuPd-BDD was performed to study its morphology. Figure 2a shows typical of unmodified crystal grains of BDD. The average size of the crystal grain of BDD was approximately 8.2 µm. After immersing the BDD in the colloidal AuNPs (figure 2b), distribution of other conductive particles was clearly shown by the presence of brighter particles on the BDD surface, indicating that AuPdNPs could be deposited on the BDD surface.

Further characterization of the modified BDD using EDS confirmed that the surface of the modified BDD composed of carbon (65.25 %), oxygen (16.28 %), nitrogen (7.97 %), gold (4.42 %) and palladium (6.09 %), confirmed that the AuPd nanoparticles was successfully deposited on the BDD surface.

3.3. Electroreduction of CO_2

Cyclic voltammetry of BDD and AuPd-modified BDD in 0.5 M H_2SO_4 was performed; figure 3a shows the difference between BDD and AuPd-modified BDD. While on the unmodified BDD surface it shows neither oxidation nor a well-defined reduction peak in the potential range of -2.0 to +1.0 V (vs. Ag/AgCl), a typical oxidation peak of gold was observed at around 0.0 V at AuPd-modified BDD. In addition, a reduction peak at around -1.3 V was also observed at AuPd-modified BDD, which probably due to the reduction of oxygen [16].

In order to study the electrochemical behavior of CO_2, cyclic voltammetry of CO_2 in 0.1 M NaCl solution was investigated at AuPd-modified electrode. Prior to the measurement, N_2 was bubbled into the solution to remove other possible impurity gas. Then, CO_2 was dissolved into the solution. Figure 3b shows the cyclic voltammograms (CVs) of the solutions with and without the presence of CO_2. The CVs with the presence of CO_2 showed a typical sharp reduction peak at around -1.5 V. This peak increased with the increase of the bubbling time of CO_2, indicating a strong relation of the peak with electrochemical reduction of CO_2. Further, the optimum time CO_2 bubbling of 60 min was fixed for the next measurements as it generates maximum current response. Investigation in various pHs was also conducted and showed that the optimum pH was 3.7 (data was not shown).

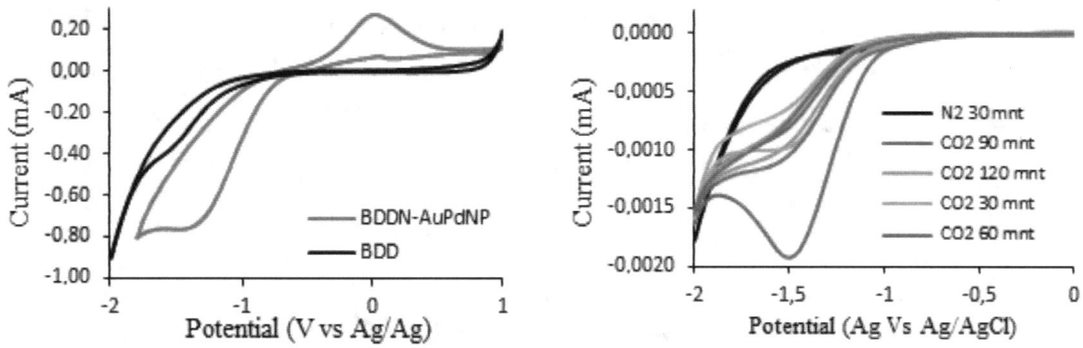

(a) (b)

Figure 3. CVs of (a) 0.5 M H_2SO_4 at unmodified BDD electrode (black line) in comparison with at AuPd-modified BDD (red line) and (b) 0.1 M NaCl solutions in the presence of dissolved CO_2 with various bubbling times

The electrochemical reduction of CO2 was performed in an electrochemical cell with two compartments separated by Nafion® membrane. A solution of 0.1 M NaCl was placed in the cathode chamber, while 0.1 M Na2SO4 solution was in the anode chamber. Electrochemical reductions were conducted by using chronoamperometric technique at the potential of -0.9 V in order to perform direct CO2 reduction on the electrode surface. The liquid products were characterized by using HPLC, while the gas products were analysed by gas chromatography. The results showed that although the products mostly contained of hydrogen gas, acetic acid and formic acid can be produced at around 9.77 % and 41.23 % of faradaic efficiency (FE), indicating that the AuPd-modified BDD is promising for an application in CO2 reduction.

4. Conclusions

Boron-doped diamond (BDD) was successfully modified with gold-palladium nanoparticles (AuPdNPs) by immersing a nitrogen-modified BDD in colloidal AuPdNPs with surface composition of 65.25 % carbon, 16.28 % oxygen, 7.97 % nitrogen, 4.42 % gold, and 6.09 % palladium. The AuPdNPs modified BDD can electrochemically convert the CO_2 at -0.9 V into acetic acid and formic acid with the faradaic efficiency of 9.77 % and 41.23 %, respectively.

Acknowledgements

This work was funded by Hibah PITTA Universitas Indonesia with contract number of 2274/UN2.R3.1/HKP.05.00/2018

References

[1] Guo X, Zhang Y, Deng Y, Li X, Xue Y, Yan Y M and Sun K 2015 *Chem. Commun.* **51** 1345–8
[2] Asai K, Ivandini T A, Falah M M and Einaga Y 2016 *Electroanalysis* **28** 177–82
[3] Hori Y, Murata A and Takasashi R 1989 *J. Chem. Sos. Faraday Trans.* I **85** 2309–26
[4] Nakata K, Ozaki T, Terashima C, Fujishima A and Einaga Y 2014 *Angew. Chem. Int. Ed.* **53** 871–4
[5] Ivandini T A 2018 *AIP Conf. Proc.* **2023** 020102
[6] Jiwanti P K, Natsui K, Nakata K and Einaga Y 2016 *RSC Adv.* **6** 102214–7
[7] Hu J W, Li J F, Ren B, Wu D Y, Sun S G and Tian Z Q 2007 *J. Phys. Chem.* C **111** 1105–12
[8] Lee Y W, Kim N H, Lee K Y, Kwon K, Kim M and Han S W 2008 *J. Phys. Chem.* C **112** 6717–22
[9] Sun D, Zhang G, Jiang X, Huang J, Jing X, Zheng Y, He J and Li Q 2014 *J. Mater. Chem.* A **2** 1767–73
[10] Nutt M O, Hughes J B and Wong M S 2005 *Environ. Sci. Technol.* **39** 1344–53
[11] Humphrey J J L, Plana D, Celorrio V, Sadasivan S, Tooze R P, Rodriguez P and Fermin D J 2015 *Chem. Cat. Chem.* **8** 952–60
[12] Kortlever R, Peters I, Balemans C, Kas R, Kwon Y, Mul G and Koper M T M 2016 *Chem. Commun.* **52** 10229–32

[13] Honda Y, Ivandini T A, Watanabe T, Murata K and Einaga Y 2013 *Diam. Relat. Mater.* **40** 7–11
[14] Luhur M S P, Ivandini T A and Khalil M 2018 *AIP Conf. Proc.* **2023** 020092
[15] Lee, K L, Wang W W and Wei P K 2008 *Plasmonics* **3** 119–25
[16] Ivandini T A, Harmesa, Saepudin E and Einaga Y 2015 *Anal. Sci.* **31** 643–49

IOP Conf. Series: Materials Science and Engineering **763** (2020) 012002 doi:10.1088/1757-899X/763/1/012002

Modified flower-like nickel oxide on carbon paste electrode (CPE) for analysis cholesterol based on non-enzymatic sensor

R T L Suropati, R Wibowo, D A Nurani and Y K Krisnandi

Department of Chemistry, Faculty of Mathematics and Natural Sciences (FMIPA), Universitas Indonesia, Kampus UI Depok, Depok 16424, Indonesia

Corresponding author's e-mail: rwibowo73@gmail.com

Abstract. The increasing circulation of unhealthy foods in various places threatens the community of malnutrition. One of which is the intake of cholesterol nutrition. When the normal limit is exceeded, it can trigger the spread of various diseases such as coronary heart disease. To anticipate the outbreak of the disease, a practical, stable, simple, and relatively inexpensive, non-enzymatic sensor device for monitoring blood cholesterol levels in some food samples such as milk and meat was developed. In this research, nickel modified catalyst on carbon paste electrodes, NiO/CPE by hydrothermal method and Ni/CPE by electrochemical method were developed. A flower-like morphology for NiO was obtained from hydrothermal method and rock-like morphology was obtained from nickel deposit. Results showed that Ni/CPE worked optimally at pH 14 with sensitivity of 0.8148 μA μM^{-1} cm^{-2} and limit of detection (LoD) of 0.1645 μM, while NiO/CPE worked optimally at pH 12 with a sensitivity of 0.1238 μA μM^{-1} cm^{-2} and LoD of 0.7804 μM. Cholesterol level measurement from the packaged milk sample showed differences of 20.42 % and 47.18 % from the nutrition table information for Ni/CPE and NiO/CPE, respectively.

1. Introduction

Cholesterol is a metabolite compound of mammals and steroid group lipid compound. Cholesterol is found in the cell membrane and circulated in the blood plasma, which is one of the important components of mammalian cell membrane and is a precursor of other biological materials, such as bile acids, steroid hormones and vitamin D [1]. The level of cholesterol in the human body is not only generated from the metabolism of the body but also from the food consumed such as fish, meat, cheese and milk [2]. Hypercholesterolemia is defined as an abnormality of cholesterol level. Hypercholesterolemia resulted from the metabolic changes in cholesterol, is a major cause of cardiovascular disorders, such as atherosclerosis and coronary heart disease [1].

Regular and continuous monitoring of cholesterol levels in the blood helps to prevent from diseases, namely hypercholesterolemia, atherosclerosis and coronary heart disease. Therefore, the development of cholesterol sensor is necessary. Recently, research on metal and metal oxide materials such as nickel, NiO and CuO are developed at a rapid pace in electrochemical systems, one of which is applied as a non-enzymatic based cholesterol sensor which is able to overcome the weaknesses in enzymatic cholesterol sensors [3]. The advantages of non-enzymatic sensors are easy to produce, low cost, high stability and not affected by its operating conditions (pH, temperature, etc.), as well as having a nano-sized structure, hence larger surface area that increases the sensitivity to cholesterol [2]. In the development of non-enzymatic sensors, carbon paste (CPE) electrodes are often used as working electrodes due to their properties like economical manufacturing, wide potential windows, and stable surface [4].

In this study, the non-enzymatic cholesterol sensors NiO/CPE and Ni/CPE were synthesized by hydrothermal and electro-deposition methods. The synthesized sensors were expected to be tested on

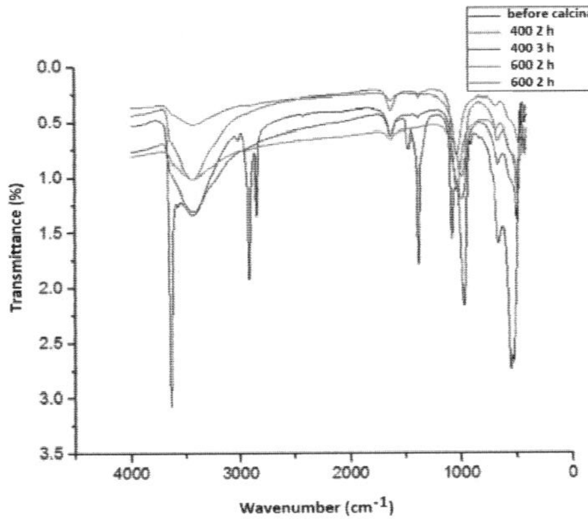

(a)

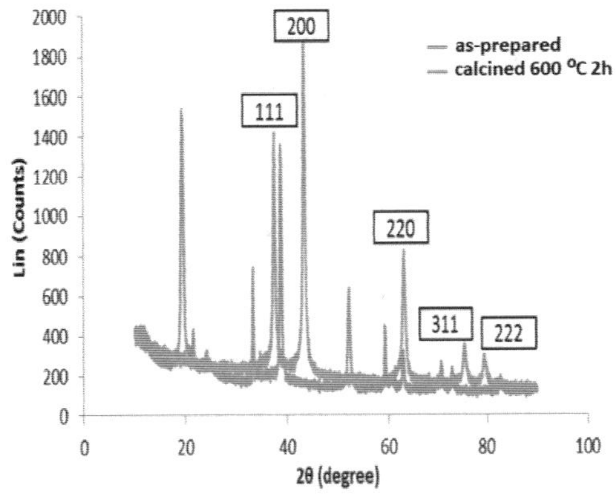

(b)

Figure 1. (a) FTIR spectra and (b) XRD pattern of NiO

milk samples and compared with the detection results with batch system. In this study, cholesterol detection, calibration, characterization and interference tests were conducted to see the usability and feasibility of the synthesized sensors.

2. Experimental

2.1. Chemicals and materials

$Ni(NO_3)_2$, Cholesterol, Cetyltrimethylammonium bromide (CTAB, ≥98%), potassium hydroxide (KOH), 1-butyl-3-methylimidazolium bis(trifluoro methylsulfonyl)imide (BMIM TFSI), graphite were purchased from Sigma Aldrich.

2.1.1. Fabrication NiO with hydrothermal method. About 25 mL of 0.3 M $Ni(NO_3)_2$ was poured into a beaker glass, and then 10 mL of CTAB solution was added stepwise under vigorous stirring for 30 minutes. Subsequently, 25 mL of 0.6 M NH_4OH was added dropwise under vigorous stirring for 2 hours [5]. The formed suspension was transferred into a 100-mL stainless steel autoclave lined with Teflon, and put in the oven at 185 °C for 24 hours [6]. The obtained precipitate was calcined at 400 °C for 2 hours. The NiO powder was characterized using Fourier Transform Infrared (FTIR), X-Ray Diffraction (XRD) and Scanning Electron Microscopy-Energy Dispersive X-Ray (SEM-EDX).

2.1.2. Fabrication carbon paste electrode (CPE) and its modified electrode. Graphite powder and Immidazole with ratio 5:1 were mixed to produce a homogenized paste, and then put into a pipette tip as the body electrode. To modify the electrode with NiO, 0.1 g as-synthesized NiO powder was mixed with the homogenized paste. Whereas, for electrode modification with Ni, the Multipulse Amperometry (MPA) technique was used, in which Ni was deposited onto CPE by reduction of Ni^{2+} (from $Ni(NO_3)_2$ solution) at potential reduction of nickel -0.326V for 90 seconds.

2.1.3. Electrochemistry measurements and sensor test. An amperometric technique was used in each modified electrode to create a calibration curve in a KOH solution with a standard cholesterol solution. Glucose, sucrose and casein were used as interferences into cholesterol solution. Milk samples were used to test the cholesterol detection sensors using a modified electrode.

3. Results and discussion

3.1. Characterization of NiO & deposit of Nickel on CPE

The infrared spectra of the NiO are presented in figure 1a. It can be found the stretching vibration of -OH (3000-3600 cm^{-1}), the vibration of the CO_2 group function (O=C=O) at 2394 cm^{-1}, the OH bending at 1400-1800 cm^{-1} and the specific vibration of the metal oxides functional group at 450-750 cm^{-1}. It can be seen that after calcination, the specific peak of metal oxides was increased and the other peak

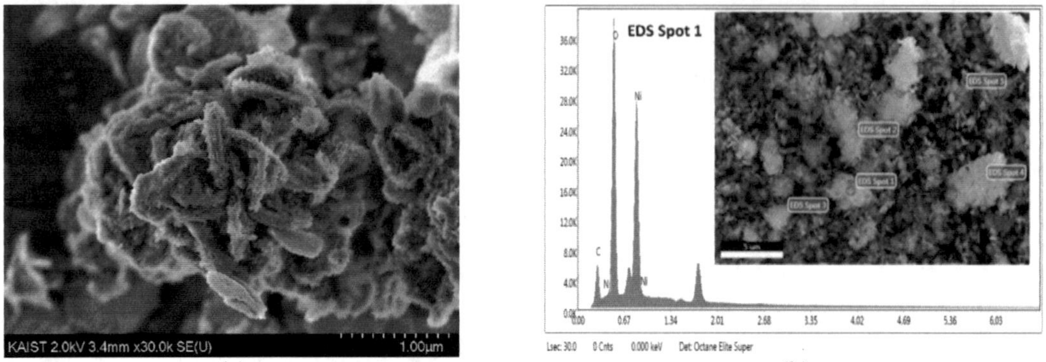

(a) (b)

Figure 2. (a) SEM image and (b) EDX of NiO

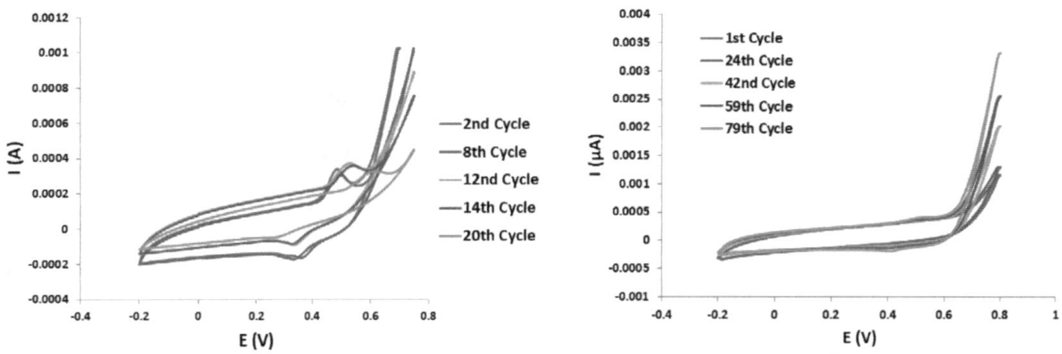

(a) (b)

Figure 3. CV of (a) Ni/CPE and (b) NiO/CPE in 1.0 M KOH solution, V range of -0.2V - +0.8V and scan rate of 100 mV/s

decreased. This indicates that the NiO became more crystalline and the impurities such as water and organic molecules have been released.

The XRD pattern of synthesized NiO by hydrothermal method before and after calcination is shown in figure 1b, which is in agreement with the diffraction patterns of NiO crystals in JCPDS 47-1049 [7,8].

Representative SEM images of the NiO powders are presented in figure 2. It can be seen the rosette-like morphology from NiO. This morphology was developed by NiOH surrounding CTAB template, so the OH^- ions will be attracted to the head of CTAB ($\sim N(CH_3)_3^+$). The concentration of alkaline solution used is responsible for the size and porosity of the as-synthesized NiO crystals, because concentration of the alkaline solution affects the diffusion process [4]. From EDX result in figure 2b, the ratio of Ni/O was ~ 2.5. Thus from XRD, infrared spectra and EDX data, it was confirmed that the NiO was successfully formed.

3.1.1. Electrode activation. Prior to use, each modified electrode is required to be activated in an alkaline solution [9]. The modified electrodes showed anodic and catodic peaks at potential -0.2V–0.8 V. The hydroxyl ions (OH^-) was involved in redox reaction. The imersion of deposited Ni or NiO in alkaline solution leads to the formation of hydrated $\alpha Ni(OH)_2$ and anhydrous $\beta Ni(OH)_2$ where the β species is more stable than α [4]. $Ni(OH)_2$ was oxidized futher and formed γ-NiOOH and β-NiOOH. In figure 3, the oxidation peak of Ni/CPE has shifted to a more anodic potential. It indicates that the stable β species has increased in KOH solution and there is a tendency of α species to β [10]. All ongoing redox reaction of Ni could be described in the following equation (1) [3]:

$$Ni^0 + 2OH^- \rightarrow \beta Ni(OH)_2 + 2e^- - \beta Ni(OH)_2 + OH^- \rightarrow NiOOH + H_2O + e^- \quad (1a)$$

$$NiO + 2OH^- \rightarrow \beta Ni(OH)_2 + \frac{1}{2}\, O_2 + 2e^- - \beta Ni(OH)_2 + OH^- \rightarrow NiOOH + H_2O + e^- \quad (1b)$$

Table 1. Comparison of calibration curve, Linearity, LoD, and Sensitivity of Ni/CPE and NiO/CPE Electrodes on variation of pH

Electrode	pH	r^2	Sensitivity (μAμM^{-1}cm^{-2})	LoD (μM)
Ni/CPE	14	0.8309	0.8148	0.1645
NiO/CPE	14	0.9585	0.0534	2.1244
Ni/CPE	13	0.9305	0.0795	1.6271
NiO/CPE	13	0.9925	0.1238	0.7804
Ni/CPE	12	0.9156	0.1559	0.6593
NiO/CPE	12	0.9312	0.1449	0.7725

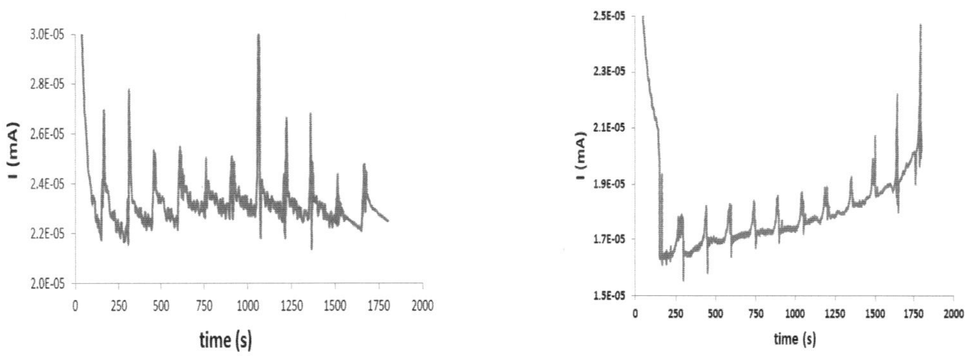

(a) (b)

Figure 4. Chrono-amperometric response of modified electrode (a) NiO/CPE and (b) Ni/CPE toward cholesterol addition in 1.0 M KOH.

3.1.2. Electrochemistry measurements and sensor test. Furthermore, Ni/CPE was being activated faster than NiO/CPE. This suggests that Ni^{0} on Ni/CPE is more easily exposed to be activated into Ni^{3+} (NiOOH) species than Ni^{2+} on NiO/CPE. Although both have FCC structures [11], when all modified electrodes were activated, Ni species will act as strong oxidizing agents for cholesterol detection sensors at specific oxidation potential (0.58 V). NiOOH was simultaneously reduced to NiO which is represented in the following equation :

$$Ni^{3+} + \text{cholesterol } Ni^{2+} + \text{cholestenone} \quad (2)$$

Figure 4 shows that the modified electrodes responded in a significant increase in peak current after each addition of cholesterol to 1.0 M KOH solution. Ni/CPE (figure 4b) exhibited a higher peak current than NiO/CPE (figure 4a) due to its high electrocatalytic activity and the large active surface of Ni/CPE. Ni/CPE can adsorb and catalyze more reactive substances and resulted in an instantaneous oxidation process [2].

The linear range, sensitivity and detection limit of modified electrode are compared at pH range 12–14 and listed in table 1. The modified electrode shows a linear relationship of electrode response with additional 0.5 μM cholesterol at range volume of 0.5 – 2.0 mL. NiO/CPE has high linearity with R^2 = 0.9925 at pH 13 whereas Ni/CPE shows a low detection limit 0.1645 μM of cholesterol sensing and high sensitivity 0.8148 μA μM^{-1} cm^{-2} at pH 14. These results indicated that Ni/CPE and NiO/CPE have a good ability for non-enzymatic detection of cholesterol.

3.1.3. Effect of other glucose and fructose as interferences. Selectivity is an important parameter of the non-enzymatic cholesterol sensor. Other substance in sample such as glucose, sucrose, and casein could interfere the detection of cholesterol. As shown in figure 5 Ni/CPE and NiO/CPE showed identical responses, the presence of glucose and sucrose interfered the cholesterol detection response, while the casein was found to be a non-interfering substance. This is thought to be due to glucose and sucrose having an oxidation potential adjacent to the cholesterol oxidation potential [9].

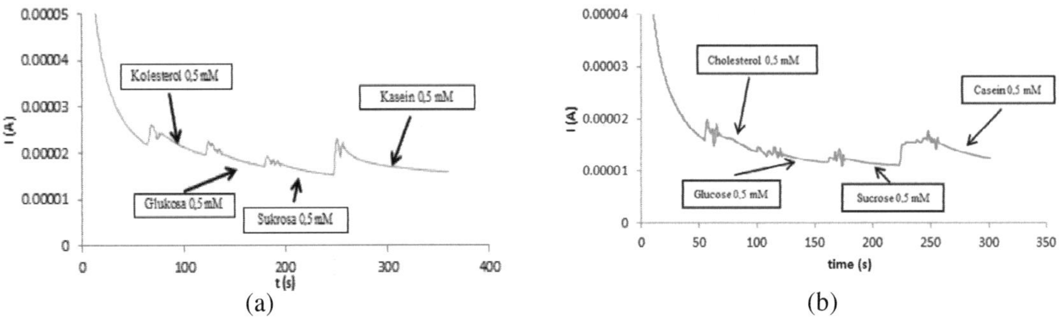

Figure 5. The interference test of modified electrodes in KOH with 0.5 mM cholesterol in the presence of 0.5 mM glucose, sucrose, and casein (a) Ni/CPE and (b) NiO/CPE

To validate the feasibility of modified cholesterol sensor, analysis of cholesterol in milk sample was performed. The test results from milk samples obtained the measured cholesterol levels using non-enzymatic sensors Ni/CPE and NiO/CPE were 86.51 μM and 37.94 μM, respectively. Compared to the calculation results of the nutrition table on the packaging milk, the calculated cholesterol level in the sample is 71.84 μM. The differences in measurement results with non-enzymatic sensors against the nutritional tables are 20.42% and 47.18% for Ni/CPE and NiO/CPE respectively. So, it can be concluded that these non-enzymatic sensors can be applied to measure cholesterol levels in milk.

4. Conclusions

Carbon Paste electrode has been successfully modified with Ni and NiO by electrodeposition and hydrothermal methods. From the above data, when using NiO/CPE does not show a strong enough response to oxidize cholesterol into ketone form because in NiO the presence of Ni bonds with O which makes Ni more difficult to be exposed to activate into Ni^{3+} species, whereas using Ni/CPE gives a higher response, because Ni is easily to be activated and the Ni^{3+} species is rich. The presence of glucose and sucrose interfered the cholesterol detection response while the casein was found to be a non-interfering substance.

Acknowledgements

This work is partially funded by PUPT BPPTN 2018 grant No. 481/UN2.R3.1/HKP.05.00/2018.

References

[1] Lewis Sr. R J 1997 *Hawley's Condensed Chemical Dictionary* 13th ed (New York: John Wiley and Sons Inc) p 265
[2] Rengaraj A, Haldorai Y, Kwak C H, Ahn S, Jeon K, Park S H and Huh Y S 2015 *J. Mater Chem. B* 3 6301–09
[3] Chen J, Sheng Q, Wang Y and Zheng J 2016 *Electroanalysis* **28** 5 979–84
[4] Yulianto E and Setiarso P 2014 *Proc. Seminar Nasional Kimia* 20 September 2014 Surabaya ed P Setiarso *et al.* (Surabaya: FMIPA Universitas Negeri Surabaya) B–14
[5] Suresh R, Ponnuswamy V, Sankar C, Manickam M, Venkatesan S and Perumal S 2017 *J. Magn. Magn. Mater.* **441** 787–94
[6] Carbone M, Bauer E M, Micheli L and Missori M 2017 *Colloids and Surf.* A **532** 178–82
[7] Rahdar A, Aliahmad M and Azizi Y 2015 *Journal of Nanostructure* **5** 145–51
[8] El-Kemary M, Nagy N and El-Mehaseb I 2013 *Mater. Sci. Semicond. Process.* **16** 1747–52
[9] Ji R, Wang L, Wang G and Zhang X 2014 *Electrochim. Acta* **130** 239–44
[10] Toghill K E, Xiao L, Stradiotto N R and Compton R G 2010 *Electroanalysis* **22** 491–500
[11] Niu X, Lan M, Xhao H and Chen C 2013 *Anal. Chem.* **85** 3561–9

Aqueous fraction of *Sapindus rarak D.C.* fruit extract mediated one-pot three component construction of 9,9-dimethyl-12-phenyl-9,10-dihydro-8H-benzo[a]xanthen-11(12H)-one

A H Cahyana, M C Maida and B Ardiansah

Department of Chemistry, Faculty of Mathematics and Natural Sciences (FMIPA), Universitas Indonesia, Kampus UI Depok, Depok 16424, Indonesia

Corresponding author's e-mail: herrykim@ui.ac.id

Abstract. In this study, we report 3-component combination of 2-naphthol, dimedone and benzaldehyde in 9,9-dimethyl-12-phenyl-9,10-dihydro-8H-benzo[a]xanthen-11(12)-one synthesis using water fraction of *Sapindus rarak D.C.* fruit extract. This extract promoted an efficient synthesis under mild reaction conditions. The compound structure was confirmed by FT-IR and Mass analysis.

1. Introduction

Benzoxanthene derivatives are noted to possess notable biological and pharmacological activities such as anti-inflammatory [1], antibacterial [2], and antiviral [3]. Three component reaction of aldehydes, 2-naphthol and cyclic 1,3-dicarbonyl compounds is one of highly interesting approaches for the preparation of benzoxanthene derivatives [4]. Recently, multicomponent synthesis of benzoxanthene derivatives have been performed in the presence of various catalysts such as $Zr(HSO_4)_4$ [5], H_2SO_4 [6], iodine [7], ZnO [8], and pyrazinium dihydrogen sulfate [9]. However, some procedures have at least one drawback, such as long reaction times, toxic solvents or high temperature needed.

In recent years, role of green chemical-based catalysts has grown tremendously. The highlighted aspect of green chemistry is reduction in the use of harmful chemical reagents [10]. Using plant or its part as a potential biocatalyst in the synthesis of organic compounds is one of the efforts that have been conducted. Organic synthesis using natural materials generates less waste than the conventional methods [11].

Sapindus rarak D.C. is a plant used as a traditional detergent. The fruits, roots, and seeds contain triterpenoids, some polyphenols and saponins that act as anti-cancer, anti-bacterial, and anti-insecticides. The content of saponin content in *Sapindus rarak D.C.* fruit reaches 10–11.5 % [12]. Saponins are group of glycoside compounds that serve as a washing agent that has properties such as soap and acts as a surfactant. Surfactants are best known to dissolve the non-polar materials. That ability has established the aqueous surfactant systems to be notably popular option for organic solvents [13] or stabilizing and capping agent in nanoparticles syntheses [14]. Therefore, with interesting properties of aqueous extract of *Sapindus rarak D.C.* fruit, we investigate its performance as an eco-friendly acidic surfactant type for the synthesis of 9,9-dimethyl-12-phenyl-9,10-dihydro-8H-benzo[a]xanthen-11(12H)-one (figure 1).

2. Materials and methods

2.1. General

All chemicals required were obtained from Merck and used without further purified. *Sapindus rarak*

IOP Conf. Series: Materials Science and Engineering **763** (2020) 012003 doi:10.1088/1757-899X/763/1/012003

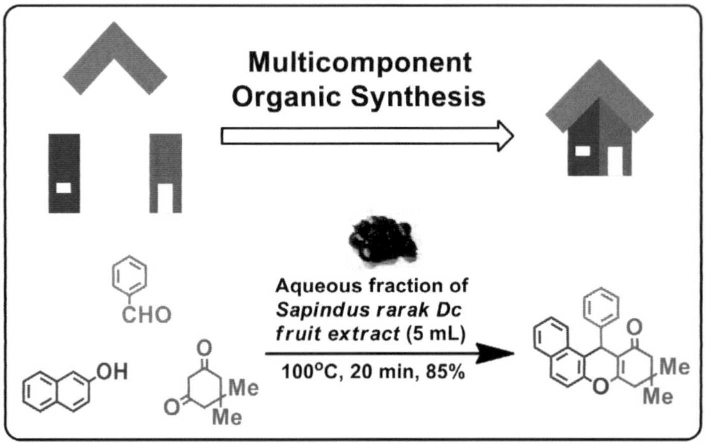

Figure 1. Multicomponent synthesis of benzo[a]xanthene using aqueous fraction of *Sapindus rarak D.C.* fruit extract

D.C. fruit was obtained from Banjarsari, Surakarta. Functional group analysis was done by Fourier Transform Infrared (FT-IR) on Shimadzu 8400 spectrometer with KBr background. Mass analysis was conducted on Shimadzu QP Mass Spectrometer 2010A having 70 eV ionization potential.

2.2. Preparation of Sapindus rarak D.C. fruit extract

Preparation of *Sapindus rarak D.C.* fruit extract (SRDFE) was carried out by maceration method. *Sapindus rarak D.C.* (50 g) dry powder was macerated in methanol (1:5 w/v) for 2 days. The extract was filtered and then centrifuged for 10 min. The filtrate was evaporated using a rotatory evaporator. Furthermore, this crude extract was separated by deionized water and hexane by a ratio of 1:1 (v / v). These fractions were phytochemically screened to determine secondary metabolites content such as saponins, alkaloids (Wagner test), flavonoids, steroids, terpenoids (Salkowski test), and tannins [15]. Further, aqueous fraction will be used to mediate multicomponent synthesis of a benzoxanthene derivative.

2.3. Synthesis of 9,9-dimethyl-12-phenyl-9,10-dihydro-8H-benzo[a]xanthen-11(12H)-one.

In 30 mL round-bottom flask, benzaldehyde (0.106 g, 1 mmol), dimedone (0.140 g, 1 mmol), 2-naphthol (0.144 g, 1 mmol) and aqueous fraction of *Sapindus rarak D.C.* fruit extract (5 mL) were stirred at 30 °C. The reaction mixture was observed by Thin-Layer Chromatography (TLC) analysis using *n*-hexane/ethyl acetate (4:1) and spots were investigated using a UV lamp (254 and 366 nm). Once the reaction was completed, the temperature was reduced to room temperature, water was added then stirred for 3 min followed by filtering. The solid residue was isolated in crystal form from aqueous ethanol (90 %) to obtain a pure product. *Sapindus rarak D.C.* extract, reaction temperature and time were varied to obtain optimized protocol.

3. Results and discussion

The phytochemical screening of water and *n*-hexane fraction of *Sapindus rarak D.C.* fruit extracts revealed the presence of several secondary metabolites. Saponins were strongly observed in water fraction but not in *n*-hexane fraction. Alkaloids were found only in *n*-hexane fraction. Meanwhile, terpenoids were detected in both fractions. The results are summarized in table 1.

FT-IR spectrum of aqueous fraction of *Sapindus rarak D.C.* fruit extract (figure 2) showed peak at 1730 cm^{-1} due to carbonyl stretching vibration. A broad peak at 3353 cm^{-1} is associated to O-H vibration. Furthermore, peaks at 2933 and 1613 cm^{-1} are ascribed to the C-H sp^3 and C=C vibrations, respectively. All those peaks may be attributable to the presence of saponins in the extract [16].

We report here the efficient and green approaches for the preparation of 9,9-dimethyl-12-phenyl-9,10-dihydro-8H-benzo[a]xanthen-11(12H)-one using one-pot, three-component condensation of

Table 1. The results of phytochemical tests in aqueous and hexane fraction of *Sapindus rarak D.C.*

Extract	Water	Hexane
Saponin	+++	-
Alkaloids	-	+
Flavonoids	-	-
Steroids	-	-
Terpenoids	++	+
Tanins	-	-

Note: + = presence, - = absence

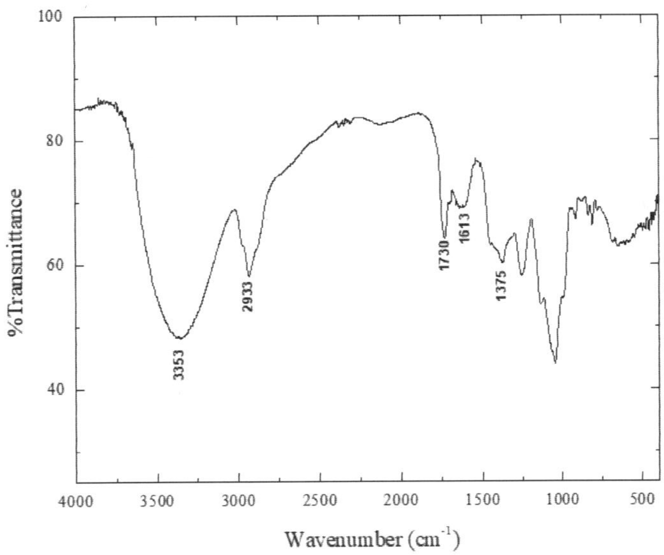

Figure 2. FT-IR spectrum of aqueous fraction of *Sapindus rarak D.C* fruit extract

benzaldehyde, 2-naphthol, and dimedone mediated by aqueous fraction of *Sapindus rarak D.C.* fruit extract.

In figure 3, FT-IR spectrum of organic product signified the presence of C=O band at 1604 cm^{-1}, whereas peaks at 2871 cm^{-1} and 2966 cm^{-1} indicated the existence of aliphatic C-H group. Meanwhile, absorption at 1371 cm^{-1} is characteristic of C-O-C group. Mass spectra data (figure 4) shows the molecular weight of benzoxanthene derivative of 354 $gmol^{-1}$, with molecular formula of $C_{25}H_{22}O_2$, fit with IUPAC name of 9,9-dimethyl-12-phenyl-9,10-dihydro-8H-benzo[a]xanthen-11(12H)-one.

To achieve optimum reaction conditions, we initially investigated the reaction with various catalysts, temperatures and times (catalyst: 5, 7 and 10 mL; reaction time: 15, 20, 25, and 30 min; temperature: 30, 50, 80 and 100 °C). The best result was obtained when using 5 mL of aqueous fraction of *Sapindus rarak D.C* fruit extract at 100 °C for 20 min (table 2 entry 7). The optimum yield of 9,9-dimethyl-12-phenyl-9,10-dihydro-8H-benzo[a]xanthen-11(12H)-one obtained was 85 %. A plausible mechanism for the production of 9,9-dimethyl-12-phenyl-9,10-dihydro-8H-benzo[a]xanthen-11(12H)-one by the existence of saponin catalysts is proposed in figure 5. The saponins, which are acidic, could strongly dissolve the reactant species through hydrogen bond formation in aqueous medium [17]. Thus as the carbonyl group will be susceptible to other nucleophilic reactant attacks. Initially, the 2-naphthol condensation with benzaldehyde generated ortho-quinone methides (o-QMs). Then, the o-QMs with dimedone produced intermediates which were cyclized and then dehydrated to generate the corresponding of 9,9-dimethyl-12-phenyl-9,10-dihydro-8H-benzo[a]xanthen-11(12H)-one [7].

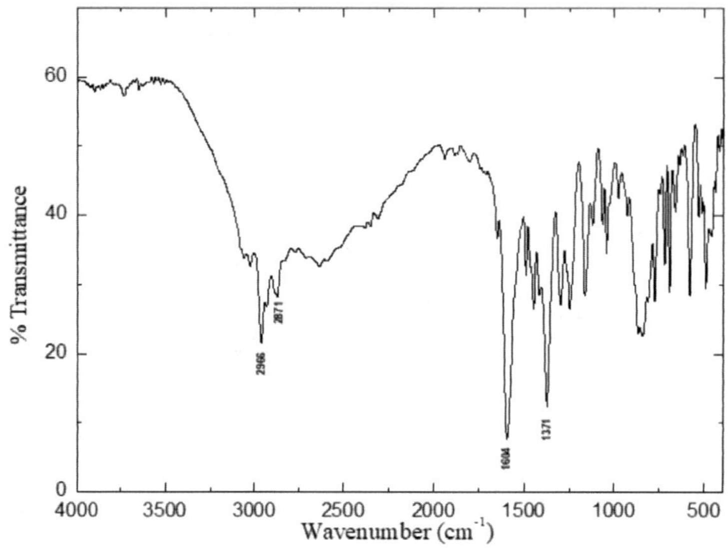

Figure 3. FT-IR spectrum of 9,9-dimethyl-12-phenyl-9,10-dihydro -8H-benzo[a]xanthen-11(12H)-one

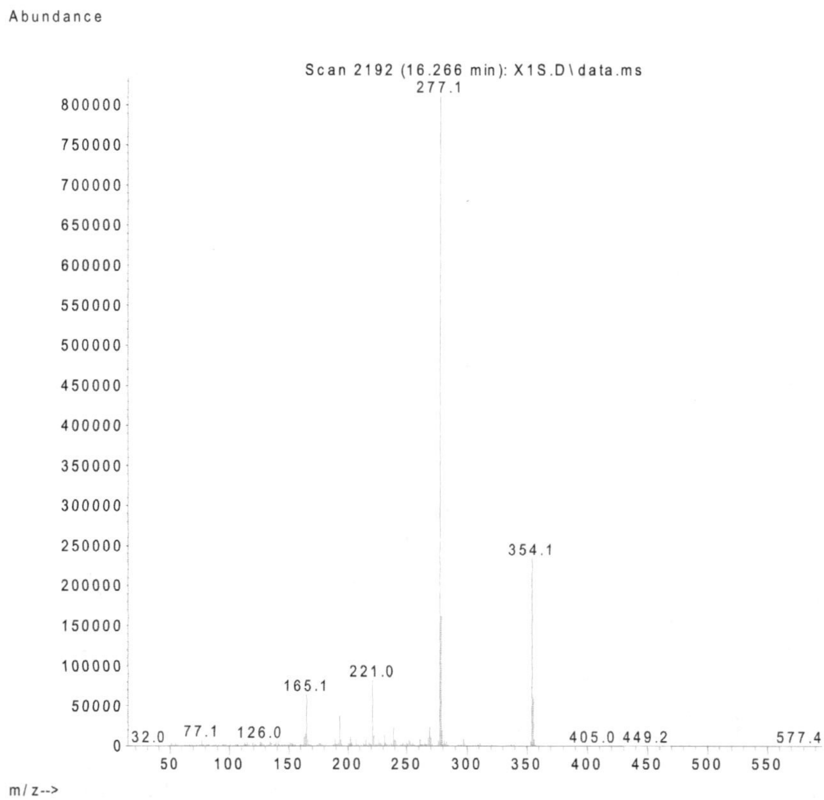

Figure 4. Mass spectrum of 9,9-dimethyl-12-phenyl-9,10-dihydro -8H-benzo[a]xanthen-11(12H)-one

Table 2. Synthesis of 9,9-dimethyl-12-phenyl-9,10-dihydro-8H-benzo[a]xanthen-11(12H)-one in different conditions using aqueous fraction of *Sapindus rarak D.C.*

Entry	fruit extract (mL)	Time (min)	Temp (°C)	Yield (%)
1	5	15	50	64
2	5	20	50	72
3	5	25	50	(i) 73
4	5	30	50	70
5	5	20	30	64
(ii) 6	5	20	80	(iii) 78
(iv) 7	5	20	100	85
(v) 8	7	20	100	(vi) 85
(vii) 9	10	20	100	85

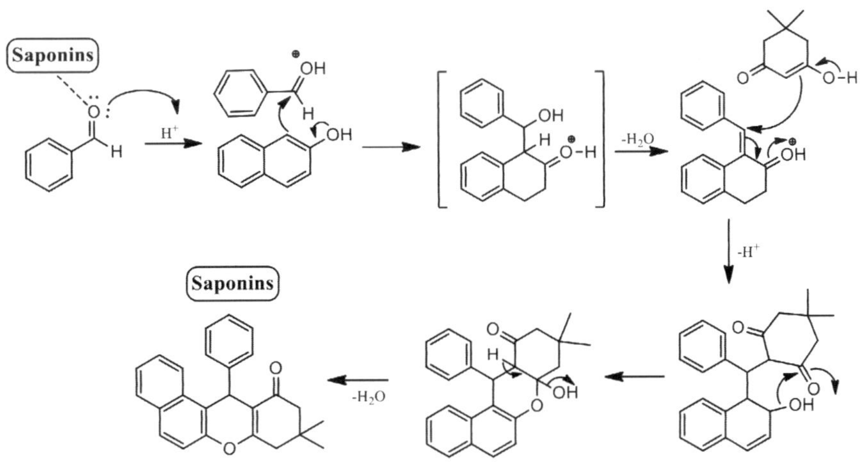

Figure 5. Proposed mechanism in the synthesis of 9,9-dimethyl-12-phenyl-9,10-dihydro-8H-benzo[a]xanthen-11(12H)-one

4. Conclusions

We successfully synthesized 9,9-dimethyl-12-phenyl-9,10-dihydro-8H-benzo[a]xanthen-11(12H)-one using aqueous extract of *Sapindus rarak D.C.* fruit extract containing saponins. The optimum conditions of the synthesis were obtained when 5 mL of extract used at 100 °C for 20 min. The present research highlighted an easy work-up, low-cost, efficient and eco-friendly manner in the synthesis of biologically potent natural product derivative with excellent yield.

Acknowledgements

This research was fully supported by Universitas Indonesia through PITTA Grant with contract number 234/UN2.R3.1/PPM.00/2018.

References

[1] Poupelin J P, Saint-Ruf G, Foussard-Blanpin O, Narcisse G, Uchida-Ernouf G and Lacroix R 1978 *Eur. J. Med. Chem.* **13** 67–71
[2] Hideo T and Teruomi J 1981 *Japan Patent* 56005480
[3] Lambert R W, Martin J A, Merrett J H, Parkes K E B and Thomas G J 1997 *PCT International. Appl. WO Patent* 9706178
[4] Nandi G C, Samai S, Kumar R and Singh M S 2009 *Tetrahedron* **65** 7129–34
[5] Foroughifar N, Mobinikhaledi A and Moghanian H 2009 *International Journal of Green Nanotechnology: Physics and Chemistry* **1** 57–63

[6] Khurana J M, Lumb A, Pandey A and Magoo D 2012 *Synth. Commun.* **42** 1796–803
[7] Wang R Z, Zhang L F and Cui Z S 2009 *Synth. Commun.* **39** 2101–7
[8] Safaei-Ghomi J and Ghasemzadeh M A 2014 *S. Afr. J. Chem.* **67** 27–32
[9] Zare A, Khanivar R, Hatami M, Mokhlesi M, Zolfigol M A, Moosavi-Zare A R, Hasaninejad, Khazaei A and Khakyzadeh V 2012 *J. Mex. Chem. Soc.* **56** 389–94
[10] Varma R S 2014 *Sustain. Chem. Process* **2** 11
[11] Alfermann A 1985 *Proc. Biocatalyst in Organic Syntheses*, ed J Tramper *et al.* (Amsterdam: Elsevier)
[12] Jassal V, Shanker U and Gahlot S 2016 *Mater. Today Proc.* **3** 1874–82
[13] Chavan H V, Sirsat D M and Mule Y B 2016 *Iran. Chem. Commun.* **4** 373–88
[14] Maryanti E, Damayanti D, Gustian I and Yudha S S 2014 *Mater. Lett.* **118** 96–8
[15] Iqbal E, Salim K A and Lim L B L 2015 *J. King Saud Univ. Sci.* **27** 224–32
[16] Almutairi M S and Ali M 2015 *Nat. Prod. Res.* **29** 1271–5
[17] Chavan H V and Bandgar B P 2013 *ACS Sustain. Chem. Eng.* **1** 929–36

Strobilanthes crispus (B.) leaf extract-assisted green synthesis of ZnO-La_2O_3 composite and preliminary study of its photocatalytic activity

R I Ashna, Y Yulizar and D O B Apriandanu

Department of Chemistry, Faculty of Mathematics and Natural Sciences (FMIPA), Universitas Indonesia, Kampus UI Depok, Depok 16424, Indonesia

Corresponding author's e-mail: yokiy@ui.ac.id

Abstract. In this study, ZnO-La_2O_3 composite as photocatalyst was synthesized via facile sol-gel route using biomaterial as a weak base source and a capping agent. ZnO-La_2O_3 composite was successfully formed through the reaction of $Zn(NO_3)_2$(aq), $La(NO_3)_3$ (aq) and *Strobilanthes crispus* (B.) aqueous leaf extract. Zn-O and La-O vibration modes of ZnO-La_2O_3 composite was analysed by Fourier Transform Infra Red (FTIR) at 400 – 700 cm^{-1} wavenumber. Optical band gap of ZnO-La_2O_3 composite determined by using Tauc plot method was about 5.21 eV. X-Ray Diffraction (XRD) pattern of ZnO-La_2O_3 composite confirmed a combination of the diffraction patterns of ZnO particle and La_2O_3 particle. Photocatalytic activity of ZnO-La_2O_3 composite and ZnO particle were observed with methylene blue using UV light irradiation. The percentage of methylene blue degradation using ZnO-La_2O_3 composite and ZnO particle were 81.75 % and 74.87 %, respectively. This result shows that La_2O_3 was improving the activity and stability of the ZnO photocatalyst.

Keywords: green synthesis, methylene blue, photocatalyst, *Strobilanthes crispus* (B.), ZnO-La_2O_3 composite

1. Introduction

Zinc oxide (ZnO) has high potential in many applications, such as nanolasers, sensors, transistors, nanogenerators and solar cells [1]. ZnO is an n-type semiconductor with a band gap of 3.20 to 3.37 eV, has been used as material for photocatalytic process [2,3]. ZnO was found as an active catalyst in photodegradation process [4]. Common procedure mostly applied for the photocatalytic activity improvement in ZnO is to encounter the doping with another semiconductors [5,6]. La_2O_3 is known to offer an efficient result for the improvement of the catalyst's activity and stability [7]. The previous reports associated with ZnO-La_2O_3 catalysts have been applied in monocrotophos degradation and phenol photodegradation [8,9].

On the other hand, there are several methods reported for the synthesis of composites, such as co-precipitation, sol-gel, and hydrothermal [10–12]. Sol-gel method has been widely used for the synthesis of metal oxide, because sol-gel is an efficient method for controlling the morphology and reactivity of solids [13]. Sol-gel method usually used in toxic hydrolysis agent like NaOH and organic solvents, and they are harmful to the environment. Green synthesis of composites is one of the promising and environmentally friendly method. In the previous report, green synthesis method was conducted using plant extract, fungi, microbes, and extracts of fruit and vegetables [14]. Green synthesis of nanoparticle and nanocomposite using plants such as *Phyllanthus embilica, Calotropis procera, Theobroma cacao, Oldenlandia corymbosa L. Physalis angulata,* and *Terminalia catappa* have been reported [14–19]. However, there was no report on ZnO-La_2O_3 composite synthesis using green synthesis method.

Strobilanthes crispus (B.) is one of medical plants in Indonesia [20]. This plant has secondary metabolite compounds that can be used to synthesize metal oxides composite. *Strobilanthes crispus*

(B.) leaf extract (SCBE) had been previously reported to contain the alkaloid, flavonoid, and saponin compounds which act as a weak base source and stabilizing agent in metal oxide composites formation [21].

In this research, ZnO-La_2O_3 composite was synthesized through a simple green route using SCBE as a weak base source and capping agent. ZnO-La_2O_3 composite was characterized by using UV-Vis Diffuse Reflectance Spectroscopy (DRS), Fourier Transform Infrared (FT-IR), and X-Ray Diffraction (XRD). Furthermore, ZnO-La_2O_3 composite was evaluated for photocatalytic degradation of methylene blue (MB)

2. Materials and methods

2.1. Materials

Strobilanthes crispus B. (SCB) was obtained from Cimanggis, Depok, West Java, Indonesia and has been determined at Indonesian Institute of Sciences (LIPI), Bogor, West Java, Indonesia. Methylene blue (MB) was obtained from Merck. Methanol and n-hexane were obtained from PT. Brataco. Analytical-grade of lanthanum nitrate hexahydrate ($La(NO_3)_3.6H_2O$) and zinc nitrate pentahydrate ($Zn(NO_3)_2.4H_2O$) were obtained from Merck, Germany.

2.2. Preparation of leaf extract

250 mL of methanol was added to the dried leaves powder of SCB and macerated for 7 days at room temperature. Methanol extract of SCB leaves was partitioned using hexane. The filtrate was then concentrated using vacuum rotatory evaporator. Distilled water was added to the concentrated methanol to obtain a stock solution of SCBE for composite synthesis.

2.3. ZnO-La_2O_3 composites synthesis

Synthesis of ZnO-La_2O_3 composite using SCBE refers to the method from the previous research [9]. $Zn(NO_3)_2$ and $La(NO_3)_3$ precursors were mixed and stirred continuously at 80 °C for 1 h. After that, SCBE was added to the mixture drop wise, then stirred simultaneously at 80 °C for 4 h. The colloidal was heated at 100 °C for 1 h and annealed at 700 °C for 3 h. ZnO-La_2O_3 crystal was obtained as white powder. ZnO particles were synthesized using the same method, but without $La(NO_3)_3$ precursor.

2.4. Characterization

Band gap measurements were determined by UV-Vis diffuse reflectance spectra (Shimadzu 2450). The functional group of compound was analyzed by FTIR (Shimadzu, Prestige 21). The diffraction pattern information was determined by XRD (Shimadzu 610) at 2θ of 20–60° with Cu Kα (λ = 0.1546 nm) radiation. The UV-Vis absorption spectrum was recorded using UV-Vis spectrophotometer (Shimadzu 2600).

2.5. Photocatalytic activity

ZnO-La_2O_3 composite was added to methylene blue of 3.0×10^{-5} M. The mixture was stirred under UV light irradiation. The reaction was observed through the absorbance change using a UV-Vis spectrophotometer for 120 min. For comparison, the photocatalytic activity of ZnO nanoparticles was evaluated using the same method.

3. Results and discussion

3.1. FT-IR analysis

Identification of functional groups of SCBE, ZnO, La_2O_3, and ZnO-La_2O_3 composite were recorded by FTIR spectroscopy at 400-4000 cm^{-1} range. Figure 1a shows broad vibrations band at 3361.1 cm^{-1} which corresponds to –OH stretching of flavonoid and saponin compounds [22, 23] and vibrations band at 1080.2 cm^{-1} which corresponds to C-N stretching of alkaloid compounds [24].

Figure 1b shows a typical Zn-O stretching vibration mode of ZnO particles at 453.0 cm^{-1} and La-O stretching vibration mode of La_2O_3 particles at 640.8 cm^{-1}. These results conform to the previous literatures [16,25]. Meanwhile, ZnO-La_2O_3 composite shows two vibration modes of ZnO particles and La_2O_3 particles at 453.0 and 640.8 cm^{-1}, respectively, indicating ZnO-La_2O_3 composite was successfully synthesized.

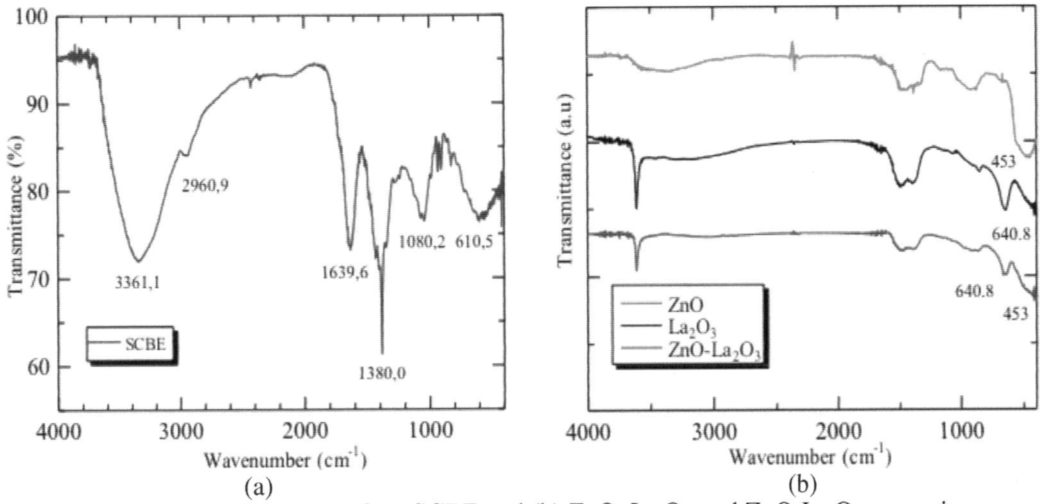

Figure 1. FTIR spectra of (a) SCBE and (b) ZnO, La_2O_3, and ZnO-La_2O_3 composite.

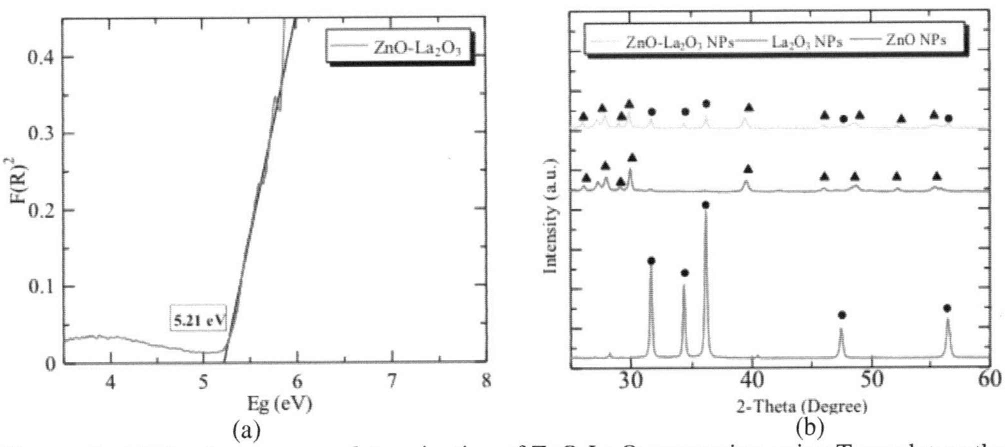

Figure 2. (a) Band gap energy determination of ZnO-La_2O_3 composites using Tauc plot method (b) XRD pattern of ZnO, La_2O_3, and ZnO-La_2O_3 composites

3.2. UV-Vis DRS analysis

The determination of band gap energy value of ZnO-La_2O_3 composite was obtained through the characterization of UV-Vis DRS. The UV-Vis DRS result presents the reflectance data which then converted into Kubelka Munk equation by plotting the curve $F(R)^2$ against Eg (eV) as shown in figure 2. According to the calculation, ZnO-La_2O_3 composite has a band gap energy (Eg) of 5.21 eV.

3.3. XRD analysis

The diffraction patterns and average crystalline size of ZnO-La_2O_3 composites are shown in figure 2 (b). The diffraction patterns of ZnO crystallites (blue line) and La_2O_3 crystallites (red line) are matched with the COD 96-230-0113 and COD 96-200-2287, respectively. The XRD results of ZnO-La_2O_3 composites (green line) shows ten diffraction patterns at 2θ of 26.06°; 27.25°; 27.94°; 29.90°; 31.71°; 34.36°; 36.20°; 39.40°; 47.48° and 56.54°, which demonstrated the combination of diffraction pattern of ZnO crystallites and La_2O_3 crystallites. The crystallite size of ZnO-La_2O_3 was calculated using the Debye-Scherer's formula. According to the calculation, ZnO-La_2O_3 composite has the crystallite size of 45.59 nm.

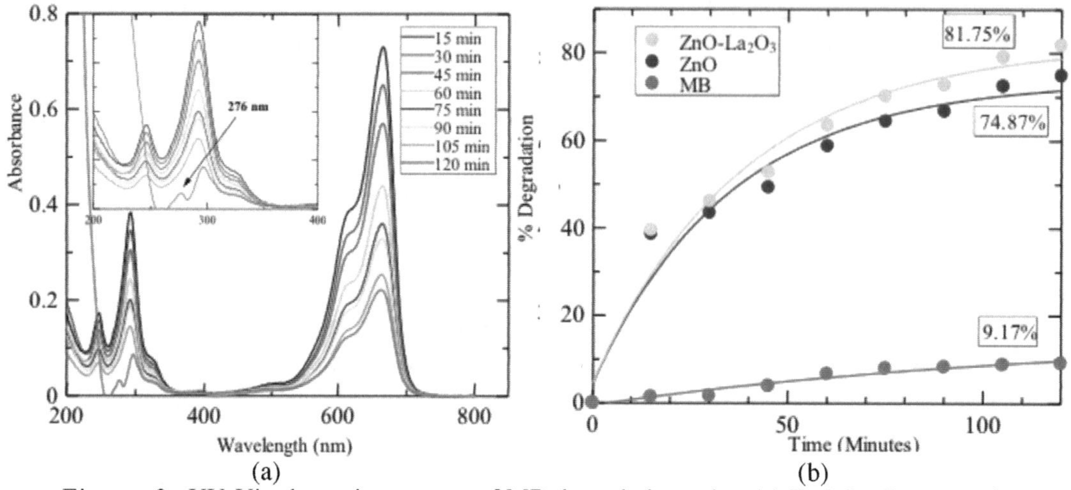

Figure 3. UV-Vis absorption spectra of MB degradation using (a) ZnO-La_2O_3 composite and (b) % degradation of MB for 120 minutes under UV light irradiation

3.4. Photocatalytic activity

The photocatalytic activity of ZnO-La_2O_3 composite was investigated by the degradation of MB under UV light irradiation. Figure 3 shows the degradation spectra of MB in 120 minutes. Investigation of MB degradation was indicated by the decreasing absorbance of UV-Vis spectra at λ_{max} of 665 nm and the decreasing colour intensity of MB. Figure 3 shows that ZnO-La_2O_3 composite and ZnO particle without photocatalyst are able to degrade MB under UV light irradiation, yielding MB degradation of 81.75 % 74.87 %, and 9.17 % respectively. This result shows that La_2O_3 was improving the activity and stability of the ZnO photocatalyst.

When a semiconductor is irradiated by light, it will produce e- (electron) on the conduction band (CB) and h^+ (hole) on the valence band (VB), which is caused by the excitation of electron from VB to its CB. The hole will then react with water (H_2O), which will cause the formation of hydroxyl radicals (OH•) and H^+ ion. Meanwhile, the electron on the CB will react with the water-dissolved O_2, which eventually will form superoxide radical, O_2•⁻. This radical ion will react with H^+ ion to form hydroxyl radicals. The hydroxyl radicals will react with one and another to form hydrogen peroxide (H_2O_2) and also react with MB to degrade it and form various intermediate products such as phenol and benzene sulfonic acid compounds, before producing H_2O and CO_2 gas [26]. Figure 3a shows the appearance of a new absorption peak at the wavelength of 277 nm. This result indicates the presence of carboxylic acid groups and aromatic chromophore groups [27]. The formed hydrogen peroxide will react with the light and cause a photolysis reaction, so the hydrogen peroxide will split into hydroxyl radicals or the hydrogen peroxide, which will react with electron on CB of the composites forming hydroxyl ions and radicals.

4. Conclusions

The ZnO-La_2O_3 composite prepared using *Strobilanthes crispus* B. leaf extract has been studied. FT-IR result showed that ZnO-La_2O_3 composite was successfully formed at 453.0 and 640.8 cm^{-1} which corresponds to the specific vibration mode of Zn-O and La-O, respectively. The optical band gap of ZnO-La_2O_3 composite was 5.21 eV. XRD result of ZnO-La_2O_3 composite showed a combination diffraction peak of ZnO and La_2O_3 particle with crystalline size of 45.59 nm. The ZnO-La_2O_3 composite and ZnO particle showed the photocatalytic activity in the methylene blue degradation at 81.75 % and 74.87 %, respectively.

Acknowledgements

The authors would like to thank Universitas Indonesia for funding this research through PITTA Grant with contract No: 2339/ UN2.R3.1/HKP.05.00/2018.

References

[1] Karunakaran C, Vinayagamoorthy P and Jayabharathi J 2014 *Powder Technol.* **254** 480–7
[2] Guo K, Liu, Z, Wang Y, Zhao Y, Xiao Y, Han J and Cui, T 2014 *Int. J. Hydrogen Energy* **39** 13408–14
[3] Ren P, Fan H and Wang X 2012 *Catal. Commun.* **25** 32–5
[4] Sakthivel S, Neppolian B, Shankar M V, Arabindoo B, Palanichamy M and Murugesan V 2003 Sol. *Energy Mater. Sol. Cells* **77** 65–82
[5] Li C, Zhang J, Yang J, Wang T, Lv X and Tang Z 2011 *Appl. Catal.* A **402** 80–6
[6] Kim C, Doh S J, Lee S G, Lee S J and Kim H Y 2007 *Appl. Catal.* A **330** 127–33
[7] Zhi G, Guo X, Wang Y, Jin G and Guo X 2011 *Catal. Commun.* **16** 56–9
[8] Anandan S, Vinu A, Lovely K L P S, Gokulakrishnan N, Srinivasu P, Mori T, Murugesan V, Sivamurugan V and Ariga K 2007 *J. Mol. Catal.* A: *Chem.* **266** 149–57
[9] Li C, Hu R, Qin L, Ding R, Li X and Wu H 2013 *Mater. Lett.* **113** 190–4
[10] Katiyar A, Kumar N and Srivastava A 2018 *Mater. Today Proc.* **5** 9144–7
[11] Mahdavi R and Talesh SSA 2017 *Ultrason. Sonochem.* **39** 504–10
[12] Bazazi S, Arsalani N, Khataee A and Tabrizi A G 2018 *J. Ind. Eng. Chem.* **62** 265–72
[13] Benhebal H, Chaib M, Salmon T, Geens J, Leonard A, Lambert S D, Crine M and Heinrichs B 2013 *Alex. Eng. J.* **52** 517–23
[14] Gawade V V, Gavade N L, Shinde H M, Babar S B, Kadam A N and Garadkar K M 2017 *J. Mater. Sci. Mater. Electron.* **28** 14033–39
[15] Joel C and Badhusha M S M 2016 *Pharm. Lett.* **8** 218–23
[16] Yulizar Y, Bakri R, Apriandanu D O B and Hidayat T 2018 *Nano Structures & Nano-Objects* **16** 300–5
[17] Yulizar Y, Latifah I, Bakri R and Apriandanu D O B 2018 *AIP Conf. Proc.* **2023** 020097
[18] Sulaiman N, Yulizar Y and Apriandanu D O B 2018 *AIP Conf. Proc.* **2023** 020105
[19] Lembang M S, Yulizar Y, Sudirman S and Apriandanu D O B 2018 *AIP Conf. Proc.* **2023** 020093
[20] Nurraihana H and Norfarizan H N A 2013 *Int. Food Res. J.* **20** 2045–56
[21] Ismail M, Manickam E, Danial A M, Rahmat A and Yahaya A 2000 *J. Nutr. Biochem.* **11** 536–42
[22] Shashank K and Pandey A K 2013 Hindawi *Sci. World J.* **2013** 16750
[23] Ustundag O G and Mazza G 2007 *Crit. Rev. Food Sci. Nutr.* **47** 231–58
[24] Sangster A W 1960 *J. Chem. Educ.* **37** 454
[25] Subhan M A, Fahim A M M, Saha P C, Rahman M M, Begum K and Azad A K 2017 *Nano-Structures & Nano-Objects* **10** 30–41
[26] Houas A, Lachheb H, Ksibi M, Elaloui E, Guillard C and Herrmann J M 2001 *Appl. Catal.* B **31** 145–57
[27] Korshin G, Chow C W K, Fabris R and Drikas M 2009 *Water Res.* **43** 1541–8

IOP Conf. Series: Materials Science and Engineering **763** (2020) 012005 doi:10.1088/1757-899X/763/1/012005

Synthesis and characterization of hydrogel NaCMC -g-poly(AA-co-AAm) modified by rice husk ash as macronutrient NPK slow-release fertilizer superabsorbent

G Andani and H Helmiyati

Department of Chemistry, Faculty of Mathematics and Natural Sciences (FMIPA), Universitas Indonesia, Kampus UI Depok, Depok 16424, Indonesia

Corresponding author's e-mail: helmiyati@sci.ui.ac.id

Abstract. In this study, sodium carboxymethyl cellulose (NaCMC) based hydrogel grafted by poly (acrylate-acrylamide) by addition of rice husk ash (RHA) and NPK fertilizer to the superabsorbent with in-situ polymerization was successfully synthesized. In this work, the effect of NaCMC concentration and mass of RHA was done. The best water swelling capacity with 3 % NaCMC, 15 % RHA is 480 g/g with percentage of water release capacity is 38 %. Then, the effect of NPK fertilizer concentration on in-situ polymerization was carried out and the best swelling capacity with 5 % NPK mass is 650.10 g/g and percentage of release capacity obtained is 25.20 % for 6 days. The successful of the synthesis of hydrogels NaCMC-g-poly(AA-co-AAm) modified by rice husk ash as macronutrient NPK slow-release fertilizer superabsorbent supported by functional group was characterized by Fourier Transform Infrared (FTIR) and surface morphology by Scanning Electron Microscopy (SEM). The hydrogels based on NaCMC biopolymers and silica from rice husk ash as macronutrient NPK slow-release fertilizer superabsorbent produced slow release can be controlled and a promising material in the future and can be applied in agriculture field.

Keywords: acrylamide, acrylic acid, hydrogel, rice husk ash, superabsorbent

1. Introduction

Superabsorbent hydrogel is a three-dimensional crosslinked polymer network, which can absorb and retain liquid within its structure [1]. In agriculture sector, especially in arid areas, the usage of superabsorbent hydrogel can reduce irrigation water consumption and plants death rate, and also increase soil's fertility [2]. In its application, adding fertilizers into superabsorbent hydrogel may increase the nutrition absorbed by plants [3].

The use of chemical fertilizers is one of the most vital factors in agricultural production [3]. To improve the products quality in agriculture, the role of chemical fertilizer is very important [4]. However, an excessive usage of chemical fertilizer may pollute the surrounding environment [5]. One of the most effective ways to overcome those issues is by modifying them into slow-release fertilizer superabsorbent in order to release nutrient in a controlled rate [6].

The utilization of polysaccharides such as cellulose has been reported as a useful material to synthesize superabsorbent hydrogel for nutrient release in plants [7]. Cellulose is an abundant, inexpensive, renewable, and biodegradable biopolymer; hence it is often used as the backbone of a hydrogel [8,9]. However, intermolecular and intramolecular hydrogen bonds among hydroxyl groups in the backbone not only limit its solubility in water, but also make cellulose less reactive. Therefore, to increase water solubility of cellulose derivatives such as carboxymethyl cellulose is used [10]. Superabsorbent that only consists of organic materials has some weaknesses, e.g. low absorption capacity, weak mechanical strength, and prone to the change of temperature. Therefore, inorganic materials such as montmorillonite, attapulgite, kaolin, and bentonite are often used in the preparation of superabsorbent [2, 11–13]. However, recently, the use of silica as composite has drawn attention because it can be produced from rice husk ash (RHA) by calcination [14]. The presence of silica nanoparticles within the hydrogel matrix improves its swelling capacity, which is an important factor in agricultural application due to hydrophilicity effect given by silica nanoparticles to enhance water retention ability of the hydrogel composite. The interactions of hydrogen bonds between functional groups of hydrogel and hydroxyl groups of silica nanoparticles create additional physical crosslink within the hydrogel network thus increases its strength [3,15]. Moreover, silica is an essential micronutrient for plants and is easily absorbed when hydrogel has degraded in the soil medium. Silica highly affects the plants growth because silica leads to the nutrient's imbalance within the plant. In addition, silica can also improve plants resistance to pests and cold weather [16,17].

In this study, a synthesis method of hydrogel superabsorbent nanocomposite by using sodium carboxymethyl cellulose, which was more soluble in water, is proposed, where our previous researches used cellulose [18,19]. The used silica nanoparticles from rice husk ash as filler in hydrogel matrix improved its swelling capacity and controlled release [20]. In this work, the effect concentration of NaCMC, RHA and NPK to superabsorbent nanocomposite fertilizers in-situ NPK was done and the swelling-release capacity for water and release NPK fertilizer were studied. The superabsorbent nanocomposite was characterized using Fourier Transform Infrared (FTIR) and Scanning Electron Microscopy (SEM).

2. Experiments

2.1. Materials
The materials used were sodium carboxy methyl cellulose, NaCMC (Merck Co.) as backbone, rice husk from Bukittinggi, HCl (Merck Co.) and H_2SO_4 (Merck Co.) for nanosilica preparation from rice husks, nitrogen gas to eliminate dissolved oxygen, acrylic acid (Merck Co.) and acrylamide (Merck Co.) as monomers, N,N'-methylenebisacrylamide (Sigma) as crosslinking agent, potassium persulfate (KPS) as initiator, ethanol (Merck Co.) and acetone (Merck Co.) to rinse the superabsorbent, and NPK fertilizer as urea, potassium phosphate and ammonium phosphate (Merck Co.)

2.2. Nanosilica preparation from rice husk
Nanosilica has been synthesized according the protocol developed by previous researcher [20]. Rice husks were washed with distilled water. Then, the solution was dipped in HCl 3 % and heated by reflux at 100 °C for 2 hours and stirred. The same process was repeated by replacing HCl 3 % with H_2SO_4 10 % and filtered, washed with distilled water. Finally, rice husks were calcinated in a furnace at 900 °C, rice husk ash (RHA) was obtained.

2.3. Superabsorbent nanocomposite synthesis
Superabsorbent nanocomposite has been synthesized according the protocol developed by previous researcher [10]. NaCMC powder was dissolved in distilled water then RHA was added while being stirred until the mixture thickened. The mixture was then flowed by nitrogen gas at 60 °C and was added by potassium persulfate (KPS). Then, acrylic acid (AA), acrylamide (AAm) and N,N'-methylenebisacrylamide (MBA) were added into the solution. The solution was heated at 70 °C for 2 hours in order for the polymerization reaction to be completed. Then, the product was immersed in acetone for 24 hours. In this work, the effect of NaCMC mass (1, 2, 3) wt.% with code; SC1, SC2, SC3 and RHA (10,15, 20) wt.% with code; SC4, SC5 and SC6, it was observed the swelling-release capacity for water, NPK fertilizer and was characterized by FTIR and SEM.

2.4. Synthesis of superabsorbent nanocomposite with in- situ NPK as slow-release fertilizer

A similar procedure from our previous work was carried out to synthesize slow-release fertilizers by adding NPK fertilizer during polymerization [13]. In this work, the effect of NPK mass (3, 4, 5) wt.% with code; SC7, SC8 and SC9, respectively with the same composition of N (urea), K (potassium phosphate) and P (ammonium phosphate).

2.5. Swelling-release capacity determination

In this work, the water swelling capacity was determined by gravimetric method, 0.1 g superabsorbent was immersed in 200 mL of distilled water at room temperature with a certain time for 8640 minutes. Superabsorbent was then removed from the absorbate to measure its weight. The calculation was done by the similar equation from our previous work [13].

To determine the release capacity, swollen superabsorbent was immersed in 200 mL of 0.9 % NaCl solution, and then weighed in a certain time. Release capacity was calculated with the following equation:

$$\textit{Water release capacity} = \frac{w_0 - w_t}{w_o} \; x \; 100 \; \% \qquad (1)$$

where w_o is the superabsorbent weight before release and w_t is the superabsorbent weight after release at a certain time

Meanwhile, to determine N, P and K release capacity, the similar procedure was the same as water release capacity, but the concentration of N, P and K was determined by the visible spectrophotometer. The release capacity of N, P and K was calculated with the following equation:

$$\textit{N, P and K release capacity} = \frac{C_1 - C_o}{C_o} \; x \; 100 \; \% \qquad (2)$$

where C_o is the superabsorbent concentration before release and C_t is the superabsorbent concentration after release at a certain time

3. Results and discussion

3.1. Swelling-release capacity

3.1.1. Water swelling capacity. Figure 1a shows the swelling capacity to time with various concentrations of NaCMC (SC1, SC2, SC3) and RHA (SC4, SC5, SC6). The optimal swelling capacity increased by the rising of NaCMC concentration. It can be seen that superabsorbent nanocomposite with NaCMC composition of 3 wt.% has the best swelling capacity in water is 420.0 g/g (SC3). This is due to the more backbone added; the more monomers are grafted so that the active groups and swelling capacity were increased. The swelling capacity increased with the increasing of RHA weight and obtained optimum RHA 15 wt.% is 480 g/g. Addition of rice husk ash (RHA) to the superabsorbent not only increases its mechanical strength, but can also increase the swelling capacity because the electrostatic repulsion force between the –COO- group of polymer chains and the negatively charged surface of SiO_2 nanoparticles causes the superabsorbent network to be easily expanded. However, in RHA 20 wt.% there is a decrease in swelling capacity. This is caused by the excessive interaction between the surface of the -OH group of silica nanoparticles and the superabsorbent hydrogel polymer component, so that the more of the physical crosslinking increases. Therefore, the cavities between the polymer chains are reduced, resulting in the decrease swelling capacity [3]. Figure 1b shows the maximum swelling capacity of superabsorbent nanocomposite at equilibrium time of 8000 minutes in water of SC1, SC2, SC3, SC 4, SC5 and SC6 are 300.60, 350.50, 420.10, 400.60, 480.30 and 450.20 g/g, respectively. It can be seen that the best swelling capacity is obtained at 480.30 with a composition of NaCMC 3 wt.% and RHA 15 wt.% (SC5).

Swelling capacity of slow release superabsorbent nanocomposite fertilizers in-situ NPK with various NPK mass; (3, 4, 5) wt.% (SC7, SC8, SC9) in water is shown in figure 2. Figure 2a shows water swelling capacity to time with various mass of NPK. The optimal swelling capacity increases by the rising of NPK mass. The presence of NPK in the superabsorbent increases the swelling capacity. This is due to the hydrophilic of the NPK group. Figure 2b shows the maximum swelling capacity of superabsorbent nanocomposite in-situ NPK at equilibrium time of 8000 minutes in water of SC7, SC8 and SC9 are

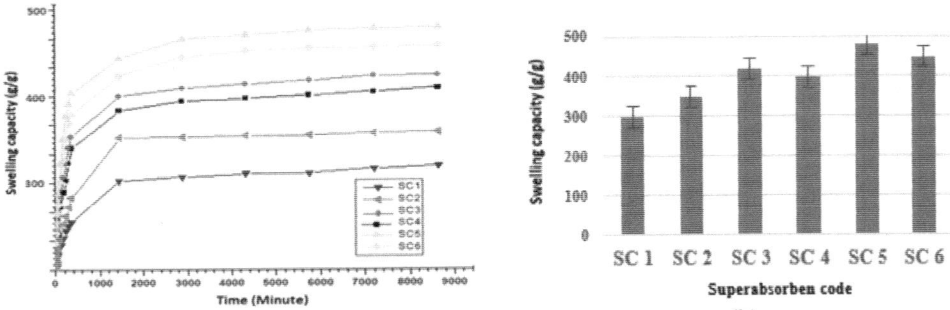

(a) (b)

Figure 1. (a) Swelling capacity to time and (b) optimum swelling capacity SC1 to SC6

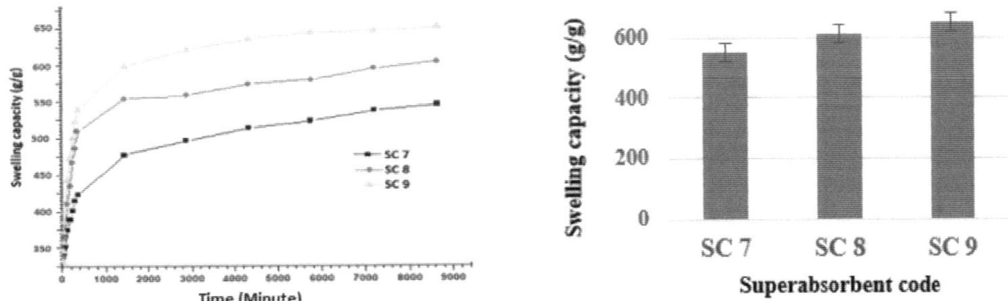

(a) (b)

Figure 2. (a) Swelling capacity to time and (b) optimum swelling capacity SC7 to SC9

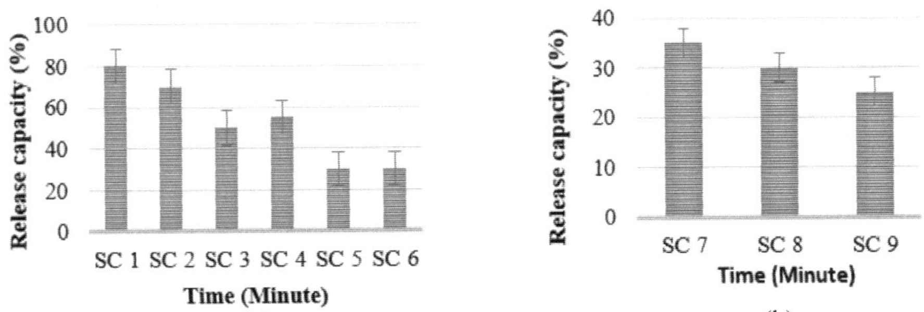

(a) (b)

Figure 3. Water release capacity of (a) SC1 to SC6 and (b) SC7 to SC9

550.30, 610.30 and 650.10 g/g, respectively. It shows that the best swelling capacity is obtained at 650.10 with composition of NaCMC 5 wt.%, RHA 15 wt.% and NPK 5 wt.% (SC9).

3.1.2. Release capacity. Nanocomposite superabsorbents that have swollen at the maximum swelling capacity are determined by the release capacity of water. Release capacity is determined to obtain a controlled slow release superabsorbent so that the superabsorbent can hold water to be released in certain time period. The maximum release capacity of SC1 to SC6 superabsorbent nanocomposite is shown in figure 3a and the superabsorbent nanocomposite SC7 to SC9 in figure 3b. Figure 3a shows the optimal

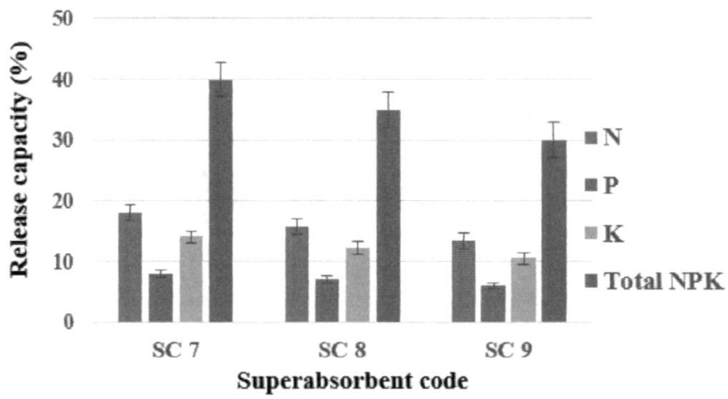

Figure 4. NPK Fertilizer Release Capacity of SC7 to SC9

release capacity at time 8640 minutes or 6 days with various concentrations of NaCMC (SC1, SC2, SC3) and RHA (SC4, SC5, SC6). The optimal release capacity decreased by the rising of NaCMC concentration. It shows that superabsorbent hydrogel with NaCMC composition 3 wt.% has the best water release capacity percentage of 50 % (SC3). The release capacity percentage decreased with the increasing of RHA weight and obtained the optimum RHA 15 wt.% of 38 %. Addition of rice husk ash (RHA) to the superabsorbent not only increases its mechanical strength, but can also decrease the release capacity. This is caused by the interaction between the surface of the -OH group of silica nanoparticles and the superabsorbent hydrogel polymer component, so that the more of the physical crosslinking, so that the ability to hold water is better and caused a decrease of water release capacity. Figure 3b shows the maximum release capacity of superabsorbent nanocomposite with various NPK concentrations of SC7, SC8, SC9 are 35.20, 30.10, 25.15 %, respectively. The release capacity percentage decreased with the increasing of NPK weight, it shows that the best release capacity percentage is obtained at 25.15 % with a composition 3 % NaCMC, 15 % RHA and 5 % NPK (SC9) at 6 days.

Figure 4a shows that the optimal NPK fertilizer release capacity at time 8640 minutes or 6 days with various NPK mass. The total NPK release capacity percentage decreased with the increasing of NPK weight of SC7, SC8 and SC9 are 40.10, 35.15 and 30.10, respectively. This is caused by interaction between the superabsorbent hydrogel polymer components with NPK fertilizer so that the ability to hold water is better so that decreases the water release capacity. The NPK fertilizer and water are released simultaneously so the total NPK fertilizer release capacity is also decreased. It shows that the best of release capacity percentage obtained is 25.20 % with the composition 3 % NaCMC, 15 % RHA and 5 % total NPK (SC9) at 6 days.

3.2. Characterisation

3.2.1. Characterisation using FTIR. Functional groups analysis is shown in figure 5. Figure 5a shows that the peak appeared at 470 cm^{-1} is related to Si-O-Si bending vibration. Meanwhile, the peaks at 805 and 1104 cm^{-1} are related to symmetric and asymmetric vibrations of Si-O-Si, respectively [15]. Absorption band that appeared in rice husk ash's spectrum in 3400-3600 cm^{-1} is related to stretching vibration of –OH of silanol [14]. In the same wavenumber range, a wide absorption band also appeared in spectrum (figure 5b) related to –OH stretching vibration [10]. A peak –COO appeared in 1597 cm^{-1} and the peaks at 1413 and 1061 cm^{-1}, the absorption bands –CH_2 scissoring vibration and CH–O–CH_2 stretching, respectively [21]. Figure 5c shows the spectrum of superabsorbent nanocomposite, a peak at 1628 cm^{-1} is related to –COO of acrylic acid (AA) and the peaks at 3200-3600 cm^{-1} weakened, it shows –OH from each of its components, the peak weakened due to polymerization reaction occurs which indicates the formation of a nanocomposite hydrogel. Figure 5d shows pure NPK fertilizer and figure 5e is superabsorbent nanocomposite in-situ NPK. From these two spectra peaks, it shows broad absorption bands at 3100-3600 cm^{-1} related to –OH and overlapped with –NH group, it shows that the spectrum of pure NPK having sharp

IOP Conf. Series: Materials Science and Engineering **763** (2020) 012005 doi:10.1088/1757-899X/763/1/012005

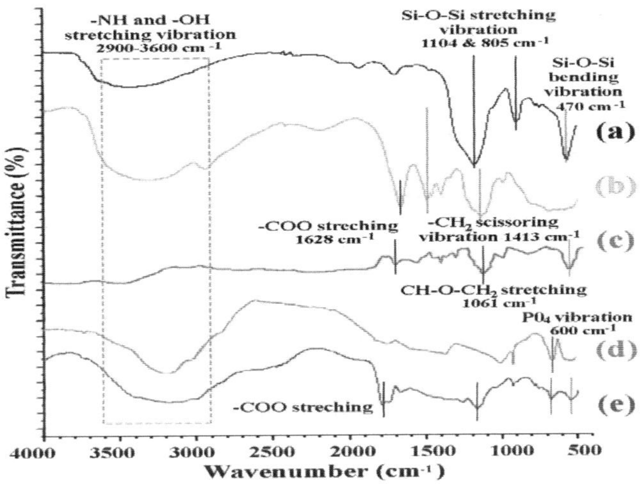

Figure 5. FTIR Spectra of (a) RHA, (b) NaCMC, (c) SC5, (d) pure NPK fertilizer and (e) SC9

(a)

(b)

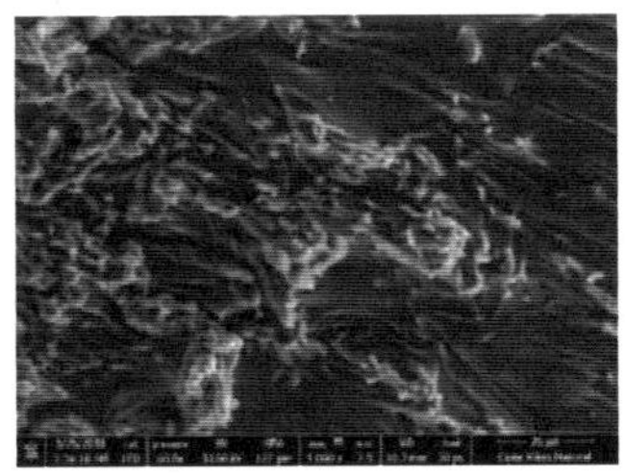
(c)

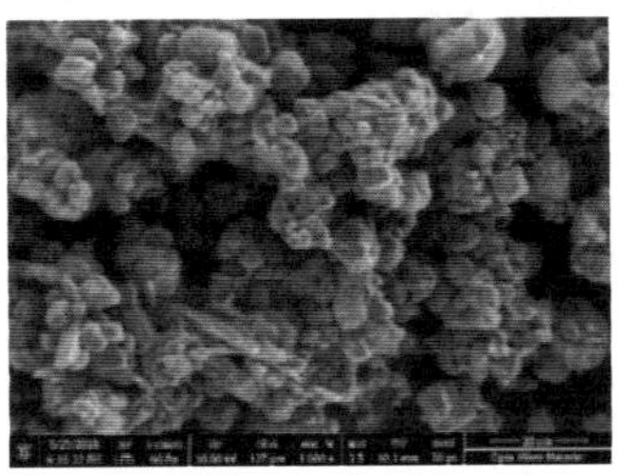
(d)

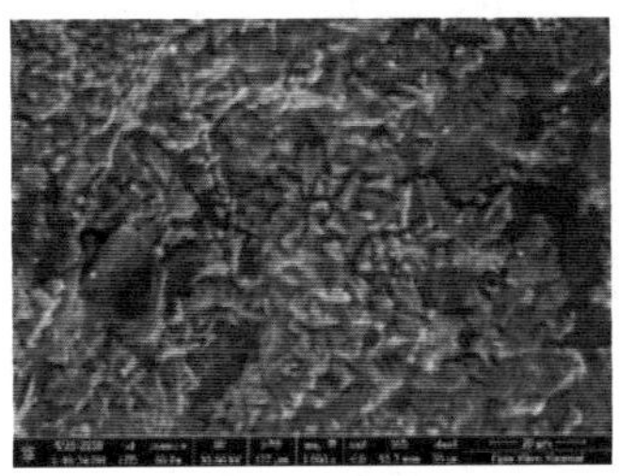
(e)

Figure 6. Micrographs of (a) RHA, (b) NaCMC, (c) SC5, (d) pure NPK fertilizer and (e) SC9

peaks. While in the superabsorbent nanocomposite in-situ NPK in figure 5e obtained the peak weaken and widen at 3100-3600 cm^{-1}, this indicates there are interaction between NPK fertilizer and components of superabsorbent. In the peak of 1719 cm^{-1} related to –COO stretching vibrations from carboxyl methyl cellulose and acrylic monomers. The peak finger print region at 550 cm^{-1} related to PO_4 group from NPK fertilizer and on superabsorbent nanocomposite in-situ NPK shows weak peak because there is interaction with the component of superabsorbent. It supported that the successful synthesis of in-situ NPK nanocomposite superabsorbents.

3.2.2. Characterization using SEM. Morphology analysis using SEM is shown in figure 6. Figure 6a shows that the RHA has particles with irregular size and shape and with rough surface with large porosity. Figure 6b shows NaCMC has smooth and uniform fiber that is evenly spread. While figure 6c shows that superabsorbent nanocomposites which are a combination of NaCMC, RHA and other components have smooth surface and it can be observed that fibers from NaCMC as backbone and RHA as fillers are evenly distributed with fine pores. Figure 6d shows NPK fertilizer has an evenly distributed granule-like surface derived from urea as N, potassium phosphate as K and ammonium phosphate as P sources. Figure 6e shows superabsorbent nanocomposite in-situ NPK has smooth pores and NPK fertilizer distributed on the surface has the similarity of pure NPK fertilizers (figure 6d). The surface morphology by SEM proves that the NPK fertilizer has been successfully encapsulated into the superabsorbent nanocomposites.

4. Conclusions

In this study, we developed superabsorbent nanocomposite and slow-release fertilizers in the form of superabsorbent through in-situ polymerisation. The result shows that the optimal NaCMC and RHA composition are 3 wt.% and 15 wt.%, respectively. Superabsorbent nanocomposite in-situ is synthesized by varying the NPK fertilizers mass, increasing of mass lead to increase the swelling capacity and decreasing release capacity. The optimal swelling capacity of superabsorbent nanocomposite with in-situ NPK in water is 650.10 g/g. The optimal total NPK release capacity of superabsorbent nanocomposite with in-situ NPK is 25.20 %. The successful of the synthesis of hydrogels nanocomposite in-situ NPK fertilizer superabsorbent supported by functional group were characterized by FTIR and surface morphology by SEM. The hydrogels NaCMC-g-poly(AA-co-AAm) modified by rice husk ash as macronutrient NPK slow-release fertilizer superabsorbent that are successfully synthesized show that NPK fertilizers slow release can be controlled and could be an interesting material for the future application in agriculture.

Acknowledgements

This research was funded by PITTA Grant 2018 from Universitas Indonesia.

References

[1] Alla S G A, Sen M and El-Naggar A W M 2012 *Carbohydr. Polym.* **89** 478–85
[2] Wu L, Liu M and Liang R 2008 *Bioresour. Technol.* **99** 547–54
[3] Olad A, Gharekhani H, Mirmohseni A and Bybordi A 2017 *Polym. Bull.* **74** 3353–77
[4] Zhang S, Guan Y, Fu G, Chen B, Peng F, Yao C and Sun R 2014 *Journal of Nanomaterials* **2014** ID 675035
[5] Rashidzadeh A and Olad A 2014 *Carbohydr. Polym.* **114** 269–78
[6] Hemvichian K, Chanthawong A and Suwanmala P 2014 *Radiat. Phys. Chem.* **103** 167–71
[7] Zhong K, Lin Z T, Zheng X L, Jiang G B, Fang Y S, Mao X Y and Liao Z W 2013 *Carbohydr. Polym.* **92** 1367–76
[8] Chang C and Zhang L 2011 *Carbohydr. Polym.* **84** 40–53
[9] Sannino A, Demitri C and Madaghiele M 2009 *Materials* **2** 353–73
[10] Bao Y, Ma J and Li N 2011 *Carbohydr. Polym.* **84** 76–82
[11] Liang R and Liu M 2006 *Ind. Eng. Chem. Res.* **45** 8610–6
[12] Schexnallder P and Schmidt G 2009 *Colloid Polym. Sci.* **287** 1–11
[13] Helmiyati and Syarifudin A 2018 *AIP Conf. Proc.* **2023** 020080
[14] Chandrasekhar S, Pramada P N and Praveen L 2005 *J. Mater. Sci.* **40** 6535–44
[15] Cândido J S, Leitão R C, Ricardo N M P S, Feitosa J P A, Muniz E C and Rodrigues F H A 2012 *J. Appl. Polym. Sci.* **123** 879–87
[16] Suriyaprabha R, Karunakaran G, Yuvakkumar R, Prabu P, Rajendran V and Kannan N 2012 *J. Nanopart. Res.* **14** 1294
[17] Yuvakkumar R, Elango V, Rajendran V, Kannan N S and Prabu P 2011 *Int. J. Green Nanotechnol.* **3** 180–90

[18] Helmiyati and Aprilliza M 2017 *IOP Conf. Ser.: Mater. Sci. Eng.* **188** 012019.
[19] Helmiyati, Fitriyani A and Meyanti F 2017 *IOP Conf. Ser.: Mater. Sci. Eng* **188** 012051
[20] Gharekhani H, Olad A, Mirmohseni A and Bybordi A 2017 *Carbohydr. Polym.* **168** 1–13
[21] Biswal D and Singh R 2004 *Carbohydr. Polym.* **57** 379–87

Synthesis of imidazoline-palmitic derivative using MAOS (microwave assisted organic synthesis) method

S Budiana, B Andreana, D U C Rahayu, D A Nurani, Y K Krisnandi and E Budianto

Department of Chemistry, Faculty of Mathematics and Natural Sciences (FMIPA), Universitas Indonesia, Kampus UI Depok, Depok 16424, Indonesia

Corresponding author's e-mail: dyahutamicr@sci.ui.ac.id

Abstract. In this work, imidazoline-palmitic derivative had been successfully synthesized by reacting triethylenetetramine (TETA) with palmitic acid (PA). Synthesis process was performed with solvent free using MAOS (Microwave Assisted Organic Synthesis) method with variation of time reactions. Dean-Stark and reflux methods using xylene as a solvent were also performed as comparisons. The percentage yield of TETA-PA imidazoline was obtained from Dean-Stark at 13 hours (95.80 %), reflux at 13 hours (88.53 %), and MAOS at 11 min (87.74 %). TETA-PA imidazoline then was identified by using thin layer chromatography and melting point analyzer followed by characterization using Fourier Transform Infrared (FTIR), UV-Vis, Proton Nuclear Magnetic Resonance (^{1}H-NMR), and Mass Spectrometry (MS) spectral data. This study provides information regarding comparison method for imidazoline derivative synthesis. For further research, TETA-PA imidazoline can be used as corrosion inhibitor towards carbon steel.

Keywords: corrosion inhibitor, imidazoline, MAOS,palmitic acid, triethylenetetramine.

1. Introduction

Corrosion is a natural process occurring in metals including carbon steel in petroleum and natural gas mining industries. The dispensing pipes are components which prone to corrosion because of corrosive media such as CO_2 and H_2S. To reduce the corrosion effect, corrosion inhibitors are given regularly to dispensing pipes. Corrosion inhibitors, which widely used in industry, are organic compound containing nitrogen, sulfur, or oxygen atom on its structure [1].

Imidazoline is a five-ring heterocyclic compound having two nitrogen atoms. According to its structure, imidazoline is often used as corrosion inhibitor due to its good adsorption properties and its ability to form layer on metal surface [2]. In the last decade, imidazoline synthesis using polyamines and fatty acid has grown rapidly since MAOS (Microwave Assisted Organic Synthesis) as novel method of imidazoline synthesis was invented. MAOS can facilitate green chemistry with solvent free, faster reaction time, and higher yield compared with reflux method [3].

Based on previous research, it is important to modify imidazoline structure in order to increase the corrosion inhibition activity [1, 4–5]. Therefore, in this research, imidazoline-palmitic derivative was synthesized by reacting triethylenetetramine (TETA) and palmitic acid (PA) using MAOS method with reflux and Dean Stark methods as comparison. PA was chosen because this compound is one of major components in tall oil, a commercial fatty acid mixture for corrosion inhibitor [2]. The aims of this study are to compare various methods on synthesizing imidazoline derivative and determine the structure of synthesized imidazoline using various spectral data. For further research, imidazoline-palmitic derivative can be developed as corrosion inhibitor towards carbon steel.

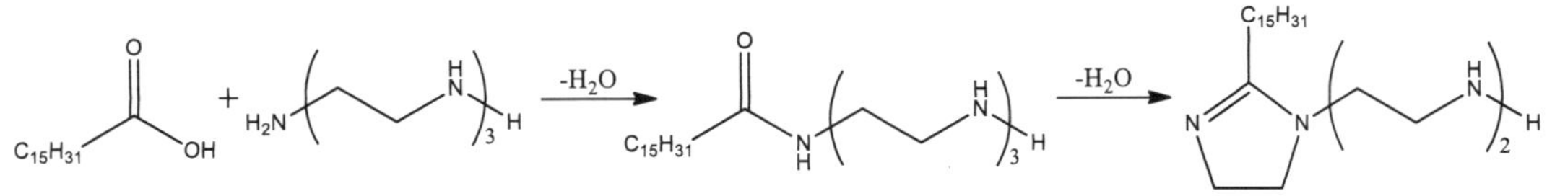

Figure 1. Reaction scheme of TETA-PA imidazoline synthesis

2. Materials and methods

2.1. General

The chemicals used were palmitic acid (PA), xylene, Thin-Layer Chromatography (TLC) plates, CH_2Cl_2, MeOH, NH_4OH 25 %, $Ce(SO_4)_2$, H_2SO_4 98 %, and Na_2SO_4 anhydrous purchased from E. Merck, while triethylenetetramine (TETA) from Sigma Aldrich. Synthesis of imidazoline-palmitic derivative, TETA-PA imidazoline, using MAOS was carried out by using microwave oven domestic Samsung ME-731K/XSE. Melting point was measured using Electrothermal Melting Temp. Apparatus. Structure elucidation of TETA-PA imidazoline was conducted using Fourier Transform Infrared (FTIR) Shimadzu IR Prestige 21, UV-Vis Shimadzu UV-2450, Proton Nuclear Magnetic Resonance (^{1}H-NMR, Agilent 500 MHz), and Mass Spectrometry (MS, Shimadzu Prominance 20-Abesciex).

2.2. Synthesis of TETA-PA imidazoline using reflux and Dean Stark methods

TETA (5 mmol) and PA (5 mmol) were mixed with 20 mL xylene in a three-neck round-bottom flask. The mixture was heated using reflux at 100–230 °C for 13 hours. Reaction was monitored by TLC with CH_2Cl_2:MeOH 8:2 and one drop of NH_4OH 25 % as an eluent, visualized under UV_{254}, and stained with $Ce(SO_4)_2/H_2SO_4$. After reaction was completed, the mixture then was evaporated to remove water as by-product and xylene as solvent. The mixture was dissolved in CH_2Cl_2 and extracted three times with aquabidest. The organic layer then was collected, dried with Na_2SO_4 anhydrous, evaporated, and stored in desiccator. TETA-PA imidazoline then was characterized using FTIR and UV-Vis. A similar reaction was also conducted using Dean Stark method by using Barret distilling receiver on the top of condenser.

TETA-PA imidazoline from reflux method was obtained as dark brown oily solid, *Rf* 0.70 in CH_2Cl_2:MeOH 8:2 with one drop of NH_4OH 25 %; FTIR with KBr ν_{max} (cm^{-1}): 3303.2, 3075.4, 2917.2, 2849.9, 1647.3, 1551.8, 1467.3; UV-Vis in MeOH λ_{max} (nm): 230, 270 (*sh*).

TETA-PA imidazoline from Dean Stark method was obtained as dark brown oily solid, *Rf* 0.70 in CH_2Cl_2:MeOH 8:2 with one drop of NH_4OH 25 %; FTIR with KBr ν_{max} (cm^{-1}): 3304.2, 3078.3, 2918.4, 2850.0, 1656.9, 1550.8, 1467.9; UV-Vis in MeOH λ_{max} (nm): 230, 270 (*sh*).

2.3. Synthesis of TETA-PA imidazoline using MAOS method

TETA (5 mmol) and PA (5 mmol) were mixed in a porcelain crucible with no solvent added. The mixture then was heated to microwave (800 watt) for 9, 11, 13, and 15 min. Similar monitoring, working up and characterization were also conducted. Furthermore, TETA-PA imidazoline from MAOS method was subjected to determine its melting point and additional characterization using ^{1}H-NMR and MS.

TETA-PA imidazoline from MAOS method was obtained as dark brown oily solid, *Rf* 0.69 in CH_2Cl_2:MeOH 8:2 with one drop of NH_4OH 25 %; mp. 77-29 °C; FTIR with KBr ν_{max} (cm^{-1}): 3303.2, 3076.5, 2915.7, 2850.0, 1638.6, 1551.8, 1467.8; UV-Vis in MeOH λ_{max} (nm): 230; ^{1}H-NMR (500 MHz, $CDCl_3$) δ_H (ppm): 3.17-3.50 (4H, *m*), 2.65-2.88 (4H, *m*), 2.31-2.59 (4H, *m*), 2.15 (2H, *t*, *J* = 7.0 Hz), 1.59 (2H, *p*, *J* = 7.0 Hz), 1.22-1.26 (24H, *m*), 0.85 (3H, *t*, *J* = 7.0 Hz); MS $(M+H)^+$: 367.3316 m/z.

3. Results and discussion

In this present work, imidazoline-palmitic derivative synthesis was performed using MAOS method and compared with other methods i.e. reflux and Dean Stark methods. In general, imidazoline synthesis consisted of two-stage reaction, i.e. amidation to form amide and cyclization to form imidazoline ring (figure 1). This two-stage reaction eliminates H_2O as by-product in both stages. To our knowledge, although this is not the first reported TETA-PA imidazoline synthesis [6], this is the first reported results on comparison methods of TETA-PA imidazoline synthesis along with its ^{1}H-NMR and MS spectral data.

IOP Conf. Series: Materials Science and Engineering **763** (2020) 012006 doi:10.1088/1757-899X/763/1/012006

Table 1. Comparison of TETA-PA imidazoline synthesis using various methods.

Parameters	MAOS	Reflux	Dean Stark
Reaction Time	11 min	13 h	13 h
Temperature (°C)	200	180	170
Solvent	-	Xylene	Xylene
% Yield	87.74	88.53	95.80

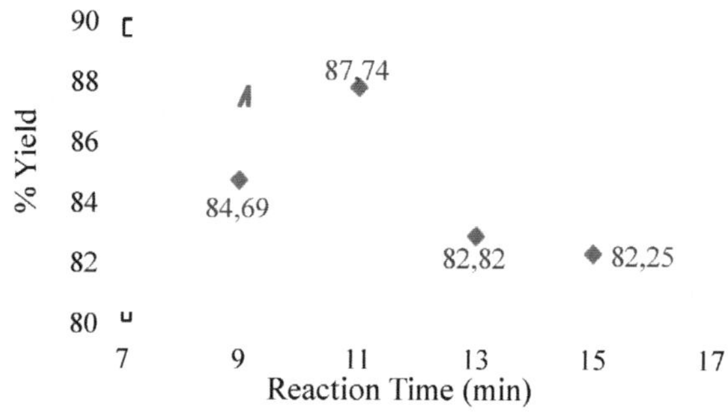

Figure 2. The percentage yield of TETA-PA imidazoline from MAOS method

Synthesis of TETA-PA imidazoline using MAOS method was carried out with variation of reaction time (9, 11, 13, and 15 min). Figure 2 showed the percentage yield of TETA-PA imidazoline as a function of reaction time. According to figure 2, the highest yield of TETA-PA imidazoline from MAOS method was obtained at 11 min with 87.74 %. TETA-PA imidazoline was yielded as dark brown oily solid with the IUPAC name of N^1-(2-(2-pentadecyl-4,5-dihydro-1H-imidazol-1-yl)-ethyl)-ethane-1,2-diamine. TETA-PA imidazoline showed melting point at 77-79 °C when measured with melting point analyzer, which indicated imidazoline derivative, was produced [2].

Comparison of TETA-PA imidazoline synthesis using MAOS, reflux, and Dean Stark methods were summarized in table 1. According to table 1, TETA-PA imidazoline from reflux and Dean Stark methods exhibited higher yield than MAOS method. According to previous research, the highest yield of synthesized imidazoline using various polyamines and fatty acids was 94.24 % for MAOS method whereas reflux method yielded a lower yield of 92.37 % [4]. The difference of experimental results may be caused from incomplete evaporation process of xylene and water as by-product. Even though MAOS method produced the lowest yield compared to others, MAOS method still revealed the percentage yield in above 80 %, facilitated shorter reaction time, reached higher temperature, and provided green chemistry with solvent free synthesis. Therefore, it can be concluded that MAOS method is the most effective and efficient method on synthesizing imidazoline derivative.

TLC analysis of TETA-PA imidazoline from MAOS, reflux and Dean Stark methods showed a similarity on *Rf* value of 0.69-0.70. Furthermore, FTIR spectra of TETA-PA imidazoline from various methods also showed the same pattern indicating similar functional groups (figure 3a). Based on figure 3a, the absorption peaks of N-H, C-H sp^3, C=N, and C-N-C bonds were appeared at 3075-3304, 2850-2918, 1639-1657, and 1467 cm^{-1}, respectively. These FTIR spectra revealed the characteristic absorption peaks for imidazoline and showed an agreement with previous research [2, 4–6]. UV-Vis spectra of TETA-PA imidazoline from various methods (figure 3b) also showed peaks appeared at the same wavelength of 230 nm which characteristic of C=N absorbance peak in imidazoline ring [2]. A shoulder absorbance peak at 270 nm on reflux and Dean Stark method indicated that xylene was still presented. Based on TLC analysis, FTIR, and UV-Vis spectra, three methods of TETA-PA imidazoline synthesis afforded the same compound.

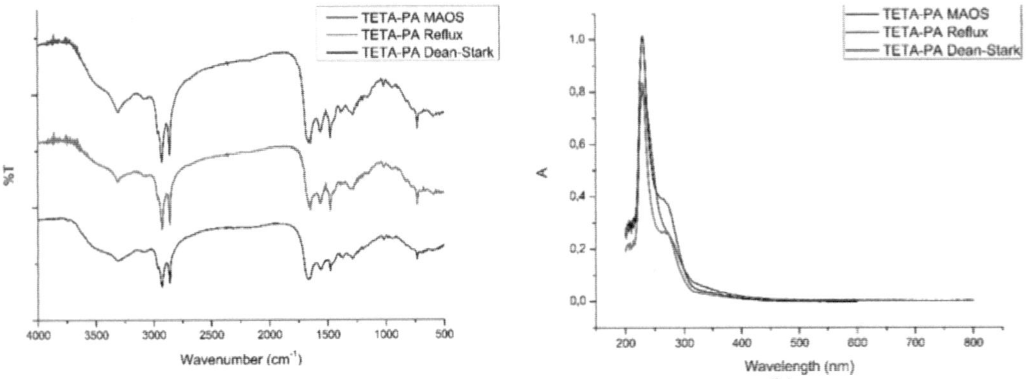

(a) (b)

Figure 3. Comparison of (a) FTIR and (b) UV-Vis spectra of TETA-PA imidazoline from MAOS, reflux, and Dean Stark methods.

(a)

(b)

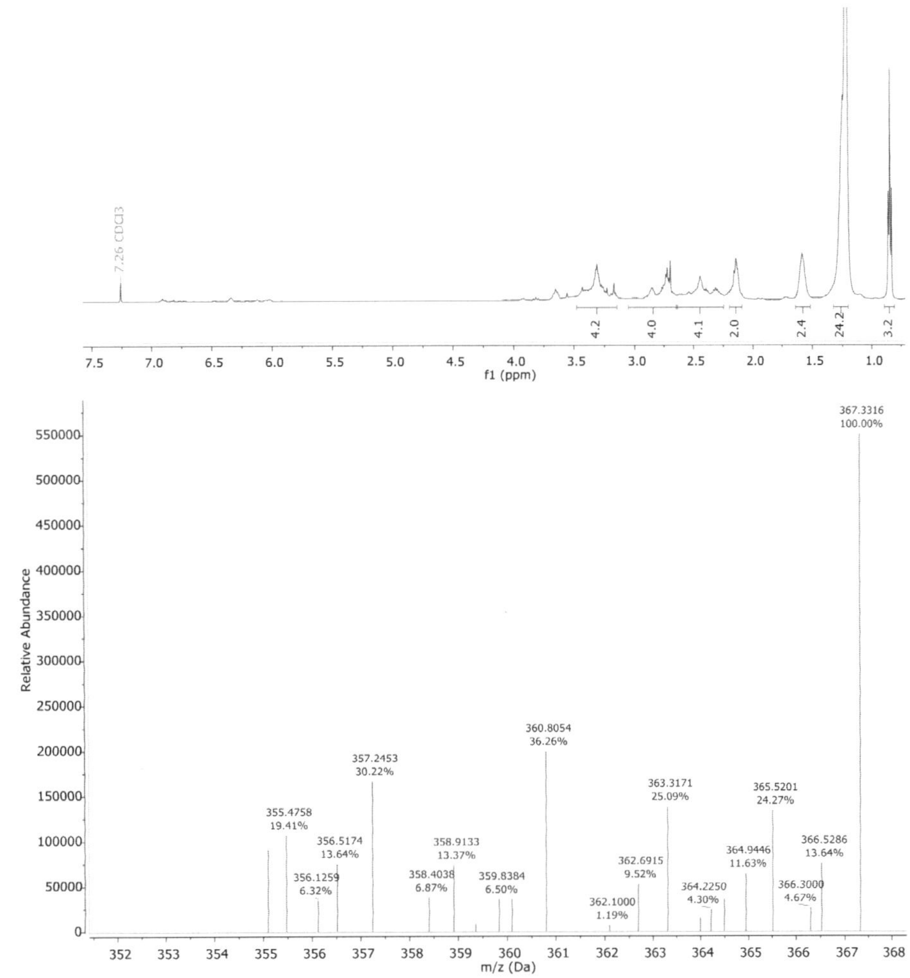

Figure 4. (a) ¹H-NMR in 500 MHz ($CDCl_3$) and (b) MS spectra of TETA-PA imidazoline from MAOS method

Since TLC analysis, FTIR, and UV-Vis spectral data revealed that TETA-PA imidazoline from MAOS, reflux, and Dean Stark methods was similar compound, therefore ^{1}H-NMR and MS spectral data to elucidate its structure were recorded only for TETA-PA imidazoline from MAOS method (Figure 4). According to ^{1}H-NMR spectra of TETA-PA imidazoline (500 MHz, $CDCl_3$) in figure 4a, the characteristic peak of proton in imidazoline ring appeared at chemical shift of 3.17-3.50 ppm. Other peaks were showed at chemical shift of 2.31-2.88 and 0.88-2.15 ppm indicated proton of pendant groups and hydrocarbon chains, respectively. ^{1}H-NMR spectra of TETA-PA imidazoline showed the similar pattern with ^{1}H-NMR spectra of imidazoline derivative from saturated fatty acid [4]. Furthermore, MS spectra of TETA-PA imidazoline in figure 4b shows $[M+H]^+$ at 367.3316 m/z which correspond to TETA-PA imidazoline, $C_{20}H_{46}N_4$. Based on all spectral data, therefore TETA-PA imidazoline had a structure as illustrated in figure 1. TETA-PA imidazoline is expected to be applied as corrosion inhibitor towards carbon steel in further research.

4. Conclusions

This work had studied imidazoline-palmitic derivative synthesis by comparing three methods i.e. MAOS, reflux, and Dean Stark. The percentage yield of TETA-PA imidazoline was obtained from Dean-Stark at 13 hours (95.80 %), reflux at 13 hours (88.53 %), and MAOS at 11 min (87.74 %). Imidazoline synthesis using MAOS method performed faster reaction time and higher reaching temperature in compare with reflux and Dean Stark methods. Structure elucidation of TETA-PA imidazoline was also conducted using various spectral data together with identification data. For further application, imidazoline-palmitic derivative can be developed as corrosion inhibitor towards carbon steel.

Acknowledgements

This work was supported by Hibah PITTA 2018 funded by Universitas Indonesia No. 2258/UN2.R3.1/HKP.05.00/2018.

References

[1] Fahrurrozie A, Sunarya Y and Mudzakir A 2010 *Jurnal Sains and Teknologi Kimia* **1** 100–11
[2] Tyagi R, Tyagi V K and Pandey S K 2007 *J. Oleo Sci.* **56** 211–22
[3] Surati M A, Jauhari S and Desai K R 2012 *Arch. Appl. Sci. Res.* **4** 645–61
[4] Wahyuningrum D, Achmad S, Syah Y M, Buchari and Ariwahjoedi B 2008 *Journal of Mathematical and Fundamental Sciences* **40** 33–48
[5] Abbasov V M, Guliyev A A, Jabrailzadeh Sh Z, Mursalov N I, Duzdaban Kh R and Mamedaliev Y 2015 *International Journal of Engineering and Innovative Technology (IJEIT)* **5** 21–3
[6] Kong X, Qian C, Fan W and Liang Z 2018 *J. Mol. Struct.* **1156** 164–71

Polyclonal antibodies production from porcine gelatin and its preliminary study for immunosensor applications

A I Astuti[1], R D Soejoedono[2], E Saepudin[1], L D Assaat[1] and T A Ivandini[1]

[1] Department of Chemistry, Faculty of Mathematics and Natural Sciences (FMIPA), Universitas Indonesia, Kampus UI Depok, Depok 16424, Indonesia

[2] Department of Animal Infectious Diseases and Veterinary Public Health, Faculty of Veterinary Medicine, Institut Pertanian Bogor, Bogor 16680, Indonesia

Corresponding author's e-mail: ivandini.tri@sci.ui.ac.id

Abstract. Gelatin was extracted from porcine skin and used as an antigen to produce the polyclonal antibodies of porcine gelatin using New Zealand white rabbits for application in the development of an ELISA-based immunosensor for porcine gelatin recognition. The extraction of porcine gelatin from porcine skin was successfully performed with 20.52 % of yield. The characterization performed by using UV and IR spectroscopy indicated that the purified gelatin has a comparable quality against the commercial standard of porcine gelatin. Further analysis of the gelatin with SDS-PAGE method observed five separation lines with molecular weight bands of 57.06; 49.50; 42.94; 33.89; and 29.40 KDa, indicating that the gelatin can be used as an antigen to produce the polyclonal antibodies of porcine gelatin. After five injections of gelatin into the rabbit, ELISA characterization of the purified serum of the blood rabbit showed that the antibody with a concentration of 0.289 mg/mL could be produced.

Keywords: ELISA, immunosensor, porcine gelatin, polyclonal antibody, rabbit

1. Introduction

Gelatin is a type of protein derived from natural collagen found in the skin and animal bones [1]. Commercial gelatin was generally produced from mammals, such as pork skin, bovine skin and fish [2-3]. The most widely used gelatin is the one from porcine, because of its superiority in gel strength, pressure resistance, water holding ability, and high melting point [4]. Gelatin is widely used in food products, such as candy, jelly, yoghurt, marshmallow, gelatin and pudding [5]. A practical method to analyze the gelatin content in food and non-food products is necessary due to its potency as allergen to some people as well as a requirement for halal labelling in Muslims' community.

The analysis of porcine gelatin has been reported by some researchers using polymerase chain reaction (PCR) [6], Fourier Transform Infrared (FTIR) spectroscopy [7], reversed-phased High Performance Liquid Chromatography (HPLC) [8], and sodium dodecyl sulphate-polyacrylamide gel electrophoresis (SDS-PAGE) [9]. However, these analysis methods were ineffective due to the requirements of special expertise in operation, long time analysis, as well as expensive instrumentations. Hence, a fast, cheap, effective, and easy method of porcine gelatin analysis is required. Immunosensor, such as ELISA type, is one of the methods that can fulfill the criteria. However, ELISA-based immunosensor requires antibodies as the sensing elements to recognize the antigen samples, while the antibodies can be prepared in an animal body, such as a rabbit.

The production of antibodies of porcine gelatin in rabbits for ELISA assays has been reported, involving the use of porcine and cow gelatin as the antigens to stimulate the production of the antibodies [5,10]. On the other hand, other report indicated that the porcine gelatin could be directly produced in animal because it contains proteins with a molecular weight of more than 10 kDa [11]. Production of antibodies using direct porcine gelatin as antigen injected into white rats has been reported [12]. However, the characterization by using the dot blot method did not give good results. In this work, the porcine gelatin was directly immunized into the body of the rabbit to stimulate the production of the antibodies. As the result, specific antibodies towards porcine gelatin can be produced as determined by ELISA method.

2. Materials and methods

2.1. Ethical approval
Approval number: 083/KEH/SKE/II/2018 was obtained as a consent to perform this research and was issued by the Animal Care and Use Committee of Research and Community Services Institution, Bogor Agricultural University.

2.2. Materials
Phosphoric acid, standard porcine gelatin (G2500) and standard bovine gelatin (G6650) were purchased form Sigma Aldrich. SDS-PAGE test materials, protein marker (Thermo), phosphate buffer saline (PBS), New Zealand female white rabbits, complete Freund's adjuvant (CFA), incomplete Freund's adjuvant (IFA), ELISA test materials, Hi-Trap protein A column, filter paper 0.45 µm nylon membrane (Whatman) and Na_3PO_4.

2.3. Extraction and characterization of gelatin from porcine skin
The raw porcine skin material was cleaned from the hair, meat, and fat, then cleansed under running water and arranged to be in size 2x2 cm^2 before soaked in water and heated at 60 ºC for 15 min. After cooling, it was immersed in 12 % phosphoric acid solution for 24 h. After washing, the gelatin was extracted by soaking in distilled water at 60 °C for ± 6 h. The liquid was filtered and kept at 4 °C to obtain gel form. In order to get the powder form, the gel was evaporated at 50 ºC and then freezes dried. Characterization was performed by using UV-Visible and FTIR spectrophotometers.

2.4. SDS-PAGE analysis of gelatin
The sodium dodecyl sulphate-polyacrylamide gel electrophoresis (SDS-PAGE) measurements were performed using 10 mg gelatin dissolved in 2 mL PBS. Briefly, the solution was added into polyacrylamide gel consists of 14 % run gel and 4 % stacking gel. The electrophoresis was performed at an applied potential of 100 V and 50 mA per gel for 4 h using Mini Protean II unit (Bio-Rad Laboratories, Inc., Richmond, CA, USA). Following the electrophoresis process, the gel was stained with coomassie blue and washed.

2.5. Production of antibodies
Production of the antibodies of porcine gelatin was performed using New Zealand female white rabbits as the tested animal. The rabbits were injected in two routes, including intravenous and subcutaneous. Intravenous injections were performed by the injection of 0.2 mL sterile PBS containing of 10 mg gelatin into rabbits for three consecutive days. The antibodies produced in the blood serum were checked after seven days. After injections through the intravenous line, the rabbit was injected with gelatin through the subcutaneous route. This injection was performed with a mixture of 1 mL sterile PBS containing 10 mg gelatin and Freund adjuvant (1: 1). Then, blood serum was checked using Agar Gel Precipitation Test (AGPT) for antibodies on the 7th and 14th days after injection.

2.6. Monitoring of the antibodies by indirect ELISA
The microplates were coated with a gelatin antigen layer (10 mg/L, 100 µL/well) in a buffer coating (carbonate bicarbonate buffer 0.05M, pH 9.6), then covered overnight at 4 ºC. After washing, the microplate was filled with the polyclonal antibody porcine gelatin antigen, incubated for 1 h, and washed. Then, 100 µL/well HRP-conjugated goat anti-rabbit IgG (1: 10,000) was added into the microplates, incubated for 1 h, and washed. The 3,3', 5,5'-tetramethylbenzidine (TMB) solution was then added into the microplates, incubated for 10 min, and added by 50 µL stop solution. The characterization was performed at the visible absorbance of 450 nm wavelengths. The microplates were washed after every coating step by using PBS Tween-20 0.05 %

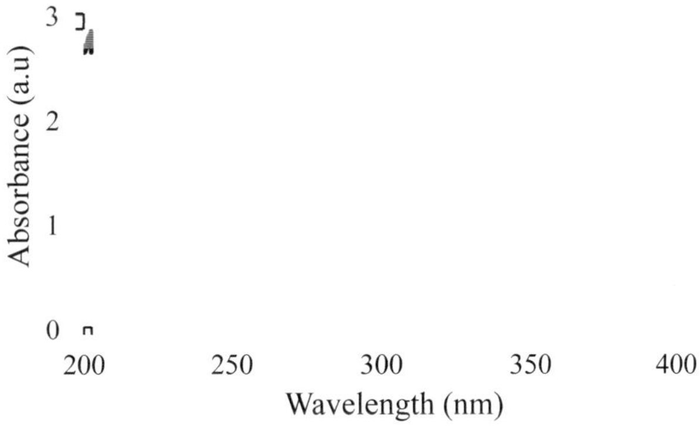

Figure 1. UV spectra of the extracted porcine gelatin (black line) in comparison with the commercial one from Sigma-Aldrich (red line)

2.7. Purification of the antibodies

Purification was performed using a commercial A Protein Kit (Sigma-Aldrich) to remove the polyclonal IgG antibodies from serum. Briefly, the blood serum was homogenized and filtered using a 0.45 μm nylon membrane. Two mL of the filtered serum was mixed with 4 mL solution of 20 mM Na_3PO_4 pH 7 and passed into the protein A column dropwise. The filtrate was then placed into a centrifuge tube (containing 0.2 mL buffer tris-HCl pH 9) to 1 mL/fraction. Each fraction was characterized by UV spectrophotometer at a 280 nm to calculate the concentration [13].

3. Results and discussion

3.1. Characterization of gelatin from porcine skin by using UV-Vis and IR spectrophotometer

The extraction of porcine gelatin was carried out by cleaning and soaking the porcine skin with warm water at around ± 60 °C. After the skin became soft, the skin was immersed in acid to convert a triple-helix collagen fiber into a single chain. The acid also served in the demineralization process to remove the minerals, such as calcium salts, due to the swelling of the materials. The last stage of the extraction process was heating, as the gelatin dissolves in warm water (T ≥ 40 °C) [14]. The effectiveness of the gelatin-preparation method was confirmed with the total yield of 20.52 %.

Comparison of the UV-Vis spectra of the extracted and standard commercial porcine gelatin is shown in figure 1. The amino acids in gelatin can be identified using the UV spectrum at the 210–240 nm which shows the absorption of the chromophore groups, including glycine, proline and arginine amino acids.

Further characterization was performed using FTIR to determine the functional groups in the gelatin. Gelatin normally consists of hydroxyl groups (O-H), carbonyl groups (C=O), and amine groups (N-H). Figure 2 shows a wide absorption band at 3100–3500 cm^{-1}, which confirmed the presence of the stretching vibration bands of O-H as well as N-H groups. The presence of O-H groups in gelatin might be also due to the use of water in the extraction process. In addition, the absorption at 2924 cm^{-1} indicated C-H stretching vibration, while three different absorption peaks at 1627 cm^{-1}, 1523 cm^{-1}, and 1235 cm^{-1} indicated the presence of the groups of amide I, amide II, and amide III, respectively.

3.2. Protein analysis in the extracted porcine gelatin using SDS PAGE

Analysis of the gelatin was also performed using SDS PAGE method by separating the protein according to the molecular weight. Comparison with the standard commercial porcine gelatin and bovine gelatin was conducted to determine the difference in protein compositions. Figure 3 shows the different patterns of SDS PAGE results of bovine and porcine gelatin. Gelatin from porcine, including the extracted one, typically generated five protein bands, indicated that the gelatin composed of five main proteins. Whereas, the standard porcine gelatin shows molecular weight bands at 59.8250; 51.9017; 45.0278; 36.5341; and 29.4020 KDa, the isolated one shows the bands at 57.0579; 49.5011; 42.9451; 33.8905; and 29.4020 KDa. On the contrary, the standard bovine gelatin generated only one

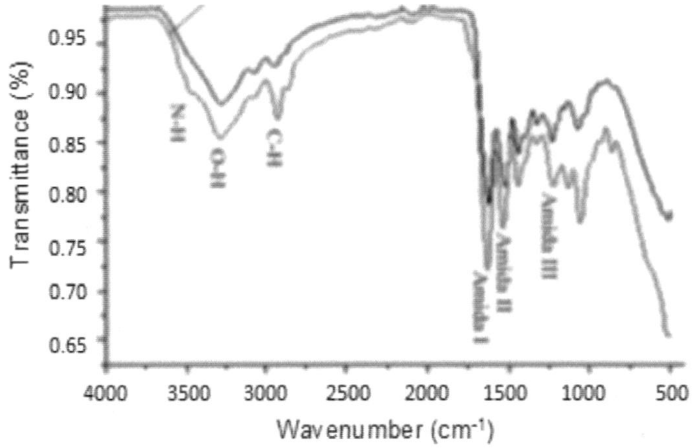

Figure 2. FTIR spectra of the extracted porcine gelatin (red line) in comparison with the commercial one (black line) from Sigma-Aldrich

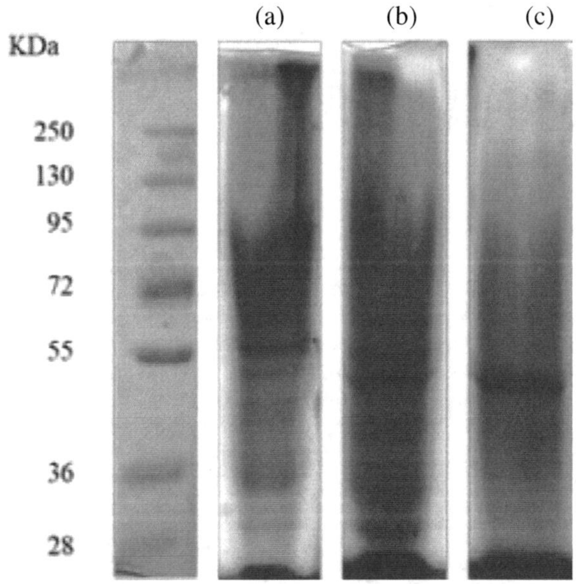

Figure 3. SDS PAGE bands of (a) the commercial and (b) the extracted porcine gelatin in comparison with (c) the commercial bovine gelatin

band at molecular weight of 47.2115 KDa. The results suggested that the porcine gelatin protein can fulfill the criteria as antigen to produce the antibody, as previously reported that protein substance must have a minimum molecular weight of 10 KDa to be effective as immunogen [11]. Moreover, the different protein compositions between bovine and porcine gelatin indicated the possibility to distinguish the gelatin from bovine and porcine.

3.3. Production of antibodies and characterization by using ELISA

The polyclonal antibodies were produced by the immunization of porcine gelatin in the New Zealand female white rabbit as the tested animal. Initially, the rabbit was immunized through intravenous, then through the subcutaneous. The first injection was performed two times to observe the rabbit's response to the gelatin antigen, while the subcutaneous injection was aimed to promote the immune responses. The last injection was repeated three times by mixing the antigen with Freund's adjuvants to induce the immune systems for long-term injections.

Table 1. Absorption of TMB during ELISA measurements of the rabbit blood serum during porcine gelatin injection

Serum day to-	Absorption of porcine gelatin extraction
0	0.165
10	0.906
14	0.870
18	0.514
21	0.653
31	0.395
38	1.313
46	1.336
52	1.242
59	1.320
65	1.328

Monitoring of the presence of antibodies was performed with analysis of the blood serum during the injection by using ELISA method. The produced antibody was used as the sensing materials, while porcine gelatin was used as the sample and TMB as the label. The amount of TMB is equivalent to the amount of the antibody. Table 1 shows the visible absorption of the blood serum from the initial stage of injection. The table shows that the absorbance has a tendency to increase and be saturated at around the 38th day. The results suggested that the rabbit has responded to the porcine gelatin antigen by producing antibodies.

The purification of the produced antibodies was performed by using Protein A filtration on the 65th days serum after totally 5 times injections. The results of the purification showed the IgG concentration of the antibody was 0.289 mg/mL.

4. Conclusions

Gelatin can be extracted from raw material of porcine skin with acid treatment with the yield up to 20.52 %. Analysis with SDS PAGE showed that the porcine gelatin has five separated bands with the molecular weights of more than 10 kDa. Injection of the extracted gelatin to rabbit Polyclonal antigens of the porcine gelatin can be produced in the rabbit after 65 days with an IgG concentration of 0.289 mg/mL.

Acknowledgements

This work was funded by Hibah PITTA Universitas Indonesia with contract number of 2274/UN2.R3.1/HKP.05.00/2018

References

[1] Yi J B, Kim Y T, Bae H J, Whiteside W S and Park H J 2006 *J. Food Sci.* **71** E376–83
[2] Gudmundsson M 2002 *J. Food Sci.* **67** 2172–6
[3] Karim A A and Bhat R 2008 *Trends Food Sci. Technol.* **19** 644–56
[4] Venugopal V 2008 *Marine Products for Healthcare: Functional and Bioactive Nutraceutical Compounds from the Ocean* (Boca Raton: CRC Press)
[5] Doi H, Watanabe E, Shibata H and Tanabe S 2009 *J. Agric. Food Chem.* **57** 1721–6
[6] Cai H, Gu X, Scanlan M S, Ramatlapeng D H and Lively C R 2012 *J. Food Compos. Anal.* **25** 83–7
[7] Hidaka S and Liu S Y 2003 *J. Food Compos. Anal.* **16** 477–83
[8] Raraswati M A, Triyana K and Rohman A 2014 *J. Food Pharm. Sci.* **2** 1–6
[9] Nur Azira T, Amin I and Che Man Y B 2012 *International Food Research Journal (IFRJ)* **19** 1175–80
[10] Venien A and Levieux D 2005 *J. Pharm. Biomed. Anal.* **39** 418–24
[11] Bellanti J A 1993 *Immunologi III*, tranlastor A S Wahab (Yogyakarta: Gadjah Mada University Press)
[12] Syamsuri A and Wardani A K 2013 *Jurnal Pangan dan Argoindustri* **1** 36–45
[13] Raja Mohd Hafidz R N, Yaakob C M, Amin I and Nurfaidzan A 2011 *International Food Research Journal (IFRJ)* **18** 813–7
[14] Ross-Murphy S B 1997 *The Imaging Sci. J.* **45** 205–9

Preliminary study on carboxylation reaction of phenylacetylene with CO_2 using nickel catalyst impregnated on mesoporous carbon

A Hadi, I Abdullah and Y K Krisnandi

Department of Chemistry, Faculty of Mathematics and Natural Sciences (FMIPA), Universitas Indonesia, Kampus UI Depok, Depok 16424, Indonesia

Corresponding author's e-mail: iman.abdullah@sci.ui.ac.id

Abstract. In this research, carboxylation reaction of phenylacetylene with CO_2 has been studied using impregnated nickel on mesoporous carbon as catalyst. Mesoporous carbon was successfully synthesized using a soft template method with Pluronic F127 as pore template, and formaldehyde and phloroglucinol as the carbon source. This material was characterized with Fourier Transform Infrared (FTIR), X-Ray Diffraction (XRD), Scanning Electron Microscopy (SEM), and Brunauer-Emmett-Teller (BET) analysis. Characterization with XRD shows the presence of two peaks at 24.26° and 42.76° which denotes the typical peak of graphite-like carbon. BET surface area analysis shows that the mesoporous carbon has a pore size distribution of 7.2 nm and surface area of 432.15 m^2/g. Modification of mesoporous carbon was carried out by impregnation of Ni metal into mesoporous carbon (Ni@MC) through reduction of Ni(II) with ethylene glycol. The Ni@MC catalyst was used as a catalyst in the carboxylation reaction of phenylacetylene with CO_2. The reactions were carried out in a batch reactor under various reaction conditions. High Performance Liquid Chromatography (HPLC) analysis of the product mixtures revealed that the highest yield of trans-cinnamic acid (2.83 %) was achieved at 85 °C for 12 h reaction time.

Keywords: mesoporous carbon, nickel, carboxylation, phenylacetylene

1. Introduction

CO_2 gas is one of the compounds that its existence in nature is very abundant. CO_2 is an inert, non-toxic, non-flammable, and the largest contributor to greenhouse gases causing the earth's surface temperature to rise. Rising earth surface temperatures is noted as Global Warming. Global Warming is a phenomenon of the earth's escalating temperature due to the increase of greenhouse gases such as CO_2, CH_4, N_2O, and CFCs, so that solar energy is trapped in the earth's atmosphere and cannot be released out of the atmosphere.

One way to reduce CO_2 emissions is by utilizing Carbon Capture and Storage (CCS) and Carbon Capture Utilization and Storage (CCUS) technologies. CCS is a technology used to reduce the increase in greenhouse gas emissions sourced from CO_2 gas. In the CCS technique, CO_2 is captured and then separated and stored. This technique is commonly applied at mining areas that produce a lot of CO_2 gas, usually used in Enhanced Oil Recovery (EOR) by injecting CO_2 into the bowels of the earth to push the oils left inside to get out. Whereas in CCUS technique, CO_2 can be reacted with another reagent to produce a product that has economic value, one of which is acrylic acid and its derivatives, which are needed in various industries as intermediate compounds in the production process. For example, Na acrylate is very important in the industry, commonly used as a monomer in a polymer super-absorber and its sales can reach 4 million tons globally [1]. Conversion of CO_2 provides many advantages, in addition to reducing CO_2 emission; it can also produce high-value products [2]. However, due to its inert character, CO_2 is difficult to react with other compounds, so a catalyst is required. One of the most commonly used catalysts in CO_2 activation is nickel [3]. Nickel is included in inexpensive metals and is usually used in the form of its complex compounds, so the catalysis system is homogeneous. There are some examples of utilization of homogeneous nickel catalyst in carboxylation reaction of some unsaturated compounds such as alkynes [4] and ynamides [5]. Thus, immobilization of the nickel homogeneous catalyst, by attaching it to a solid support, could provide various advantages provided by heterogeneous catalyst, especially for industrial scale applications, such as ease of separation, thermal and another durability.

There are some reports on the use of heterogeneous metal catalyst on carboxylation reaction of unsaturated hydrocarbons with CO_2. The carboxylation reaction of the terminal alkynes with CO_2 has been successfully performed by using Cu nanoparticle catalyst with Al_2O_3 support to produce alkynyl carboxylic acid derivates (R-COOtBu) but it requires a high pressure [6]. Furthermore, the carboxylation reaction of the terminal alkyne with CO_2/H_2 has also been successfully carried out using dual catalyst NHC-Ag/Pd to produce R-COOH compounds [7]. However, it is well known that Pd catalyst is expensive. Activated carbon has also been reported as catalyst support for some metals such as copper [8] and silver [9] to produce alkynyl carboxylic acids and carbonates, respectively. However, the product was different from the Ni-catalyzed carboxylation product.

In this work, the carboxylation reaction of terminal alkyne was conducted using impregnated Ni catalyst on mesoporous carbon as heterogeneous catalyst. To the best of our knowledge, this kind of work has not been reported. The carboxylation reaction was carried out at atmospheric pressure and low temperature, using phenylacetylene as the alkyne, and as results, similar compound to cinnamic acid was obtained as the product.

2. Experimental

2.1. Materials

All chemicals used in this experiment were analytical grade and used without further purification. Phloroglucinol, Pluronic F127, $Ni(NO_3)_2.6H_2O$, $MgCl_2$, zinc powder, formaldehyde, DMF 99.8 %, phenylacetylene 98 %, and trans-cinnamic acid 97 % were purchased from Sigma Aldrich. Ultra-high purity CO_2 gas was purchased from CV Retno Gas.

2.2. Synthesis of mesoporous carbon (MC)

Synthesis of mesoporous carbon was carried out using a previously reported procedure [10]. In a typical synthesis, dissolution of 1.25 g of phloroglucinol and 1.25 g of poly(ethylene oxide)–poly(propylene oxide)–poly(ethylene oxide) triblock copolymer (Pluronic F127) was conducted in the mixture of 9.7 g of 10:9 weight ratio of ethanol and water. After the solid completely dissolved, 0.08 mL of 37 % HCl was inserted into the solution and stirred for 30 min until a light pink color is observed. The solution was then added with 1.25 mL of 37 % formaldehyde. After 30 min, the solution turned cloudy and separated into two layers after 1–2 h. The polymer-rich bottom layer was taken and continued stirring overnight (12 h) to form an elastic, non-sticky monolith. The obtained monolith was then displaced into an autoclave and cured in an oven at 100 °C for 24 h. The material was carbonized in a tubular furnace under nitrogen flow with the heat rate of 1 °C/min from 100 to 400 °C and 5 °C/min from 400 to 850 °C and finally maintained at 850 °C for 2 h.

2.3. Impregnation of Ni into mesoporous carbon (Ni@MC)

Impregnation of Ni into mesoporous carbon was conducted using previously reported method [11]. 0.5 g of MC was dispersed into 10 mL of water. 1.23 g of $Ni(NO_3).6H_2O$ was dissolved into 5 mL of water and then added drop wise into the dispersed carbon. The mixture was stirred for 6 h and dried in air. The obtained material and 20 mL of ethylene glycol were transferred into a round-bottom flask and the compound was heated under reflux at 170 °C for 6 h. The final solid was filtered, washed by water and ethanol, and dried at room temperature.

2.4. Carboxylation reaction

Phenylacetylene (0.8 mmol), Ni@MC (0.04 mmol), $MgCl_2$ (0.6 mmol), Zn (0.6 mmol) and DMF (4 mL) were mixed in the reactor at room temperature. Then, balloon filled with CO_2 gas was fitted to the reactor. The air in the reactor was removed by flowing the CO_2 gas into the reactor and then closed. The mixture was stirred for 12 h at 50 °C and 1 atm. The time and temperature of reaction were varied: (8 h, 12 h and 16 h) and (25 °C, 50 °C, 85 °C). To obtain the targeted product i.e. cinnamic acid, the mixture was multi-extracted with CH_2Cl_2 (3 x 5 mL). The organic layer was collected, and the

Figure 1. SEM images of (a) MC, (b) Ni@MC (x2000).

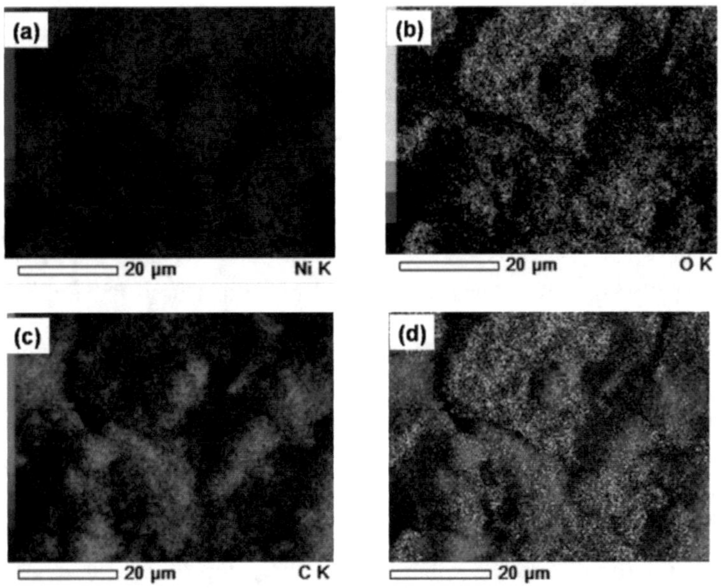

Figure 2. Mapping images of Ni@MC: (a) Ni, (b) O, (c) C, and (d) overlay of Ni, O & C.

aqueous layer was acidified with 10% HCl (2 mL), then extracted with diethyl ether (3 x 5 mL). A collection of organic layers was taken and then dried with sodium sulfate (Na_2SO_4) then decanted. The remaining solvent was evaporated to obtain a dry white powder (cinnamic acid). The product was then analyzed and identified using HPLC. The product analysis was performed using the Diamonsil C18 column (200 mm x 4.6 mm, 5 µm particle size). The detector used was UV-Vis, which operated at the wavelength (λ) of 278 nm. The mobile phase used was methanol and 1 % acetic acid with a ratio of 55:45 (v/v). The flow rate of the mobile phase was 1 mL/min. All analysis was performed under isothermal conditions at 25 °C [12].

3. Results and discussion

Morphology of MC and Ni@MC were examined with scanning electron microscope (SEM) and the results are shown in figure 1. All materials seem to have rough surface with various sizes. Ni@MC has a coarser morphology when compared to mesoporous carbon morphology (MC). This may be due to the presence of NiO and Ni(0) attached to the mesoporous carbon thus altering the surface. This surface characteristic might increase the surface area of Ni@MC, which could be useful for catalysis application.

From figure 2, it is shown that Ni particles were evenly distributed although there are some spots that do not contain Ni. The images also indicate that nickel species existed on mesoporous carbon

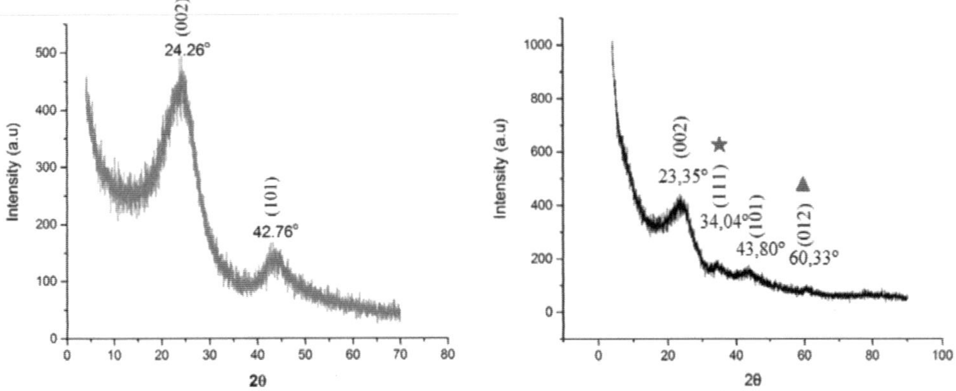

(a) (b)

Figure 3. XRD patterns of (a) MC and (b) Ni@MC

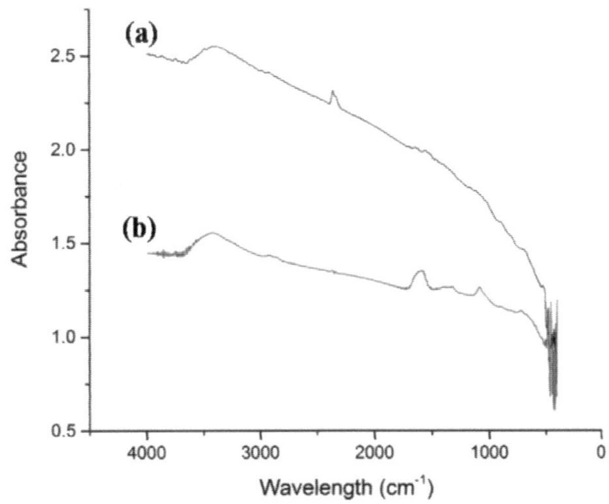

Figure 4. FTIR spectrum of (a) MC and (b) Ni@MC

were mixture of both Ni(0) and NiOx, since oxygen atoms appeared at almost the same location to that of nickel, figure 2d. The presence of Ni(0) metal is shown as blue area in overlay image. The existence of this zero valent nickel is important since it could be the active catalyst site for carboxylation reaction. The existence of NiO in the materials could be explained by two reasons: it might be due to incomplete reduction during reflux with ethylene glycol, or there were some Ni(0) being re-oxidized by air during material characterization.

The diffraction pattern of MC and Ni@MC are shown in figure 3. In XRD pattern of both materials, there are two peaks at around 24° and 43° which shows the typical peak of lattice planes (002) and (101), respectively, in carbon material. This is in agreement with JCPDS File Index for carbon-based materials, No. 75-1621f.

In the diffraction pattern of Ni@MC, there are new peaks located at 34.04° (111) indicated by the star symbol and at 60.33° (012) indicated by the triangle symbol. These peaks are the peak of NiO (111) and Ni (012), according to the Ni-NiO diffraction pattern reported by Yung *et al.* [14].

Figure 4 shows that the FTIR spectra of Ni@MC look almost identical to that of MC. The absence of C-H stretching peak at 2900-3000 cm^{-1} indicates no organic compounds remain in the materials. However, there are some new peaks in Ni@MC FTIR spectra at 1600 and 1100 cm^{-1}. This could be due to the presence of Ni-NiO species impregnated into mesoporous carbon.

IOP Conf. Series: Materials Science and Engineering **763** (2020) 012008 doi:10.1088/1757-899X/763/1/012008

Table 1. Results of carboxylation reaction.

Entry	Reaction conditions	Yield of trans-cinnamic acid (%)
1	25 °C, 8 h	Not detected
2	25 °C, 12 h	Not detected
3	25 °C, 16 h	Not detected
4	50 °C, 8 h	Not detected
5	50 °C, 12 h	Not detected
6	50 °C, 16 h	Not detected
7	85 °C, 8 h	1.52
8	85 °C, 12 h	2.83
9	85 °C, 16 h	0.62

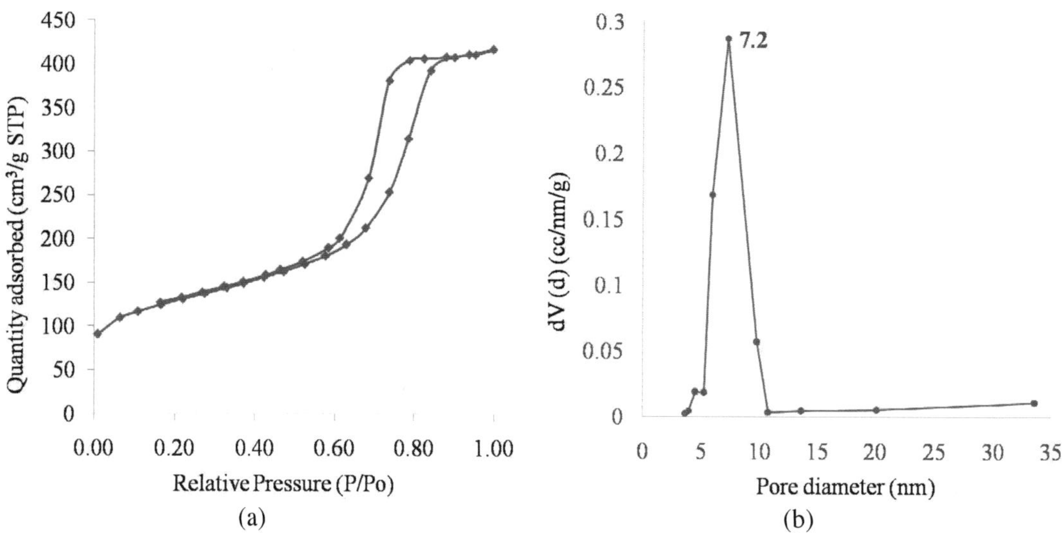

Figure 5. BET N_2-surface area analysis: (a) Adsorption desorption isotherm and (b) BJH desorption MC.

The N_2 adsorption-desorption isotherm curve of mesoporous carbon is shown in figure 5. In the N_2 desorption adsorption process, a hysteresis loop is generated which can indicate the type of isotherm curve. The resulting MC isotherm curve indicates an IV type isotherm curve corresponding to the IUPAC type of adsorption curve. This type of isotherm curve is also supported by the distribution of pore size diameter. From the mesoporous carbon desorption BJH curve, it is shown that the pore size diameter is concentrated at 7.2 nm. Thus, the pore size diameter is in the range of mesoporous pore size (2–50 nm).

The results of the carboxylation reaction of phenylacetylene with CO_2 are shown in table 1. Based on HPLC analysis of reaction mixtures, trans-cinnamic acid ($C_6H_5CH{=}CHCOOH$) was detected as product when the reactions were conducted at 85 °C (entries 7-9). Conducting the reactions at lower temperatures gave no desired product (entries 1–6). The reaction of carboxylation of phenylacetylene with CO_2 gave highest yield at 85 °C for over 12 h (entry 8). These results indicate that high temperature is required for carboxylation of phenylacetylene using heterogeneous Ni catalyst.

4. Conclusions

Modification of mesoporous carbon has been conducted by impregnation of Ni metal into mesoporous carbon (Ni@MC). The Ni@MC catalyst was used as a catalyst in the carboxylation reaction of phenylacetylene with CO_2. The highest yield of cinnamic acid (2.83 %) was achieved at 85 °C for 12 h reaction time.

Acknowledgements
The authors gratefully acknowledge PITTA Grant 2018 (2272/UN2.R3.1/HKP05.00/2018) from Universitas Indonesia for funding this project. Prof. Sung Min Choi is greatly acknowledged for his generosity providing access for SEM characterization at KAIST.

References
[1] Lejkowski M L *et al.* 2012 *Chem. Eur. J.* **18** 14017–25
[2] Maeda C, Miyazaki Y and Ema T 2014 *Catal. Sci. Technol.* **4** 1482–97
[3] Tsuji Y and Fujihara T 2012 *Chem. Commun.* **48** 9956–64
[4] Wang X, Nakajima M and Martin R 2015 *J. Am. Chem. Soc.* **137** 8924–7
[5] Doi R, Abdullah I, Taniguchi T, Saito N and Sato Y 2017 *Chem. Commun.* **53**(55) 7720–3
[6] Bondarenko G N, Dvurechenskaya E G, Magommedov E S, and Beletskaya I P 2017 *Catal. Lett.* **147** 2570–80
[7] Yu D, Zhou F, Lim D S W, Su H and Zhang Y 2017 *Chem. Sus. Chem.* **10** 836–41
[8] Xie J N, Yu B, Zhou Z H, Fu H C, Wang N and He L N 2015 *Tetrahedron Lett.* **56** 7059–62
[9] Qiu J, Zhao Y, Wang H, Cui G and Wang J 2016 *RSC Adv.* **6** 54020–6
[10] Górka J, Zawislak A, Choma J and Jaroniec M 2008 *Carbon* **46** 1159–61
[11] Pamungkas A Z, Abdullah I and Krisnandi Y K 2019 *IOP Conf. Ser.: Mater. Sci. Eng.* **496** 012003
[12] Huang T, Chen N, Lai Y, Wang D, Yan J and Gu J 2011 *J. Med. Plants Res.* **5** 1313–20
[13] Yuan D, Chen J, Hu X, Zeng J, Tan S and Liu Y 2008 *Int. J. Electrochem. Sci.* **3** 1268–76
[14] Yung T Y, Huang L Y, Chan T Y, Wang K S, Liu T Y, Chen P T, Chao C Y and Liu L K 2014 *Nanoscale Res. Lett.* **9** 444

IOP Conf. Series: Materials Science and Engineering **763** (2020) 012009 doi:10.1088/1757-899X/763/1/012009

Synthesis of V_2O_5 nanoparticles using *Calophyllum inophyllum* leaf extract as photocatalyst under visible light irradiation

W Megawati[1], Y Yulizar[1], S Sudirman[1,2] and D O B Apriandanu[1]

[1] Department of Chemistry, Faculty of Mathematics and Natural Sciences (FMIPA) Universitas Indonesia, Kampus UI Depok, Depok 16424, Indonesia

[2] Center for Science and Technology of Advanced Materials (PSTBM Batan, Gedung 40 Kawasan Puspitek Serpong, Tangerang Selatan, Banten 15314, Indonesia

Corresponding authour's e-mail : yokiy@ui.ac.id

Abstract. Synthesis of V_2O_5 nanoparticles (NPs) using *Calophyllum inophyllum* leaf extract (CILE) as an eco-friendly material has been investigated. V_2O_5 NPs were prepared using NH_4VO_3 as precursor. The typical absorption peak of V_2O_5 NPs has been characterized using UV–Vis spectrophotometer at 204 nm. V_2O_5NPs have the band gap value of 2.42 eV confirmed by UV-Visible Diffuse Reflectance Spectroscopy (UV-vis DRS). X-ray diffraction (XRD) analysis shows that an average crystal size of V_2O_5NPs was 68 nm. Identification of functional groups was characterized by Fourier-Transform Infrared (FTIR) spectroscopy. The absorption band at 1023 cm^{-1} shows the V=O bond and the V-O-V bond at 849 and 578 cm^{-1}, respectively. Size distribution of V_2O_5 NPs was characterized by particle size analyzer (PSA). Scanning Electron Microscopy-Energy Dispersive X-Ray (SEM-EDX) Spectroscopy analysis was used to determine the morphological and the element composition of V_2O_5 NPs. The degradation percentage of methylene blue (MB) using V_2O_5NPs was 74.83 % under visible light irradiation for 120 min.

1. Introduction

Nanomaterials are the pillar of nanoscience and nanotechnology. One form of nanomaterials is the metal oxide V-O-V. Metal oxide nanoparticles have significant contributions in many application fields, namely in chemistry, physics and materials. In technological applications, metal oxide nanoparticles are used in water treatment photocatalysts [1]. On the other hand, photocatalyst has become a new technology for solving the environmental pollution problems. Photocatalyst has been used for degradation of organic pollutants in wastewater by the presence of sunlight or visible light sources [2].

Among the various oxide-based semiconductors, V_2O_5 is especially interesting one due to its widely applications like electrochemical devices and catalyst [3]. In addition, V_2O_5 has narrow band gap (Eg), which is susceptible to photoactivation in visible light area, and has been used in semiconductor-type photocatalyst for degradation of water pollutants under visible light area.

Several methods have been used for the synthesis of nanomaterials in a relatively short time. However, conventional nanomaterial synthesis such as top-down approach and bottom-up approach leads to the adsorption of toxic chemicals on the material surface that may have adverse the effects on various applications [4]. The use of biological materials such as bacteria [5], mushrooms [6], enzymes [7], and plant leaf extracts [8] for the synthesis of nanomaterials offers many benefits such as being environmentally friendly, and are suitable for pharmaceutical and other biomedical use as well as does not involve harmful substance in synthesis procedures. In our previous reports, some metal oxide nanoparticles were synthesized using plant extract such as ZnO-CuO, Nd_2O_3, CuO, and La_2O_3 [9–12].

Table 1. Phytochemical screening result of CILE

Secondary metabolites	Flavonoid	Terpenoid	Steroid	Alkaloid	Tannin	Polyphenol	Saponin
Results	+	-	-	+	-	+	+

+ : positive result, - : negative result

The potential of plants in Indonesia as biological materials for synthesis of nanomaterials has not been fully explored yet. *Calophyllum inophyllum* is one of the most extensively distributed plants in Indonesia and has many benefits. Susanto *et al.* [13] have reported that phytochemical compounds in *Calophyllum inophyllum* leaves contain alkaloids, tannins, polyphenols, triterpenoid, flavonoid, and sapononin.

In this research, synthesis of V_2O_5 NPs was done using *Calophyllum inophyllum* leaves extract as a weak base source and a capping agent. Furthermore, the synthesized nanoparticles were evaluated in the photocatalytic activity for degradation of methylene blue (MB) as model pollutants under visible light irradiation.

2. Materials and experimental procedures

2.1. Preparation of Calophyllum inophyllum leaves extract
Calophyllum inophyllum leaves were washed and grounded to form powders. Powders were macerated using distilled methanol (1:5) (m/v) for 7x24 hours by stirring for 20 min per day. The compound was filtered and partitioned with n-hexane (1:1) (v/v). The methanol fraction was evaporated and dissolved in distilled water. The result was then filtered to obtain a stock solution of *Calophyllum inophyllum* leaves extract (CILE) for nanoparticles synthesis [14]. In addition, phytochemical screening was carried out for qualitative determination of flavonoid, terpenoid, steroid, alkaloid, tannin, polyphenol, and saponin in CILE using the method as reported by Saraswathi and Santhakumar [15].

2.2. Synthesis of V_2O_5 NPs
10 mL of 1.31% CILE was added into 100 mL of 0.025M NH_4VO_3 (*Merck*) and stirred at 80 °C for 4 hours. The solutions were then aged for 24 hours to form colloids. Finally, the result was dried at 100 °C then calcined at 500 °C for 4 hours to obtain the powder of V_2O_5NPs.

2.3. Characterization
The diffraction pattern of V_2O_5NPs was identified by X-Ray Diffraction (XRD, Shimadzu) at 2θ range of 0–80°. The morphology of V_2O_5 NPs was investigated by Scanning Electron Microscopy (SEM, Zeiss). Elemental analysis of V_2O_5 NPs was confirmed by Energy Dispersive X-Ray (EDX, Bruker). UV-Vis absorption spectra were recorded by UV-Vis spectrophotometer (Shimadzu 2600) at the wavelength between 200–800 nm. Size distribution of V_2O_5NPs was characterized by Particle Size Analyzer (PSA, Malvern Zetasizer 1600). Identification of functional groups was characterized by Fourier Transform Infrared (FTIR, Prestige 21 Shimadzu) spectroscopy and the band gap value measurement was determined by UV-Vis Diffuse Reflectance Spectroscopy (DRS, Shimadzu 2600).

2.4. Photocatalytic activity analysis
Photocatalytic activity was evaluated in MB degradation. 5.0 mg of V_2O_5 NPs were added into 25 mL of $2.0x10^{-5}$ M MB solution. The mixture was sonicated for 30 min and further stirred for 2 hours under sodium lamp irradiation. UV-Vis absorption spectra were recorded every 30 min to observe the decrease of MB absorbance.

3. Results and discussion
Based on phytochemical screening result as shown in table 1, CILE has positive results for alkaloids, saponin, polyphenols, and flavonoid, which can act as hydrolyzing and capping agent in nanoparticles synthesis.

3.1. Characterization of V_2O_5NPs

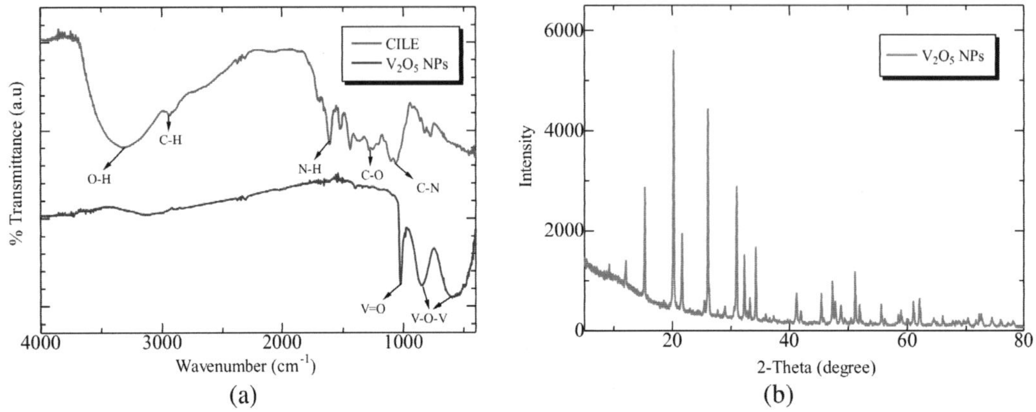

Figure 1. (a) FTIR spectra and (b) XRD patterns of V_2O_5NPs

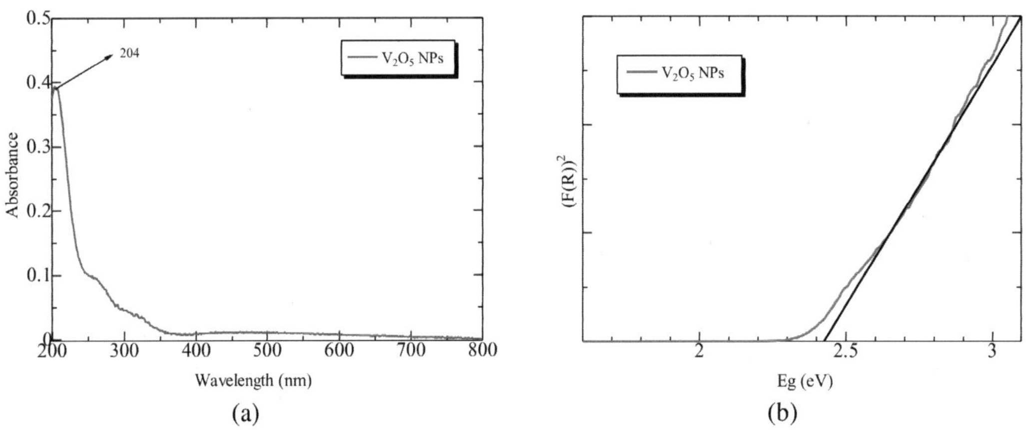

Figure 2. (a) UV-Vis absorption spectrum and (b) Tauc's plot of V_2O_5NPs

FTIR spectra of CILE and V_2O_5 NPs are shown in Figure 1a. CILE shows the vibration of C-N stretching and N-H bending at wavenumber of 1061 and 1607 cm^{-1}, respectively. These two vibrations indicate the presence of alkaloid. The presence of polyphenol and flavonoid are shown by the vibration of -OH stretching at 3316 cm^{-1}. The vibration of C-O stretching at 1284 cm^{-1} and the vibration of C-H stretching at 2939 cm^{-1} indicates the presence of saponin [16]. In fact, the vibration at 1023 cm^{-1} is attributed to the V=O stretching (vanadyl oxygen). Both vibrations at 849 and 578 cm^{-1} are corresponded to V-O-V deformation modes.

Figure 1b shows diffraction patterns of V_2O_5. The 2θ value of V_2O_5 has conformity with COD of V_2O_5 No. 96-101-1292. V_2O_5 has diffraction peaks at 2θ value of 15.31°, 20.22°, 21.65°, 26.09°, 30.96°, 32.31°, 32.80°, 34.25°, 51.15°, and 62.05°. The average of crystal size was calculated using the Debye-Scherrer equation:

$$D = \frac{K\lambda}{\sqrt{\beta_s^2 - \beta_e^2 \cos\theta}} \quad (1)$$

where D is crystalline size, λ is wavelength of the X-Ray radiation (0.1541 nm), K is generally taken as 0.89, β_s and β_e are the peak widths at half-maximum height of sample and the broadening of equipment. According to the calculation, the average crystals size of V_2O_5 is about 68 nm.

Optical properties of V_2O_5NPs are characterized by UV-Vis spectrophotometer as shown in figure 2a. The strong absorption peak occurs at maximum wavelength of 204 nm due to the transition of

Table 2. Elemental composition of V_2O_5NPs.

Element	C	V	O
Atomic %	17.72	39.95	42.29

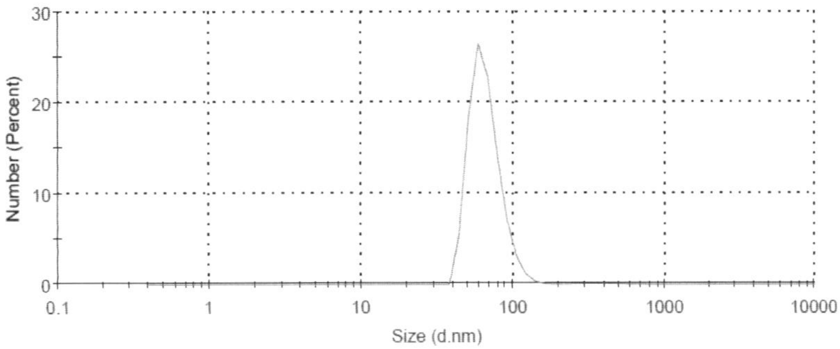

Figure 3. PSA result of V_2O_5NPs

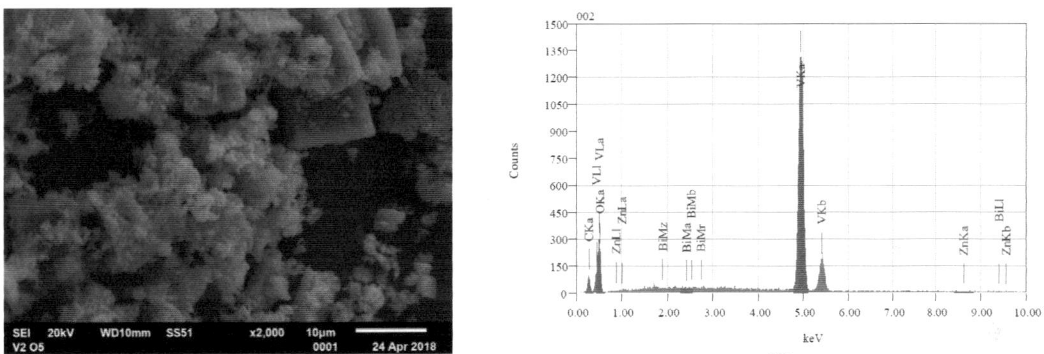

(a) (b)

Figure 4. (a) SEM image and (b) EDX spectra of V_2O_5NPs

holes between vanadium and oxygen [17]. Figure 2b shows the Tauc's plot of V_2O_5 NPs as resulted from UV-Vis DRS characterization to find out the band gap value. The band gap value of V_2O_5 NPs was obtained by calculating from the Kubelka-Munk equation. The calculation result shows the band gap value of V_2O_5NPs was 2.42 eV.

PSA characterization aims to know the average of particles size distribution of V_2O_5. Figure 3 shows the average of particles size distribution of V_2O_5NPs is 66.55 nm.

Figure 4a shows that the SEM image of V_2O_5 NPs morphology has an agglomerate shape. Further analysis about elemental composition on V_2O_5NPs is illustrated in figure 4b. The EDX spectra of V_2O_5 NPs clearly confirm the presence of vanadium and oxygen. The elemental composition is shown in details in table 2. The presence of C happens due to the remains of organic compounds from CILE.

3.2. Photocatalytic performances of V_2O_5NPs

UV-Vis absorption spectra of MB degradation were observed in the decrease of absorbance at maximum wavelength of 664 nm as shown in both Figure 5a and 5b. In addition, Figure 5c illustrates that MB degradation percentage with and without V_2O_5NPs were 18.65 and 74.83 %, respectively after 120 min under visible irradiation. This result demonstrates that V_2O_5NPs has an efficient photocatalytic activity to degrade organic pollutant such as MB.

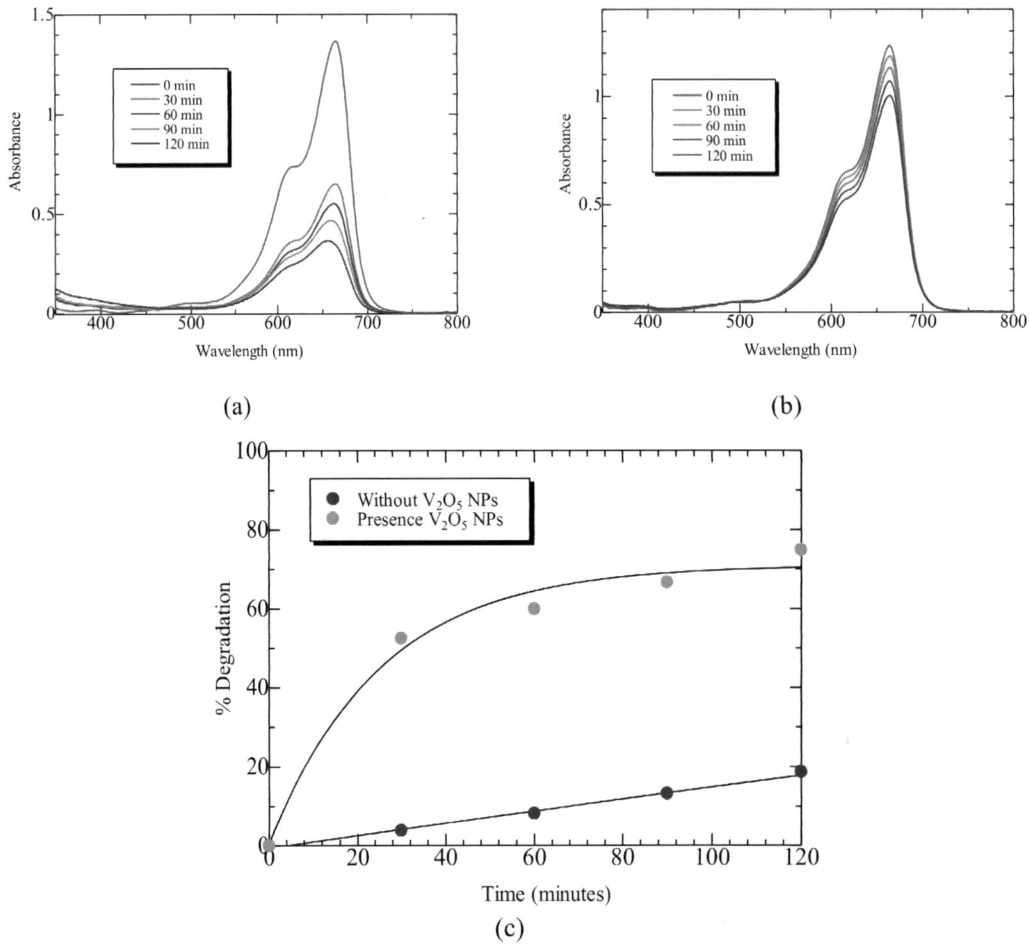

Figure 5. UV-Vis absorption spectra of MB degradation (a) with and (b) without V_2O_5 NPs and (c) Degradation percentage of MB

4. Conclusions

V_2O_5 NPs were successfully synthesized using *Calopyhllum inophylum* leaf extract (CILE) with average particles size distribution of 66.55 nm. V_2O_5 NPs have a typical absorption peak at maximum wavelength of 204 nm and an average crystals size of 68 nm. V_2O_5 NPs show V=O bond at 1023 cm^{-1} and the V-O-V bond at 849 and 578 cm^{-1}. The band gap value of V_2O_5 NPs was 2.42 eV. The degradation percentage of MB was 74.83 % for 120 min. According to the results and discussions, we report that an eco-friendly based V_2O_5 NPs has good photocatalytic activity in MB degradation under visible light irradiation.

Acknowledgements

The authors would like to thank Universitas Indonesia for funding this research through PITTA Grant with contract No. 2338/UN2.R3.1/HKP.05.00/2018 and PT DKSH (Malvern) for supporting characterization of particle size analyzer.

References

[1] Scott-Emuakpor E O, Kruth A, Todd M J, Raab A, Paton G I and Macphee D E 2012 *Appl. Catal. B: Environ.* **123–124** 433–9
[2] Jose J, Paliwal M, Gupta N, Punjabi P B and Sharma V K 2008 *Int. J. Chem. Sci.* **6** 111–21
[3] Li B, Xu Y, Rong G, Jing M and Xie Y 2006 *Nanotechnology* **17** 2560–6

[4] Sangeetha G, Rajeshwari S, and Venckatesh R 2011 *Mater. Res. Bulletin* **46** 2560–6
[5] Saifuddin N, Wong C W and Yasumira N 2009 *J. Chem.* **6** 61–70
[6] Bhainsa K C and D'Souza S F 2006 *Colloids Surf.* B*: Biointerfaces* **47** 160–4
[7] Willner I, Basnar B and Willner B 2007 *FEBS J.* **274** 302–9
[8] Parashar V, Parashar R, Sharma B and Pandey A C 2009 *Digest. J. Nanomater. Biostruct.* **4** 45–50
[9] Yulizar Y, Bakri R, Apriandanu D O B and Hidayat T 2018 *Nano Struct. Nano-Objects* **6** 300–5
[10] Lembang M S, Yulizar Y, Sudirman S and Apriandanu D O B 2018 *AIP Conf. Proc.* **2023** 020093
[11] Yulizar Y, Latifah I, Bakri R and Apriandanu D O B 2018 *AIP Conf. Proc.* **2023** 020097
[12] Sulaiman N, Yulizar Y and Apriandanu D O B 2018 *AIP Conf. Proc.* **2023** 020105
[13] Susanto D F, Aparamarta H W, Widjaja A and Gunawan S 2017 *Asian Pac. J. Trop Biomed.* **7** 773–81
[14] Yulizar Y and Ayun Q 2017 *IOP Conf. Ser.: Earth Environ. Sci.* **60** 012006.
[15] Saraswathi V S and Santhakumar K 2017 *J. Photochem. Photobiol.* B*: Biol.* **169** 47–55
[16] Maobe M A G and Nyarango 2013. *Global J. Pharm.* **7** 61–8
[17] Raj A T, Ramanujan K, Thangavel S, Gopalakrishan S, Raghavan N and Venugopal G 2015 *J. Nanosci. Nanotech.* **15** 3802–8

Imidazoline-lauric derivative synthesis using MAOS (microwave assisted organic synthesis) method and its activity as corrosion inhibitor towards carbon steel

B Andreana, D U C Rahayu, D A Nurani, Y K Krisnandi and E Budianto

Department of Chemistry, Faculty of Mathematics and Natural Sciences (FMIPA), Universitas Indonesia, Kampus UI Depok, Depok 16424, Indonesia

Corresponding author's e-mail: dyahutamicr@sci.ui.ac.id

Abstract. Imidazoline is a heterocyclic compound containing nitrogen atom, which is well-known in industry, especially in petroleum field as a corrosion inhibitor of oil refineries in sea environment. In searching of potential organic compounds for corrosion inhibitor, imidazoline-lauric derivative had been successfully synthesized by reacting triethylenetetramine (TETA) and lauric acid (LA) with solvent free and variation of reaction time using MAOS (Microwave Assisted Organic Synthesis) method. The optimum yield of TETA-LA imidazoline was 72.6 % at 7 min reaction time in 800 watts. TETA-LA imidazoline was identified using TLC and melting point analyzer while its structure had been confirmed using Fourier Transform Infrared (FTIR), UV-Vis, Proton Nuclear Magnetic Resonance (^{1}H-NMR), and Mass Spectrometry (MS) spectral data. The highest percentage of inhibition efficiency (% IE) using Tafel polarization method was 74.5 % at 500 ppm of TETA-LA imidazoline. It can be concluded that imidazoline-lauric derivative is potent to be developed as corrosion inhibitor towards carbon steel.

Keywords: imidazoline, lauric acid, triethylenetetramine, MAOS, corrosion inhibitor.

1. Introduction

In petroleum industry, to fulfil the increasing oil consumption, oil refineries are placed in deep sea to find new resource of oil and gas. Oil refineries are made of carbon steel, which is easily to have corrosion due to salinity of seawater. Therefore, corrosion inhibitor such as imidazoline should be added to the sea since it is the most efficient and economical ways to prevent the corrosion [1–2]. Imidazoline is a heterocyclic compound having three carbons and two nitrogen arranged at 1 and 3 positions on its five-member cyclic and has carbon chain attached to $C(2)$ and $N(3)$. Imidazoline can be synthesized by reacting carboxylic acid derivatives with polyamine having more than two amino groups [3].

In recent years, research on imidazoline derivatives as corrosion inhibitor has been rapidly growing. By varying side chain on carboxyl-derived or amino group, many scholars found how imidazoline structure impacts its activity [1,4–6]. Others also discovered that by modifying imidazoline to its quaternary ammonium salt improves its performance [2–3,7–10]. In this study, lauric acid (LA) was used as a source of carboxylic acid derivative and triethylenetetramine (TETA) as polyamines to produce imidazoline with different pendant group from previous study [1]. Therefore, the aim of this study is to synthesize imidazoline-lauric derivative and evaluate its activity as corrosion inhibitor towards carbon steel. Imidazoline was synthesized using MAOS (Microwave Assisted Organic Synthesis) method since it provides the most effective and efficient method for organic synthesis, as well as to support green chemistry [11].

2. Materials and methods

2.1. General

MAOS method was performed using domestic microwave oven Samsung ME-731K/XSE. Imidazoline-lauric derivative was characterized using Fourier Transform Infrared (FTIR) Shimadzu IR Prestige 21, UV-Vis Shimadzu UV-2450, Proton Nuclear Magnetic Resonance (^{1}H-NMR) Agilent 500 MHz, and Mass Spectrometry (MS) Shimadzu Prominance 20-Abesciex. Melting point was measured using Electrothermal Melting Temp. Apparatus. Corrosion inhibition activity was determined using eDAQ 450 potentiostat and Versasatat II with low carbon steel JIS G3123 grade SGD 400D (C 0.08 %, Mn 0.40 %, Si 0.19 %, P 0.033 %, S 0.021 %, and Fe 99.276 %) from PT. Citra Tanamas as working electrode (WE), SCE as reference electrode (RE), and platinum as auxiliary electrode (AE). The chemicals used for synthesis and corrosion activity test were triethylenetetramine (TETA) from Sigma Aldrich, lauric acid (LA) from PT. Evyap, while TLC plates, CH_2Cl_2, MeOH, NH_4OH 25 %, $Ce(SO_4)_2$, H_2SO_4 98 %, Na_2SO_4, acetone, and NaCl form Merck.

2.2. MAOS (microwave assisted organic synthesis) method

TETA (5 mmol) and 5 mmol of LA were heated in microwave (800 watts) with variation of time (7, 8, 9, and 11 min) at 200-240 °C. After reaction was completed, the mixture was cooled and dissolved in CH_2Cl_2. The organic layer was extracted with water three times to remove unreacted precursors. The organic layer was then dried with Na_2SO_4 and concentrated to obtain imidazoline-lauric derivative, TETA-LA imidazoline. TETA-LA imidazoline was monitored by Thin-Layer Chromatography (TLC) using CH_2Cl_2:MeOH 8:2 with one drop NH_4OH 25 % as an eluent, visualized under UV lamp (λ = 254 nm), and sprayed with $Ce(SO_4)_2$ in H_2SO_4. TETA-LA imidazoline then was identified using melting point analyzer and characterized using FTIR, UV-Vis, ^{1}H-NMR, and MS.

TETA-LA imidazoline was obtained as dark-brown oily solid; *Rf* 0.88 in CH_2Cl_2:MeOH (8:2) with one drop NH_4OH 25 %; mp. 63–64 °C; FTIR with KBr ν_{max} (cm^{-1}): 3304.2, 3071.1, 2913.2, 2851.9, 1648.2, 1554.7, 1467.9; UV-Vis in MeOH λ_{max} (nm): 233; ^{1}H-NMR (500 MHz, $CDCl_3$) δ_H (ppm): 3.41 (2H, *t*, 10 Hz); 3.32 (2H, *t*, 10 Hz); 2.79–3.14 (4H, *m*); 2.73 (2H, *t*, *J* = 8.0); 2.39–2.55 (2H, *m*); 2.15 (2H, *t*, 7.5 Hz); 1.58 (2H, *p*, 7.5 Hz); 1.22–1.27 (16H, *m*); 0.84 (3H, *t*, *J* = 7.5 Hz); MS $(M+H)^+$: 311.3515 m/z.

2.3. Corrosion activity test

The WE was abraded with SiC abrasive paper grade 600 and rinsed with acetone and bidistilled water. NaCl 1 % solution was saturated with CO_2 gas for 30 min as corrosive medium and blank solution. Test solution was made by adding various concentration of TETA-LA imidazoline (100, 200, 300, 400, and 500 ppm) to NaCl 1 % solution. WE, RE, and AE were then arranged in NaCl 1 % as an electrochemical cell. Potentiodynamic polarization was performed from -2000 to +500 mV at 50 mV/s scanning rate. The electrochemical parameters, i.e. corrosion potensial (E_{corr}) and corrosion density (I_{corr}), and also percentage of inhibition efficiency (%IE) were obtained by using Tafel polarization method and equation (1).

$$\%IE = \frac{I\ blank\ (mA/cm^3) - I\ sample\ (mA/cm^3)}{I\ blank\ (mA/cm^3)} \times 100\% \qquad (1)$$

3. Results and discussion

In this study, imidazoline-lauric derivative synthesis was conducted by reacting TETA with LA using MAOS method yielding TETA-LA imidazoline with the IUPAC name of *N*^{1}-(2-(2-undecyl-4,5-dihydro-1*H*-imidazole-1yl)ethyl)ethane-1,2-diamine. TETA-LA imidazoline had main structure of imidazoline ring containing *N*(1)-*C*(2)-*N*(3) bonds with long hydrocarbon chain attached to C(2) and pendant group attached to N(3), with a physical form of dark-brown oily solid. Figure 1 shows the reaction scheme of TETA-LA imidazoline synthesis. According to figure 1, TETA-LA imidazoline synthesis consisted of two-stages reaction, i.e. formation of amide followed by cyclization. In both stages, H_2O as by product was produced [1]. To our knowledge, although this is not the first reported of the synthesis of TETA-LA imidazoline [6], this is the first reported of ^{1}H-NMR and MS spectral data of TETA-LA imidazoline as well as its activity as corrosion inhibitor towards carbon steel in NaCl 1 % solution.

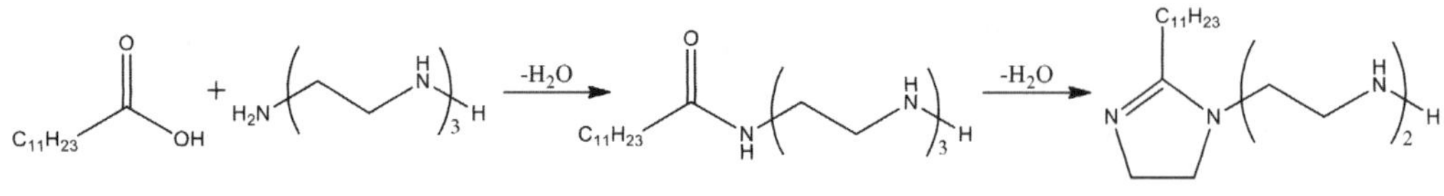

Figure 1. Reaction scheme of imidazoline-lauric derivative (TETA-LA) synthesis.

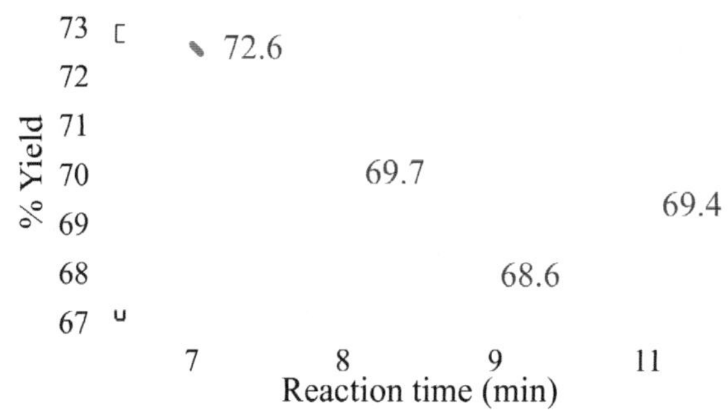

Figure 2. The percentage yield of TETA-LA imidazoline.

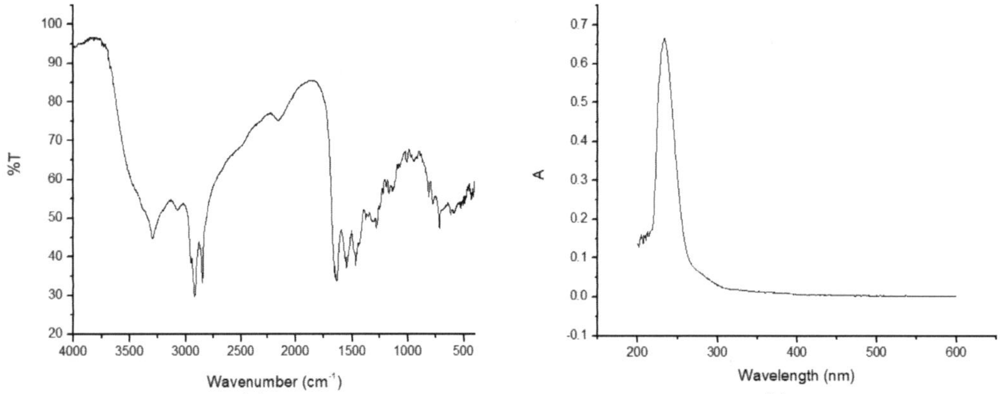

(a) (b)

Figure 3. (a) FTIR and (b) UV-Vis spectra of TETA-LA imidazoline

Synthesis of TETA-LA imidazoline was carried out by varying reaction times in 7, 8, 9, and 11 min (figure 2), where the optimum yield was obtained at 7 min (72.6 %). According to figure 2, the shorter reaction time, the higher yield of TETA-LA imidazoline was obtained. However, the minimum time of imidazoline synthesis using MAOS method with domestic microwave oven (800 watts) was 7 min. As reported before, in shorter than 7 min reaction time, there was a possibility that amide product had not completely continued into imidazoline [12].

FTIR spectra of TETA-LA imidazoline are shown in figure 3a. According to figure 3a, the wavenumber of 3050–3300 cm^{-1} was the absorption peaks of N-H bonds on the ethylene groups while 2800–2950 cm^{-1} was the shifting vibration of C-H bonds on the alkyl and ethylene groups. The specific absorption of imidazoline ring was appeared at around 1600 and 1400 cm^{-1} which belong to absorption peaks of C=N and C-N-C bonds on the imidazoline ring, respectively [3, 5–6]. Figure 3b shows UV-Vis spectra of TETA-LA imidazoline, where the maximum absorbance peak appeared at the wavelength of 233 nm. It indicated that imidazoline product had been formed, where the characteristic peak that appeared at around 230 nm indicated C=N bond of imidazoline [3].

^{1}H-NMR spectra of imidazoline derived from saturated fatty acid commonly consist of three regions of peak, i.e. peaks from hydrocarbon chain (δ_H 0.80-2.20 ppm), pendant group (δ_H 2.40-3.40 ppm), and imidazoline ring (δ_H 3.30–3.70 ppm) [4, 12-13]. Figure 4a shows ^{1}H-NMR spectra of TETA-

(a)

(b)

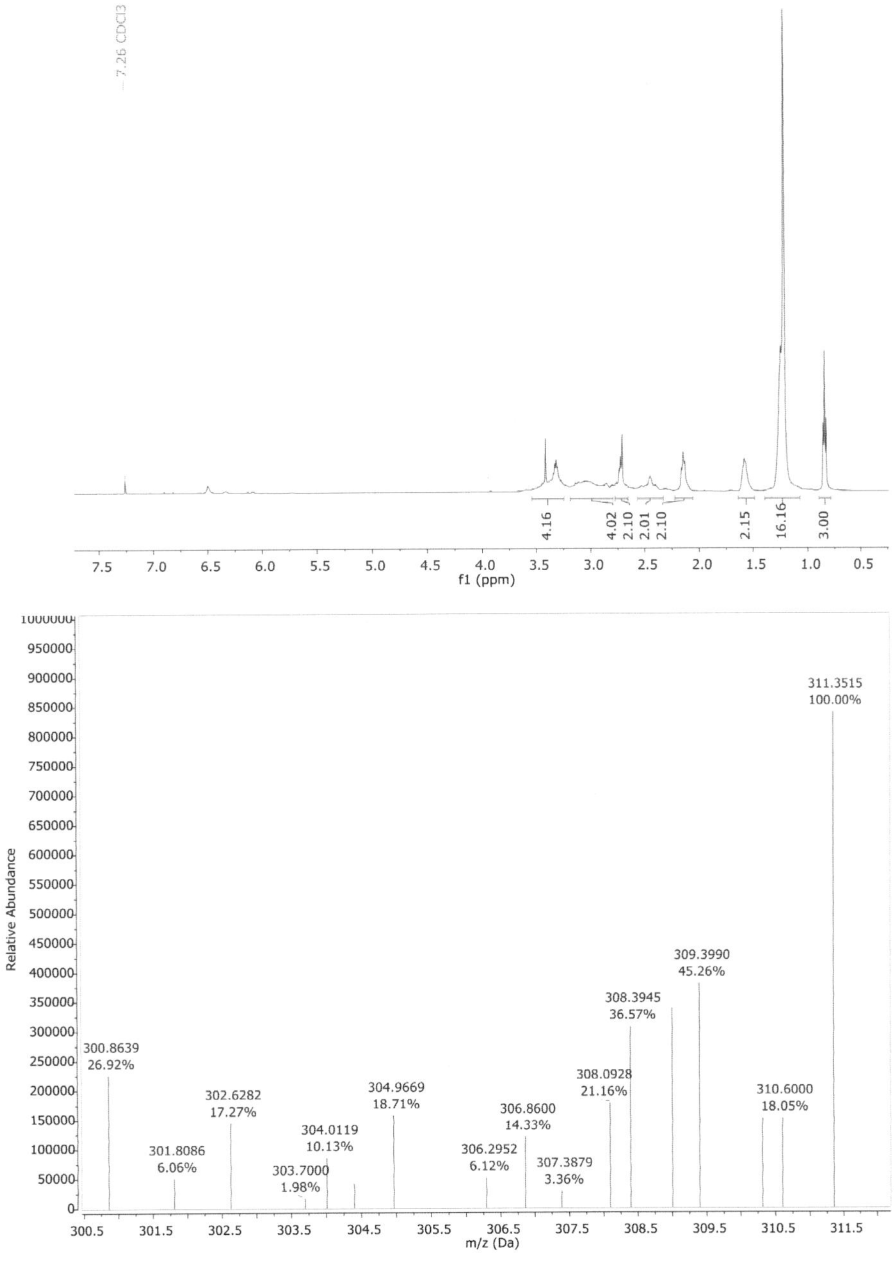

Figure 4. (a) ^{1}H-NMR and (b) MS spectra of TETA-LA imidazoline

LA imidazoline (500 MHz, $CDCl_3$). Based on figure 4a, proton signal at chemical shifts of 3.32 and 3.41 ppm indicated proton signal of imidazoline ring. Proton signal of imidazoline ring usually appears as deshielding peak than others since it is near the electronegative atoms in cyclic. Proton signal at chemical shifts of 2.39–3.14 ppm pointed to proton signal of pendant group with 8 proton integrations belongs to 2 ethyl-amino groups from TETA. The remaining proton signal at chemical shift 0.84–2.15 ppm is specified to proton signal of hydrocarbon chain with 23 proton integrations belongs to -$C_{11}H_{23}$ groups from LA. By elucidating ^{1}H-NMR spectra, TETA-LA imidazoline had the structure as illustrated in figure 1. Furthermore, MS characterization was also conducted to confirm the imidazoline structure by adding HCOOH and resulting m/z value in $[M+H]^+$ (figure 4b). The $[M+H]^+$ value for TETA-LA imidazoline was 311.3515 m/z which correspond to $C_{18}H_{38}N_4$.

Table 1. The corrosion inhibition activitiy using Tafel polarization method.

TETA-LA imidazoline compound	I_{corr} (mA)	E_{corr} (mV)	% IE
Blank	3.119	315	-
100 ppm	2.774	298	11.1
200 ppm	2.284	369	26.8
300 ppm	2.235	435	28.3
400 ppm	1.047	228	66.4
500 ppm	0.797	192	74.5

Corrosion inhibition by using saturated fatty acid imidazolines was the common method because these compounds is able to be adsorbed on metal surface through π-electron of imidazoline ring and lone pair of electrons of N atoms [1, 14–15]. Table 1 shows the corrosion inhibitor activity of TETA-LA imidazoline on low carbon steel in NaCl 1 %. According to table 1, the higher concentration of imidazoline addition, the higher number of inhibition efficiency percentage (% IE). Presumably, this data can be described because the higher imidazoline concentration in test solution, the more imidazoline compound adsorbed and formed an inhibition layer on the metal surface. Therefore, the metal was protected from corrosion medium. This result revealed that the presence of hydrocarbon chain had an important role to inhibit the corrosion since corrosion medium will be difficult to interact with metal surface. The head (imidazoline ring) and pendant group as hydrophilic sites will promote bonding of the molecule to the metal surface while hydrocarbon chain as hydrophobic site will protect the metal surface to keep the corrosion medium away from the metal surface [14–15].

4. Conclusions

Imidazoline-lauric derivative, TETA-LA imidazoline, had been successfully synthesized using MAOS method with the optimum yield of 72.6 % at 800 watts at 7 min reaction time. The highest corrosion inhibition activity was achieved at 500 ppm of TETA-LA imidazoline with % IE value of 74.5 %. Imidazoline-lauric derivative can be used as an alternative corrosion inhibitor derived from organic compounds, since it gives high activity of anticorrosion in this work.

Acknowledgements

This work was supported by Hibah PITTA 2018 funded by Universitas Indonesia with contract number 2258/UN2.R3.1/HKP.05.00/2018.

References

[1] Quraishi M A, Rafiquee M Z A, Saxena N and Khan S 2008 *Prot. Met.* **44** 91–8
[2] Fei F, Hu J, Wei J, Yu Q and Chen Z 2014 *Constr. Build. Mater.* **70** 43–53
[3] Tyagi R, Tyagi V K and Pandey S K 2007 *J. Oleo Sci.* **56** 211–22
[4] Jawich M W S, Oweimreen G A and Ali S A 2012 *Corros. Sci.* **65** 104–12
[5] Abbasov V M, Guliyev A A, Jabrailzadeh Sh Z, Mursalov N I, Duzdaban Kh R and Mamedaliev Y 2015 *International Journal of Engineering and Innovative Technology (IJEIT)* **5** 21–3
[6] Kong X, Qian C, Fan W and Liang Z 2018 *J. Mol. Struct.* **1156** 164–71
[7] Fahrurrozie A, Sunarya Y and Mudzakir A 2010 *Jurnal Sains and Teknologi Kimia* **1** 100–11
[8] Karadağ A and Destegül A 2013 *J. Mol. Liq.* **177** 369–75
[9] Zhao M, He H, Dai C, Wu X, Zhang Y, Huang Y and Gu C 2018 *J. Mol. Liq.* **268** 875–81
[10] Bajpai D and Tyagi V K 2008 *J. Surfact. Deterg.* **11** 79–87
[11] Chawla A, Sharma A and Sharma A K 2012 *Der Pharma Chem.* **4** 116–40
[12] Wahyuningrum D, Achmad S, Syah Y M, Buchari and Ariwahjoedi B 2008 *Journal of Mathematical and Fundamental Sciences* **40** 33–48
[13] Yoo S H, Kim Y W, Chung K, Baik S Y and Kim J S 2012 *Corros. Sci.* **59** 42–54
[14] Bajpai D and Tyagi V K 2006 *J. Oleo Sci.* **55** 319–29
[15] Rustandi A, Soedarsono J W, Febriary M C and Barrinaya M A 2017 *Proc. Seminar Nasional Teknologi (Senatek) 2015* 17 Januari 2015 Malang (Malang: Institut Teknologi Nasional Malang) 851–9

IOP Conf. Series: Materials Science and Engineering **763** (2020) 012011 doi:10.1088/1757-899X/763/1/012011

Synthesis and antibacterial activity of 2-((1,4-dioxo-1,4-dihydronaphthalen-2-yl)amino)ethyl acetate

A H Cahyana, B Ardiansah and A C Widodo

Department of Chemistry, Faculty of Mathematics and Natural Sciences (FMIPA), Universitas Indonesia, Kampus UI Depok, Depok 16424, Indonesia

Corresponding author's e-mail: herrykim@ui.ac.id

Abstract. Derivatization of lawsone compound (1) was done by reacting it with ethanolamine to give 2-((2-hydroxyethyl)amino)-1,4-naphthoquinone (2), and then treating with acetic anhydride to produce the desired compound 2-((1,4-dioxo-1,4-dihydronaphthalen-2-yl)amino) ethyl acetate (3). The target compound possesses the antibacterial activity against E. coli and S. aureus.

1. Introduction

Antibacterial activity is one of the natural product properties for killing or inhibiting the growth of bacteria [1]. The necessity of antibacterial in Indonesia was high since 2008 until 2010 [2]. Moreover, there are many cases of drugs resistancy caused by bacterial mutation [3]. Nowadays, researchers have been searching for new natural products or their derivatives with potent antibacterial activities.

2-hydroxy-1,4-naphthoquinone (lawsone) is one of natural products isolated from henna leaf (*Lawsonia inermis*) [4]. It is usually utilized as hair dye for woman. Lawsone compound has been known to show many bioactivities, such as antifungal, anti-inflammatory, antioxidant, and antibacterial [5]. The antibacterial activity is affected by the presence of two carbonyl groups at C-1 and C-4 position of naphthoquinone ring. Meanwhile, functional group at C-2 affect hydrophilicity of naphthoquinone [6]. Herein, we report two-step synthesis of 2-((1,4-dioxo-1,4-dihydronaphthalen-2-yl)amino)ethyl acetate (3) from lawsone (figure 1). Then, antibacterial activity will be tested using disc diffusion method [7] against the gram-positive bacteria of *S. aureus* and gram-negative bacteria of *E. coli*.

2. Materials and methods

2.1. Materials

All chemicals were analytical grade and purchased from some chemical companies. *Staphylococcus aureus and Escherichia coli* were obtained from Biochemistry Laboratory, Department of Chemistry, Universitas Indonesia. The growth of bacterial was supported by nutrient broth (NB) and nutrient agar (NA) from Merck, Germany. Dimethyl Sulfoxide (DMSO) was used as solvent in antibacterial test and also as negative control, whereas chloramphenicol was used as positive control. Fourier Transform-Infrared (FT-IR) spectra were recorded on Shimadzu Prestige 21 spectrophotometer. Mass spectra were obtained from Shimadzu QP Mass Spectrometer (MS) 2010A with 70 eV electron ionization potential.

2.2. Methods

2.2.1. Synthesis of 2-((2-hydroxyethyl)amino)-1,4-naphthoquinone (compound 2). Synthesis of 2-((2-hydroxyethyl)amino)-1,4-naphthoquinone (2) was initiated by mixing of lawsone (1), ethanol, and

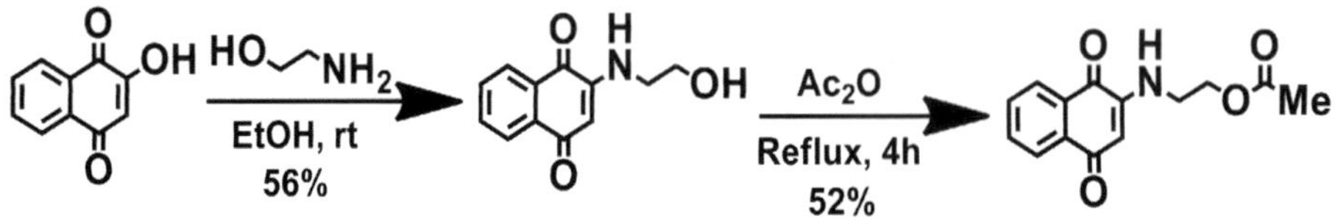

Figure 1. Synthesis of lawsone derivative

ethanolamine as reagent. Lawsone (1 mmol) was dissolved in 20 mL ethanol in 50 mL round-bottom flask. Ethanolamine (1 mmol) was inserted into the flask. Reaction was performed at room temperature for 2 h, and product formed was monitored by Thin-Layer Chromatography (TLC). Solvent was volatilized, and then crude product was purified by column chromatography to obtain compound 2.

2.2.2. Synthesis of 2-((1,4-dioxo-1,4-dihydronaphthalene-2-yl)amino)ethyl acetate (compound 3). Compound 2 (0.25 mmol) and acetic anhydride (5 mL) was reacted under reflux system at 115 °C for 4h. After completion, solvent was volatilized with vacuum and crude product was refined by column chromatography to get compound 3 as red solid.

2.2.3. Antibacterial activity assay. In antibacterial activity assay, samples were dissolved in DMSO (2000 μg/mL). Chloroamphenicol and DMSO were used as positive and negative control, respectively. Nutrient agar (20 mL) was mixed with nutrient broth (NB) bacterial suspension (200 μL) in which absorbance was measured within range 0.08 to 0.10 [8]. The mixture was put in the petri dish. A 6 mm disc paper was immersed in one antibacterial sample for 30 seconds and put on the hard agar. Petri dish was put inside an incubator for 48 h. Diameter of growth inhibition zone was measured to determine antibacterial activity.

3. Results and discussion

3.1. Synthesis proccess
Compound 2 was synthesized by utilizing the resonance properties of naphthoquinone ring. Resonance in naphthoquinone ring affected the stabilization of strucutre. Amine group which is good nucleophilic attacked and bonded with carbon atom at C-2 [9]. Hydroxyl group of lawsone bonded with a proton and naphthoquinone ring and released hydroxyl group as water molecule, while electron stabilized the stucture immediately. Reaction yielded at 56% of 2 as red solid. We synthesized compound 2 again with water as solvent with the isolated yield of 46%.

Compound 3 was synthesized by acetylation of hydroxyl group. Resonance of anhydride acetic gave a place of carbon atom to be attacked by hydroxyl group. This affected the oxygen atom to withdraw electron and release as acetic acid immediately, while electron stabilized the structure [10]. Yield percentage of the reaction was found to be 52% as red solid.

3.2. Characterization of products
In FT-IR spectrum of 2-((2-hydroxyethyl)amino)-1,4-naphthoquinone (2), a medium broad peak appeared from 3200 cm^{-1} to 3600 cm^{-1} was interpreted as overlapping of secondary amine (-NH-) and hydroxyl groups (-OH). A sharp peak appeared at 1725 cm^{-1} and a weak peak located at 1650 cm^{-1} were assumed as identity of carbonyl (C=O) and C=C double bond, respectively. FT-IR spectrum of 2-((1,4-dioxo-1,4-dihydronaphthalene-2-yl)amino)ethyl acetate (3) showed a different functional group than the former product. A medium sharp peak appeared at 3333 cm^{-1} was assumed as secondary amine group (-NH) [11]. Moreover, at wavenumber of 1050 cm^{-1}, a weak sharp peak attributed to C-O-C functional group. The FT-IR spectra of compound 2 and 3 can be seen in figure 2.

In MS characterization, both products were measured to ensure the relative molecular mass corresponding to structure of compound. Figure 3a shows MS spectrum of 2-((2-hydroxyethyl)amino)-1,4-naphthoquinone (compound 2). The m/z value of compound 2 was 217. This number was matched with Mr of the desired product having molecular formula of $C_{12}H_{11}NO_3$. For the MS characterization of 2-((1,4-dioxo-1,4-dihydronaphthalene-2-yl)ethyl acetate (compound 3), the m/z value of this compound was found to be 259. The figure 3b shows MS spectrum of 2-((1,4-dioxo-1,4-

IOP Conf. Series: Materials Science and Engineering **763** (2020) 012011 doi:10.1088/1757-899X/763/1/012011

Table 1. Result of antibacterial activity test

Antibacterial Sample	*S. aureus* (mm)	*E.coli* (mm)
Positive control (Chloroamphenicol)	16.1	12.2
Negative control (DMSO)	No activity	No activity
Compound 1	11.0	20.0
Compound 2	9.8	17.1
Compound 3	10.0	10.8

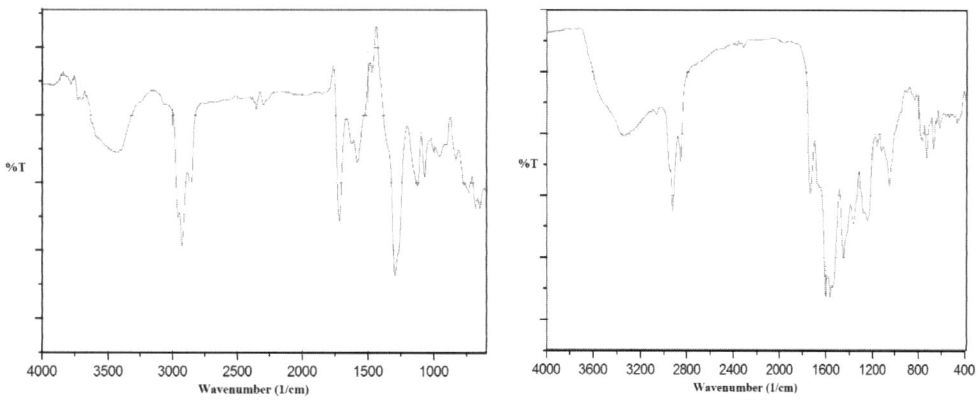

(a) (b)

Figure 2. FT-IR Spectrum of lawsone derivates: (a) 2-(2-hydroxyethyl)amino)-1,4-naphthoquinone; (b) 2-((1,4-dioxo-1,4-dihydronaphthalene-2-yl)amino)ethyl acetate

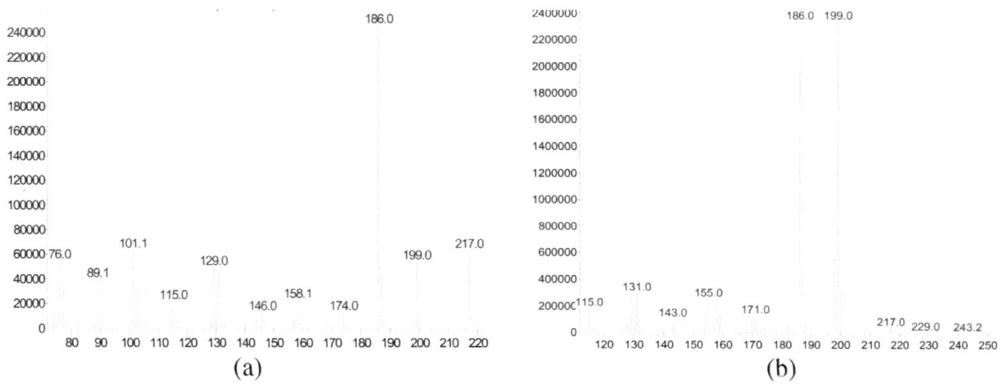

(a) (b)

Figure 3. Mass spectra of (a) 2-((2-hydroxyethyl)amino)-1,4-naphthoquinone (compound 2) and (b) 2-((1,4-dioxo-1,4-dihydronaphthalene)amino)ethyl acetate (compound 3)

dihydronaphthalene-2-yl)amino)ethyl acetate (3). This number was matched with Mr of product having formula of $C_{14}H_{13}NO_4$. Both products have an odd m/z number which indicate the presence of nitrogen atom within the structures [13].

3.3. Antibacterial activity test

Disc diffusion method is one of the antibacterial tests where the measurement of growth inhibition zone is the indicator of antibacterial activity. Therefore, antibacterial activity is measured by the ability to inhibit or kill a bacterial which form an inhibition zone [7,8]. Classification of antibacterial activity ability listed by diameter of growth inhibition zone. Very strong: ≥ 30 mm; strong: 21–29 mm; medium: 16–20 mm; weak: 6–15 mm; and no activity ≤ 6 mm [6].

In antibacterial test against *S. aureus*, all tested compounds showed weak activity (table 1). Compound (1) was able to form 11 mm inhibition zone, while (2) and (3) were able to form 9.8 and 10 mm, respectively. In another antibacterial test against *E. coli*, the activity of compound (1) and (2) were stronger than compound (3). Lawsone (1) was able to form 20 mm inhibition zone, while compound (2) and (3) were able to form 17.1 and 10.8 mm, respectively. These results prove that lawsone derivates have antibacterial activity against *S. aureus* and *E. coli.*

4. Conclusions

We have successfully synthesized lawsone derivatives (compound 2 and compound 3) in medium yields. These compounds showed antibacterial activity against *S. aureus* and *E. coli* at 2000 µg/mL sample concentration. However, derivative products have weaker antibacterial activity than lawsone (1). It may be due to the solubility effect of the derivatives that could limit the interaction with bacteria.

Acknowledgements

We would like to thank Universitas Indonesia for funding this project through PITTA Grant 2018 with contract number 234/UN2.R3.1/PPM.00/2018.

References

[1] Leekha S, Terrell C L and Edson R S 2011 *Mayo Clin. Proc.* **86** 156–67
[2] Pradipta I S, Ronasih E, Kartikawati A D, Hartanto H, Amelia R, Febrina E and Abdulah R 2015 *J. Family and Community Med.* **22** 101–5
[3] Ventola C L 2015 *Pharm. Ther.* **40** 277–83
[4] Ashnagar A and Shiri A 2011 *Int. J. ChemTech Res.* **3** 1941–4
[5] Jordao A K, Vargas M D, Pinto A C, da Silva F C and Ferreira V F 2015 *RSC Adv.* **83** 67909–43
[6] Rahmoun N M, Boucherit-Atmani Z, Benabdallah M, Boucherit K, Villemin D and Choukchou-Braham N 2013 *Am. J. Med. Biol. Res.* **1** 16–22
[7] British Society for Antimicrobial Chemotherapy 2003 *BASC Method for Antimicrobial Susceptibility Testing* (Birmingham: BSAC)
[8] National Committee for Clinical Laboratory Standards 2003 Performance Standard for Antimicrobial Disk Susceptibility Tests: Approval Standard 8th ed (Wayne: National Committee for Clinical Laboratory Standards) Document M2-A8
[9] Elavarasan S, Bhakiaraj D and Gopalakrishnan M 2013 International Journal of Pharmaceutical Science and Health Care (IJPHC) **6** 131–44
[10] Fritz J S and Schenk G H 1959 *Anal. Chem.* **31** 1808–12
[11] Hijji Y M, Barare B and Zhang Y 2012 *Sens. Actuators* B*: Chem.* **169** 106–12
[12] Williams D and Fleming I 1995 Spectroscopic Methods in Organic Chemistry 5th ed (London: McGraw-Hill)
[13] Gross J H 2010 Mass Spectrometry 2nd ed (Berlin: Springer) p 265

Cellulose-Fe_3O_4 nanocomposite based on rice husk as catalyst for synthesis of methyl ester from waste cooking oil

L Trisnawati and H Helmiyati

Department of Chemistry, Faculty of Mathematics and Natural Sciences (FMIPA), Universitas Indonesia, Kampus UI Depok, Depok 16424, Indonesia

Corresponding author's e-mail: helmiyati@sci.ui.ac.id

Abstract. Cellulose-Fe_3O_4 nanocomposite has been successfully synthesized by modifying Fe_3O_4 onto nanocellulose acetate surface derived cellulose from rice husk. It was supported by Fourier Tranform-Infrared (FT-IR) characterization, where there was absorption at 900 cm^{-1}, which was a characteristic for cellulose absorption, and at 572 cm^{-1}, which was the absorption of Fe-O. The morphology of cellulose-Fe_3O_4 nanocomposite by Scanning Electron Microscopy (SEM) having fibers shape were the characteristic of cellulose and Fe_3O_4 nanoparticle attached on the surface of nanocellulose and supported by SEM mapping. The characterization of Tunneling Electron Microscopy (TEM) obtained nano Fe_3O_4 on the nanocellulose surface with average size around 55 nm. The cellulose-Fe_3O_4 nanocomposites have the capability, as catalyst for the synthesis of methyl esters from waste cooking oil with the optimal conditions at 60 °C for 120 minutes with the obtained conversion is 77.56 %. The nanocomposite of based biopolymer with nanomagnetite active side is potential for catalysts from the used cooking oil to biodiesel as renewable energy and can replace limited fossil energy.

Keywords: cellulose, catalyst, nanocomposite, rice husk, waste cooking oil

1. Introduction

Rice husk is one of the agricultural wastes, which is by-product in the process of rice mill into rice [1,2]. Rice husk contains organic material about 40 % cellulose, 30 % hemicellulose, 10 % lignin and other mineral components, therefore rice husk is considered as a potential source of cellulose from agricultural plant waste [3]. Cellulose is one of the most abundant biopolymers in nature, which is biodegradable, non-toxic, and environmentally friendly materials [4,5].

In the recent years, nanocomposite materials have been evolved as heterogeneous catalysts for various applications. Cellulose can be utilized as supporting materials for catalyst, cellulose-based magnetite nanocomposite biopolymers as a catalyst for the synthesis of methyl oleate [6].

Nanocomposite polymer is a combination of polymer matrix and nanoparticle inorganic. When an inorganic nanoparticle is added to the nanocellulose polymer matrix it will be affected to the strength of the resulting nanocomposite [7,8]. The resulting nanocomposite will provide superior properties derived from the properties of nanocellulose and inorganic nanoparticles, which will form better properties [9,10]. Waste cooking oil (WCO) can be utilized as the main ingredient for the production of methyl ester by oil transesterification and short-chained alcohol reaction in the presence of a catalyst [11]. Methyl ester biodiesel can be utilized as alternative fuels that are environmentally friendly, non-toxic and produce low emission [12]. Waste cooking oil contains triglycerides and free fatty acids. It can be an effective and economical solution related to the disposal of WCO into the environment [13].

In this study, the synthesis of Fe_3O_4 nanocomposite based on cellulose biopolymer isolated from rice husk has been done and nanocomposite have been synthesized to be used as a catalyst for the synthesis of methyl ester from WCO derived from household waste. The several stages carried out in this study were cellulose isolation from rice husk, nanocellulose synthesis, Fe_3O_4 synthesis, cellulose- Fe_3O_4 nanocomposite synthesis, and nanocomposite application as a catalyst for synthesis of methyl ester from WCO. Characterizations of synthesis products were done using Fourier Transform-Infrared (FT-IR), X-Ray Diffraction (XRD), Scanning Electron Microscope (SEM) and Tunneling Electron Microscopy (TEM).

2. Experimental

2.1. Materials

Rice husk from Bukittinggi as the source material for cellulose; toluene and ethanol (Merck), KOH (Merck), sodium chlorite (Merck) for cellulose extraction of rice husk; anhydrous acetate (Merck) for the synthesis of nanocellulose; iron (III) chloride ($FeCl_3.6H_2O$) (Sigma Aldrich) and iron (II) chloride ($FeCl_2.4H_2O$) (Sigma Aldrich) for the synthesis of iron oxide; methanol (Sigma Aldrich) and waste cooking oil (WCO) as material for methyl ester production.

2.2. Synthesis of nanocellulose from rice husk

In the first step, cellulose was isolated from rice husks refer to our previous studies [14]. Then, nanocellulose acetate synthesis from cellulose refers to the previous reported [6]. Second step, cellulose soaked in HCl with stirring at 60 °C for 2 hours, after washing and filtering the nanocellulose will be obtained. The last step, the nanocellulose was functionalized with anhydrous acetic acid and nanocellulose acetate was obtained and characterized by FTIR, XRD and SEM.

2.3. Synthesis of cellulose-Fe_3O_4 nanocomposite

The first stage was conducted the synthesis of magnetite (Fe_3O_4) by co-precipitation method refers to previous reported [8]. The second stage was synthesis of cellulose-Fe_3O_4 nanocomposite refers to the reported method [16]. The nanocomposite formed were filtered and rinsed with distilled water, ethanol, and acetone. Then, they were dried at 60 °C for 24 hours, and the cellulose-Fe_3O_4 nanocomposite was obtained and characterized by FTIR, XRD, SEM and TEM.

2.4. Catalytic activity of cellulose-Fe_3O_4 nanocomposite

Catalytic activity of cellulose-Fe_3O_4 nanocomposites as catalysts was applied to waste cooking oil to produce biodiesel-using methanol with a transesterification reaction. The effect of temperature and reaction time was carried out to observe the conversions.

3. Results and discussion

3.1. Characterization

Characterization using FTIR for determination of the functional groups, FTIR analysis used IR-Prestige-21 Shidmazu IR-type instrumentation with a wavelength of 4000-400 cm^{-1}. The results of the spectrum can be seen in figure 1. Figure 1a is the spectrum of rice husks, which the characteristic absorption band around 3200-3450 cm^{-1} is O-H stretching and the absorption band at 2920 cm^{-1} shows C-H stretching. The peaks at 1516 cm^{-1} is lignin, and at 500 cm^{-1} is Si-O-Si group. These two peaks only appear for rice husks and did not appear at the spectrum of nanocellulose rice husk. Figure 1b shows the spectrum of pure nanocellulose, there is an appearance of characteristic band absorption of cellulose at 900 cm^{-1}, 1060 cm^{-1} and 1680 cm^{-1} is β-glycosidic bond. The peak around 3200-3450 cm^{-1} is the vibration of the O-H group and at 2913 cm^{-1} shows the C-H stretch. Figure 1c is nanocellulose rice husk, it shows that lignin, hemicellulose, and silica have disappeared. The characteristic band absorption of cellulose appears at 900 cm^{-1}, 1050 cm^{-1} and 1690 cm^{-1} is β-glycosidic bond, which is similar to the pure nanocellulose spectrum in figure 1b. Figure 1d is the spectrum of the nanocellulose acetate, in which this spectrum is similar to the nanocellulose rice husk in figure 1c. There is no significant change between nanocellulose rice husk with nanocellulose acetate, which proves that anhydrous acetic acid did not change the structure or functional groups of nanocellulose. It only changed the size of the nanocellulose to be smaller. Figure 1e shows the spectrum of magnetite (Fe_3O_4), there is appearance of a typical peak of Fe-O at 572 cm^{-1}. Figure 1f shows the spectrum of the cellulose- Fe_3O_4 nanocomposite. It can be observed that the identical absorption band of cellulose appear at 850 cm^{-1}, 1020 cm^{-1} and 1690 cm^{-1} is the β-glycosidic bond and there is absorption peak of

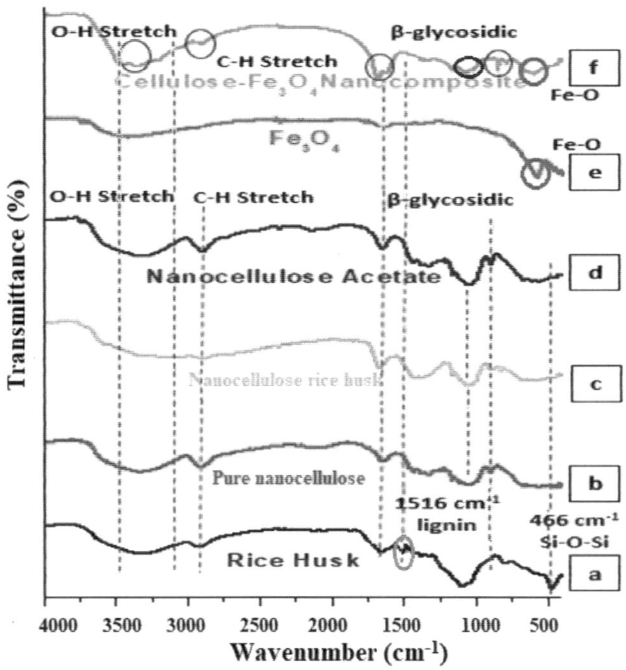

Figure 1. FTIR spectra of (a) rice husk, (b) pure nanocellulose, (c) nanocellulose rice husk, (d) nanocellulose acetate, (e) magnetite (Fe_3O_4), and (f) cellulose-Fe_3O_4 nanocomposite

Fe-O at 572 cm^{-1}. The results of the spectrum indicated that cellulose-Fe_3O_4 nanocomposites have been successfully synthesized.

The characterization of XRD was used to analyze the crystal structure of the synthesized sample. XRD pattern is shown in figure 2. Figure 2a is the diffraction pattern of nanocellulose acetate, there are three peaks appear at 2θ = 18.10°, 22.4° and 34.5° with the sharpest peak appeared at 2θ = 22.8°. These peaks are characteristics for cellulose [2,15]. Figure 2b is the nano Fe_3O_4 diffraction pattern, it can be observed that the peaks at 2θ = 30.1°, 35.2°, 43.4°, 53.2°, 57.2°, 60.5° and 64.5° with the highest and sharpest peak appeared at 2θ = 35°. This result is similar to the previous report [8]. In figure 2c, cellulose-Fe_3O_4 nanocomposite have typical peaks derived from nanocellulose at 2θ = 18.0°, 22.9° and the peaks of nano Fe_3O_4 are at 2θ = 30.5°, 35.6°, 43.8°, 53.9°, 61.1° and 65.5°. The peaks are dominated by Fe_3O_4, due to the nano Fe_3O_4 attached on the surface of the nanocellulose. The crystallinity index of cellulose-Fe_3O_4 nanocomposite obtained is 64.06 % and the crystal size calculated using the Debye Scherrer equation is 26.89 nm. The results of XRD-pattern supported that the cellulose-Fe_3O_4 nanocomposite has been synthesized successfully.

The SEM micrograph is shown in figure 3. Figure 3a shows that the surface morphology of cellulose rice husk with magnification of 1500 times have the form of fibers with its surface quite smooth. Figure 3b is the surface morphology of nanocellulose acetate with magnification of 1500 times has smoother fibers and evenly distributed. This shows that the addition of anhydrous acetate did not alter the structure of the cellulose, but it has capability to modify the nanocellulose into a smaller size. Figure 3c shows that the morphological results of Fe_3O_4 nanoparticles with a magnification of 2000 times have granular shape. Figure 3d shows that the morphology of cellulose-Fe_3O_4 nanocomposite with magnification of 2000 times have fibers that were the characteristic of cellulose and Fe_3O_4 nanoparticle attached on the surface of nanocellulose. This supports the results of the diffraction pattern by XRD. This result shows that nanocomposite of cellulose-Fe_3O_4 have been successfully synthesized.

The SEM-mapping is shown in the figure 4, for observing the distribution of the elements on the nanocomposite. Figure 4a shows the overall distribution of elements of the nanocomposite, the largest parts are the cellulose support material colored in green. Figure 4b is the distribution of C elements

IOP Conf. Series: Materials Science and Engineering **763** (2020) 012012 doi:10.1088/1757-899X/763/1/012012

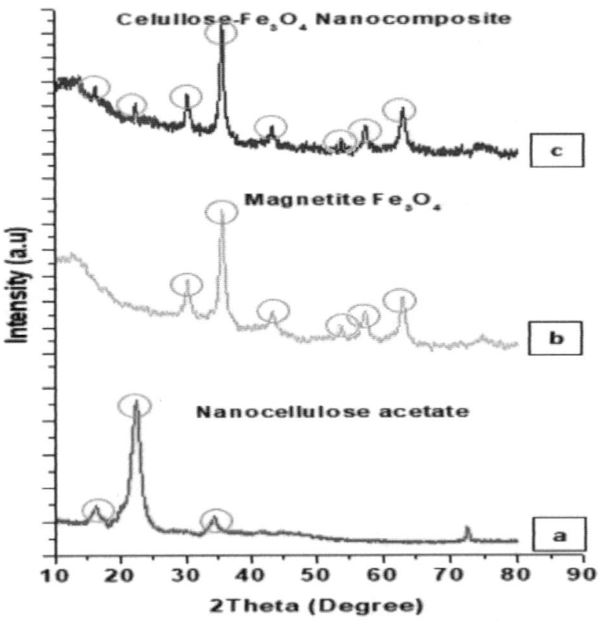

Figure 2. The XRD-patterns of (a) nanocellulose acetate, (b) nanoparticles Fe_3O_4 and (c) cellulose- Fe_3O_4 nanocomposites

Figure 3. SEM micrograph of (a) nanocellulose rice husk, (b) nanocellulose acetate, (c) nanoparticles Fe_3O_4 and (d) cellulose-Fe_3O_4 nanocomposite

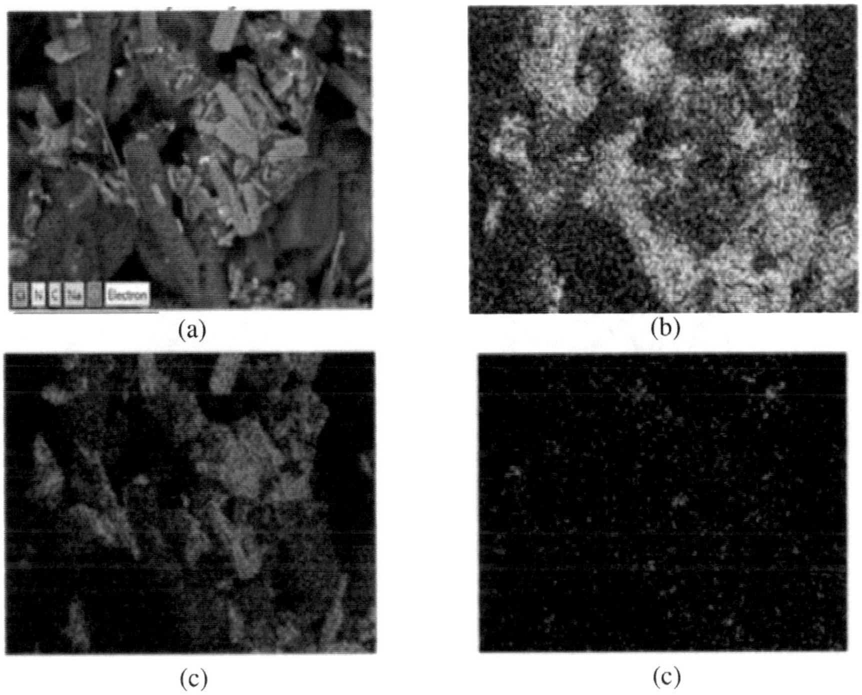

Figure 4. SEM-Mapping of (a) the overall of elements (b) C elements, (c) O elements, and (d) Fe elements in nanocomposite

colored in light blue evenly because C element is the main component of nanocellulose. Figure 4c is the distribution of element O colored in red evenly, due to O element is the main component of nanocellulose and nano Fe_3O_4. Figure 4d is the distribution of Fe elements colored in bright purple derived from nano Fe_3O_4, but its element was not so much within the nanocomposites as compared to the distribution of C and O elements. It is due to nanocellulose as support meanwhile the nano Fe_3O_4 as the active side of the catalyst [6]. From the result of SEM-Mapping, it shows that nanocomposite of cellulose-Fe_3O_4 have been successfully synthesized.

Characterization using TEM was performed to determine the morphology and size of the nanocomposite. The result of TEM characterization can be seen in figure 5. Figure 5 shows the surface morphology of cellulose-Fe_3O_4 nanocomposite with a scalebar of 100 nm. It can be observed gray colored is nanocellulose in the form of fiber and the distributed nano Fe_3O_4 on the surface nanocellulose with granular shape small dark-colored have an average size of 55 nm. This indicates that nano Fe_3O_4 was successfully composited on the surface of nanocellulose. This result supported the previous characterization. The nanocomposite used as a catalyst of waste cooking oil with methanol to biodiesel methyl ester.

3.2. Catalytic activity of cellulose-Fe_3O_4 nanocomposite

The catalytic activity of synthesized cellulose-Fe_3O_4 nanocomposite used for conversion of waste cooking oil with methanol to methyl ester and evaluated by the effect of temperature, and time reaction. The reaction temperature has an effect to the trans esterification reaction, it is caused by the temperature effect on the reaction rate [17]. The effect of temperature was conducted at 50 °C, 55 °C, 60 °C, and 65 °C as shown in figure 6. The temperatures rise will increase connversion of methyl esters. The conversion of optimum methyl ester is obtained about 77.56 % with reaction temperature at 60 °C. However, the conversion of methyl esters decreased at 65 °C, because at high temperature before reacting with WCO, methanol may evaporate because the boiling point of methanol about 64.5 °C so that the conversion of methyl ester decreased. The effect of time reaction to conversion of waste cooking oil to methyl esters was conducted at 30, 60, 90, 120 and 150 minutes as shown in figure 7. As the reaction time increased, the conversion of methyl ester was also increased, it can be

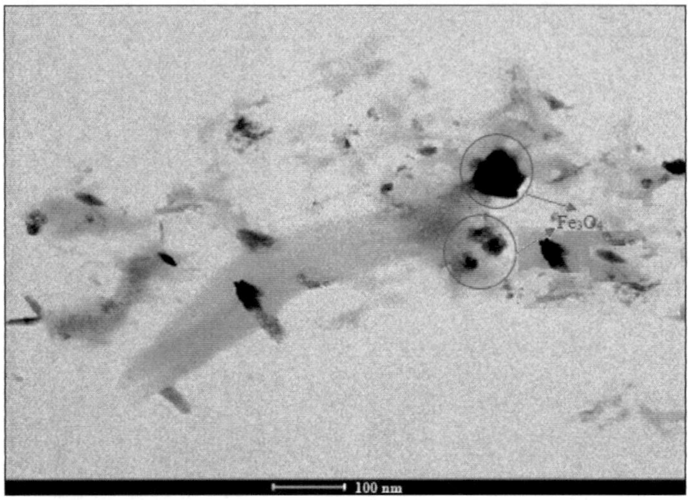

Figure 5. The TEM results of cellulose-Fe_3O_4 nanocomposite

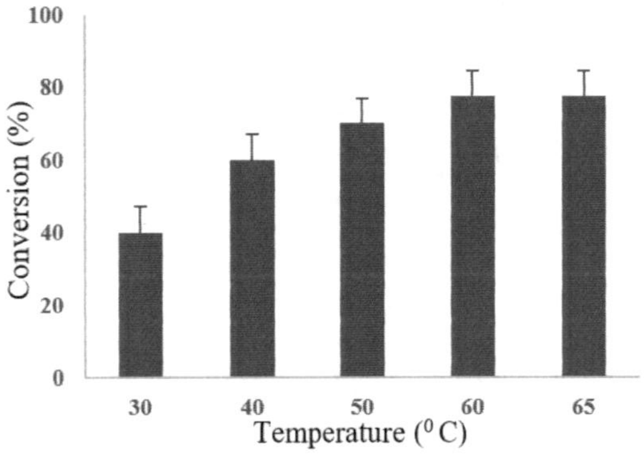

Figure 6. Effect of reaction temperature to conversion

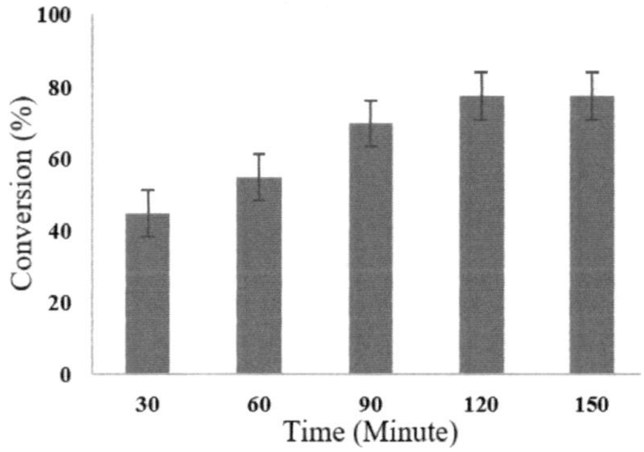

Figure 7. Effect of reaction time to conversion

observed from reaction time 30 to 120 minutes, the conversion percentage increased, but it decreased after 120 minutes. It was probably because at that condition the equilibrium had been reached. The optimal conversion of methyl ester was obtained at reaction time of 120 minutes is 77.56 %.

4. Conclusions

Cellulose-Fe_3O_4 nanocomposite has been successfully synthesized, which was supported by FT-IR characterization, where there was absorption at 900 cm^{-1}, which was a characteristic for cellulose absorption, and at 570 cm^{-1} which was the absorption of Fe-O. The morphology of cellulose-Fe_3O_4 nanocomposite by SEM having fibers shape that were the characteristic of cellulose and Fe_3O_4 nanoparticle attached on the surface of nanocellulose and supported by SEM mapping. The characterization of TEM obtained nano Fe_3O_4 on the surface nanocellulose of the average size around 55 nm. The cellulose-Fe_3O_4 nanocomposites have the capability, as catalyst for the synthesis of methyl esters from waste cooking oil with the optimal conditions at 60 °C for 120 minutes obtained conversion is 78.56 %.

Acknowledgements

The authors would like to thank Universitas Indonesia for funding this research through PITTA Grant 2018 with contract No: 2268/UN2.R3.1/HKP.05.00/2018

References

[1] Pode R 2016 *Renew. Sustain. Energy Rev.* **53** 1468–85
[2] Helmiyati, Abbas G H and Kurniawan S 2017 IOP Conf. Ser.: Mater. Sci. Eng. **188** 012024
[3] Battegazzore D, Bocchini S, Alongi J, Frache A and Marino F 2014 *Cellulose* **21** 1813–21
[4] Azad S and Mirjalili B B F 2016 *RSC Adv* **6** 96928–34
[5] Ma J, Li X and Bao Y 2015 *RSC Adv* **5** 59745–57
[6] El-Nahas A M, Salaheldin T A, Zaki T, El-Maghrabi H H, Marie A M, Morsy S M and Allam N K 2017 *Chem. Eng. J.* **322** 167–80
[7] Müller K *et al.* 2017 *Nanomaterials* **7** E74
[8] Zarei S, Niad M and Raanaei H 2018 *J. Hazard. Mater.* **344** 258–73
[9] Cai J, Liu S, Feng J, Kimura S, Wada M, Kuga S and Zhang L 2012 *Angew. Chem. Int. Ed. Engl.* **51** 2076–9
[10] Helmiyati and Dini F W 2018 *AIP Conf. Proc.* **2023** 020101
[11] Arumugam A and Ponnusami V 2017 *Heliyon* **3** e00486
[12] Gashaw A, Getachew T and Teshita A 2015 *Journal of Forest Products and Industries (JFPI)* **4** 80–5
[13] Farooq M, Ramli A and Subbarao D 2013 *J. Clean Prod.* **59** 131–40
[14] Helmiyati, Saefumillah A and Yulianti W 2014 *Asian J Chem* **26** 7337–42
[15] Helmiyati, Fitriyani A and Meyanti F 2017 *IOP Conf. Ser.: Mater. Sci. Eng.* **188** 012051
[16] Maleki A, Ghamari N and Kamalzare M 2014 *RSC Adv.* **4** 9416–23
[17] Alsharifi M, Znad H, Hena S and Ang M 2017 *Renew. Energy* **114** 1077–89

IOP Conf. Series: Materials Science and Engineering **763** (2020) 012013 doi:10.1088/1757-899X/763/1/012013

Arsenic and arsenate detection with anodic stripping voltammetry technique using boron-doped diamond modified by platinum

R Sitorus, A Saefumillah and T A Ivandini

Department of Chemistry, Faculty of Mathematics and Natural Sciences (FMIPA), Universitas Indonesia, Kampus UI Depok, Depok 16424, Indonesia

Corresponding author's e-mail: ivandini.tri@sci.ui.ac.id

Abstract. High homogeneity and stability of platinum particles modified at boron-doped diamond (BDD) electrode was successfully performed using chemical seeding, followed by electrodeposition technique. The platinum particles on the surface of BDD characterized using Scanning Electron Microscopy- Energy Dispersive X-Ray Spectroscopy (SEM-EDS) showed that the coverage of Pt particles was 1.54 % on the BDD surface. These modified electrodes were used for As^{3+} and As^{5+} sensors using the anodic stripping voltammetry technique in 0.1 M phosphate buffer solution pH 6. Prior to the detection, pretreatment was required by the addition of 0.1 M $NaBH_4$ to change As^{5+} to As^{3+}. The optimum parameters were at a deposition potential of -500 mV, a deposition time of 150 s, and a scan rate of 200 mV/s. The voltammograms were linear for As^{3+} and As^{5+} (R^2 = 0.9797 and 0.9903, respectively) in the concentration range of 0 to 100 ppb. Estimated detection limits of 16.50 and 8.19 ppb for As^{3+} and As^{5+} were achieved, respectively, suggesting that the method is promising for the specification detection of As^{3+} and As^{5+}.

Keywords: arsenic, arsenate, anodic stripping voltammetry, boron-doped diamond, platinum

1. Introduction

Arsenic contamination in water has become a serious problem for human health and the ecosystem. Arsenic, mainly accumulated in sewage and industrial waste, is a hazardous substance [1]. It is poisoning and can cause death in high doses [2]. It is also carcinogenic [3] and easily distributed in the environment [4], therefore, a detection method to monitor its contamination in the environment is necessary.

Various methods and techniques were generally employed in the detection of arsenic, such as hydride generation atomic fluorescence spectrometry, chemiluminescence, inductively coupled plasma atomic emission spectrometry, and inductively coupled plasma mass spectrometry [5]. However, these methods require relatively complex procedures and expensive instrumentations as well [6]. On the opposite, electrochemistry method offers the same efficiency with cheaper instrument. It can also develop to be portable, therefore, making it an ideal method for field analysis. Various types of working electrodes can be used, including gold [7], iridium [8], silver and platinum [9]. However, their high background currents make their modification on the surface of less conductivity substrates to be interesting [10]. Stripping voltammetry technique particularly has an advantage in pre-concentration step, which can provide high concentration of analyte on the surface [6]. Boron-doped diamond (BDD), popular with its low background current and wide potential window, is one of the interesting materials for substrates, but it is not very easy to be modified due to high stability of its sp^3 configuration [11–12].

(a) (b) (c)

Figure 1. SEM image of (a) Bare BDD (b) after wet seeding and (c) after electrodeposition step.

In this work, BDD electrodes were modified with platinum for an application as arsenic and arsenate detection. The modification was performed by amperometry with an addition step, i.e. wet chemical seeding, to provide a better distribution of the deposited particles. High homogeneity and stability of Pt particles could be achieved, which increased the sensitivity and stability of measurements.

2. Materials and methods

2.1. Apparatus and chemicals
All apparatus and chemicals used in this research were BDD (supplied by Einaga's laboratory, Keio University, Japan), an electrochemical cell made from Telfon® completed with magnetic stirrer, platinum and Ag/AgCl (saturated KCl) electrodes, $NaAsO_2$, $Na_2HAsSO_4.7H_2O$, $H_2PtCl_6.6H_2O$, K_2HPO_4, KH_2PO_4, and $NaBH_4$ were supplied from Wako, Japan.

2.2. Modification of the BDD with Pt
Modification using wet chemical seeding was performed by dropping subsequently $NaBH_4$ and H_2PtCl_6 solutions on the BDD surface, then washed and dried after 24 h. This treatment was modified from Gao *et al.* [12] and repeated three times. Next, the electrodeposition was conducted at a potential of -0.2 V for 15 minutes using H_2PtCl_6 solution. After calcinated the electrode at 700 °C in N_2 atmosphere for 5 minutes, the materials were resulted by using Scanning Electron Microscopy-Energy Dispersive X-Ray Spectroscopy (SEM-EDS) and then used as working electrode to detect arsenic trace.

2.3. Detection procedure of As^{3+} and As^{5+}
Examination of the Pt-BDD for arsenic sensor was performed in phosphate buffer solution (PBS) pH 6 containing various concentrations of As^{3+} and As^{5+}. Parameters of anodic stripping voltammetry (ASV) were fixed at a deposition potential of -500 mV and deposition time of 150 s with a stripping rate of 200 mV/s. For As^{5+} detection, 0.1 M $NaBH_4$ that added before was stirred for 1 min to reduce As^{5+} to As^{3+} before ASV technique was performed. For the detection of As^{3+} and As^{5+} mixture, initially ASV was conducted to determine As^{3+} concentration, then 0.1 M $NaBH_4$ that added before, was stirred for 1 min before the second ASV technique was performed.

3. Results and discussion

3.1. Modification of the BDD with Pt
Modification of the BDD surface was performed by combining the wet seeding to grow the crystal nuclei of Pt on the BD surface and electrodeposition to grow the Pt crystal on the BDD surface [12]. Figure 1 shows the SEM images of the BDD before and after modification with Pt with magnification of 2000 × (figure 1a) and 10000 × (figure 1b and figure 1c). After wet seeding step, smaller particles were deposited on the surface of the BDD, while after electrodeposition step the particles were grown larger and distributed more homogeneously.

Characterization was also performed by using Energy Dispersive X-Ray Spectroscopy (EDS) to know the coverage of Pt on the surface of BDD. Table 1 displays the summary of atomic comparison

Table 1. Summary of EDS measurements of the Pt-modified BDD after the wet seeding and electrodeposition steps

Element	Surface composition of BDD after wet seeding step	Surface composition of BDD after electrodeposition step
Carbon	96.74	94.91
Oxygen	02.32	02.74
Platinum	00.94	01.54

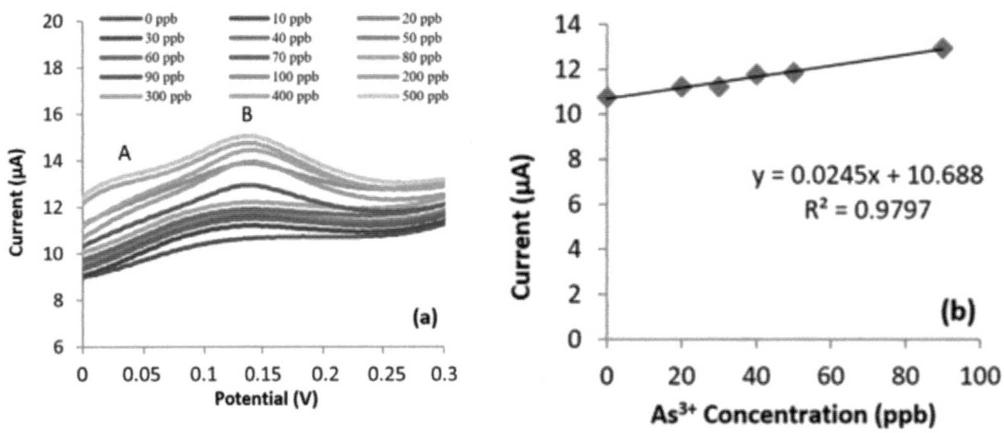

(a) (b)

Figure 2. (a) Stripping voltammogram of various concentrations of As^{3+} together with (b) the current plots at the potentials +0.15 V vs. As^{3+} concentrations. The stripping voltammetry parameters were at deposition potential of -0.5 V for 150 s and scan rate of 200 mV/s

on the modified BDD surface, showing that after wet seeding step 0.94 % surface was covered. This amount increased to be 1.54 % after the electrodeposition step. On the surface of BDD, oxygen was also observed, which probably due to the oxygen sites of the BDD grown when it was prepared and stored [7,11]. The Pt sites on the surface will be used for arsenic detection to increase the interaction between As and Pt particles ASV measurements. Therefore, platinum distribution on the BDD surface has a prominent effect on the sensitivity of the electrode towards arsenic.

3.2. As^{3+} detection

The detection of As^{3+}using Pt-modified BDD was conducted in the concentration range of 0 ppb to 500 ppb as shown in figure 2. The deposition process during ASV was performed at -500 mV. The purpose of this step was to concentrate As^{3+} in the solution on the BDD surface, therefore, the amount of As^{3+} can be reduced in the stripping step to As^{0} will be higher, resulting in a higher oxidation peak of As^{0} converting back into As^{3+}. The voltammograms in figure 2a shows an oxidation peak at around +0.15 V (B), which increased with the increasing of As^{3+} concentration, indicating that the oxidation peak of As^{0} to As^{3+} occurred in this potential. Another oxidation peak at around +0.05 V was proposed due to the release of H_2 gas generated in the deposition step.

3.3. As^{5+} detection

The detection of As^{5+} using Pt-modified BDD was conducted in various concentrations of 0 to 200 ppb as shown in figure 3. In this detection, it is necessary to add $NaBH_4$ as a chemical reducing agent as to reduce As^{5+} needs a very negative potential that can cause a competition with hydrogen evolution. As the results, the reduction As^{5+} on the surface of electrode was imperfect. Figure 3a shows that the oxidation peak observed at the same potential at around 0.15 V (A), confirmed that the peak belongs to

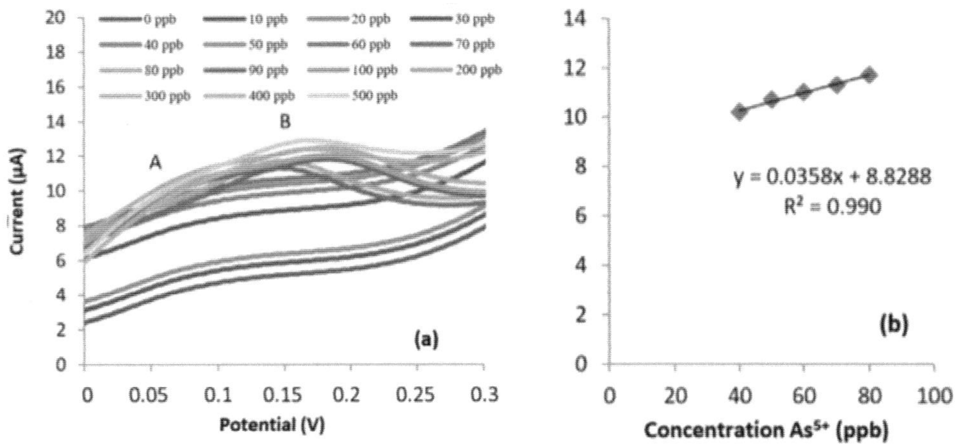

(a) (b)

Figure 3. (a) Stripping voltammogram of various concentrations of As^{5+} together with (b) the current plots at the potentials +0.15 V vs. As^{5+} concentrations. Stripping voltammetry parameters were the same as in figure 2.

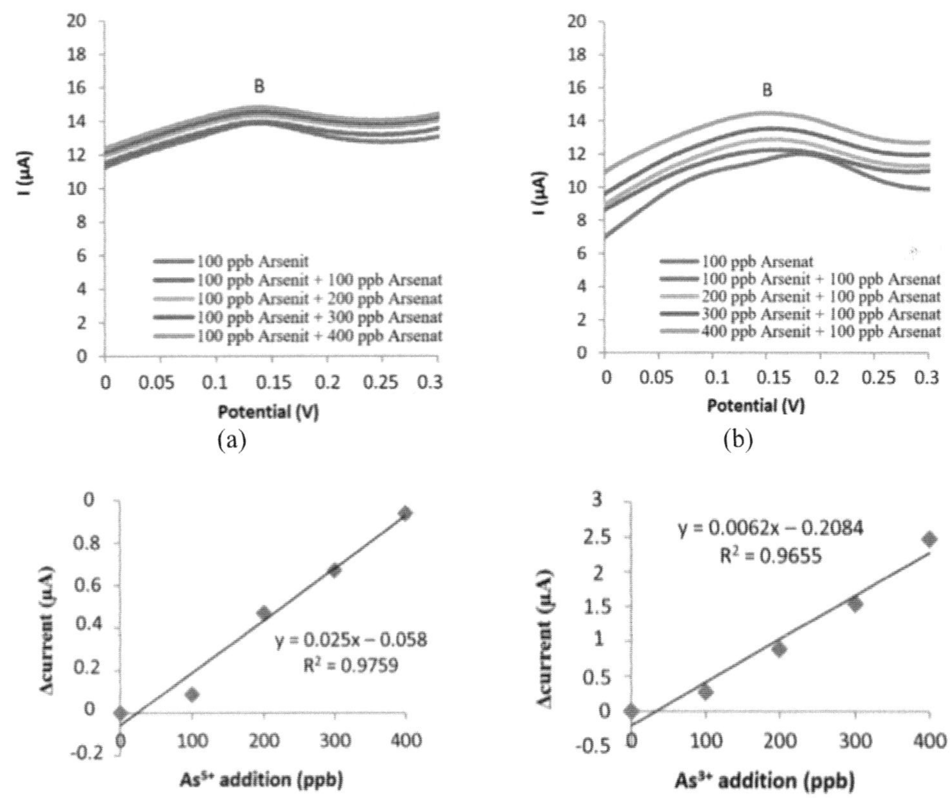

(c) (d)

Figure 4. Stripping voltammograms of 100 ppb (a) As^{3+} solutions with the addition of various concentrations of As^{5+} (b) As^{5+} solutions with various concentrations of As^{3+} together with (c) and (d) which showed the plots of the related currents at the potentials +0.15 V of (a) and (b) respectively. The stripping voltammetry parameters were the same as figure 2.

the oxidation of As^0 to As^{3+}. Other peak at around 0.05 V was not detected due to less formation of the hydrogen using this procedure. The plots of current responses vs. As^{5+} concentrations in figure 3b show the linearity from 40–100 ppb.

3.4. As^{3+} and As^{5+} mixture detection

Next, the detection of As^{3+} and As^{5+} mixture solution was conducted using ASV in two steps. First, ASV was performed to determine As^{3+}. Then, the same solution was added with $NaBH_4$ before the ASV measurement to measure the total arsenic concentration of As^{3+} and As^{5+}. The concentration of As^{5+} can be determined by subtracting As^{3+} from the total arsenic concentration. Two series of testing was examined. In the first one, solution of 100 ppb of As^{5+} was used with the addition of various concentrations of As^{3+} from 100 to 400 ppb, whereas in the second one, 100 ppb As^{3+} solution was used as the constant variable with addition of 100 to 400 ppb As^{5+}. The voltammograms of the first and the second series are shown in figure 4a and figure 4b, respectively. In both figures, the increase of current responses was observed with the increase of As^{3+} (figure 4a) and As^{5+} (figure 4b) concentrations. The linear calibration curves are shown in figure 4c and 4d, respectively, for As^{3+} and As^{5+} with a correlation value (R^2) of 0.98 and 0.979, indicating that the method can be applied for real applications. Furthermore, the stability of the Pt-modified BDD was examined during the detection of As^{3+} and As^{5+} in five consecutive days. The % RSD of under 3 % was observed for both, suggesting that the Pt-modified BDD was stable enough to be employed for the detection of both As^{3+} and As^{5+} species.

4. Conclusions

Modification of BDD with Pt particles has been successfully examined using combination of the wet seeding and electrodeposition process, resulting in 1.54 % coverage of Pt on the surface of BDD. Applications of the modified BDD as the working electrode to detect As^3 and As^{5+} using anodic stripping voltammetry showed a good linearity in the measurements of individual solutions of As^{3+} and As^{5+} as well as of the mixture solution of As^{3+} and As^{5+} with good linearity and stability.

Acknowledgements

This work was funded by Hibah Kompetensi Publikasi Internasional Terindeks Universitas Indonesia, with contract number 2273/UN2.R3.1/HKP.05.00/2018

References

[1] Agency for Toxic Substances and Disease Registry 2017 *Support Document to The 2017 Subtance Priority List (Candidates for Toxicological Profiles)* (Atlanta: Division of Toxicology and Human Health Sciences ATSDR)
[2] Levin-Scherz J K, Patrick J D, Weber F H, Garabedian Jr C 1987 *Ann. Emerg. Med.* **16** 702–4
[3] Lee A M, Fraumeni J F 1969 *J. Natl. Cancer I.* **42** 1045–52
[4] Wedepohl K H 1969 *Handbook of Geochemistry* (New York: Springer)
[5] Bagherian G, Chamjangali M A, Amin A H, Ameri S 2012 *Eurasian J. Anal. Chem.* **7** 78–88
[6] de Carvalho L M, do Nascimento P C, Koschinsky A, Bau M, Stefanello R F, Spengler C, Bohrer D and Jost C 2007 *Electroanal.* **19** 1719–26
[7] Ivandini T A, Yamada D, Watanabe T, Matsuura H, Nakano N, Fujishima A and Einaga Y 2010 *J. Electroanal. Chem.* **645** 58–63
[8] Ivandini T A, Sato R, Makide Y, Fujishima A and Einaga Y 2006 *Anal. Chem.* **78** 6291–98
[9] Hrapovic S, Liu Y, Loung J H T 2007 *Anal. Chem.* **79** 500–7
[10] Nagaoka Y, Ivandini T A, Yamada D, Fujita S, Yamanuki M and Einaga Y 2010 *Chem. Lett.* **39** 1055–7
[11] Macpherson J V 2015 *Phys. Chem. Chem. Phys.* **17** 2935–49
[12] Gao F, Yang N and Nebel C E 2013 *Electrochim. Acta* **112** 493–9

Synthesis of impregnated Ni(0) zeolite ZSM-5 as catalyst for carboxylation reaction of acetylene with CO_2

F Mujiantono, I Abdullah and Y K Krisnandi

Department of Chemistry, Faculty of Mathematics and Natural Sciences (FMIPA), Universitas Indonesia, Kampus UI Depok, Depok 16424, Indonesia

Corresponding author's e-mail: yuni.krisnandi@sci.ui.ac.id

Abstract. Conversion of carbon dioxide (CO_2) into other compounds has become very advantageous because of the abundance in the atmosphere. However, carbon dioxide (CO_2) is thermodynamically and kinetically stable so it needs low valence metal such as Ni(0) to be activated and react with other compounds. In this study, nickel metal-impregnated ZSM-5 was prepared as heterogeneous catalyst for carboxylation reaction of acetylene with carbon dioxide. Microporous and hierarchical ZSM-5 were synthesized using single template and double template, respectively. X-Ray Diffraction (XRD) characterization results show that ZSM-5 material with high crytallinity was successfully synthesized. Scanning Electron Microscopy (SEM) images show ZSM-5 has hexagonal coffin-like morphology. Brunauer-Emmett-Teller (BET) surface area analysis proves that the ZSM-5 has combination of micro and meso-sized pores (hierarchical pores). The elemental analysis using Atomic Absorption Spectroscopy (AAS) gives nickel content of 1.9 % in microporous ZSM-5 and 2.1 % in hierarchical ZSM-5. HPLC analysis show new peak is observed at retention time of 3.6 minutes for carboxylation reaction of acetylene (CO_2: acetylene of 0.2:0.8) with optimum condition at 12 h and 80 °C when using microporous Ni/ZSM-5 or hierarchical counterpart as catalyst.

Keywords: carboxylation, microporous Ni(0)/ZSM-5, hierarchical Ni(0)/ZSM-5, acetylene, CO_2 transformation

1. Introduction

Global warming is a problem that arises mainly due to excess of greenhouse gases (e.g. carbon dioxide, methane, nitrogen oxide) in the atmosphere, covering the earth and reflecting heat radiation back to the surface of the earth. This leads to problems in the environment such as rising sea level due to polar ice melting, increasing atmospheric storms, increasing species and organisms that cause disease [1]. In recent years, emissions from carbon dioxide gases generated by human activities such as industrial have become very high [2]. Accumulation of carbon dioxide in the atmosphere has huge risk because carbon dioxide can last longer than other greenhouse gases in the atmosphere. Therefore, some methods are required to reduce the acumulation of carbon dioxide in the atmosphere, such as CO_2 storage and utilization of CO_2 [3].

Conversion of carbon dioxide into other compounds becomes very advantageous yet challenging. Because carbon dioxide has a thermodynamic and kinetic stability so it needs low valence metal to react [4]. Currently there has been several carbon dioxide conversion using metal complexes catalyst [5]. However, most of the catalysts are homogeneous catalysts which are efficient in reaction but difficult to be separated from the products. For example, homogeneous nickel catalyst has been employed as good catalyst in carboxylation of various unsaturated hydrocarbons such as cyanamides for production of amino acid derivatives [6], and ethylene for production of acrylic acid [7]. ZSM-5 has been used widely as catalysts and catalyst support in many reactions. For example, Co/ZSM-5 as catalyst in partial oxidation of methane to methanol [8,9], conversion of biomass to levulinic acid [10], cracking of hexadecane [11], and conversion of CO_2 to cyclic carbonates [12]. ZSM-5 modified with nickel is considered as an efficient heterogeneous catalyst for carboxylation reaction [13] because ZSM-5 has large pore size and high surface area makes it possible to be used as catalyst system that can convert carbon dioxide that is absorbed efficiently.

This work reports carboxylation reaction of acetylene with carbon dioxide, using nickel impregnated ZSM-5 zeolite. Hierarchical ZSM-5, synthesized using double template method, was used and its activity was compared with that from micoporous ZSM-5.

2. Experimental

2.1. Materials
The materials used were tetraethyl orthosilicate (TEOS, 98%, Sigma Aldrich), sodium aluminate (Sigma Aldrich), tetra propylammonium hydroxide (TPAOH, 1M, Sigma Aldrich), poly(diallyldimethylammonium chloride) (PDDAM, 35 %, Aldrich), nickel(II) nitrate ($Ni(NO_3)_2 \cdot 6H_2O$, Sigma Aldrich), ethylene gas (99.99 % UHP from BOC), N_2 gas (99.99 % UHP), H_2 gas (99,99 % UHP), CO_2 gas (99.99 % UHP), magnesium chloride ($MgCl_2$ Merck), dimethylformamide (DMF, merck), and deionized water. All materials in this work were used without further purification.

2.2. ZSM-5 synthesis and Ni/ZSM-5 preparation
The synthesis method is similar to that reported by Wang *et al.* [14] with some modifications [8]. Hierarchical ZSM-5 was synthesized using a molar ratio of 1 Al_2O_3: 64 SiO_2 : 10 $(TPA)_2O$: 3571H_2O. The water contained in the obtained zeolite was removed by drying at room temperature and then calcined at 823 K. Metal insertion into the zeolite was carried out using a wet impregnation method with reduction via hydrogen gas; 1 g of ZSM-5 was added to 0.2495M metal solution prepared from nickel nitrate hydrate and stirred for 24 h at room temperature and was calcined at 823 K under flowing hydrogen gas. The ZSM-5 and Ni/ZSM-5 zeolites were characterized with X-Ray diffraction (XRD), Fourier-transform infrared spectroscopy (FTIR) Shimadzu Prestige 21, Scanning Electron Microscopy with Energy Dispersive X-Ray Spectroscopy (SEM-EDX) SEM FEI quanta FEG 450, and Atomic Absorption Spectroscopy (AAS) Shimadzu AA 13000.

2.3. Catalytic test: carboxylation of acetylene
Into a 100 mL-round bottom flask reactor, 0.05 g Ni/ZSM-5 catalyst, 0.032 g promoter $MgCl_2$, 0.011g Zn, 0.15 mL H_2O and 10 mL DMF were added. Then, flowed gas acetylene with a pressure of 0.2 atm and gas CO_2 with a pressure of 0.8 atm for 90 sec of each gas. The reactor was heated with temperature variation of (298, 313, 353) K with time variation of (3, 6, 12) h. After the reaction was completed, the autoclave was cooled to room temperature and the resulting product was collected for analysis with High Performance Liquid Chromatography (HPLC) Ultimate 3000, in which the column used was C18 3.5 μm, 4.6 x 150 mm, temperature 313 K, UV spectrum 210 nm and eluents used were acetonitril and 0.1 %H_3PO_4.

3. Results and discussion

3.1. Characterization of ZSM-5
Microporous ZSM-5 was prepared with single template tetrapropylammonium hydroxide (TPAOH) as mordenite inverted framework (MFI) structure directing agent while hierarchical ZSM-5 was produced with double template TPAOH and PDDAM as mesoporous structure template using hydrothermal method. During aging, the positive charge of PDDAM interacts with the negative charge of ZSM-5 crystal to be structured, so the ZSM-5 MFI framework could be grown around the PDDAM polication [14]. The XRD pattern of as-synthesized both microporous and hierarchical ZSM-5 (figure 1a) shows certain peaks in the 2θ range from 7° to 9° and from 22° to 25°, which is consistent with the standard ZSM-5 pattern [14], indicating that the ZSM-5 structure was successfully synthesized. Figure 1b shows the XRD patterns of modified ZSM-5 catalysts. From both microporous and hierarchical ZMS-5, some peaks were identified at position 2θ about 44.5° and 53° indicating the presence of nickel(0) species [15].

Figure 2 presents the FTIR spectra of microporous and hierarchical ZSM-5 before (a) and after (b) calcination. The spectrum of both ZSM-5 before calcination shows several extra peaks at 2960–

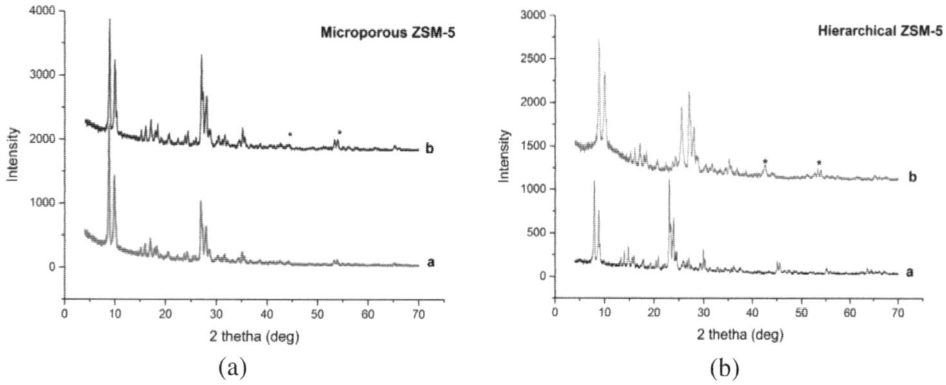

Figure 1. XRD patterns of (a) microporous and (b) hierarchical ZSM-5 for as-synthesized ZSM-5 (red lines) and Nickel/ZSM-5 (brown and black lines)

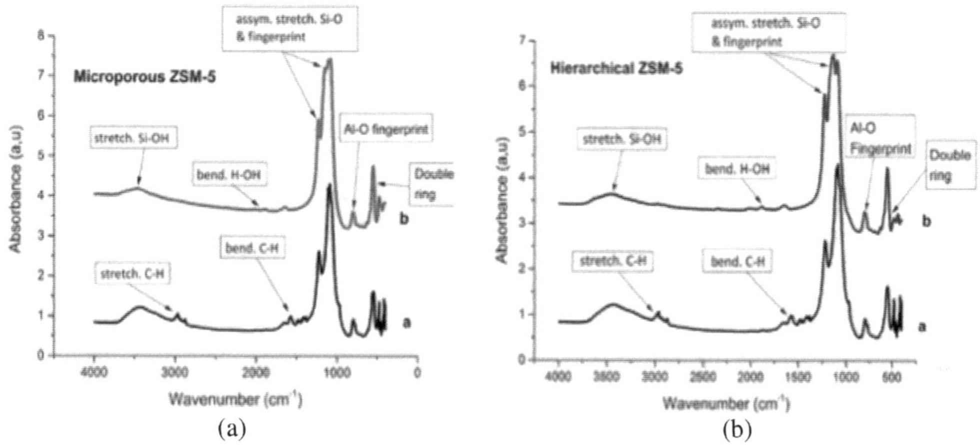

Figure 2. FTIR spectra of (a) microporous and (b) hierarchical ZSM-5 zeolites before (black lines) and after (red and blue lines) calcination at 823 K

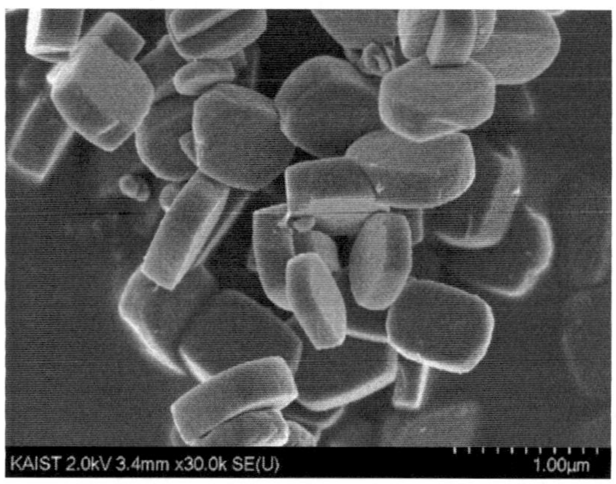

Figure 3. SEM image of hierarchical ZSM-5, 30000 magnification

Table 1. Results of acetylene carboxylation using Ni(0)/ZSM-5 (area below the peak of the product)

Reaction temperature (K)	Ni/ZSM-5 (hierarchical)			Ni/ZSM-5 (microporous)		
	3h	6h	12h	3h	6h	12h
298	34.9968	36.8503	42.0126	0.1430	32.2852	29.0601
313	27.7825	27.7568	41.6828	42.8235	34.2039	46.9155
353	40.4486	21.7988	61.3878	10.6264	27.2937	49.6533

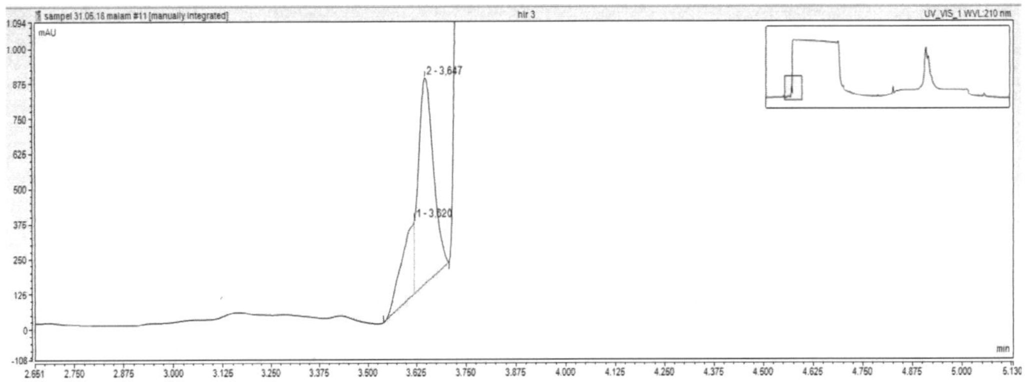

Figure 4. Chromatograph of product from carboxylation reaction of acetylene on hierarchical Ni(0)/ZSM-5 (condition: CO_2: acetylene: 0.2 : 0.8, 298 K, 12 h, catalyst 50 mg)

2850 cm^{-1} (C–H stretching) and at 1470–1350 cm^{-1} (C–H bending) compared with both ZSM-5 after calcination. These peaks show the presence of templates used in the synthesis of zeolite (TPAOH and PDDAM). Calcination was used to remove the templates used in the synthesis of both microporous and hierarchical zeolites. From figure 2b it can be seen that the extra peaks are gone and indicated the template were fully removed.

Figure 3 shows the SEM image of hierarchical ZSM-5. It can be seen that the ZSM-5 crystals have hexagonal shape and high crystalinity. Calculated from EDX data, the Si/Al ratio was 27.

Ni/ZSM-5 heterogeneous catalysts were synthesized using a wet impregnation method. The elemental analysis using AAS indicates that the loading of nickel on hierarchical ZSM-5 zeolites was about 2.1 wt.% and microporous ZSM-5 was 1.9 wt.%. This is close to the target loading of 2.0 wt.%.

3.2. Carboxylation of acetylene

Figure 4 shows a typical chromatograph of products from carboxylation reaction of acetylene on hierarchical Ni(0)/ZSM-5. It can be seen that there is a new peak with retention time (Rt) of 3.647 min, after reaction took place at 353 K and 12 h, while using hierarchical or microporous Ni(0)/ZSM-5 catalyst. This new peak should be more polar than the expected acrylic acid (a published Rt for acrylic is 6.9 min [16]), when using C18 column with acetonitrile and phosphoric acid (0.1 % H_3PO_4) as eluent. While table 1 summarizes the results obtained in various reaction conditions, where the optimum condition took place at 353 K and 12 h.

Figure 5 is the FTIR spectra of the reaction products. A broad band assigned to vibration of O-H appeared at 3600–3400 cm^{-1}, while sharp peaks attributed to vibration of C-H at 2950–2840 cm^{-1}, vibration of C=O at 1680 cm^{-1}, bending vibration of –C–H at 1480–1440 cm^{-1} and vibration of C-OH at 1200–1020 cm^{-1} were also observed. The peaks from products obtained from the reaction using either hierarchical Ni(0)/ZSM-5 or microporous Ni(0)/ZSM-5 as catalyst were looked similar. From the FTIR spectra, it could be suggested that the product is fumaric acid (HOOC-CH=CH-COOH), which is in accordance with the results observed in HPLC analysis. This product might be formed from a double carboxylation reaction of acetylene in the presence of magnesium halide ($MgCl_2$) as promoter [17].

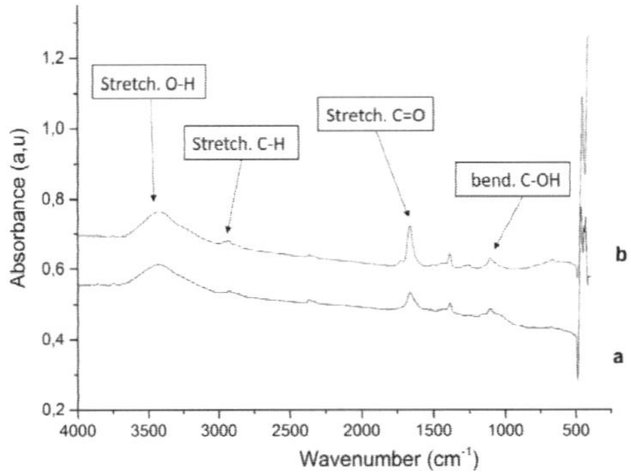

Figure 5. FTIR spectra of product from carboxylation of acetylene using (a) hierarchical, and (b) microporous Ni/ZSM-5 catalysts.

4. Conclusions

The synthesis of microporous and hierarchical ZSM-5 zeolites followed by modification with nickel metal has been carried out. The catalysts were further used in the reaction of acetylene carboxylation with carbon dioxide targeted acrylic acid as product. HPLC analysis did not show the presence of acrylic acid in the reaction mixture. Alternatively, more polar compound such as fumaric acid may be produced.

Acknowledgements

This work is supported by PITTA Grant 2018 funded by Universitas Indonesia No.2272/UN2R3.1 /HKP.05.00/2018.

References

[1] Intergovernmental Panel on Climate Change (IPCC) 2014 *Climate Change 2014: Synthesis Report* ed R K Pachauri and L Meyer (Geneva: IPCC)
[2] Olah G A, Goeppert A and Prakash G K S 2009 *Beyond Oil and Gas: The Methanol Economy*, 2nd ed (Darmstadt: Wiley-VCH)
[3] Leung D Y C, Caramanna G and Maroto-Valer M M 2014 *Renew. Sustain. Energy Rev.* **39** 426–43
[4] Sakakura T, Choi J C and Yasuda H 2007 *Chem. Rev.* **107** 2365–87
[5] Wang X, Wang H and Sun Y 2017 *Chem.* **3** 211–28
[6] Doi R, Abdullah I, Taniguchi T, Saito N and Sato Y 2017 *Chem. Commun.* **53** 7720–3
[7] Al-Ghamdi M, Vummaleti S V C, Falivene L, Pasha F A, Beetstra D J and Cavallo L 2017 *Organometallics* **36** 1107–12
[8] Krisnandi Y K, Putra B A P, Bahtiar M, Zahara, Abdullah I and Howe R F 2015 *Procedia Chem.* **14** 508–15
[9] Krisnandi Y K, Samodro B A, Sihombing R and Howe R F 2015 *Indonesian Journal of Chemistry* **15** 263–8
[10] Krisnandi Y K, Nurani D A, Agnes A, Pertiwi R, Antra N F, Anggraeni A R, Azaria A P and Howe R F 2019 *Indonesian Journal of Chemistry* **19** 115–23
[11] Zahara Z, Krisnandi Y K, Wibowo W, Nurani D A, Rahayu D U C and Haerudin H 2018 *AIP Conf. Proc.* 2023 020088
[12] Zhao Q N, Song Q W, Liu P, Zhang Q X, Gao J H and Zhang K 2017 *Chinese J. Chem.* **36** 187–93
[13] Xie H, Lin T, Shi L and Meng X 2016 *RSC Adv.* **6** 97285–92
[14] Wang L, Zhang Z, Yin C, Shan Z and Xiao F S 2010 *Microporous Mesoporous Mater.* **131** 58–67
[15] Ziebro J, Łukasiewicz I, Borowiak-Palen E and Michalkiewicz B 2010 *Nanotechnology* **21** 4960–3
[16] Yang J, Gioia A J D 2006 HPLC Determination of *Free Acrylic Acid and Monomers in Polyacrylates* (Milford: Waters Corporation) Application Note
[17] Fujihara T, Horimoto Y, Mizoe T, Sayyed F B, Tani Y, Terao J, Sakaki S and Tsuji Y 2014 *Org. Lett.* **16** 4960–3

Effect of surfactants mixing of Tween 80 and Nonidet RK-18 in preparation of microspheres based on polyblend of poly(lactic acid) and polycaprolactone

A Ferrariski and E Budianto

Department of Chemistry, Faculty of Mathematics and Natural Sciences (FMIPA), Universitas Indonesia, Kampus UI Depok, Depok 16424, Indonesia

Corresponding author's e-mail: emilb@ui.ac.id

Abstract. In preparation of microspheres process with solvent evaporation method, microspheres are formed from the oil in water (O/W) emulsion droplets as a template in which the type and concentration of surfactant will affect the stability of the emulsion. This research aims to investigate the effect of composition and concentration of a mixture of Tween 80 and Nonidet RK-18 as a mixed emulsifier in the preparation of microspheres based on polyblend of PLA/PCL. The solvent evaporation method involves emulsification and dispersing the polyblend emulsion in water. The microspheres obtained were characterized using FTIR, stereomicroscopes, and PSA instrumentations. The result of this research shows that the mixed emulsifier produces smaller microspheres compared to the result of using Tween 80 or Nonidet RK-18 individually because of the presence of synergistic effect. When the surfactants mixture is used, the larger Tween 80 ratio in the mixture will increase the ratio of smaller microspheres obtained because of the high hydrophilicity of Tween 80 that can stabilize O/W emulsion. On the other hand, the addition of Nonidet RK-18 in an increasing manner of ratio or concentration may destabilize the emulsion droplets through Ostwald ripening that can occur because Nonidet RK-18 has a low cloud point (17 °C).

Keywords: lactic acid, solvent evaporation, microspheres

1. Introduction

When microspheres are used as a controlled drug delivery carrier, before we evaluate the microencapsulation efficiency and drug loading of drug loaded microcapsules, a research to observe the size and particle size distribution of hollow microcapsules (microspheres) is needed to be done. This is due to the fact that if the required size and size distribution of microspheres as drug delivery device are not attained, the utilization of microspheres as a carrier will not be optimum [1].

According to Jain [2], the size of microspheres as drug delivery device is required to be less than 250 μm. Moreover, Kim and Pack [3] stated that the appropriate size of microspheres as controlled drug delivery device falls between the ranges of 1-100 μm with a narrow particle size distribution.

One of the important parameters that affect microspheres size obtained is the surfactant used [4].It stems from the fact that in microspheres making process with solvent evaporation method, there is an emulsification process before dispersing the emulsion in water. This emulsification process is a stage of oil forming in water (o/w) emulsion droplets as a template in which the type and concentration of surfactant will affect the stability of the emulsion. A dual surfactants system is utilized in this research because of the synergistic interface stabilising effect between two surfactants with different properties found in previous studies.

Based on the explanation above, this research is conducted to investigate the effect of composition and concentration of a mixture of Tween 80 and Nonidet RK-18 as a mixed emulsifier in microspheres

Table 1. Variations of mixed surfactants compositions

Fixed variable	Tween 80:Nonidet RK-18 ratio
Tween 80 and Nonidet RK-18 concentration (0.5 %)	40:60 50:50 60:40 65:35 70:30

Table 2. Variations of mixed surfactants concentrations

Fixed variable	Tween 80:Nonidet RK-18 ratio
Tween 80 and Nonidet RK-18 composition (60:40 Tween 80: Nonidet RK-18 ratio)	40:60 50:50 60:40 65:35 70:30

making process based on blend of Poly(Lactic Acid) (PLA) and Polycaprolactone (PCL) with solvent evaporation method. The result of this research is expected to show the correlations between variations of composition and concentration of surfactants mixture and the microspheres size obtained.

2. Materials and methods

2.1. Materials
The subsequent compounds were employed as obtained: L-(+)-lactic acid (purity of ≥ 98 %) was obtained from Sigma-Aldrich, Polycaprolactone CAPA-6800 (glass transition temperature of -60 °C) was obtained from Perstorp, dichloromethane and Tween 80 from Merck, Nonidet RK-18 was a gift from Evonik.

2.2. Instruments
Thermo Scientific Cimarec SP88857105 Stirring Hot Plate, Fourier Transform Infra Red (FTIR) IRPrestige-21 Shimadzu, and Particle Size Analyzer Beckman Coulter Light Scattering-100 (PSA LS-100).

2.3. Lactic acid polycondensation
Polymerizations were conducted in a three-neck flask completed with magnetic stirrer, thermometer, and a reflux condenser. Fifty millilitres of L-LA were added to the flask and then heated gradually to 120 °C, after which the temperature was maintained at 120 °C for an hour. After one hour, the temperature was then brought to 150 °C with N_2 flowed to the reactor. After that, this condition was kept for 22 hours. After the reaction finished, the reactor was cooled down to room temperature for two hours. The product was then removed from the three-neck flask and characterized with FT-IR.

2.4. Preparation of polyblend solutions
PLA and PCL polyblend was prepared with 60:40 (mole ratio) compositions. The polymers were wholly mixed in dichloromethane with the addition of Nonidet RK-18 and stirred until entirely dissolved to obtain the 10 % polyblend solution.

2.5. Preparation of microspheres at different dual surfactants compositions and concentrations
Preparation of PLA/PCL microspheres was attained by a modified solvent evaporation technique [1]. Five millilitres of 10 % polyblend solution were emulsified in Tween 80 with different compositions and concentrations. The emulsification process was carried out by stirring at 700 rpm for 1 hour. The emulsion was then dispersed in 400 mL distilled water and stirred for another hour. The microspheres formed were acquired by filtration, washed with distilled water, and dried in oven (≤ 40°C) overnight. The variations of mixed surfactants compositions and concentrations used in this research are listed in table 1 and table 2.

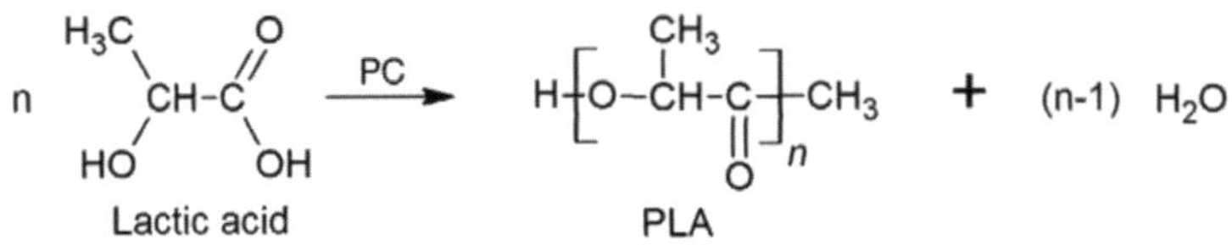

Figure 1. Reaction scheme of lactic acid polycondensation reaction to produce PLA.

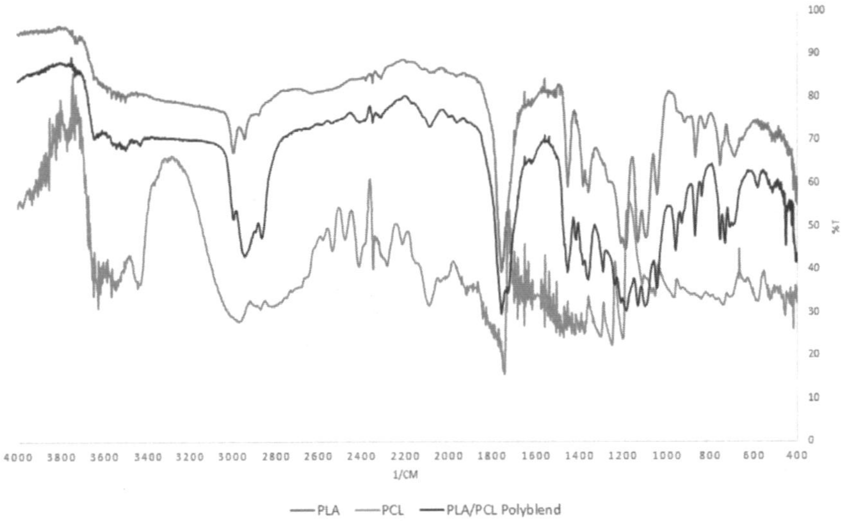

Figure 2. IR spectra of PLA, PCL, and PLA/PCL polyblend

2.6. Microspheres characterization

After microspheres were obtained by utilizing the mixture of Tween 80 and Nonidet RK-18 as the emulsifier, it was characterized using FTIR to ensure that microspheres were successfully formed from polyblend of PLA and PCL. The microspheres were also characterized using Particle Size Analyzer (PSA) to measure the particle size distribution of the microspheres produced.

3. Results and discussion

3.1. Lactic acid polycondensation

Condensation polymerization is used in this research as a PLA synthesis method from lactic acid (figure 1), which this polymerization occurs without the presence of catalyst and solvent thus classified as an economic reaction. However, this simple polycondensation reaction has some downsides such as difficult to obtain the desired polymer with high molecular weight (M_w), produces PLA with inferior mechanical strength, and it manifests PLA degradation reaction simultaneously at high temperature [5–6].

When the reaction takes place, water molecules are also produced as the by-product, which is necessary to be removed because it can diffuse into the amorphous fragments of PLA followed by random hydrolysis of the ester bonds. Fragmented PLA will produce Oligomer Lactic Acid (OLA) and reduce molecular weight of PLA subsequently. Hence, N_2 is flowed to the reactor to drive the water vapor away.

3.2. Preparation of microspheres based on polyblend of PLA/PCL

Polyblend is a mixture of two or more different polymers or copolymers, which are associated physically (without the presence of covalent bonds) [7]. The absence of chemical bonds implies that there will be no new vibrations observed in the IR spectra of the polyblend compared to the polymers (figure 2). The mentioned vibrations identified from the IR spectra are C-H sp^3 stretching on a region of 2850-3030 cm^{-1}, C=O stretching on 1725 cm^{-1},and stretching of saturated ester (C-O) on 1130 cm^{-1}.

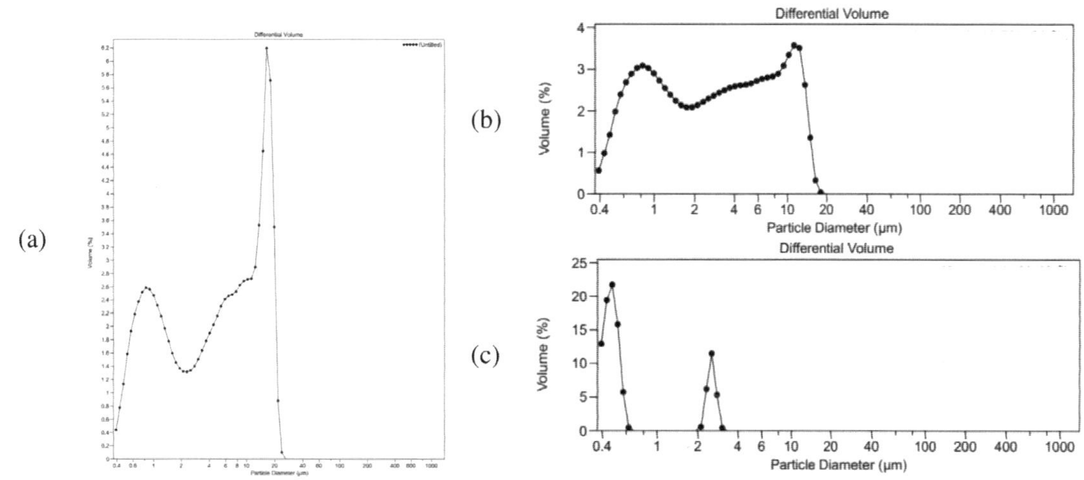

Figure 3. Particle size distribution of microspheres based on polyblend of PLA/PCL with (a) 40:60 (b) 50:50 and (c) 60:40 Tween 80/Nonidet RK-18 ratio.

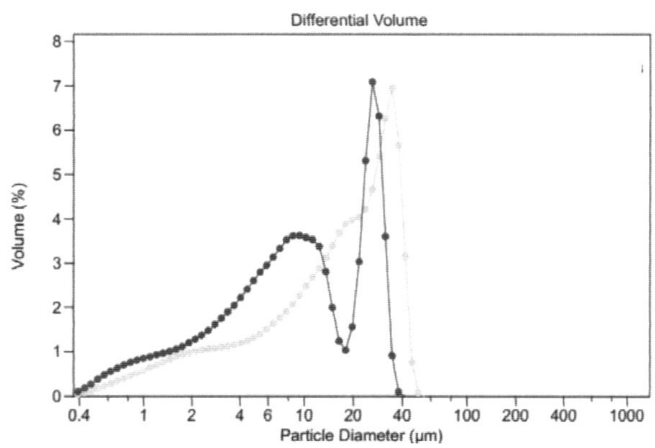

Figure 4. Particle size distribution of microspheres based on polyblend of PLA/PCL with Tween 80 (blue) and Nonidet RK-18 (green) as sole emulsifier.

3.3. Preparation of microspheres at different dual surfactants compositions

This research studies particle size distribution of microspheres prepared at various compositions of Tween 80 and Nonidet RK-18 as mixed emulsifier with 40:60; 50:50; and 60:40 Tween 80:Nonidet RK-18 ratios (figure 3), which compared to the size distribution of microspheres prepared using sole surfactant (Tween 80 or Nonidet RK-18) (figure 4). Microspheres prepared with dual surfactants displays smaller particle size of microspheres (< 25 μm) compared to the utilization of sole surfactant (a peak is observed at 40 μm). This might be owing to the presence of synergistic effect between Tween 80 and Nonidet RK-18 that produces an interfacial layer, which protects emulsion droplets during emulsification process [4]. The synergistic effect itself is defined by Rosen as a condition where the properties of the mixture are exceeding the properties obtainable from the single surfactant system [8].

To get a better understanding on the roles of each surfactant constituents in dual surfactants system, particle size distributions of microspheres prepared with Tween 80:Nonidet RK-18 ratios of 60:40; 65:35 and 73:30 are observed in figure 5. The particle size distributions indicate that larger Tween 80 ratio in the surfactants mixture will produce higher quantity of the smaller microspheres obtained (0.4–1 μm) whereas an increase in Nonidet RK-18 will decrease the quantity of smaller microspheres (0.4–1 μm) and increase the quantity of bigger microspheres obtained (2–4 μm). This might be for the

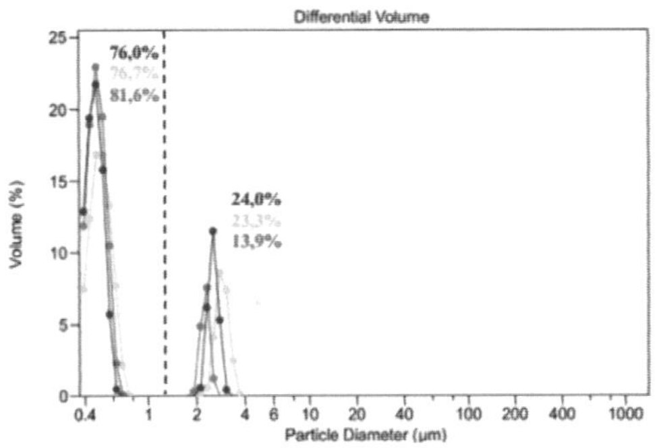

Figure 5. Particle size distribution of microspheres based on polyblend of PLA/PCL with Tween 80: Nonidet RK-18 ratios of 60:40 (blue); 65:35 (green); and 73:30 (red).

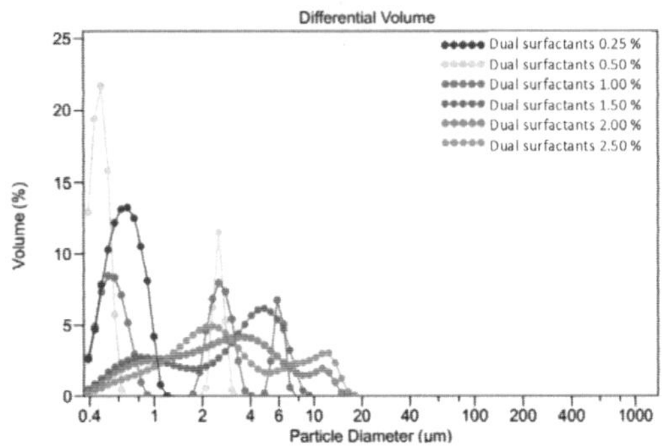

Figure 6. Particle size distribution of microspheres based on polyblend of PLA/PCL at different dual surfactants concentration.

reason that Tween 80 has high hydrophilicity (HLB 15.0 ± 1.0) while Nonidet RK-18 has low hydrophilicity (HLB 6). Tween 80 with its hydrophilic property will dissolve readily in water and according to Bancroft rule will produce a stable emulsion when water serves as the dispersing phase. To put it another way, higher Tween 80 ratio will stabilize O/W emulsion and will produce smaller microspheres.

3.4. Preparation of microspheres at different dual surfactants concentrations
After the investigation of surfactants compositions effect, the microspheres produced with various surfactant concentrations (0.25 %; 0.5 %; 1.0 %; 1.5 %; 2.0 % and 2.5 %) were studied. In figure 6, dual surfactants 0.25 % produced microspheres with monomodal size distribution with a single peak at 0.688 μm, dual surfactants 0.5 % exhibited bimodal size distribution with peaks at 0.474 μm and 2.539 μm, dual surfactants 1.0 % exhibited trimodal size distribution with peaks at 0.520 μm, 2.539 μm, and 5.878 μm, respectively. For dual surfactants with concentrations from 1.5 % to 2.5 %, microspheres size distributions produced broad distributions compared to the others.

The formations of new peaks in bigger size from the utilization of dual surfactants 0.25 % to 0.5 % and 1.0 % implied that there were droplets destabilization that might be due to the fact that Nonidet RK-18 has a low cloud point (17 °C) whilst the cloud point of Tween 80 is higher at 65 °C. Cloud point itself defines as a temperature above which the surfactant solution becomes turbid due to phase separation and destabilization. When exceeding its cloud point, Nonidet RK-18 that consists PEO copolymer block will undergo an excessive dehydration resulting in the domination of hydrophobic attraction to hydrophilic repulsion and prompt the micelles aggregation. This micelles aggregation or flocculation leads a direct contact between oil droplets and drives an Ostwald ripening by decreasing the diffusion path length. Ostwald ripening itself covers a diffusive transfer of the dispersed phase from smaller to larger droplets [9–10].

4. Conclusions
Microspheres based on polyblend of PLA/PCL has been successfully prepared with solvent evaporation method implied from the synergic IR spectra of the microspheres and the polymer constituents. The mixed emulsifier produces smaller microspheres compared to the result of using Tween 80 or Nonidet RK-18 individually because of the presence of synergistic effect. When the surfactants mixture is used, the larger Tween 80 ratio in the mixture will increase the ratio of smaller microspheres obtained because of the high hydrophilicity of Tween 80 that can stabilize O/W emulsion. On the other hand, the addition of Nonidet RK-18 in an increasing manner of ratio or concentration may destabilize the emulsion droplets through Ostwald ripening that can occur because Nonidet RK-18 has a low cloud point (17 °C).

Acknowledgements
The authors wish to acknowledge the support of Universitas Indonesia for funding this research through PITTA Grant with contract No:233/UN2.R3.1/PPM.00/2018 and PT. Evonik Indonesia.

References

[1] Kemala T, Budianto E and Soegiyono B 2012 *Arab. J. Chem.* **5** 103–8
[2] Jain R A 2000 *Biomaterials* **21** 2475–90
[3] Kim K K and Pack D W 2006 *BioMEMS and Biomedical Nanotechnology* ed M Ferrari *et al.* (New York: Springer) pp 19–50
[4] Thakare M, Israel B, Garner S T, Ahmed H, Garner P, Elder D, Price J C and Capomacchia A C 2013 *Pharm. Dev. Technol.* **18** 1213–9
[5] Laonuad P, Chaiyut N and Ksapabutr B 2010 *Optoelectron. Adv. Mat.* **4** 1200–2
[6] Maharana T, Mohanty B and Negi Y S 2009 *Prog. Polym. Sci.* **34** 99–124
[7] Kim B K, Hwang S J, Park J B and Park H J 2003 *J. Microencapsul.* **22** 193–203
[8] Rosen M J 2004 *Surfactants and Interfacial Phenomena* 3rd ed (New Jersey: John Wiley and Sons)
[9] Santos J, Calero N, Trujillo-Cayado L A, Garcia M C and Muñoz J 2017 *Colloids Surf. B Biointerfaces* **159** 405–11
[10] Sharma M K, Aswal V K and Hassan P A 2007 *BARC Newsletter* **285** 84–7

Crude glycerol from waste cooking oil treatment as an alternative solvent in synthesis of diethyl-2,6-dimethyl-4-phenyl-1,4-dihydropyridine-3,5-dicarboxylate

A H Cahyana, B Ardiansah and H Hanifah

Department of Chemistry, Faculty of Mathematics and Natural Sciences (FMIPA), Universitas Indonesia, Kampus UI Depok, Depok 16424, Indonesia

Corresponding author's e-mail: herrykim@ui.ac.id

Abstract. Diethyl-2,6-dimethyl-4-phenyl-1,4-dihydropyridine-3,5-dicarboxylate, a 1,4-DHP derivative, has been synthesized from benzaldehyde, ethyl acetoacetate and ammonium acetate via a one-pot three component reaction in crude glycerol as solvent. Crude glycerol was obtained from the treatment of waste cooking oil. The product was obtained in 69 % yield when the reaction was conducted at 90 °C for 1 h under reflux in 10 mL glycerol.

Keywords: waste cooking oil, hydrolysis, glycerol, solvent, 1,4-dihydropyridine.

1. Introduction

Expeditious and sustainable processes in producing organic compounds are concerning numerous synthetic chemists, where the structural complexity of products frequently requisites multistep synthetic pathways, manifolded reagents, sumptuous reaction media and tedious post-reaction work-up, leading to highly exhausting experimental protocols [1]. Currently, organic chemists have concentrated considerable attention towards more environmentally friendly synthetic methodologies that are applicable in laboratory as well as industrial scale [2]. In this status, multicomponent reactions (MCRs) have become an efficient strategy in contemporary synthesis of complex organic molecules from simple precursors because they can achieve conversion in one-step reaction without isolation of any intermediate [3]. MCRs provide a great benefit of time-saving one-pot syntheses, satisfactory yields, high atom economy, and energy-saving, thus allowing a rapid investigation of a working hypothesis [4]. Consequently, developing new multicomponent protocol is an interesting field of the study to generate molecular diversities which have a widespread of applications.

1,4-Dihydropyridine (1,4-DHP) skeleton is an integral part of pharmacophores in medicinal chemistry [5]. Commercially available drugs having this moiety, such as amlodipine and nivaldipine (figure 1) are evidence of the importance of 1,4-DHP [6]. Derivatives of 1,4-DHP have exhibited a broad spectrum of pharmacological properties, such as cocaine-dependent regulators [7], hepatoprotective [8], HIV-1 protease inhibitors [9], vasodilators [10] and antitumor [11] activities. First introduced by Arthur Hantzsch in 1882, recently many multicomponent reactions in 1,4-DHPs syntheses have been developed, such as using $Au(PPh_3)Cl/AgOTf$ [12], montmorillonite [13], magnetic guanidinylated chitosan nanobiocomposite [14], V_2O_5/ZrO_2 [15], nano-silica supported tin tetrachloride [16] and [3,6-DOMDA]OTf ionic liquid [17]. Despite good results were obtained, some reported methods have at least one limitation, such as usage of excess and toxic organic solvents, expensive metal-based catalysts and the use of high reaction temperatures. Hence, development of

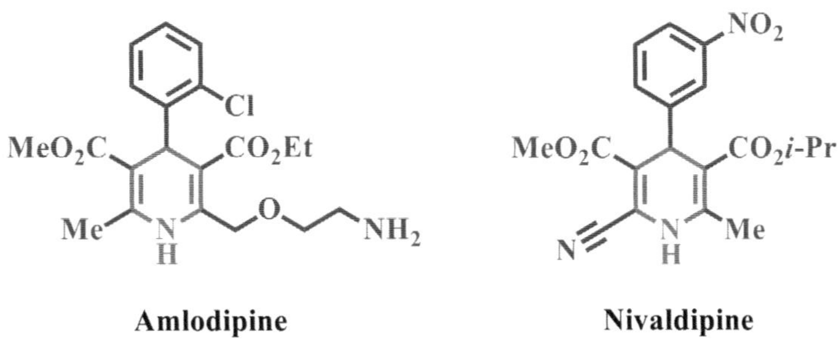

Figure 1. Two commercial drugs containing 1,4-DHP scaffold

new reaction conditions for synthesis of 1,4-DHP derivatives is still an interesting research area. In the present work, we report preparation of crude glycerol from the treatment of waste cooking oil and its use as an alternative solvent for catalyst-free one-pot multicomponent reaction in 1,4-DHP synthesis.

2. Materials and methods

2.1. Materials
Waste cooking oil was collected from the cafeteria in the faculty. All chemicals were synthesis grade and used without purification, such as benzaldehyde, cinnamaldeyde, ethyl acetoacetate and ammonium acetate. Fourier Transform Infrared (FTIR) analysis was performed on Shimadzu Prestige-21 spectrophotometer. UV-Vis spectra were noted on Shimadzu 2450 double beam spectrophotometer. Gas Chromatography and Mass Spectrometry (GC-MS) analysis was conducted on Shimadzu GC-MS QP Mass Spectrometer 2010A.

2.2. Methods

2.2.1. Preparation of glycerol from waste cooking oil. Waste cooking oil was filtered using filter paper prior to use. The filtered waste cooking oil (20 g) and potassium hydroxide 1N in ethanol (100 mL) were stirred at 60 °C for 60 min. After the reaction mixture was cooled down to room temperature, hydrochloric acid 3N (50 mL) was slowly inserted and stirred. The mixture will separate into two phases. Crude glycerol was found at lower phase. Finally, glycerol was collected and ready to be used as solvent in 1,4-DHP synthesis.

2.2.2. Synthesis of 1,4-DHP in glycerol. In a 50 mL round-bottom flask equipped with condenser, benzaldehyde (1 mmol), ammonium acetate (1 mmol) and ethyl acetoacetate (2 mmol) in prepared glycerol (10 mL) were stirred and refluxed (80–100) °C. The reaction progress was observed using TLC. After the process, the temperature of the compound was reduced to room temperature, and then cold water (50 mL) was added. The obtained product was separated by filtration. Then, crude product was recrystallized from ethanol to get pure compound, diethyl-2,6-dimethyl-4-phenyl-1,4-dihydropyridine-3,5-dicarboxylate (compound 1).

3. Results and discussion
Glycerol was prepared from hydrolysis of waste cooking oil using potassium hydroxide. Base-catalyzed hydrolysis of triglyceride in waste cooking oil was performed under reflux system at 80-100·C in ethanol to produce glycerol and soaps. Then, acidic work-up to the mixture produces two separated phases, glycerol and free fatty acids (figure 2). Glycerol was then recovered and ready to be used as solvent.

Screening of best reaction conditions was performed by using benzaldehyde, ethyl acetoacetate and ammonium acetate via a one-pot multicomponent reaction in crude glycerol (10 mL) as solvent. Reaction time was varied from 1 to 3 h, whereas temperature from 80 to 100 °C (table 1). At 80 °C for 1 h, yield of product was found to be 25 %. Surprisingly, when temperature was increased to 90 °C, the product was obtained in 69 %. However, increasing temperature to 100 °C cannot improve the product yield. Unfortunately, prolong reaction time (2 or 3h) at any temperature gave 1,4-DHP in the range of 26–40 % yield only.

Table 1. Optimization of reaction conditions[a] for MCR

Entry	Time (h)	T (°C)	Yield (%)[b]
1	1	80	25
2	1	90	69
3	1	100	30
4	2	80	28
5	2	90	40
6	2	100	28
7	3	80	27
8	3	90	38
9	3	100	26

[a] Ph-CHO (1 mmol), ethyl acetoacetate (2 mmol), NH4OAc (1 mmol)
[b] isolated yield

(a)

(b)

Figure 2. (a) reaction mixture (b) glycerol after separation

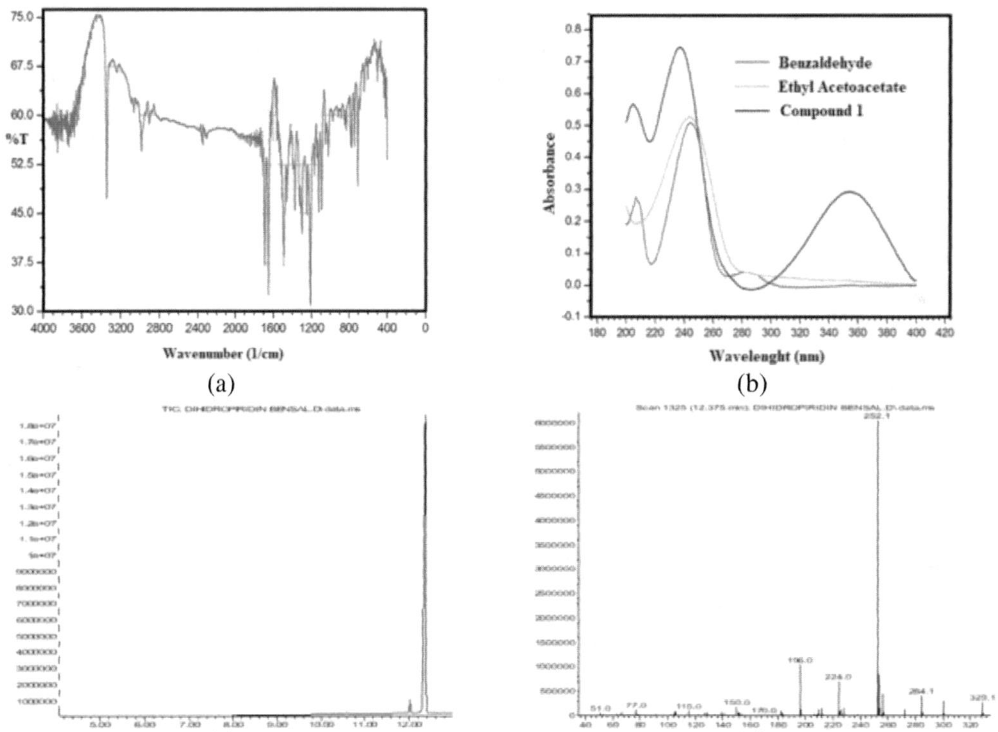

(a) (b) (c) (d)

Figure 3. (a) FTIR (b) UV-Vis (c) GC and (d) Mass spectra of diethyl -2,6-dimethyl-4-phenyl-1,4-dihydropyridine-3,5-dicarboxylate

A 1,4-DHP derivative, diethyl-2,6-dimethyl-4-phenyl-1,4-dihydropyridine-3,5-dicarboxylate, was characterized by means of FTIR, UV, Vis and GC-MS (figure 3). In FTIR, a sharp peak appeared at 3337 cm^{-1} is due to N-H stretching vibration from secondary amine. Peaks around 2900-2980 cm^{-1} indicate the presence of C-H sp^3 vibration in the molecule (figure 3a). From UV-Vis analysis, it was found that the molecule has maximum lambda at 237 and 354 nm (figure 3b), as characteristic for n to π* and π to π*, respectively. The product has retention time at 12.37 min (figure 3c), and has m/z value of 329.1 (figure 3d).

4. Conclusions

Glycerol obtained from hydrolysis of waste cooking oil can be used as an alternative solvent in synthesis of diethyl-2,6-dimethyl-4-phenyl-1,4-dihydropyridine-3,5-dicarboxylate, a derivative of 1,4-DHP. The mildness of the synthesis and experimental simplicity using non-toxic solvent make this protocol efficient for multicomponent reaction in 1,4-dihydropyridine synthesis.

Acknowledgements

This work was supported by PITTA Grant 2018 from Universitas Indonesia with contract No. 234/UN2.R3.1/PPM.00/2018.

References

[1] Filho J FA, Lemos B C, de Souza A S, Pinheiro S and Greco S J 2017 *Tetrahedron* **73** 6977–7004
[2] Goel V, Bajwan A, Chauhan S and Goel S 2018 *Chem. Sci. Trans.* **7** 343–7
[3] Patil S, Pawar P B, Jadhav S D and Deshmukh M B 2013 *Asian J. Chem.* **25** 9442–6
[4] Dömling A 2006 *Chem. Rev.* **106** 17–89
[5] Velena A, Zarkovic N, Troselj K G, Bisenieks E, Krauze A, Poikans J and Duburs G 2016 *Oxidative Medicine and Cellular Longevity* **2016** ID 1892412
[6] Khedkar S A and Auti P B 2014 *Mini-Rev. Med. Chem.* **14** 282–90
[7] Johnson B A, Devous Sr M D, Ruiz P and Ait-Daoud N 2001 *Am. J. Psychiatry* **158** 1191–8
[8] Wei Y, Lu Y, Zhu Y, Zheng W, Guo F, Yao B, Xu S, Wang Y, Jin L and Li Y 2018 *Biochim. Biophys. Acta* **1862** 2261–70
[9] Hilgeroth A, Dressler C, Neuhoff S, Spahn-Langguth H and Langguth P 2000 *Pharmazie* **55** 784–5
[10] Di Stilo A, Visentin S, Cena C, Gasco A M, Ermondi G and Gasco A 1998 *J. Med. Chem.* **41** 5393–401
[11] Mohamed M F, Darweesh A F, Elwahy A H M and Abdelhamid I A 2016 *RSC Adv.* **6** 40900–10
[12] Wang S, Chen H, Zhao H, Cao H, Li Y and Liu Q 2013 *Eur. J. Org. Chem.* **2013** 7300–4
[13] Liu Y P, Liu J M, Wang X, Cheng T M and Li R T 2013 *Tetrahedron* **69** 5242–7
[14] Maleki A, Firouzi-Haji R and Hajizadeh Z 2018 *Int. J. Biol. Macromol.* **116** 320–6
[15] Bhaskaruni S V H S, Maddila S, van Zyl W E and Jonnalagadda S B 2018 *Catal. Today* **309** 276–81
[16] Bamoniri A, Mirjalili B B F and Fouladgar S 2016 *J. Taiwan Inst. Chem. Eng.* **63** 396–403
[17] Baghery S and Zolfigol M A 2017 *J. Mol. Liq.* **232** 174–81

IOP Conf. Series: Materials Science and Engineering **763** (2020) 012017 doi:10.1088/1757-899X/763/1/012017

Investigation of metalloporphyrins in Duri crude oil asphaltene fraction

T Affendi[1], Y Yulizar[1], I Kurniawaty[2], H Haerudin[2] and D O B Apriandanu[1]

[1] Department of Chemistry, Faculty of Mathematics and Natural Sciences (FMIPA), Universitas Indonesia, Kampus UI Depok, Depok 16424, Indonesia

[2] Pertamina Research and Development, Jalan Raya Bekasi Km 20, Jakarta 13920, Indonesia

Corresponding author's e-mail: yokiy@sci.ui.ac.id

Abstract. Metalloporphyrins are a compound that leads the contaminant in catalytic cracking of crude oil. This study aims to investigate the metalloporphyrin in crude oil from Duri, Riau, Indonesia. The extraction process yielded maltene and asphaltene up to 71.87 % and 1.07%, respectively. Asphaltene fraction was extracted by soxhlet using methanol solvent. The separation of asphaltene fraction was perfomed by column chromatography using mixing of toluene: chloroform: methanol (5:5:1) eluents. The Fourier Transform Infrared (FTIR) spectrum of asphaltene indicated the presence of a pyrolle ring at 800 cm^{-1} in the fingerprint region as a porphyrin-formed group. The UV-Vis absorption spectra of methanol separation showed free porphyrin at 580 nm and metalloporphyrins at 500–700 nm. The Microwave Plasma-Atomic Emission Spectrometer (MP-AES) qualitative analysis of asphaltene fraction showed the presence of vanadium, nickel, and iron. The Liquid Chromatography-Mass Spectrometry (LC-MS) analysis on asphaltene15th fraction showed the presence of $C_{34}H_{39}N_4VO_2$ compound.

1. Introduction

Crude oil is produced from the natural process of microorganism, animals, and plants and occurs slowly over a period of thousands of years. The main constituents of crude oil are hydrocarbons and organic compounds that contain sulfur, nitrogen, oxygen, and organometallics [1]. Hydrocarbons are classified into four main groups based on their polarity and solubilty. This classification is known as SARA, which includes saturated, aromatic, resins, and asphaltenes [2].

Almost 50% of the crude oil fraction can be used as renewable oil. The renewal oil from heavy fraction and remediation of soil pollution interested a lot of researchers to investigate, due to the presence of asphaltene [3]. Asphaltene fraction contains metal complexes, which is difficult to investigate and also responsible for the catalyst contaminant [4,5]. The metal complexes are mostly vanadium and nickel [6]. Vanadium and nickel are usually from naphtenic acids in the form of organometallic complex such as metalloporphyrins [7].

As the result, in this study, metalloporphyrins were investigated from crude oil, and their structure and content in the oil were determined. Asphaltene was mixed with silica gel and extracted by soxhlet method with methanol. A column chromatography method was perfomed to separate the crude oil fraction using silica gel and the mixture of toluene: chloroform: methanol as mobile phase. A series of further soxhlet extractions and purification produced the target compounds. The separation results were characterized using Fourier Transform Infrared (FTIR) spetroscopy, UV-Vis spectrophotometer, Liquid Chromatography-Mass Spectrometry (LC-MS), and Microwave Plasma Atomic Emission Spectrometer (MPAES) to determine the metal content and metalloporphyrin structure in the crude oil.

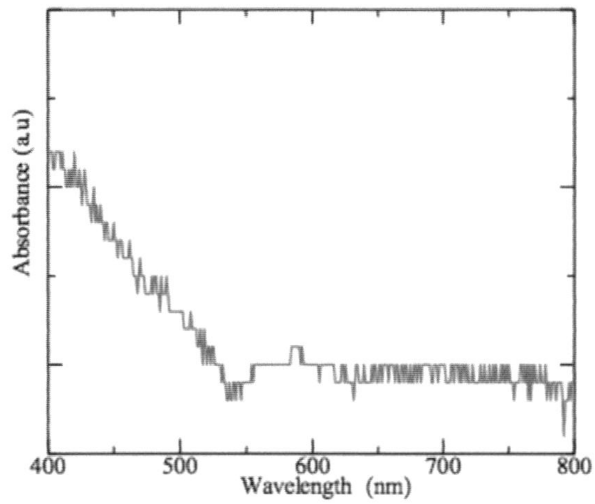

Figure 1. UV-Vis absorption spectrum of the asphaltene extract in methanol

2. Materials and methods

2.1. Materials
Crude oil was obtained from Duri, Riau, Indonesia. Methanol, n-heptane, toluene, silica gel 70–230 mesh were purchased from Merck (Germany), whereas double distilled water was used as a solvent.

2.2. Distillation of Duri crude oil
The crude oil was distilled at 260°C. The fraction with a boiling point greater than 260 °C was filtered and marked as the crude oil residue, whereas the metal content was determined [8].

2.3. Reflux of Duri crude oil residue
5 g of Duri crude oil residue was dissolved into 150 mL n-heptane and refluxed for an hour as reported by Keshmirizadeh [9]. The result of reflux was kept in the dark condition for 2 h, and then it was filtered and washed with warm n-heptane. The obtained residue and filtrate were used for further soxhlet extraction.

2.4. Soxhlet extraction of the Crude oil residue using methanol
Soxhlet extraction was perfomed by the method as reported in previous research [10]. Filter paper containing residues was placed in the extraction chamber. Round squash was filled with methanol solvent. Meanwhile, the round pumpkin containing the soxhlet extraction in n-heptane was evaporated at 60 °C.

2.5. Column chromatography
Column chromatography method was adapted from the previous reports [11]. The asphaltene fraction was mixed with pure silica to form dispersion. The mixture was injected into a silica gel-filled coloumn chromatography with the eluen composition of toluene: chloroform: methanol (5:5:1). The obtained fractions were characterized by UV-Vis spectrophotometer (Shimadzu 2600) to determine the constituent compounds, FTIR spectroscopy (IR Prestige-21 Shimadzu) to identify the functional groups, MP-AES (Agilent 4210) to determine the elemental content and composition of the fractions, LC-MS (Agilent 1200) to investigate the molecular formula and chemical structure of the compound in the fractions.

3. Results and discussion
The distillation process was conducted to separate the oil from the crude oil to the each fraction based on the boiling point. The percentages of maltene and asphaltene fractions in crude oil were 71.87 % and 1.07%, respectively. The methanol soxhlet extraction of asphaltene was investigated using UV-Vis spectrophotometer. The use of methanol is highly considered due to its inability to mix with oil and is an appropriate solvent for extraction of metallophorphyrin [11]. Figure 1 shows a weak peak in

IOP Conf. Series: Materials Science and Engineering **763** (2020) 012017 doi:10.1088/1757-899X/763/1/012017

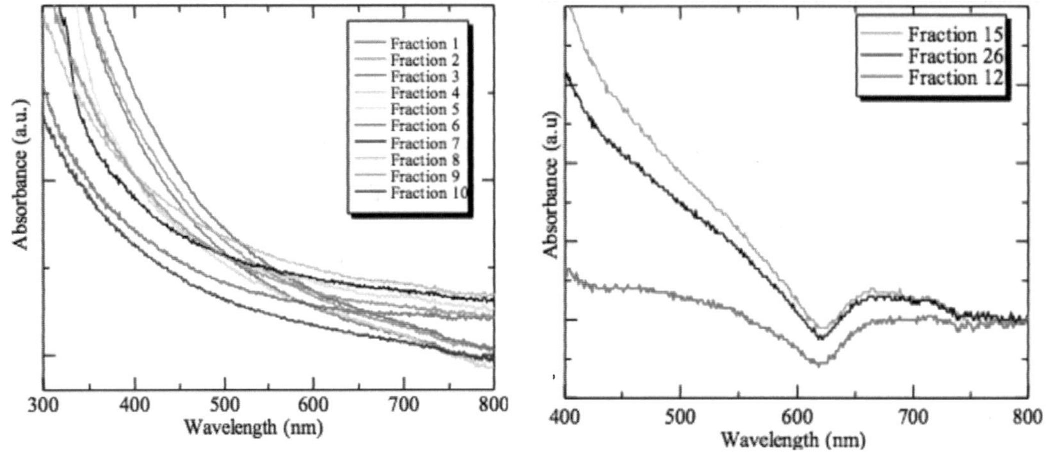

(a) (b)

Figure 2. UV-Vis Absorption Spectra of Metalloporphyrin

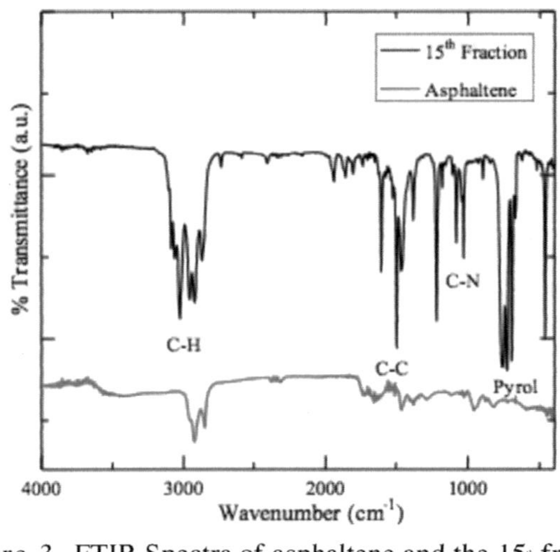

Figure 3. FTIR Spectra of asphaltene and the 15th fraction

the visible range of 480–700 nm that indicated the presence of porphyrin in the methanol fraction as described in the previous report [3].

The asphaltene fraction was further separated by column chromatography and each fraction was characterized by UV-Vis spectrophotometer as shown in figure 2. Figure 2a shows there was no peak at all in fractions at the visible range, whereas figure 2b shows absorption peak around 666 nm that indicated the presence of metalloporphyrins at fraction 12, 15 and 26. Metalloporphyrin has typical absorption around 480–700 nm [12]. Fraction 15 has the highest absorbance and lowest maximum wavelength. This result assumes that the presence of metalloporphyrins could be significant at fraction 15. Moreover, it was further characterized by FT-IR, MP-AES, and LC-MS.

The comparison of FTIR spectra of the asphaltene and the 15th fractions after column chromatography is presented in figure 3. The presence of C-H groups was indicated by a band at 2865 cm^{-1}. The bands at 2865, 1500, 1116, and 800 cm^{-1} corresponds to the presence of C-H, C = C aromatic, C-N, and pyrrole groups, respectively. The presence of pyrrole group illustrates the typical vibration mode of porphyrin at 800 cm^{-1} as reported by Mishra *et al.* [13]. Therefore, the presence of a porphyrin in the fractions is clearly revealed.

Table 1. Elemental composition of asphaltene fractions after coloumn chromatography

Sample	Concentration (ppm)		
	Ni	Fe	V
12nd Fraction	0.042	0.015	0.011
15th Fraction	0.026	0.010	0.074
26th Fraction	0.034	0.012	0.110

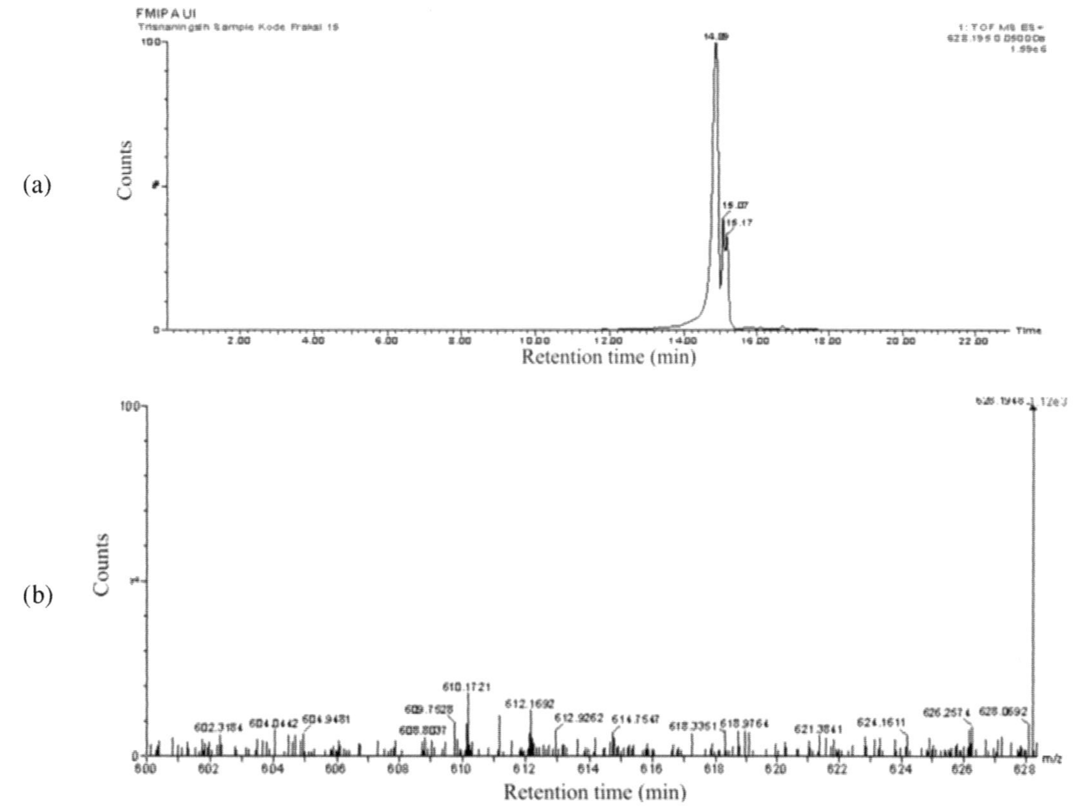

Figure 4. (a) Liquid chromatogram and (b) Mass spectrogram of 15th fraction

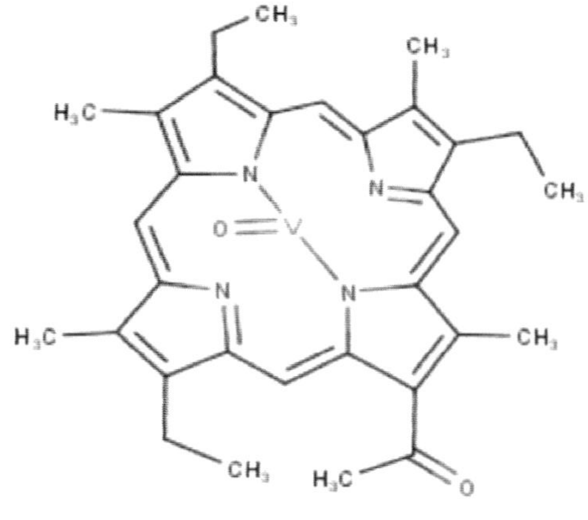

Figure 5. Molecular structure of $C_{34}H_{30}N_4VO_2$

Table 1 shows the MP-AES data that reveals the presence of elements commonly observed in asphaltene fractions. LC analysis shows a peak at retention time of 14.89 min as presented in figure 4a. Figure 4b shows the value of *m/z* represented by the pseudomolecular peak ion (M + H + ACN) is 628. Therefore, the compound has a molecular weight of 586. This indicates compatibility with the previous research [14] that the molecular weight of 586 is $C_{34}H_{39}N_4VO_2$ as a porphyrin compound bound to vanadium metals as shown in figure 5. This result was supported by the UV-Vis absorption spectrum, which shows metal bonds to porphyrin.

4. Conclusions

Metalloporphyrin has been investigated in asphaltene fraction of Duri crude oil with the chemical structure of $C_{34}H_{39}N_4VO_2$ confirmed by LC-MS analysis. FTIR spectrum confirmed a pyrrole ring appears at 800 cm^{-1}, which corresponds, to porphyrin groups. UV-Vis absorption spectrum also showed a vanadium bond porphyrin at 666 nm.

Acknowledgements

This work was funded by Pertamina Join Research Grant between Departement of Chemistry Faculty of Mathematics and Natural Science Universitas Indonesia and Research & Technology Center PT. Pertamina (Persero) 2018.

References

[1] Ali M F and Abbas S 2006 *Fuel Proc. Technol.* **87** 573–84
[2] Vale G M R, Silva M M, Damin I C F, Filho P J S and Welz B 2008 *Talanta* 74 1385–91
[3] Ghannam M T, Hasan S W, Abu-Jdayil B and Esmail N 2012 *J. Petrol. Sci. Eng.* **81** 122–8
[4] Abbas S, Maqsood Z T and Ali M F 2014 *Pet. Sci. Technol.* **28** 1770–7
[5] González G, Sousa M A and Lucas E F 2006 *Energy Fuels* **20** 2544–51
[6] Hirschberg A, de Jong L N J, Schipper B A and Meijer J G 1984 *Soc. Pet. Eng. J.* **24** 283–93
[7] Liu H, Mu J, Wang Z, Ji S, Shi Q, Guo A, Chen K and Lu J 2015 *Energy Fuels* **29** 4803–13
[8] Kurniawaty I, Yulizar Y, Gamar I, Haerudin H and Apriandanu D O B 2018 *AIP Conf. Proc.* **2023** 020087
[9] Keshmirizadeh E, Shobeirian S and Memariani M 2013 *J. Appl. Chem. Res.* **7** 15–24
[10] Zhao X, Liu Y, Xu C, Yan Y, Zhang Y, Zhang Q, Zhao S, Chung K, Grays M R and Shi Q 2013 *Energy Fuels* **27** 2874–82
[11] Yakubov M R, Abilova G R, Sinyashin K O, Milordov D V, Tazeeva E G, Yakubova S G, Borisov D N, Gryaznov P I, Mironov N A and Borisova Y Y 2017 *Phthalocyanines and Some Current Applications* ed Y Yilmaz (London: IntechOpen)
[12] Rio Y, Rodríguez-Morgade M S and Torres T 2008 *Org. Biomol. Chem.* **6** 1877–94
[13] Mishra S, Kaur S, Tripathi S K, Mahajan C G and Saini G S S 2006 *J. Chem. Sci.* **118** 361–9
[14] Dechaine G P and Gray M R 2010 *Energy Fuels* **24** 2795–808

IOP Conf. Series: Materials Science and Engineering **763** (2020) 012018 doi:10.1088/1757-899X/763/1/012018

Sodium alginate-TiO_2-bentonite nanocomposite synthesis for photocatalysis of methylene blue dye removal

R Subekti and H Helmiyati

Department of Chemistry, Faculty of Mathematics and Natural Sciences, Universitas Indonesia, Kampus UI Depok 16424, Indonesia

Corresponding author's e-mail: helmiyati@sci.ui.ac.id

Abstract. In this study, nanocomposite based on biopolymer using sodium alginate (NaAlg) and bentonite intercalated by TiO_2 was synthesized. Bentonite in the nanocomposite was used as a template for TiO_2 nanoparticles formation through in-situ method and NaAlg was used as catalyst supporter in order to obtain nanocomposite material, which was expected as dye waste removal. The nanocomposite was characterized using Fourier Transform Infrared (FTIR), Scanning Electron Microscopy (SEM), Energy Dispersive X-Ray (EDX) and Tunneling Electron Microscopy (TEM). The obtained nanocomposite had TiO_2 average nanoparticles size around 35 nm. Nanocomposite was applied for photocatalysis to remove the dye waste of methylene blue (MB). Degradation percentage was achieved at 90.50 % with the optimum condition at pH 8 with photocatalysis time at 30 min and 30 mg nanocomposite mass. Biopolymer-based sodium alginate and TiO_2-bentonite nanocomposite can be applied to reduce dyestuff that is biodegradable and environmentally friendly waste.

Keywords: biopolymers, bentonite, catalyst, nanocomposite, sodium alginate

1. Introduction

Society's high demand of textile causes the textile industry to grow massively and carries negative impacts for human beings and environment. Recently, one of the utmost concerns is how to handle the environmental pollution, especially in water, due to the wastewater from textile industry. Wastewater that contains dye is very dangerous because most dyes are hard to be degraded, resistant, and toxic [1]. One of the most used dyes is methylene blue (MB), which is a non-biodegradable organic pollutant caused by the presence of benzene groups that are difficult to be degraded, in which its exposure will cause asphyxiation, vomiting, diarrhea, and nausea [2]. Therefore, to prevent the risk, some efforts to remove dye in environment should be taken.

Some techniques can be done to treat dye waste, such as adsorption and photocatalysis. These techniques are widely used because adsorption is easier to be applied, low-cost and have a high uptake. However, the adsorption process only absorbs waste toxic and it will not be degraded into non-toxic compounds. Meanwhile, photocatalysis technique is widely used because it can decompose toxic into non-toxic compounds. This is due to in photocatalytic process a lot of hydroxyl radical that attacks dye organic into CO_2, H_2O, and other non-toxic substances will be formed, so it will be environmentally friendly [3].

In the photocatalyst process, the used materials have the ability to absorb photons and usually owned by semiconductor materials. Metal semiconductor materials such as TiO_2, ZnO, Fe_2O_3, CdS, and ZnS can be utilized as catalysts in a photocatalytic process. The most widely used semiconductor for photocatalysis is titanium dioxide (TiO_2), due to its stability under UV-light. In this study, bentonite as filler is used due to the high adsorption capacity of its porous structure [4–6].

Utilization of the biopolymer as nanocomposite has caught attention of many researchers around the globe; the cause is its biodegradable properties and decreases the needs for synthetic polymer. One of the biopolymers that have many advantages is alginate, which extracted from brown algae. Alginate is a renewable and biodegradable natural polymer, which has high capacity to remove toxic pollutants such as dye waste [7–9]. Some researches have made some attempts to treat waste using composite such as Harikumar *et al.* [7], who has succeeded to degrade some dyes using calcium alginate-TiO_2 composite through adsorption and photocatalytic process. Laysandra *et al.* [10] has successfully reduced methylene blue and rhodamine B dye wastes through photocatalytic and adsorption process.

In this study, nanocomposite was done by combining sodium alginate, bentonite with inorganic substances to obtain nanocomposites that have superior property, derived from sodium alginate and bentonite having large surface area, high pores and good degradation properties of TiO_2, so that obtained a maximum photocatalytic power in reducing methylene blue dyestuff. In this work, the effect of nanocomposite mass (10, 20, 30, and 40 mg), photocatalysis time (1 h with 10 min interval), and the effect of pH (3, 7, 8, and 9) of methylene blue solution have been studied.

2. Experimental

2.1. Materials

Materials used were titanium tetraisopropoxide, TTIP (Merck Co) as the precursor of TiO_2; ethanol p.a. (Merck Co) as solvents; HCl (Merck Co) as a catalyst in TiO_2 synthesis; sodium alginate (Sigma) and bentonite (Sigma Aldrich) as nanocomposite components; HNO_3 (Merck Co) as a catalyst for TiO_2-bentonite composite synthesis; $CaCl_2$ (Merck Co) to form beads on nanocomposite; H_2SO_4 (Merck Co) as bentonite activator and dye acidity controller, NaOH (Merck Co) to control dye basicity; methylene blue (Merck Co) as the dye to be tested.

2.2. TiO_2 synthesis

TiO_2 has been synthesized according to previous reports [11]. The precursor of titanium tetraisopropoxide (TTIP) was added into distilled water, ethanol, HCl and then stirred at room temperature and dried. The solid was calcinated at 450 °C for 2 h. The nano TiO_2 was characterized by Fourier Transform Infrared (FTIR) and Scanning Electron Microscopy (SEM).

2.3. Sodium alginate-TiO_2 synthesis

The synthesis of sodium alginate-TiO_2 was performed using ex-situ method [12]. TiO_2 (2 wt.%) and sodium alginate (2 wt.%) was stirred, then added with 0.5 M $CaCl_2$ solution to form beads of NaAlg-TiO_2. The beads were dried in an oven at 30 °C for 1 h. The sodium alginate-TiO_2 was characterized by FTIR and SEM.

2.4. Bentonite-TiO_2 synthesis

The synthesis of TiO_2-bentonite was performed using in-situ method [12,13]. TTIP was dropped into ethanol and stirred for 30 minutes, and then it was added into HNO_3 until TiO_2 sol was formed. Furthermore, bentonite was added to the solution and stirred, the solid was dried and calcinated at 500 °C. The TiO_2-bentonite was characterized by FTIR and SEM

2.5. Synthesis of sodium alginate-TiO_2-bentonite nanocmposite

The synthesis of sodium alginate-TiO_2-bentonite referred to previous reports [5]. The sodium alginate was dissolved in distilled water and added with TiO_2-bentonite, then stirred until the solution homogenous, and it was injected into $CaCl_2$ solution. The obtained solid was dried at 50 °C for 1 h and the nanocomposite of sodium alginate-TiO_2-bentonite was characterized by FTIR, Scanning Electron Miscroscopy-Energy Dispersive X-Ray (SEM-EDX) and Tunneling Electron Microscopy (TEM).

2.6. Photocatalysis Activity

The activity of photocatalysis was tested against methylene blue solution, performed by the three types of nanocomposite (sodium alginate-TiO_2; bentonite-TiO_2; sodium alginate-TiO_2-bentonite) by referring to previous reports [5]. Photocatalysis process was conducted for 1 h with UV light, and analyzed for every 10 min using UV-Vis spectrophotometer. The degradation capacity or removal capacity of methylene blue catalytic process by nanocomposite was calculated using the following equation:

$$The\ degradation\ capacity\ (\%) = \left(\frac{C_i - C_e}{C_i}\right) \times 100 \quad (1)$$

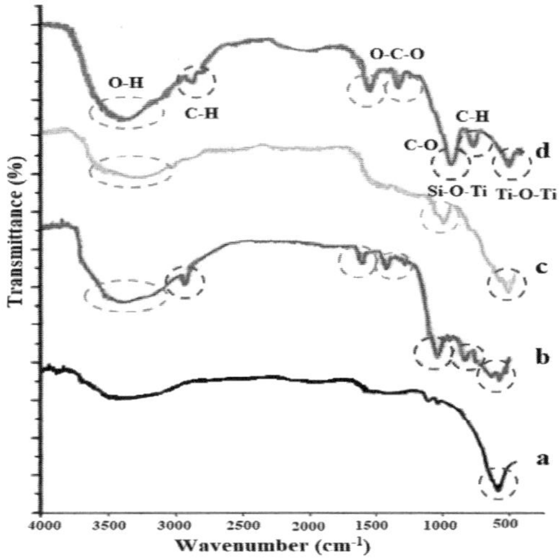

Figure 1. FTIR analysis of (a) TiO_2; (b) sodium alginate-TiO_2; (c) bentonite-TiO_2; (d) sodium alginate-TiO_2-bentonite nanocomposite

where C_i is initial concentration of methylene blue (ppm) and C_e is methylene blue concentration at equilibrium (ppm).

3. Results and discussion

3.1. FTIR Analysis

Figure 1a shows the absorption of infrared of TiO_2, sodium alginate-TiO_2 (figure 1b) bentonite-TiO_2 (figure 1c) and sodium alginate-TiO_2-bentonite nanocomposite (figure 1d). In figure 1a, we can see the absorption peak of TiO_2 at 590 cm^{-1} indicating the stretching and bending vibrations of Ti-O-Ti [14]. Figure 1b shows that absorption peak around 3150-3600 cm^{-1} which is O-H stretching vibration of hydroxyl groups, weak absorption band at 2937 cm^{-1} shows C-H stretching. The peaks at 1400 and 1628 cm^{-1}, shows symmetric and asymmetric O-C-O stretching vibrations and at 500-700 cm^{-1} is characteristic for Ti-O-Ti from TiO_2 (b). Figure 1c shows the absorption band at 3150-3600 cm^{-1} indicating the O-H stretching vibration of Ti-OH and Si-OH interacted with water molecules to form hydrogen bond. It also appears a peak of Si-O-Ti at 965 cm^{-1}. In figure 1d, we can see the absorption band at 3250-3600 cm^{-1} indicating O-H stretching vibration, and the weak absorption band at 2937 cm^{-1} which is characteristic for the C-H stretching. The peaks at 1475 and 1645 cm^{-1} indicating the asymmetric and symmetric stretching of carboxylate O-C-O [15]. There is also a peak at 1035 cm^{-1} refers to C-O stretching of uronic acid and the peak at 820 cm^{-1} is the absorption band of C-H vibration mannuronic acid and peaks at 550 cm^{-1} indicating the presence of Ti-O-Ti bond and we can see peak of TiO_2 at 500 cm^{-1}, whereas at pure TiO_2 showed at 590 cm^{-1}, shift of peak shows that the TiO_2 have been successfully impregnated in nanocomposite.

3.2. SEM and EDX analysis

Surface characterization using SEM is shown in figure 2. Figure 2a shows the morphology of sodium alginate, we can see short fibers are distributed evenly, which is the characteristic of sodium alginate. Figure 2b shows the morphology of TiO_2, and have small granules shape. Figure 2c is the morphology of bentonite, there are pores so that causes a good adsorption ability. Meanwhile, figure 2d shows the morphology of NaAlg-TiO_2-Bentonite, it can be observed the presence of fibrous derived from sodium alginate and the pores from bentonite. There are TiO_2 as small granules particles on the surface of sodium alginate and in the pores bentonite. This shows the successful synthesis of the nanocomposite.

Figure 3 shows EDX analysis of NaAlg-TiO_2-Bentonite. It can be observed the content of elements in nanocomposite. There are O, C, Si, Ti, Al, etc. (41.7, 19.3, 11.2, 9.2, 5.1 and 13.5 wt.%), it shows at most of O content derived from sodium alginates, monomers and bentonite. While, Si and Al are from

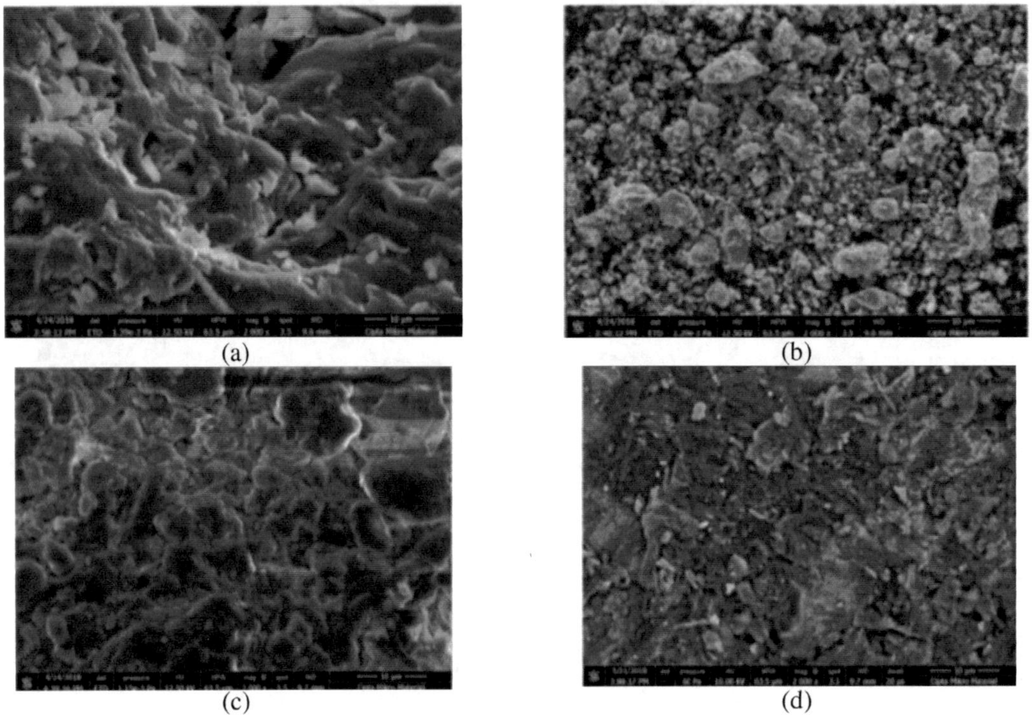

Figure 2. Surface morphology of (a) sodium alginate; (b) TiO_2; (c) bentonite and (d) sodium alginate-TiO_2- bentonite nanocomposite

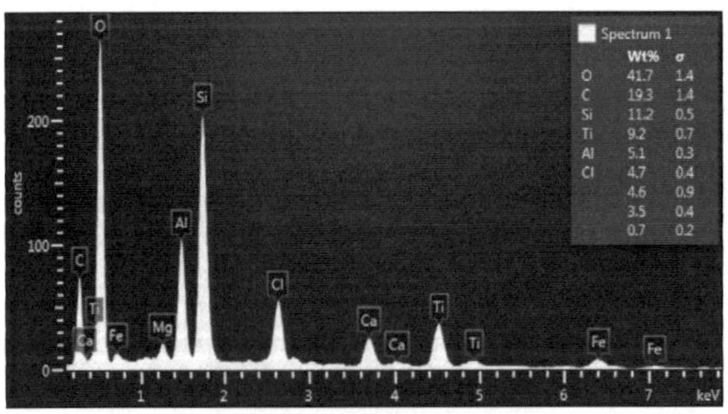

Figure 3. EDX analysis of sodium alginate-TiO_2- bentonite nanocomposite

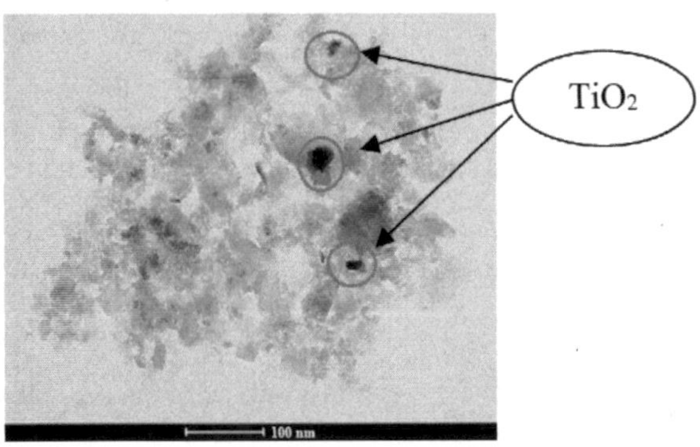

Figure 4. Micrograph of NaAlg-TiO_2-Bentonite nanocomposite at resolution of 100 nm

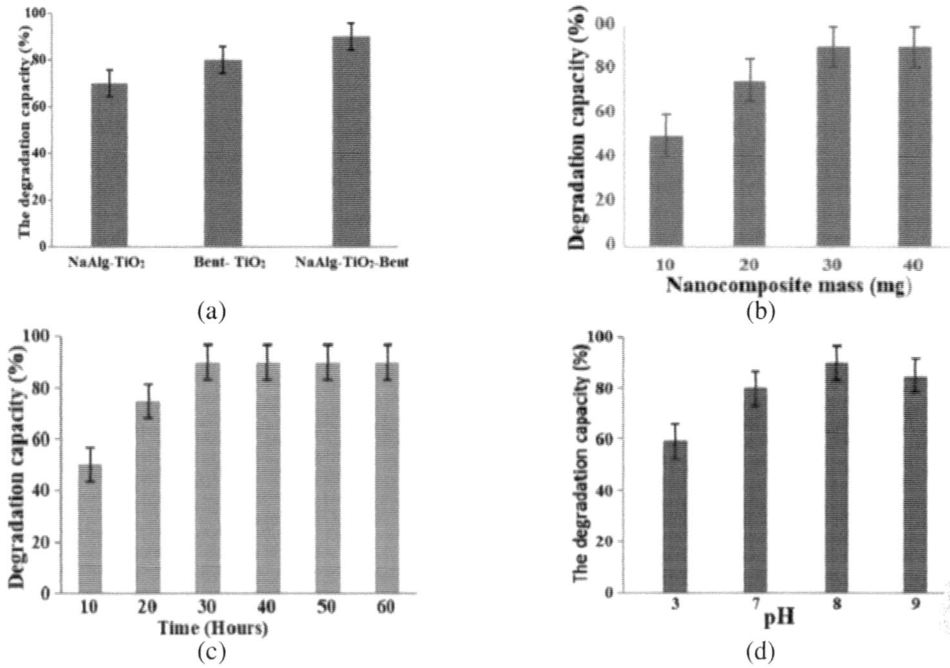

Figure 5. The variation of (a) nanocomposite types; (b) nanocomposite mass; (c) time and (d) pH

bentonite. The content of Ti is 9.2 wt.% in peak appears in three peaks at 0.2, 4.3 and 4.9 keV. The EDX data supports the successful synthesis of NaAlg-TiO_2-bentonite nanocomposite.

3.3. TEM analysis
Characterization using transmission electron microscopy (TEM) with 100 nm resolutions can be seen in sodium alginate-TiO_2-bentonite nanocomposite in figure 4. We can see the surface morphology of the TiO_2 nanoparticles are small, black particles with TiO_2 in average size around 35 nm and bentonite as templates shows gray color and sodium alginate as a supporting agent with brighter color can be observed. This supports the results of the previous characterization showing the success of nanocomposite synthesis.

3.4. Photocatalysis activity
The photocatalysis activity was done by determining the best composite. There are three types of nanocomposites have been synthesized (NaAlg-TiO_2, bentonite- TiO_2 and NaAlg-TiO_2-Bentonite). Each of nanocomposite was performed the degradation process against methylene blue (MB). Degradation capacity of MB in three types of nanocomposites is shown in figure 5a. The methylene blue before the degradation process have dark blue color and after the photocatalysis process, it got paler. We could see that the percentage of degradation sequence of methylene blue for the three types of nanocomposites is NaAg-TiO_2 (69.05 %) < Bentonite-TiO_2 (79.10 %) < NaAlg-TiO_2-Bentonite (90.50 %), respectively. The greater the percent of degradation, the solution will be clearer. This is caused by the nanocomposite of NaAlg-TiO_2-Bentonite have strong adsorption ability derived from sodium alginate and bentonite and combination of TiO_2 have good degradation ability and obtained the best degradation capacity. For the next optimization, the best nanocomposite of NaAlg-TiO_2-Bentonite was used. Figure 5b shows the effect of NaAlg-TiO_2-Bentonite nanocomposite mass to MB degradation capacity in the photocatalysis process. We can see the percentage of degradation capacity MB increased with rise of the nanocomposite mass and reached an optimum at 30 mg in 90.50 %. Furthermore, the increase of mass more than 30 mg, caused no significant change to degradation capacity.

Figure 5c shows the time effect to the degradation capacity of MB on photocatalysis process. We can see the percentage of degradation capacity MB increased with the rise of time and reached an optimum time at 30 min, which obtained 90.10 %, and then with the increase of time, there is no significant change to degradation capacity. Figure 5d shows the degradation capacity of MB increased with the rise of pH. In acidic condition, the excessive H^+ ion in solution causes the active site of nanocomposites surface creates electrostatic repulsion between H^+ ions and methylene blue cationic with the nanocomposites surface. Therefore, the amount of adsorbed methylene blue will decrease. The reaction estimation of the active site of TiO_2 in acidic condition can be seen in equation (2). In base condition or the pH of the solution increases, the photocatalytic process will increase with the rise of pH. This is due to the change of charge of TiO_2, in a high pH, TiO_2 would be negatively charged. Furthermore, the increase of OH group would increase OH radicals. Therefore, since methylene blue is a cationic dye, there will be attraction between the negatively charged nanocomposite surfaces with positively charged of methylene blue, so that at high pH, photodegradation effectivity would be increased [16]. The optimum degradation capacity was obtained with condition of pH 8. The reaction estimation of the active site of TiO_2 in base condition is shown in equation (3).

$$pH\ (acid) \quad : Ti - OH + H^+ \quad \rightarrow TiOH_2^+ \tag{2}$$

$$pH\ (bacic) \quad : Ti - OH + OH^- \quad \rightarrow TiO^- + H_2O \tag{3}$$

The best degradation capacity using sodium-alginate-TiO_2-bentonite nanocomposite for the degradation of methylene blue obtained at nanocomposite mass 30 mg, photocatalysis time 30 min and pH 8 was 90.10 %. The results of this degradation showed better results than our previous work with 51.68 % degradation [16].

4. Conclusions

Among some composites that have been synthesized, NaAlg/TiO_2/bentonite composite has the best ability to degrade methylene blue. The success of nanocomposite synthesis was supported by characterization of FTIR, SEM, EDX and TEM. EDX characterization shows that TiO_2 has the composition of 9.2 wt.%, with particle size 35 nm according to TEM characterization. The percentage of degradation sequence of methylene blue for the three types of nanocomposites is NaAg-TiO_2 (69.05 %) < Bentonite-TiO_2 (79.10 %) < NaAlg-TiO_2-Bentonite (90.50 %), respectively. NaAlg-TiO_2-bentonite nanocomposite has the best ability to degrade methylene blue. The optimal photocatalysis conditions obtained at 30 mg nanocomposite mass for 30 min at pH 8 is 90.50 % of methylene blue have been degraded.

Acknowledgements

This research was supported by Research Fund from University of Indonesia for fiscal year 2018.

References

[1] Oladipo A A and Gazi M 2014 *J. Water Process Eng.* **2** 43–52
[2] Tarawou T and Horsfall Jr M 2007 *Bioremediation J.* **11** 77–84
[3] Ahmad M, Ahmed S, Swami B L and Ikram S 2015 *International Journal of Pharmacognosy (IJP)* **2** 280–9
[4] Shen D, Fan J, Zhou W, Gao B, Yue Q and Kang Q 2009 *J. Hazard. Mater.* **172** 99–1107
[5] Vahidhabanu S, Karuppasamy D, Adeogun A I and Babu R B 2017 *RSC Adv.* **7** 5669–78
[6] Chen X, Wu Z, Liu D and Gao Z 2017 *Nanoscale Res. Lett.* **12** 143
[7] Harikumar P S, Joseph L and Dhanya A 2013 *J Environ. Eng. Ecol. Sci.* **2** 2
[8] Papageorgiou S K, Katsaros F K, Kouvelos E P and Kanellopoulos N K 2009 *J. Hazard. Mater.* **162** 1347–54
[9] Mahmoud G A, Mohamed S F and Hassan H M 2015 *Desalin. Water Treat.* **54** 2765–76
[10] Laysandra L, Sari M W M K, Soetaredjo F E, Foe K, Putro J N, Kurniawan A, Ju Y H and Ismadji S 2017 *Heliyon* **3** e00488
[11] Mehrizad A, Gharbani P and Tabatabii S M 2009 *J. Iran. Chem. Res.* **2** 145–9
[12] Lam W H, Chong M N, Horri B A, Tey B T and Chan E S 2017 *J. Appl. Polym. Sci.* **134** 45002
[13] Hadjltaief H B, Zina M B, Galvez M E and Da Costa P 2016 *J. Photochem. Photobiol.* A *Chem.* **315** 25–33
[14] Sun Z, Chen Y, Ke Q, Yang Y and Yuan J 2002 *J. Photochem. Photobiol.* A *Chem.* **149** 169–74
[15] Helmiyati H and Dini F W 2018 *AIP Conf. Proc.* **2023** 020101
[16] Helmiyati and Aprilliza M 2017 *IOP Conf. Ser.: Mater. Sci. Eng.* **188** 012019

[17] Karamanis D, Ökte A N, Vardoulakis E and Vaimakis T 2011 *Appl. Clay Sci.* **53** 181–7
[18] Helmiyati H and Wahyuningrum K D 2018 *AIP Conf. Proc.* **2023** 020107

IOP Conf. Series: Materials Science and Engineering **763** (2020) 012019 doi:10.1088/1757-899X/763/1/012019

Modification of boron-doped diamond with gold through wet-chemical seeding and electrodeposition techniques for the application of acrylamide biosensor

L E Anggraini, E Saepudin and T A Ivandini

Department of Chemistry, Faculty of Mathematics and Natural Sciences (FMIPA), Universitas Indonesia, Kampus UI, Depok, 16424, Indonesia

Corresponding author's e-mail: ivandini.tri@sci.ui.ac.id

Abstract. Acrylamide is known as carcinogenic and neurotoxin substrates. An alternative method for acrylamide detection is by using hemoglobin-based biosensors as it can provide a simple, rapid, and sensitive method. In this research, the electrodes for biosensor of acrylamide were prepared by modifying boron-doped diamond (BDD) with gold and hemoglobin (Hb). Wet-chemical seeding followed by electrochemical overgrowth of seeds technique was performed to modify the BDD electrodes with gold. The characterization with Scanning Electron Microscopy-Energy Dispersive Spectroscopy X-Ray (SEM-EDS) showed the gold coverage of 12.74 % on the BDD surface. The gold-modified BDD was then modified with 0.25 mM Hb. The prepared sensor can be successfully applied in the concentration range of acrylamide from 0.6 to 6 μM with high linearity (R^2 = 0.9901) and an estimated limit of detection of 0.845 μM.

Keywords: acrylamide, biosensor, boron-doped diamond, gold, hemoglobin

1. Introduction

Carbohydrate-rich foods, such as potatoes, bread, coffee, cigarettes, and cereal products, can stimulate the formation of carcinogenic compounds called acrylamide through the frying or roasting process at high temperatures [1–3]. The acrylamide formation occurs through Maillard reaction, involving the heating of asparagines and reducing sugars [1,3–7]. Research over the years proves that acrylamide is a neurotoxin and toxic compound and also rapidly absorbed by humans and distributed in brain, thymus, heart and liver [1,2,4,5,8]. Accordingly, this substance is categorized as "probably carcinogenic to humans" by the International Agency on Research on Cancer (IARC) [2].

Various analytical methods, such as chromatography and mass spectroscopy, are available for acrylamide detection. But these methods are not applicable for common people. On the contrary, biosensor methods can overcome these problems due to their simple, rapid, sensitive and spesific applications [8, 9–13].

In this work, hemoglobin (Hb) is used as the biosensing since Hb and acrylamide interacts each other through N-terminal valine of Hb. This interaction decreases the electrochemical signal of acrylamide. On the other hand, boron-doped diamond (BDD) is widely applied in electrochemical applications because it has stable surface morphology and microstructures, low background currents, and wide potential windows [9,14–16]. However, BDD has low sensitivity to hemoglobin, therefore, modification with gold is required to facilitate the electron transfer between Hb and BDD [4,16–18].

Furthermore, a common method to modify gold onto BDD surface, i.e. electrodeposition, often produces poor electrode stability [9,18,19]. Therefore, in this work, the method to prepare gold-modified BDD (Au-BDD) electrode was performed through wet-chemical seeding and electrochemical overgrowth of seeds techiques to increase the electrode's stability.

2. Materials and methods

2.1. Reagents and instruments

The polycrystalline BDD films (B:C ratio of the precursor gas was 1:1000) were supplied from Einaga's Laboratory, Keio University, Japan. The films were prepared in chemical vapor deposition method by using methane as the carbon source and trymethoxy borane as the boron source. Detail of the preparation was described here [20]. Human hemoglobin H7379 and acrylamide were obtained from Sigma-Aldrich, while sodium borohydrate, sodium hydroxide, isopropanol, n-hexsane, chloroauric acid tetrahydrate ($HAuCl_4.4H_2O$) and other chemicals were purchased from Merck. Acetate buffer solution (ABS) pH 5 was made by mixing acetic acid and sodium acetate. Ultra-pure deionized water was obtained from Millipore Direct® Q5-UV at resistivity of 18.2 MΩcm.

Potentiostat (Metrohm DropSens) with three electrodes system was used, served as working electrode was Hb/Au-BDD, while counter and reference electrodes were Pt wire and Ag/AgCl, respectively. Rapid Thermal Annealing (RTA), Scanning Electron Microscopy-Energy Dispersive X-Ray Spectroscopy (SEM-EDS) analysis (JEOL JCM-6000PLUS NeoScope), Raman Spectrophotometer (Olympus), and High Performance Liquid Chromatography (LC210) were utilized in this research.

2.2. Modification of BDD with gold by using wet-chemical seeding and electrochemical overgrowth of seeds techniques

To modify the BDD, solutions of 1M $NaBH_4$ in 0.1M NaOH and 1mM $HAuCl_4$ were subsequently dropped onto the BDD surface. After 24 h, the BDD was washed and dried. The deposition procedure was repeated 3 times. Then, the amperometric deposition was conducted using 1 mM $HAuCl_4$ solution at a constant potential of -0.2 V. Next, rapid thermal annealing at 700 °C in N_2 atmosphere for 5 min was conducted. After annealing, the electrode was refreshed by using cyclic voltammetry (CV) in 0.1M H_2SO_4 for 50 cycles to activate the electrode, followed by amperometric deposition at -0.2 V. [8]. Stability of Au-BDD electrode was examined with 30 min sonicated and 100 times CV, while characterization of the electrode was performed by using CV and SEM-EDS.

2.3. Preparation of hemoglobin modified Au-BDD (Hb/Au-BDD) electrode

Hb powder was dissolved in 0.1 M acetate buffer solution (ABS) pH 5. Volume of 15 μL Hb solution was dropped onto Au-BDD electrode, then stored at 4 °C for 24 h to keep the Hb condition from being damaged and remain stable to the surface of the electrode. Characterization was performed by SEM EDS.

2.4. Response measurements of Hb/Au-BDD electrode

The measurements of acrylamide biosensor were performed by using CV technique in the potential range of -0.2 to +0.8 V, served as working electrode was Hb/Au-BDD, while counter and reference electrodes were Pt wire and Ag/AgCl, respectively. The scan rate was 0.1 V/s. Acrylamide standard solutions were prepared by dissolution of acrylamide powder in ABS pH 5.

3. Results and discussion

3.1. Preparation of Au-BDD electrode by using wet-chemical seeding and electrochemical overgrowth of seeds technique

Theoretically, there will be a reaction between $NaBH_4$ and $HAuCl_4$ on the surface of the BDD electrodes according to the following reaction:

$$HAuCl_4 + 4\,NaBH_4 + 12\,H_2O \rightleftharpoons Au + 4\,B(OH)_3 + 4\,NaCl + 29\,H^+$$

This process deposited gold particles on the surface of the BDD by physical adsorption. Then, the electrodeposition was carried out at a potential range of -0.5 to +1.5 V, where a reduction reaction from Au^{3+} to Au^0 occurred to form gold metal particles on the surface of BDD, particularly at the gold particles. Figure 1 shows the comparison of the cyclic voltammogram of BDD before modification (red line), after wet chemical seeding (green line), and after electrodeposition (black line). The calcination by using rapid thermal annealing (RTA) at 700 °C for 5 min in N_2 atmosphere was then conducted to improve the growth of gold particles on the BDD surface. N_2 atmosphere was used to prevent the formation of metal oxides when heated at this high temperature. However, because of this treatment, the electrode surface was inactive as shown by lower gold oxidation-reduction peak currents after annealing (light blue line).

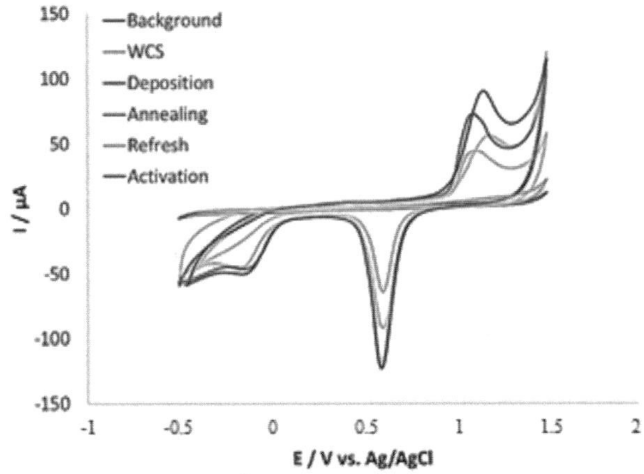

Figure 1. Cyclic voltammograms performed in 0.1 M acetate buffer solution pH 5 at scan rate 100 mV/s of at BDD before modification (red line), after wet chemical seeding of gold (green line), after first electrodeposition black line), after annealing (light blue), after refreshing treatment (orange line), and after activation by using 2nd electrodeposition (blue line).

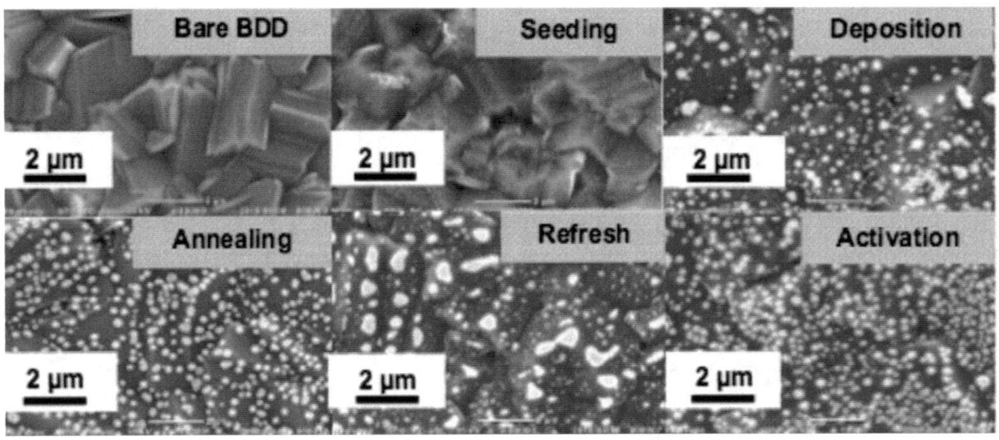

Figure 2. SEM images every steps Au-BDD electrodes.

CV was performed to refresh the Au-BDD surface by repeating the oxidation-reduction in a 0.1M H_2SO_4 solution for 50 times. After the treatment, recovery of the peak was observed as shown in figure 1 (orange line), indicating the electrode was electrochemically active. However, the electrochemical redeposition of gold using $HAuCl_4$ solution was necessary to recover the current response of gold (blue line).

SEM images were conducted to study the change on the BDD surface after modification. Figure 2 shows the SEM images of the BDD surface before and at every step after modification. The figure shows that at the seeding step, less gold particles were deposited, while the number and size of gold particles increased after the first electrodeposition. Next, the annealing treatment homogenized the particle size and distribution, which was then randomly spread after refreshing treatment. However, after the 2nd electrodeposition the particles were more homogeneously deposited.

Sonication of the Au-BDD was performed for 30 min to examine the stability of the deposited gold. The CV performed in ABS pH 5 showed no change of the peak in the voltammograms before and after sonication (figure 3a). The current decrease was around 1.95 %. Furthermore, CV was also applied for repetitive 100 cycles in a solution of ABS pH 5, where a slightly decrease of the peak current was observed with an RSD of 1.35 % (figure 3b). The results indicated a good stability of the Au-BDD.

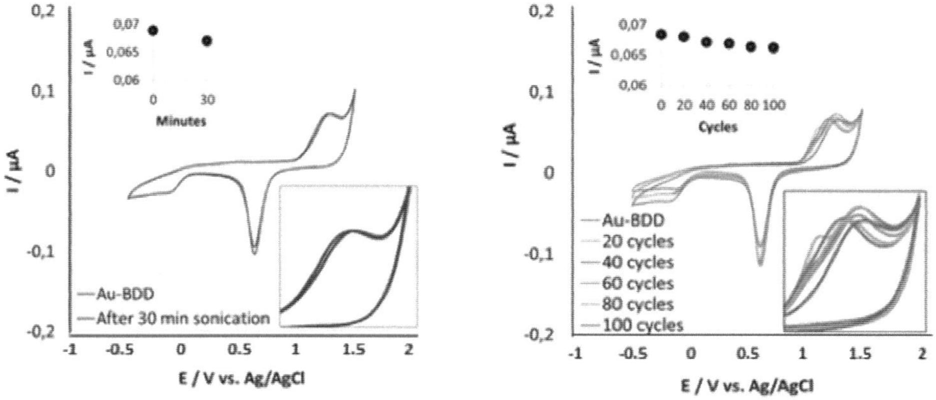

(a) (b)

Figure 3. Cyclic voltammograms of the prepared Au-BDD (a) before and after sonication for 30 min in double distilled water; (b) its 100 repetitive cycles in 0.2 M ABS pH 5.

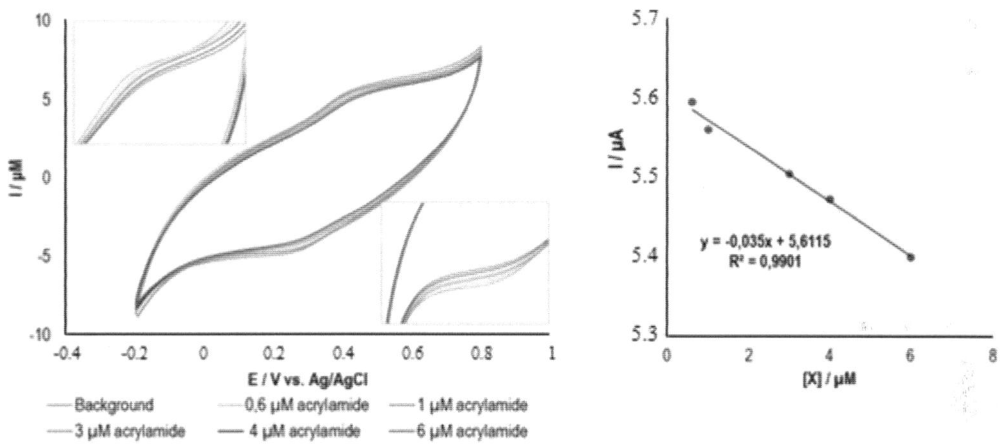

(a) (b)

Figure 4. (a) Voltammograms of 0.2 M ABS in the presence of various acrylamide concentrations at Hb/Au-BDD electrode. Scan rate was 100 mV/s. (b) the oxidation currents

3.2. Preparation of hemoglobin modified Au-BDD (Hb/Au-BDD) electrode

After obtaining Au-BDD electrode with homogenous gold distribution and high stability, Hb was modified at the Au-BDD. The interaction between Hb and Au-BDD electrode occurrs between the amino acid cysteine or NH_4^+-lysine residues of Hb structure and gold surface through the amino and thiol groups [3,4,18]. In the presence of acrylamide, the interaction was disturbed, resulting in the decrease of Hb currents due to the stronger attraction of amine groups in Hb toward acrylamide. This interaction causes a stunted redox system that occurs in Fe atoms in hemoglobin, so that the oxidation current response and the reduction of Fe (III)/(II) decreases.

Figure 4a shows the voltammograms of 0.2M ABS pH 5 in the presence of various concentrations (0.6–6.0μM) of acrylamide. Couple of oxidation reduction peaks were observed at the potentials of +0.5 and +0.3 V, respectively, reflecting the oxidation-reduction of Fe(III)/Fe(II) in Hb structure. Furthermore, the decrease of the oxidation currents peak was linear (R^2= 0.991) with the equation of y = -0.035x + 5.6115 (figure 4b) suggested that the sensors can be used for detection of acrylamide. The estimated detection limit of 0.845 μM was achieved. The repetitive measurements on 1μM acrylamide show an RSD of 1.073 %. The 5 times measurements, suggested the excellent stability of the electrode.

Table 1. Comparison of various electrochemical biosensors for acrylamide detection

Electrode	Optimum pH	Linear Range (M)	LOD (M)	R^2	Sensitivity ($\mu A/\mu M.cm^2$)	RSD repeatability (%)	Reference
Hemoglobin/Au modified BDD (seeding method)	5	6×10^{-7} to 6×10^{-6}	8.45×10^{-7}	0.99	0.0350	1.073	This paper
Hemoglobin/Au modified BDD (electrodeposition method)	5	5×10^{-6} to 5×10^{-5}	5.14×10^{-6}	0.97	0.0016	-	[9]
HbNPs modified Au Polycrystalline	5	1×10^{-10} to 1×10^{-7}	1×10^{-10}	0.99	-	3.85	[8]
Polyethersulphone double membrane	7.2	1×10^{-3} to 1×10^{-1}	6.31×10^{-4}	0.99	-	-	[10]
Polycarbonate double membrane	7.2	5×10^{-4} to 1×10^{-1}	9.12×10^{-5}	0.99	-	-	[10]
Multi-walled carbon nanotube/ CoNPs/Polyanilin	5.5	5×10^{-9} to 75×10^{-3}	2×10^{-8}	0.99	-	2.35	[11]
Screen printed carboxyl single-walled carbo nanotubes	-	1×10^{-5} to 2×10^{-4}	3×10^{-8}	0.99	0.009	-	[12]
Multi-walled carbon nanotube/ FeONPs/chitosan composite	5	3×10^{-9} to 9×10^{-8}	2×10^{-11}	0.99	0.0024	2.2	[13]

Table 1 shows comparison of various biosensor electrodes for detecting acrylamide compound. The results show that current electrode (hemoglobin/gold modified BDD through seeding method) has a better sensitivity (0.035 $\mu A/\mu M.cm^2$) and stability (1.073 %) compared to other electrodes.

Although the detection limit of the current sensor was a bit higher than previous biosensor based on Au Polycrystalline modified HbNPs [8], multi-walled carbon nanotube/CoNPs/Polyaniline [10], screen printed carboxyl single-walled carbon nanotubes [12], and multi-walled carbon nanotubes/FeONPs/chitosan composite [13], but slightly lower than previous biosensor based on Hb/Au modified BDD [9], polyethersulphone double membrane [10], and polycarbonate double membrane [10].

4. Conclusions

Gold-modified boron-doped diamod (Au-BDD) was prepared by wet-chemical seeding and electrodeposition technique in order to produce excellent particles stability and homogenity. Stability was demonstrated by the repeated sonication and voltammetry cyclic, while the homogenity of the particles was confirmed by SEM characterization. Hemoglobin-modified on the surface of the Au-BDD electrode was then applied for acrylamide sensors by using cyclic voltammetry. The results showed that Hb-modified Au-BDD electrode can be applied as acrylamide biosensor and able to detect acrylamide with linear corelation of 0.901 in the concentration range of 0.6 to 6.0μM.

Acknowledgements

This work was funded by Hibah Kompetensi Publikasi Internasional Terindeks Universitas Indonesia, Contract No. 2273/UN2.R3.1/HKP.05.00/2018

IOP Conf. Series: Materials Science and Engineering **763** (2020) 012019 doi:10.1088/1757-899X/763/1/012019

References

[1] Tareke E, Rydberg P, Karlsson P, Eriksson S and Törnqvist M 2002 *J. Agric. Food. Chem.* **50** 4998–5006
[2] Lineback D R, Coughlin J R and Stadler R H 2012 *Annu. Rev. Food Sci. Technol* **3** 15–35
[3] Stobiecka A, Radecka H and Radecki J 2007 *Biosens. Bioelectron.* **22** 2165–70
[4] Hu Q, Xu X, Li Z, Zhang Y, Wang J, Fu Y and Li Y 2014 *Biosens. Bioelectron.* **54** 64–71
[5] Stadler R H, Blank I, Varga N, Robert F, Hau J, Guy P A, Robert M C and Riediker S 2002 *Nature* **419** 449–50
[6] Mottram D S, Wedzicha B L and Dodson A T 2002 *Nature* **419** 448–9
[7] Zyzak D V *et al.* 2003 *J. Agric. Food Chem.* **51** 4782–7
[8] Yadav N, Chillar A K and Pundir C S 2018 *Int. J. Biol. Macromol.* **107** 1000–13
[9] Umam K, Saepudin E and Ivandini T A 2017 *IOP Conf. Ser.: Mater. Sci. Eng.* **188** 012006
[10] Silva N A F, Matos M J, Karmali A and Rocha M M 2011 *Port. Electrochim. Acta* **29** 361–73
[11] Batra B, Lata S, Sharma M and Pundir C S 2013 *Anal. Biochem.* **433** 210–7
[12] Gonzáles-Fuentes F J, Maríquez J, Godínez L A, Escarpa A and Mendoza S 2014 *Electroanalysis* **26** 1039–44
[13] Batra B, Lata S, Sharma M and Pundir C S 2013 *Bioprocess. Biosyst. Eng.* **36** 1591–9
[14] Suzuki A, Ivandini T A, Kamiya A, Nomura S, Yamanuki M, Matsumoto K, Fujishima A and Einaga Y 2007 *Sens. Actuators* B*: Chem.* **120** 500–7
[15] Gao F and Yang N 2013 *Electrochim. Acta* **112** 493–9
[16] Li M, Zhao G, Geng R dan Hu H 2008 *Bioelectrochem.* **74** 217–21
[17] Friedman M 2003 *J. Agric. Food Chem.* **51** 4504–26
[18] Garabagiu S and Mihailescu G 2011 *J. Electroanal. Chem.* **659** 196–200
[19] Wulandari R, Saepudin E and Ivandini T A 2019 *Sens. Mater.* **31** 1105–17
[20] Ivandini T A and Einaga Y 2017 Heavy metal sensing based on diamond electrodes *Carbon-Based Nanosensor Technology (Springer Series on Chemical Sensors: Methods and Applications)* vol 17 ed. C Kranz (Cham: Springer)

Synthesis of Mn_3O_4/ZSM-5 zeolite from natural aluminosilicate minerals and its application as heterogeneous catalyst for α-cellulose conversion to levulinic acid

S F Fhaza, Y K Krisnandi and D U C Rahayu

Department of Chemistry, Faculty of Mathematics and Natural Sciences (FMIPA) Universitas Indonesia, Kampus UI Depok, Depok 16424, Indonesia

Corresponding author's e-mail: yuni.krisnandi@sci.ui.ac.id

Abstract. Microporous ZSM-5 was successfully synthesized using Bayat-Klaten natural zeolite and Belitung kaolin as its mineral source through 2 stages of treatment, i.e. pre-treatment of natural zeolite and kaolin followed by synthesis of ZSM-5 zeolite. Bayat-Klaten natural zeolite as source of Si and Al was first treated as follows: activation, purification, fragmentation, and dealumination, whereas Belitung kaolin as the source of Si, to cover Si deficiency in the synthesis, was treated as follows: activation, purification, and Si extraction. Based on Energy Dispersive X-Ray (EDX) analysis, the Si/Al ratio was 24.8 in dealuminated natural zeolite and 98 % of Si after kaolin extraction. ZSM-5 zeolite was synthesized and characterized using X-Ray Diffraction (XRD), Fourier Transform Infrared (FTIR), Scanning Electron Microscopy-Energy Dispersive X-Ray (SEM-EDX), and Brunauer-Emmett-Teller (BET) analysis. The XRD pattern showed a similarity with typical ZSM-5 zeolites. Impregnation with Mn(II) followed by calcination at 550 °C giving Mn_3O_4/ZSM-5 (2.2 wt.% of Mn), that then employed as catalyst for α-cellulose conversion to levulinic acid. High Performance Liquid Chromatography (HPLC) analysis in product reaction showed that the conversion reaction for 6 h resulting in the highest yield of levulinic acid (7.2 %). Therefore, it was revealed that the natural mineral-based ZSM-5 has potential to be used as heterogeneous catalyst.

Keywords: heterogeneous catalyst, levulinic acid, Mn_3O_4/ZSM-5, natural aluminosilicate, zeolite.

1. Introduction

Nowadays, research about zeolites has been raised due to the uniqueness and benefits of zeolites that can be used in some applications [1]. One of the zeolites that can be used is natural zeolite. However, the utilization of natural zeolite as a catalyst is still limited, because the natural zeolite still has many pollutants that will cause interference in catalyst performance [2–3]. On the other hand, Indonesia has abundant mineral resources such as natural zeolite and kaolin [4]. Therefore, it is necessary to conduct research regarding the appropriate modification and treatment for these minerals so that these natural zeolites can compete with the synthetic zeolites [4–5].

In this study, ZSM-5 zeolite was synthesized from Bayat-Klaten natural zeolite and the silica shortage will be covered from Belitung kaolin, which was previously extracted. ZSM-5 zeolite was then used as catalyst in α-cellulose conversion to levulinic acid [6]. Levulinic acid is a platform chemical derived from glucose degradation from cellulose hydrolysis. Levulinic acid's production is still captured the scholar's attention since it has wide range applications [7-9].

2. Materials and methods

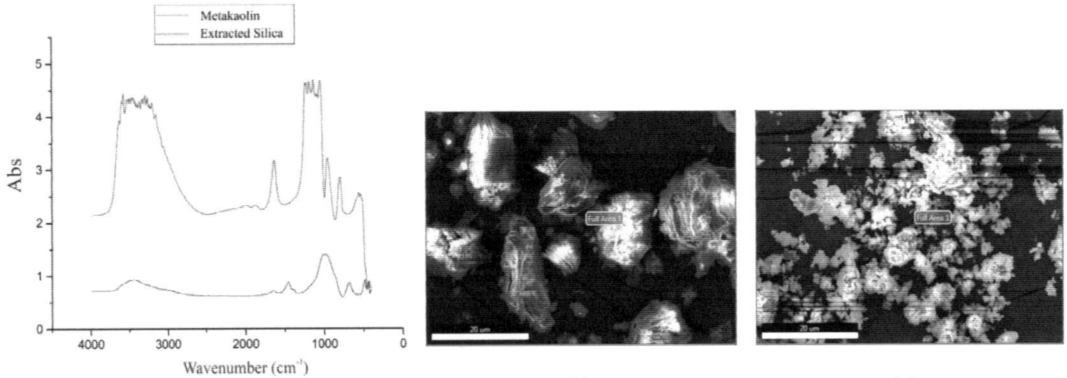

(a) (b) (c)

Figure 1. (a) FTIR spectra of metakaolin and extracted silica and SEM image of (b) raw kaolin and (c) extracted silica.

2.1. Materials

Natural zeolite was obtained from Bayat-Klaten, Central Java, Indonesia whereas kaolin was obtained from Belitung, Indonesia. The chemicals used were α-cellulose and tetrapropylammonium hydroxide (TPAOH) purchased from Sigma-Aldrich, levulinic acid (97 %) from TCI, HCl from Smart Lab Indonesia, while NaOH pellets, glacial CH_3COOH, H_2O_2 30 %, $Na_2S_2O_4$, $Na_3C_6H_5O_7$, NH_4Cl, H_2SO_4(96 % p.a.), PDDA (34 % p.a.), NaOAc buffer pH 5, Na-citric-bicarbonate, and Mn(II) from E. Merck.

2.2. Pretreatment process of Bayat-Klaten natural zeolite and Belitung kaolin

Belitung kaolin was purified prior extracted, follow published procedure to yield metakaolin [10]. Bayat-Klaten natural zeolite was pretreated in four stages: activation, purification, fragmentation, and dealumination by following the published method [2–3]. Kaolin, metakaolin, and all pretreated natural zeolites in each step were characterized using Fourier Transform Infrared (FTIR), X-Ray Diffraction (XRD) and Energy Dispersive X-Ray (EDX).

2.3. Synthesis of ZSM-5 micropore and impregnation of Mn_3O_4/ZSM-5

ZSM-5 micropore was synthesized by following the published procedures [11]. The synthesized ZSM-5 micropore was then characterized using Scanning Electron Microscopy (SEM), XRD, Brunauer-Emmett-Teller (BET) particle analytical, and FTIR. Impregnation of Mn_3O_4 to ZSM-5 was conducted with Mn(II) 2 % (w/w). Homogenized solution with 1 g of ZSM-5 was made and stirred for 24 h at room temperature until paste formed. The paste was then dried at 60 °C for 1 night, then calcinated for 3 h. Mn_3O_4/ZSM-5 then was characterized using XRD, FTIR and Atomic Absorption Spectroscopy (AAS).

2.4. α-Cellulose conversion to levulinic acid

α-Cellulose (0.1 g) was mixed with Mn_3O_4/ZSM-5 (0.01 g), H_3PO_4 40 % (2 mL), and H_2O_2 30 % (0.05 mL) in a three-neck round-bottom flask. The mixture was heated under reflux at 130 °C for variation of reaction time (0, 2, 4, 6, 8, and 10 h). A similar reaction was also conducted without the presence of Mn_3O_4/ZSM-5.

3. Results and discussion

In this study, Belitung kaolin was successfully purified to yield metakaolin. FTIR spectra for the extracted kaolin and metakaolin (figure 1a) showed that raw kaolin tended to have sharp peak because it has an independent SiOH which is not binding to water. Therefore, the peak intensity appeared as wide peak. On the other hand, the extracted silica showed wider peak than raw kaolin because it was interacted with water, resulting a higher peak intensity of H-OH vibration. SEM-EDX result in figure 1b, figure 1c and table 1 proved that after silica extraction, the percentage of silica instantly turned up, while the mole percentage of others were close to zero which caused the Si/Al mole ratio to be infinite.

IOP Conf. Series: Materials Science and Engineering **763** (2020) 012020 doi:10.1088/1757-899X/763/1/012020

Table 1. EDX result of kaolin and extracted silica

Elements	Kaolin	Extracted Silica
Na	0.03	0
Si	17.01	39.71
Al	17.74	0.07
Ratio Si/Al	0.92	∞

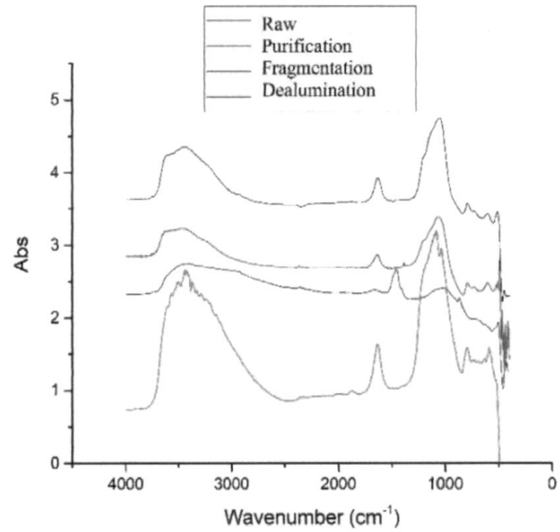

Figure 2. FTIR spectra for Bayat-Klaten natural zeolite before and after pretreatment process.

(a) (b) (c)

Figure 3. SEM image of Bayat-Klaten natural zeolite
(a) raw, (b) fragmented, and (c) dealuminated.

Figure 2 shows FTIR spectra of Bayat-Klaten natural zeolite in each pretreatment process. Because Bayat-Klaten natural zeolite is an aluminosilicate, it appears vibration for SiO and AlO at wavenumbers of 800-1300 cm^{-1}. Peak at wavenumber of 500 cm^{-1} was subjected to double ring signified to zeolite structure. Compared with raw Bayat-Klaten natural zeolite, the purified zeolite did not show the peak difference, indicating that purification process did not alter its structure. In fragmentated zeolite, it showed new peak at wavenumber of 1500 cm^{-1} indicating the presence of $AlONa^{-}$ and 1000-1200 cm^{-1} belongs to $SiONa^{-}$. At wavenumber of 500-600 cm^{-1}, peak of double ring decayed significantly, indicating that structure was successfully decomposed to the monomer form. For dealuminated zeolite, the FTIR spectra showed the similar pattern with raw zeolite because the existing zeolite remain the same, but the intensity of the peak was higher because the purity of zeolite.

Figure 3 and table 2 exhibited SEM-EDX result of each step of natural zeolite pretreatment. According to figure 3 and table 2, dealumination process was successfully applied in reducing the mole of Al giving the final mole ratio of Si/Al was suitable for ZSM-5 synthesis.

Table 2. EDX results of Bayat-Klaten natural zeolite before and after pretreatment process.

Elements	Raw	Fragmented	Dealuminated
Na	0.66	23.76	1.39
Si	34.37	19.96	34.74
Al	6.69	3.52	1.34
Ratio Si/Al	4.91	5.46	24.8

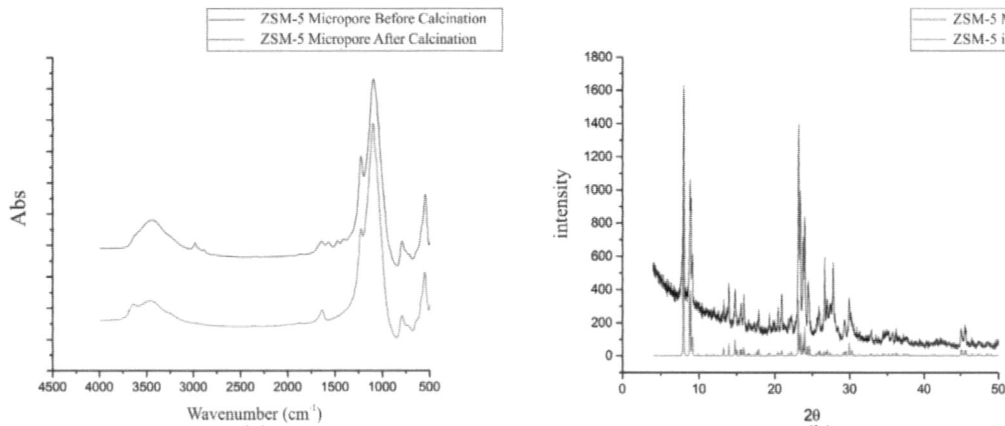

(a) (b)

Figure 4. (a) FTIR spectra of ZSM-5 before and after calcination (b) XRD result of ZSM-5 micropore.

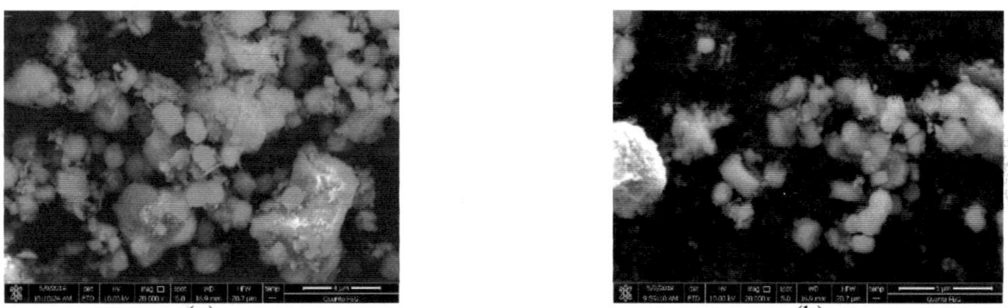

(a) (b)

Figure 5. SEM image of (a) ZSM-5 micropore and (b) Mn_3O_4/ZSM-5.

FTIR spectra of ZSM-5 before and after calcination are shown in figure 4a. FTIR spectra of ZSM-5 before calcination show a peak at wavenumber of 2960-2850 cm^{-1} and 1470-1350 cm^{-1} belongs to vibration of C-H stretching and C-H bending, respectively. These absorption bands come from the template used in ZSM-5 synthesis. This calcination process was performed to remove the organic template. Thus, FTIR spectra of ZSM-5 after calcination showed the loss of template which indicated by the loss of C-H stretching and C-H bending vibration. With calcination, the absorption band at wavenumber of 3600 cm^{-1} belongs to the absorption band of silanol group (Si-OH) on the surface of the zeolite as well as located within the zeolite (as shown in SEM image in figure 5). The percentage of mass loss after calcination was 14.67 %. In the wavenumber of 700-1100 and 950-1250 cm^{-1}, there were peaks belong to the vibration of Si-O and Al-O which is the framework of aluminosilicate. Furthermore, XRD result of the synthesized ZSM-5 zeolite (figure 4b) proved that this zeolite was successfully synthesized.

To prove that Mn_3O_4 was successfully impregnated on ZSM-5 zeolite, FTIR spectra was measured (figure 6). According to figure 6, the difference between ZSM-5 and Mn_3O_4/ZSM-5 can be observed at wavenumber of 3500 cm^{-1} indicating the presence of internal Si-OH. Mn_3O_4/ZSM-5 zeolite showed a decrease in intensity of Si-OH group compared with ZSM-5 zeolite.

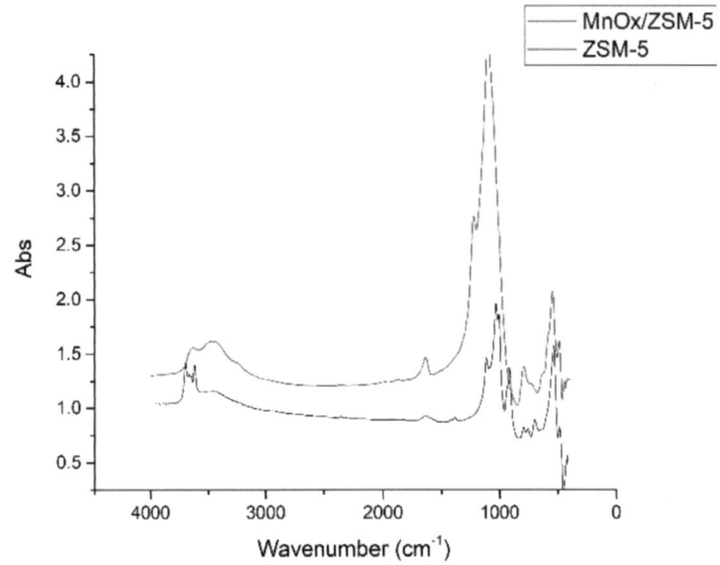

Figure 6. FTIR spectra of MnOx/ZSM-5 zeolite compared with ZSM-5 zeolite.

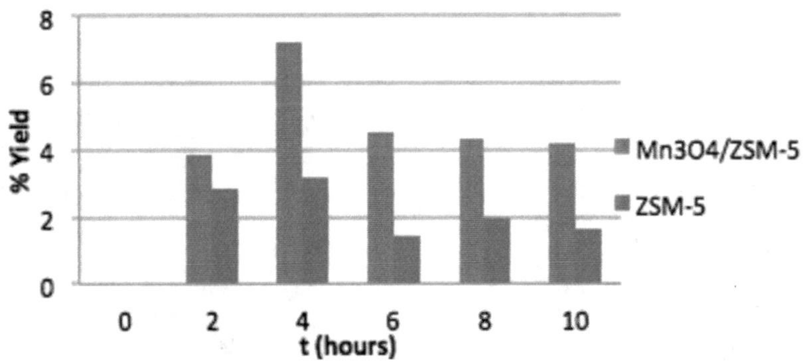

Figure 7. The percentage yield of levulinic acid analyzed using HPLC

Figure 7 shows the percentage yield of levulinic acid from α-cellulose conversion. Based on figure 7, Mn_3O_4/ZSM-5 is able to produce better yield (7.2 %) compared to unimpregnated ZSM-5 catalyst, with the optimum reaction time of 4 h. The Mn impregnated catalyst is better because the metal itself has an active side, so indirectly for the Mn_3O_4/ZSM-5 has more active side rather than ZSM-5. Therefore, reaction using Mn_3O_4/ZSM-5 catalyst produced more levulinic acid than the ZSM-5 catalyst. The plausible mechanism for cellulose conversion can be described via hydrolysis followed by isomerization of cellulose to yield fructose then further dehydration to give 5-HMF as intermediate of this conversion [12].

4. Conclusions

Belitung kaolin can be used as Si source to cover the Si deficiency in Bayat-Klaten natural zeolite in order to synthesize the ZSM-5 zeolite, refers to its EDX, SEM, FTIR, and XRD analysis. ZSM-5, zeolite with Si and Al source from kaolin and natural zeolite is capable to replace the synthetic ZSM-5 zeolite and can be used to catalyze α-cellulose conversion into levulinic acid. HPLC analysis showed the highest α-cellulose conversion to levulinic acid was 7.2 % with 4 h of reaction time.

Acknowledgements
This work was funded with PUPT BPPTN research grant no. 481/UN2.R3.1/HKP.05.00/2018. Prof. Sun Min Choi was greatly acknowledged for giving access to SEM measurement in KAIST.

References

[1] Tao Y, Kanoh H, Abrams L and Kaneko K 2006 *Chem. Rev.* **106** 896–910
[2] Rohayati, Krisnandi Y K and Sihombing R 2017 *AIP Conf. Proc.* **1862** 030094
[3] Krisnandi Y K, Mahmuda I, D U C Rahayu and Sihombing R 2018 *J. Phys: Conf. Ser.* **1095** 012044
[4] Kusdarto 2008 *J of Indonesia Zeolites* **7** 78–87
[5] Yue Y, Liu H, Yuan P, Li T, Yu C, Bi H and Bao X 2014 *J. Catal.* **319** 200–10
[6] Chen H, Yu B and Jin S 2011 *Bioresour. Technol.* **102** 3568–70
[7] Chen Y, Li G, Yang F and Zhang S M 2011 *Polym. Degrad. Stabil.* **96** 863–9
[8] Bevilaqua D B, Rambo M K D, Rizzetti T M, Cardoso A L and Martins A F 2013 *J. Clean. Prod.* **47** 96–101
[9] Cha J Y and Hanna M A 2002 *Ind. Crops Prod.* **16** 109–18
[10] Ming D W and Dixon J B 1987 *Clays Clay. Miner.* **35** 463-8.
[11] Wang L, Zhang Z, Yin C, Shan Z and Xiao F S 2010 *Micropor. Mesopor. Mater.* **131** 58–67
[12] Girisuta B, Janssen L P B M and Heeres H J 2006 *Green Chem.* **8** 701–9

Effect of nonionic surfactant in the preparation of microsphere based on blend of poly(lactic acid) and polycaprolactone

E Budianto and R Dhafiningtia

Department of Chemistry, Faculty of Mathematics and Natural Sciences (FMIPA), Universitas Indonesia, Kampus UI Depok, Depok 16424, Indonesia

Corresponding author's e-mail: emilb@ui.ac.id

Abstract. Microspheres of poly(lactic acid) (PLA) and poly(ε-caprolactone) (PCL) were blended and prepared using the water in oil (w/o) emulsion solvent evaporation method. The PLA/PCL blend was formulated with the composition of 60:40 (w/w), nonionic surfactant was utilized as the emulsifier. This study observed the distribution of the microspheres particle size by varying the nonionic surfactant volumes, emulsion stirring speed, and dispersion stirring time. The microspheres were characterized using Fourier Transform Infrared (FTIR) and Particle Size Analyzer (PSA). Physical forms of microspheres were observed using an optical microscope as well. The IR spectra of PLA/PCL blend showed that only physical interaction was occured between them. Moreover, the result of this study showed that when 2.0 mL nonionic surfactant was added, the uniform size distribution of the formed microspheres was observed at 31.50 μm. Furthermore, the microspheres that formed through emulsion stirring speed at 900 rpm revealed that the formed microspheres have a uniform size distribution at 31.50 μm, while the uniform size distribution at 34.58 μm was observed in the microspheres that formed during the dispersion stirring time at 90 minutes.

1. Introduction

Microspheres are commonly used in the drug delivery systems because they have small size and efficient properties as particles carrier. Microspheres are small spherical particles with diameters 1–1000 μm [1]. The size of microspheres for drug delivery systems should not be greater than 250 μm, ideally less than 125 μm [2]. Biodegradable polymers are used as coatings or agent (drug or protein) in the form of microspheres or microcapsules [3]. The notable examples of biodegradable polymers have been widely used for the drug delivery systems are poly (lactic acid) (PLA), polycaprolactone (PCL), poly (glycolic acid), and poly (lactic-co-glycolic acid) [4]. Meanwhile, PLA has low permeability compared to PCL, although the degradation time is shorter than PCL [5]. The PCL and PLA's mixture is supposed to generate a suitable polymer with higher speed degradation period as well as preferable permeability [6].

According to results of experiments, it showed that the performance of the microspheres is highly affected by the amount and types of surfactants [7]. Nonionic surfactant is frequently used in the preparation of microsphere using solvent evaporation method. They are generally imparted with good stability and compatibility [8–10]. The microsphere yielded from PLA-PCL was produced using Nonidet RK-18 as nonionic surfactant. Nonidet RK-18 is a low foaming nonionic surfactant with high emulsification and has biocompatible properties, which can control microsphere size distribution. This study aims to determine the effect of nonionic surfactants by varying the Nonidet RK-18' (nonionic surfactant) volumes, emulsion stirring speed and dispersion stirring time on the microsphere size distribution.

2. Materials and methods

Table 1. The variations were used for prepared microsphere using 10 % polyblend solution

Composition formulation	Variations		
	Volumes of Surfactant (mL)	Stirring Speed (rpm)	Dispersion Stirring Time (min)
60 PLA : 40 PCL	1.0 1.5 2.0 2.5	700 800 900	30 90 120

2.1. Materials
Lactic acid (90 %) and dichloromethane from Merck, polycaprolactone Perstop CAPA-6800, and Nonidet RK-18 were obtained from Evonik Indonesia: Specialty chemicals.

2.2. Synthesis of poly(lactic acid)(PLA)
50 mL of lactic acid was added into 250 mL glass reactor equipped with a mechanical stirrer and thermometer [11]. The reactor was heated in an oil bath until the temperature reached 120 °C and was made it constant for 1 hour. The reaction was continued at 150 °C for 22 hours under the flow of nitrogen gas. After 22 h, the sample was cooled and characterized by Fourier Transform Infrared (FTIR) Shimadzu Prestige-21 spectrometer.

2.3. Preparation of 10 % polyblend solution
The PLA and PCL blend were prepared with the ratio of 60:40 (w/w). The polymers were blended in dichloromethane and stirred for 15 min to obtain the 10 % polyblend solutions [12].

2.4. Various Paratemeters in preparation of microsphere
Microspheres were prepared using water in oil (w/o) emulsion solvent evaporation method. 5 mL of 10 % polyblend solution was emulsified in 2 mL surfactant and 1 mL distilled water with stirring speed of 700 rpm for 1 hour. Furthermore, the organic phase was dispersed into 300 mL of the aqueous phase and stirred at 900 rpm for 1 h to completely evaporate dichloromethane. The produced microspheres was filtrated with distilled water and dried in an oven [13]. Variations of nonionic surfactant are listed in table 1. The particle size distribution and appearance of the produced microspheres were determined by an optical microscope (OM) and particle size analyzer (PSA). The functional groups of microspheres were analyzed by FTIR.

3. Results and disscussion
Microsphere was prepared by using single water in oil (w/o) emulsion the solvent evaporation method. Initially, the aqueous external phase was added with organic phase and emulsified [13]. The organic phase employed in the research is dichloromethane, because it has a low boiling point at 39 °C [12]. This study utilized Nonidet RK-18 as the surfactant to stabilize the emulsion. Nonidet RK-18 is a low foaming sufactant with high emulsification properties and has a HLB value of 6.

3.1. Characterization by FTIR
Poly(lactic acid) was synthesized through the direct polycondensation of lactic acid under vacuum without catalyst using nitrogen gases. Analysis of FTIR aims to determine the functional group of PLA [12]. Figure 1 shows the FTIR spectrum of the PLA obtained from synthesis that C=O bond was visible at 1758 cm^{-1} which resembles the characteristic of carbonyl ester, 1992 cm^{-1} is characteristic of C-O bond and at 2996 cm^{-1} which shows the existence of C-H sp^{3}. The result of PLA spectrum was compared with the LA spectrum obtained from the previous research [14]. The formation of poly(lactic acid) was characterized by the band shift at 1730 to 1758 cm^{-1} show exist C=O stretch in the carboxylic acid to ester.

The purpose of FTIR analysis is to investigate and match the functional group of the polyblend's constituent components [12]. According to figure 1, the C=O, C–O–C, and C–C peaks were clearly observed at 1754, 1175 and 1200 cm^{-1}, respectively, in the IR spectra of PLA and PCL [14]. Based on the analysis of functional groups, the PLA and PCL spectra reappeared in the microspheres spectra. It shows that the functional groups of polyblend as microspheres were the combination of their constituent components. There was a physical interaction between PLA and PCL because there were no new peaks [12].

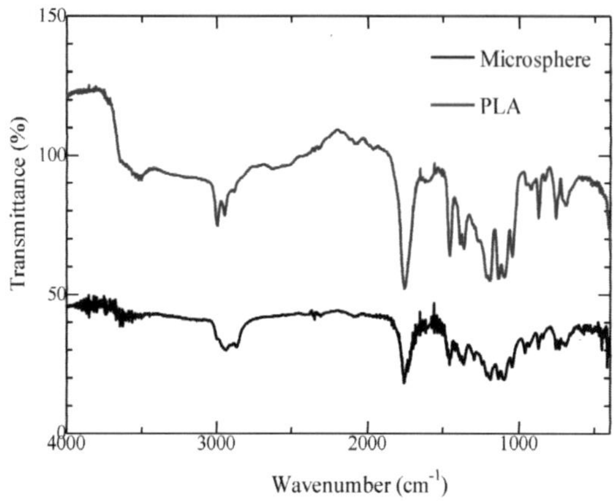

Figure 1. FTIR spectrum of microspheres

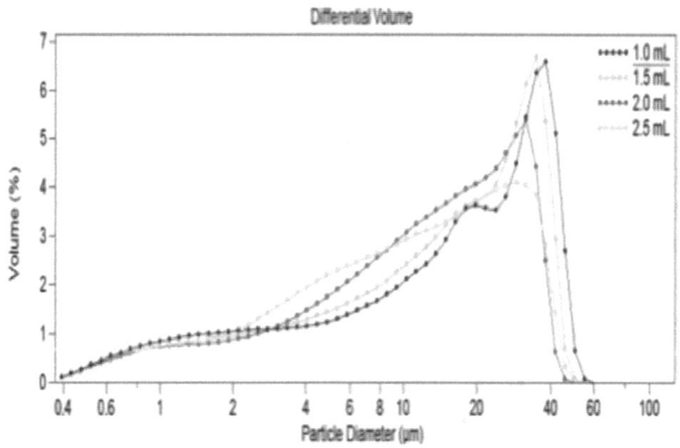

Figure 2. Variation of nonionic surfactant volumes - particle size analysis of microsphere

3.2. Particle size distribution of microsphere prepared with variation of nonionic surfactant volumes

The distribution of microspheres particle size was performed by varying the nonionic surfactant volumes at 1.0 mL, 1.5 mL, 2.0 mL, and 2.5 mL. Figure 2 and figure 3 shows that the increase of surfactant volumes will affect the stability of emulsion and resulted in smaller particle size with uniform size distribution [15]. The addition of surfactant will reduce the continuous phase surface and will make the diffusion layer thicker around the microdroplet, intended to prevent the coalescence [16,17]. This phenomenon can be seen in figure 2 and figure 3, nonionic surfactant volumes of 1.0 mL and 1,5 mL resulted in the size distribution of formed microspheres at 37.97 μm, and 34.58 μm. When 2.0 mL nonionic surfactant was added, the uniform size distribution of the formed microspheres was observed at 31.50 μm. However, the use of nonionic surfactant volumes of 2.5 mL revealed that the distribution was not uniform. The size of the formed microsphere is depended on the size and the stability of the emulsion droplets [18].

3.3. Particle size distribution of microsphere prepared with variation of emulsion stirring speed

The emulsion stirring speed was varied to determine the effect of nonionic surfactant on microsphere particle size distribution. As shown in figure 4 and figure 5, the particle size distribution decreases with an increase in stirring speed [15]. The lower stirring speed will produce larger size microspheres. While at low stirring speed 700 rpm the particle size distribution was obtained at 37.97 μm. As the stirring speed is increased to 800 rpm, the particle size distribution is reduced to 34.58 μm. A further increase in the stirring speed to 900 rpm formed particle size distribution with a uniform size distribution at 31.50 μm. The high stirring speed of emulsion can easily break down the emulsion into small droplets where there is a high impact between the stirrer and emulsion droplets [19].

IOP Conf. Series: Materials Science and Engineering **763** (2020) 012021 doi:10.1088/1757-899X/763/1/012021

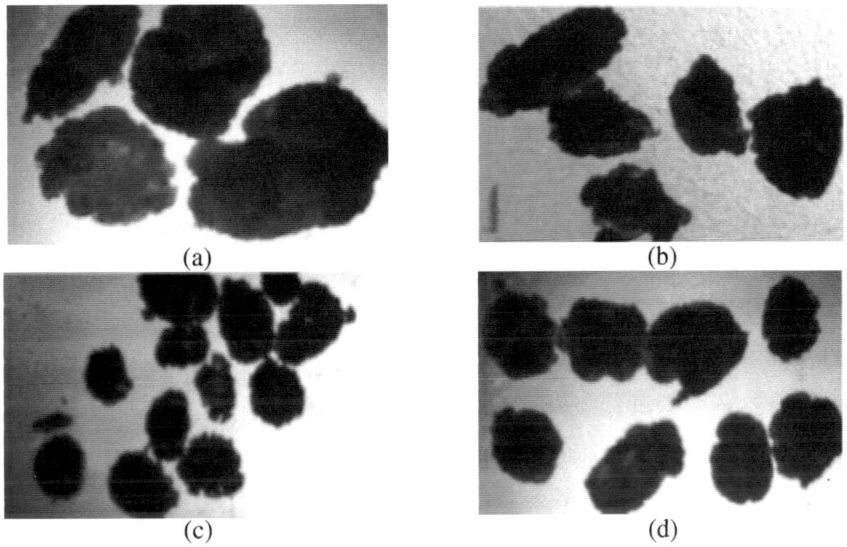

Figure 3. Variation of nonionic surfactant volumes - Optical Microscope (6.75 ×)
(a) 1.0 mL, (b) 1.5 mL, (c) 2.0 mL and (d) 2.5 mL

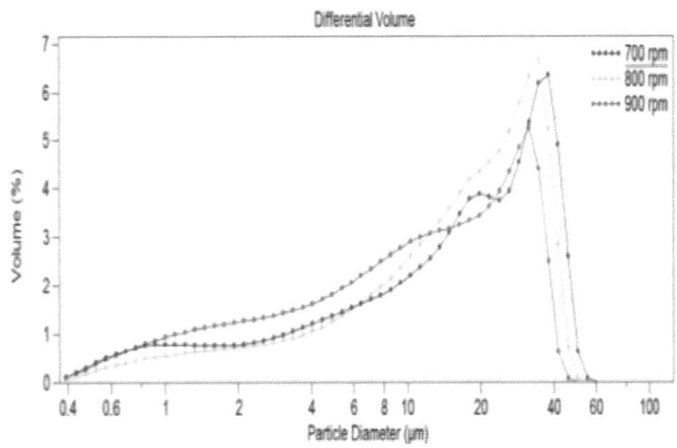

Figure 4. Variation of emulsion stirring speed - particle size analysis of microsphere

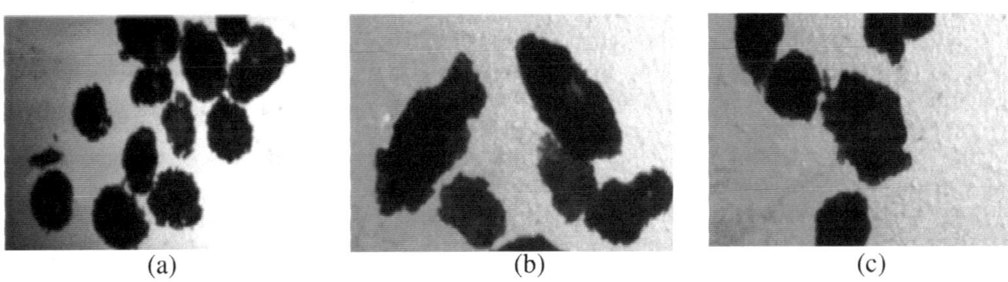

Figure 5. Variation of emulsion stirring speed (a) 700 rpm, (b) 800 rpm and (c) 900 rpm
Particle Size - Optical Microscope (6.75 ×)

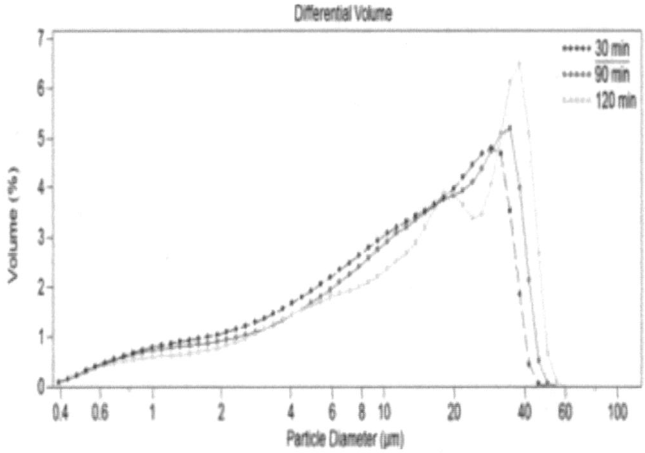

Figure 6. Variation of dispersion stirring time - particle size analysis of microsphere

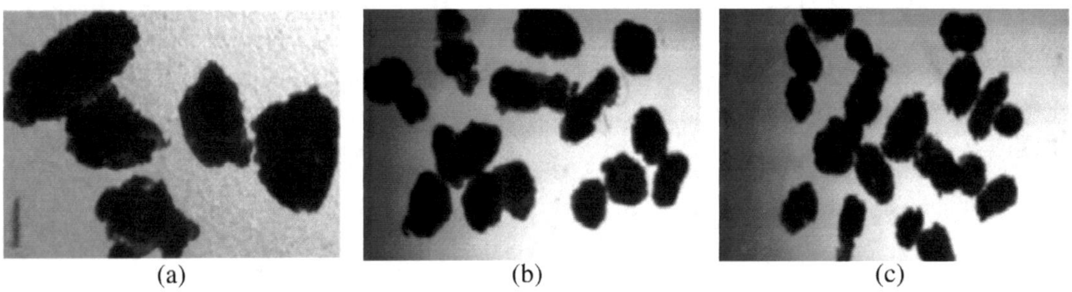

Figure 7. Variation of dispersion stirring time (a) 30 min, (b) 90 min, and (c) 120 min Particle Size - Optical Microscope (6.75 ×)

3.4. Particle size distribution of microsphere prepared with variation of dispersion stirring speed

Dispersion stirring time will affect the particle size distribution of microspheres as well. Longer stirring time will produce a collision between the particles, resulting in a smoother dispersion. In addition, stirring time will determine whether the dichloromethane has completely evaporated or not. The result from figure 6 and figure 7 showed that the uniform size distribution at 34.58 µm was observed in the microspheres, which formed during the dispersion stirring time at 90 minutes. Whereas, the sample with 30 minutes and 120 minutes resulted the size distribution of the formed microsphere at 28.70 µm and 37.97 µm, respectively.

4. Conclusions

Microspheres of polyblend PLA and PCL with nonionic surfactants have been successfully carried out by a solvent evaporation method. The interaction occurs between PLA and PCL in polyblend of microspheres was a physical interaction based on the FTIR spectra. In this research, the effect of nonionic surfactant on the particle size distribution was found to be decreased with the increasing volume, stirring speed and dispersion stirring time. The variation of nonionic surfactant volumes, emulsion stirring speed and dispersion stirring speed resulted in the size of microspheres with a uniform distribution at 31.50 µm with a volume of 2.0 mL, 31.50 µm with the stirring speed at 900 rpm and dispersion stirring time at of 90 minutes at 34.58 µm.

Acknowledgements

The authors would like to thank Universitas Indonesia for financially supporting this research through PITTA Grant Univeristas Indonesia with contract No. 233/UN2.R31/PPM.00/2018.

References

[1] Betala S, Varma M M and Abbulu K 2017 *International Research Journal of Pharmacy (IRJP)* **8** 103–8
[2] Jain R A 2000 *Biomaterials* **21** 2475–90
[3] Mehta R, Kumar V, Bhunia H and Upadhyay S N 2005 *Journal Macromolecular Science* C **45** 325–49
[4] Lu Y and Chen S C 2004 *Adv. Drug Deliv. Rev.* **56** 1621–33
[5] Porjazoska A, Karal-Yilmaz O, Kayaman-Apohan N, Cvetkovska M and Baysal B M 2004 *Croat. Chem. Acta* **77** 545–51
[6] Gunatillake P A and Adikari R 2003 *Eur. Cell. Mater.* **5** 1–16
[7] Pachuau L and Mazumber B 2009 *Int. J. PharmTech Res.* **1** 966–71
[8] Jiao J 2008 *Adv. Drug. Deliv. Rev.* **60** 1663–73
[9] Schick M J 1987 *Nonionic Surfactants: Physical Chemistry* (New York: Marcel Dekker Inc)
[10] Wang B, Siahaan T and Soltero R 2005 *Drug Delivery: Principles and Applications* (Hoboken: John Willey and Sons)
[11] Hyon S H 2000 *Yonsei Med. J.* **41** 720–34
[12] Kemala T, Budianto E and Soegiyono B 2012 *Arab. J. Chem.* **5** 103–8
[13] Choksi N and Desai H 2017 *International Journal of Applied Chemistry (IJAC)* **13** 377–84
[14] Chen C C, Chueh J Y, Tseng H, Huang H M and Lee S Y 2003 *Biomaterials* **24** 1167–73
[15] Nilkumhang S and Basit A W 2009 *Int. J. Pharm.* **377** 135–41
[16] Li M, Rouaud O and Poncelet D 2008 *Int. J. Pharm.* **363** 26–39
[17] Jones D S and Pearce K J 1995 *Int. J. Pharm.* **118** 199–205
[18] Berchane N S, Jebrail F F, Carson K H, Rice-Ficht A C and Andrews M J 2006 *J. Microencapsul.* **23** 539–52
[19] Ahlin P, Kristl J, Kristl A and Vrecer F 2002 *Int. J. Pharm.* **239** 113–20

IOP Conf. Series: Materials Science and Engineering **763** (2020) 012022 doi:10.1088/1757-899X/763/1/012022

Preparation of ZnO nanoparticles in n-hexane-water system using *Moringa oliefera* leaf extract and its photodegradation activity

Y Pratiwi, Y Yulizar and D O B Apriandanu

Departement of Chemistry, Faculty of Mathematics and Natural Sciences (FMIPA), Universitas Indonesia, Kampus UI Depok, Depok 16424, Indonesia

Corresponding author's e-mail: yokiy@sci.ui.ac.id

Abstract. In this study, ZnO nanoparticles were successfully synthesized using *Moringa oliefera* leaf extract in two phases system of n-hexane-water. The presence of secondary metabolites in n-hexane acts as a weak base source and precursor of $Zn(NO_3)_2$ in water were needed for the formation of ZnO nanoparticles. ZnO nanoparticles were further characterized using UV-Vis Diffuse Reflectance Spectroscopy(DRS), Fourier Transform Infrared (FTIR), X-Ray Diffraction (XRD), Scanning Electron Microscopy-Energy Dispersive X-Ray (SEM-EDX), and Tunneling Electron Microscopy (TEM). Based on TEM characterization, the particles size of ZnO nanoparticles was approximately 67 nm. The band gap energy of ZnO was 3.2 eV. The photocatalytic activities of ZnO nanoparticles were observed through the methylene blue (MB) degradation under UV-light irradiation. The percentage of MB degradation was 99 % for 2 h.

Keywords: ZnO nanoparticles, *Moringa oleifera*, n-hexane-water system, photodegradation, methylene blue

1. Introduction

The preparation method of metal oxide nanoparticle has been done by green synthesis. One of them is green synthesis with liquid-liquid extraction [1]. Green synthesis method using plant extract is more environmentally friendly and harmless. The shape and size of nanoparticles can be controlled [2]. Green synthesis of nanoparticle has been mostly reported for metal oxide like zinc oxide (ZnO). ZnO is a semiconductor with a wide band gap (3.37 eV) and has good photocatalytic properties in the UV-Visible region [3]. ZnO can absorb ultraviolet (UV) radiation from sunlight or fluorescent lamps. Moreover, it will produce oxidation reactions where electrons are released and captured. The photocatalytic reaction is activated by photon absorption with the same or higher energy levels of the band gap energy in catalyst [4]. Various types of photocatalysts from metal oxide nanoparticles have been reported to degrade some organic compounds [5–7].

On the other hand, some of Indonesian plants for green synthesis of metal and metal oxides nanoparticles are *Oldenlandia corymbosa L, Graptophyllum pictum, Terminalia catappa, Physalis angulata, Theobroma cacao* and *Polyscias fruticosa* [8–12]. One of the plants used for green synthesis of ZnO nanoparticles is *Theobroma cacao* [11]. *Moringa oleifera* is a plant species of *moringaceae* that has many functions due to the presence of secondary metabolites. The secondary metabolites of *Moringa oleifera* consist of alkaloids, flavonoids, tannins, saponins, and steroids [13].

To the best of our knowledges, we reported the synthesis of ZnO nanoparticles using moringa leaf extract on n-hexane-water system and the test of its photocatalytic activity for the degradation of methylene blue.

2. Experimental

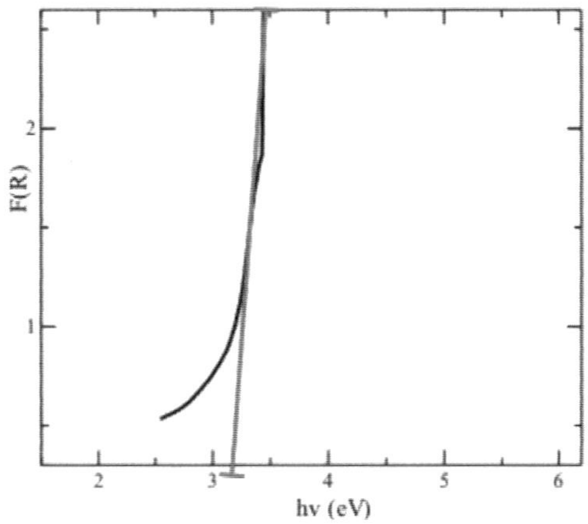

Figure 1. The band gap energy of ZnO nanoparticles

2.1. Apparatus and chemicals

Moringa oliefera was obtained from Study Center of Tropical Biopharmaca, Bogor, Indonesia. Zn $(NO_3)_2.4H_2O$ as precursor and methylene blue as a model of dyes was purchased from Merck, Germany. All reagents and solvents were in analytical grade.

2.2. Preparation of Moringa leaf extract

Moringa leaves were washed, dried and grinded to form powders. Moringa leaves powders were macerated in hexane for 7 x 24 h. The results were partitioned in aquabidest to separate the polar and non-polar metabolites compounds. The hexane fraction contains non-polar secondary metabolite and the aqueous fraction contains polar secondary metabolite. The hexane fraction was used as Moringa leaf extract for ZnO nanoparticles synthesis.

2.3. Synthesis of ZnO Nanoparticles

2.0 x 10^{-2} M $Zn(NO_3)_2$ solutions was reacted with the hexane fraction of moringa leaf extract. The mixture was extracted for 1.5 h to get the hexane phase and water phase of $Zn(OH)_2$ sol. The water phase of $Zn(OH)_2$ sol was heated at 80 °C for 4 h to get the gel system. The result was calcined at 500 °C for 4 h.

2.4. Characterization of ZnO nanoparticles

UV-Vis Diffuse Reflectance Spectroscopy (DRS), Shimadzu 2400 was used to calculate the band gap energy. Fourier Transform Infrared (FT-IR) spectrophotometry, IR Prestige-21 Shimadzu, was used to determine the functional group. Scanning electron microscopy (SEM), Zeiss Evo MA10, was used to analyze the surface morphology. Transmission electron microscopy (TEM), JEOL JEM 1400, was used to know the particle size. Energy dispersive x-ray spectroscopy (EDX), Zeiss Evo MA10, was used to determine the elemental composition. X-Ray Diffraction (XRD), Shimadzu 610 was used to determine the crystallinity. The methylene blue photodegradation was studied by UV-Vis spectrophotometer (Shimadzu 2600).

2.5. Photodegradation activity test

ZnO powder was added into methylene blue 2.0 x 10^{-5} M. The mixture was stirred under UV irradiation every 30 min for 2 h. The photodegradation activity was performed by observing the absorbance of methylene blue by UV-Vis spectrophotometer.

3. Results and discussion

ZnO nanoparticles were characterized by UV-Vis DRS at the wavelength of 200–800 nm. The band gap energy value was determined by the Tauc Plot method as shown in figure 1. The result shows a linear regression equation of straight line as the band gap energy value. The band gap energy of ZnO nanoparticles was 3.2 eV.

IOP Conf. Series: Materials Science and Engineering **763** (2020) 012022 doi:10.1088/1757-899X/763/1/012022

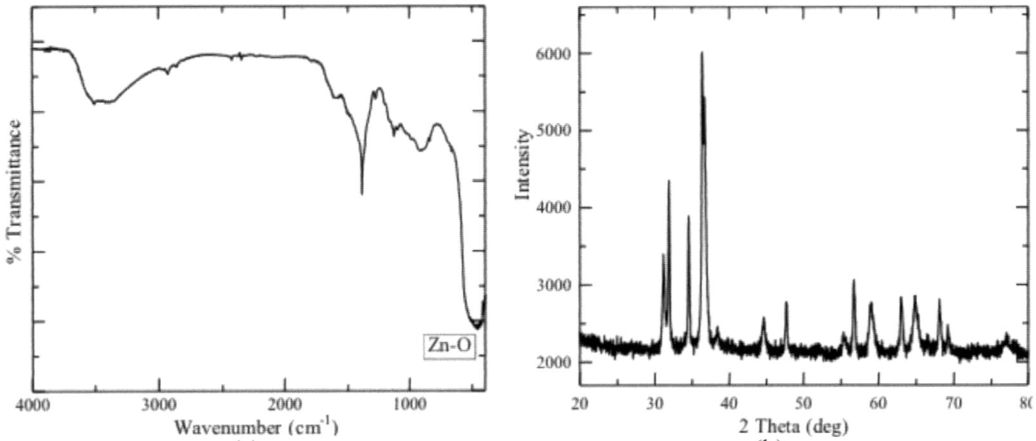

(a) (b)

Figure 2. (a) FTIR spectrum and (b) XRD pattern of ZnO nanoparticles

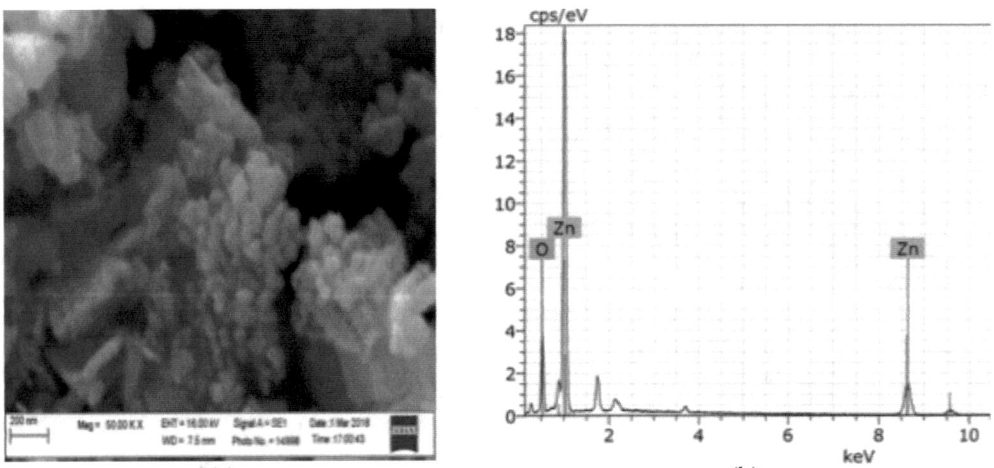

(a) (b)

Figure 3. (a) SEM image and (b) EDX spectrum of ZnO nanoparticles

The FT-IR spectrum of ZnO nanoparticles is shown in figure 2a. The absorption band at 474 cm^{-1} is assigned to the presence of a Zn-O vibration bond, indicating that ZnO has been successfully formed. The C-C vibration bond at 1407 cm^{-1} assumes that there are still organic compounds due to the remain of secondary metabolites compounds on the ZnO surface.

XRD characterization was performed to find the typical peak, diffraction pattern, and crystallinity of ZnO nanoparticles. As can be seen in figure 2b, ZnO nanoparticles have the diffraction pattern of 2θ at 31.37°; 34.41°; 36.24°; 47.53°; 56.57°; 56.90°; 59.72°; 63.03° and 66.75°. According to this pattern, the crystal phase of ZnO was theoretically in accordance with hexagonal wurtzite. This result conforms with the data of ZnO from AMCSD No. 96-900-4181. The SEM results describe that the morphology of ZnO nanoparticles synthesized using moringa leaf extract on the n-hexane-water system have the aggregate form and small spheres as shown in figure 3a. EDX result clearly indicates that ZnO nanoparticles contain the atoms of zinc and oxygen as shown in figure 3b.

TEM characterization was conducted to determine the shape and particle size of ZnO nanoparticles. As can be seen in figure 4a, the particle sizes of ZnO nanoparticles were approximately 67 nm in agglomerated form. The photocatalytic activity showed that ZnO nanoparticles have a role as photocatalyst for MB degradation at 99 % for 2 h as shown in figure 4b. The results also showed that the longer of irradiation time can cause much light absorption and a greater decrease of methylene

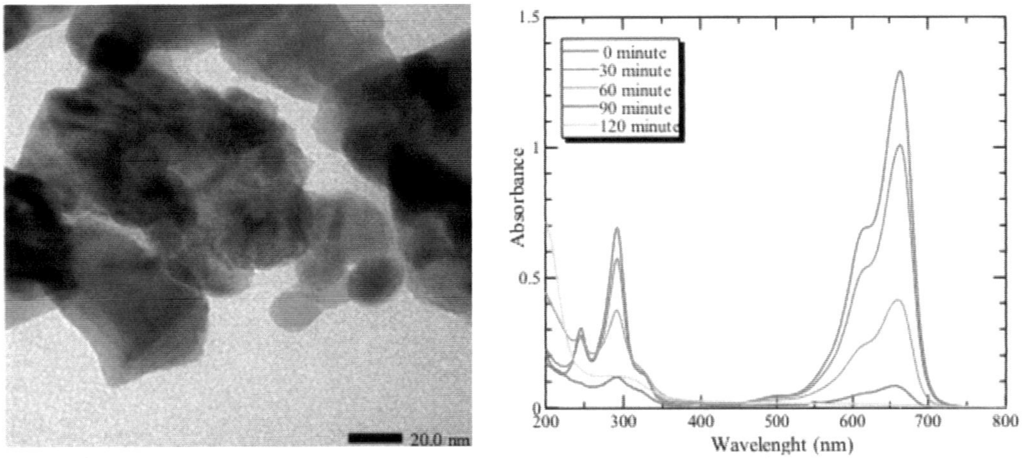

(a) (b)

Figure 4. (a) TEM image of (a) ZnO nanoparticles and
(b) UV-Vis absorption spectra of MB degradation

Blue concentration. This is due to the increasing number of electrons (e^-) from the valence band to the conduction band on catalyst, which produces a hole (h^+) in the valence band. The resulting electrons are capable to reduce the color by reacting with electron acceptor from O_2 adsorbed on the catalyst surface to form superoxide radical anions (O_2^- •). The result hole is able to oxidize the organic molecules by reacting with OH^- or H_2O to produce OH •. O_2^- •. Finally, OH• is a very strong species, which is capable to degrade almost all azo dyes into the simpler products [14].

4. Conclusions

ZnO nanoparticles were successfully synthesized in n-hexane-water system using *Moringa oliefera* leaf extract. The formation of ZnO nanoparticles can be described by the presence of secondary metabolites in n-hexane phase that acts as a weak base source and $Zn(NO_3)_2$ in water phase. Based on TEM characterization, the particles size of ZnO nanoparticles was approximately 67 nm. The band gap energy of ZnO was 3.2 eV. The photocatalytic activity of ZnO nanoparticles were observed for the MB degradation under UV light irradiation. The percentage of MB degradation was 99 % for 2 h.

Acknowledgements

The authors would like to thank Universitas Indonesia for funding this research through PITTA Grant Universitas Indonesia with contract No.2338/UN2.R3.1/HKP.05.00/2018.

References

[1] Yulizar Y, Wahyuningsih N, Asri N D and Watarai H 2012 *Makara J. Sci.* 16 169–77
[2] Vidya C, Hiremath S, Chandraprabha M N, Antonyraj M A L, Gopal I V, Jain A and Bansal K 2013 *Int. J. Curr. Eng. Technol.* 82 118–20
[3] Sadjadi S and Eskandari M 2013 *Ultrason. Sonochem.* 20 640–3
[4] Konstantinou I K and Albanis T A 2004 *Appl. Catal.* B*: Environ.* 49 1–14
[5] Scott-Emuakpor E O, Kruth A, Todd M J, Raab A, Paton G I and Macphee D E 2012 *Appl. Catal.* B*: Environ.* 123–124 433–9
[6] Teng F, Liu Z, Zhang A and Li M 2015 *Environ. Sci. Technol.* 49 9489–94
[7] Gao S, Gu B, Jiao X, Sun Y, Zu X, Yang F, Zhu W, Wang C, Feng Z, Ye B and Xie Y 2017 *J. Am. Chem. Soc.* 139 3438–45
[8] Yulizar Y, Latifah I, Bakri R and Apriandanu DOB 2018 *AIP Conf. Proc.* 2023 020097
[9] Lembang M S, Yulizar Y, Sudirman S and Apriandanu D O B 2018 *AIP Conf. Proc.* 2023 020093
[10] Sulaiman N, Yulizar Y and Apriandanu D O B 2018 *AIP Conf. Proc.* 2023 020105
[11] Yulizar Y, Bakri R, Apriandanu D O B and Hidayat T 2018 *Nano-Structures and Nano-Objects* 16 300–5

[12] Yulizar Y and Ayun Q 2017 *IOP Conf. Ser.: Earth Environ. Sci.* **60** 012006
[13] Ikalinus R, Widyastuti S K and Setiasih N L E 2015 *Indonesia Medicus Veterinus* **4** 71–9
[14] Akpan U G and Hameed B H 2009 *J. Hazard. Mater.* **170** 520–9

IOP Conf. Series: Materials Science and Engineering **763** (2020) 012023 doi:10.1088/1757-899X/763/1/012023

Influence of pyrolyzed temperature on the formation of silicon carbide fiber from polycarbosilane in electrospinning method

M Masripah[1,2], **A Zulys**[1] **and J Setiawan**[2]

[1] Departement of Chemistry, Faculty of Mathematics and Natural Sciences (FMIPA), Universitas Indonesia, Kampus UI Depok, Depok 16424, Indonesia

[2] The Center for Nuclear Fuel Technology (PTBBN) Batan, Kawasan Puspitek Serpong, Tangerang Selatan, Banten 15314, Indonesia

Corresponding author's e-mail: zulys@ui.ac.id

Abstract. The synthesis of silicon carbide (SiC) fibers by electrospinning method was done using polycarbosylane (PCS) as precursor and dimethylformamide (DMF)/toluene as solvent. Knowing the heat treatment of fibers consist of curing and pyrolisis, there were two steps in the major reaction. The curing process is by heating in air to make an oxidative cross-linking of the PCS. The pyrolysis process causes chemical decomposition from organic to inorganic compounds. These composition changes were described through Fourier Transform Infrared FTIR, Thermagravimetric analysis (TGA), X-Ray Diffraction (XRD) and Energy Dispersive X-Ray Spectroscopy (EDS) analysis. The products of electrospinning and curing were characterized by FTIR and TGA. The conversion to ceramic of PCS under various pyrolysis temperatures (1200, 1300, 1400, 1500) °C has been investigated and the products of this process were characterized by FTIR and XRD. Elemental analysis of the fiber as received and after cured showed that the ratio of H and C to Si decreased with holding time at the cure-temperature while the amount of oxygen increased. Meanwhile, the pyrolyzed generally produced silicon oxycarbide SiC_xO_y. Crystalline SiC β-SiC was found at temperature 1500 °C.

Keywords: SiC fibers, polycarbosilane, electrospinning, curing, pyrolysis

1. Introduction

Silicon carbide (SiC) is known as one of highly significant semiconductor substances generated through the covalent bond amongst Si and C. This substance is a perfect preference for the applications where tough environment is involved, such as high-power, high frequency as well as high temperature. PCS conversion into ceramic is normally obtained through a heating process of the polymer precursor to the temperature higher than 1000 °C in inert atmosphere condition. Numbers of stages in SiC fibers were observed to be involved in the PCS pyrolytic conversion and oxidation-cured methods. The SiC fibers derived from the pyrolysis process of polymer carry beneficial properties, namely the flexibility and fine-diameter. Polycarbolsilane's synthesis as precursor, polymer-derived C fibers performed by Yajima *et al.* [1] has been widely investigated. The said method involves four processes: synthesis, spinning, curing and pyrolysis.

Electrospinning is an exceptional multipurpose system -enables the production of fine-fibers in tiny diameters, from couple of microns to tens of nanometer- by the means of processing or melting the polymeric solution [2]. Electrospinning is a fiber fabrication method using electrical fields to stretch spinnable jets from polymer solutions. It has been used as a versatile technique of fiber fabrication such as polymers and ceramic fibers. The working principle is to conduct electric spinning using an electric field to stretch spinnable jets from polymer solutions.

The synthesis of SiC fibers by electrospinning of PCS solutions is followed by thermal curing and pyrolysis. PCS melts before decomposition. Thus, it cannot be directly heated up to obtain SiC. During crosslinking, oxygen helps to bridge the PCS molecules to form larger ones. The curing process allows PCS to have a more stable structure. The PCS after crosslinking does not melt before decomposition during the subsequent heat treatment in pyrolysis process [3–4].

During the preparation of SiC fibers, in the curing process, oxidation takes place in PCS fibers due to the heat treatment in the air. This heat treatment is intended to provide the infusible article and to maintain its form, as well as to generate a high ceramic results afterward [5]. This stage is a crucial process, since considerable mass of fibers could be lost throughout the pyrolysis process.

As SiC-forming precursor, PCS has been widely utilized by many researchers, in which the process of producing SiC fibers from PCS involves few steps of method, i.e. synthesis, spinning, curing and pyrolysis [1,6].

The temperature's influence on SiC formation process was studied in this research. The fibers were pyrolyzed at temperature 1200°, 1300°, 1400 °C and 1500 °C. The fibers were analyzed in order to describe the structural change during the heat treatment. The characterization was done by Fourier Transform Infrared (FTIR), Thermagravimetric analysis (TGA), X-Ray Diffraction (XRD) and Energy Dispersive X-Ray Spectroscopy (EDS).

2. Experimental

Polycarbosilane (PCS) was utilized as a precursor through dissolution of toluene and N, N-dimethylformamide (DMF) as solvent. PCS at 1.2 g/mL was dissolved with solvent with a composition of 30 % (% DMF/toluene). The solution was gently stirred for 12 hours minimum to make sure the PCS solution is homogeneous. This solution served as feeding solution for electrospinning process. During the electrospinning process, the voltage was set at 17 kV and the tip to collector distance was set at 10 cm. In each process, the feeding solution was set at 2 mL and feed rate at 1 mL/hour. The obtained fibers was cured at 200 °C for 1 h at the rate of 2–3 °C/min. For the conversion of the cured fibers into the SiC fibers, the fibers were pyrolyzed at (1200,1300, 1400 and 1500) °C for 1 h at the rate of 2–3 °C/min.

The characterization of the generated fibers from electrospinning was done with Fourier Transform Infra Red (FTIR). The cured fibers were characterized with FTIR and Thermogravimetric Analyzer (TGA). The pyrolyzed fibers were characterized by FTIR, X-ray Diffraction (XRD) and Energy Dispersive X-ray Spectroscopy (EDS).

3. Results and discussion

In this chapter we discuss the structural changes of PCS, starting from the original form until pyrolysis took place.

3.1. The crosslinking in curing process

The crosslinking behaviors in curing process of PCS fibers will be discussed. PCS melts before decomposition. Thus, it cannot be directly heated up to obtain SiC. During curing process, oxygen helps to bridge the PCS molecules to form the larger ones. The PCS after crosslinking does not melt before decomposition 1. The effect of the temperature and the heat-treatment duration on the crosslinking of the PCS was studied.

The as-received PCS used in this work had three major structural elements, they are: Si-C bonds, Si-CH_2-Si bonds and SiCH_3 bonds. The three of them, Si-CH_3 and Si-CH_2-Si bonds are crucial in consequence of the relation to the oxygen control included during the next curing process. C-H bonds and Si-H bonds are the side groups of PCS. The curing process allowed PCS to have a more stable structure so that they would not melt during the subsequent heat treatment. FTIR spectra for PCS before curing was similar to what Setiawan *et al.* [7] have done for several DMF concentrations. Comparison of FTIR results for samples of before and after curing, is shown in figure 1 where the absorption peak appears in the sample after curing where Si-H strain (2100 cm^{-1}), Si-CH_3 (1250 cm^{-1}) and CH (2850 cm^{-1}) represent the Si-CH_3 group.

There is an increase peak of Si-CH_2-Si (1000 cm^{-1}). There are two peaks that appears on the FTIR graph of curing fiber, Si-O group (470 cm^{-1}) and Si-OH (3200 cm^{-1}) with a very dominant peak, strain CH (1450-1350 cm^{-1}), cluster C = C (1640-1680 cm^{-1}), other than that, there is a CH 2850 cm^{-1} group. In the curing process there was a condensation reaction between the residual CH groups in the structure, where it was released H_2 and CH_4 and formed Si-C-Si bridges with the consequence of increasing

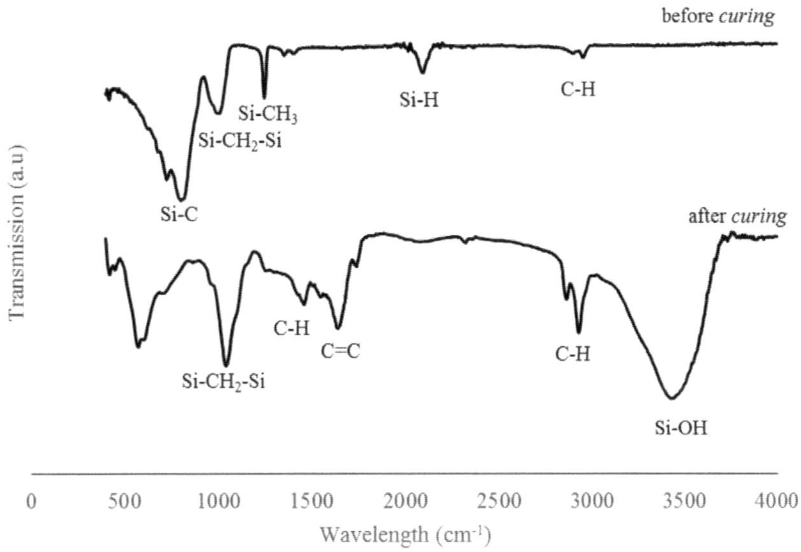

Figure 1. The FTIR result before and after curing

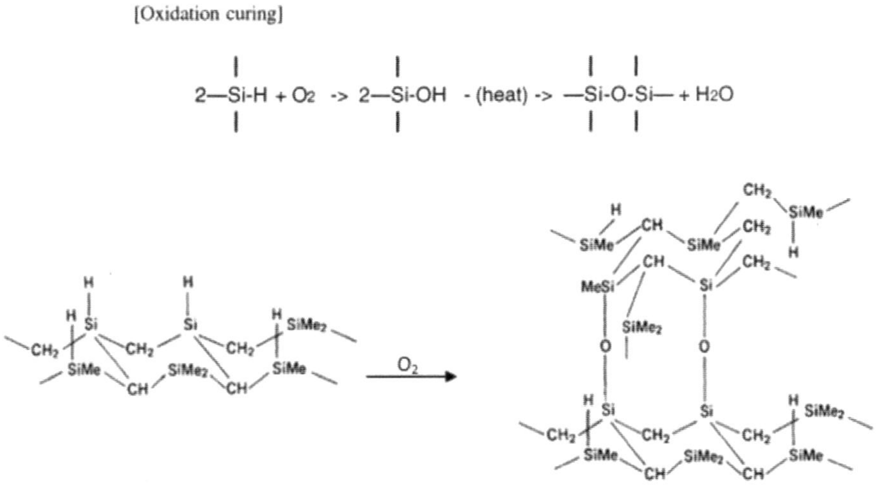

Figure 2. Crosslinking in polycarbosilane during curing in air [6, 8].

crosslinking network. Si-CH$_3$, Si-H and C-H groups can react with each other to produce hydrogen and methane gas.

The presence of oxygen during the curing process causes a reaction between oxygen with Si and C groups, causing the decomposition of PCS so that the Si-H group decreases and the Si-O and Si-OH groups increase, indicating the reaction of the oxidation bridge, C=C also appears because of the dehydrogenation or the release of hydrogen. The results of this study indicate that they are not much different from the research of [4, 6, 8–9]. The phenomenon of crosslinking on PCS fibers during the curing with air conditions is described in figure 2.

The influence of the heat-treatment temperatures on the PCS decomposition was observed with TGA, measurements, which was made to a temperature of 1000 °C, with air conditions and a rate of 10 °C / minute of heating. Figure 3 shows the PCS fibers prior to curing at room temperature up to 200 °C. The mass decreased of about 5.7 % due to evaporation in the PCS solvent, it was possible to occur due to the release of oligomers with low molar masses and also possible due to the evaporation of water contained in the fibers due to air humidity during storage [10].

IOP Conf. Series: Materials Science and Engineering **763** (2020) 012023 doi:10.1088/1757-899X/763/1/012023

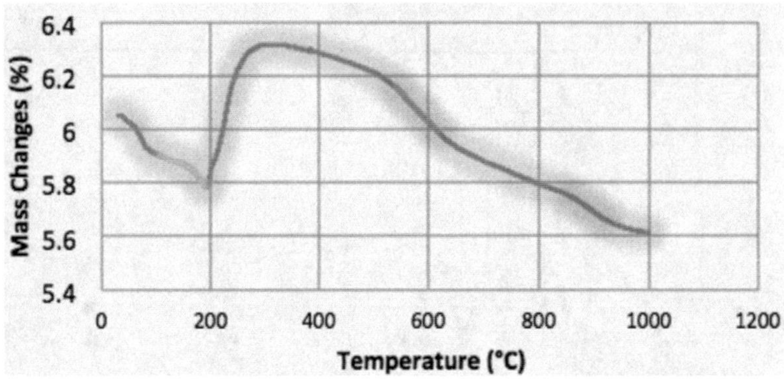

Figure 3. TGA Result of PCS fiber after electrospinning.

The PCS gained was about 5.8 % by weight at temperature ~ 200 °C, and then peaked to 6.3 % at temperature ~ 300 °C. This would be due to the interaction of oxygen with Si, which disintegrated from Si-H bond. So, the addition of 0.5 % weight is due to the entry of oxygen into the PCS structure to form of what called as crosslinking. Then, the weight decreased at temperature ~ 400 °C and furthermore the higher the heating temperature decreased the fiber weight. The weight change curve started to dropping down when the temperature was above 300 °C. This suggests that the early decomposition of the side groups in PCS such as methyl silicone bonds, silicon hydrogen bonds and carbon hydrogen bonds, lost the mass due to hydrogen and low molecular weight hydrocarbons' lost during the dehydrogenation and dehydrocarbonation condensation reactions. PCS fibers are highly sensitive against oxygen in the heat treatment process in the air, so thermal decomposition in the fibers mainly occurs on the PCS fiber side chain [9]. The curve pattern decreases from 300 °C to 1000 °C has some steep drop point, indicating the decomposition takes place through several stages. There are two parts of steep decrease at ambient temperature 530 °C to 620 °C, then at a temperature of about 855 °C to 920 °C, indicating at the temperature decomposition occurs so much loss of mass.

The TGA results were also supported by FTIR fiber samples prior to curing and after curing as shown in figure 1. There were peaks that appear on the FTIR graph, Si-O (470 cm^{-1}) and Si-OH (3200 cm^{-1}) with a very dominant peak. The oxygen that enters the molecule causes weight increase. When the temperature reached around 260 °C to 300 °C there was no more weight increase. The oxygen during the process reacted with Si and C groups to form Si-O and O-H groups. After the temperature reached 400 °C, the PCS backbone begins to decompose. The Si-C and Si-CH$_2$-Si bonds become very weak and at that time it is believed that the breaking of Si-CH$_3$, Si-H and CH groups decreases, so the PCS weight decreases [4,11–12].

3.2. Pyrolisis

Figure 4 shows the FTIR spectra for SiC fiber that pyrolyzed at different temperatures. The organic parts of PCS were transformed into a more inorganic state during pyrolysis. High weight lost and gaseous evolution of the most of H_2 and CH_4 were related with these reactions. Si-H and C-H bonds were immediately broken. FTIR indicates that the major bonds available were Si-CH$_3$ (1250 cm^{-1}) and Si-CH$_2$-Si (1020 cm^{-1}) and Si-OH (3400 cm^{-1}). The band that broadening at about 3700 cm^{-1} probably because of O-H stretching of Si-OH. Along with the temperature increase, C-H bond decreased while the Si-H, and Si-CH$_3$ bonds disappeared. At the temperature of 1400 °C the connectivity of Si-C-Si bonds increased and a network structure can be formed through condensation reaction amongst CH_4 and CH_3 units, then, Si-C (820 cm^{-1}) bonds started to appear and also C-O (1200 cm^{-1}) bonds which was caused by chemical decomposition with reaction [4] as follows:

$$SiC_xO_y \rightarrow SiC_{(s)} + SiO_{(g)} + CO_{(g)} + C_{(s)} \quad (1)$$

Oxygen was introduced in oxidation cured; the oxygen actually will be lost as SiO gas and CO gas during heating in pyrolysis process. However, CO gas evaporates at temperature no less than 1500 °C, so to remove all oxygen content it needs to be heated to more than 1500 °C. The crystals of SiC were formed when the temperature raised to 1400 ºC or above, it was probably because of the crystal growth of β-SiC [4,13–14]. During the decomposition process, the evolution of H_2 gas and Si-H bond was interrupted and C-H was broken, while Si-C-Si bond increased. During pyrolysis temperature of 1400 °C and 1500 °C the C-O bonds appeared.

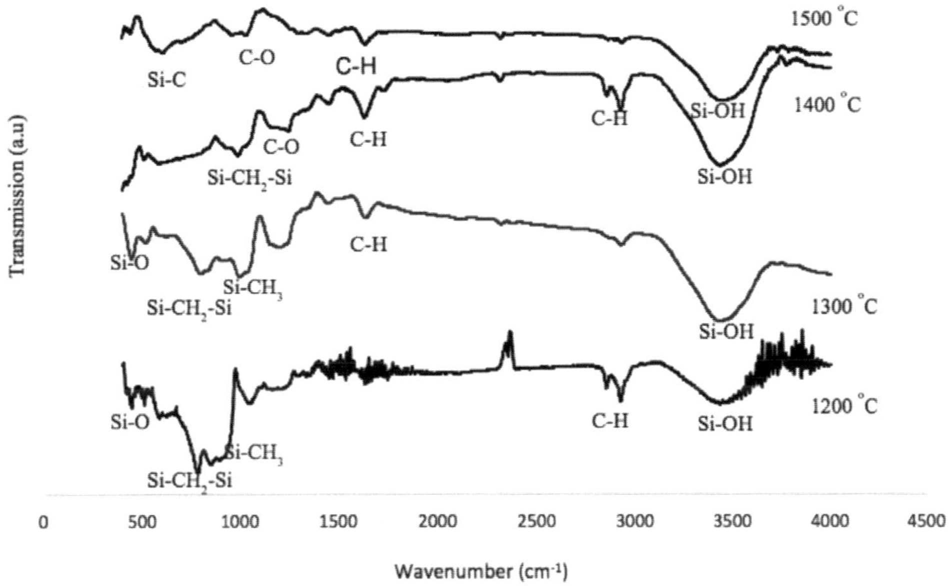

Figure 4. FTIR spectra of SiC fibers

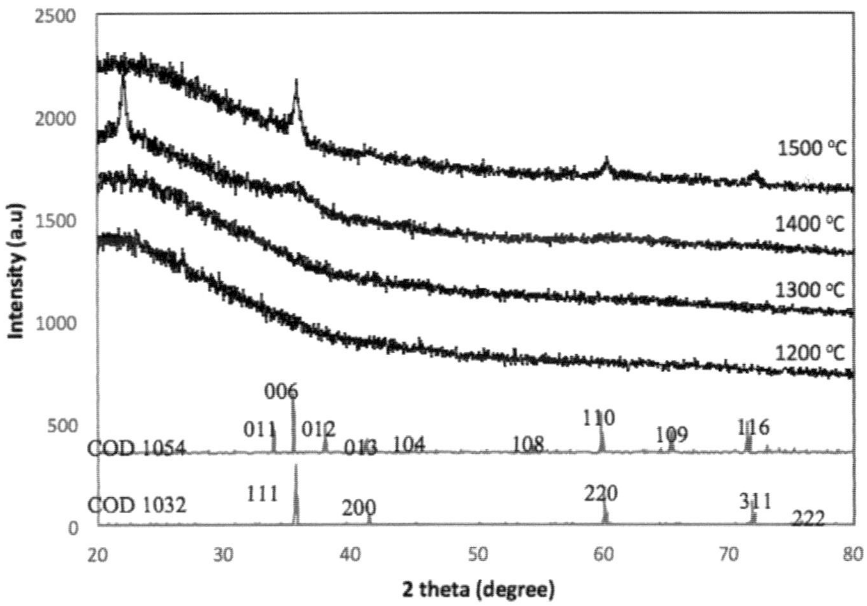

Figure 5. XRD spectra of SiC fibers

In the XRD pattern of figure 5, during the pyrolysis at 1400 °C and 1500 °C, the SiC crystals begin to grow, showed by the peak between 30-40 °, indicating the formation of SiC crystals at hkl (111) and a low peak between 60-70 ° which also indicating the SiC (220). The fibers of pyrolysis at 1200 °C and 1300 °C are still amorphous, so there is no peak in the XRD graph. Crystallinity increases with the increasing temperature; crystal growth begins at temperatures above 1300 °C.

It has been described in introduction, that SiC has a varied crystal form (polytypes) XRD data shows only 1500 °C pyrolysis sample fibers and 1400 °C pyrolysis sample fibers from which have a crystal lattice but are still in small amounts, marked by gentle peaks and many noise-like spectra that

Table 1. The results of EDS for fiber composition.

Elements	Temperature (°C)			
	1200	1300	1400	1500
Si (%)	27.31	20.78	30.75	41.11
O (%)	55.71	35.13	51.63	44.78
C (%)	16.67	43.53	17.62	14.11
Fiber composition	$SiO_{2.04}C_{0.61}$	$SiO_{1.70}C_{2.09}$	$SiO_{1.69}C_{0.57}$	$SiO_{1.09}C_{0.34}$

indicating many structures of SiC are still in amorphous form. The SiC crystal structure that may grow is β-SiC or 3C-SiC, where the β-SiC crystal structure is formed on the surface of the fiber when the temperature is increased to1300 °C or higher [4,15–16]. So, only SiC fibers in the pyrolysis at 1400 °C and 1500 °C only appear on the XRD graph, since the SiC fiber phase at temperatures below 1300 °C is still amorphous, when the temperature rises to 1300 °C or greater, SiC_xO_y decomposition occurs, and forms SiO gas, SiC crystals and free carbon, then SiC crystal precursors occur because of the reaction between free carbon and SiO gas forms large crystals can be seen on fibers' surface. SiC crystal formation reaction can be seen in the following equation:

$$SiO_{(g)} + 2C_{(s)} = SiC_{(s)} + CO_{(g)} \quad (2)$$

3C-SiC is the only polytype with a cubic crystal structure also called β-SiC. 3C-SiC has a sequence of arrangements of a double atomic layer is ABCABC. The "C" in '3C-SiC' indicates the cubic crystal structure and' '3' represents the amount of layers of two atoms in a loop unit (ABC). A layer of Si (or C) atoms in a stacking order might have a cubic (k) or hexagonal (h) shape, but 3C-SiC only has a type-k atom.

The element content in pyrolysis fibers was analyzed using EDS. The results stated that the chemical structure was Si-O-C, with SiC_xO_y ratio shown in table 1. The oxygen contained in the fibers comes from the curing process because the curing process was carried out under air conditions causing an increase in oxygen content due to oxidation and replacement of Si-CH_3 bonds with Si-O-Si and Si-O-C during cross linking, thereby increasing the percentage of oxygen content. Table 1 shows the phenomenon that by heat treatment during pyrolysis, the oxygen content decreased, it is because during the pyrolysis process there is formation of gas product (SiO and CO) [4, 6, 17, 18].

4. Conclusions

SiC fibers have been successfully synthesized by electrospinning method with polycarbosilane precursors and DMF / Toluene solvents, followed by heat treatment in curing and pyrolysis processes. The curing process was performed with air condition at temperature of 200 °C. The heat treatment analysis was conducted using TGA which proved that the process of fiber decomposition to ceramic form occurred from about 300 °C. The conversian of organic parts in PCS resulted in a more inorganic state during pyrolysis. High weight lost and gaseous evolution of the most of H_2 and CH_4 were related with these reactions. Si-H and C-H bonds were immediately broken. The formation of SiC fibers occurred at the pyrolysis stage that is the heat treatment with temperature variation 1200 °C, 1300 °C, 1400 °C and 1500 °C. SiC fibers obtained showed an amorphous and slightly crystalline phase in pyrolysis fiber of 1400 °C. Crystalline SiC β-SiC was found at temperature 1500 °C with elements composition of $SiO_{1.09}C_{0.34}$.

Acknowledgements

This research supported by Center for Nuclear Fuel Technology – Nuclear Energy Agency and cooperation with Department of Chemistry, Faculty of Science and Mathematics, University of Indonesia.

References

[1] Yajima S, Hasegawa Y, Hayashi J and Iimura M 1978 *J. Mater. Sci.* **13** 2569–76
[2] Kurban Z 2011 *Electrospun Nanostructured Composite Fibres for Hydrogen Storage Applications* (London: University College London) Doctoral Thesis
[3] Chen Y, Li S, Luo Y and Xu C 2011 *Solid State Sci.* **13** 1664–7
[4] Yue Y 2014 *Synthesis of Silicon Carbide Fibers from Polycarbosilane by Electrospinning Method* (Clemson: Clemson University) Master Thesis

[5] Setiawan J, Pribadi S, Prajono, Poertadji S and Sigit 2015 *Adv. Mater. Res.* **1123** 223–6.
[6] Takeda M, Saeki A, Sakamoto J, Imai Y and Ichikawa H 1999 *Compos. Sci. Technol.* **59** 787–92.
[7] Setiawan J, Pranjono, Saptaaji R, Poertadji S and Sigit 2015 *Proc. 17th Int. Conf. Physics and Its Applications (ICOPIA-14)* 16-17 September 2014 Solo ed R Suryana and Nuryani (Paris: Atlantis Press) pp 127–130
[8] Okamura K 1987 *Composites* **18** 107–20
[9] Zhaeng C, Li X, Wang H and Zhu B 2006 *Transactions of Nonferrous Metals Society of China* **16** 44-8.
[10] Mustika D, Sihombing R, Pribadi S, Langenati R, Sujatno A, Dimyati A, Setiawan J and Indarto E 2015 *Urania* **21** 29–38
[11] Hong J, Cho K Y, Shin D G, Kim S H and Riu D H 2018 *J. Inorg. Organomet. Polym. Mater.* **28** 2221–30
[12] Lu P, Huang Q, Mukherjee A and Hsieh Y L 2011 *J. Mater. Chem.* **21** 1005–12
[13] Hasegawa Y 1989 *J. Mater. Sci.* **24** 1177–90
[14] Ly H Q, Taylor R, Day R J and Heatley F 2001 *J. Mater. Sci.* **36** 4037–43
[15] Houyem A and Hmida S B H 2011 Silicon carbide : synthesis and properties *Properties and Applications of Silicon Carbide* ed R Gerhardt (London: IntechOpen) pp 361–89
[16] Happe M *et al.* 2016 *Bioresour. Technol.* **200** 435–43
[17] Sugimoto M, Shimoo T, Okamura K and Seguchi T 1995 *J. Am. Ceram. Soc.* **78** 1013-7.
[18] Takeda M, Imai Y, Ichikawa H, Kasai N, Seguchi T and Okamura K 1999 *Compos. Sci. Technol.* **59** 793–9

Study of urinary 8-hydroxy-2'-deoxyguanosine (8-OHdG) in rats (*Rattus norvegicus*) due to bisphenol A (BPA) exposure based on fenton-like reaction

H Franklyn, B Budiawan, S Handayani and I C Dani

Department of Chemistry, Faculty of Mathematics and Natural Sciences (FMIPA), Universitas Indonesia, Kampus UI Depok, Depok 16424, Indonesia.

Corresponding author's e-mail: dr.budiawan@gmail.com

Abstract. This study was conducted to analyze the formation of DNA adduct 8-OHdG due to oxidative damage of DNA caused by Bisphenol (BPA) and chromium (VI) exposure. The in vivo study was conducted in the rat (Rattus norvegicus) group with exposure of BPA (2 mg/kg BW), the mixture of BPA (2 mg/kg BW) and Cr (VI) (1.28 µg/kg BW) for 28 days, respectively. Urine of rats was collected every week as samples for tracking the 8-OHdG compounds. The formation of 8-OHdG in urine samples was analyzed using LC-MS/MS with reverse-phase chromatography. Mixture of ammonium acetate pH 4.0 20 mM and acetonitrile were used as mobile phase in LC-MS/MS for an elution gradient. The results showed that the exposure of BPA, the mixture of BPA and Cr (VI) might cause the formation of urinary 8-OHdG in rats. The longer exposure time causes the increase level of urinary 8-OHdG formations. The study also showed that urinary 8-OHdG concentration increases due to synergistic effect between BPA and Cr (VI) exposure.

1. Introduction

Bisphenol A (BPA; 2,2-bis (4-hydroxy-phenyl) propane) is a chemical compound produced in large quantities in the industrial field. This compound is used as a monomer for producing polycarbonate (PC) in plastic and epoxy resin, or as an antioxidant and stabilizer in various plastics [1]. With so many benefits, the use of BPA are currently widespread in the production of water-drain pipes, children's toys, medical devices, dental products, electronic devices, and CD/DVDs. Along with the development of industries in many countries, BPA as a chemical is commonly used in many industries in the product also continues to grow to this day.

Based on its main function as a plastic monomer and epoxy resin, BPA can enter the human body through two main lines, food and non-food intake. If it exposes to heat and in contact with acidic or alkaline solutions, the ester in BPA as polycarbonate or epoxy resins will rapidly hydrolyze [2]. This causes the migration of BPA into food or drink from containers that use BPA as polycarbonate monomer or epoxy resin. Many countries have banned or restricted the use of BPA due to the ease of BPA release and the adverse health effects. The side effects caused by BPA exposure are toxicity, teratogenic, carcinogenic, and especially estrogenic effects [3]. In the human body, BPA can also trigger the production of reactive oxygen species (ROS) [4]. BPA in the human body cannot directly undergo the reaction that causes the hydroxyl radical to form. However, BPA is notable to reduce the activity of antioxidant enzymes in mitochondria, such as superoxide dismutase enzymes, catalase, glutathione reductase, and glutathione peroxidase, accompanied by elevating levels of hydrogen peroxide and lipid peroxidation [5]. Cr (VI) is known to be easily passing through the cell membrane and get in the body cells [6]. Cr (VI) will undergo a reduction process to a lower valence state, Cr (V), Cr (IV), and Cr (III). In this process, the reduction reaction with Cr (VI) will produce metabolites in the form of a large number of reactive oxygen species [7]. This study have analyzed the DNA adduct of 8-hydroxy-2'-deoxyguanosine (8-OHdG) in rat urine after 28 days of exposure to BPA and Cr (VI) chemical compounds.

2. Materials and methods

2.1. *Chemicals*
The BPA, 8-OHdG and SPE cartridge DSC-18LT tube were purchased from Sigma-Aldrich (SG). HPLC grade acetonitrile and methanol, potassium dichromate, ammonium acetate, dimethyl sulfoxide and diethyl ether were from Merck (ID). All other chemicals and solvents were an analytical grade.

2.2. *Ethical approval*
Experimental protocol was approved by the ethics committee of the Faculty of Medicine, Universitas Indonesia (Indonesia).

2.3. *Animal preparation*
The present study was conducted on 8 weeks old healthy Sprague-Dawley rats and housed in cages. The animals were first acclimatized for 21 days to adapt to the new environments and gained the appropriate weight (130–200 grams). Animals were provided with standard feed in the cage and allowed water ad libitum throughout the experimental period. Rats were divided into three groups, group A is a control group without chemical treatment, group B is a group with BPA exposure, and group C is a group with the chemical mixture exposure of BPA and Cr(VI).

2.4. *Formulation*
BPA was dissolved into dimethyl sulfoxide and water. BPA solution was administered orally in Group B with dose volume of 10 mL/kg [8]. The daily administration was continued for 28 days. The mixture of BPA and potassium dichromate were administrated orally in Group C with dose volume of 5 mL/kg respectively [9]. The daily administration was continued for 28 days.

2.5. *Experimental design*
Experimental groups consist of three groups. Each group has 5 male rats. Rats of Group A were kept as negative control and were given water orally. Rats of group B were given BPA orally; rats of group C were given the mixture of BPA and potassium dichromate. The dose for group B was 2 mg/kg body weight for 28 days. The dose of BPA and potassium dichromate for group C was 4 mg/kg and 1.28 μg/kg body weight, respectively, for 28 days. The doses in this experiment were based on sub-chronic toxicity study.

2.6. *Sample collection and preparation*
Rats urine samples were collected using metabolic cages and frozen immediately for analysis. The samples were thawed at room temperature. Before analysis, 5 mL of methanol and 5 mL of 20 mM ammonium acetate pH 4.0 were used to precondition the SPE cartridge. The mixture of 20 mM ammonium acetate pH 4.0 and acetonitrile (97:3) (v/v) were used to wash the SPE cartridge after loading 500 μL of urine sample, and 8-OHdG was eluted into the mixture of 20 mM ammonium acetate and acetonitrile (80:20)(v/v).

2.7. *Calibration curves, LOD and LOQ*
Calibration curve was obtained using 8-OHdG standard solutions in the 1–50 ng/mL. The limit of detection (LOD) and limit of quantification (LOQ) were determined statistically using the linear regression equation from the calibration curves. The measurement value will be equal to the value of b on the linear regression equation of $y = a+bx$, whereas the standard deviation of the blank is equal to the standard deviation of residual $S(y/x)$.

2.8. *Instrumental analysis*
Samples were inspected in a 20A liquid chromatography system (Shimadzu, Japan) coupled with an AB Sciex Q-Trap 3200 mass spectrometer (LC-MS/MS). All analytes were separated in a Hypersil Gold C_{18} column (50 x 2.1 mm, 5 μm, Thermo Scientific). Acetonitrile and 20 mM ammonium acetate pH 4.0 were used as mobile phase. The gradient elution program is displayed in table 1. The following conditions were applied to the instrument, 0.5mL/min of flow rate, 40 °C of column temperature, and 10μL of injection volume.

The initial temperature was fixed at 550 °C, and the ionization voltage was 5500 V. The analyte was analyzed in multiple reactions monitoring (MRM) mode with a dwell time of 300 ms. Both

IOP Conf. Series: Materials Science and Engineering **763** (2020) 012024 doi:10.1088/1757-899X/763/1/012024

Table 1. The gradient mobile phases for 8-OHdG analysis

Time (min)	Ammonium acetate (%)	Acetonitrile (%)
0.5	95	5
1.0	5	95
2.0	5	95
2.5	95	5
3.0	Stop	

Table 2. Optimization of MS/MS parameters for 8-OHdG analysis

Precursor ion (Q1)	Product ion (Q3)	Retention time (min)	DP (eV)	EP (eV)	CE (eV)	CXP (eV)
284.1	168.2	0.738	43.0	4.0	17.0	3.0
	140.1	0.738	43.0	4.0	40.0	3.0

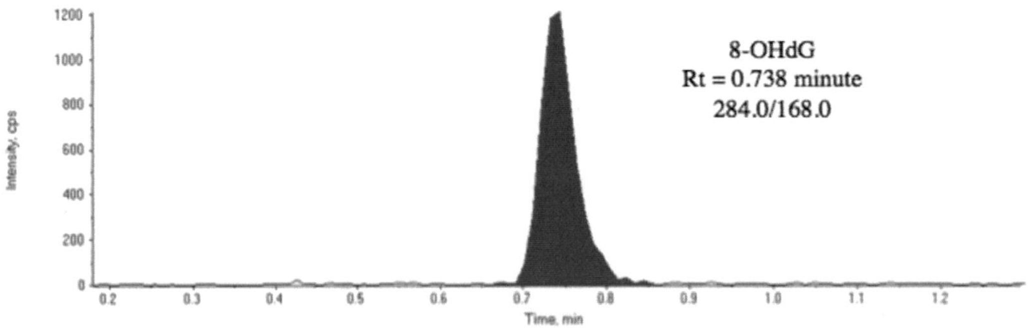

Figure 1. Chromatogram Standard of 8-OHdG 100 ppb LC-MS/MS

nebulizer gas (GS1) pressure and the turbo heater gas (GS2) pressure were 55 psi. The curtain gas (CUR) was 20 psi and the collision gas (CAD) pressure was medium, respectively. The MS/MS parameters, containing parent ion (Q1), product ion (Q3), collision energies (CE), declustering potential (DP), entrance potential (EP) and collision exit potential (CXP) were optimized. The optimized parameters are listed in table 2.

3. Results and discussion

3.1. Calibration curves, LOD and LOQ
Figure 1 shows chromatogram of standard solution of 8-OHdg. The calibration curves for the standard compound was linear with correlation coefficient (≥ 0.999) in the range of 1–50 ng/mL, which covers the lowest concentration of 8-OHdG in rats urine, and the slope was 34.321. The limit of detection and limit of quantification was 1.350 ng/mL and 4.499 ng/mL, respectively, according to the calculation of the linear regression equation from the calibration curves.

3.2. Detection of 8-OHdG in urine samples

3.2.1. The effect of BPA exposure on the formation of 8-OHdG. In this experiment, in vivo study was conducted to see the formation of DNA adduct 8-OHdG in the urine sample due to BPA exposure for 28 days with a dose of 2 mg/kg body weight. BPA exposure in rats could generate reactive oxygen species such as •OH radicals and leads to the formation of DNA adduct 8-OHdG, as shown in figure 2. The lowest and highest of 8-OHdG concentrations are equal to 27.597 ng/mL and 31.683 ng/mL, respectively. The mean of 8-OHdG concentrations due to BPA exposure is 29.258 ng/mL.

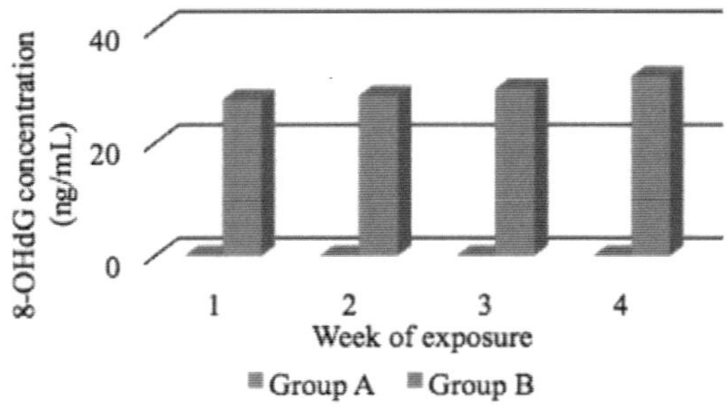

Figure 2. The effect of Group A and Group B in the formation of 8-OHdG.

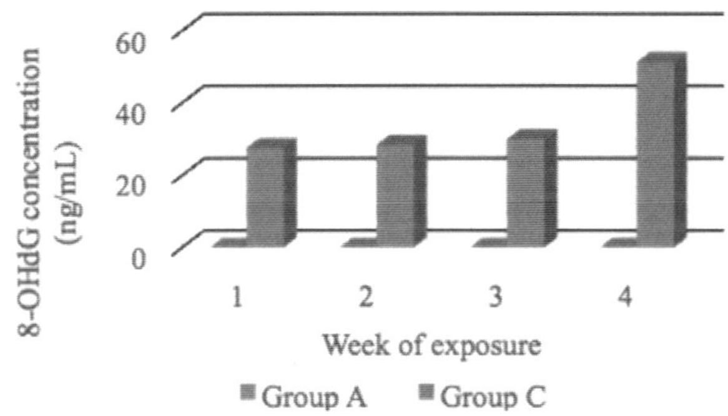

Figure 3. The effect of Group A and Group C in the formation of 8-OHdG.

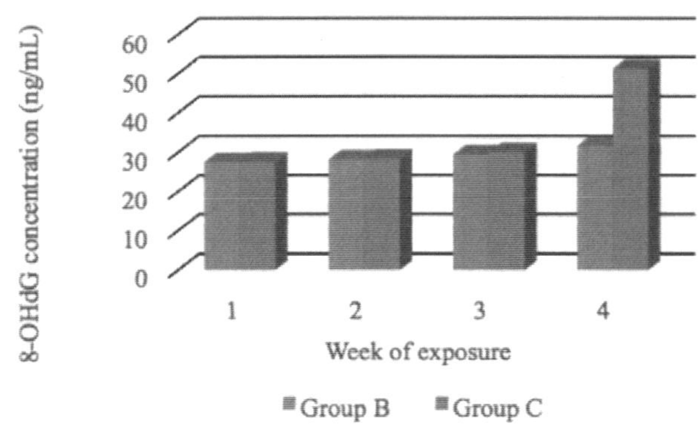

Figure 4. The synergistic effect of Group B and Group C in the formation of 8-OHdG.

3.2.2. The effect of BPA and Cr(VI) exposure on the formation of 8-OHdG. This experiment was conducted to see the synergistic effect of BPA and Cr(VI) exposure for 28 days with the dose of 4 mg/kg and 1.28 μg/kg body weight, respectively. The exposure of BPA and Cr(VI) metal of Group C may cause the formation of DNA adduct 8-OHdG and its concentration is higher than Group A, as shown in figure 3. From the data results, it is proved that the presence of Cr(VI) metal will enhance the production of reactive oxygen species such as •OH radicals.

The results of urine samples measurements in group C showed that Cr(VI) exposure had a synergistic effect on the formation of DNA adduct 8-OHdG as seen from the increase of 8-OHdG levels compared to group B, as shown in figure 4. Cr(VI) metal will undergo reduction process in the presence of NAD(P)H to a lower and more stable valence state, Cr(V), Cr(IV), and Cr(III), the

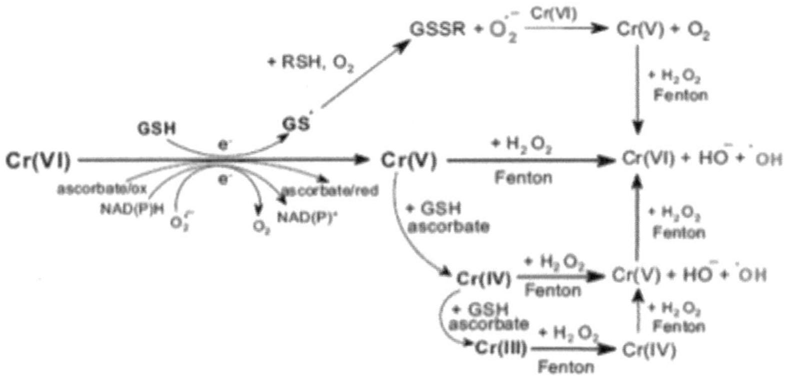

Figure 5. Fenton-like reaction of Cr (VI) metal

reaction will produce a large number of reactive oxygen species as shown in figure 5 (Fenton-like reaction) [6]. The lowest and highest of 8-OHdG concentrations is equal to 27.638 ng/mL and 51.290 ng/mL, respectively. The mean of 8-OHdG concentrations due to BPA and Cr(VI) exposure is 34.356 ng/mL.

4. Conclusions

BPA and Cr(VI) exposure to rats can lead to the formation of DNA adduct 8-OHdG. Due to BPA exposure for 28 days, the formation of urinary 8-OHdG reaches 31.683 ng/mL. The Cr(VI) exposure may increase the urinary 8-OHdG concentration due to its synergistic effect with BPA exposure. BPA and Cr(VI) exposure can lead to urinary 8-OHdG formation up to 27.638 ng/mL.

References

[1] Geens T *et al.* 2012 *Food Chem. Toxicol.* 50 3725–40
[2] vom Saal F S and Hughes C 2005 *Environ. Health Perspect.* 113 926–33
[3] Michałowicz J 2014 *Environ. Toxicol. Pharmacol.* 37 738–58
[4] Bindhumol V, Chitra K C and Mathur P P 2003 *Toxicology* 188 117–24
[5] Kabuto H, Hasuike S, Minagawa N and Shishibori T 2003 *Environ. Res.* 93 31–5
[6] Jomova K and Valko M 2011 *Toxicology* 283 65–87
[7] Patlolla A K, Barnes C, Yedjou C, Velma V R and Tchounwou P B 2009 *Environ. Toxicol.* 24 66–73
[8] Karnam S S, Ghosh R C, Mondal S and Mondal M 2015 *Vet. World* 8 738–44
[9] Dolai D P, Dey S K, Dash S K and Roy S 2016 *American Journal of Applied Scientific Research (AJASR)* 2 59–64

Synthesis and characterization of Au-TiO_2 Janus nanostructures

M F Mulyantono, J Gunlazuardi and M Khalil

Department of Chemistry, Faculty of Mathematics and Natural Sciences (FMIPA), Universitas Indonesia, Kampus UI Depok, Depok 16424, Indonesia

Corresponding author's email: mkhalil@sci.ui.ac.id

Abstract. This work reports a comprehensive study on the synthesis and characterization of Au-TiO_2 Janus nanostructures from pre-synthesized Au nanoparticles and anatase TiO_2 nanoparticles. Here, well-distributed Au nanoparticles with mean particle diameter of ~35 nm were successfully prepared using seed-mediated method, which exhibited a unique surface plasmon resonance (SPR) properties at 525.5 nm. Meanwhile, anatase TiO_2 nanoparticles with perfectly spherical morphology were synthesized *via* sol-gel method, which could easily be integrated by the as-prepared Au nanoparticles. Based on the result, it is found that the resultant nanostructured exhibited a Janus heterostructure feature, where Au nanoparticles were efficiently found to be attached on the surface TiO_2 nanoparticles. Furthermore, spectroscopic analyses also revealed that the integration of Au nanoparticles could widen the absorption spectrum of TiO_2 to visible range due to the presence of SPR phenomenon.

Keywords: TiO_2 nanoparticles, Au nanoparticles, Janus nanostructures, plasmonic, HR-TEM

1. Introduction

During the last several decades, research on the development of new nanostructures has become a hot topic in the fields of materials, physics, and chemistry. Furthermore, many interests have also been widely extended from one dimension to two or three-dimensional spaces, as well as the integration of multiple nanocomposites structures with controllable morphologies to achieve various new properties [1–3]. Owing to the rapid advances in synthetic nanochemistry, which enables one to fabricate numerous nanostructured materials with well-defined sizes, shapes, crystal facets, and compositions have carved the pathway for developing a new type of materials with excellent physicochemical properties [4]. Among many kinds of heterostructured nanomaterials, Janus nanostructure has been recently attracted much attention due to their exceptional abilities to generate synergetic and tunable physical and chemical properties which would not be accessible in their individual counterparts [5].

According to literature, colloidal semiconductor nanoparticles such as anatase TiO_2 nanoparticles have been widely considered as one of the most active catalytic materials for various photocatalytic reactions due to their high photo-activity, stability, low cost and corrosion resistance [6]. However, the application of colloidal TiO_2 nanoparticles as a photocatalyst for solar-driven photocatalytic reactions where the excitation energy is mostly originated from visible light can often be limited by two major issues, i.e. (1) large band gap, and (2) rapid electron-hole recombination [7]. Recently, many efforts have been carried out to circumvent these two significant issues. One of the conventional means to solve these limitations is *via* the utilization of heterostructured nanomaterials as the catalyst for instance by integrating TiO_2 nanocatalyst with photosensitive materials such as dyes or quantum dots nanoparticles to enhance the photocatalytic activity of TiO_2 in a broader light spectrum, or with plasmonic particles to reduce electron-hole recombination process [8–9].

Recently, tremendous efforts have been focused on the development of highly photoactive Janus Au-TiO_2 nanostructures due to its remarkable enhanced photocatalytic performances and power conversion efficiencies [8]. It is believed that the integration of plasmonic particles such as Au nanoparticles to TiO_2 would introduce two new prominent features. i.e., (a) surface plasmon resonance (SPR) which enables the strong absorption at the visible area, and (b) Schottky junction which could enhance the charge separation efficiency. Furthermore, it is also reported that, upon excitation, this integration would also shift the Fermi level of TiO_2 nanoparticles to a more negative potential, which enables one to reduce the frequent possibility of electron-hole recombination process. Nevertheless, these features are also reported to be significantly affected by various factors, such as morphology and crystal phase of the semiconductor itself, morphology of the plasmonic particles, and more importantly how stable the integration between each component in the Janus heterostructure is. In this study, an investigation of the synthesis and characterization of Au-TiO_2 Janus nanostructure was presented. Here, plasmonic Au nanoparticles were prepared *via* seed-mediated approach and integrated at acidic condition onto the surface of anatase TiO_2 nanoparticles. In this present study, anatase TiO_2 was synthesized by a sol-gel method where titanium(IV) isopropoxide (TTIP) was used as the titanium precursor and hydrolyzed at basic condition. Furthermore, the as-prepared Au-TiO_2 Janus nanostructure was characterized using X-Ray Diffraction (XRD), UV-Vis Spectrophotometer, Diffuse Reflectance Spectroscopy (DRS) and High-Resolution Transmission Electron Microscope (HR-TEM).

2. Materials and methods

2.1. Materials

Gold(III) chloride hydrate ($HAuCl_4.xH_2O$) (purity: 99.95 %), hydroxylamine hydrochloride and sodium citrate were used for the synthesis of Au nanoparticles and purchased from Sigma-Aldrich. Titanium(IV) isopropoxide (TTIP) (purity: 97 %) was obtained from Sigma-Aldrich and used as titanium precursor. NH_4OH solution (28-30 % NH_3 in H_2O) and absolute ethanol were also purchased from Sigma-Aldrich and used in the sol-gel reaction. Finally, hydrochloric acid was also used for the synthesis of Au-TiO_2 Janus nanostructure.

2.2. Synthesis of Au nanoparticles

In this study, Au nanoparticles were synthesized via seed-mediated technique according to a method reported by Haiss and co-workers [10]. Accordingly, Au seeds were firstly synthesized by rapidly adding 12.5 mL of 40 mM sodium citrate solution into 125 mL of 0.254 mM $HAuCl_4xH_2O$ under vigorous stirring until the color of the solution becomes deep red. To obtain Au nanoparticle seeds, the mixture was then boiled for 15 minutes and then cooled to room temperature. To prepare Au nanoparticles with bigger particle size, 20 mL of Au seeds solution was diluted with 180 mL of water. The solution was then mixed with 7 mL of 2 M hydroxylamine hydrochloride (NH_2OH•HCl) followed by the dropwise addition of 6 mL of 25.4 mM of $HAuCl_4{\cdot}xH_2O$ under vigorous stirring at room temperature. The resulting Au nanoparticles solution was then stored at 4 °C in darkness and used for further investigations.

2.3. Synthesis of anatase TiO_2 nanoparticles

In this work, a sol-gel method reported by Li and co-workers was used to synthesize anatase TiO_2 nanoparticles [11]. However, TTIP was used as titanium source instead of tetrabutyl titanate (TBOT). Typically, 0.75 mL of TTIP was added into a mixture of 100 mL of absolute ethanol and 0.3 mL NH_4OH which was then vigorously stirred at 4 °C for 24 hours. The result solution was then centrifuged at 6000 rpm, and the obtained white precipitates were collected, washed with deionized water and ethanol, respectively. The precipitate was dried at 60 °C overnight and then calcined at 450 °C for 2 hours and used for further investigations.

2.4. Preparation of Au-TiO_2 Janus nanostructure

Au-TiO_2 Janus nanostructure was fabricated by integrating Au nanoparticles onto the surface of anatase TiO_2 nanoparticles at acidic condition. Here, 0.05 g of the as-synthesized nanoparticle TiO_2 powder was added into 25 mL of the colloidal as-synthesized Au nanoparticles solution. The mixture was then vigorously stirred at room temperature while the pH of the solution was adjusted to pH 3 by slowly adding 0.01M of HCl solution. The mixture was then stirred for 30 minutes at room temperature, and the precipitate was recovered by centrifugation, followed by washing with deionized water and ethanol, respectively. The powder was then dried in a vacuum oven at 60 °C overnight and used for further investigations.

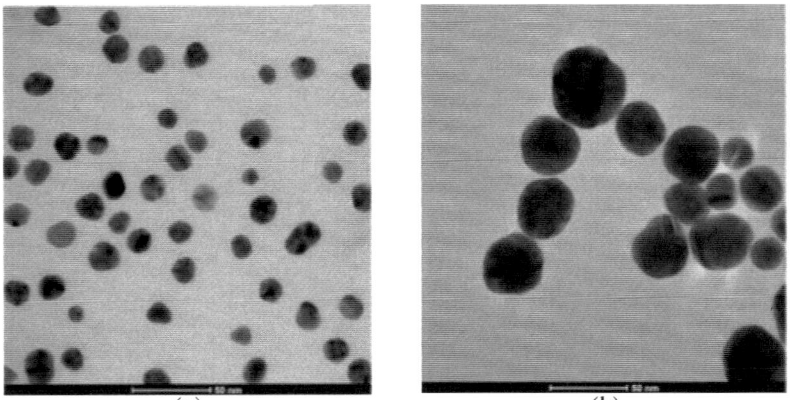

(a) (b)

Figure 1. TEM images of (a) Au seed particles, and (b) final Au nanoparticles (Scale bar: 50 nm).

2.5. Characterization

To study the physical and chemical properties of the as-prepared nanoparticles, several characterization techniques were carried out in this work. To investigate the crystal structure of the as-synthesized nanoparticles, X-Ray Diffraction (XRD) analysis was carried out using PANanalytical X'Pert Pro MPD (PANanalytical B.V., Almelo, Netherlands) equipped with fast detector X'Celerator and CuKα radiation as the source. Meanwhile, micrographic images of the nanoparticles were collected using TECNAI G2 Spirit Twin High-Resolution Transmission Electron Microscope (HR-TEM). To further determine the spectroscopic properties of the as-prepared nanoparticles, both spectroscopic UV-Vis and DRS analyses were carried out using Shimadzu UV-2450.

3. Results and discussion

3.1. Synthesis of Au nanoparticles

In general, tremendous efforts have been conducted to develop a facile method for the synthesis of Au nanoparticles with full control of size and morphology. In this work, Au nanoparticles were prepared using a seed-mediated method where a smaller Au nanoparticles seeds were firstly made using sodium citrate as strong reducing and capping agent. These Au seeds were then used to fabricate bigger Au nanoparticles using a milder reducing agent, i.e., hydroxylamine hydrochloride. Figure 1 presents the TEM images of Au seed particles and the final Au nanoparticles. Based on the result, it is shown that monodisperse polyhedral Au seed particles with average particle diameter ~13 nm were successfully synthesized (figure 1a). Furthermore, bigger polyhedral Au nanoparticles with an average particle diameter of around ~35 nm could easily be fabricated from these seed particles (figure 1b).

In general, the mechanism of Au nanoparticle growth in a seed-mediated method could be described using LaMer's nucleation theory [12]. Typically, the nanoparticles growth could be divided into three consecutive stages, i.e. (i) atom production from reduction reaction of the metal precursor, (ii) nucleation of seed particles, and (iii) nanocrystal growth due to the addition of atoms into the particle seeds. During the first and second stages, fast reduction of metal precursor using strong reducing such as sodium citrate is typically preferred to induce more self-nucleation atom events, which ultimately lead to the formation of many small nanoparticle seeds. Meanwhile, mild or weak reducing agent often renders a fewer self-nucleation process which often leads to the creation of less and bigger particles. It is believed that this phenomenon occurs due to the rapid build-up of the atom saturation. Finally, the particles size would gradually increase due to the continuous addition of metal atoms until the activation energy for fluctuation of the crystal structure becomes very high. At this point, mild or weak reduction agent such as hydroxylamine is typically preferred to avoid the fast addition of metal atoms into the particle clusters creating large particles with wide size distribution [13].

Furthermore, UV-Vis absorption was measured to investigate the spectroscopic properties of the as-synthesized Au nanoparticles. Figure 2 shows the UV-Vis spectrum of the as-synthesized Au nanoparticles. Based on the result, it could be seen that the as-synthesized Au nanoparticles absorbed strongly at 525.5 nm due to the surface plasmon resonance (SPR) phenomenon (λ_{SPR}). It is believed that

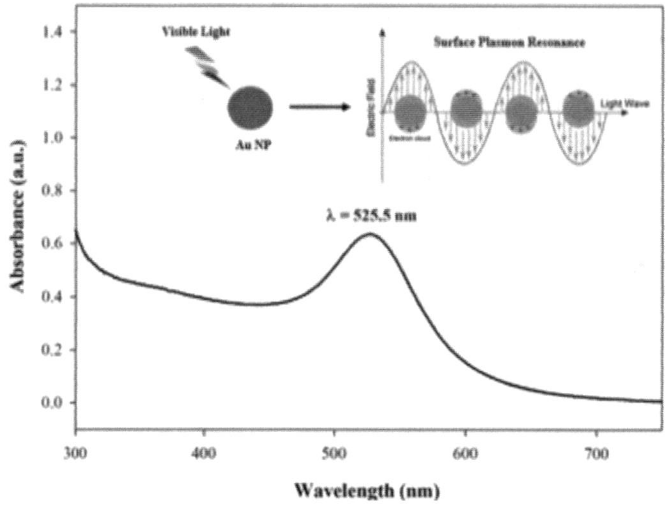

Figure 2. UV-Vis spectrum of the as-synthesized final Au nanoparticles synthesized using seed-mediated method.

the position of this peak is strongly affected by the size and shape of the nanoparticles. Reports have shown that the position of λ_{SPR} is typically shifted to higher wavelength as the particle size increases. Recently, Haiss and co-workers have accurately estimated the diameter of Au nanoparticles using the following equation [10]:

$$d = \frac{\ln\left(\frac{\lambda_{SPR} - \lambda_0}{L_1}\right)}{L_2} \quad (1)$$

where, d is the particle diameter (nm), λ_{SPR} represents the experimental peak position for SPR. Meanwhile, λ_0, L_1, and L_2 are the fit parameters determined from the theoretical values. The values for λ_0, L_1, and L_2 are 512, 6.53 and 0.0216, respectively [10]. Based on the calculation using equation (1), it can, therefore, be estimated that the calculated particle diameter of the as-synthesized was 33 nm, which is very close to the experimental average particle diameter obtained from TEM analysis (~35 nm) (figure 1b).

3.2. Synthesis of anatase TiO_2 nanoparticles

Anatase TiO_2 nanoparticles were synthesized *via a* sol-gel method where TTIP was used as the titanium precursor in basic condition. In this work, XRD analysis was carried out to investigate the crystalline structure of the as-synthesized TiO_2 nanoparticles. Figure 3a shows the XRD diffractogram of the reaction products before and after calcination. Based on the result, it can be observed that an amorphous phase was initially obtained after the sol-gel reaction. Nevertheless, the crystalline structure transformed into anatase after calcination at 450°C for 2 hours. Results demonstrated that all the Bragg peaks suggest that the primary crystal phase of the reaction product after calcination could be unambiguously indexed as anatase phase and no other peaks were obtained for both rutile or brookite phases (JCPDS, No. 21-1272). Furthermore, the broadening of major peak for (101) anatase phase could also be used as an indication for the formation of nano-scale TiO_2 particles grains.

Besides, TEM analysis was also carried out to obtain micrographic information about the morphology, particle size and particle size distribution of the as-synthesized anatase TiO_2 nanoparticles. According to the result (figure 3b), the sol-gel reaction of TTIP at basic condition followed by calcination was able to produce a highly monodisperse spherical anatase TiO_2 nanoparticles with an average particle size diameter of around ~520 nm. . According to literature, the crystal growth mechanism of these nanoparticles was initiated by the formation of small TiO_2 seed particles which cluster together to form a spherical TiO_2 nanocluster *via* LaMer's growth and Ostwald ripening process [11].

IOP Conf. Series: Materials Science and Engineering **763** (2020) 012025 doi:10.1088/1757-899X/763/1/012025

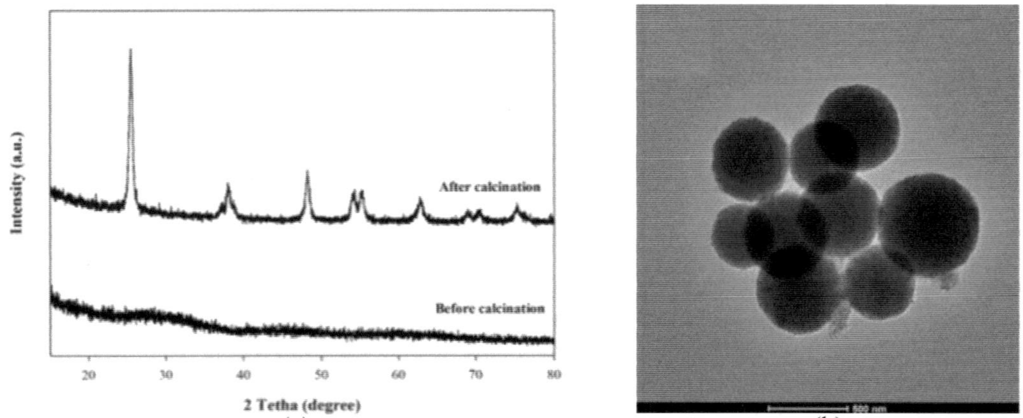

(a) (b)

Figure 3. (a) XRD pattern of the as-synthesized TiO_2 nanoparticles before and after calcination; (b) TEM images of as-synthesized TiO_2 nanoparticles after calcination (scale bar: 500 nm).

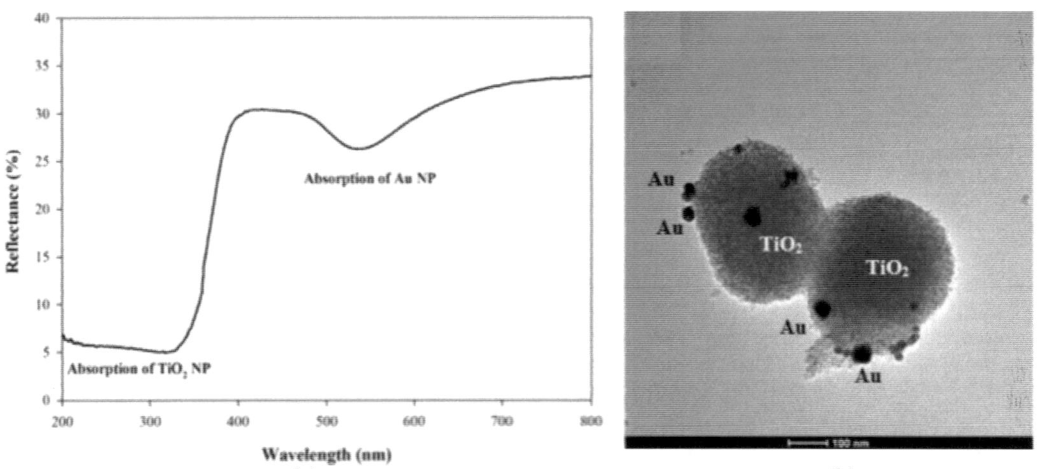

(a) (b)

Figure 4. (a) UV-Vis DRS reflectance and (b) TEM image of the as-synthesized Au-TiO_2 Janus nanostructure (scale bar: 100 nm).

3.3. Synthesis of Au-TiO_2 Janus nanostructure
In this work, the formation of Au-TiO_2 Janus nanostructures was firstly investigated by determining the UV-Vis reflectance of the reaction products obtained from the integration of Au and TiO_2 nanoparticles reaction. Based on the result (figure 4a), it could be observed that Au nanoparticles could easily be integrated onto the surface of the as-prepared TiO_2 nanoparticles at low pH. This is proven by the strong absorption of both UV (200-350 nm) and visible light (525.5 nm). Strong absorption at UV region is believed due to the presence of the energy band gap between TiO_2 valance and conduction band (3.2 eV). Meanwhile, the strong absorption of energy at visible region can be ascribed due to the presence of SPR phenomenon of Au nanoparticles. The successful integration of Au nanoparticles onto the surface of TiO_2 was also supported by the result obtained from TEM analysis. The result from micrographic image analysis of the reaction product shows that few Au nanoparticles (ø = ~ 35 nm) were attached on the surface of TiO_2 nanoparticles (ø = ~ 520 nm) suggesting the successful formation of Janus nanostructure (figure 4b). Besides, TEM image also shows that during the integration of Au nanoparticles on the surface of TiO_2 at low pH, the morphological properties and size of both nanoparticles were pertained and remain unchanged.

4. Conclusions

The successful formation of Au-TiO_2 Janus nanostructure was reported in this work. Here, the as-prepared Au nanoparticles (ø = ~ 35 nm) synthesized *via* seed-mediated method could easily be attached to the surface of anatase TiO_2 nanoparticles (ø = ~ 520 nm) prepared by sol-gel method at low pH. The result from UV-Vis DRS analysis demonstrated that the as-synthesized Au-TiO_2 Janus nanostructures were strongly absorbing at both UV and visible region, i.e., 200–350 and 525.5 nm, respectively. The strong absorption at UV region is believed due to the presence of TiO_2 band gap, whereas SPR phenomenon of Au nanoparticles renders a strong absorption visible region. Moreover, results also show that the morphology and particle size of both Au and TiO_2 nanoparticles remained unchanged during the integration process.

Acknowledgements

This work is financially supported by Universitas Indonesia through Hibah PITTA (No. 2287/UN2/R3.1/HKP.05.00/2018).

References

[1] Lemire C, Meyer R, Shaikhutdinov S K and Freund H J 2004 *Surf. Sci.* **552** 27–34
[2] van Aert S, Batenburg K J, Rossell M D, Erni R and van Tendeloo G 2001 *Nature* **470** 374–7
[3] Yi L, Liu Y, Yang N, Tang Z, Zhao H, Ma G, Su Z and Wang D 2013 *Energy Environ. Sci.* **6** 835–40
[4] Yan C, Nikolova L, Dadvand A, Harnagea C, Sarkissian A, Perepichka D F, Xue D and Rosei F 2010 *Adv. Mater.* **22** 1741–5
[5] Schick I, Lorenz S, Gehrig D, Tenzer S, Storck W, Fischer K, Strand D, Laquai F and Tremel W 2014 *Beilstein J. Nanotechnol.* **5** 2346–62
[6] Tahir M and Amin N S 2015 *Appl. Catal.* B *Environ.* **162** 98–109
[7] Ola O and Maroto-Valer M M 2015 *J. Photochem. Photobiol.* C **24** 16–42
[8] Li H, Bian Z, Zhu J, Huo Y, Li H and Lu Y 2007 *J. Am. Chem. Soc.* **129** 4538–9
[9] Bessekhouad Y, Chaoui N, Trzpit M, Ghazzal N, Robert D and Weber J V 2006 *J. Photochem. Photobiol.* A **183** 218–24
[10] Haiss W, Thanh N T K, Aveyard J and Fernig D G 2007 *Anal. Chem.* **79** 4215–21
[11] Li W, Yang J, Wu Z, Wang J, Li B, Feng S, Deng Y, Zhang F and Zhao D 2012 *J. Am. Chem. Soc.* **134** 11864–7
[12] Polte J 2015 *Cryst. Eng. Comm.* **17** 6809–30
[13] You H and Fang J 2016 *Nano Today* **11** 145–67

The effects of PMSF and cysteine addition into partially purified bromelain from pineapple (*Ananas comosus [L] Merr*) cores to its kinetics behaviour

S Setiasih, M Putri, S Handayani and S Hudiyono

Departement of Chemistry, Faculty of Mathematics and Natural Sciences (FMIPA), Universitas Indonesia, Kampus UI Depok, Depok 16424, Indonesia

Corresponding author's e-mail: setiasih@ui.ac.id

Abstract. Bromelain found in various tissues in the pineapple plant (Ananas Comusus [L] Merr. The bromelain will be taken from pineapple cores for kinetic study. The effect of addition organic compounds, such as phenylmethylsulfonyl fluoride (PMSF) and cysteine on its kinetics behaviour will be determined by proteolytic activities of the bromelain. The stages of this study were started from isolation of crude enzyme, fractionation with ammonium sulphate salt, dialysis, and ion-exchange column chromatography with CM sephadex C-50 for enzyme purification. The obtained enzyme fractions from crude enzyme, ammonium sulphate, dialysis, and CM-sephadex C-50 ion exchange chromatography have specific activities of 45.39, 238.31, 311.16, and 427.13 U/mg, respectively. The purification stages result in increasing of the bromelain specific activities by 9 times compared to the crude enzyme. The Michaelis-Menten constant (Km) value for the purified bromelain was 0.85 % casein and the Vmax value was 0.41 U/min. Addition of PMSF and cysteine compound effect on its Vmax and Km with various trends. The Vmax value will be decreased by addition of PMSF and increased by addition of cysteine. Thus, proteolytic activity of bromelain from pineapple cores can be significantly increased by the addition of cysteine compounds and inhibited by the addition of PMSF.

Keywords: enzymes, bromelain, pineapple, kinetic study enzymes, chromatography

1. Introduction

Protease enzyme is an enzyme that works to hydrolyze the peptide bonds in proteins into oligopeptides and amino acids. Enzyme is a protein for catalyzing chemical reaction. Enzymes will convert the arrangement of molecules into simpler structures, in particular, and works like keys and locks, including in context only from certain molecular arrangements. The work activity is affected by several factors such as inhibitor, cofactor, coenzyme, temperature, pH, concentration of enzyme and substrate [1].

Pineapple fruit (Ananas comosus) is the largest component containing bromelain protease enzyme, which is 95 % cysteine protease mixture. For the people of Indonesia, Pineapple (Ananas comosus (L) Merr) is one type of fruits commonly known and eat freshly. In the process production, pineapple always produces big enough waste around 60 % and 7 % among them is pineapple core which contains bromelain [2]. Meanwhile, bromelain shows therapeutic effect for human body [3].

In this study, the pineapple part of the weed will bind proteolytic activity and its specific activity after bromelain is fractionated using ammonium sulfate to precipitate the protein, then dialyzed to remove it and purified by ion exchange chromatography column. Furthermore, to determine its stability, the bromelain fraction obtained will be determined the proteolytic activity and bromelain kinetics studies of organic compound addition, such as cysteine and phenylmethylsulfonyl fluoride PMSF.

Table 1. Purification of the crude enzyme by using ammonium sulphate and dialysis on bromelain enzyme

Fraction	Volume (mL)	Total		Specific Activity (U/mg)	Purification Factor (times)
		Proteolytic Activity (U)	Protein (mg)		
Crude Enzyme (CE)	150	2171.50	47.87	45.39	-
Fraction 1 (0 %-20 %)	3	54.65	0.62	88.13	1.94
Fraction 2 (20 %-50 %)	10	195.4	0.71	275.21	6.06
Fraction 3 (50 %-80 %)	8	107.47	2.83	38.03	0.84
Residue Enzyme (RE)	171	1234.05	34.61	35.66	0.79
Fraction Dialysis (FD)	8	174.00	0.56	311.16	6.86

2. Experimental methods

2.1. Isolation, ammonium sulphate precipitation, and dialysis

The isolation, ammonium sulfate precipitation, and dialysis stages were conducted according to the method that has been done with modification [4]. Different circumstance is the centrifuge condition, which was operated at 8000 rpm for 20 minutes. The obtained bromelain fractions were kept for the determination of proteolytic activity and total protein content.

2.2. Specific activity determination

The protein content was determined by Lowry Method. Meanwhile, the proteolytic activities of bromelain fractions were performed by Kunitz method using casein as substrate with modification. The specific activity of the enzyme was obtained by dividing the total units of enzyme activity (U) with the total protein content (mg) [5].

2.3. Enzyme purification by ion-exchange chromatography

The enzyme fraction obtained from the dialysis process was purified by an ion-exchange column chromatography method using CM sephadex C-50. The enzyme sample was fed into the chromatography column and eluted using HCl 0.05M pH 8.0. Further elution was then eluted using Tris-HCl 0.05M buffer pH 8.0, which contains NaCl salt solution with gradual concentration of 0.1M, 0.2M, and 0.3M [6]. The elution results were collected in a fractional reaction tube with a volume of 5 mL/tube. Each collected fraction was determined for its protein content by measuring its absorbance using a UV-Vis spectrophotometer at a wavelength of 280 nm and its proteolytic activity by Kunitz method. Fractions with the highest proteolytic activity are classified to specify their specific activity.

2.4. The effect of various compounds on bromelain activity

The compounds used in this study were PMSF and cysteine. The enzyme solutions, which contain the compounds, were tested for their proteolytic activities with variation concentration of casein substrate to determine Michaelis-Menten constant (Km) value and maximum velocity (Vmax) value. The values were determined by plotting its concentration substrate and velocity data to Lineweaver-Burk plot method.

3. Results and discussion

3.1. Isolation, ammonium sulphate precipitation, and dialysis

The bromelain fraction of ammonium sulfate 20 %-50 % has the highest specific activity with the purity level of 6.06 times compared to the crude enzyme (table 1). The fraction 2 still contain ammonium sulfate salt to be purified, one of the best methods to remove the salt is by dialysis method. The salt and small non bromelain protein will diffuse through cellophane membrane. From the table 1, dialysis fraction has higher purity level than the fraction 2 with specific activity of 311.16 U/mg.

Table 2. Specific activity-specific data of bromelain fraction in the purification step by ion-exchange column chromatography

Fraction	Total		Specific Activity (U/mg)	Purification Factor (times)
	Proteolytic Activity (U)	Protein (mg)		
AP 1 (Fraction 9-16)	6.87	0.02	325.88	7.18
AP 2 (Fraction 25-39)	4.47	0.01	379.41	8.36
AP 3 (Fraction 47-55)	5.46	0.01	461.03	10.15

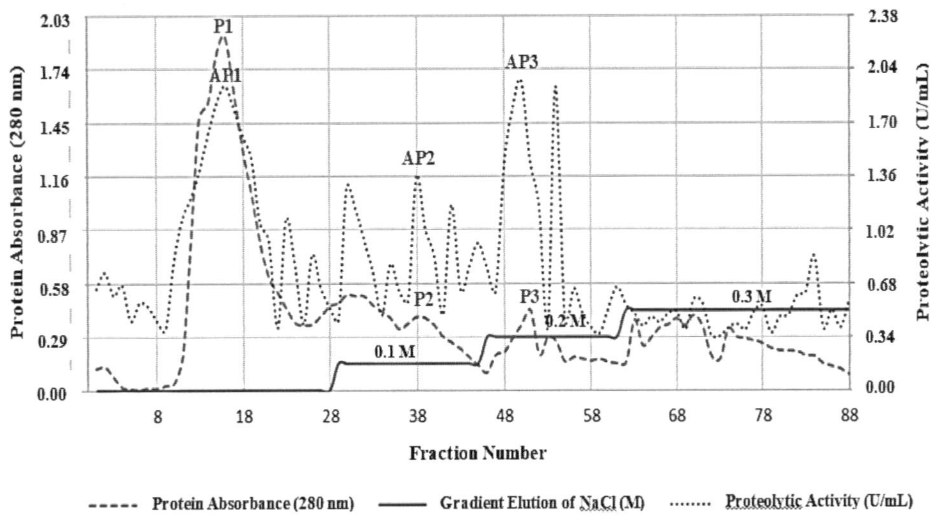

Figure 1. Results of chromatogram of CM sephadex C-50 chromatography column. Separation Conditions: column size 3 x 20 cm, matrix volume 141.3 mL, elution with buffer solution, flow rate 1.5 mL/minute, eluent volume 5 mL/tube.

3.2. Ion exchanger column chromatography using CM sephadex C-50

This purification step is carried out on the fraction of the dialysis result. The stationary phase used in this study is the CM sephadex C-50 resin matrix, which is a weak cation exchange chromatography column (negatively charged resin). From this purification obtained as much as 88 fractions of result of column. Furthermore, the fractions were measured at maximum protein absorbance 280 nm. Furthermore, each fraction is measured for its proteolytic activity based on Kunitz's method. The results of both measurements are described in the form of chromatograms and obtained 3 protein peaks (figure 1).

All fractions around the peaks are combined and determined for its specific activity. Specific activity data of mixture fraction is shown in table 2. The measurement results showed the highest specific activity at the peak fraction of AP3 protein of 461.03 U/mL with purity level of 10.6 times compared to the crude enzyme.

3.3. Enzymatic kinetics

The determination of kinetics parameters was performed on purification isolation enzymes based on Kunitz method under optimum conditions by varying casein substrates (0.25, 0.5, 0.75, 1.0, 1.25, and 1.5) %. In figure 2, we can see the value of proteolytic activity each concentration of casein substrate. The data showed that 1% of the substrate concentration had the highest activity, so that each active side of the enzyme had bound to the substrate optimally. Determined the enzymatic rate of the Michaelis-Menten constant to determine the hyperparabolic relationship between substrate

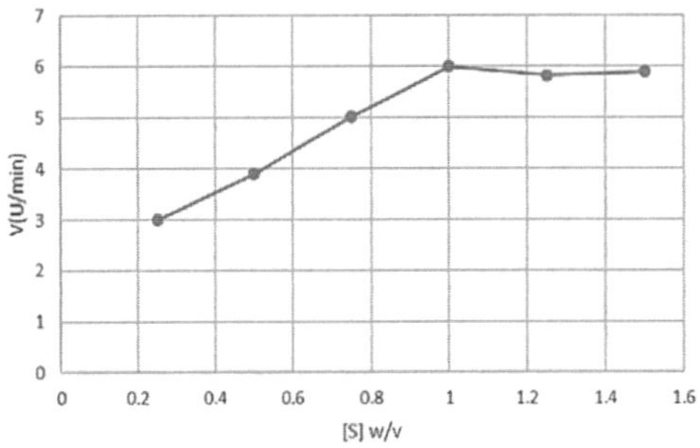

Figure 2. Graph of enzymatic reaction rate to substrate concentration

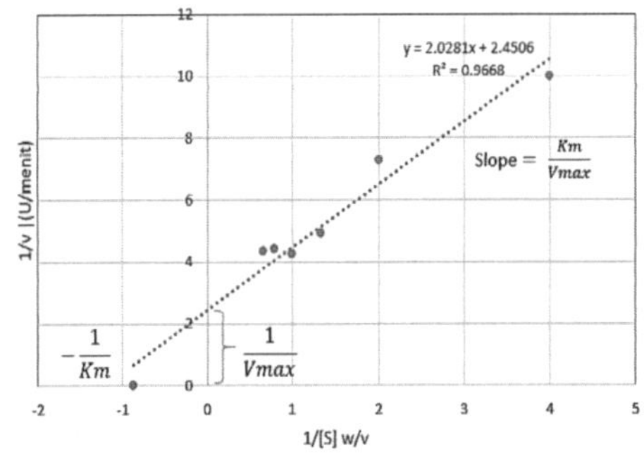

Figure 3. Graph of Lineweaver-Burk plot for the hydrolysis of the casein substrate

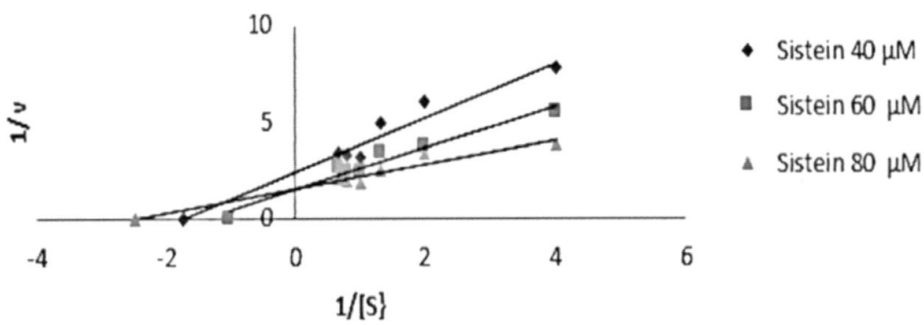

Figure 4. Graph of Lineweaver-Burk Plot for addition cysteine

concentration and enzymatic rate. In addition, (0–1.0) % of the substrate is accompanied by an enzymatic rate rise, whereas in the addition of 1.25 % to 1.50 % the enzymatic rate is relatively constant, since the active side of the enzyme has been saturated by the substrate.

The Km value of an enzyme can be calculated by the Lineweaver-Burk equation obtained from the Michaelis-Menten equation which then produced a Lineweaver-Burk diagram which can be seen in figure 3, the Lineweaver-Burk Plot for the hydrolysis of the casein substrate by the result of the refractionation bromelain dialysis. The Michaelis-Menten (Km) constant value for the purified enzyme was 0.85 % Casein and the Vmax value was 0.41 U / min.

3.4. Enzymatic kinetics after addition of cysteine and PMSF compounds

Proteolytic activity of bromelain after the addition of cysteine solution showed an increasing trend along with the increase of activator concentration. This is due to alteration of bromelain conformation consisting of sulfhydryl group at its active center establishing an easier way for the substrate to interact with enzyme molecules. Based on the Line-WeaverBurk plot in figure 4, bromelain with addition of 40 μM, 60 μM, and 80 μM have Km value of 0.59, 0.96, and 0.40 % casein and Vmax 0.65, 0.41, 0.64 U/min.

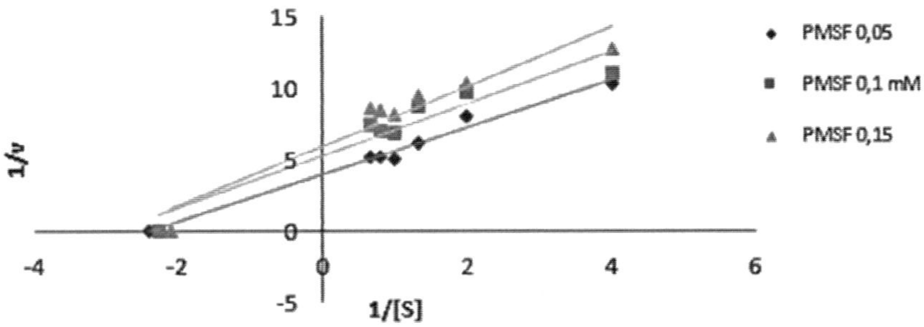

Figure 5. Graph of Lineweaver-Burk Plot for addition PMSF

In spite of the effect of PMSF on bromelain activity showed a decrease of activity along with increasing concentration of PMSF compound with variation 0.05 mM, 0.10 mM and 0.15 mM (figure 5). This decrease in bromelain activity may be due to a conformational change that causes the enzyme's center of activation to undergo a modification so that the enzyme interaction with the substrate is inhibited. The results of this test indicate that PMSF is a strong enough inhibitor against the enzyme bromelain, where at a concentration of 0.15 mM PMSF has been able to provide a relatively large inhibitory effect on proteolytic activity of enzymes. Based on the data, the Michaelis-Menten (Km) constant value for enzymes with the addition of 0.05, 0.10 and 0.15 mM PMSF compound were obtained by 0.42, 0.45 and 0.48 % casein and Vmax value of 0.25, 0.12 and 0.17 U/min, respectively.

4. Conclusions

The bromelain from pineapple cores has been successfully isolated and partially purified by ammonium sulfate precipitation, dialysis, and ion-exchange chromatography using CM-sephadex The specific activities were sequentially increased by these purification stages and the highest purity level of the bromelain was obtained in AP3 fraction from ion-exchange chromatography. The optimum concentration of casein substrate for bromelain's proteolytic activity determination is in 1% of casein. The proteolytic activity of bromelain from pineapple cores can be significantly increased by the addition of cysteine compounds and inhibited by the addition of PMSF.

Acknowledgements

This project was funded by Biaya Operasional Perguruan Tinggi Negeri (BOPTN) DIKTI 2016.

References

[1] Lehningher A L 1982 *Principles of Biochemistry* (New York: Worth Publisher)
[2] Oktaviani R, Rahayu K and Suhartatik N 2016 *Jurnal Teknologi dan Industri Pangan* **1** 134–43
[3] Costa H B, Fernandesa P M B, Romão W and Ventura J A 2014 *Ind. Crops Prod.* **59** 163–8
[4] Setiasih S, Prabowo H A, Budianto E and Hudiyono S 2018 *J. Appl. Pharm. Sci.* **8(10)** 017–24
[5] Musfiroh F F, Setiasih S, Handayani S, Hudiyono S and Ilyas N M 2018 *IOP Conf. Ser.: Mater. Sci. Eng.* **299** 012017
[6] Gautam S S, Mishra S K, Dash V, Goyal A K and Rath G 2010 *Thai J. Pharm. Sci.* **34** 67–76

Study of orthophosphate detection method using DGT (diffusive gradients in thin films) device with modified binding layer Fe-chitosan-bentonite biocomposite in aquatic system

S Andini[1], A Saefumillah[1], M A E Hafizah[2] and R T Yunarti[1]

[1] Department of Chemistry, Faculty of Mathematics and Natural Sciences (FMIPA), Universitas Indonesia, Kampus UI Depok, Depok 16424, Indonesia

[2] PT Evonik Sumi Asih, Tambun, Bekasi 17510, Indonesia

Corresponding author's e-mail: asep.saefumillah@sci.ui.ac.id

Abstract. In this work, a study of orthophosphate detection technique in the aquatic system has been conducted. The detection technique used in this study was Diffusive Gradients in Thin Films (DGT) with Fe-Chitosan-Bentonite (Fe-CSBent) biocomposite as the binding agent. The phosphate species diffused through the polyacrylamide gel and then bound by Fe-CSBent as the binding agent on the binding layer. The adsorption of phosphate by Fe-CSBent is through electrostatic attraction, inner sphere complexation, and ion-exchanged. The synthesized Fe-CSBent biocomposite was characterized using Fourier Transform Infrared (FTIR), Scanning Electron Microscopy (SEM), Energy Dispersive X-Ray Spectroscopy (EDS) and X-Ray Diffraction (XRD) analysis, and the synthesized diffusive gel and binding gel were characterized using FTIR analysis. In this experiment, we examined the DGT-Fe-CSBent capability in orthophosphate binding with parameters such as the deployment time, pH of the solution and the presence of co-existent anion i.e. sodium tripolyphosphate (STPP). The mass of bound orthophosphate can be calculated after eluted the binding gel with acid and then measured with the molybdenum blue method using UV-Vis spectrophotometry. Based on the experiment, DGT-Fe-CSBent bound the orthophosphates more efficiently in concentration of 2.2970 μg/mL with 90.208 % efficiency, compared to DGT-CSBent in concentration of 1.7333 μg/mL with 79.874 % efficiency. It was also known that the presence of STPP influenced the amount of orthophosphate concentration bound by Fe-CSBent in the binding gel. DGT-Fe-CSBent device that was made to be selective in predicting bioavailable phosphate, but the presence of other phosphate species such as sodium tripolyphosphate may be influenced orthophosphate bound by DGT.

Keywords: phosphate, diffusive gradients in thin films (DGT), biocomposite, chitosan, bentonite

1. Introduction

Phosphate is an important compound for aquatic ecosystems because it is one of the nutrients needed for the growth of aquatic plants [1]. Eutrophication is a high nutrient level in aquatic systems, such as nitrogen (N) and phosphate (P) causing excessive algal blooms. This can cause the low content of dissolved oxygen in the water resulting in the demise of life in the aquatic ecosystem. The high phosphates concentration in aquatic ecosystems can come from agricultural fertilizers, detergents, soaps, and industrial activity [2]. The limit of phosphate tolerance in water is <0.5 mg/L [3]. That is why an effective technique for the detection of phosphates in aquatic systems is needed.

IOP Conf. Series: Materials Science and Engineering **763** (2020) 012027 doi:10.1088/1757-899X/763/1/012027

Various technologies of phosphate detection have been found and used, such as microbial remediation, adsorption, ion exchange, and others. The adsorption technique is more desirable because it is more effective, easy to operate, economical and selective against pollutants in aquatic systems [4]. Another approach for phosphate detection in the environment is the diffusive gradients in thin films (DGT) system. DGT is a system made for the measurement of available phosphate (bioavailable) where the DGT can absorb bioavailable phosphate in water in accordance with phosphate uptake carried out by biota such as algae. The DGT system has been used for the measurement of the dissolved phosphate in-situ in waters. This is due to the interaction of phosphate species in water, which can cause the concentration of phosphate species to change when the sample is stored [5]. A DGT device comprises three layers: cellulose nitrate membrane, diffusive layer in the form of polyacrylamide gel, and binding layer containing a binding agent or adsorbent in a polyacrylamide gel. The mechanism of DGT system, i.e. phosphate diffused through cellulose nitrate membrane and diffusive gel, then bound by the binding agent or adsorbent in the binding gel. The success of the DGT system is determined by the performance of the binding gel containing the adsorbent according to the analyte studied [6].

In this study, we used DGT system for phosphate binding with Fe-Chitosan-Bentonite (Fe-CSBent) biocomposite as binding agent in binding layer. Bentonite is a type of smectite clay, which has an isomorphic replacement of Si^{4+} with Al^{3+} on tetrahedral sites and Al^{3+} with Fe^{3+} and Mg^{2+} on octahedral sites, which produces net electrical charge on its surface [7]. Chitosan is a biopolymer derived from natural crustacean shells. Chitosan is a derivative product of chitin compounds obtained from the process of deacetylation of chitin under alkaline conditions (NaOH) [8]. To increase the capacity of phosphate uptake in acidic medium, modification of active amine group ($-NH_2$) was modified in chitosan such as functionalization, grafting, cross-linking and biocomposite forms [4]. In this study, Chitosan-Bentonite biocomposite adsorbents were used and loaded with Fe^{3+} to improve the stability of biocomposite and phosphate uptake capacity in water. The synthesized Fe-CSBent biocomposite were characterized using Fourier Transform Infrared (FTIR), Scanning Electron Microscopy (SEM), Energy Dispersive X-Ray Spectroscopy (EDS) and X-Ray Diffraction (XRD) analysis, and the synthesized diffusive gel and binding gel were characterized using FTIR analysis. Phosphate binding test was conducted on DGT-Fe-CSBent with various parameters, i.e. pH and presence of co-existent anion i.e. sodium tripolyphosphate (STPP). The mass of bound phosphate can be calculated after eluted the binding gel with acid and then measured with molybdenum blue method using UV-Vis spectrophotometry.

2. Materials and methods

2.1. Materials

Chitosan, bentonite (PT. Madani Alam Lestari), cellulose nitrate membrane (GE Healthcare Life Sciences), $FeCl_3.6H_2O$, glacial acetic acid, N,N'-methylenebisacrylamide, acrylamide, ammonium persulphate, N,N,N',N'-Tetramethylethylenediamine (TEMED) 99 %, potassium antimony tartrate, ammonium molybdate, H_2SO_4 (p), and ascorbic acid.

2.2. Synthesis of Fe-CSBent biocomposites

The preparation of Fe-CSBent biocomposites is based on the procedure of Kumar and Viswanathan [4]. Chitosan (CS) was dissolved in 2 % acetic acid solution and stirred for 2 hours at 40 °C. Na-Bentonite (Bent) was dispersed with double distilled water then stirred for 2 hours. Slowly added the bentonite dispersion into the chitosan solution, then heated at 60 °C for 24 hours with high intensity stirring. After 24 hours, the metal ion was loaded by adding $FeCl_3.6H_2O$ 3 % (w/v) and stirred for 2 hours. The obtained Fe-CSBent biocomposite slurry was kept for 24 hours for the aging process, followed by filtered the slurry and washed with double distilled water. The biocomposite was dried and mashed into powder and sieved to obtain a uniform biocomposite size.

2.3. Gel preparation and mounting

Preparation of gel solution was based on Zhang *et al.* [5] procedure. Combined acrylamide and crosslinkers N, N'-methylenebisacrylamide until homogeneous then added initiator ammonium persulfate and catalyst TEMED. The gel solution was immediately casted between two glass plates separated by 0.1 mm wide glass and heated at 42 ± 2 °C for 45 minutes. The binding gel was made by mixing the Fe-CSBent biocomposite and gel solution and the same procedure was performed as

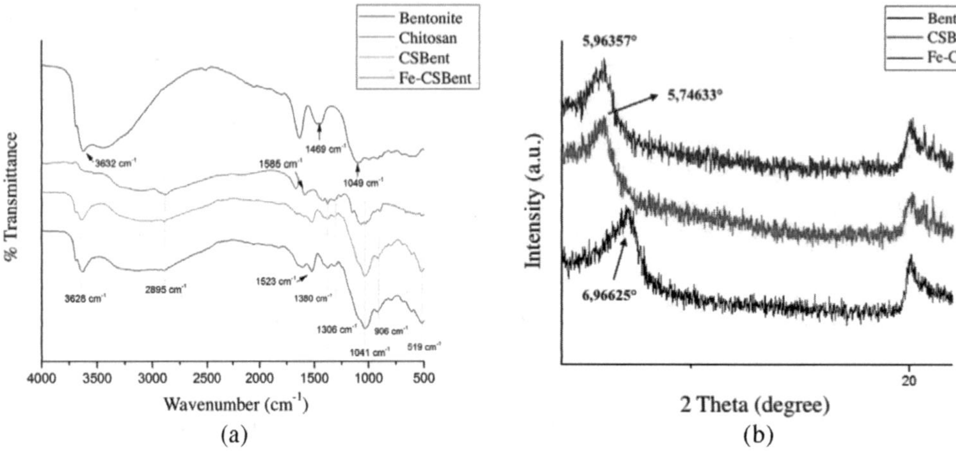

Figure 1. (a) FTIR characterization of Bentonite, Chitosan, CSBent and Fe-CSBent
(b) XRD characterization of Bentonite, CSBent and Fe-CSBent

diffusive gel. Gel diffusive and binding gel were hydrated with double distilled water for at least 24 hours before use. Throughout this hydration, the gel will be expanded to a steady shape and also washed the impurities on the gel. The diffusive gel and binding gel were stored in 0.1M $NaNO_3$ and the double distilled water, respectively.

Prior to the installation of the DGT component, the gels (diffusive gel and binding gel) was cut with a 25 mm diameter DGT cutter and the cellulose nitrate membrane was soaked in double distilled water. The binding gel was placed first on the DGT device, followed by diffusive gel and cellulose nitrate membrane. The DGT device was then closed properly.

2.4. Measurement of DGT-Fe-CSBent performance and response

Orthophosphate uptake by DGT-Fe-CSBent and DGT-CSBent used 2 ppm KH_2PO_4 solution. The DGT device was inserted into the solution for 24 hours (optimum time). DGT-Fe-CSBent performance test included 4 parameters, homogeneous of the binding gel, deployment time, pH of the solution, and presence of co-existent anion i.e. sodium tripolyphosphate (STPP).

2.5. Analytical methods

After immersion of the DGT device, the binding gel was eluted in H_2SO_4 0.25M for 16 hours. Initial phosphate concentrations and eluted concentrations were measured with the molybdenum blue method using UV-Vis spectrophotometry. The concentration of phosphate in the DGT can be calculated using the equation derived from Fick's First Law [5]:

$$C = \frac{M\,\Delta g}{D\,A\,t}$$

2.6. Characterization

The synthesized biocomposites were characterized by Fourier Transform Infrared (FTIR), X-Ray Diffraction (XRD), Scanning Electron Microscopy (SEM) and Energy Dispersive X-Ray Spectroscopy (EDS).

3. Results and discussion

3.1. Characterization of biocomposites

FTIR spectra of bentonite, chitosan, CSBent and Fe-CSBent are shown in figure 1a. The band observed around 3600 cm^{-1} indicated the presence of O–H stretch of bentonite, CSBent and Fe-CSBent. At 2895 cm^{-1} indicated the presence of C–H stretch band of chitosan, CSBent, and Fe-CSBent. In the chitosan spectra, it showed that N–H bending band at 1585 cm^{-1}, but it showed in the CSBent and Fe-CSBent spectra that the N–H bending band is at 1523 cm^{-1}. This is due to $-NH_2$ in chitosan protonated into $-NH_3^+$ during biocomposite synthesis [9]. At 1041 cm^{-1}, it showed the presence

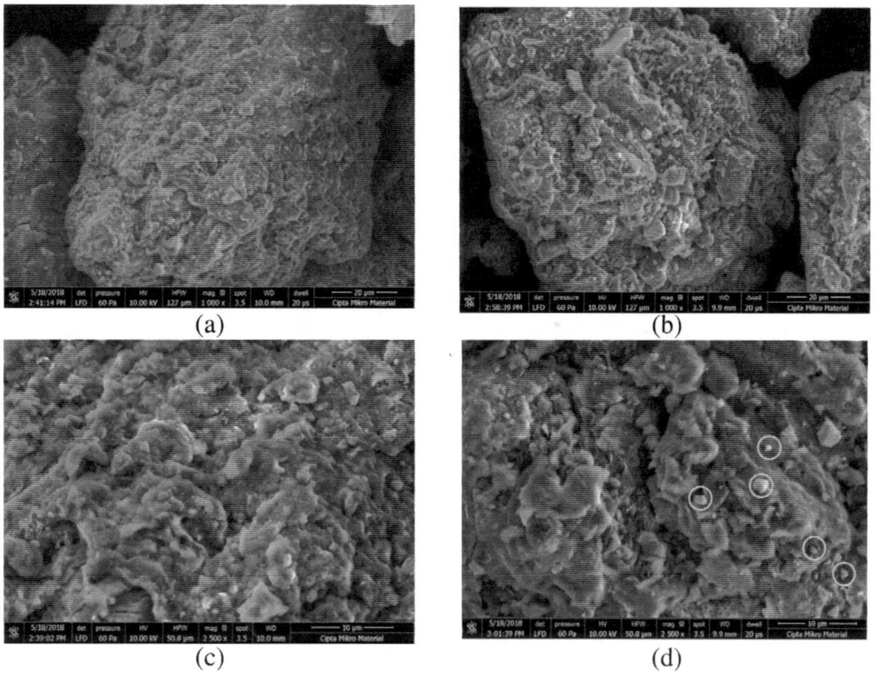

Figure 2. SEM Image of (a) Fe-CSBent before adsorption phosphate 1000 x magnification, (b) Fe-CSBent after adsorption phosphate 1000 x magnification, (c) Fe-CSBent before adsorption phosphate 2500x magnification, and (d) Fe-CSBent after adsorption phosphate 2500 x magnification.

of C–O–C band in chitosan, CSBent, and Fe-CSBent [10]. The presence of Al–Al–OH (stretching deformation) and Al–O(OH)–Al (stretching vibration) bands are found in 906 cm^{-1} [11] and Si–O–Al (vibration bands) and Si–O– (bending and stretching vibration) bands are found in 519 cm^{-1} in CSBent and Fe-CSBent [12].

Based on diffraction pattern that are shown in figure 1b, it showed that the peak of bentonite shifted between bentonite before and after synthesized into biocomposite. According to Bragg's law, the smaller of θ (degree) the greater the value of d. This explains the shift in the diffraction degree of bentonite at $2\theta = 6.96625°$ and synthesized CSBent at $2\theta = 5.74633°$ which caused by the increasing interlayer distance between atoms in bentonite after being synthesized into biocomposite. This proves that the expansion of bentonite structure after synthesis.

The SEM image in figure 2a and figure 2c show the presence of pores on the surface morphology of Fe-CSBent before phosphate adsorption with 1000x and 2500x magnification, respectively. Figure 2b and figure 2d show that on the surface of Fe-CSBent after phosphate adsorption is asymmetrical and believed to be succeeded in phosphate adsorption. The SEM-EDS results (figure 3a and figure 3b) show an increase in element O and the presence of element P in the Fe-CSBent sample indicated that the phosphate was successfully adsorbed by Fe-CSBent.

3.2. Characterization of diffusive and binding gel

The FTIR spectra of acrylamide, diffusive gel, CSBent gel, and Fe-CSBent gel are shown in figure 4. The FTIR spectra of acrylamide show the characteristic bands at 3180–3350 cm^{-1}, 1673 cm^{-1}, 1612 cm^{-1}, and 1299 cm^{-1} indicating the presence of N–H stretch, C=O stretch, C=C stretch, and C–N stretch, respectively. The FTIR spectra of diffusive gel, CSBent gel, and Fe-CSBent gel show the characteristic bands at 3180 cm^{-1}, 2900 cm^{-1}, 1673 cm^{-1}, and 1296 cm^{-1}, indicating the N–H stretch, C–H stretch, C=O stretch, and C–N stretch, respectively. In the acrylamide FTIR spectra, it is shown the existence of the C=C band but no existence of the C=C band in the diffusive gel and binding gel. This is because polymerization of polyacrylamide gel has been successfully performed.

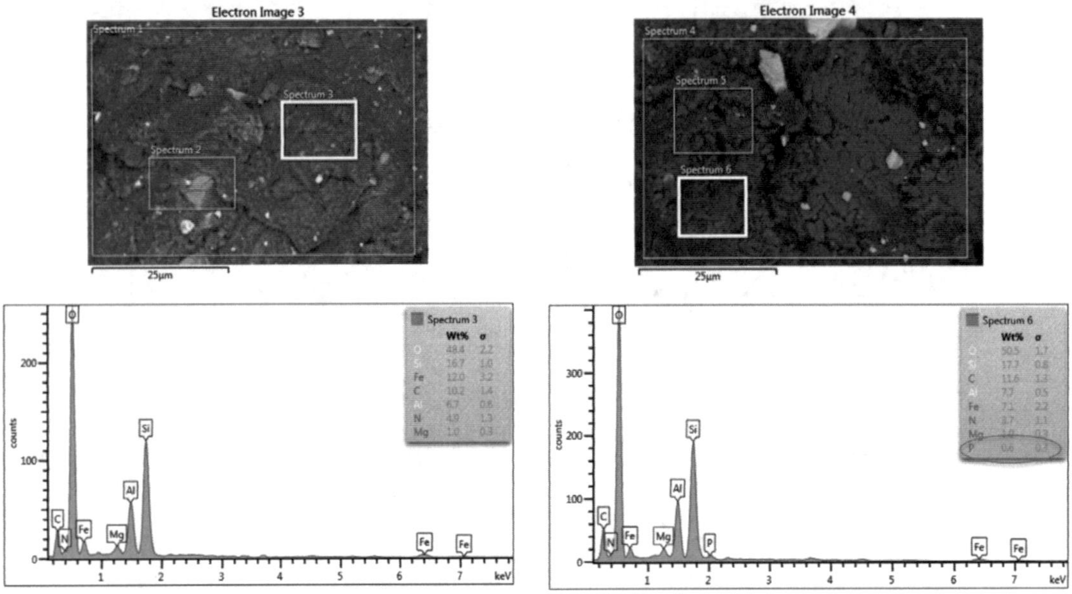

(a) (b)

Figure 3. SEM-EDS of (a) Fe-CSBent before adsorption phosphate and (b) Fe-CSBent after adsorption phosphate.

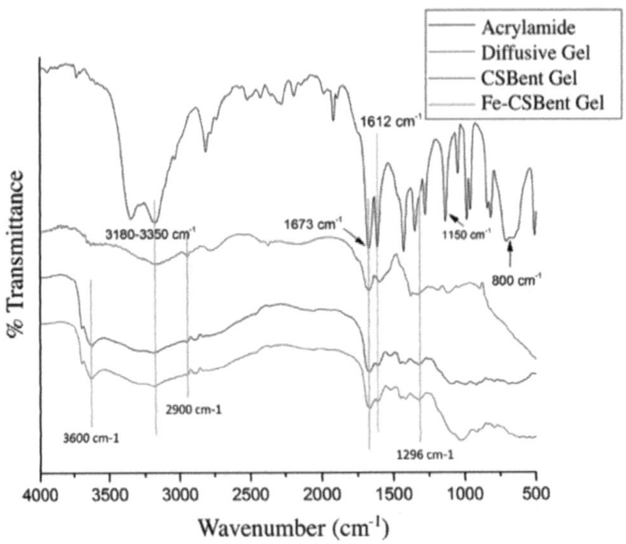

Figure 4. FTIR characterization of acrylamide, diffusive gel, CSBent gel, and Fe-CSBent gel

3.3. Mechanism of phosphate binding by DGT-Fe-CSBent

The mechanism of phosphate binding by DGT is the phosphate species diffused through the diffusive layer and then bound by the adsorbent in the binding gel. The use of cross linker N, N'-methylenebisacrylamide is to create gel that has pores, so only orthophosphates can be diffused through the gel. The diffused orthophosphate is then bound by Fe-CSBent. Figure 5 shows the binding of phosphates by Fe-CSBent through three mechanisms: electrostatic interaction between the adsorbent surface and the phosphate, complexation of Lewis metal ion with the phosphate, and ion exchange. The chitosan surface ($-NH_2$) at the low pH condition (pH<5.0) will be protonated ($-NH_3^+$),

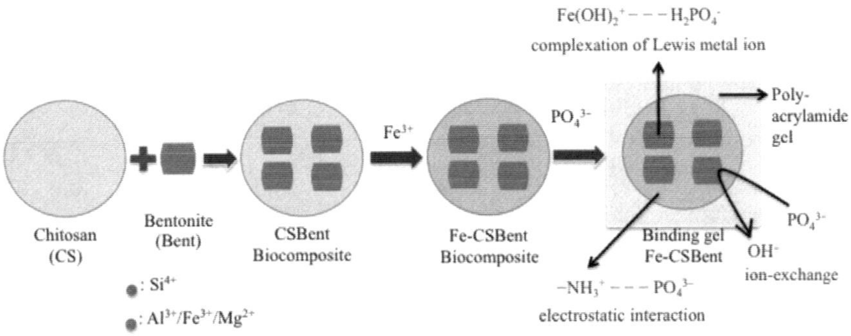

Figure 5. Mechanism of phosphate binding by DGT-Fe-CSBent.

which causes the chance of electrostatic interaction between phosphate ions and the surface of the adsorbent [10,13]. The metal ion, in this case Fe^{3+} is considered to act as Lewis acid that will combine with negatively charged phosphate species that are considered as strong Lewis bases. The inner sphere complex formation between $H_2PO_4^-$ and the hydrated metal oxide in the synthesized biocomposite, $Fe(OH)_2^+$ could be occurred [14]. The hydroxyl group (–OH) in bentonite will be replaced with phosphate ion by ion-exchange in Fe-CSBent biocomposite, the possible mechanism of phosphate binding by DGT-Fe-CSBent [4].

3.4. DGT-Fe-CSBent performance and response
Homogeneity tests were performed to determine whether the binding gel used in the test was homogeneous or not. Based on the experiment, C_{DGT} values from the four repetitions on both gels obtained have almost the same results (figure 6a). The average orthophosphate concentration bound by DGT-CSBent was 1.7333 µg/mL with standard deviation of 0.1784 and 79.874 % efficiency. The average orthophosphate concentration bound by DGT-Fe-CSBent was 2.2970 µg/mL with a standard deviation of 0.1105 and 90.208 % efficiency. DGT-Fe-CSBent binds orthophosphates more efficiently compared to DGT-CSBent. It can be concluded that both CSBent and Fe-CSBent gels were homogeneous enough.

The optimum deployment time of DGT obtained in this experiment was at 24 hours, after 24 hours the C_{DGT} of orthophosphate values were almost stagnant because of the lack of empty sites available on the Fe-CSBent binding gels so that have reached the equilibrium stage (figure 6b). The influence of pH solution was done to determine the effect of pH on phosphate binding and to determine the tolerable pH range in phosphate binding by DGT. Phosphate binding test done at pH 3–6 showed good results. At pH>6, there was a decrease in phosphate binding efficiency by DGT due to the electrostatic repulsion between negatively charged phosphate ions and hydroxyl ions (OH^-). At pH< 6, electrostatic attraction occurred between the negatively charged phosphate ions and the protonated biocomposite surface resulting in maximum phosphate binding [4].

3.5. Effect of co-existing anion (sodium tripolyphosphate) on orthophosphate binding by DGT
Phosphates in nature are available in the form of orthophosphates, polyphosphates (pyro-, meta-, etc.), as well as organic phosphates. Phosphate analysis can be classified into two methods, first is the total phosphate analysis, i.e. conversion of phosphate form to orthophosphate by oxidation destruction process and second is soluble phosphate analysis, i.e. measuring the phosphate from orthophosphate without destruction [15]. The destruction process used strong acid that is H_2SO_4 and $HClO_4$. The destruction process will hydrolyze the polyphosphate into orthophosphate, which its concentration as the total phosphate can be determined by the molybdenum blue method [16]. Research using DGT method is a method made for the measurement of bioavailable phosphate where DGT can absorb the bioavailable phosphate in water in accordance to the phosphate uptake by biota such as algae.

It can be seen that C_{DGT} of orthophosphate (without destruction) increases by the addition of STPP concentration (figure 7). This is due to the STPP which may have diffused through the diffusive gel and bound by the Fe-CSBent on the binding gel. To prove that STPP diffused through the diffusive gel and bound by Fe-CSBent on the binding gel, a comparison of C_{DGT} of orthophosphate (without

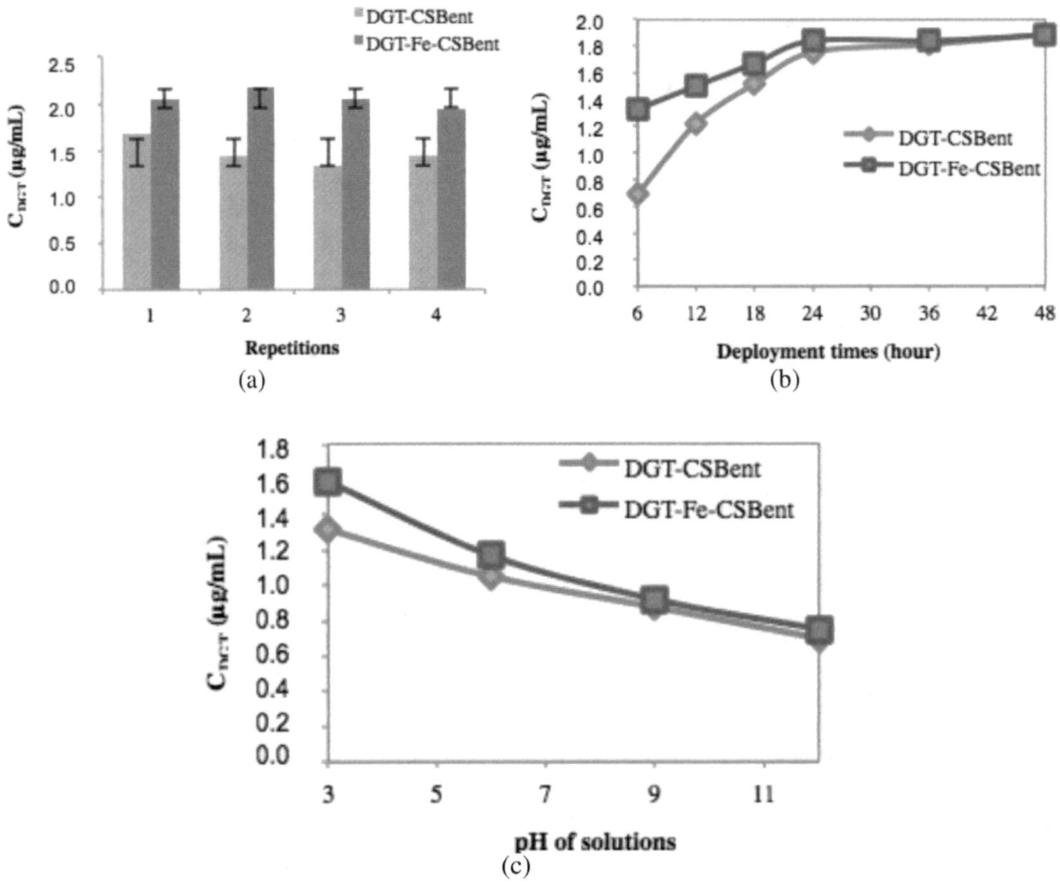

Figure 6. DGT performance and response with various parameters
(a) homogeneous of the binding gel, (b) deployment times, and (c) pH of solutions.

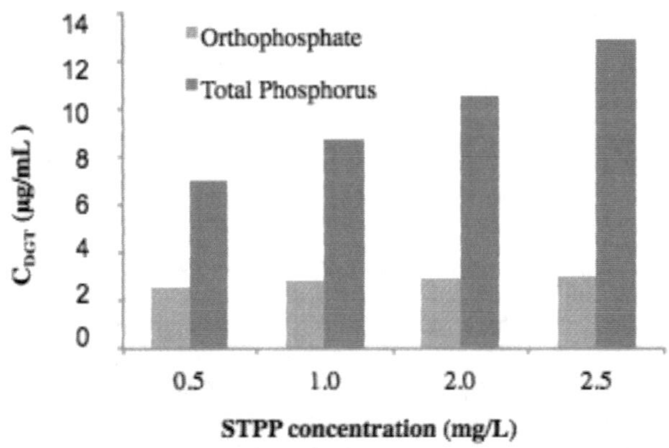

Figure 7. Effect of co-existing anion (STPP) on orthophosphate binding by DGT

destruction) and total phosphate (with destruction) was performed. It appears that the total phosphate concentration (orthophosphate and polyphosphate of STPP) obtained from the destruction analysis is greater than the orthophosphate concentration (without destruction). It can be concluded that in the elution solution there is also STPP bound to Fe-CSBent in the binding gel. Although STPP diffuses into binding gel, orthophosphate is the bioavailable phosphate in the aquatic system in accordance to phosphate uptake by biota such as algae.

4. Conclusions

Modification of binding gel with biocomposite in DGT can bind with phosphate. DGT-Fe-CSBent binds with orthophosphates more efficiently compared to DGT-CSBent. DGT-Fe-CSBent device, which was made to be selective in predicting the bioavailable phosphate, but the presence of other phosphate species such as polyphosphate, may influence the orthophosphate binding by DGT.

Acknowledgements

The authors are thankful to Hibah PITTA Universitas Indonesia 2018 for providing monetary support to carry out this research work.

References

[1] Lu J, Liu H, Zhao X, Jefferson W, Cheng F and Qu J 2014 *Colloids Surf. A* **455** 11–8
[2] Wang Z, Xing M, Fang W and Wu D 2016 *Appl. Surf. Sci.* **366** 67–77
[3] Xu Y, Dai Y, Zhou J, Xu Z P, Qian G and Lu G Q M 2010 *J. Mater. Chem.* **20** 4684–91
[4] Kumar I A and Viswanathan N 2017 *Int. J. Biol. Macromol.* **104** 1539–47
[5] Zhang H, Davidson W, Gadi R and Kobayashi T 1998 *Anal. Chim. Acta* **370** 29–38
[6] Rumhayati B 2010 *Jurnal Ilmu Dasar* **11** 160–6
[7] Onal M and Sarikaya Y 2007 *Power Technol.* **172** 14–8
[8] Rinaudo M 2006 *Prog. Polym. Sci.* **31** 603–32
[9] Moussout H, Ahlafi H, Aazza M and Amechrouq A 2018 *Thermochim. Acta* **659** 191–202
[10] Anirudhan T S, Rijith S and Tharun A R 2010 *Colloids Surf.* A *Physicochem. Eng. Asp.* **368** 13–22
[11] Zhao D L, Feng S J, Chen C L, Chen S H, Xu D and Wang X K 2008 *Appl. Clay Sci.* **41** 17–23
[12] Yuan P, He H, Bergaya F, Wu D, Zhou Q and Zhu J 2006 *Microporous Mesoporous Mater.* **88** 8–15
[13] Chen P H, Hwang Y H, Kuo T Y, Liu F H, Lai J Y and Hsieh H J 2007 *J. Med. Biol. Eng.* **27** 23–8
[14] Blaney L M, Cinar S and SenGupta A K 2007 *Water Res.* **41** 1603–13
[15] Clesceri L S, Greenberg A E and Eaton A D 1998 *Standard Methods for the Examination of Water and Wastewater* 20th ed (Washington DC: American Public Health Association)
[16] Nagul E A, McKelvie I D, Worsfold P and Kolev S D 2015 *Anal. Chim. Acta* **890** 60–82

The effect of sorbitan monooleate as surfactant in preparation of polyblend polylactic acid and polycaprolactone microspheres

A N Ryofi and E Budianto

Department of Chemistry, Faculty of Mathematics and Natural Sciences (FMIPA), Universitas Indonesia, Kampus UI Depok, Depok 16424, Indonesia

Corresponding author's e-mail: emilb@ui.ac.id

Abstract. Polyblend of polylactic acid (PLA) and polycaprolactone (PCL) microspheres with polymodal size distribution were synthesized through an emulsification-solvent evaporation method in water-in-oil (W/O) emulsions. Sorbitan monooleate was used as emulsion stabilizer and size controller in preparation of the microspheres. The effect of sorbitan monooleate addition in preparation on the size of the microspheres was studied by varying the volumes of sorbitan monooleate, the stirring speeds of emulsion formation, and dispersion times. The microspheres were characterized using Fourier Transform Infrared (FT-IR) and Particle Size Analyzer (PSA). The results show that there is no chemical interaction between PLA and PCL. This phenomenon can be seen from the FT-IR spectrum, which shows both PLA and PCL component. The particle size of microspheres decreased with the increasing volume of surfactant, emulsion stirring speed, and dispersion time. The size of microspheres obtained ranges from 0.4 μm to 10 μm. Physical forms of microspheres were observed using optical microscope. The result shows that some microspheres are spherical and others are in aggregates form.

Keywords: solvent evaporation, polyblend, polylactic acid, polycaprolactone, sorbitan monooleate

1. Introduction

Microspheres are very important in drug delivery system due to their tiny size as well as other superior characteristics. Microspheres are well-known as a perfect solution to formalize a potent dose for a drug control release simulation [1]. Microsphere is a term used for small spherical particles with diameter 1–1000 μm [2]. Microspheres are typically free flowing solid powders, which consist of proteins or synthetic polymer, known to be biodegradable in nature and can be delivered by several routes like oral, parenteral, nasal, ophthalmic, transdermal, colonal etc. [3].

Polylactic acid (PLA), originating from lactic acid and polycaprolactone (PCL) has biocompatibility and biodegradability characteristics and are excellent materials to use in the medical sciences [4]. PLA is aliphatic polyester, thermoplastic, hydrophobic, and has been used in medical application such as cell transplantation scaffold and PCL has been applied in drug delivery system due to its great permeability property [5,6]. PCL holds a high crystallinity characteristic; yet, its degradation rate is low [7]. D,L-PLA forms of PLA experience a more rapid degradation rate compared to PCL, however, it carries a 3-5% lower strain compared to PCL, thus PLA is more brittle than PCL [8].

The PCL mixture together with other polymers with lower molecular weight can be applied in the numbers of application, namely, polymer modification for drug delivery system [2]. The PCL mixture and PLA are counted to generate a suitable polymer with a more rapid degradation rate and better permeability. The application of this biodegradable polymer offers many benefits, since it can be conveniently absorbed by the body once the hydrolysis process is completed, hence, will not cause any poisoning in the body [9].

The requirements of microspheres for drug delivery systems must have a size not greater than 250 μm [2]. The size of the microsphere is the main determinant of the rate of drug release [10]. Thus, the rate and drug release duration depends on the surface area, thickness of the walls, and pores of the drug introducing microspheres, so that the size distribution of microspheres is a significant aspect in the microspheres characterization. In addition, particle size is vital parameter in the drug dosage preparation form, such as oral, parenteral and others [11]. By changing the size to be simpler it enables the localization of microspheres to specific organs or tissues. Microspheres that have a size in the range of 40-50 μm will produce efficiency of around 70-90% so that a uniform microsphere size will increase the efficiency and rate of drug release [12].

Recently, many researchers have investigated the preparation of microspheres with various surfactants. However, there have been few reports on research on the control of polyblend PLA-PCL microsphere size [13] using a non-ionic hydrophobic surfactant, so in this study, the preparation of the blend of polylactic acid and polycaprolactone microspheres with sorbitan monooleate as surfactants has been studied by the method of water-in-oil (w/o) emulsion solvent evaporation. The use of sorbitan monooleate as an emulsion stabilizer was selected because sorbitan monooleate has biocompatibility properties and has been used in biomedical applications, oral drug pharmaceutical formulations, and food industries. In addition, the sorbitan monooleate acts as a microsphere size controller with a uniform size distribution. The effect of sorbitan monooleate in the control of size in the formation of microspheres is studied through variations in volume sorbitan monooleate, stirring speeds of emulsion formation, and dispersion times. The size distribution is affected by the agitation process throughout the emulsion process in the w/o system [12–14].

2. Materials and methods

2.1. Materials

Lactic acid (90%) and dichloromethane were purchased from Merck, while polycaprolactone was purchased from Pestorp CAPA-6800, and sorbitan monooleate was purchased from Evonic with the commercial name is span 80.

2.2. Synthesis of polylactic acid

The reaction was performed in a 250 mL, three-necked flask reactor completed with a magnetic stirrer, thermometer, and a water-cooling condenser. A volume of 50 mL lactic acid was added in to the reactor. The temperature was gradually increased to 120 °C and kept constant for 1 hour. During the reaction, the solution was stirred continuously. After that, the reactor was flowed with nitrogen gas for 15 min. At the same time, the temperature was raised up to 150 °C and kept at constant temperature for 22 h. After the reaction finished, it was cooled down at room temperature for 2 h, while stirring. White-yellow sticky solids were obtained.

2.3. Preparation of 10% polyblend solutions

The amount of 0.5 g polyblend of PLA and PCL with 60:40 ratio was prepared. The polymers were perfectly mixed in dichloromethane and stirred for 30 min to acquire the 10% polyblend solution.

2.4. Preparation of microspheres

5mL 10% polyblend solution and sorbitan monooleate with different volumes were added as an emulsifier; this was stirred at 700 rpm for 1 h. The emulsion product was then dispersed in 250 mL distilled water, and stirred (900 rpm) for 1 h. At the same time, in dispersion phase was occurred evaporation of dichloromethane. The produced microsphere was obtained by filtration, cleaned with distilled water, and dried in an oven at temperature maximum 40 °C or at room temperature. The particle size of the microspheres products generated from each formula were studied by Particle Size Analyzer (PSA), optical microscope and the microspheres' functional group was analyzed by Fourier Transform Infrared (FT-IR). Details of the various preparation conditions are described below.

3. Results and discussion

In this study, polylactic acid was obtained with direct polycondensation without catalyst and solvent. In this reaction lactic acid acts as self-catalyzed polymerization. Without exogenous strong acid, the

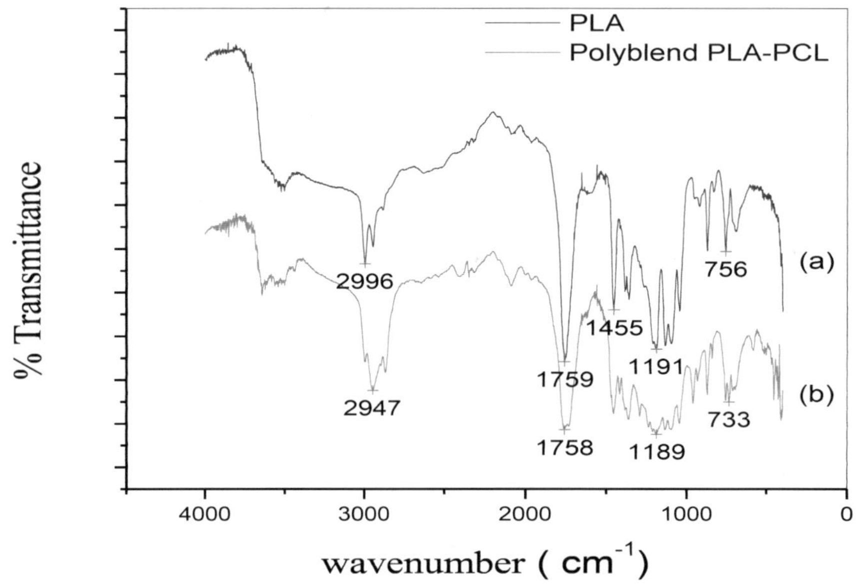

Figure 1. FT-IR spectra of (a) PLA and (b) polyblend of PLA-PCL.

acid monomer has an important part to serve as self-catalyst in the esterification process [15]. The polycondensation process without catalyst did note generate a higher molecular weight of PLA, since the lactic acid dehydrative polycondensation is promoted a self-catalysis from carbonyl group, the condensation progress declined as the number of carboxylic group in the lactic acid decreased in the polycondensation reaction [16,17].

3.1. Characterization of polylactic acid by FT-IR

Figure 1a shows FT-IR spectra of lactic acid and polylactic acid synthesized at 150 °C for 22 h. In Figure 1a IR spectra of PLA presented absorption peaks characteristic of ester (1759 cm^{-1} and 1191 cm^{-1} for –COO– and C–O–C respectively) and CH_3 groups (2996 cm^{-1}). According to Laonuad *et al.* [18] as lactic acid was polymerized, the hydroxyl group will react with carboxylic group from another molecule, therefore the broad –OH stretch peak at 3428 cm^{-1} assigned for PLA will be reduced. The hydroxyl absorption peak (in case of PLA) is almost vanished, suggesting that the amount of hydroxyl groups was decreased and the sharper absorption peak of C=O stretching (1759 cm^{-1}) was presented as the polymer was continuously synthesized.

3.2. Characterization polyblend of PLA-PCL by FT-IR

FT-IR spectroscopy was used to ensure that no chemical interaction between the polymers had occurred. From the inspection of FT-IR spectra of blend of PLA and PCL in figure 1b, it showed a characteristic of ester group (C=O) at 1758 cm^{-1} with stretching vibration and C-O-C at 1189 cm^{-1}. The C-H stretching of methyl occurs in the 2947 cm^{-1} region. According to Jung *et al.* [19] polycaprolactone has ester carbonyl group at 1729 cm^{-1} with stretching vibration and C-O-C at 1185 cm^{-1}. The FT-IR spectrum corresponds to microspheres formulation was almost identical to polymer spectra. The interaction between PLA and PCL is only physical interaction [2]. The proposed physics interaction is the van der Waals force that resulted from the presence of momentary dipoles or London forces in non-polar molecules of PLA and PCL.

3.3. Effect of emulsifier volume on the size of the PLA and PCL microspheres

For this experiment, the volume of sorbitan monooleate was varied from 0.5 to 1.5 mL, the W/O ratio was 1:5, the emulsion stirring speed was 700 rpm for 1 h, and the stirring time of dispersion was 900 rpm for 1 h. Figure 2 shows the effect of emulsifier volume on the size of the PLA and PCL microspheres. Bimodal size distributions are obtained and can be seen in table 1. The results indicated

Table 1. Blend of PLA-PCL microsphere size distribution obtained with different variations

Variations		Microsphere size distribution (µm)		
		1	2	3
Volumes of sorbitan monooleate (mL)	0.5	0.393-0.755	1.748-2.787	
	1.0	0.393-0.829	1.748-3.358	
	1.5	0.393-0.755	1.748-2.539	
Stirring speeds of emulsion (rpm)	700	0.393-0.755	1.748-2.787	
	800	0.393-0.910	1.451-3.059	
	900	0.393-0.829	1.321-2.539	
Dispersion times (hour)	0.5	0.393-1.592	1.748-4.443	4.877-7.370
	1.5	0.393-1.592	1.748-4.877	5.354-10.290
	2.0	0.393-0.829	1.592-2.787	

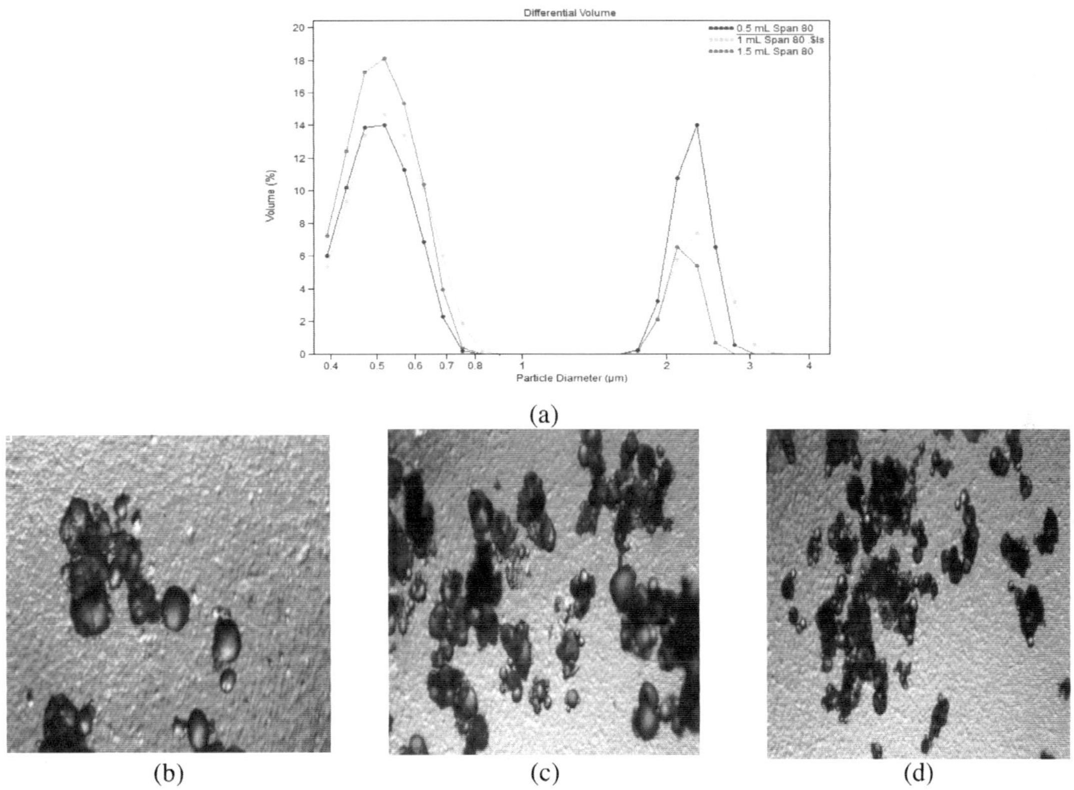

Figure 2. (a) Size distribution of PLA and PCL microspheres of variation of volume surfactant and optical microscope images (11.25×) of microspheres with volume of sorbitan monooleate: (b) 0.5, (c) 1 and (d) 1.5 mL.

that as the volume of the emulsifier increased, the average size decreased. This can be explained on the basis that the primary function of the emulsifier in stabilization of droplets is to generate a thin film around the droplets to avoid their coalescence. Emulsifier is one of the critical aspects that can influence the microspheres' particle size and its shape. Emulsifier can reduce the interfacial tension of W/O in order to produce a smaller and more stable droplets in the dispersed phase. So, when the emulsifier (sorbitan monooleate) volume was increased from 0.5 mL to 1.5 mL, the size of the microspheres decreased from 1.128 µm to 0.764 µm.

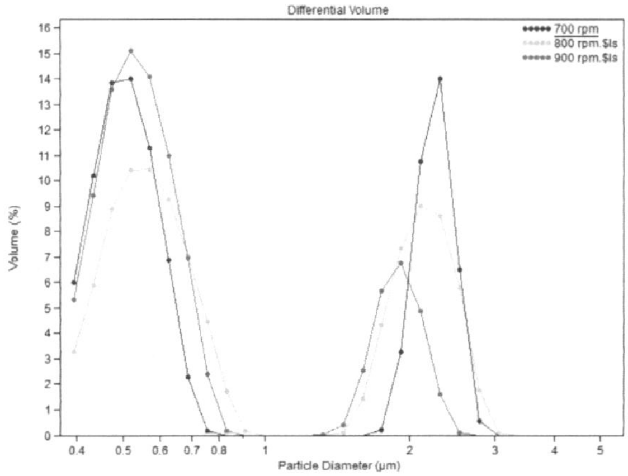

(a)

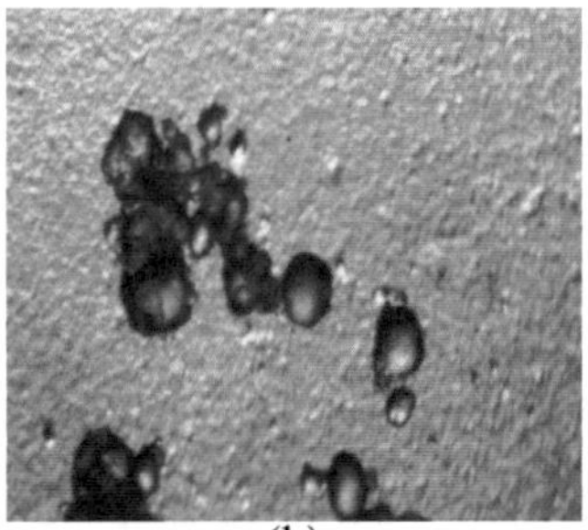

(b)

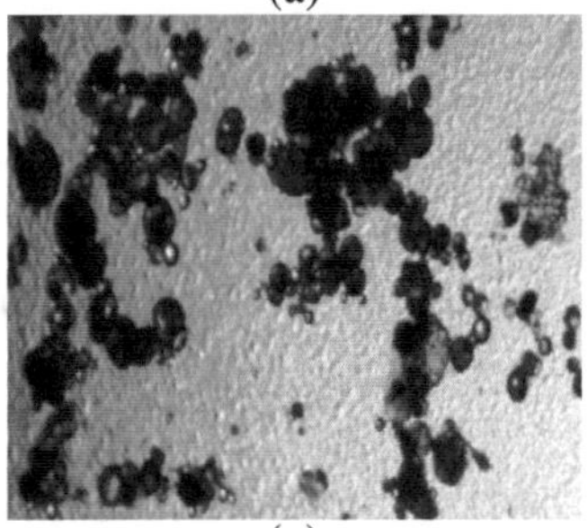

(c)

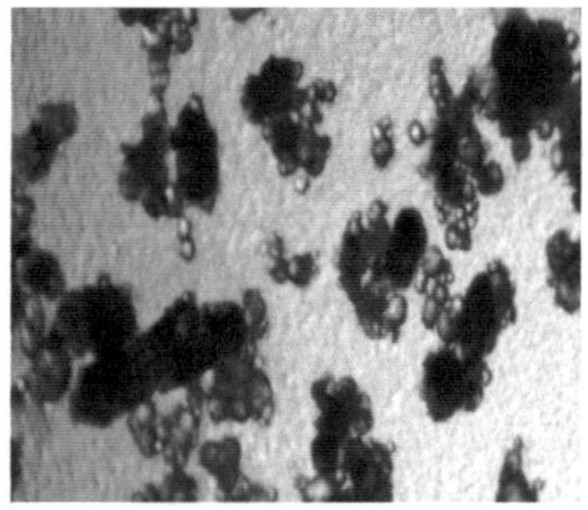

(d)

Figure 3. (a) Size distribution of microspheres with different emulsion stirring speed and optical microscope images (11.25×) of microspheres with emulsion stirring speed of: (b) 700, (c) 800 and (d) 900 rpm.

The formation of a non-uniform particle size can be caused by the properties of sorbitan monooleate, which has a cloud point far below room temperature. The non-ionic surfactant's solubility decreases as the temperatures increases, and these surfactants are probably start to lose the surface-active properties at a higher transition temperature – known as cloud point [20]. This phenomenon takes place due to at a higher temperature of the cloud point, the surfactant-rich phase of the swollen micelles will be separated and transition can be observed by the increase of dispersion turbidity. The system consists of an almost micelle-free dilute solution and a surfactant-rich micellar phase [20,21]. This separation is due to a significant increase in aggregation amount and a decrease in intermicellar repulsions [21,22] that generates a different density of the micelle-rich and micelle-poor phases so that the more surfactant volume added the more agglomeration formed and the solution looks more turbid.

3.4. Effect of emulsion stirring speed on the size of the PLA and PCL microspheres

Variations of emulsion stirring speeds were performed at speeds of 800 and 900 rpm. Other conditions were made constant (0.5 mL volume of sorbitan monooleate, 1 h emulsion stirring time, and dispersion stirring speed 900 rpm for 1 h). The emulsion stage is the step that determines the size of the microsphere because at this stage droplet formation will occur which will become solid microspheres.

The effect of stirring speed of emulsion formation on the size of the PLA-PCL microsphere has been characterized using PSA and can be seen in figure 3a. The result data of the variation was compared with the initial stirring speed of emulsion (700 rpm). Based on figure 3a, bimodal size distributions were obtained. There are two main size groups of microspheres. The details size distribution can be seen at table 1.

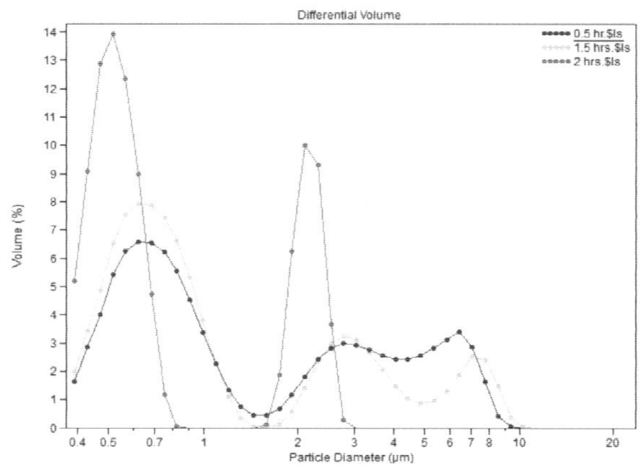

(a)

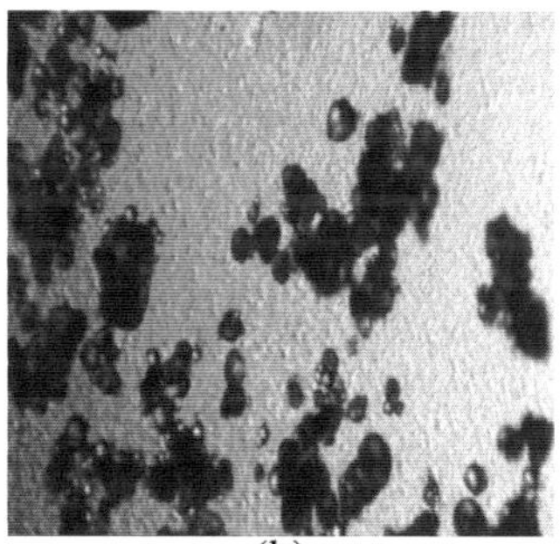

(b)

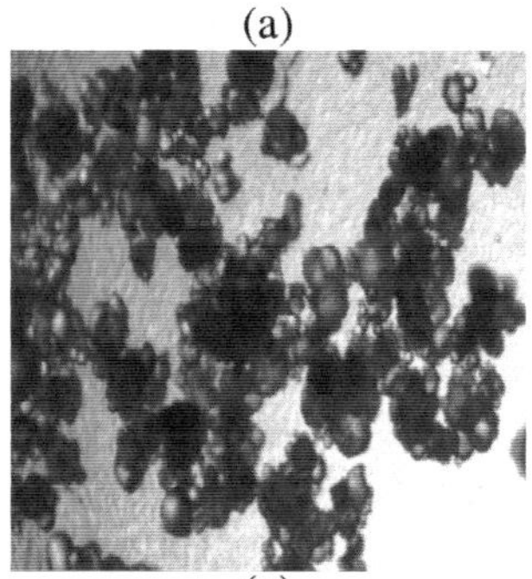

(c)

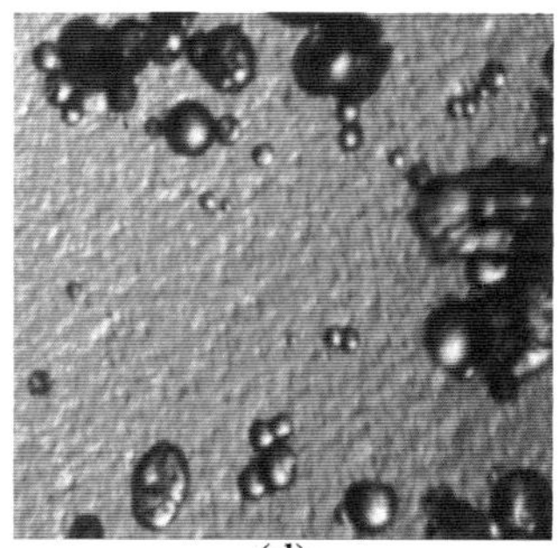

(d)

Figure 4. (a) Size distribution of microspheres with different stirring time dispersion and optical microscope images (11.25×) of microspheres with stirring time of dispersion of: (b) 0.5, (c) 1.5 and (d) 2 hours.

The average size of the microspheres at a variation in stirring speed of 800 rpm is 1.178 µm and the stirring speed of 900 rpm is 0.839 µm. On the other hand, sample with a speed of 700 rpm has an average size of 1.128 µm. At a speed of 700 rpm it appears that the distribution size is smaller than the speed of 800 rpm so that the average size becomes smaller. The effect of variations in the speed of the emulsion stirring on the average size of the microspheres shows that as the emulsion-stirring rate increases (800 rpm and 900 rpm) the resulting microsphere size becomes smaller. The decrease in the average size of the microsphere is due to an increase in stirring rate, the energy applied to the water phase increases, and the collision increases so that it can break the droplet into smaller droplets. In addition, droplets that are not easily to be flocculated or precipitated are supported in the presence of strong stirring centrifugal forces resulting in smaller microspheres. Whereas, with slower stirring the force applied to the water phase is less so that the microspheres are divided into droplets and some aggregates [10]. Droplets can join one another and obtained agglomeration.

3.5. Effect of dispersion time on the size of the PLA and PCL microspheres

With the other parameters is kept at constant (sorbitan monooleate volume 0.5 mL, emulsion stirring speed 700 rpm for 1 h, varying of stirring time of dispersion on (0.5, 1.5 and 2 h) stirring speed of dispersion 900 rpm, and the ratio of W/O 1:5), the effect of stirring time of dispersion on size of the PLA and PCL microspheres was studied. As shown in table 1 and figure 4a, bimodal and polymodal size distributions are obtained. Based on figure 4, the average size seemed as the dispersion time increased, the average size decreased. Stirring time of dispersion for 0.5 to 2 hours obtained the size of PLA and PCL microspheres 2.233, 1.918, and 1.045 µm respectively.

The reason is unclear but may be related to the dispersion step there is evaporation of solvent from droplet to continuous phase, then from continuous phase to free air [23]. A decrease in the average size of the microspheres may occur as the longer contact time between the droplets and the continuous phase allows the more complete solvent evaporation of the droplets diffusing into the continuous phase and the deposition of droplets into solid microspheres is also better so that the resulting average microsphere size is smaller. However, when droplets are in continuous phase in a short time, the solidification of the microspheres is yet to be perfect, and the residual solvent is constantly present in the droplets [24].

4. Conclusions

Polyblend of polylactic acid and polycaprolactone microspheres with sorbitan monooleate as surfactant were successfully synthesized. There is no chemical interaction between blend of PLA and PCL, its only physical interaction. The FT-IR spectra of polyblend of PLA and PCL are combinations of PLA and PCL. The effect of sorbitan monooleate as surfactant on the size of microspheres by increasing the amount of surfactant volume, the stirring speed of emulsion formation, dispersion time obtained the microsphere size decrease. The size of microspheres obtained in range from 0.4 µm to 10 µm. Physical form of microspheres was observed using optical microscope. The result shows that some microspheres are spherical and others in aggregates form.

Acknowledgements

The authors would like to thank Universitas Indonesia for funding this research through PITTA Grant Universitas Indonesia with contract No:233/UN2.R3.1/PPM.00/2018.

References

[1] Sharma N, Purwar N and Gupta P C *2015 International Journal of Pharmaceutical Sciences and Research (IJPSR)* **6** 4579–87
[2] Ramteke K H, Jadhav V B and Dhole S N 2012 *IOSR Journal of Pharmacy (IOSRPHR)* **2** 44–8
[3] Kemala T, Budianto E and Soegiyono B 2012 *Arab. J. Chem.* **5** 103–8
[4] Ramot Y, Haim-Zada M, Domb A J and Nyska A 2016 *Adv. Drug Deliv. Rev.* **107** 153–62
[5] Thomson R C, Wake M C, Yaszemski M J and Mikos A G 1995 Biodegradable polymer scaffolds to regenerate organs *Biopolymers II* (*Advances in Polymer Science* vol 122) ed N A Peppas and R S Langer (Berlin: Springer) pp 245–74
[6] Pillai C K S and Sharma C P 2010 *J. Biomater. Appl.* **25** 291–366
[7] Woodruff M A and Hutmacher D W 2010 *Prog. Polym. Sci.* **35** 1217–56
[8] Porjazoska A, Karal-Yilmaz O, Kayaman-Apohan N, Cvetkovska M and Baysal B M 2004 *Croat. Chem. Acta* **77** 545–51
[9] Gunatillake P A and Adhikari R 2003 *Eur. Cell. Mater.* **5** 1–16
[10] Peng Z Y, Shen Y Q and Li Z P 2012 *J. Macromol. Sci.* B **51** 12–21
[11] Sun R, Shi J, Guo Y and Chen L 2009 *Front. Chem. Chin.* **4** 222–8
[12] Zhu K J, Li Y, Jiang H L, Yasuda H, Ichimaru A, Yamamoto K, Lecomte P and Jerome R 2005 *J. Microencapsul.* **22** 25–36
[13] Achmad F, Yamane K, Quan S and Kokugan T 2009 *Chem. Eng. J.* **151** 342–50
[14] Esposito E, Cortesi R and Nastruzzi C 1996 *Biomaterials* **17** 2009–20
[15] Berchane N S, Jebrail F F, Carson K H, Rice-Ficht A C and Andrews M J 2006 *J. Microencapsul.* **23** 539–52
[16] Odian G 2004 *Principles of Polymerization* 4th ed (NewYork: John Wiley & Sons)
[17] Moon S I, Lee C W, Miyamoto M and Kimura Y 2000 *J. Polym. Sci.* A **38** 1673–9
[18] Laonuad P, Chaiyut N and Ksapabutr B 2010 *Optoelectron. Adv. Mater.* **4** 1200–02
[19] Jung J W, Lee H, Hong J M, Park J H, Shim J H, Choi T H and Cho D W 2015 *Biofabrication* **7** 045003
[20] Schramm L L, Stasiuk E N and Marangoni D G 2003 *Annu. Rep. Prog. Chem. Sect.* C **99** 3–48
[21] Staples E J and Tiddy G J T 1978 *J. Chem. Soc. Faraday Trans. 1* **74** 2530–41
[22] Tiddy G J T 1980 *Phys. Reports* **57** 1–46
[23] Li M, Rouaud O and Poncelet D 2008 *Int. J. Pharm.* **363** 26–39
[24] Tiwari S and Verma P 2011 *International Journa of Pharmacy and Life Science* **2** 998–1005

Synthesis of metal oxide-hierarchical ZSM-5 (Co_3O_4/ZSM-5 and Fe_2O_3/ZSM-5) as catalysts for partial oxidation of biomethane to methanol

A T Putrananda, D A Nurani and Y K Krisnandi

Department of Chemistry, Faculty of Mathematics and Natural Sciences (FMIPA), Universitas Indonesia, Kampus UI Depok, Depok 16424, Indonesia

Corresponding author's e-mail: yuni.krisnandi@sci.ui.ac.id

Abstract. This study presents a partial oxidation of biomethane from anaerobic organic waste of cow dung (83.7 % of methane and 16.3 % of oxygen) to methanol using modified cobalt and iron metal oxide-hierarchical ZSM-5 catalysts. Hierarchical ZSM-5 was synthesized using double template methods with tetrapropylammonium hydroxide (TPAOH) and dimethyl diallyl ammonium chloride acrylamide copolymer (PDDAM) as the first and secondary templates, respectively. The as-synthesized ZSM-5 materials were characterized with Scanning Electron Microscopy-Energy Dispersive X-Ray Spectroscopy (SEM-EDS), Fourier Transform Infrared (FTIR), X-Ray Diffraction (XRD), and the results show the similarities of physicochemical properties with standard ZSM-5 materials. The hierarchical ZSM-5 was then modified with cobalt or iron oxides through impregnation with incipient wetness. The results of XPS analysis show that the impregnated cobalt and iron metal oxide are Co_3O_4 and Fe_2O_3, respectively. While the AAS analysis indicates that the metal content in Co_3O_4/ZSM-5 and Fe_2O_3/ZSM-5 are 2.1 % (Co) and 2.4 % (Fe), respectively. The partial oxidation of bio-methane to methanol was carried out in an atmospheric fixed batch reactor, c.a. 200 mL, with biomethane to N_2 ratio of 0,2:2 (in bar). The results of application using biomethane show the hierarchical Fe_2O_3/ZSM-5 catalyst has higher methanol yield (17.61 %), compared to the yield by using hierarchical Co_3O_4/ZSM-5 (10.99 %). In addition, no side product is detected after reaction. Thus, it can be concluded that Fe_2O_3 is more active than Co_3O_4 as catalyst and biomethane can be utilized as source in this partial oxidation reaction.

Keywords: bio-methane, metal oxide, methanol, partial oxidation, hierarchical ZSM-5

1. Introduction

Nowadays, many of the world's interest is environmentally friendly renewable energy sources as a result of the limited fossil fuel resources, given the increasing energy needs. Biogas is one of the most promising renewable energy sources in terms of environmental and economical factors [1]. Biogas is generally produced from anaerobic organic waste derived from the environment such as livestock's manure. Biogas contains methane and carbon dioxide as the largest component, and small amounts of nitrogen, oxygen, hydrogen sulfide, halogen, and aromatic compounds.

Utilization of methane gas can be done by methane reforming process to get syngas, and converting to methanol with partial oxidation reaction. Methane conversion through partial oxidation reactions can be done directly. However, this direct conversion has drawbacks, given that methane is a stable compound and requires high temperatures and is suitable for making methane to be unstable in order to be converted into other compounds with high percent conversion and high selectivity rates [2]. Thus, a catalyst, which can be used in the partial oxidation reaction of methane, is required. Catalytic oxidation of methane with N_2O using Fe-zeolite systems, in order to optimize methane conversion activity was reported [3]. However, N_2O gas is rather difficult to handle, compared to molecular O_2 or even air. Beznis *et al.* [2] reported the methane oxidation into methanol and formaldehyde through Co–ZSM-5 zeolite agglomerates, by adjusting the reactivity and selectivity using alkaline and acid treatments on the zeolite. Krisnandi *et al.* [4] used a hierarchical Co-oxides/ZSM-5 catalyst, and the reaction showed a high selectivity toward methanol conversion. Furthermore, partial oxidation of methane over hierarchical H-ZSM-5 modified with cobalt oxide species has also been carried out [5]. In this work, we used a biogas (biomethane) as a methane source for partial oxidation of biomethane to methanol using Co_3O_4/ZSM-5 and Fe_2O_3/ZSM-5 catalysts.

2. Experimental

2.1. Materials

For synthesis of hierarchical ZSM-5 the materials needed were tetraethyl orthosilicate (TEOS, 98%, Sigma Aldrich), sodium aluminate (Sigma Aldrich), tetrapropylammonium hydroxide (TPAOH, 1 M, Sigma Aldrich), dimethyl diallyl ammonium chloride acrylamide copolymer (PDDAM, 10 wt%, Aldrich), and deionized water. For modification with metals oxide the materials needed were cobalt (II) nitrate ($Co(NO_3)_2\cdot 6H_2O$, Sigma Aldrich), iron(III) nitrate ($Fe(NO_3)_3\cdot 9H_2O$, Sigma Aldrich). For the catalytic test the materials needed were bio-methane (anaerobic organic waste derived from livestock's manure, PT SWEN IT Ciomas-Bogor), methane gas (99.99% UHP from BOC), and N_2 gas (99.99 % UHP).

2.2. Synthesis of hierarchical ZSM-5

The hierarchical ZSM-5 was prepared by hydrothermal reaction using Wang *et al.* [6] methods with some modifications. All the materials with molar ratio of 1 Al_2O_3 : 64.34 SiO_2 : 10.07 $(TPA)_2O$: 3571.70 H_2O were mixed under stirring at 373 K for 3 h. The PDDAM was added slowly and stirred at the ambient temperature for 48 h. The hydrothermal reactions were carried out in an autoclave at 443 K for 144 h. The product was filtered and dried at the ambient temperature. Activation ZSM-5 by calcination at 823 K. The hierarchical ZSM-5 was characterized by X-ray diffraction (XRD), Fourier transform infrared (FTIR, Shimadzu Prestige 21) and Scanning Electron Microscopy with Energy Dispersive X-ray (SEM-EDX, SEM FEI quanta FEG 450).

2.3. Preparation of M_xO_y/ZSM-5

Metal oxide-hierarchical ZSM-5 (M_xO_y/ZSM-5) was prepared using wet impregnation method. 1 g of ZSM-5 was mixed with 0.2495M metal nitrate solution and stirred at the ambient temperature for 24 h. The activation of M_xO_y/ZSM-5 by calcination at 823 K. The catalysts were characterized by XRD and atomic absorption spectroscopy (AAS) instrument.

2.4. Biomethane analysis

The biomethane used in this research was from anaerobic organic waste derived from livestock's manure (PT SWEN IT Ciomas-Bogor). The composition of bio-methane was analyzed using Gas Chromatography – Mass Spectrometry (GC-MS Shimadzu QP 2010 Ultra).

2.5. Catalytic test: partial oxidation of bio-methane

The partial oxidation of bio-methane tested for its catalytic property was performed in atmospheric fixed batch reactor. The amount of Co_3O_4/ZSM-5 and Fe_2O_3/ZSM-5 catalysts used was 0.5 g. The reaction of catalytic test was performed at 423 K for 2 h with composition of bio-methane and nitrogen pressures of 0.2 and 2 bar, respectively. The product was extracted with ethanol and analyzed using gas chromatography (GC, Shimadzu 2010 FID Detector).

3. Results and discussion

3.1. Characterization of hierarchical ZSM-5

In this method the TPAOH template was used as a directing structure of MFI, which is a structural type of ZSM-5. The alumina source in synthesis of ZSM-5 was obtained from $NaAlO_2$, while the silica source was obtained from TEOS, and the PDDAM was used as a mesopore template. The positively charged of PDDAM can interact with the negatively charged of ZSM-5 materials, so the ZSM-5 framework could be established around the PDDAM [6]. The standard ZSM-5 material has a typical diffraction pattern on 2θ between 7–10° and 22–25° [7]. The XRD pattern of as-synthesized

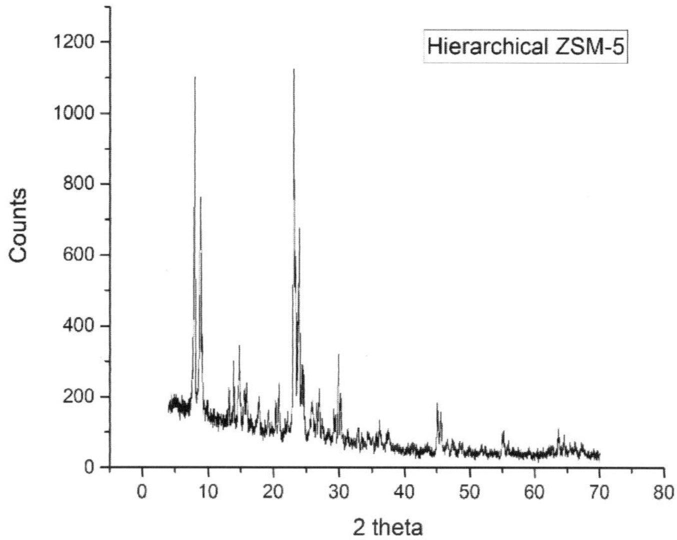

Figure 1. XRD patterns of synthesized hierarchical ZSM-5

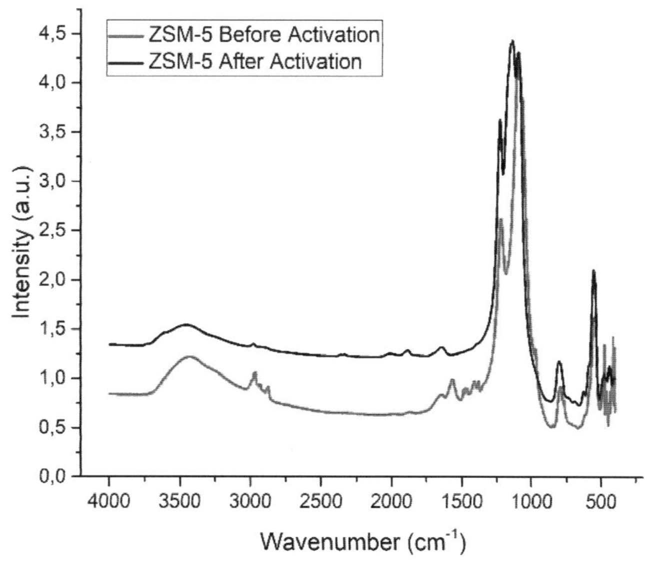

Figure 2. FTIR spectra of ZSM-5 before and after activation with calcination at 823 K

hierarchical ZSM-5 (figure 1) shows certain peaks in the 2θ range from 7° to 10° and from 22° to 25°, which is consistent with the standard ZSM-5 pattern, indicating that the ZSM-5 structure was successfully synthesized.

The FTIR spectra of ZSM-5 in before and after activation by calcination at 823 K are shown in figure 2. The spectrum of ZSM-5 before activation shows several extra peaks at 2960–2850 cm^{-1} which is C–H stretching, and at 1475–1350 cm^{-1} is C–H bending, compared with the ZSM-5 after activation, which can be attributed to the templates used in the synthesis of ZSM-5 material, i.e., TPAOH and PDDAM. Calcination process was used to remove the organic templates in the synthesis of ZSM-5 material. It can be seen that the template was removed as indicated by the loss of the C–H stretching and C–H bending vibration bands. Moreover, the pores and channels in the zeolite became empty after calcination.

Figure 3. SEM analysis for Hierarchical ZSM-5

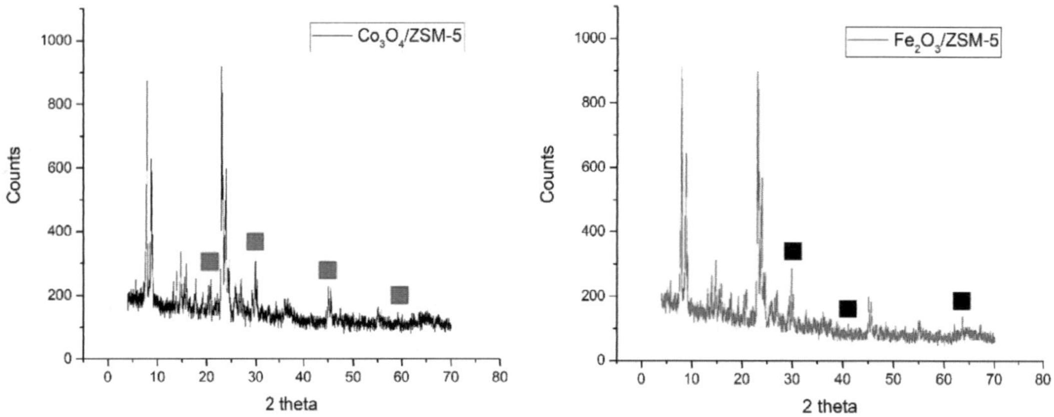

(a) (b)

Figure 4. XRD patterns of (a) Co_3O_4/ZSM-5, and (b) Fe_2O_3/ZSM-5

SEM-EDX was used to determine the crystal structure by the morphology and the Si/Al ratio of the synthesized ZSM-5. In agreement with the standard ZSM-5 morphology [8], figure 3 presents the morphology of as-synthesized hierarchical ZSM-5 has hexagonal or coffin-shaped. Rough surface observed in ZSM-5 indicates the loss of PDDAM template used in the hierarchical ZSM-5 synthesized. The EDX analysis was performed to determine the composition and quantity of elements of a material. It can also be used to determine the Si/Al ratio of ZSM-5 material. By the results, the Si/Al ratio on as-synthesized ZSM-5 material was 27.

3.2. Preparation of M_xO_y/ZSM-5

M_xO_y/ZSM-5 heterogeneous catalysts were synthesized using a wet impregnation method, then the activation of catalysts with calcination up to 823 K. The elemental analysis with AAS indicates the distribution of metals (% weight loading) in ZSM-5 material. It was about 2.1 % and 2.5 % for Co_3O_4/ZSM-5 and Fe_2O_3/ZSM-5, respectively. Figure 4 shows the XRD patterns of modified ZSM-5 catalysts. The results of XRD analysis also show the similarity of diffraction patterns for the modification of cobalt and iron metal oxides in the hierarchical ZSM-5 which is in the diffraction pattern with high intensity at 2θ of 7–10° and 22–25°. The similarity indicates that the process of metal oxide impregnation in ZSM-5 material does not damage the material structure of ZSM-5. However, the modification of the metal oxide into the ZSM-5 material causes a decrease in the crystallinity of the ZSM-5 material. This is due to the interaction between metals with the ZSM-5 framework. Figure 4a shows the diffraction patterns at 2θ = 31°, 45°, 55° and 60° attributed to Co_3O_4 [9]. Moreover, at figure 4b the diffraction pattern at 2θ = 30°, 42° and 64° are attributed to Fe_2O_3 [10].

Table 1. The Bio-Methane Component

Component	Concentration
Methane	83.7%
Oxygen	16.3%

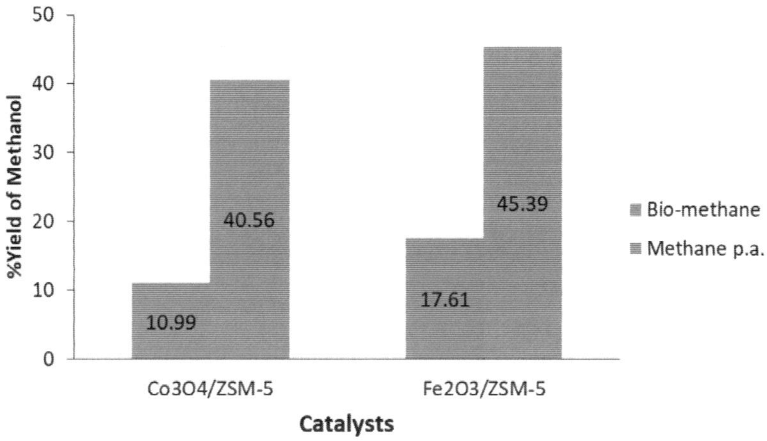

Figure 5. Results of partial oxidation of methane using metal oxide-modified ZSM-5

3.3. Biomethane component

Biogas is one of the most attractive sources of environmentally friendly renewable energy and its exploitation is very economically beneficial [1]. Biogas used in the application of partial oxidation of methane to methanol, produced from anaerobic organic waste derived from livestock's manure. Biomethane was transferred from the supplier using biogas bag made of thick synthetic rubber. The biomethane was analysed by using Gas Chromatography-Mass Spectrometry (GC-MS) instrument to determine the composition of the biomethane. Table 1 shows the content of the compounds in the biomethane sample derived from livestock's manure, i.e., methane and oxygen gas. Based on this result, biomehtane contains only methane and oxygen, which can be used as feed without further purification.

3.4. Methane partial oxidation catalytic test

The results of biomethane partial oxidation using as-synthesized ZSM-5 catalysts are shown in figure 5. From those results, it can be observed that the catalysts modified in the hierarchical ZSM-5 modified with the cobalt and iron metal oxide was selective to methanol production. This is consistent with our previous results [4,5]. However, the amount of methanol yield obtained in the biomethane partial oxidation reaction is smaller than that of methane p.a. This may be due to the amount of biomethane feed into the reactor was smaller, carried out manually by pressing the bio-methane from poly-bag. Based on the process, the maximum pressure of bio-methane gas that could enter into the reactor is only 0.2 bar and the optimum reaction condition (pressure = 0.75 bar) could not be reached. The oxygen content in the biomethane also increases the yield of methanol, which also in consistency with the previous results [5]. The metals oxides impregnated in ZSM-5 are capable of being the active site of the catalyst for oxidizing methane to methanol. The finding in this work is the Fe_2O_3/ZSM-5 has the higher % yield of methanol than Co_3O_4/ZSM-5 catalyst.

4. Conclusions

Hierarchical ZSM-5 modified by cobalt and iron oxides could be used as catalysts in partial oxidation of biomethane. Biogas, which derived by anaerobic organic waste, could be used as methane source in partial oxidation biomethane. Modified cobalt and iron oxides-hierarchical ZSM-5 catalysts were selective to methanol production. The catalytic test using biomethane shows that the hierarchical Fe_2O_3/ZSM-5 catalyst gives the highest methanol yield i.e. 17.61 %.

Acknowledgements
This research was funded by PITTA grant No. 2253/UN2.R3.1/KKP/05.00/2018 from Universitas Indonesia.

References
[1] Vita A, Italiano C, Previtali D, Fabiano C, Palella A, Freni F, Bozzano G, Pino L and Manenti F 2018 *Renew. Energy* **118** 673–84
[2] Beznis N V, van Laak A N C, Weckhuysen B M and Bitter J H 2011 *Micropor. Mesopor. Mater.* **138** 176–83
[3] Knops-Gerrits P P and Goddard III W A 2001 *J. Mol. Catal.* A **166** 135–45
[4] Krisnandi Y K, Putra B A P, Bahtiar M, Abdullah I and Howe R F 2015 *Procedia Chem.* **14** 508–15
[5] Krisnandi Y K, Samodro B A, Sihombing R and Howe R F 2015 *Indonesian Journal of Chemistry* **15** 263–8
[6] Wang L, Zhang Z, Yin C, Shan Z and Xiao F S 2010 *Micropor. Mesopor. Mater.* **131** 58–67
[7] Treacy M M J and Higgins J B 1986 *Collection of Simulated XRD Powder Patterns for Zeolites* 5th ed (Amsterdam: Elsevier)
[8] Lu J, Bartholomeeusen E, Sels B F and Shyrvers D 2017 *J. Microsc.* **265** 27–33
[9] Yue W and Zhou W 2007 *J. Mater. Chem.* **17** 4947–52
[10] Ganachari S V, Joshi V K, Bhat R, Deshpande R, Salimath B, Rao N V S and Venkataraman A 2012 *International Journal of Science Research (IJSR)* **1** 77–9

NaAlg-PVA-g-AAm/ZnO nanocomposite hydrogel as material of wound dressing: synthesis and characterization

H Helmiyati and U Hartina

Department of Chemistry, Faculty of Mathematics and Natural Sciences (FMIPA), University of Indonesia, Kampus UI Depok, Depok 16424, Indonesia

Corresponding author's e-mail: helmiyati@sci.ui.ac.id

Abstract. In this study, sodium alginate and polyvinyl alcohol grafted acrylamide with modified ZnO nanoparticles to nanocomposite hydrogel NaAlg-PVA-g-AAm/ ZnO have been successfully synthesized. The characterizations of hydrogels were supported by Fourier Transform Infrared (FTIR), X-Ray Diffraction (XRD), Scanning Electron Microscopy (SEM) and Tunneling Electron Microscopy and Energy Dispersive X-Ray (TEM-EDX). The image of TEM obtained shows that ZnO nanoparticle are distributed evenly with about 70 nm diameter in hydrogel nanocomposite. The results for NaAlg-PVA-g-AAm nanocomposite hydrogel modified ZnO nanoparticles was obtained the best nanocomposite at 32.21 ppm Zn^{2+} concentration. It has Zn^{2+} ion maximum loading capacity at 80 %, maximum release capacity of 25 % measured by AAS, and its water maximum swelling capacity is 230.20 (g/g). While the maximum swelling capacity for NaAlg-PVA-g-AAm nanocomposite hydrogel is 106.30 (g/g). From in-vitro antibacterial activity test, *S.aureus* (gram positive bacteria) has more resistance than *P. Aeruginosa* (gram negative bacteria) in which *S. aureus* inhibition percentage is 60.22 % and *P. Aeruginosa* is 40.10 % with the minimum inhibitory concentration of 31.25 ppm.

Keywords: acrylamide, hydrogel, nanocomposite, polyvinyl alcohol, sodium alginate

1. Introduction

Hydrogel is a network of hydrophilic polymers that can absorb substance in large amount [1]. One of the biopolymers is sodium alginate (NaAlg) isolated from brown algae with concentration about 40 % of its dry weight [2]. This biopolymer has attracted wide attention because its abundant in nature makes this biopolymer renewable and biodegradable linear polysaccharide and also non-toxic. Other unique natural properties of sodium alginate are its hydrophilicity, biocompatibility and low cost, make it suitable for many biomedical applications [3–5].

Polyvinyl alcohol (PVA) has been widely used to be mixed with alginate because it has good mechanical properties and chemical stability, low cost, non-carcinogenic, and biodegradable. PVA was composited with sodium alginate to form a nanocomposite hydrogel using grafting method with hydrophilic vinyl monomer such as acrylamide (AAm) which can increase its swelling capacity and crosslinking with N, N'-methylene-bis-acrylamide (MBA) to form a hydrophilic polymer network, insoluble in water [6].

In some previous studies, it has been known that ZnO has antibacterial properties that can inhibit gram-negative or gram-positive bacteria. Antibacterial properties are important for wound dressing materials, considering fluid wound is a medium easily overgrown with bacteria. Using a hydrogel matrix as a nanoreactor allows the release of ZnO nanoparticles and Zn^{2+} ions to be more controlled [7–9]. Zinc oxide is element which effective against microorganisms such as *Staphylococcus aureus*, *Escherichia coli*, *Pseudomonas aeruginosa*, and *Aspergillus niger*. Mechanism of action is with penetrating the bacterial cell wall via diffusion leading to cell death [10].

In this study, we have blended sodium alginate used as the back bone polymer with polyvinyl alcohol loaded with ZnO nanoparticle to form polymeric nanocomposites, based on zinc oxide, an element which effective against microorganisms as has been described previously [10]. The synthesized of hydrogel nanocomposite by free radicals graft polymerization of acrylamide (AAm) using methylenebisacrylamide (MBA) as crosslinking agent and potassium persulfate (KPS) as an initiator. The grafting of acrylamide onto sodium alginate and PVA has been reported previously [3,6]. The effects of zinc nitrate concentration on capacity of Zn^{2+} loading, the water swelling and the release Zn^{2+} the superabsorbent hydrogel were investigated. The formation of silver nanoparticles on nanocomposite was confirmed by Fourier Transform Infrared (FTIR), X-Ray Diffraction (XRD), Scanning Electron Microscopy (SEM) and Tunneling Electron Microscopy-Energy Dispersive X-Ray (TEM-EDX). The best nanocomposite hydrogels tested to antibacterial activity was done using two types of cultural bacteria, there are *P. aeruginosa* negative bacteria and *S. aureus* positive bacteria.

2. Experimental

2.1. Material

Sodium alginate (Sigma Aldrich) as backbone, polyvinyl alcohol (PT. Brataco Chemical) as an additive polymer, acrylamide (Sigma Aldrich) as a monomer, potassium persulfate (Merck co), N, N'-methylene-bis-acrylamide (Merck Co), $Zn(NO_3)_2$ (Sigma Aldrich), NaOH (Merck co), nutrient agar for microbiology (Merck Co.), peptone ISO 6579 (Merck Co) and beef extract Lab-Lemco Powder (Oxoid Ltd.) for bacterial growth medium, chloramphenicol (Bernofarm Pharmaceutical) as a positive control on antibacterial tests, culture of *Staphylococcus aureus* bacteria and culture of *Pseudomonas aeruginosa* bacteria.

2.2. Synthesis of NaAlg-PVA-g-AAm/ZnO hydrogel

The first step was the NaAlg-PVA-g-AAm hydrogel synthesis, by referring to the reported method [6] with modification. The second step was synthesis of NaAlg-PVA-g-AAm modified with ZnO nanoparticles using the Zn $(NO_3)_2$ precursor referred to the reported method [7] with modification. The NaAlg-PVA-g-AAm was put in the zinc nitrate precursor solution for 24 hours. Hydrogels that had been loaded with Zn^{2+} ions were washed with distilled water to remove the zinc ions attached on the surface of the hydrogel. Furthermore, hydrogel was put in 0.2M NaOH for 24 hours, then washed with distilled water, and dried in an oven at 50 °C.

In this work, the effect of the $Zn(NO_3)_2$ concentration of precursor have been studied at 0, 6.25, 12.5, 31.25 and 62.5 ppm by code H0, H1, H2, H3 and H4, respectively. The nanocomposite hydrogels that were successfully synthesized were characterized by FTIR, XRD, SEM and TEM-EDX.

2.3. Capacity of loading-swelling-release

The swelling capacity of water was determined using the gravimetric method, based on our previous study [11]. The swelling capacity is calculated using equation (1):

$$\textit{Swelling capacity (g/g)} = \frac{m_t - m_o}{m_o} \quad (1)$$

where m_o is the initial mass of dry hydrogel and m_t is the mass of hydrogel after swelling at a given time.

The loading capacity of Zn^{2+} ionic was determined using atomic absorption spectroscopy (AAS) instrumentation. The loading capacity is calculated using equation (2):

$$\text{Loading capacity (\%)} = \frac{C_t - C_o}{C_o} x\ 100 \quad (2)$$

where C_o is the initial concentration of $Zn(NO_3)_2$ and C_t is the concentration of Zn $(NO_3)_2$ precursor after loading at a given time.

The release capacity of Zn^{2+} ionic was determined using atomic absorption spectroscopy (AAS). The release capacity is calculated using equation (3):

$$\textit{Release capacity (\%)} = \frac{C_t - C_0}{C_0} x\ 100 \quad (3)$$

where C_o is the loaded concentration of $Zn(NO_3)_2$ in hydrogel and C_t is the released concentration of Zn $(NO_3)_2$ at a given time.

2.4. Antibacterial activity

An antibacterial activity test was performed on a modified ZnO nanocomposite hydrogel with a

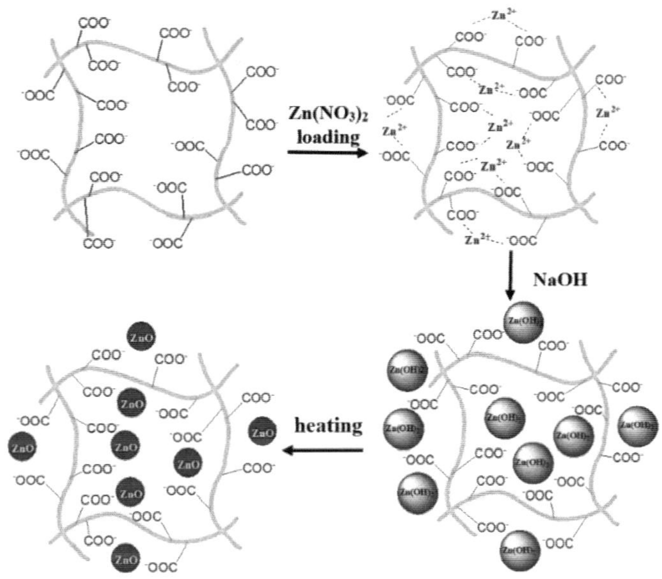

Figure 1. The scheme of formation ZnO nanoparticle in hydrogel

variation of $Zn(NO_3)_2$ concentrations. The antibacterial activity in-vitro test was carried out by dilution method referred to previous study [12]. Percentage of inhibition is the percentage of bacterial cells that can be inhibited the growth in 24 hours compared to positive and negative controls. The positive controls are bacteria culture with hydrogel of chloramphenicol antibiotic loaded; meanwhile negative controls are bacteria culture without hydrogel. The measurements of turbidity are performed using optical density (OD) at a maximum wavelength 625 nm. The bacteria used were two types, *Staphylococcus aureus* positive bacteria and *Pseudomonas aeruginosa* negative bacteria. Percentage of inhibition is calculated using the equation (4):

$$\textit{Percentage of Inhibition (\%)} = \frac{OD\ negative\ control - OD\ sample}{OD\ negative\ control} \times 100 \qquad (4)$$

3. Results and discussion

3.1. *Hydrogel of NaAlg-PVA-g-AAm/ZnO*

In this study, NaAlg-PVA-g-AAm hydrogel was utilized as nanoreactor to form ZnO nanoparticle where the product was known as hydrogel nanocomposite. Polymerization process for hydrogel synthesis was free-radical polymerization by ex-situ method. Alginate macroradical formation was initiated using potassium persulfate. Afterwards, the active site of polymeric chain would initiate acrylamide so that it causes the grafting reaction. The addition of N, N'-methylene-bis-acrylamide as crosslinking agent would create three-dimensional network called hydrogel [6,11]. The method of ex-situ was used to obtain evenly distributed ZnO nanoparticles within the hydrogel matrix. ZnO nanoparticles formation within the hydrogel matrix is explained in figure 1.

Figure 1 shows formation of ZnO nanoparticle in hydrogel. When hydrogel was swollen by $Zn(NO_3)_2$ aqueous solution, Zn^{2+} ions entered the hydrogel matrix and filled in the carboxylate and hydroxyl groups along the alginate, polyvinyl alcohol and acrylamide chain. Then, Zn^{2+} ion-loaded hydrogel was added into 0.2 M NaOH aqueous solution, OH^- ions entered the hydrogel matrix and reacted with Zn^{2+} ions to form zinc hydroxide. Furthermore, it was heated at 50 °C for a few hours to obtain ZnO nanoparticles; the ZnO nanoparticles were formed along with the loss of water [7].

3.1.1. Capacity of Zn^{2+} loading. The amount of Zn^{2+} ions loaded into the NaAlg-PVA-g-AAm hydrogel was calculated by measuring the difference in concentration of the remaining $Zn(NO_3)_2$, with the initial solution $Zn(NO_3)_2$ concentration. The effect of $Zn(NO_3)_2$ concentration on the loading capacity of Zn^{2+} ions absorbed into the hydrogel is shown in figure 2.

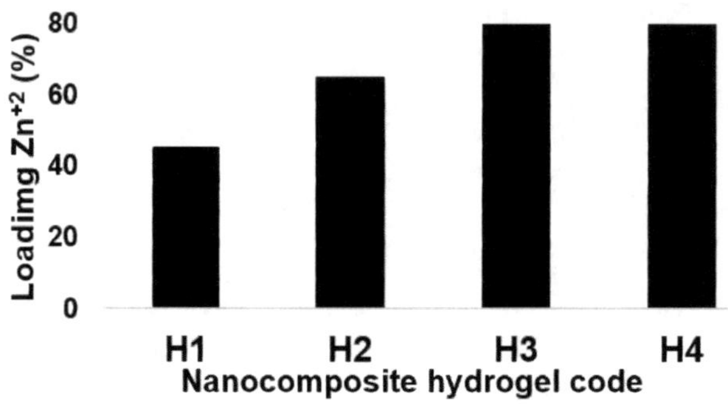

Figure 2. Effect of $Zn(NO_3)_2$ concentration on the loading of Zn^{2+}

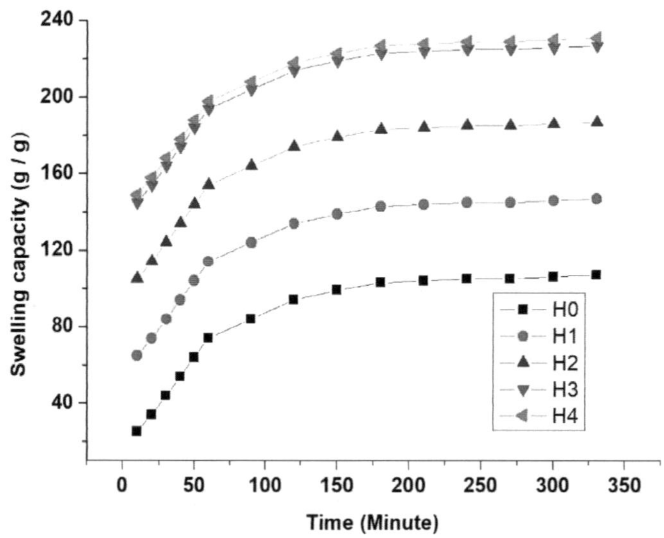

Figure 3. Swelling capacity of nanocomposite hydrogel

Figure 2 shows that the greater the concentration of $Zn(NO_3)_2$ used as a precursor, the Zn^{2+} ion loaded in matrix hydrogel will be increased. It caused the rise concentration of a solution, the greater the collision between particles that occurs so that Zn^{2+} ions the more diffuse into the hydrogel matrix. It shows that at the 31.25 ppm concentration of $Zn(NO_3)_2$ with H3 code obtained the maximum loading capacity 80 %.

3.1.2. Swelling capacity. Water swelling capacity of nanocomposite NaAlg-PVA-g-AAm hydrogel modified ZnO nanoparticles with various concentrations of $Zn(NO_3)_2$ is shown in figure 3. Swelling capacity is measured against time, until a balance time about 300 minutes. It can be seen that the hydrogel without ZnO (H0) has the smallest swelling capacity; this is due to the absence of ZnO nanoparticles so that the hydrogel matrix pore is not expanded. Figure 3 shows that the higher the precursor concentration of zinc nitrate, the greater the swelling capacity of NaAlg-PVA-g-AAm nanocomposite hydrogels modified ZnO nanoparticles. This is due to the more ZnO nanoparticle formed in the hydrogel matrix so that the pore will expand as explained in figure 1, so that its ability increases to absorb water [13,14]. The effect of the $Zn(NO_3)_2$ concentration to maximum swelling capacity at 300 minutes are 0 (H0), 6.25 (H1), 12.5 (H2), 31.25 (H3) and 62.5 ppm (H4) obtained 106.30, 146.30, 186.20, 230.20, 226.00 g/g, respectively. It shows that swelling maximum obtained on hydrogel at 31.25 ppm concentration is 230.20 g/g.

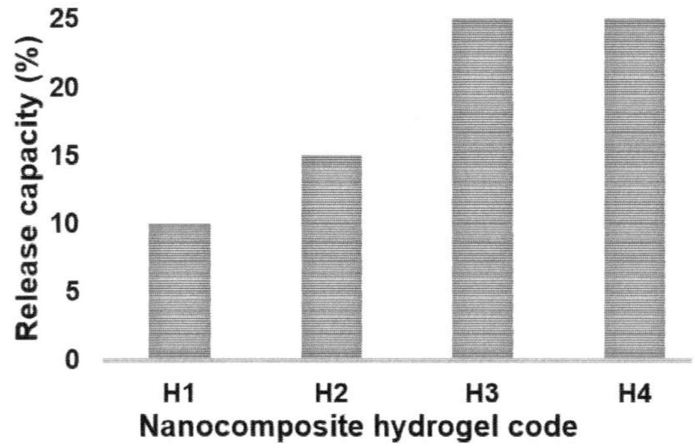

Figure 4. Maximum release capacity of Zn^{2+} ions from varieous of hydrogels

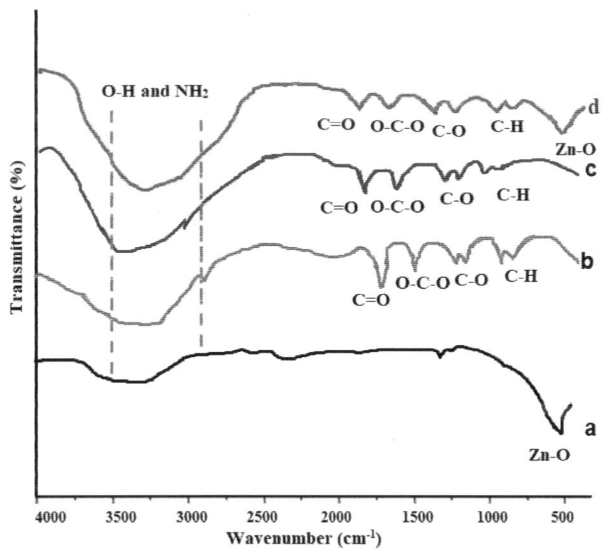

Figure 5. FTIR spectra of (a) ZnO, (b) natrium alginate, (c) NaAlg-PVA-g-AAm, and (d) NaAlg-PVA-g-AAm/ZnO

3.1.3. Release capacity. Nanocomposite hydrogel swollen at the maximum swelling capacity tested against the release capacity test of Zn^{2+} ions was determined by the atomic absorption spectroscopy (AAS) as shown in figure 4. It shows that the higher the precursor concentration of zinc nitrate, the greater the release capacity of Zn^{2+} ions. Release of Zn^{2+} ions from the hydrogel matrix was related to its maximum swelling capacity. If maximum swelling capacity is higher, it will cause more Zn^{2+} ions to be released from the hydrogel matrix. It shows that maximum release obtained at 31.25 ppm $Zn(NO_3)_2$ concentration is 25 % at 300 minutes.

3.2. Characterization

The result shows that loading capacity of Zn^{2+}, water swelling and the Zn^{2+} release of NaAlg-PVA-g-AAm modified with ZnO nanoparticles obtained the best hydrogel nanocomposite at 32.21 ppm Zn^{2+} concentration is 80%, 230 g/g and 25 %, respectively. The best nanocomposite hydrogel was characterized by FTIR, XRD, SEM and TEM-EDX.

3.2.1. FTIR analysis. Figure 5 shows FTIR spectra, in figure 5a, it shows the absorption spectra of the nano ZnO at 475 cm^{-1} peak, that is the identical absorption peak of the Zn-O bond, similar to our

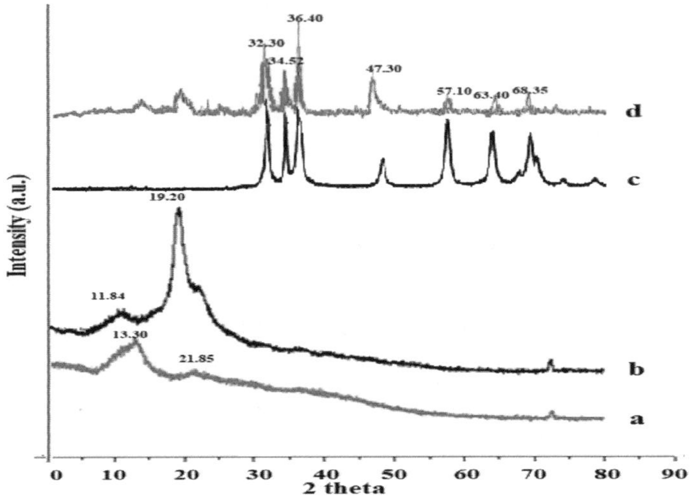

Figure 6. XRD pattern of NaAlg (a), NaAlg-PVA-g-AAm (b), ZnO (c), andNaAlg-PVA-g-AAm/ZnO (d)

previous study [5]. Figure 5b shows the spectra of the sodium alginate, the main peak at 752-855 cm^{-1} is mannuronic acid and 990 cm^{-1} and 1100 cm^{-1} are the C-O peaks of uronic acid, both of these peaks are identical band of sodium alginate. The peaks at 1480 and 1700 cm^{-1} which indicated the asymmetric and symmetric stretching of carboxylate O-C-O [2] and it can be observed in the group of C=O at 1650 cm^{-1} and C- O at 1300 cm^{-1} and O-H peak at 3000-3500 cm^{-1}. Figure 5c shows the spectra of the NaAlg-PVA-g-AAm hydrogel, there is a characteristic peak of alginate, PVA grafted with AAm identifies the C=O carboxylic group shifts to 1700 cm^{-1} and C-O 1350 cm^{-1}, this is due to the interaction between the groups and the OH peak becomes wide and sharp at 2900-3500 cm^{-1}; this is caused by a combination of OH groups from alginate, PVA, and overlap with acrylamide amine ($-NH_2$) monomers. Figure 5d shows the spectra of hydrogel nanocomposite modified by ZnO nanoparticles. Loading of Zn^{2+} into the NaAlg-PVA-g-AAm hydrogel network caused electrostatic interaction occur between Zn^{2+} ions and hydrophilic groups in the hydrogel pores and network. The interaction of hydrophilic groups such as O-H, N-H and carboxylate (C=O and C-O) intensity decrease [6,15]. As for O-H bond peak, its broaden and sharp band, it due to combination O-H of alginate, PVA, overlap amine ($-NH_2$) acrylamide monomer and zinc hydroxide remain. It can also be observed that the Zn-O peak shifts to 500 nm, this supports ZnO nanoparticles formed on the hydrogel and indicates the hydrogel has been successfully modified by ZnO.

3.2.2. XRD analysis. To ensure that ZnO nanoparticles were formed within the hydrogel matrix through ex-situ method, the sample was analyzed with X-Ray diffraction. XRD patterns hydrogels can be seen in figure 6. Figure 6a shows diffraction pattern of sodium alginate with 2θ =13.30° and 21.85° peaks [2]. Figure 6b shows diffraction pattern of NaAlg-PVA-g-AAm hydrogel can be seen 2θ peaks; 19.20° and 11.84°, it indicated that the peaks derived from PVA and overlap with sodium alginate peaks. Figure 6c shows diffraction pattern of ZnO nanoparticle has sharp peaks, there are 32.30°, 34.52°, 36.40°, 47.30°, 57.10°, 63.40° and 68.35°, which are the characteristic of ZnO nanoparticle. Figure 6d shows diffraction pattern of NaAlg-PVA-g-AAm/ZnO, it can be observed that all peaks derived from sodium alginate, PVA and ZnO nanoparticle are appeared. However, there is shift of the peaks due to interaction between groups, this supports the results of previous characterizations.

3.2.3. Characterization by SEM. The morphology of surface by SEM is shown in Figure 7. Figure 7b shows the surface morphology of NaAlg-PVA-g-AAm hydrogel has a smooth surface and smaller pores. It shows that graft polymerization reaction had occurred among sodium alginate, PVA and acrylamide, so it caused the pores on hydrogel. Figure 7c shows the surface morphology of NaAlg-PVA-g-AAm modified hydrogel have larger pores. It is due to ZnO nanoparticle have entered into hydrogel so that the pores will be bigger, it can be seen the ZnO nanoparticle have rod-shaped. This supports the swelling capacity to rise after the hydrogel is modified with ZnO nanoparticle.

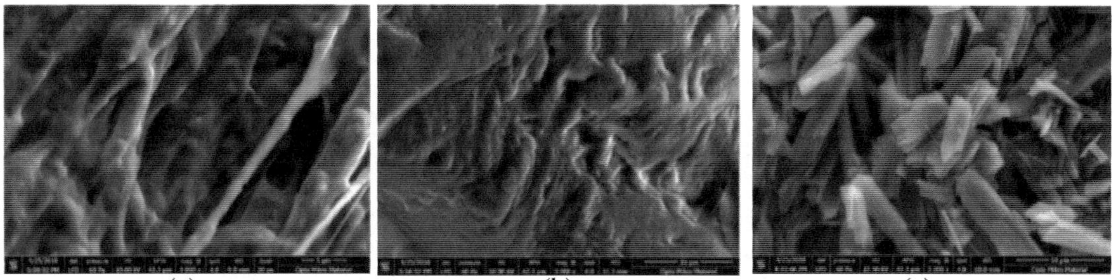

(a) (b) (c)

Figure 7. SEM images of NaAlg (a), NaAlg-PVA-g-Aam (b), and NaAlg-PVA-g-AAm/ZnO (c) with 3000x magnification

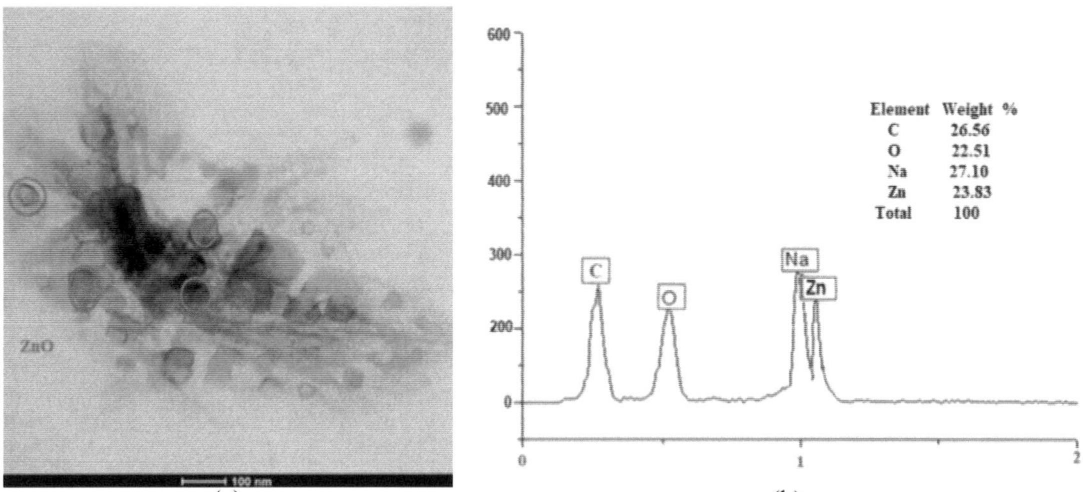

(a) (b)

Figure 8. TEM images (a), and EDX analysis (b) of NaAlg-PVA-g-AAm/ZnO

3.2.4. Characterization by TEM. Figure 8 shows TEM image of NaAlg-PVA-g-AAm/ZnO hydrogel nanocomposite. Figure 8a shows that the ZnO nanoparticle are evenly distributed with about 70 nm diameter in hydrogel nanocomposite. Figure 8b shows EDX spectrum NaAlg-PVA-g-AAm/ZnO hydrogel have C, O, Na and Zn elements are 26.56, 22.51, 27.10, and 23.83 %.wt, respectively.

From the characterization of FTIR, XRD, SEM and TEM-EDX can support the formation of ZnO nanoparticles in the hydrogel matrix.

3.3. Antibacterial activity
Antibacterial activity test was done using two types of cultural bacteria, which were *Staphylococcus aureus* positive bacteria and *Pseudomonas aeruginosa* negative bacteria. Figure 9 shows the results of antibacterial activity test of NaAlg-PVA-g-AAm/ZnO hydrogel. Antibacterial activity caused by Zn^{2+} ions and ZnO nanoparticles were released from the hydrogel. In figure 9, it can be seen that positive gram *S. aureous* bacteria was more resistant compared to negative gram *P. aeruginosa* bacteria. Maximum percent inhibition for *S. aureus* and *P. aeruginosa* was seen at concentration of $Zn(NO_3)_2$ of 31.25 ppm with inhibition percentage of 60.22 % and 40.10 %, respectively. NaAlg-PVA-g-AAm hydrogel nanocomposite modified with ZnO nanoparticles that have been successfully synthesized is more effective to *S. aureous* positive gram compare to *P. aeruginosa* gram negative bacteria.

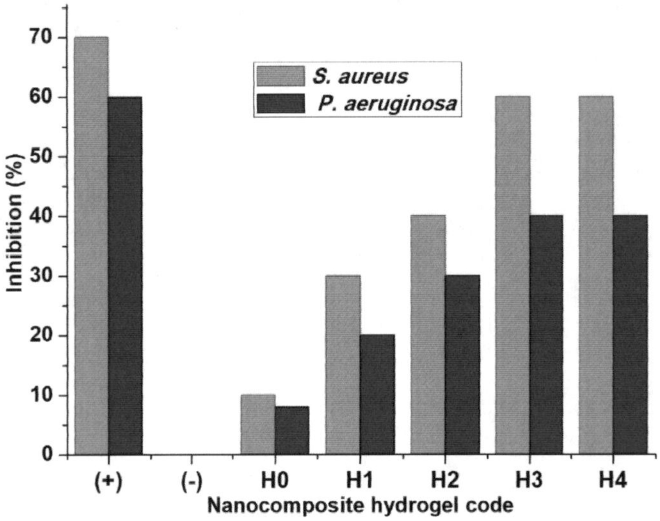

Figure 9. Antibacterial activity

4. Conclusions

Nanocomposite of NaAlg-PVA-g-AAm hydrogel modified with ZnO nanoparticles has been successfully synthesized. This study has obtained the best hydrogel nanocomposite with loading capacity of Zn^{2+} is 80 %, water swelling capacity is 230.20 g/g and the Zn^{2+} release is 25 % at 32.21 ppm Zn^{2+}concentration. The results were supported by FTIR, XRD, SEM and TEM-EDX characterization. Maximum percent inhibition for *S. aureus* and *P. aeruginosa* with 32.21 ppm Zn^{2+} concentration are 60.22 % and 40.10 %, respectively. Hydrogel nanocomposite modified with ZnO nanoparticles that have been successfully synthesized is more effective for *S. aureous* positive gram compared to *P. aeruginosa* negative gram bacteria.

Acknowledgements

The authors gratefully acknowledge the financial supports from Hibah PITTA Universitas Indonesia Grant for fiscal year 2018.

References

[1] Bhowmick S and Koul V 2016 *Mater. Sci. Eng.* C **59** 109–19
[2] Helmiyati and Aprilliza M 2017 *IOP Conf. Ser.: Mater. Sci. Eng.* **188** 012019
[3] Shalumon K T, Anulekha K H, Nair S V, Nair S V, Chennazhi K P and Jayakumar R 2011 *Int. J. Biol. Macromol.* **49** 247–54
[4] Motshekga S C, Ray S S and Maity A 2018 *J. Colloid Interface Sci.* **512** 686–92
[5] Helmiyati H and Wahyuningrum K D 2018 *AIP Conf. Proc.* **2023** 020107
[6] Ghasemzadeh H and Ghanaat F 2014 *J. Polym. Res.* **21** 355
[7] Yadollahi M, Gholamali I, Namazi H, Aghazadeh M 2015 Int. J. Biol. Macromol. **74** 136–41
[8] Hashem M, Sharaf S, Abd El-Hady M M and Hebeish A 2013 *Carbohydr. Polym.* **95** 421–7
[9] Wahid F, Yin J J, Xue D D, Xue H, Lu Y S, Zhong C and Chu L Q 2016 *Int. J. Biol. Macromol.* **88** 273–9
[10] Siddiqi K S, Rahman A, Tajuddin and Husen A 2018 *Nanoscale Res. Lett.* **13** 141
[11] Helmiyati, Saefumillah A and Yulianti W 2014 *Asian Journal of Chemistry* **26** 7337–42
[12] Goy R C, Morais S T B and Assis O B G 2016 *Rev. Bras. Farmacogn.* **26** 122–7
[13] Pandey S and Ramontja J 2016 *Int. J. Biol. Macromol.* **93** 712–23
[14] Basu S, Samanta H S and Ganguly J 2018 *Soft Mater.* **16** 7–19
[15] Deen G R and Chua V 2015 *Gels* **1** 117–34

IOP Conf. Series: Materials Science and Engineering **763** (2020) 012031 doi:10.1088/1757-899X/763/1/012031

Synthesis, characterisation and catalytic activity of V_2O_5 nanoparticles using *Foeniculum vulgare* stem extract

J L Al Jabbar[1], D O B Apriandanu[1], Y Yulizar[1] and S Sudirman[2]

[1] Department of Chemistry, Faculty of Mathematics and Natural Sciences (FMIPA), Universitas Indonesia, Kampus UI Depok, Depok 16424, Indonesia

[2] Center for Science and Technology of Advanced Materials (PSTBM) Batan, Kawasan Puspitek Serpong, Tangerang Selatan, Banten 15314, Indonesia

Corresponding author's e-mail: yokiy@ui.ac.id

Abstract. A green synthesis route of metal oxide has been developed to synthesize the V_2O_5 nanoparticles. V_2O_5 was synthesized using *Foeniculum vulgare* stems extract (FSE) and ammonium monovanadate as precursors. The role of secondary metabolite compounds affect the particle size of V_2O_5. The crystalline phase and crystallite size of V_2O_5 were investigated by X-Ray Diffraction (XRD) with the orthorhombic crystalline phase and the average crystallite size around 78.6 nm. The presence of functional groups was evaluated by Fourier Transform Infrared (FT-IR). The morphology, particle size, and chemical composition of V_2O_5 were analyzed by Scanning Electron Microscopy (SEM), Tunneling Electron Microscopy (TEM) and Energy Dispersive X-Ray (EDX). V_2O_5 nanoparticles show a good catalytic activity for the reduction of methylene blue (MB).

Keywords: V_2O_5, *Foeniculum vulgare* stems, catalytic activity, methylene blue

1. Introduction

Metal oxide nanoparticles have become an interesting research subject due to their potential application as catalyst [1–4]. Vanadium pentoxide (V_2O_5) is a transition metal oxide that has an important role in the development of science and technology [5], for instance as a catalyst [6,7], electronic devices [8] and biomedical application[9].

Vanadium generally has different oxidation numbers that give some oxide structures such as VO, V_2O_3, VO_2 and V_2O_5. Among of them, V_2O_5 is probably the most studied one, due to its thermodynamic stability [10]. Synthesis of V_2O_5 nanoparticles has been successfully carried out by conventional chemical methods using hazardous compounds [11] as base source and capping agent. Along with the advancement in science and technology, V_2O_5 nanoparticles can be synthesized by green chemical methods. This method usually uses extracts from microorganism or parts of plant [12]. Conceptually, this method has an approach such as physico-chemical synthesis method [13]. The development of this method in synthesis of V_2O_5 nanoparticles is still limited, for example by using saccharomyces *cerevisiae* [14] and rambutan peel [9]. Unfortunately, synthesis of nanoparticles using plants from Indonesia has not been widely reported.

Foeniculum vulgare grows wildly on the plateau with an alkaloids content of 119.37 μg/mg [15] and the other compounds such as flavonoids, steroids, and saponins. Secondary metabolite compounds plays a vital role as an alternative base sources, stabilizer, and capping agent in the synthesis of metal oxide nanoparticles [16]. Phytochemical tests were used to confirm the secondary metabolite compounds in *Foeniculum vulgare* qualitatively.

IOP Conf. Series: Materials Science and Engineering **763** (2020) 012031 doi:10.1088/1757-899X/763/1/012031

In this study, synthesis of V_2O_5 nanoparticles was performed using *Foeniculum vulgare* stems extract. In addition, V_2O_5 nanoparticles have been characterized by UV-Vis spectrophotometer, Fourier Transform Infrared (FT-IR) spectroscopy, X-Ray Diffraction (XRD), Scanning Electron Microscope-Energy Dispersive X-Ray (SEM-EDX), and Transmission Electron Microscopy (TEM). The catalytic activity of V_2O_5 nanoparticles was investigated for reduction of methylene blue (MB).

2. Experimental

2.1. Materials

Foeniculum vulgare stems were obtained from Pasuruan, East Java, Indonesia. Ammonium monovanadate (NH_4VO_3) was purchased from Merck chemicals. All reagents and solvents were analytical grade.

2.2. Preparation of stems extract

The *Foeniculum vulgare* stems extraction was adapted from the previous report with some modifications [17,18]. 50.0 g of dried *Foeniculum vulgare* stems was macerated in 250 mL of methanol for a week. The filtrate was partitioned by *n*-hexane. The methanol fraction was evaporated with a rotary vacuum evaporator. The concentrated methanol fraction was dissolved in distilled water to obtain aqueous *Foeniculum vulgare* stems extract (FSE). The presence of secondary metabolite compounds was identified qualitatively by phytochemical test.

2.3. Phytochemical test

Preliminary phytochemical test was conducted to determine the presence of alkaloids, flavonoids, terpenoids, steroids, tannins, saponins, and phenolic as described by Beyazen *et al.* [19]. For alkaloids identification test: 1.0 mL of FSE was reacted with 3.0 mL of 2N HCl and 4.5 mL of Wagner's reagent sequentially. Brownish red solution indicates the presence of alkaloids. For flavonoids test: 2.0 mL of FSE was reacted with 2.0 mL of 0.1M NaOH and then added H_2SO_4 dropwise. The fading color indicates the presence of flavonoids. For phenolic test: 2.0 mL of FSE was reacted with 1.0 mL distilled water and then added 5 % $FeCl_3$ dropwise. Dark green solution indicates the presence of phenolic. For saponins test: 1.0 mL of FSE was mixed and shaken with 3.0 mL of distilled water. The appearing bubbles indicate the presence of saponins.

2.4. Synthesis of V_2O_5 nanoparticles

1M NH_4VO_3 solution was reacted with FSE by stirring for 2 h at 80 °C and then incubated at room temperature for 7 days. The orange powder was obtained after the calcination at 600 °C for 4 h.

2.5. Characterization of V_2O_5 nanoparticles

The UV-Vis absorption of V_2O_5 nanoparticles was analyzed using UV-Vis spectrophotometer (Shimadzu 2600) at wavelength range 200–800 nm. The functional groups of FSE and V_2O_5 nanoparticles were recorded by FT-IR (Prestige-21 Shimadzu). The structural properties of V_2O_5 nanoparticles were identified by XRD (PANalytical Empyrean) using Cu Kα as radiation (1.5406 Å). The morphology and composition information of the V_2O_5 nanoparticles were characterized by SEM-EDX (Zeiss Evo MA10). The particle sizes of V_2O_5 nanoparticles were analyzed by TEM (Tecnai D2360 S-Twin).

2.6. Catalytic activity

The catalytic activity of V_2O_5 nanoparticles was examined by monitoring of methylene blue (MB) reduction using $NaBH_4$ as a reducing agent. The amount of V_2O_5 and 1 mM MB was mixed with 0.1M $NaBH_4$ and then stirred for 5 min. The MB's concentration was determined through observation of the decreasing absorbance at 664 nm using UV-Vis spectrophotometer.

3. Results and discussion

The phytochemical test was performed to investigate the presence of alkaloids, flavonoids, phenolic, and saponins in FSE. The methanol filtrate showed the positive result for all secondary metabolites. However, FSE only showed the presence of alkaloids, flavonoids, and phenolic (table 1).

3.1. UV-Vis absorption spectra analysis

Table 1. Phytochemical test of FSE

Secondary metabolite compound	Fraction	
	Methanol	Aqueous
Alkaloids	+	+
Flavonoids	+	+
Phenolic	+	+
Saponins	+	-

+ to indicate presence, - to indicate absence

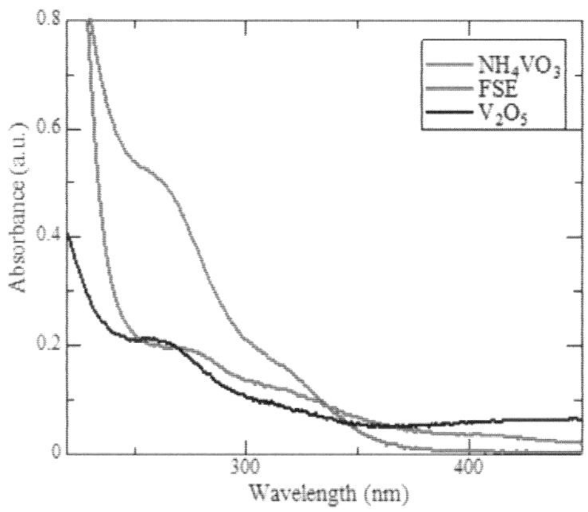

Figure 1. UV-Vis absorption spectra of NH_4VO_3 solution, FSE, and V_2O_5 nanoparticles

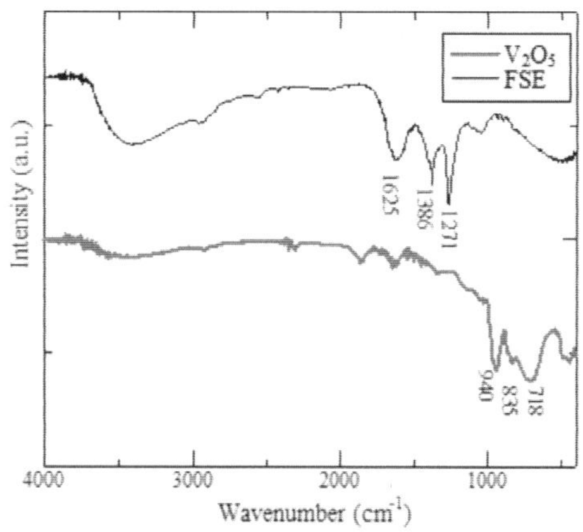

Figure 2. FT-IR spectra of FSE and V_2O_5 nanoparticles

The UV-Vis absorption spectra of V_2O_5 nanoparticles (black line) are shown in figure 1. For the comparison, the spectrum of NH_4VO_3 solution (blue line) and FSE (red line) were also analyzed. The maximum wavelength of NH_4VO_3 solution, FSE, and V_2O_5 nanoparticles were appeared at 276, 265, and 452 nm, respectively. The maximum wavelength of V_2O_5 nanoparticles is in accordance with the previous report [20].

3.2. FT-IR analysis

FT-IR analysis aims to reveal the functional groups of FSE and V_2O_5 nanoparticles. Figure 2 presents the wavenumber changes of FSE and V_2O_5 nanoparticles. The characteristic bands of alkaloids occur at 1625 cm^{-1}, 1386 cm^{-1} and 1271 cm^{-1} at the spectrum of FSE (black line) and these results were in accordance with previous report [21]. However, the band of alkaloids was not appeared at the spectrum of V_2O_5 nanoparticles (blue line). The characteristic vibration of V=O and V-O-V bond is shown at 940 cm^{-1}, 835 cm^{-1}, and 718 cm^{-1}. These results have the conformity with the previous report [22].

3.3. Structural and morphology

The crystalline properties of V_2O_5 nanoparticles were examined by XRD. As can be seen from figure 3, the diffraction patterns of V_2O_5 nanoparticles show the orthorhombic crystalline phase indexed to (ICDD No. 96-101-1292, space group Pmn21). There are 2 peaks at 2θ = 20° and 26° with the structural plane (001) and (101) indicating the presence of V_2O_5 nanoparticles [9]. The crystallite size of V_2O_5 nanoparticles was estimated by Debye-Scherrer equation obtained 78.6 nm.

The morphology of V_2O_5 nanoparticles prepared by a green synthesis method was characterized by SEM, as depicted in figure 4a. V_2O_5 nanoparticles reveal a form of cubic agglomeration with particle size on a scale bar of 200 nm. The corresponding EDX spectrum is shown in figure 4b indicating the elemental composition of vanadium and oxygen with the mass percentage of 11.77 % and 25.80 %, respectively.

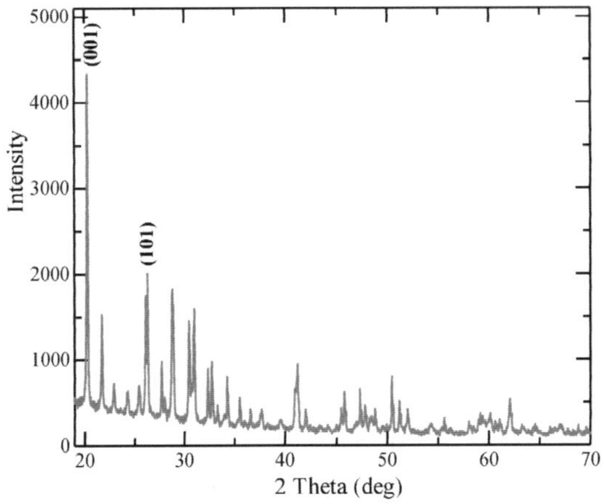

Figure 3. XRD pattern of V_2O_5 nanoparticles

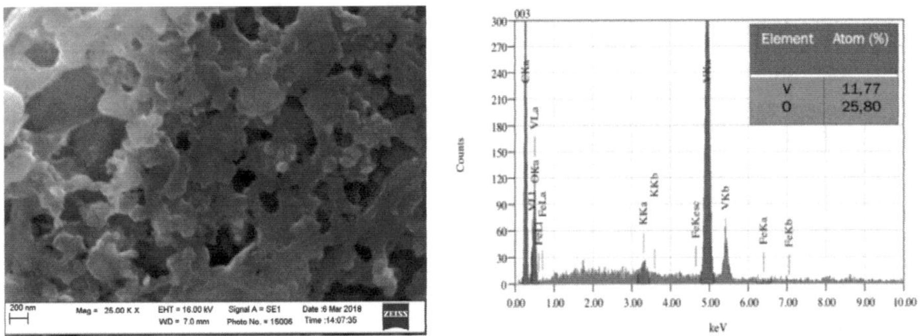

(a) (b)

Figure 4. (a) SEM image and (b) EDX spectrum of V_2O_5 nanoparticles

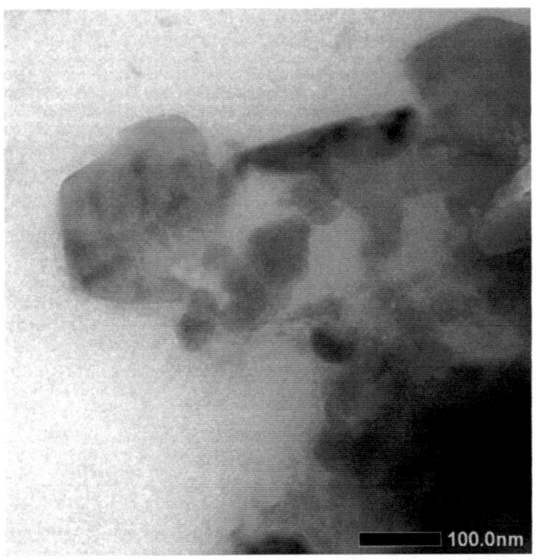

Figure 5. TEM image of V_2O_5 nanoparticles

For further characterization on the morphology of V_2O_5 nanoparticles, the TEM image exhibits cubic-like agglomerated morphology, which has an average diameter of around 70 nm as shown in figure 5.

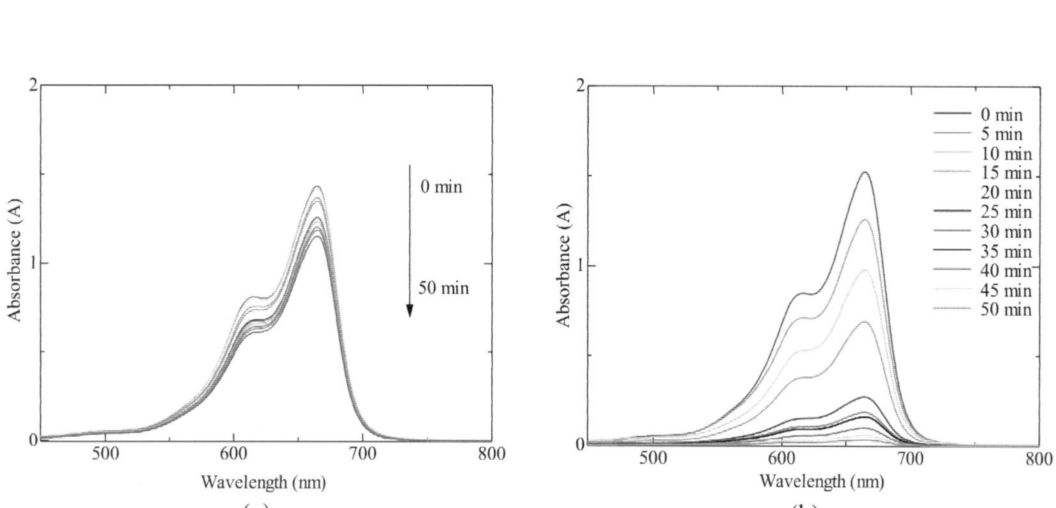

(a) (b)

Figure 6. UV-Vis absorption spectra of MB reduction (a) without and (b) with V_2O_5 nanoparticles catalyst

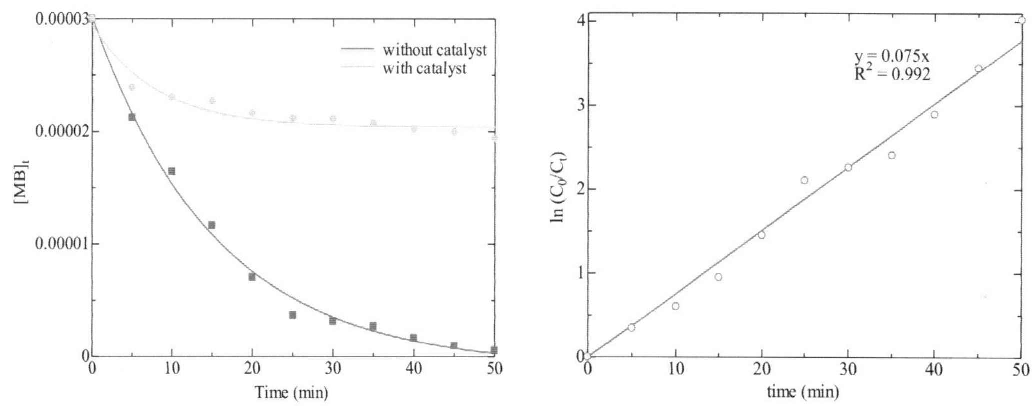

(a) (b)

Figure 7. (a) Plot of MB concentration as a function of time and (b) plot of ln (C_0/C_t) against time for MB reduction

3.4. Catalytic activity of V_2O_5 nanoparticles

The catalytic performance of synthesized V_2O_5 nanoparticles reduced MB to leuco methylene blue (LMB) using $NaBH_4$. The reaction was observed by UV-Vis spectrophotometer in wavelength range of 450-800 nm at room temperature as shown in figure 6. MB shows the UV-Vis absorption peak at maximum wavelength of 664 nm. Figure 6a shows slightly decrease of MB absorbance.

Figure 6b exhibits the decreasing absorbance of MB peak at 664 nm using V_2O_5 nanoparticles catalyst. The decreasing of MB concentration using V_2O_5 nanoparticles catalyst is faster than without catalyst as shown in figure 7a. V_2O_5 nanoparticles help the electron transfer from BH_4^- to MB. A linier correlation between ln (C_0/C_t), where C_t was concentration at various times and C_0 was concentration at the initial time, against reduction time (t, min) indicates that the MB reduction follows pseudo-first order kinetics. The observed rate constant (k_{obs}) was calculated from the slope of the straight line and obtained 0.075 min^{-1} as shown in figure 7b.

4. Conclusions

The V_2O_5 nanoparticles have been successfully prepared using *Foeniculum vulgare* stems extracts by green synthesis methods. According to XRD characterization, the crystalline phase of V_2O_5

nanoparticles shows an orthorhombic shape with the crystallite size of 78.6 nm. This result was strengthened by the supporting data from SEM, TEM, and EDX. The obtained V_2O_5 nanoparticles show the catalytic activity for MB reduction in 50 min. The reduction of MB followed pseudo-first order kinetics with k_{obs} value of 0.075 min^{-1}.

Acknowledgements
The authors would like to thank Universitas Indonesia for funding this research through PIT-9 Grant 2019.

References

[1] Zhang P, Zhan Y, Cai B, Hao C, Wang J, Liu C, Meng Z, Yin Z and Chen Q 2010 *Nano Res.* **3** 235–43
[2] Wu S, He Q, Zhou C, Qi X, Huang X, Yin Z, Yang Y and Zhang H 2012 *Nanoscale* **4** 2478–83
[3] Xie Y, Yan B, Xu H, Chen J, Liu Q, Deng Y and Zeng H 2014 *ACS Appl. Mater. Interfaces* **6** 8845–52
[4] Sulaiman N and Yulizar Y 2018 *Mater. Sci. Forum* **917** 167–71
[5] Menezes W G, Reis D M, Benedetti T M, Oliveira M M, Soares J F, Torresi R M and Zarbin A J 2009 *J. Colloid Interface Sci.* **337** 586–93
[6] Chin S, Jurng J, Lee J H and Moon S J 2009 *Chemosphere* 75 1206–9
[7] Kristensen S B, Kunov-Kruse A J, Riisager A, Rasmussen S B and Fehrmann R 2011 *J. Catal.* 284 60–7
[8] Wang Y and Cao G 2006 *Chem. Mater.* **18** 2787–804
[9] Yuvakkumar R and Hong S I 2016 *Mater. Sci. Eng.* C **65** 419–24
[10] Asim N, Radiman S, Yarmo M A and Banaye Golriz M S 2009 *Micropor. Mesopor. Mater.* **120** 397–401
[11] Uchaker E, Zhou N, Li Y and Cao G 2013 *J. Phys. Chem.* C **117** 1621–6
[12] Annu, Ali A and Ahmed S 2019. Green synthesis of metal, metal oxide nanoparticle and heir various application *Handbook of Ecomaterial* ed L M Torres-Martínez *et al.* (Cham: Springer) chapter 25 pp 2281-300
[13] Hussain I, Singh N B, Singh A, Singh H and Singh S C 2016 *Biotechnol. Lett.* **38** 545–60
[14] Talavera N, Navarro M, Sifontes A B, Díaz Y, Villalobos H, Nino-Vega G, Boada-Sucre A and Gonzalez I 2013 Green synthesis of nanosized vanadium pentoxide using Saccharomyces cervisiae as biotemplate *Recent Research Developments in Materials Science* vol 10, ed S G Padalai (Kerala: Research Signpost) chapter 5 pp 89–102
[15] Madhu M, Sailaja V, Satyadev T N V S S and Satyanarayana M V 2016 *J. Pharmacogn. Phytochem.* **5** 25–9
[16] Marslin G, Siram K, Maqbool Q, Selvakesavan R K, Kruszka D, Kachlicki P and Franklin G 2018 *Materials (Basel)* **11** 940
[17] Yulizar Y, Bakri R, Apriandanu D O B and Hidayat T 2018 *Nano Structure & Nano-Objects* **6** 300–5
[18] Haque M M, Begum S, Sohrab M H, Ahsan M, Hasan C M, Ahmed N and Haque R 2013 *Pharmacogn. Mag.* **9** 76–80
[19] Beyazen A, Dessalegn E and Mamo W 2017 *International Journal of Chemical Studies (IJCS)* **5** 18–25
[20] Sreedhara M B, Vasu K and Rao C N R 2014 *Z. Anorg. Allg. Chem.* **640** 2737–41
[21] Strekal' N D, Motevich I G, Nowicky J W and Maskevich S A 2007 *J. Appl. Spectrosc.* **74** 31–7
[22] Karim Z, Ansari M A, Shah M R, Karim A and Khan K M 2015 *Russ. J. Gen. Chem.* **85** 936–8

IOP Conf. Series: Materials Science and Engineering **763** (2020) 012032 doi:10.1088/1757-899X/763/1/012032

Evaluation of dissolution of purified bromelain from pineapple core extract (*Ananas comosus [L.] Merr*) in the form of chitosan-coated microspheres

T K Zahrah, S Setiasih and S Hudiyono

Department of Chemistry, Faculty of Mathematics and Natural Sciences (FMIPA), Universitas Indonesia, Kampus UI Depok, Depok 16424, Indonesia.

Corresponding author's e-mail: siswati.setiasih@sci.ui.ac.id

Abstract. The aim of this study is to coat the bromelain from pineapple's core into chitosan microspheres crosslinked by glutaraldehyde to maintain its activity until it reaches the intestines. The crude enzyme was first purified by ammonium sulfate precipitation method, followed by the dialysis. The bromelain fraction of the dialysis has a specific activity of 294.44 U/mg with a purity level of 15.68 fold compared to the crude enzyme. Coating bromelain using glutaraldehyde crosslinked microsphere showed an efficiency of 87.14 %. The dissolution test results showed the release rate of bromelain in artificial gastric fluid is 9.35 % and in artificial intestinal environment is 79.92 %. The result of this study indicates that chitosan-coated microspheres are good enough as bromelain coatings for slow-release matrices.

1. Introduction

Pineapple (*Ananas comosus* [L.] Merr) is the one of the tropical plant cultivated in Indonesia. In the food industry, usually pineapple core are not used and becomes waste. Bromelain is a proteolytic enzyme contained in almost all pineapple's part such as skin, stems, and core [1]. Purified bromelain purified has a therapeutic effect such as fibrinolytic, antithrombotic, and anti-inflammatory properties and has been used in animal and human studies [2]. In the stomach fluid, bromelain will experience a decrease in activity after four first hours [3]. This is due to the deactivation and degradation of bromelain by gastric enzymes and acidic gastric acid pH [4]. In previous studies, the activity of bromelain experiences disruption in the stomach fluid and has activity in the intestinal environment [5]. Therefore, in application as an oral medication, bromelain needs to be coated in order to be released on target. In addition to hydrogel, microspheres could also be other ingredient of the drug encapsulation material [6].

2. Experimental

2.1. Materials

Tools needed in this research are beaker glass, measuring gourd, measuring glass, cooler box, refrigerated centrifuge, pH meter, blender, UV-Vis spectrophotometer, thermometer, oven, micropipette, tip, vial bottle, cuvette, FTIR spectrometer, analytical balance, SEM, refrigerator, magnetic stirrer, hot plate, centrifuge tube, and dissolution apparatus. While, for necessary materials for doing the research were pineapple core, buffer phosphate, ammonium sulfate, Folin-Ciocalteu reagent, bovine serum albumin, $CuSO_4.5H_2O$, K-Na tartrate, Na_2CO_3, NaOH, Aquadest, buffer tris-HCl, casein TCA, chitosan powder, magnesium silicate, Al powder, glutaraldehyde, EDTA, aquademin, glacial acetic acid, BaCl, and HCl.

2.2. Crude enzymes preparation

In this study, bromelain was isolated from the Palembang pineapple core obtained from the Pasar Induk Kramat Jati, East Jakarta. The core pineapple was cut into small dice-shape and destroyed with blender. Core enzyme solution was filtered and centrifuged at 6000 rpm, 4 °C for 20 minutes. This filtrate is called crude enzyme. Set aside some ml of core pineapple solution and crude enzyme. Protein content test was performed with Lowry method and proteolytic activity test with Kunitz method [7-9].

2.3. Purification of crude enzymes
The crude enzyme filtrate in a cold state was then purified by multilevel fractionation using ammonium sulfate with interval concentrations variation of 0–20 %; 20–50 %; and 50–80 % [9]. Furthermore, the highest specific activity fraction, followed by dialysis purification method was conducted to remove ammonium sulfate.

2.4. Synthesis of chitosan microspheres with crosslinker glutaraldehyde
Chitosan 4 % (w/v) in acetic acid solution was mixed with magnesium silicate and aluminum powder to synthesize the chitosan microspheres with simple coacervation method [10].

2.5. Determination of crosslinking degree and swelling ratio
The crosslinking degree of the microspheres can be determined [11]. Cross-linked swelling microsphere behavior was explored by placing at 5 mL resevoir containing gastric fluid artificial (pH 1.2) and intestinal environment artificial (pH 7.4) [10].

2.6. Encapsulation bromelain to chitosan microspheres with post loading method
Immersed the microspheres into a bromelain solution for 24 hours. Furthermore, the microspheres which has been soaked overnight was washed using a phosphate buffer pH 7. After that, it was allowed to dry and became encapsulated chitosan microspheres. To determine the % efficiency, the encapsulated chitosan microspheres was rinsed with phosphate buffer 0.2 M pH 7 to determine the untrapped bromelain.

2.7. Dissolution of bromelain test
The dissolution test was carried out by loading the encapsulated microsphere into a dissolution apparatus. Sample was taken with interval of 30 minutes for 2 hours in artificial gastric fluid. Meanwhile, the filtrate was taken 10 mL. After two hours, the microspheres were transferred into the artificial intestinal environment with an interval of 30 minutes in the first two hours and 60 minutes in the next hour until eight hours. Each sample was tested to Kunitz and Lowry methods.

2.8. Microspheres characterization
Chitosan, non-covalent chitosan microspheres, crosslinked glutaraldehyde chitosan microspheres and encapsulated chitosan microspheres were characterized using an instrument Fourier Transform Infrared (FTIR) to determine the functional group changes. Furthermore, an analysis to see the morphology was performed using Scanning Electron Microscopy (SEM) for crosslinked glutaraldehyde microspheres before encapsulation, after encapsulation, and dissolution.

3. Results and discussion

3.1. Synthesis of chitosan microspheres with crosslinker glutaraldehyde
Chitosan was dissolved in acid condition using 5 % acetic acid because it is an amide with pKa 6.5. It can be protonated at low pH and dissolve in acidic atmosphere. The addition of aluminum powder to form artificial pores supports the impregnation of crosslinked chitosan microspheres with an amine crosslinking group [10]. This process is a way of making microspheres using a simple coacervation method, which utilizes the physicochemical properties of insoluble chitosan in basic pH medium, but precipitates when has a contact with an alkaline solution.

Furthermore, the immersion water was removed and the microspheres were re-immersed using a 2.5 % glutaraldehyde crosslinking agent for 24 hours. Also, the synthesized chitosan microspheres without addition of aluminum powder and magnesium silicate (chitosan microspheres variation b) and non-covalent chitosan microspheres was used as comparison. The dried microspheres can be seen in figure 1.

Figure 2a shows that chitosan powder has an absorption-uptake characteristic of a particular wave number. The characteristic lies in the widespread uptake of the wave number 3550-3200 cm^{-1} for the

Figure 1. Microspheres synthesized with (a) variation a (with addition of magnesium silicate and aluminum powder), (b) variation b (without addition of magnesium silicate and aluminium powder) and (c) noncovalent microspheres.

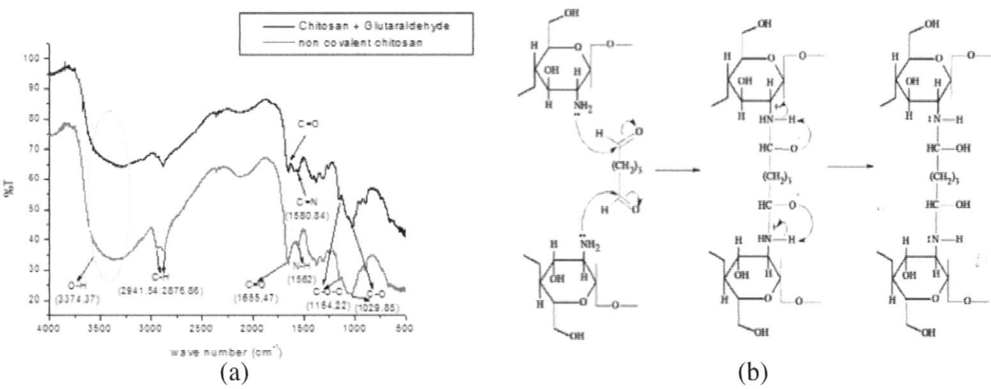

Figure 2. (a) FTIR spectra with (b) reaction scheme

vibration of the hydroxyl group (O-H) stretching and the vibration of the overlapping N-H amine group and the wave number 1570-1515 cm^{-1} for the N-H bending vibration. In the wave number 1649 cm^{-1} is the absorption resulting from the vibration of the remaining C=O acetyl group remaining on the chitin chain. Chitin has C=O chain and N-H chain. This shows that the chitosan used was not in 100 % purity level. Then, the wavenumber 1154.22 cm^{-1} shows the vibration of the bridge C-O-C, which is the typical absorption of the saccharide structure [12].

3.2. Reaction of crosslinking

Chitosan has several functional groups in the structure i.e. the primary amine group ($-NH_2$) and the hydroxyl group (-OH). In crosslinking reaction, chitosan with glutaraldehyde is reacted to an acidic atmosphere resulting in a nucleophilic addition reaction. It occurs when the free electron pair of the chitosan primary amine attacks the carbonyl group on glutaraldehyde, which has a positive partial. Then, there will be tetrahedral dipolar intermediates. The proton on N will be transferred to O and produce the neutral compound of carbinoamine. The crosslinking scheme can be seen in figure 2b. 5% acetic acid as solvent will make the tetrahedral carbinolamine to undergo protonation of the acid catalyst to the hydroxyl group. So, causing the electron pair on nitrogen to push the release of water to

Table1. The degree of crosslinking and swelling ratio microspheres with glutaraldehyde variety a, b, and non covalent.

Microspheres	Average % crosslinking degree ± SD (n = 5)	Average % Swelling Ratio pH 1.2 ± SD (n = 5)	Average % Swelling Ratio pH 7.4 ± SD (n = 5)
variety a	94.53 ± 0.67	116.45 ± 4.08	53.60 ± 2.18
variety b	96.79 ± 0.36	72.34 ± 3.80	10.43 ± 0.57
non covalent	-	122.71 ± 12.08	141.13 ± 7.03

Table 2. Specific activity of bromelain each purification

Sample	Volume (mL)	Total of proteolytic activity (U)	Total of protein content (mg)	Specific activity (U/mg)	Level of purity (fold)
Core Solution	550	467.5	169.4	2.760	-
Crude Enzyme	100	326.7	17.4	18.776	1
Ammonium sulfate fractionation method					
F1 (0-20)%	4	25.532	0.552	46.2536	2.46
F2(20-50)%	6	99.798	0.594	168.01	8.95
F3 (50-80)%	1.8	3.089	0.0648	47.669	2.53
F4	100	25	2.1	11.905	0.63
Dialysis method					
F5 Dialysis	5.1	94.605	0.3213	294.44	15.68

produce iminium ion. Then, the H^+ ion will be released from nitrogen into a neutral imine product with a new double bond of C=N. A series of crosslinking processes between chitosan and glutaraldehyde is a formation reaction of schiff bases [13].

3.3. Crosslinking degree and swelling ratio

The degree of crosslinking and swelling ratio are the parameters to test the physical properties of the microspheres. Microspheres immersed in a crosslinking agent have more rigid physical properties than non-covalent microspheres. The process of determining the degree of crosslinking is by soaking the microspheres with 5 % acetic acid for 24 hours. The indicator of crosslinking is the ratio of the dry mass of the microsphere after being immersed by the solvent to the dry mass of the microsphere before being immersed by the solvent.

Based on the results of cross-linked determination, the microspheres chitosan variety *a* has a lower crosslinked degree than variety *b* due to the presence of magnesium silicate on the surface which blocks the chitosan part for cross-linking (table 1).

The ability of microspheres in absorbing liquid phase to coat can be known by conducting swelling ratio test. The microspheres immersion is related to the ability of amine groups present in protonated chitosan due to the presence of H^+ concentration in the solution. The amine chitosan group will form the $-NH_3^+$ group, causing repulsion in the polymer chain and there is stretching on the polymer chain. Microspheres that have a lower degree of crosslinking will have a higher swelling ratio and will be considered by the selected microspheres.

3.4. Isolation of crude enzyme

The temperature in this process was 4 °C to inactivate the activity of other protease enzymes that can damage bromelain. The pineapple core solution was then centrifuged using a refrigerated centrifuge at 6000 rpm for 20 minutes.

Determination of protein content was done by Lowry method. The principle of this method is to sharpen the color produced in the biuret method, where Cu (II) -protein will be formed and in the alkaline atmosphere Cu (II) will be reduced to Cu (I). Furthermore, the determination of proteolytic activity was performed using Kunitz method. Proteolytic activity is the ability of a bromelain to hydrolyze or break down casein the substrate under its optimum condition [14]. The obtained specific activity based on test results of protein enzyme levels and proteolytic activity test is shown in table 2.

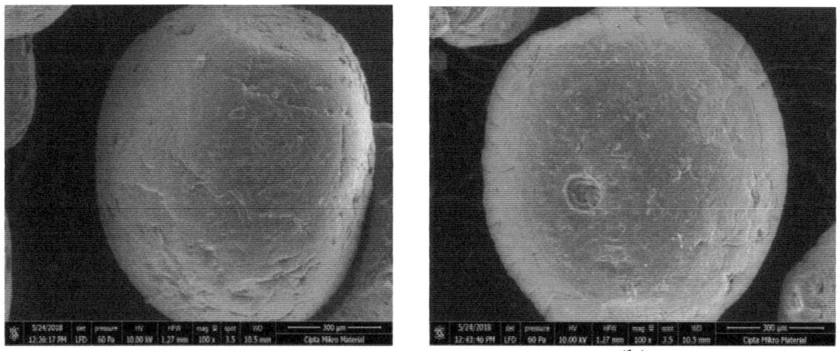

(a) (b)

Figure 3. The microsphere characterization 100x magnification with SEM instrument (a) before and (b) after encapsulation

Pineapple core and crude enzyme solutions have a specific activity of 2.76 U/mg and 18.78 U/mg. Table 2 shows an increase of enzyme activity after centrifugation.

3.5. Fractionation enzyme with ammonium sulfate

The principle of fractionation ammonium sulfate is salting out. Interactions between proteins are stronger than protein interactions with the solvent and makes the proteins can be separated from high salt concentrations [15]. The protein molecules are triggered to interact and form an aggregate in lower solubility. The variation in saturation level aims to determine the appropriate ammonium sulfate concentration to precipitate bromelain.

The solution, which has been reserved for ± 18 hours was centrifuged (6000 rpm, 20 min, ± 3 °C). The obtained pellets were then reconstituted at 0.2M phosphate buffer pH 7.0. The second fraction level of ammonium sulfate (20–50 %) showed the highest fraction with specific activity of 168.01 U/mg and the purity level 8.95 times compared with the crude enzyme. Then, carried out further purification step that is dialysis method.

Dialysis causes a decrease in bromelain protein content from 0.396 mg to 0.321 mg as a result of the release of a small non-bromelain protein from the cellophane membrane bag. This leads to an increase in the specific activity of bromelain to 294.44 with a purity rate of 15.68 times compared to crude enzymes. Furthermore, the results of this dialysis will be coated using chitosan microspheres.

3.6. Encapsulation of purified bromelain into chitosan microspheres

The chitosan microspheres variety *a* was used to coat the purified bromelain. The selection was based on swelling test results at pH 7.4 for 30 minutes. Variety *a* gave a larger swelling ratio i.e. 53.60 % compared to variation *b* i.e. 10.43 % (see table 1). A larger swelling ratio will give the effectiveness of coating bromelain into a microsphere. Meanwhile, although the noncovalent chitosan swelling was the largest among the variation i.e. 141.13 %, the microspheres cannot be used. This was decided after the assay was applied into artificial gastric orientation (pH 1.2) for two hours, which all microspheres were completely destroyed.

Bromelain was coated for 24 hours then washed with 2 ml of 0.2 M phosphate buffer pH 7.0 for measuring the amount of enzymes un-entangled in the chitosan microspheres. Thus, the percentage efficiency of bromelain coating of post loading method i.e. 87.14 ± 0.59 %. According to figure 3, a fairly obvious difference is observed between the microspheres before and after encapsulation. Microsphere that have not yet been incorporated in bromelain have a more subtle morphology than microspheres that have been incorporated in bromelain. In addition, on the microsphere surface after insertion of bromelain there are many streaks are arranged uniformly. It indicates the presence of bromelain entering the microsphere with the occurrence of swelling on the microsphere.

Furthermore, characterization was done using FTIR. Spectrum is presented in figure 4. Meanwhile, the spectra between microspheres of chitosan cross-linked glutaraldehyde before and after insertion of bromelain showed the same absorption peaks. This information indicated that there is no chemical interaction between microspheres and purified bromelain. However, there is a physical interaction in the form of hydrogen bonds shown in sharper intensity changes at about 3500-3080 cm^{-1}. The

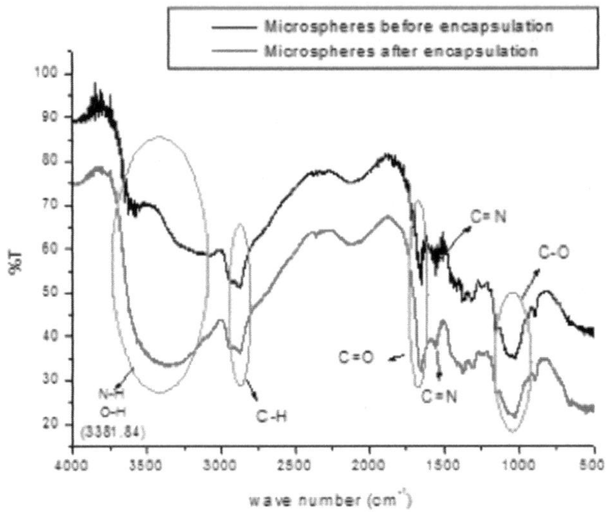

Figure 4. Spektra IR microspheres before and after encapsulation.

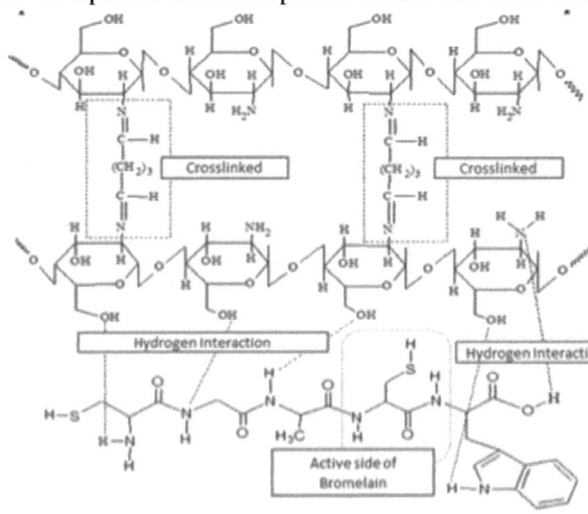

Figure 5. Proposed interactions chitosan-coated microspheres application with amino acid residues from bromelain

following proposed interactions between chitosan and amino acid residue fragments from bromelain by previous studies in figure 5.

3.7. In vitro dissolution test of bromelain

Dissolution is a process of the release of a drug, which aims to determine protein levels and proteolytic activity of bromelain escapes from chitosan microspheres. The profile can be seen from taking sample every 30 minutes for two hours in gastric fluid artificial and every 30 minutes then continue every 60 minutes up to 480 minutes in an intestinal environment artificial. This study used the basket dissolution method. The release rate of bromelain from the cross-linked glutaraldehyde chitosan microsphere is presented in figure 6. The figure shows that bromelain successfully escaped from the microspheres with the release rate i.e. 9.53 % in artificial gastric fluid and the release rate in the artificial intestinal environment i.e. 79.92 % made the total release of bromelain after ten hours was 89.45 %.

The proteolytic activity of each pH based on the sampling time interval was also measured and presented in figure 7 for artificial gastric (pH 1.2) and transferred to the artificial intestinal environment (pH 7.4). It can be concluded that bromelain has the highest proteolytic activity when it was in the intestinal environment after eight hours of in-vitro test process of 1.083 U/mL The

IOP Conf. Series: Materials Science and Engineering **763** (2020) 012032 doi:10.1088/1757-899X/763/1/012032

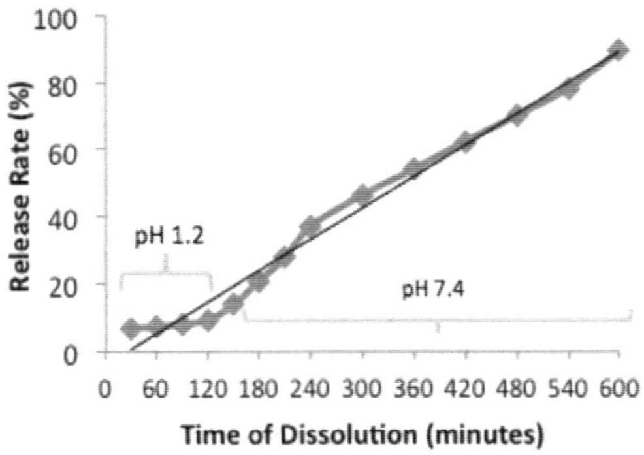

Figure 6. Graph of profile dissolution test of bromelain in the form of chitosan-coated microspheres

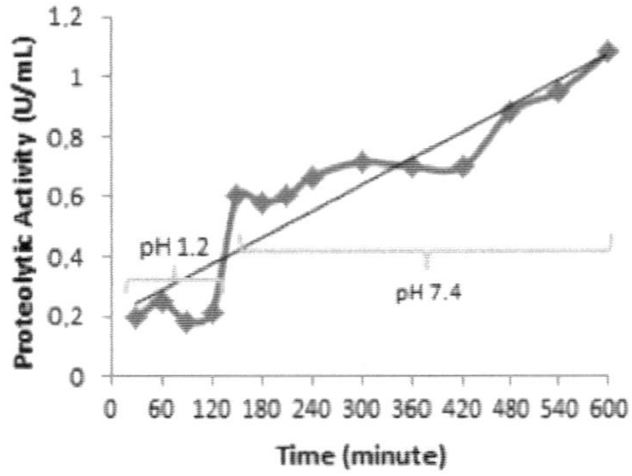

Figure 7. Graph of proteolytic activity of each sampling

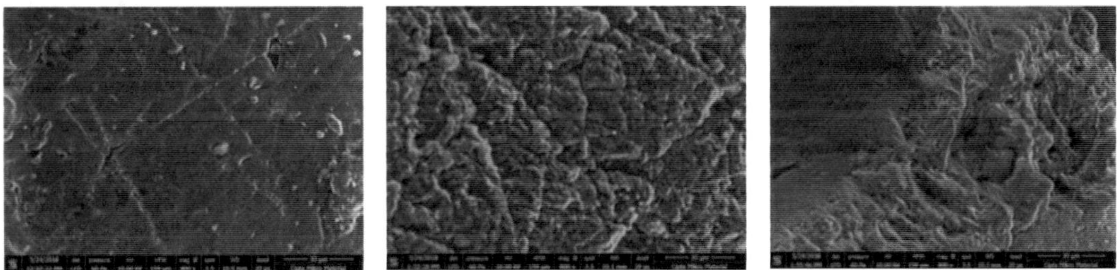

(a) (b) (c)

Figure 8. Microspheres surface morphological of dissolution test: (a) before dissolution, (b) after dissolution in artificial gastric fluid, and (c) after dissolution in artificial intestinal environment (SEM 800 times magnification).

fluctuative of proteolytic activity of enzyme is caused by the general properties of the enzyme, which works at the optimum pH and optimum temperature. The concentration of the enzyme also causes proteolytic activity. The destruction of the microspheres is marked by the large number of holes in which it causes enzymes to escape from the microspheres. The longer the microspheres are immersed in the artificial intestinal environment (pH 7.4) with the aid of a speed of 100 rpm, will cause the microspheres to be further damaged (figure 8). Accordingly, the more bromelain would be released in the intestines.

4. Conclusions

The crude bromelain enzyme from the pineapple core was successfully isolated with specific activity of 18.78 U/mg. Purification by fractionation method yielded the bromelain with purity level 8.95 fold i.e. 168.01 U/mg and dialysis method yielded bromelain with purity 15.68 fold i.e. 294.44 U/mg (ammonium sulfate concentration 20–50 %) compared with crude enzyme. Bromelain coated chitosan microsphere with an efficiency of 87.14 ± 0.59 %. Chitosan microspheres used as coatings have a crosslinking degree of 94.53 % and the swelling ratio at pH 7.4 of 53.60 %. A dissolution profile in artificial gastric fluid is 9.35 % for two hours and in artificial intestinal environment is 79.92 % for eight hours. The highest proteolytic activity is performed in 600 minutes in an artificial intestinal environment (pH 7.4). The results of this study indicate that chitosan-coated microspheres are good enough as bromelain coatings for slow-release matrices.

References

[1] Wan J, Guo J, Miao Z and Guo X 2016 *Food Chem.* **197A** 450–6
[2] Pavan R, Jain S, Shraddha and Kumar A 2012 *Biotechnology Research International* **2012** *ID* 976203
[3] Setiasih S, Adimas A C D, Dzikria V and Hudiyono S 2017 *IOP Conf. Ser.: Mater. Sci. Eng.* **299** 012016

[4] Shiew P S, Fang Y L and Majid F A A 2010 *Proc. 3rd Int. Conf. Biotechnology for the Wellness Industry* 8-9 October 2010 Kuala Lumpur (Kuala Lumpur: Universiti Teknologi Malaysia)
[5] Hale L P 2004 *Int. Immunopharmacol.* **4** 255–64
[6] Agnihotri S A, Mallikarjuna N N and Aminabhavi T M 2004 *J. Control. Release* **100** 5–28
[7] Febriani K, Wahyuni I, Setiasih S and Hudiyono S 2017 *AIP Conf. Proc.* **1862** 030095
[8] Soares P A G, Vaz A F M, Correia M T S, Pessoa A and Carneiro-da-Cunha M G 2012 *Sep. Purif. Technol.* **98** 389–95
[9] Soares P A G, Coelho D, Mazzola P, Silveira E, Carneiro-da-Cunha M G, Pessoa Jr A and Tambourgi E B 2011 *Chem. Eng. Trans.* **24** 979–84
[10] Gonçalves V L, Laranjeira M C M, Fávere V T and Pedrosa R C 2005 *Polímeros* **15** 6–12
[11] Abdel-Mohsen A M, Aly A S, Hrdina R, Montaser A S and Hebeish A 2011 *J. Polym. Environ.* **19** 1005–12
[12] Silverstein R M,Webster F X and Kiemle D J 2005 *Spectromectric Identification of Organic Compound* 7th ed (Hoboken: John Wiley and Sons)
[13] Singh A, Narvi S S, Dutta P K and Pandey N D 2006 *Bull. Mater. Sci.* **29** 233–8
[14] Gautam S S, Mishra S K, Dash V, Goyal A K and Rath G 2010 *Thai Journal of Pharmaceutical Sciences (TJPS)* **34** 67–76
[15] Reece J B and Campbell N A 2011 *Campbell Biology* 9th ed (Boston: Benjamin Cummings)

Synthesis and characterization of Cu^{2+} imprinted polymer-tannin extract from mango leaf (*Mangifera indica L.*) for selective separation of Cu^{2+}

F A Aribah, D A Nurani, D U C Rahayu and Y K Krisnandi

Department of Chemistry, Faculty of Mathematics and Natural Sciences (FMIPA), Universitas Indonesia, Kampus UI Depok, Depok 16424, Indonesia

Corresponding author's e-mail: d.arifa@sci.ui.ac.id

Abstract. Tannins are secondary metabolites in Mango (*Mangifera indica L.*). This compound has a cluster of polyphenols as functional ligands for chelating Cu(II). Cu(II)-imprinted polymer-tannin (Cu-IP-tannin) was successfully synthesized by polymerization of phenol and formaldehyde with the complex of Cu(II)-tannins in acidic medium. Non-imprinted polymer (NIP) was similarly prepared without the presence of Cu(II) ions. The polymers were characterized by Scanning electron microscopy energy dispersive x-ray spectroscopy (SEM-EDS) and Fourier transform infra-red (FTIR). Cu(II) ions were completely removed by leaching the dried and powder of imprinted polymer with HNO_3 0.5 M. The optimum pH of maximum adsorption of Cu(II)-IP-tannin was 7.0 for 120 minutes of contact time. The adsorption of Cu(II)-IP-tannin followed the Freundlich isotherm model with a maximum adsorption capacity of 99.08 mg/g. The relative values of the selectivity factor (α_r) of Cu(II)/Ni(II), Cu(II)/Pb(II) and Cu(II)/Fe(II) on a single metal ion selectivity test were 23.89, 55.71 and 26.25, respectively.

Keywords: copper, tannins, ion imprinted polymer, selective separation, mango (*Mangifera indica L.*)

1. Introduction

Pollution of heavy metals like copper (Cu) especially in the waters brought attention of many parties due to its accumulation in organisms, toxicity and long-term effects. Solid phase extraction currently has been widely used for the separation of metal ions before quantitative analysis [1–3]. This technique was developed by the use of ion exchangers or chelating agent absorbent [4]. The previous research reported tannins as an effective functional ligand used for chelating agent for metal Cu(II) [5-6].

Absorbent selectivity can be improved by Ion Imprinted Polymer (IIP) [7]. In addition, IIP has a high affinity, a good selectivity, an easy preparation, and economical. IIP can be prepared from the reaction between templates, functional monomer/ligand binding agents, crosslinking, and initiator [8].

In this research, we describe the synthesis of a new ion imprinted polymer based on tannins from mango leaf as a chelating agent and metals ion, Cu(II), as a template. The complex is subsequently crosslinked with phenol and formaldehyde using sulfuric acid as a catalyst and initiator. The synthesized polymer was then tested for their adsorption ability in various pH and contact times. The maximum adsorption capacity, reusability and adsorption isotherm were also determined. In addition, the effect of interfering cations (Ni^{2+}, Pb^{2+}, and Fe^{2+}) was investigated to study the IIPs selectivity. The main object of this research is to investigate the selective analysis of Cu(II) in the presence of competing species by IIP synthesized using natural chelating agent.

2. Experimental

2.1. Materials
All chemicals used in this research were mango leaves (*Mangifera indica L.*) (Jakarta, Indonesia); Phenol ≥ 99.99 % (Merck), formaldehyde 37 % (Merck), Na_2CO_3, dichloromethane, HNO_3 65 % (Merck), H_2SO_4 97 % (Merck), acetic anhydride, chloroform, *n*-hexane, ethyl acetate (EtOAc), NaOH 99 % (Merck,), $FeCl_3 6H_2O$ (Merck), Wagner's reagent, iodine, KI, HCl 37 % (Merck), double distilled water, and nitrogen gas with high purity. Analytical grade $Pb(NO_3)_2$, $Cu(NO_3)_2.3H_2O$, $Ni(NO_3)_2.6H_2O$, $Fe(NO_3)_2.9H_2O$ were purchased from Merck.

2.2. Instrumentation
The polymer was specified by Fourier Transform Infra-red (FT-IR) spectroscopy (Prestige 21 Shimadzu). Scanning Electron Microscope (SEM) was used to obtain the surface morphology of synthesized polymer, Energy Dispersive X-ray Spectroscopy (EDS) to characterize the composition of atoms in polymers before and after leaching, Spectrophotometer UV-Vis Shimadzu 2460 to study the complex of metal-tannins and Atomic Absorption Spectroscopy (AAS) Shimadzu AA 6800 to analyze metal ions.

2.3. Isolation of tannin from Mango leaf (Mangifera indica L.)
5 kg of wet mango leaves were dried under the sunlight for 3-7 days. The dried leaves were mashed to get the powder of mango leaves. 1 kg powder of dried mango leaves were macerated three times with 5 L of distilled water for 24 h at room temperature. The extract was filtered using Whatman filter paper. The filtrates were extracted in dichloromethane (4:1) for 5 min. The aqueous fraction was then dried by freeze dryer.

2.4. Phytochemical screening
The phytochemical analysis was carried out on water extract using standard procedure to identify the constituents as described by Harborne [9]. Wagner's reagent was used to identify alkaloids (formation of brownish precipitate). Tannins identification was tested by two drops of 2 % $FeCl_3$. Blue dark color and precipitate indicated the presence of hydrolysable tannins [9]. 0.5 mL extract was diluted in 5 mL distilled water and heated in water bath. Formation of froth indicated the presence of saponins [9]. 0.5 mL extract was added to the mixture of 0.5 mL chloroform and 0.5 mL acetic anhydride. The reddish brown coloration showed positive result for terpenoids [9]. Whereas, the presence of steroid was marked by the color change from violet to blue or green [9]. Flavonoids were identified by adding 1 mL extract in 0.1 M NaOH (1:1) followed by the drop wise addition of 10 % HCl. A yellow solution that turned colorless indicates the presence of flavonoids [9].

2.5. Synthesis Cu-IP-tannin
The tannin extract was complexed with Cu(II) ions (1:1) in demineralized water as solvent. The solution was stirred for 120 min at room temperature. Then, phenol (50 mmol), HCHO (50 mmol), and 50 mL H_2SO_4 0.2 M 50 mmol were added to the mixture. The mixture was stirred until homogeneous and purged with nitrogen gas for 10 minutes to release the dissolved oxygen. Polymerization was performed in a temperature-controlled water bath reflux at 65 °C for 5 hours. The resulting product was washed with demineralized water repeatedly to get a neutral pH. The entrapped Cu(II) ions were leached by stirring the polymer in HNO_3 0.5 M for 27 h in batch experiments. The Cu(II) contents of the filtrate were determined by AAS. The polymer was then washed with demineralized water and dried at 50 °C for 24 h. The procedure of non-imprinted polymers synthesis was similar with the procedure of imprinted polymer synthesis without Cu(II) as the template.

2.6. Copper adsorption study

Table 1. Phytochemical Screening of water extract of the Mango Leaf

Test	Result
Flavonoid	+
Terpenoid	-
Steroid	-
Tannins	+
Saponins	-
Alkaloid	+

Notes : (+) presence, (-) absence

50 mg of imprinted polymer was added into 10 mL of Cu(II) solution at various concentrations of 10, 20, 100, 250, 500 and 1000 mg/L with adjusted pH (2–12) at room temperature for several times (30–150 min) in stirred batch experiments. The remaining copper concentration in the solution was measured by AAS. The adsorption properties were investigated by Langmuir isotherm using equation (1) and Freundlich isotherm using equation (2) [10–11].

$$\frac{C_e}{q_e} = \frac{1}{q_e K_e} + \frac{C_e}{q_m} \quad (1)$$

where q_e is the amount of adsorbed Cu(II) in imprinted polymer (μmol/g), C_e is the equilibrium concentration of Cu(II) ions (μg/mL), q_m is the Langmuir constant associated to adsorption capacity, and K_e is the Langmuir constant associated to the energy of adsorption.

$$\log\frac{x}{m} = \log K + \frac{1}{n}\log C \quad (2)$$

where K and n are Freundlich constants attributed to adsorption capacity and adsorption intensity.

2.7. Selectivity experiments

50 mg of Cu-IP-Tannin and NIP were added into mixed metal solutions with metal concentration of 5 mg/L Cu(II), Pb(II), Ni(II) and Fe(II). All selected competitor ions have similar charge with Cu(II). The experiments were occurred at room temperature with optimum pH and contact time. The amount of each metal was calculated from the initial and equilibrium solution concentration analyzed by AAS. The adsorption capacity, distribution ratio, relative selectivity factor of Cu(II) in regards to Pb(II), Ni(II) and Fe(II) were calculated using equation (3) [10]:

$$Q = \frac{(C_0 - C_e)V}{W}, \quad D = \frac{Q}{C_e}, \quad \alpha = \frac{D_{Cu}}{D_{M\cdot}} \quad (3)$$

where Q = adsorption capacity (g/g), C_0 = initial concentration of each metal (g/L), C_e = equilibrium concentration of each metal (g/L), V = volume of ion metal solution (L), W = mass of imprinted polymer (g), D = the distribution ration (mL/g) and α = relative selectivity factor of Cu(II) and other metals.

3. Results and discussion

3.1. Phytochemical analysis

The presence of flavonoid, terpenoid, steroid, tannins, saponins and alkaloid in water extract of mango leaf was observed by phytochemical screening following the protocol described in Harborne [9]. In table 1, the result shows that tannins were successfully isolated from mango leaf. This compound contained flavonoid and alkaloid that was indicated by the fading color from yellow to colorless solution [9]. The result also showed the brownish precipitate that indicates the presence of alkaloid [9]. It is suggested that the detected alkaloids can be due to the alkaloid compound binding to the tannin compound.

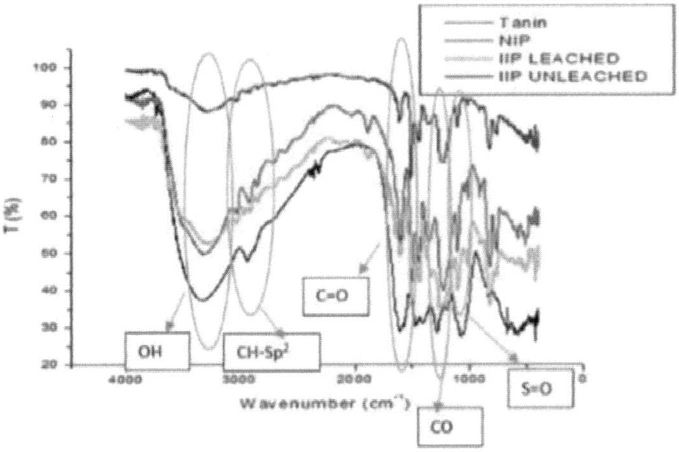

Figure 1. FT-IR Spectrum of Tannin Extract from Mango Leaf, Cu-IP-Tannin unleached, Cu-IP-Tannin leached and NIP

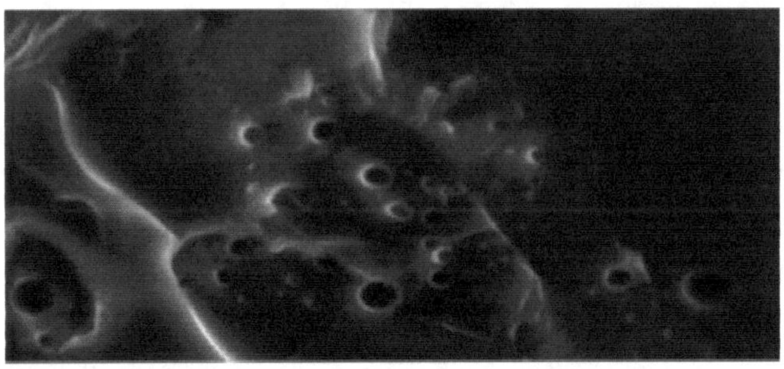

Figure 2. SEM Image of Cu-IP-Tannin Leached (5000x Magnification)

3.2. Characterization study

The FTIR spectra of isolated tannin compound showed the characteristic of stretching vibration of –OH at 3326 cm^{-1}, the typical C=O and C-O ester stretching bands at 1611 cm^{-1} and 1310 cm^{-1} respectively.

FTIR spectra of NIP and IIP show similar backbones due to their identical components (figure 1). The strong absorption peak at 1611 cm^{-1} is specified for the stretching vibration of C=O and a peak at 1310 cm^{-1} attributed to C-O. The typical sulphonic acid absorption band S=O at 1096 cm^{-1} appeared in imprinted polymer. The shifting of –OH peak to lower wavenumber was observed in IIP due to intermolecular H-bonding [12]. The intensity of –OH stretching band of imprinted polymer was lower compared to that of tannin. Those peaks confirmed that tannin has been copolymerized with phenol-formaldehyde in H_2SO_4 solution. Furthermore, the similarities of FTIR spectra of IIP leached and unleached indicated that the polymer network was not affected by the leaching process.

The morphology of Cu-IP tannin after leaching was characterized by Scanning Electron Microscopy (SEM). Cu-IP-Tannin leached has formed a circular mold as shown in figure 2. It indicated that all of Cu metal in Cu-IP-Tannin leached has released all and left a mold. EDS (Energy Dispersive X-Ray Spectroscopy) characterization of Cu- IP-Tannin leached and Cu-IP-Tannin unleached was performed to ensure that all of Cu ions metal in the Cu-IP-Tannin process was released. There are some elements of Cu on Cu-IP-Tannin before leaching can be seen in figure 3. In addition to Cu elements, there are also some

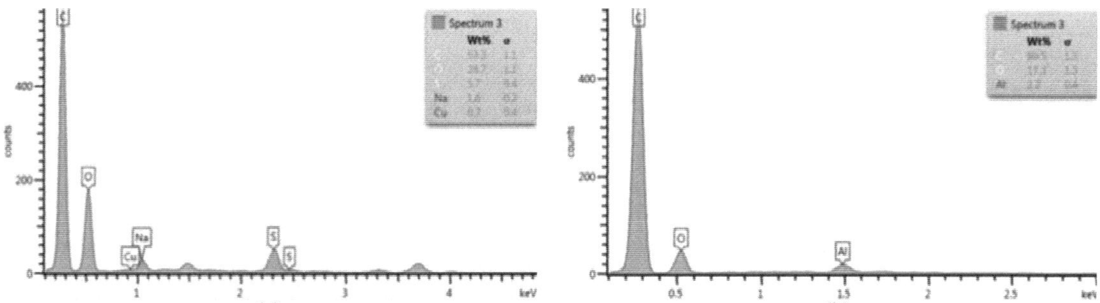

(a) (b)

Figure 3. EDS Spectrum of Cu-IP-Tannin unleached (a) and Cu-IP-tannin leached (b)

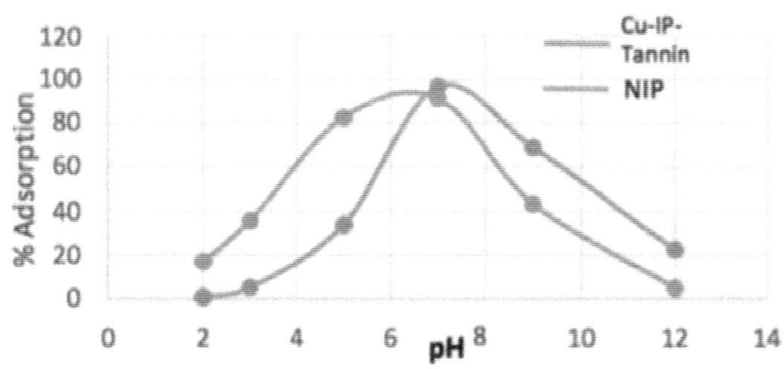

Figure 4. Effect of pH on the adsorption of Cu(II) on imprinted polymer; 50 mg of IIP and NIP; 5 mg/L Cu(II); room temperature

elements of C, S, Na and O from tannins and crosslinkers. There is no Cu ion on Cu-IP-Tannin leached as shown in figure 3. It is indicated that Cu-ions on Cu-IP-Tannin leached have removed.

3.3. Effect of pH
The pH influence on Cu(II) ions adsorption onto the synthesized polymer (IIP and NIP) was observed in batch experiments by altering pH of Cu(II) solution (5 mg/L) from pH 2 to 12. Figure 4 shows similar properties of Cu(II) adsorption on IIP and NIP. Copper ions adsorption increase with increasing pH from 2 to 7 and decrease at higher pH (figure 4). At the acidic atmosphere (pH 2 and 3) there was a competition between Cu^{2+} ions and H^{+} ions to occupy the template formed in the polymer matrix. In addition, H^{+} ions have a greater affinity for binding ligands than Cu^{2+} ions. The solubility of Cu(II) ions is governed by hydroxide concentration and precipitated as copper hydroxide at pH > 7. This also depends on the Cu(II) ions concentration in the medium [13]. Hydroxide ions have a high affinity to Cu^{2+} ions, so that Cu^{2+} ions prefer to bind with OH^{-} ions in solution compared to adsorbents, which have a specific cavity to Cu^{2+} ions. This condition leads to the fact that the optimum adsorption of Cu(II) was reached at pH = 7.

3.4. Equilibrium adsorption time
The equilibrium adsorption time of Cu(II) ions from aqueous solution onto the imprinted polymer was investigated. Based on the figure, at the contact time of 30, 60, 90, and 120 minutes, the adsorption percent increased and tended to be constant at 150 min (figure 5). The interaction of Cu-IIP with Cu(II) could strengthen the ion-dipole interaction. The equilibrium contact time was obtained at 120 minutes with Cu absorption of 92 %. This equilibrium adsorption time was influenced by the complexation rate such as affinity of Cu(II) ions and template groups in polymer structure [13].

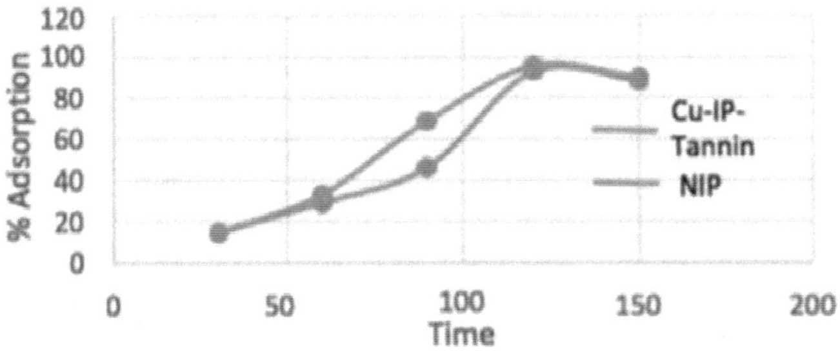

Figure 5. Equilibrium adsorption time of Cu(II) ions on to imprinted polymer (IIP and NIP); 50 mg imprinted polymer; 5 mg/L Cu(II); room temperature

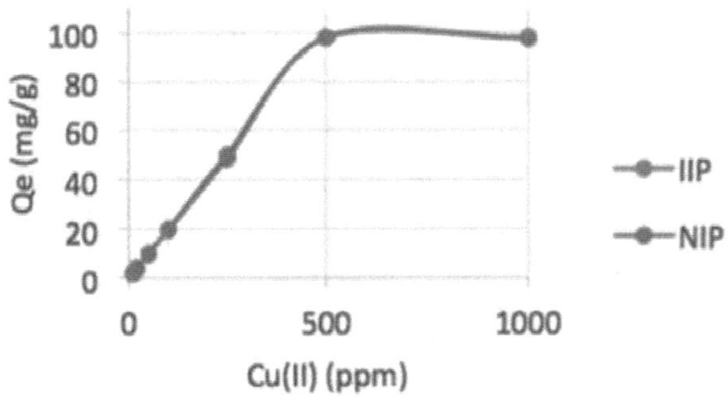

Figure 6. Adsorption capacity of Cu-IP-Tannin and NIP

3.5. Adsorption capacity

The adsorption profile of Cu(II) in the concentration range 0–1000 mg/L on to imprinted polymer was investigated by batch experiments. 50 mg of synthesized polymer was added to 10 mL Cu(II) solution at optimum pH (pH 7) and equilibrium contact time (120 min). The number of Cu(II) ions adsorbed per unit mass of synthesized polymer escalated with the increase of the initial Cu(II) concentration (figure 6). The initial Cu (II) concentration was escalated to reach saturation (adsorption capacity values). The maximum capacity of synthesized polymer was 99.08 mg/g of Cu-IIP and 99.6 mg/g of NIP. These results indicate that Cu-IIP has as high adsorption capacity as NIP. However IIP has fixed functional group due to imprinting effect, while the functional monomers are randomly distributed in polymer matrix of NIP [14].

The adsorption isotherm was studied to understand the mechanism of adsorption of Cu^{2+} metal ions by Cu-IP-Tannin and NIP, so that it can be known how the molecular distribution was adsorbed between the liquid phase and the solid phase. The type of adsorption isotherm, which often used in the determining of the adsorption isotherm pattern of an adsorbent, is the Langmuir in equation (1) and Freundlich in equation (2). It can be seen in figure 7 that Freundlich model is more suitable in the fitting of isothermal data adsorption of Cu-IP-tannin, while Langmuir model does not have good linearity. The adsorption profile of NIP towards Cu(II) ions follows Langmuir adsorption model ($R^2 = 0.9826$). Meanwhile, the plot of log Q_e vs log C_e exhibits that the adsorption profile of NIP does not obey the Freundlich isotherm adsorption model in the concentration range studied (figure 7). These results show that Cu-IP-tannin has heterogeneous binding sites [10]. The adsorption chamber holds more than one molecular layer and not all adsorbed molecules come into contact with the surface of the adsorbent (multilayer) surface. Langmuir's adsorption isotherm model assumed that the adsorption process of NIP occurs only on the certain homogeneous sites. It is suggested that Cu-IP-tannin has both specific binding sites and unspecific binding sites while NIP only has unspecific binding sites. The specific binding sites are related to the selectivity of adsorbent due to better separation process [15].

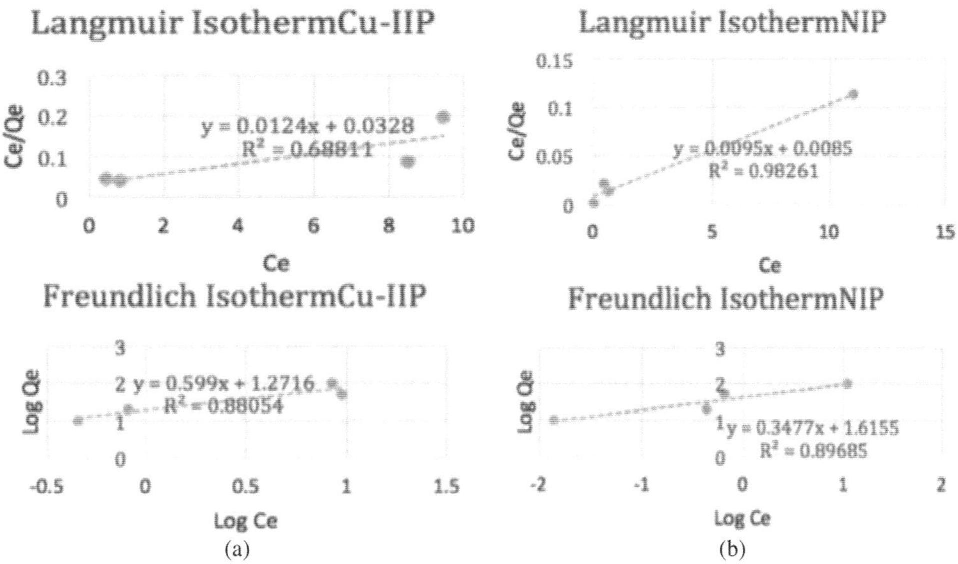

Figure 7. Langmuir and Freundlich Isotherm model of (a) Cu-IP-Tannin and (b) NIP

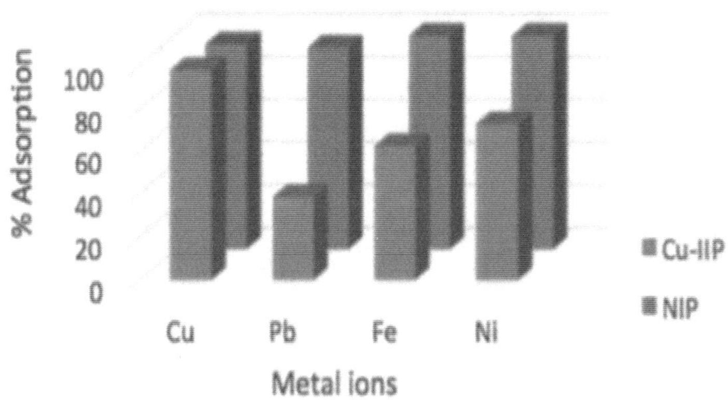

Figure 8. Competitive adsorption of Cu(II) and other metal ions; 5 mg/L of Cu(II), Pb(II),Fe(II), and Ni(II); 50 mg sythesized polymer; room temperature

3.6. Selectivity study

The adsorption of Cu(II) in the presence of competing divalent ion Pb(II), Fe(II), Ni(II) onto the surface of IIP was studied to be compared with that of NIP. The adsorption of competing ions was lower than Cu(II) adsorption (figure 8). It can be influenced by the size and shape of the cavity sites and affinity of ligand towards imprint metal ion [16]. The relative values of the selectivity factor (αr) of Cu (II)/Ni (II), Cu (II)/Pb (II) and Cu (II)/Fe (II) on a single metal ion selectivity test were 23.89, 55.71, and 26.25, respectively. This result indicated that the synthesized Cu-IP-Tannin was selective to those metals (αr > 1). Therefore, Cu(II) could be detected by the existence of Ni(II), Pb(II) and Fe(II), while NIP had the αr value of < 1.

3.7. Reusability
To study the reusability of imprinted polymer, five identical measurements were performed using 50 mg prepared polymer in 10 mL Cu(II) 5 mg/L.H_2SO_4 0.5 M was used for Cu(II) elution. From the calculation results, %RSD (Relative Standard Deviation) for Cu-IP-Tannin was 0.52 %, while CV (Coefficient Variation) Horwitz for Cu-IIP was 1.02 %. The precision was good, if RSD value was lower than CV Horwitz [17]. It can be concluded that Cu-IIP has good reusability in 5 repetitions analysis.

4. Conclusions
A new Cu-imprinted sorbent based on natural chelating agent (tannins) was developed by polymerization of phenol-formaldehyde as crosslinkers in acidic medium using sulfuric acid for selective adsorption of Cu ions. The imprinted polymer has high adsorption capacity towards Cu(II) ions. The Langmuir-Freundlich model results showed that Cu-IP-tannins has heterogeneous surface. In comparison with NIP, Cu-IP-tannins also showed high selectivity towards Cu(II) in the presence of competing metal ions. Furthermore, the Cu-IP-Tannin can be reused for five times analysis.

Acknowledgements
The authors would like to thank Universitas Indonesia for funding this research through PITTA Grant Universitas Indonesia with contract No. 2253/UN2.R3.1/KKP/05.00/2018

References
[1] Meouche W, Laatikainen K, Margaillan A, Silvonen T, Siren H, Sainio T, Beurroies I, Denoyel R and Branger C 2017 *Eur. Polym J.* **87** 124–35
[2] Poole C F 2003 *Trends Anal. Chem.* **22** 362–73
[3] Rao T P, Daniel S and Gladis J M 2004 *Trends Anal. Chem.* **23** 28–35
[4] Camel V 2003 *Spectrochim. Acta* B **58** 1177–233
[5] Ross A R S, Ikonomou M G and Orians K J 2000 *Anal. Chim. Acta* **411** 91–102
[6] Karamać M 2007 *Pol. J. Food Nutr. Sci.* **57** 357–62
[7] Branger C, Meouche W and Margaillan A 2013 *React. Funct. Polym.* **73** 859–75
[8] Rao T P, Kala R and Daniel S 2006 *Anal. Chim. Acta* **578** 105–16
[9] Harborne J B 1984 Phenolic compounds *Phytochemical Methods* (Dordrecht: Springer) Chapter 2 pp 37–99
[10] Shamsipur M, Besharati-Seidani A, Fasihi J and Sharghi H 2010 *Talanta* **83** 674–81
[11] Li Y H, Di Z, Ding J, Wu D, Luan Z and Zhu Y 2005 *Water Res.* **39** 605–9
[12] Singh D K and Mishra S 2009 *J. Hazard. Mater.* **164** 1547–51
[13] Say R, Birlik E, Ersöz A, Yilmaz F, Gedikbey T and Denizli A 2003 *Anal. Chim. Acta* **480** 251–8
[14] Roushani M, Abbasi S, Khani H and Sahraei R 2015 *Food Chem.* **173** 266–73
[15] Liang H, Chen Q, Ma J, Huang Y and Shen X 2017 *RSC Adv.* **7** 35394–402
[16] Shampsipur M and Besharati-Seidani A 2011 *React. Funct. Polym.* **71** 131–9
[17] Albert R and Horwitz W 1997 *Anal. Chem.* **69** 789–90

Oil palm empty fruit bunch (OPEFB) (*Elaeis guineensis Jacq.*) cellulose conversion into levulinic acid using hierarchical Mn/ZSM-5 heterogeneous catalyst

D U C Rahayu, J E Hutauruk, Y K Krisnandi and E Saepudin

Department of Chemistry, Faculty of Mathematics and Natural Sciences (FMIPA), Universitas Indonesia, Kampus UI Depok, Depok 16424, Indonesia

Corresponding author's e-mail: yuni.krisnandi@sci.ui.ac.id

Abstract. Oil palm empty fruit bunch (OPEFB) is one of the potential biomasses containing cellulose that can be used as a source for renewable energy. However, the cellulose is tightly connected with lignin in lignocellulose system, which needs to be separated, so called delignification, before being used. To investigate the effect of delignification on conversion reaction, the pretreated cellulose was treated with different delignification processes, i.e. alkali delignification using 10 % NaOH (w/v) and oxidative delignification using 2 % NaOCl (v/v). Furthermore, cellulose conversion from OPEFB into levulinic acid was performed using hierarchical Mn/ZSM-5 (Mn/Hi_ZSM-5) heterogeneous catalyst with various reaction times and concentration of H_3PO_4. The cellulose conversion into levulinic acid in the absence of catalyst was also conducted to study the effect of catalyst. High Performance Liquid Chromatography (HPLC) analysis showed the highest cellulose conversion into levulinic acid was achieved using cellulose from alkali delignification (28.08 % yield) in optimum reaction condition of 40 % H_3PO_4 (v/v), 30 % H_2O_2 (v/v), and 0.1 g of Mn/Hi_ZSM-5 at 100 °C for 10 h. Mn/Hi_ZSM-5 revealed good selectivity for levulinic acid with the absence of 5-HMF as an intermediate compound in conversion reaction.

Keywords: cellulose, levulinic acid, hierarchical Mn/ZSM-5, oil palm empty fruit bunch (OPEFB)

1. Introduction

Increasing petroleum oil prices has forced chemical industry to discover raw materials for producing chemicals that can be used as alternative energy [1]. Lignocellulose contained in biomass has been well known as a promising alternative energy source for crude oil [2]. Indonesia has a large palm oil area with a number of palm oil refining companies. From the palm oil refining process, oil palm empty fruit bunch (OPEFB) are produced which consist of 44.2 % cellulose, 33.5 % hemicellulose, and 20.4 % lignin [3]. Because of its constituents, OPEFB becomes a potential biomass to produce levulinic acid, an important product derived from glucose degradation as a result from cellulose hydrolysis. Levulinic acid is a chemical platform with wide range of application such as pharmaceutical uses, textile dye, antifreeze, animal feed, coating material, solvent, plasticizer, food flavoring agent, and resin [4–6].

In this research, OPEFB was used as raw material to produce cellulose, which then converted into levulinic acid. Prior to conversion process, hemicellulose and lignin content in OPEFB should be reduced by pretreatment process. Pretreatment process was divided into two parts, i.e. alkaline delignification with NaOH 10 % (w/v) and oxidative delignification with NaOCl 2 % (v/v). Chen *et al.* [17] reported the conversion of cotton cellulose into levulinic acid using Mn/ZSM-5 catalyst under phosporic acid media. In this paper, conversion of OPEFB sourced cellulose into levulinic acid using hierarchical Mn/ZSM-5 heterogeneous catalyst was reported. Formerly, this catalyst had been used in our research to convert delignified rice husk into levulinic acid [8].

2. Materials and methods

2.1. Materials
OPEFB was obtained from Adolina Palm Oil Refining Company at Perbaungan Medan, North Sumatera, Indonesia. The chemical used were in pro-analytical grade i.e. H_2O_2 30 %, H_3PO_4 85 %, NaOH 99 %, HCOOH 98–100 %, H_2SO_4 98 %, NaOCl 2 %, CH_3COOH 100 %, and ion exchanger resin Amberlite from Merck, while levulinic acid > 97 % and 5-HMF > 95 % from TCI. Hierarchical 2 % Mn/ZSM-5 (Mn/Hi_ZSM-5) was synthesized according to Krisnandi *et al.* [8].

2.2. Preparation of OPEFB
OPEFB was cut into small pieces with size of 4–5 cm then washed with hot water to remove the impurities and other substances from the palm oil refining process. OPEFB was then dried at 50 °C overnight. The dry sample was ground and sieved into 315 μm particle size.

2.3. Pre-treatment processes of OPEFB
OPEFB was subjected into two pretreatment processes i.e. alkali delignification with NaOH 10 % (w/v) and oxidative delignification with NaOCl 2 % (v/v). Alkali delignification process was conducted using NaOH 10 % (w/v) at 55 °C for 90 min (cellulose A), while oxidative delignification was carried out after NaOH 10 % (w/v) delignification at the same condition (cellulose B). The chemical contents in OPEFB before and after pretreatment were determined according to Krisnandi *et al.* [8].

2.4. Cellulose conversion from delignified OPEFB into levulinic acid
Pretreated cellulose A and B (1 g) were mixed with 0.5 mL H_2O_2 30 % (v/v), 20 mL H_3PO_4 with concentration of 30 or 40 % (v/v), and Mn/Hi_ZSM-5 (0.1 g) in three-neck round bottom flask. The mixture was heated at 100 °C in reflux for 10 h while stirring continuously until the completion process. During the reaction, sampling was carried out every 2 h for analysis. Similar reaction was also conducted without addition of Mn/Hi_ZSM-5.

2.5. Product and residue analysis
Sample from every 2 h reaction was centrifuged for 15 min. Supernatant was then separated by decantation. The filtrate solution was mixed with ion exchanger resin Amberlite and filtered. The residue was then washed and neutralized with distilled water for Fourier Transform Infrared (FTIR) analysis with KBr. High Performance Liquid Chromatography (HPLC) analysis was performed using Shimadzu Prominance 20 HPLC System (Shimadzu Japan) under following conditions: RezexTM ROA-Organic Acid H+ (8 %) column, flow rate of H_2SO_4 0.005 N as a mobile phase was 0.6 mL/min, detector UV wavelength was 220 nm, and retention time was 45 min. The injection volume was 20 μL for each analytes. The identification for each unknown sample was conducted by comparison to the retention time of its standard. The yield of cellulose conversion was calculated on weight basis according to equation (1).

$$\%yield(wt\%) = \frac{product\ weight\ (g)}{feedstock\ initial\ weight\ (9)} x100\% \qquad (1)$$

3. Results and discussion
Pretreatment with NaOH allowed the swelling of materials, resulted easy dissolution of lignin and hemicellulose. The delignification reactions involved cleavage of non-phenolic β-O-4-linkage, phenolic α-O-4-linkage, and release of lignin [9]. Therefore, it will lead to decrease crystallinity and polymerization degree of cellulose [10]. Chemical analysis revealed that pretreatment process could reduce the lignin and hemicellulose content, compared to the raw OPEFB (table 1).

FTIR spectra of cellulose before and after pretreatment are illustrated in figure 1. FTIR spectra revealed broad absorption band located from 3000-3500 cm^{-1} belongs to stretching mode of hydroxyl groups (ν–OH). The intensity of the broad band was decreased after pretreatment, due to decrease of phenolic compounds from lignin. Besides, absorption band located at 2900 cm^{-1} showed ν–CH_2 of vinyl

Table 1. Chemical constitution of OPEFB before and after pretreatment.

Chemical Composition	Before Pretreatment (wt. %)	After Pretreatment (wt. %)
Cellulose	50.24	39.79
Hemicellulose	14.28	11.97
Lignin	37.44	5.41

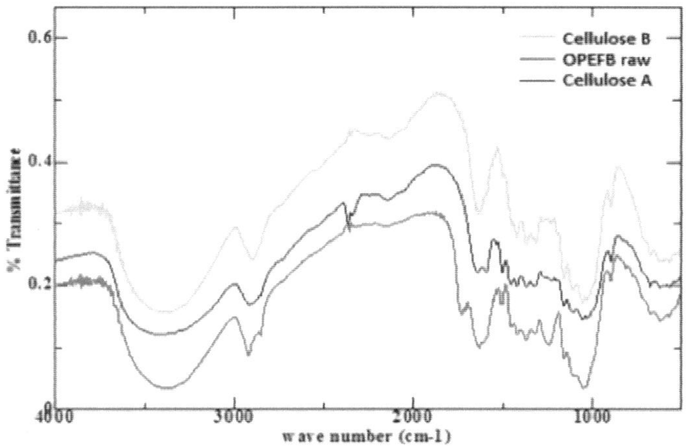

Figure 1. FTIR spectra of cellulose before and after pretreatment.

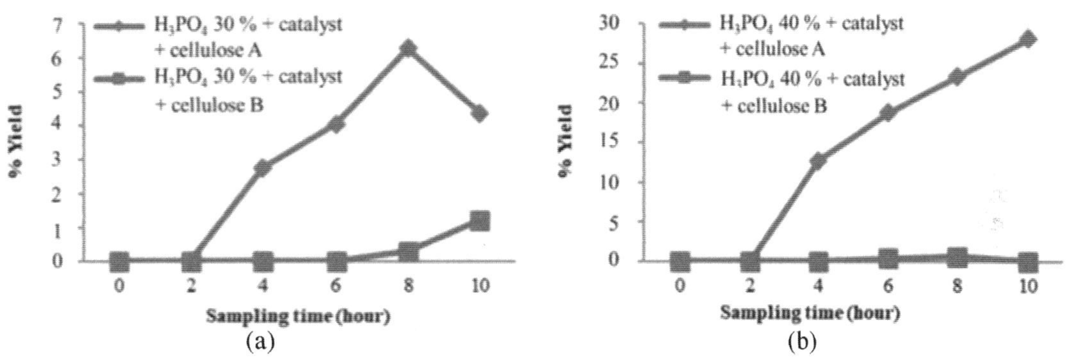

Figure 2. The percentage yield of levulinic acid with reaction condition of 0.1 g Mn/Hi_ZSM-5, 0.5 mL H_2O_2 30 % (v/v), refluxed at 100 °C with 20 mL H_3PO_4 (a) 30 % and (b) 40 % (v/v).

groups [11]. Furthermore, absorption band for C=O groups hemicellulose at 1700-1740 cm^{-1} was absent indicating the removal of hemicellulose [12]. Moreover, the peak showed the C-H rock vibration of cellulose was observed at 896 cm^{-1} [13].

The results of conversion of cellulose-rich OPEFB using two different concentrations of H_3PO_4 (v/v) are shown in figure 2. According to figure 2a, the optimum yield of levulinic acid using H_3PO_4 30 % (v/v) were 6.28 % and 1.22 % in the same reaction condition for cellulose A (8 h) and cellulose B (10 h), respectively. Compared with using H_3PO_4 40 % (v/v) (figure 2b), the optimum yields of levulinic acid were 28.08 % and 0.61 % in the same reaction condition for cellulose A and B at 8 h, respectively. Phosporic acid was added to maintain the conversion reaction to degrade and depolymerize the structure of cellulose [14]. During reaction, there was an interaction between hydroxyl groups from cellulose to H_3PO_4 by forming phosphate ion from auto-protolysis. In this reaction, the role of H_3PO_4 was to degrade the inter- and intra-molecular hydrogen bond of macromolecule structure of cellulose to become a simple molecule of glucose [15]. Based on figure 2, the best H_3PO_4 concentration to obtain the optimum yield of levulinic acid (28.08 %) was achieved from the reaction that used H_3PO_4 40 % (v/v) in cellulose A.

IOP Conf. Series: Materials Science and Engineering **763** (2020) 012034 doi:10.1088/1757-899X/763/1/012034

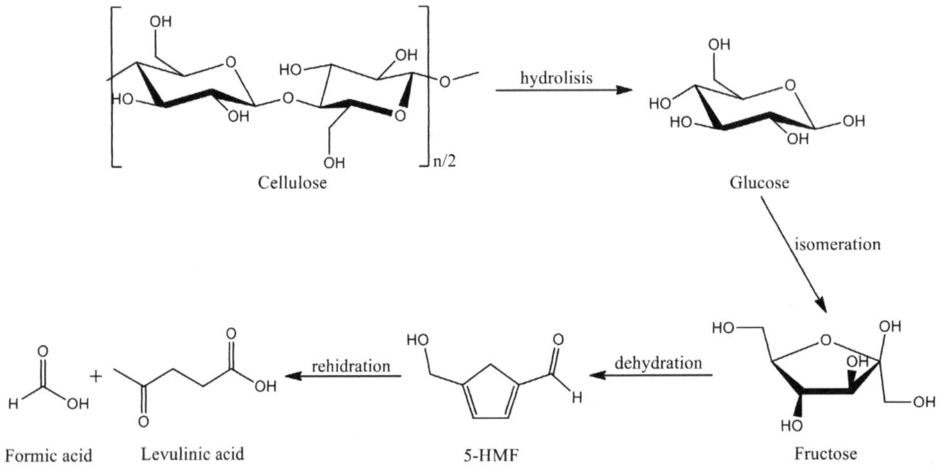

Figure 3. Proposed mechanism for cellulose conversion into levulinic acid

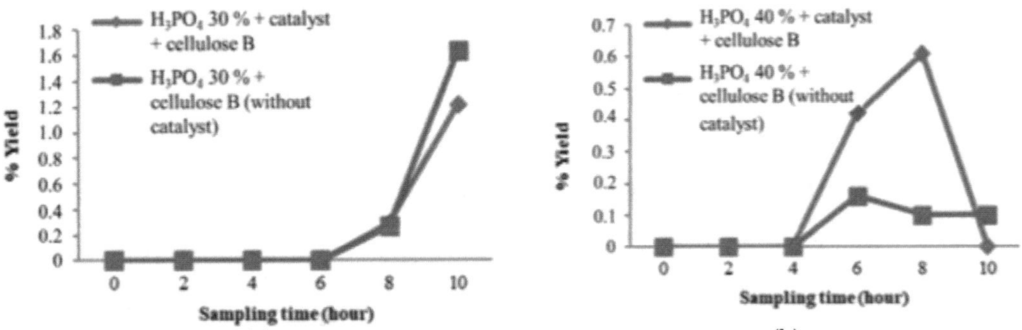

(a) (b)

Figure 4. The percentage yield of levulinic acid with reaction condition of 0.1 g Mn/Hi-ZSM-5, 0.5 mL H_2O_2 30 % (v/v), refluxed at 100 °C with 20 mL H_3PO_4 (a) 30 % and (b) 40 % (v/v).

The role of Mn/Hi_ZSM-5 as a catalyst in conversion reaction was very important. In cellulose conversion into levulinic acid, Mn/Hi_ZSM-5 with combination of Fenton-like reagent are able to increase the molar yield of 5-HMF as intermediate compound [16]. This phenomenon can be attributed to the chemical reaction occurred (equation (2) and equation (3)) in cellulose conversion. Proposed mechanism for cellulose conversion into levulinic acid is shown in figure 3.

$$+H_2O_2 \rightarrow Mn^{3+} + HO^- + HO^{\bullet} \quad (2)$$

$$Mn^{3+} + H_2O_2 \rightarrow Mn^{2+} + H_2O^{\bullet} + H^{+\bullet} \quad (3)$$

To study the role of Mn/Hi_ZSM-5 catalyst, the conversion reaction without catalyst was also conducted and the effect thereof was studied. Based on figure 4a, the optimum yield of levulinic acids from cellulose B conversion without and with catalyst in the same condition were 1.64 % and 1.22 %, respectively. Compared with figure 4b, the optimum yield of levulinic acid with various concentrations of H_3PO_4 were 0.61 % and 0.16 % using H_3PO_4 40 % (v/v) at 8 h and 30 % (v/v) at 6 h, respectively. During reaction, depolymerization of cellulose was occurred by breaking the β-1-4-glycoside bond which was activated by Fenton like reagent [7].

Furthermore, the results also show that different treatment to OPEFB gave different % yield of levulinic acid. The cellulose B, treated with oxidative delignification, resulted in lower % yield of levulinic acid. The results depicted in figure 4a and figure 4b shows that the performance of Mn/Hi_ZSM-5 was not as effective, compared to its activity in similar reaction with delignified rice husk [8]. It is suggested that the catalyst was deactivated by the OCl^- ion that remains in the cellulose B, which may interfere the reaction.

4. Conclusions

OPEFB has a great potential to be used as raw material to yield levulinic acid due to its cellulose content. Cellulose from alkaline delignification was recommended to be used as substrate, since it gives high % yield of levulinic acid at optimum reaction condition in this work.

Acknowledgements

This work was supported by PUPT BOPTN 2015 research grant No. 1086/H2.R12/HKP.05.00/2015.

References

[1] Fang Q and Hanna M A 2002 *Bioresour. Technol.* **81** 187–92
[2] Sun Y and Cheng J 2002 *Bioresour. Technol.* **83** 1-11.
[3] Chin S X, Chia C H and Zakaria S 2013 *Bioresources* **8** 447–60
[4] Chang C, Cen P and Ma X 2007 *Bioresour. Technol.* **98** 1448–53
[5] Bevilaqua D B, Rambo M K D, Rizzetti T M, Cardoso A L and Martins A F 2013 *J. Clean. Prod.* **47** 96–101
[6] Cha J Y and Hanna M A 2002 *Ind. Crops Prod.* **16** 109–118
[7] Chen Y, Li G, Yang F and Zhang S M 2011 *Polym. Degrad. Stab.* **96** 863–9
[8] Krisnandi Y K, Nurani D A, Agnes A, Pertiwi R, Antra N F, Anggraeni A R, Azaria A P and Howe R F 2019 *Indonesian Journal of Chemistry* **19** 115–23
[9] Misson M, Haron R, Kamaroddin M F A and Amin N A S 2009 *Bioresour. Technol.* **100** 2867–73
[10] Damisa D, Ameh J B and Umoh V J 2008 *Afr. J. Biotechnol.* **7** 2444–50
[11] Jahan M S, Saeed A, He Z and Ni Y 2011 *Cellulose* **18** 451–9
[12] Fahma F, Iwamoto S, Hori N, Iwata T and Takemura A 2010 *Cellulose* **17** 977–85
[13] Alemdar A and Sain M 2008 *Bioresour. Technol.* **99** 1664–71
[14] Akzo Nobel N V, Boerstoel Hanneke and Y P M A Marco 1996 *Process for Making Cellulose Extrudates Patent* WO 9606207 (Geneva: WIPO)
[15] Fadeeva J, Shmukler L and Safonova L 2003 *J. Mol. Liq.* **103-104** 339–47
[16] Dükkancı M, Gündüz G, Yılmaz S and Prihod'ko R V 2010 *J. Hazard. Mater.* **181** 343–50

Three-component reaction in 1,4-dihydropyridine synthesis catalyzed by Copper(I) Iodide nanoparticles

A H Cahyana, B Ardiansah and N S Aisyah

Department of Chemistry, Faculty of Mathematics and Natural Sciences (FMIPA), Universitas Indonesia, Kampus UI Depok, Depok 16424, Indonesia

Corresponding author's e-mail: herrykim@ui.ac.id

Abstract. We have synthesized two 1,4-dihydropyridine derivatives via one-pot three-component reaction. The reactions were catalyzed by copper(I) iodide nanoparticles with crystallite size of 78.4 nm. Formations of cyclocondensation products were confirmed by Fourier Transform Infrared (FTIR), UV-Vis and Gas Chromatography-Mass Spectrometry (GC-MS) spectra.

Keywords: 1,4-dihydropyridine, multicomponent reaction, CuI.

1. Introduction

Nitrogen-containing six-membered heterocycles are valuable in the field of pharmaceutical and biological chemistry [1]. Specifically, 1,4-dihydropyridines (DHPs) are known to show various pharmacological activities such as calcium-channel blockers [2], antihypertensive [3], antianginal [4], antitumor [5], anticancer [6] and anti-inflammatory [7] activities. Other importances in medicinal chemistry are found to be reported such as antitubercular [8], analgesic [9], and platelet aggregation inhibitory [10] actions. Flashing back to more than one hundred years ago, the core structure of DHP was first synthesized by Hantzsch in 1882, which involves multicomponent cyclocondensation of aldehyde, ammonia and β-ketoester [11]. The Hantzsch method generally performed either in acetic acid or under reflux in alcohols for lengthy reaction times [11]. Due to the interesting and significant biological activities, considerable efforts have been developed to provide more efficient protocols for the synthesis of DHPs [12].

In the modern era of organic synthesis, there is an elevating awareness of clamant necessity to confine the use of harmful chemicals as well as harsh reaction protocols, also generation of by-products, which can decrease atom economy [13]. Multicomponent reactions (MCRs) are one-pot syntheses employing three or more reactants and incorporating all their skeletons or sub-skeletons to furnish products with interesting structural motifs [14,15]. They are powerful tools for efficient production of molecular diversity in a single reaction [16]. In the literature, many organic/ inorganic materials have been reported to catalyze MCRs in DHPs synthesis, such as various ionic liquids [17–20], $CuSO_4$/chitosan [21], bismuth nitrate [22] and cellulose/sulfuric acid [23]. Although most of them showed appreciable performance, some reported methods still have one or more drawbacks like dangerous solvents, expensive reagents, tedious work-up after reaction, long time and high temperature [24]. Hence, the search of new method or catalyst that could be superior in this reaction is still highly required. In the present work, we report preparation of copper(I) iodide nanoparticles and its ability in synthesis of two DHP derivatives.

2. Materials and methods

2.1. Materials

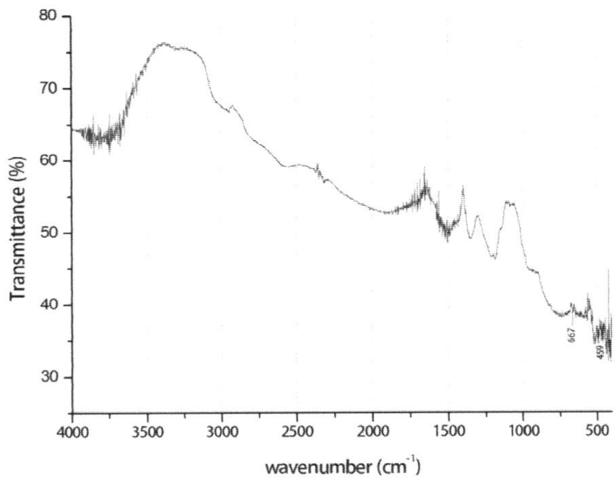

Figure 1. IR spectrum of CuI nanoparticles

Cinnamaldehyde, benzaldehyde, and other chemicals including solvents were analytical grade, used without further purification. TLC was performed on silica gel 60 F_{254}. Catalyst morphology and crystallinity were studied using Scanning Electron Microscope-Energy Dispersive X-Ray Analysis (SEM-EDX) and X-ray Diffractometer (XRD), respectively. Presence of functional groups was analyzed on Shimadzu Prestige-21 Fourier Transform Infrared (FTIR) spectrophotometer. Beside FTIR, organic products were also identified by UV-Vis and Gas Chromatography and Mass Spectrometry (GC-MS) instruments.

2.2. Methods

2.2.1. Preparation of CuI nanocatalyst. The catalyst was prepared by ultrasonic irradiation approach. Copper(II) sulfate was used as precursor. First, copper(II) sulfate (1 mmol) was ultrasonically cleaned for 20 sec in acetone, then in 2M hydrochloric acid, followed by repeated rinsing with distilled water. After drying, precursor was added into potassium iodide solution (2 mmol in 40 mL distilled water). The mixture was sonicated for 30 min obtained grey precipitates were filtered, washed with distilled water and dried in vacuum between room temperature to 40 °C [25].

2.2.2. Synthesis of 1,4-DHPs. In a 20 mL round-bottom flask integrated with cooling system, a mixture of benzaldehyde (2 mmol), ethyl acetoacetate (4 mmol), ammonium acetate (2 mmol) and 5 mL ethanol as solvent was stirred in the presence of various amount of CuI nanoparticles (varying from 0–10 mol%). The reaction mixture was cooled to room temperature then extracted with ethyl acetate. Organic layer was washed with water and brine, and dried over anhydrous sodium sulfate. After evaporation, crude product was purified by recrystallization from hot ethanol to afford compound 1a, diethyl-2,6-dimethyl-4-phenyl-1,4-dihydropyridine-3,5-dicarboxylate). Compound 1b, (E)-diethyl-2,6-dimethyl-4-styryl-1,4-dihydropyridine-3,5-dicarboxylate was synthesized by employing cinnamaldehyde instead of benzaldehyde, applying the best condition obtained from optimization of reaction conditions in synthesis of compound 1a.

3. Results and discussion

3.1. Characterization of CuI nanoparticles

Copper(II) sulfate and potassium iodide react in oxidation-reduction manner. Iodide acts as reducing agent, whereas Cu(II) as oxidizing agent. After mixing the precursors, grey to brown powder as product was expected to be the mixture of CuI, I_2 and potassium sulfate [26]. The solid was washed several times with distilled water and ethanol. Copper(I) iodide nanoparticles were first characterized by FTIR (figure 1). Two bands at 667 and 459 cm^{-1} are characteristics of CuI samples. Broadband around 3600-3400 cm^{-1} is attributed to intermolecular hydrogen bonding presents in residual water. These observations are in agreement with previous work by Humud *et al.* [27].

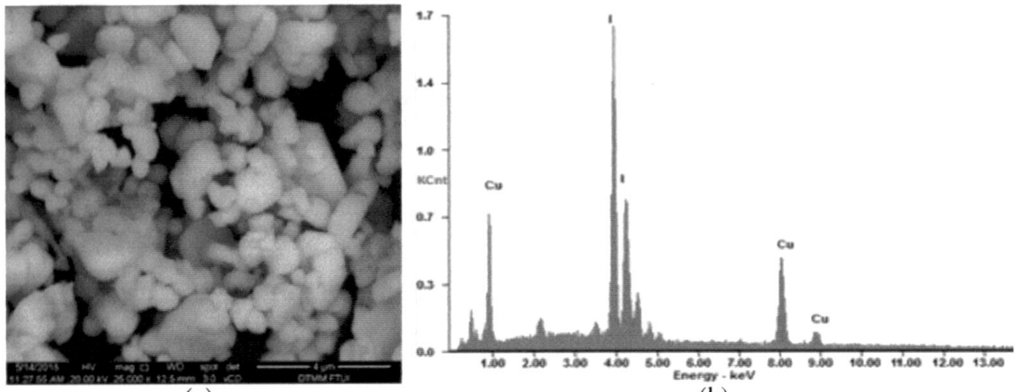

(a) (b)

Figure 2. (a) SEM image and (b) EDX spectrum of CuI nanoparticles

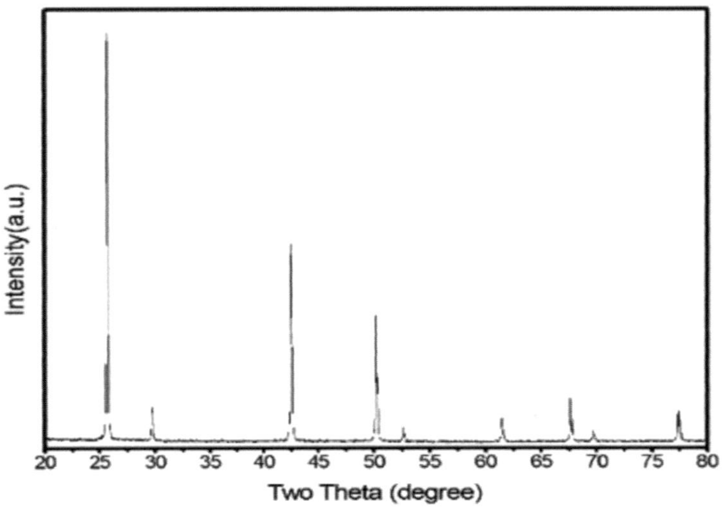

Figure 3. (a) XRD patterns of CuI nanoparticles

Characterization using SEM shows surface morphology of the CuI nanoparticles (figure 2a). It can be seen that most of them were found to have spherical shape. The particles are well dispersed and distributed equally in all directions. However, little agglomeration of the particles was also observed, resulting larger size irregular shape of some copper(I) iodide particles. EDX result in figure 2b shows elemental composition of the sample. Atom percentage of Cu and I was found to be 49.06 and 50.94 %, respectively. In the spectrum, it was found only peaks of Cu and I, absent of any other elements. This result signifies that copper(I) iodide nanoparticles was obtained in pure.

XRD pattern of CuI sample is depicted in figure 3. The crystal structure of the sample was determined by comparing its diffraction degree with literature (JCPDS Card No. 06-0246). The sample has diffraction peaks at 2θ of 25.7°, 29.7°, 42.4°, 50.2°, 52.5°, 61.4°, 67.6°, 69.6°, 77.6° and 83.0°. No other diffraction peaks arising in the XRD pattern, so it indicates the high phase purity of the sample [28]. The crystallite size diameter (D) of the sample was calculated by Debye-Scherrer equation ($D = K\lambda/\beta\cos\theta$) and found to be 78.4 nm.

3.2. Synthesis of 1,4-DHPs

Benzaldehyde was chosen as model substrate for the optimization study and subjected for three-component reaction in the presence of different CuI nanoparticles amount. Initially, control reaction was conducted in the absence of catalyst, which provides product in 27% yield only (table 1). It is

Table 1. Optimization of reaction conditions[a] in synthesis of compound 1a

Entry	CuI catalyst (mol%)	Solvent	Time (min)	Yield[b] (%)
1	None	Ethanol	40	27
2	2	Ethanol	40	46
3	5	Ethanol	40	59
4	10	Ethanol	40	70
5	10	Water	40	67
6	10	Solvent-free	40	27
7	10	Ethanol	80	49
8	**10**	**Ethanol**	**120**	**85**
9	10	Ethanol	180	73

[a] benzaldehyde (1 mmol), ethyl acetoacetate (4 mmol), NH_4OAc (2 mmol), solvent (5 mL)
[b] isolated yield

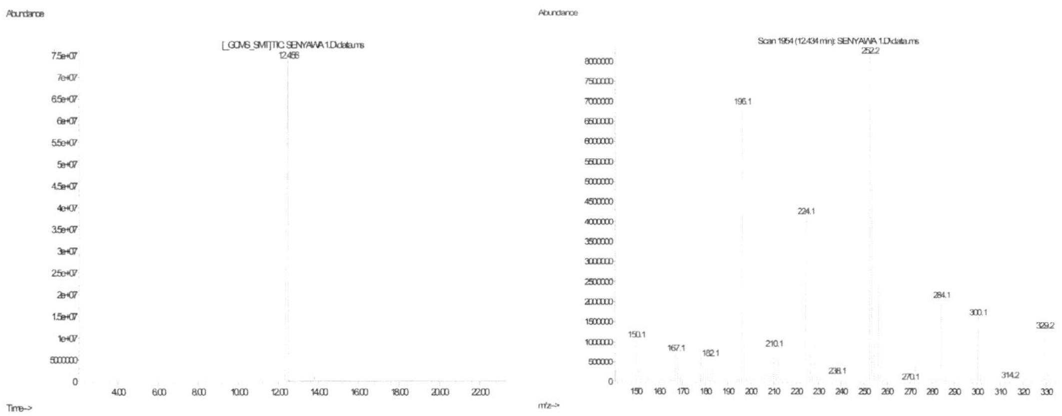

Figure 4. GC-MS spectra of compound 1a

noteworthy that the yield of compound 1a was increased up to 70% with higher catalyst amount (table 1, entry 2–4). Employing water as solvent instead of ethanol yielded 67% of product, slightly lower than that of ethanol. Unfortunately, the yield obtained only 27% when reaction was performed under solvent-less condition. Prolonging time up to 120 min gave the best result, product 1a was found to be 85% (table 1, entry 8). However, after 180 min reaction, the product was slightly decreased to 73%. It can be presumed that the product is slightly less stable under this condition. Having established the best condition, we further synthesized compound 1b under same condition. Yield of 1b was found to be 90%.

Organic products were first characterized by FTIR. Compound 1a has some important absorption. A sharp peak at 3342 cm^{-1} indicating the presence of N-H vibration (secondary amine). Peaks at 1652 and 1213 cm^{-1} are attributed to the C=O and C-O vibration, respectively. Compound 1b has maximum wavelength at 354 nm. GC-MS spectra of compound 1a are depicted in figure 4. It confirmed the formation of desired product with molecular cation value of 329.2. Meanwhile, like 1a, compound 1b shows N-H absorption band at 3337 cm^{-1}, C=O vibration at 1651 cm^{-1} and C-O at 1223 cm^{-1}. Lambda maximum for 1b was found to be 353 nm. From GC-MS, molecular cation of 355.1 was observed, indicating the formation of desired product (figure 5). To test the reusability of catalyst, we have also performed reactions using recycled catalyst. For first run, 85% yield of product was obtained. After separation, copper(I) iodide nanocatalyst was washed with ethanol and dried under vacuum. Then, it is used for next three consecutive reactions under same condition. Yields were found to be 73, 61 and 49% for second, third and fourth run, respectively. The result indicated that the catalyst could be reused for 4 consecutive experiments with partial loss of catalytic activity.

IOP Conf. Series: Materials Science and Engineering **763** (2020) 012035 doi:10.1088/1757-899X/763/1/012035

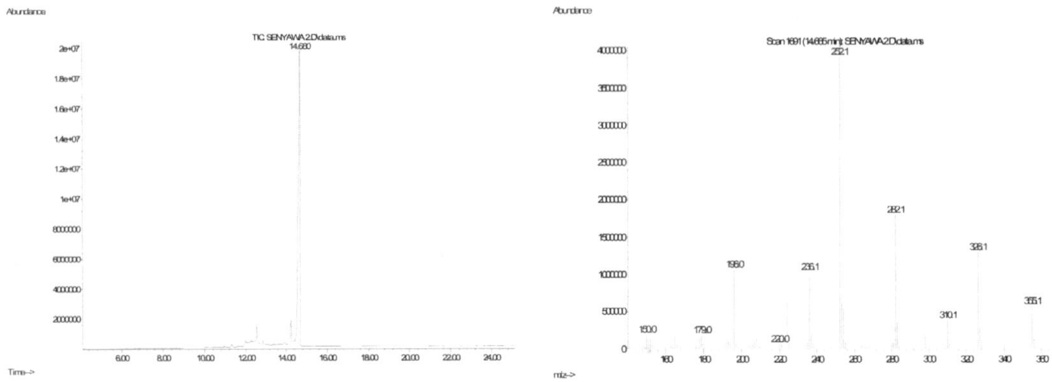

Figure 5. GC-MS spectra of compound **1b**

4. Conclusions

We have successfully prepared copper(I) iodide nanoparticles by oxidation-reduction method with the help of ultrasonic irradiation. The material was successfully used to catalyze three-component reaction in synthesis of two derivatives of 1,4-DHPs.

Acknowledgements

We thank to Universitas Indonesia for funding this research through PITTA Grant 2018 with contract No. 234/UN2.R3.1/PPM.00/2018.

References

[1] Kaur N and Kishore D 2014 *Synth. Commun.* **44** 1173–211
[2] Toal C B, Meredith P A and Elliott H L 2012 *Blood Press.* **21** S3–10
[3] Fard S R H, Miri R and Nekooeian A A 2016 *Res. Pharm. Sci.* **11** 497–504
[4] Loev B, Goodman M M, Snader K M, Tedeschi R and Macko E 1974 *J. Med. Chem.* **17** 956–65
[5] Abbas H A, El Sayed W A and Fathy N M 2010 *Eur. J. Med. Chem.* **45** 973–82
[6] Kumar R S, Manilal A, Nasser A J A, Merdekios B, Chen X and Idhayadhulla A 2014 *Journal of Pharmacology and Toxicology (JPT)* **9** 119–28
[7] Idhayadhulla A, Kumar R S, Nasser A J A, Kavimani S and Indhumathy S 2015 *Pharm. Chem. J.* **49** 463–6
[8] Iman M, Davood A, Dehqani G, Lotfinia M, Sardari S, Azerang P and Amini M 2015 *Iran J. Pharm. Res.* **14** 1067–75
[9] Shojaii A, Motaghinejad M, Norouzi S and Motevalian M 2015 *Iran J. Pharm. Res.* **14** 263–9
[10] Sunkel C E, Fau de Casa-Juana M, Cillero F J, Priego J G and Ortega M P 1988 *J. Med. Chem.* **31** 1886–90
[11] Hantzsch A 1882 *Justus Liebigs Ann. Chem.* **215** 1–7
[12] Aswin K, Logaiya K, Sudhan P N and Mansoor S S 2012 *J. Taibah Univ. Sci.* **6** 1–9
[13] Zonouz A M and Sahranavard N 2010 *E- J. Chem.* **7** S372–6
[14] Azzam R A and Mohareb R M 2015 *Chem. Pharm. Bull.* **63** 1055–64
[15] Wan J P and Liu Y 2012 *RSC Adv.* **2** 9763–77
[16] Varádi A, Palmer T C, Notis Dardashti R and Majumdar S 2016 *Molecules* **21** 19
[17] Sharma S D, Hazarika P and Konwar D 2008 *Catal. Commun.* **9** 709–14
[18] Wang L M, Sheng J, Zhang L, Han J W, Fan Z Y, Tian H and Qian C T 2005 *Tetrahedron* **61** 1539–43
[19] Ko S and Yao C F 2006 *Tetrahedron* **62** 7293–9
[20] Sharma M G, Rajani D P and Patel H M 2017 *R. Soc. Open Sci.* **4** 1–8
[21] Dekamin M G, Kazemi E, Karimi Z, Mohammadalipoor M and Naimi-Jamal M R 2016 *Int. J. Biol. Macromol.* **93** 767–74
[22] Bandyopadhyay D, Maldonado S and Banik B K 2012 *Molecules* **17** 2643–62
[23] Murthy Y L, Rajack A, Taraka Ramji M, Jeson Babu J, Praveen C and Aruna Lakshmi K 2012 *Bioorg. Med. Chem. Lett.* **22** 6016–23
[24] Mahinpour R, Moradi L, Zahraei Z and Pahlevanzadeh N 2018 *J. Saudi Chem. Soc.* **22** 876–85

[25] Ziarati A, Safaei-Ghomi J and Rohani S 2013 *Ultrason. Sonochem.* **20** 1069–75
[26] Shahbazi S and Afshar S 2014 *Mater. Lett.* **115** 190–3
[27] Humud H R, Kadhem S J and Khudhair D M 2017 *Journal of Chemical and Pharmaceutical Research (JCPR)* **9(1)** 31–6
[28] Johan M R, Si-Wen K, Hawari N and Aznan N A K 2012 *Int. J. Electrochem. Sci.* **7** 4942–50

Bi_2O_3 nanoparticles: synthesis, characterizations, and photocatalytic activity

N Nurmalasari, Y Yulizar and D O B Apriandanu

Department of Chemistry, Faculty of Mathematics and Natural Sciences (FMIPA), Universitas Indonesia, Kampus UI Depok, Depok 16424, Indonesia

Corresponding author's e-mail: yokiy@ui.ac.id

Abstract. In this study, Bi_2O_3 nanoparticles have been successfully synthesized using *Jatropha multifida* L. leaf extract (JMLE). Bi_2O_3 nanoparticles were characterized by Fourier Transform Infrared (FT-IR) spectroscopy, X-Ray Diffraction (XRD), UV-Vis Diffuse Reflectance Spectroscopy (UV-Vis DRS), Scanning Electron Microscope-Energy Dispersive X-Ray (SEM-EDX) and Transmission Electron Microscope (TEM). These various characterizations were performed to analyze the functional groups, structural, optical, morphological, composition, and photocatalytic properties. The synthesized Bi_2O_3 have monoclinic structure, optical band gap of 3.34 eV, agglomerated morphology, and particle size of 17.26 nm. Photocatalytic activity of Bi_2O_3 nanoparticles was investigated for the degradation of methylene blue (MB) under UV light irradiation with degradation percentage of 92.53 %.

Keywords: Bi_2O_3 *Jatropha multifida*, photocatalytic activity, methylene blue

1. Introduction

Bi_2O_3 has many advantages as anode, semiconductor, catalyst, and photodegradation of organic dyes [1–4]. Bi_2O_3 nanoparticles have been synthesized by sol gel, successive ionic layer adsorption and reaction (SILAR), redox reaction, and green synthesis methods [1, 5–7]. Bismuth oxide has less toxic, good conductivity, large energy band gap, and good photocatalyst activity [4]. Therefore, Bi_2O_3 nanoparticles are synthesized as photocatalyst to decompose some organic dyes.

Currently, the synthesis of nanoparticles is mostly by green synthesis. It aims to reduce the use of hazardous chemicals, which pollute the environment. Green synthesis methods are environmentally friendly, low cost, and the used material is easy to obtain. Metal oxide nanoparticles that has been synthesized with green synthesis method are ZnO/CuO, Fe_3O_4, CdO, and La_2O_3 [8–11]. One of the raw materials is plant leaf extract.

Jatropha multifida is a kind of traditional plants in Indonesia. It is used as antiseptic, antioxidant, antibacterial, antimicrobial, and others [12–17]. *Jatropha multifida* also contains secondary metabolites such as alkaloid, flavonoid, tannin, steroid, phenol, and saponin [16,18].

Synthesis of Bi_2O_3 nanoparticles with aqueous fraction of plant extracts has not been studied. In this study, we reported the synthesis of Bi_2O_3 nanoparticles with *Jatropha multifida* leaf extract (JMLE) and their photocatalytic activity under UV light irradiation. Photocatalytic activity of Bi_2O_3 nanoparticles was performed for methylene blue (MB) degradation.

2. Materials and methods

2.1. Materials

Bismuth nitrate alkaline $Bi_5O(OH)_9(NO_3)_4$ (Merck, Germany) was used as precursor. *Jatropha multifida* leaves were collected from Conservation Unit of Biopharmaceutical Cultivation (UKBB) Study Center of Tropical Biopharmaca, Bogor Agriculture Institute, Bogor, Indonesia. All reagents and solvents, which used in the investigation, were in analytical grade.

2.2. Preparation of Jatropha multifida leaf extract
Jatropha multifida leaf was washed, dried and mashed. 50 mg of leaf powder was macerated in 250 mL of methanol by stirring every day for 7 days. The methanol fraction was partitioned with hexane, and it was evaporated]. The extract was added with deionized water as *Jatropha multifida* leaf extract (JMLE) stock solution.

2.3. Phytochemical screening
Each fraction of leaf extract was phytochemically tested to determine the secondary metabolites content such as alkaloids, flavonoids, steroids, terpenoids, tannins, polyphenols, and saponins.

2.4. Synthesis of Bi_2O_3
0.1 mmol of $Bi_5O(OH)_9(NO_3)_4$ was added into JMLE solution, then stirred at 80 °C for 2 h. The results were centrifuged for 30 min. The formed particles were rinsed with deionized water and recentrifuged for 15 min. The centrifuged result was dried and calcined at 500 °C for 3 h to obtain Bi_2O_3 powder.

2.5. Characterization of Bi_2O_3
The Fourier Transform Infrared (FT-IR) spectrophotometer (IR Prestige-21 Shimadzu) was performed to determine functional groups in JMLE and Bi_2O_3 nanoparticles. UV-Vis Diffuse Reflectance Spectroscopy (DRS) Shimadzu 2400 was used to investigate the optical studies at the wavelength range of 200–800 nm. Scanning Electron Microscopy (SEM) Zeiss Evo MA10 and Tunneling Electron Microscopy (TEM) JEOL JEM 1400 were used to know the surface morphology and the particle size of nanoparticles. Energy Dispersive X-Ray (EDX) was used to determine the composition of nanoparticles. X-Ray Diffraction (XRD) characterization (PANalytical EMPYREAN) was used to determine the crystal structure at 2θ range of 10°–90°. The photodegradation of methylene blue was investigated using UV-Vis spectrophotometer (Shimadzu 2600) at the wavelength of 200–800 nm.

2.6. Photocatalytic activity of Bi_2O_3
The photocatalytic activity test of Bi_2O_3 nanoparticle was carried out for the degradation of methylene blue under UV light irradiation (Gold star PLB 9 W-BLB, 170-240 V). All experiments were performed at room temperature for 150 min.

3. Results and discussion

3.1. Phytochemical screening
The phytochemical screening was performed to determine the content of secondary metabolites such as alkaloids, flavonoids, tannins, polyphenols, steroids, terpenoids, and saponins. Methanol fraction and aqueous fraction of JMLE contain alkaloids, flavonoids, tannins, polyphenols, and saponins. While, hexane fraction contains alkaloid and saponin. Hence, phytochemical screening of JMLE aqueous fraction shows the positive results for alkaloids as the main secondary metabolite in Bi_2O_3 nanoparticle synthesis. This result is demonstrated in further analysis by FT-IR spectroscopy.

3.2. FT-IR analysis
FT-IR spectroscopy was performed to determine the functional group of JMLE and Bi_2O_3 nanoparticles at the wavenumber 400–4000 cm^{-1}. Figure 1 shows FT-IR spectra of Bi_2O_3 nanoparticles and JMLE. The sharp absorption peaks at 1383 cm^{-1} is assigned to the C-O vibration and 3420 cm^{-1} as -OH stretching from peak shift at JMLE that may still be left on the surface of Bi_2O_3 nanoparticles. The vibration of Bi-O is shown at wavenumber 825, 586, and 495 cm^{-1}. It has the conformity with C–O vibration modes at 1388 cm^{-1} [4]. The wavenumber of 400-650 cm^{-1} and 3433 cm^{-1} also correspond to BiO_6 octahedron and –OH vibration, respectively as reported by Yuvakkumar and Hong [20]. The absorption peaks of JMLE assign to the stretching vibration modes of O-H at 3320 cm^{-1} and the stretching vibration of C-H at 2930 cm^{-1}. These two vibrations demonstrate the presence of flavonoid as reported by Venkatesan *et al.* [21]. C=O group vibration at 1610 cm^{-1} and C-N stretching vibration at 1076 cm^{-1} show the presence of saponin and alkaloid, respectively. The vibration mode of C-C stretching is assigned at 1406 cm^{-1}. These results are in line with the previous studies that the vibration

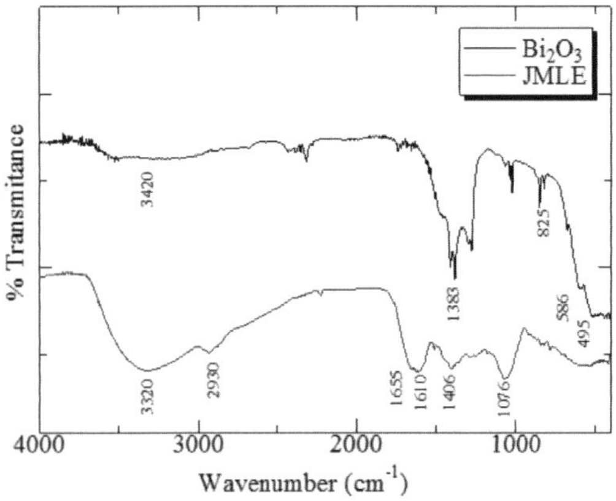

Figure 1. FT-IR spectra of Bi_2O_3 nanoparticles and JMLE.

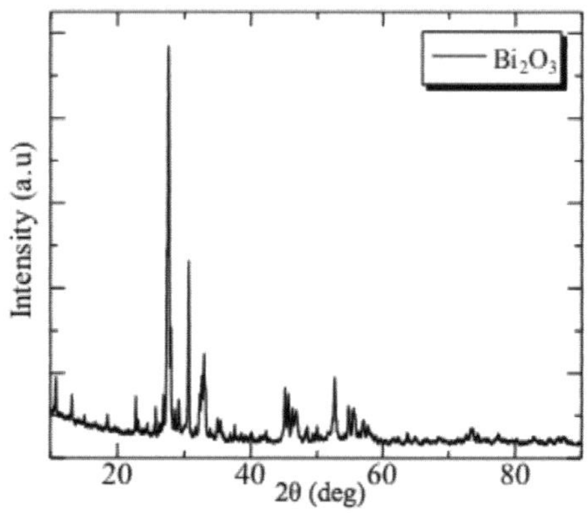

Figure 2. XRD pattern of Bi_2O_3 nanoparticles.

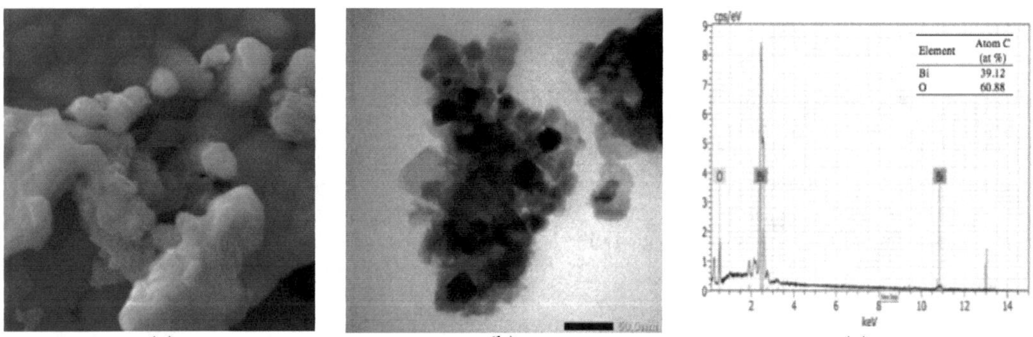

(a) (b) (c)

Figure 3. (a) SEM image, (b) TEM image, and (c) EDX spectrum of Bi_2O_3 nanoparticles.

of C=O, C-N stretching, and C-C stretching are assigned at the wavenumber of 1622, 1401 and 1076 cm^{-1}, respectively [21–25].

3.3. XRD analysis

XRD pattern of Bi_2O_3 nanoparticles gives a strong and sharp peak, indicating that Bi_2O_3 nanoparticles are crystals phase as shown in figure 2. The peaks are related to the monoclinic crystals phase of Bi_2O_3 suitable with COD No. 96-152-6459 at the 2θ value of 24.55°, 25.75°, 26.92°, 27.34°, 33.18°, 35.00°, 35.42°, and 37.56°. The average crystallite size of Bi_2O_3 nanoparticles can be calculated by Debye Scherrer equation and obtained the average crystal size of 69.24 nm.

3.4. SEM-EDX and TEM analysis

Figure 3a shows the SEM images of Bi_2O_3 nanoparticles with the magnification of 25,000 times, indicating the morphology of Bi_2O_3 nanoparticles are agglomerated. Figure 3b shows TEM images with the scale bar of 50 nm with the morphology of Bi_2O_3 nanoparticles are cubic like and agglomerated. The particle size of Bi_2O_3 is 17.26 nm. According to the EDX spectrum as shown in figure 3c, the quantitative EDX analysis indicates the atomic ratios of Bi and O are 2:3, which is confirmed as Bi_2O_3.

3.5. UV-Vis DRS analysis

UV-Vis DRS was performed to investigate the optical studies and to determine the band gap energy of semiconductor materials. The reflectance value is converted by the Kubelka Munk equation. The band gap energy can be determined by plot Tauc method. Figure 4 shows Bi_2O_3 nanoparticles with the band

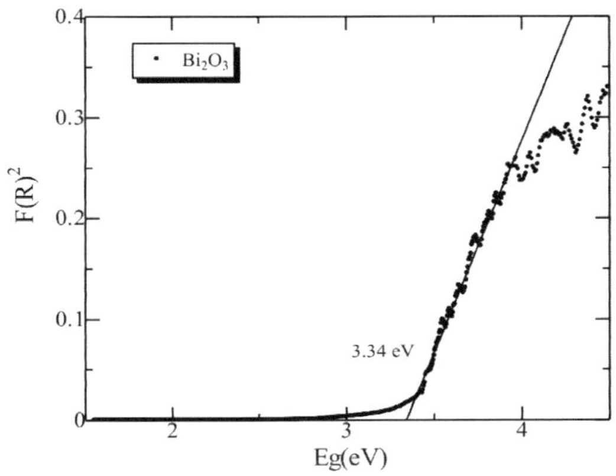

Figure 4. The band gap energy (E_g) of Bi_2O_3 nanoparticles.

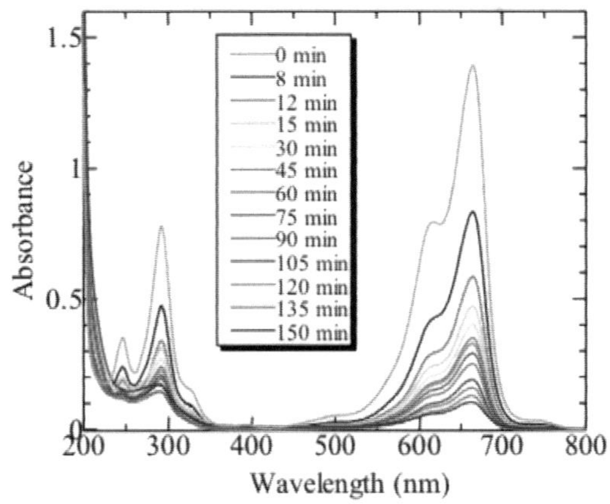

Figure 5. The UV-Vis absorption spectra of methylene blue degradation.

gap energy of 3.34 eV. Based on the band gap energy value, Bi_2O_3 nanoparticles can work as photocatalyst under UV light irradiation [4].

3.6. Photocatalytic activity of Bi_2O_3

The photocatalytic activity of Bi_2O_3 nanoparticles was observed for the degradation of methylene blue under UV light irradiation. The UV-Vis absorption peak decreases at λ_{max} of 664 nm for 150 min as shown in figure 5. The degradation percentage of methylene blue was obtained at 92.53% using Bi_2O_3 nanoparticles under UV light irradiation.

The UV light irradiation activates Bi_2O_3 nanoparticle to form electron-hole pairs. The catalyst, which is exposed by UV light irradiation will produce hole in the valence band and electron in the conduction band. It causes the excitation of electrons from the valence band to the conduction band. The hole which reacts with H_2O will form •OH, while the reaction with O_2 will produce superoxide radicals (O_2•$^-$) [26]. The reaction will degrade methylene blue to the smaller molecules.

4. Conclusions

Bi_2O_3 nanoparticles have been successfully synthesized by green synthesis method using *Jatropha multifida* leaf extract (JMLE) as a weak base source and capping agent. The phytochemical screening shows that JMLE contains alkaloid, flavonoid, tannin, polyphenol, and saponin. FT-IR characterization shows the vibration of Bi-O at the wavenumber 825, 586, and 495 cm^{-1}. The band gap energy of Bi_2O_3 nanoparticles is 3.34 eV and its morphology is agglomerated with the particle size of 17.26 nm. The photocatalytic activity test of Bi_2O_3 nanoparticles has been conducted for degradation of methylene blue at 92.53 % for 150 min.

Acknowledgements

The authors would like to thank University of Indonesia for funding this research through PITTA Grant University of Indonesia with contract No. 2339/UN2.R3.1/HKP.05.00/2018.

References

[1] Fang W, Fan L, Zhang Y, Zhang Q, Yin Y, Zhang N and Sun K 2017 *Ceram. Int.* **43** 8819–23
[2] Li L, Meng L, Wang F and Wang Y 2017 *Mater. Sci. Semicond. Process.* **68** 48–52
[3] Masui T, Furukawa S and Nobuhito I 2006 *Chem. Lett.* **35** 1032–3
[4] Abu-Dief A M and Mohamed W S 2017 *Mater. Res. Express* **4** 035039
[5] Sudrajat H and Sujaridworakun P 2017 *Mater. Des.* **130** 501–11
[6] Raut S S, Bisen O, and Sankapal B R 2017 *Ionics (Kiel)* **23** 1831–7
[7] Karnan T and Samuel S 2016 *Ceram. Int.* **42** 4779–87
[8] Yulizar Y, Bakri R, Apriandanu D O B and Hidayat T 2018 *Nano-Structure and Nano-Objects* **16** 300–5
[9] Sari I P and Yulizar Y 2017 *IOP Conf. Ser.: Mater. Sci. Eng.* **191** 012014
[10] Permana Y N and Yulizar Y 2017 *IOP Conf. Ser.: Mater. Sci. Eng.* **188** 012018

[11] Sulaiman N, Yulizar Y and Apriandanu D O B 2018 *AIP Conf. Proc.* **2023** 020105
[12] Aiyelaagbe O O 2001 *Fitoterapia* **72** 544–6
[13] Anani K, Adjrah Y, Améyapoh Y, Karou S D, Agbonon A, de Souza C, and Gbeassor M 2016 *Pharmacogn. Res.* **8** 142–6
[14] Aransiola M N, Ehikhase C, Mmegwa J C and Wahab I O 2014 *IOSR Journal of Pharmacy and Biological Sciences (IOSR-JPBS)* **9(4)** 53–7
[15] Falodun A, Imieje V, Erharuyi O, Joy A, Langer P, Jacob M, Khan S, Abaldry M and Hamann M 2014 *Asian Pac. J. Trop. Biomed.* **4** 374–8
[16] Rampadarath S, Puchooa D and Ranghoo-Sanmukhiya V M 2014 *Asian Pac. J. Trop. Med.* **7** S380–3
[17] Shoji M, Woo S Y, Masuda A, Win N N, Ngwe H, Takahashi E, Kido H, Morita H, Ito T and Kuzuhara T 2017 *BMC Complement. Altern. Med.* **17** 96
[18] Nwokocha, Blessing A, Agbagwa I O and Okoli B E 2011 *Research Journal of Phytochemistry* **5** 107–14
[19] Apriandanu D O B and Yulizar Y 2017 *IOP Conf. Ser.: Mater. Sci. Eng.* **188** 012013
[20] Yuvakkumar R and Hong S I 2015 *Spectrochim. Acta A Mol. Biomol. Spectrosc.* **144** 281–6
[21] Venkatesan S, Pugazhendy K, Sangeetha D, Vasantharaja C, Prabakaran S and Meenambal M 2012 *Int. J. Pharm. Biol. Sci. Arch.* **3** 969–72
[22] Khalil A S, Rahim A A, Taha K K and Abdallah K B 2013 *Journal of Applied Industrial Sciences (JAIS)* **1** 78–88
[23] Sangster A W 1960 *J. Chem. Educ.* **37** 454–9
[24] Kime B, Barminas J T, Nkafamiya I I and Akinterinwa A 2015 *International Journal of Innovative Science, Engineering and Technology (IJISET)* **2** 733–7
[25] Kayalvizhi K, Cathrine L and Banu K S 2015 *International Journal PharmTech Research* **8** 166–70
[26] Adhikari S P, Dean H, Hood Z D, Peng R, More K L, Ivanov I, Wu Z and Lachgar A 2015 *RSC Adv.* **5** 91094–102

Calcium alginate-TiO_2/SiO_2 nanocomposite for glucose conversion to 5-hydroxymethylfurfural

I Lestari and H Helmiyati

Department of Chemistry, Faculty of Mathematics and Natural Sciences (FMIPA), Universitas Indonesia, Kampus UI Depok, Depok 16424, Indonesia

Corresponding author's e-mail: helmiyati@sci.ui.ac.id

Abstract. The sodium alginate biopolymer used to form calcium alginate-TiO2/SiO2 nanocomposite has been successfully synthesized. The first stage is synthesis of TiO2/SiO2 composites by the sol-gel method using tetraethyl orthosilicate (TEOS) and titanium isopropoxide (TTIP) precursors. Calcium alginate nanocomposite is formed from the crosslinking process between Ca^{2+} ions from $CaCl_2.2H_2O$ with sodium alginate combined with TiO_2/SiO_2 composites, characterized by Fourier Transform Infrared (FTIR), Scanning Electron Microscopy (SEM) and Tunneling Electron Microscopy (TEM). The average particle size of TiO_2/SiO_2 composite on the surface of calcium alginate obtained by TEM is about 50 nm. The application of calcium alginate-TiO_2/SiO_2 nanocomposite as catalyst is used for conversion of glucose into 5-hydroxymethylfurfural (HMF) using dimethyl sulfoxide (DMSO) solvents. The best glucose conversion obtained at 140 °C for 4 hours is 97 % and optimum HMF yield obtained at 140 °C and 5 hours is 40 %. Nanocomposites based on sodium alginate biopolymers combined with TiO_2/SiO_2 composites can be developed as new superior materials and promising catalysts because it is easily obtained and biodegradable.

Keywords: biodegradable, glucose, HMF, nanocomposite, sodium alginate

1. Introduction

The most widely used biopolymer in several industrial sectors is alginate based on brown algae in the form of sodium alginate. Indonesia is the second largest country to produce 50 % of seaweed, which is widely found in the waters. A species of seaweed is *Sargassum sp*, so that Indonesia has great potential in producing and using the alginate compound [1]. Calcium alginate nanoparticles can be formed from sodium alginate by ionic gelation method with the addition of calcium chloride to form a calcium alginate gel, which is crosslinked with Ca^{2+} ion. Nano calcium alginate has a large surface area, mechanical strength and high stability [2].

Recently, heterogeneous catalysts are used as substitute for homogeneous catalysts, due to the use of heterogeneous catalysts have advantages, such as the properties of acids or bases can be controlled, have stability to heat and environmentally friendly. The heterogeneous catalysts mostly used are metal oxides namely, SiO_2, TiO_2, ZrO_2 and so on, because it has the acid and base properties [3]. The formation of metal oxide nanocomposite with biopolymers is one alternative to increase the catalyst activity. The incorporation of alginate with metal oxides such as TiO_2 and SiO_2 will produce a porous catalyst, thermal stability, mechanical strength and better catalytic activity [4].

Glucose and fructose can be converted into furan derivatives, namely 5-hydroxymethylfurfural (HMF) and 5-ethoxymethylfurfural (EMF). The source of glucose in nature is easier to obtain than fructose [5], so the synthesis of 5-hydroxymethylfurfural is more economical using glucose as raw material [6]. 5-hydroxymethylfurfural (HMF) is a compound that is widely used for the production of chemicals and fuels. HMF compounds are taken from petroleum sources, with limited availability [7]. The synthesis process of 5-hydroxymethylfurfural uses solvents; high boiling point of organic solvents is preferable than water, such as dimethyl sulfoxide (DMSO) [8]. Some researchers have conducted studies of HMF synthesis from cellulose, glucose and fructose. Glucose conversion to 5-hydroxymethylfurfural using TiO_2 catalyst in water, the glucose conversion obtained 30 % HMF [9]. The catalysts of SO_4 /TiO_2/SiO_2, which have a large number of Bronsted acid sites, produce 77 % HMF from fructose in DMSO solvents [10].

In this study, calcium alginate-TiO_2/SiO_2 nanocomposite was synthesized in several stages; the first stage was the synthesis of TiO_2/SiO_2 composites, which referred to the previously reported study [10]. The second stage is the formation of TiO_2/SiO_2 calcium alginate nanocomposite which referred to the study reported [11,12] in which composite TiO_2/SiO_2 is combined into sodium alginate solution with crosslinked Ca^{2+} ion. The success of nanocomposite synthesis is supported by characterization with Fourier Transform Infrared (FTIR), Scanning Electron Microscopy (SEM) and Tunneling Electron Microscopy (TEM). The nanocomposite is used as a catalyst for conversion of glucose to HMF by referring to the previously reported study [3,8].

2. Materials and methods

2.1. Materials

Sodium alginate from brown algae according to our previous study [13,14]; $CaCl_2.2H_2O$ (Merck Co) as calcium alginate forming material; tetraethyl orthosilicate, TEOS (Merck Co); titanium isopropoxide, TTIP (Sigma Aldrich) as a precursor of TiO_2 and SiO_2; glucose anhydrous (Merck Co) as reactants to form 5-hydroxymethylfurfural, HMF and dimethyl sulfoxide, DMSO (Merck Co) as glucose solvents.

2.2. Synthesis of TiO_2/SiO_2 composites

The synthesis of TiO_2/SiO_2 refers to previous studies [10]. The precursors of tetraethyl orthosilicate and titanium isopropoxide were dissolved with ethanol separately. Then, the tetraethyl orthosilicate solution was added with HCl solution until pH 3 is reached. The solution of titanium isopropoxide was mixed into a tetraethyl orthosilicate solution and stirred for 2 hours at 80 °C. The mixture was calcined at temperature 550 °C for 6 hours.

2.3. Synthesis of TiO_2/SiO_2 calcium alginate nanocomposite

The synthesis of calcium alginate-TiO_2/SiO_2 nanocomposite referred to previous studies [11,12]. TiO_2/SiO_2 composites were added to sodium alginate solution which has been mixed with calcium chloride and stirred until the suspension was formed and stirred for 1 hour, furthermore dried by freeze drying, to obtain calcium alginate-TiO_2/SiO_2 nanocomposite and characterized by FTIR, SEM and TEM.

2.4. Nanocomposite application as catalyst of glucose to HMF

The conversion of glucose to HMF referred to previous studies [3,8]. Anhydrous glucose was dissolved in DMSO, and then calcium alginate-TiO_2/SiO_2 nanocomposite catalyst was added to the glucose solution. The unreacted glucose and HMF concentrations formed were analyzed by High Performance Liquid Chromatography (HPLC). Glucose analysis was used EC-C18 columns, refractive index detectors, and water mobile phases. HMF analysis was used C18 reversed phase column with a UV-Vis detector at a wavelength of 285 nm. A mixture of methanol and water was used as the mobile phase. Glucose and HMF concentrations were obtained using a standard curve.

3. Results and discussion

3.1. Calcium alginate - TiO_2 / SiO_2 nanocomposite

The first stage of TiO_2/SiO_2 composites synthesis was using sol-gel method with tetraethyl orthosilicate precursors dissolved in ethanol $Si(OEt)_4$ and titanium isopropoxide dissolved with ethanol $Ti(OPr)_4$. The hydrolysis process occurs when adding water to tetraethyl orthosilicate solution produces $Si(OH)(OEt)_3$ silica colloid as can be seen in equation (1). The addition of $Ti(OPr)_4$ produces $Ti[OSi(OEt)_3]_4$ titanium-silica colloidal in equation (2). Furthermore, it was heated to form TiO_2/SiO_2 composite, where the remaining water and ethanol will evaporate; the estimated reaction can be seen in equation (3) by referring to Kibombo *et al.* [15] with modification.

$$Si(OEt)_4 + H_2O \rightarrow Si(OH)(OEt)_3 + EtOH \quad (1)$$

$$4\,Si(OH)(OEt)_3 + Ti(OPr)_4 \rightarrow Ti[OSi(OEt)_3]_4 + 4\,PrOH \quad (2)$$

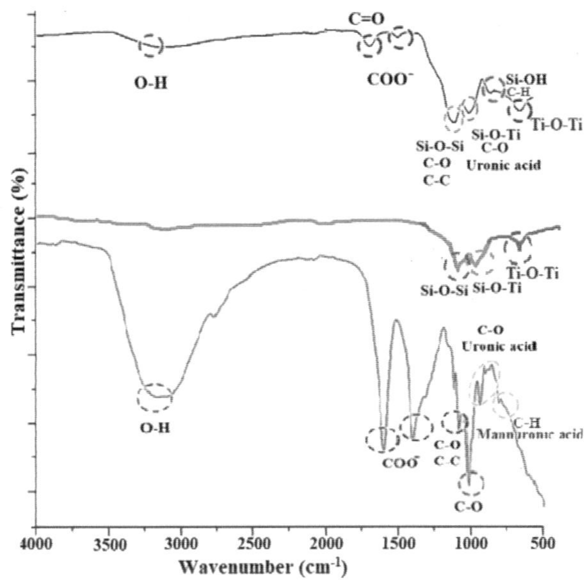

Figure 1. FTIR spectrum of (a) sodium alginate (b) TiO_2/SiO_2 and (c) nanocomposite of calcium alginate-TiO_2/SiO_2

$$Ti[OSi(OEt)_3]_4 + n\,H_2O \rightarrow O - Si - O - Ti - O - Si - O + n\,EtOH \quad (3)$$

The second stage was done for the synthesis of calcium alginate- TiO_2/SiO_2 nanocomposite, by combining sodium alginate, which has been mixed with calcium chloride, then added to TiO_2/SiO_2 composites. The nanocomposite formed will provide pore properties, thermal stability and mechanical strength so that catalytic activity will be better [16].

3.2. Characterization

3.2.1. Analysis of FTIR. Figure 1a shows the sodium alginate FTIR spectrum around 2900–3300 cm^{-1} is stretching O-H vibration with high intensity; it causes sodium alginate to have many hydroxyl groups. The asymmetric and symmetric stretching vibrations of COO- at 1604.1 cm^{-1} and 1407.1 cm^{-1}, respectively. The C-O stretching vibration of the pyranose ring at 1135.19 $cm^{-1}$and 1025.02 cm^{-1}. The alginate typical vibrations of C-O uronic acid at 946.8 and 885.2 cm^{-1} and vibrations of C-H mannuronic acid at 805.6 cm^{-1}. Figure 1b shows spectrum of TiO_2/SiO_2 composite, the peak Si-O-Si at 1075.9 cm^{-1} derived from SiO_2 compound and the peak of Si-O-Ti at 942.7 cm^{-1}, this indicates that Si-O-Ti bonds was formed. The peak of Ti-O-Ti appears at 641.00 cm^{-1} derived from TiO_2 compound. Figure 1c shows calcium alginate-TiO_2/SiO_2 spectrum consists of O-H vibration around 2925–3275 cm^{-1}, its intensity decreases compared to sodium alginate in figure 1a; it is caused O-H group to bind with TiO_2/SiO_2. The asymmetric of COO- at 1691.5 cm^{-1} which overlaps to vibration of C = O and the symmetric of COO- at 1499.6 cm^{-1}. The Si-O-Si vibration at 1025.7 cm^{-1} overlaps with vibration of C-O and C-C. The Si-O-Ti vibration at 1075.4 cm^{-1}, which overlaps with the vibration of C-O uronic acid at 995.4 cm^{-1}. There is Si-OH vibration at 840.9 cm^{-1}, which overlaps with the vibration of C-H mannuronic acid. The Ti-O-Ti vibration is shifted to 600 cm^{-1}. The characterization with FTIR showed the successful formation of $TiO_2/_{SiO_2}$ calcium alginate nanocomposite.

3.2.2. Morphology with SEM. The morphology using SEM is shown in figure 2. Figure 2a is the morphology of sodium alginate having fibril layer with uniform surface. Figure 2b is composite of TiO_2/SiO_2 shows the small granular of TiO_2 compounds and the SiO_2 silica in larger granular and non-uniform. Figure 2c is calcium alginate- TiO_2/SiO_2 nanocomposite shows the presence of evenly distributed TiO_2/SiO_2 composites on the surface of sodium alginate. These results support the successful synthesis of TiO_2/SiO_2 nanocomposites.

IOP Conf. Series: Materials Science and Engineering **763** (2020) 012037 doi:10.1088/1757-899X/763/1/012037

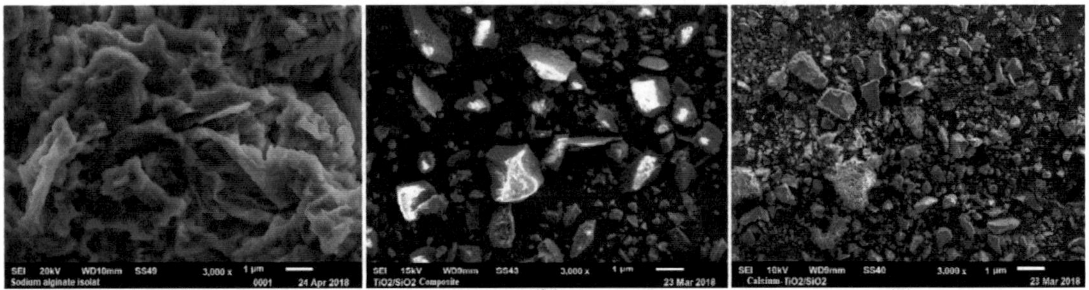

(a) (b) (c)

Figure 2. SEM Morphology of (a) sodium alginate (b) TiO_2/SiO_2 (c) nanocomposite of calcium alginate-TiO_2/SiO_2

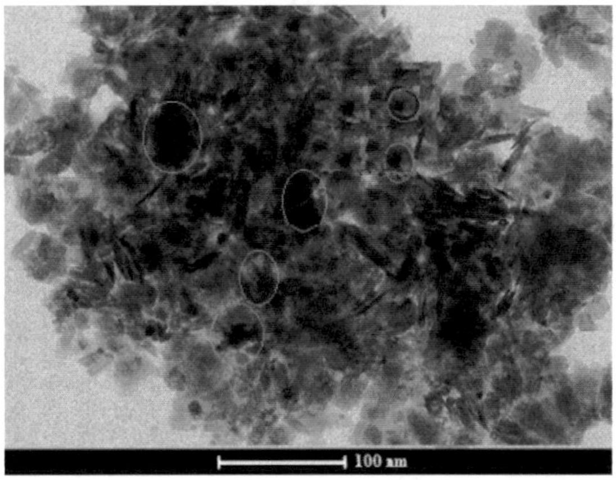

Figure 3. TEM morphology of calcium alginate-TiO_2/SiO_2 nanocomposite

3.2.3. TEM morphology. The morphology of calcium alginate-TiO_2/SiO_2 nanocomposite analyzed by TEM is shown in figure 3. It can be seen the grey colored of calcium alginate, while the TiO_2/SiO_2 composite is black, it was caused by TiO_2/SiO_2 composite on the surface of calcium alginate, so that can be observed clearly. The average particle size of TiO_2/SiO_2 composite at calcium alginate surface obtained by TEM is about 50 nm.

3.2.4. Application nanocomposite as catalyst of HMF from glucose. The formation of HMF from glucose using catalyst of calcium alginate-TiO_2/SiO_2 nanocomposite was done with reaction temperature variations of 80°, 100°, 120°, 140° and 160 °C and time variations of 1, 2, 3, 4 and 5 hours. The unreacted glucose and formed HMF were analyzed by HPLC. Figure 4 shows the effect of temperature of glucose conversion, it was observed the unreacted glucose or conversion of glucose to HMF. It was seen that glucose conversion increases as the temperature rises and optimal conversion was obtained at 140 °C, and then constant up to 160 °C. This is because at low temperatures the reaction rate is slow and at high temperature at 140 °C the optimum reaction rate is obtained, so that glucose conversion is 97 %. Figure 5 shows the effect of time to glucose conversion. It can be seen that as the reaction time rises, the glucose reacted will increase and the obtained optimum time is 4 hours and constant up to 5 hours is 97 %.

After obtaining the optimum temperature and time for glucose conversion to HMF by measurement of unreacted glucose, it was obtained at 140 °C and 4 hours. Furthermore, the concentration of HMF formed was determined against the reaction time as can be seen in figure 6. It shows that the formation of HMF can be observed at 4 hours and maximum at 5 hours is 40 %.

The reaction profile of glucose concentration consumption and HMF formation at temperature 140 °C by time variation 1, 2, 3, 4, 5 and 6 hours are illustrated in figure 7. In figure 7 it can be

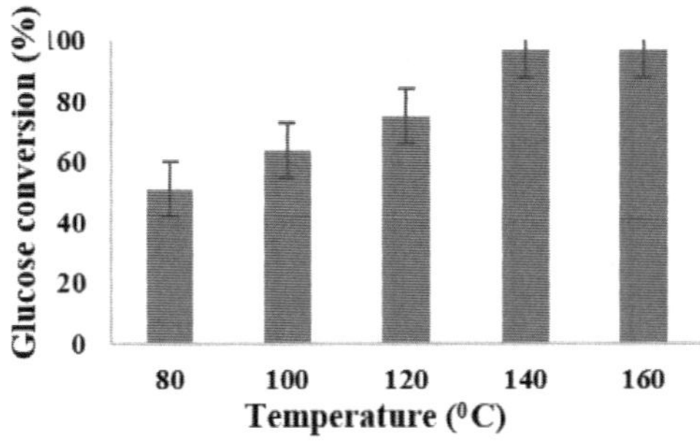

Figure 4. Effect of temperature to glucose conversion

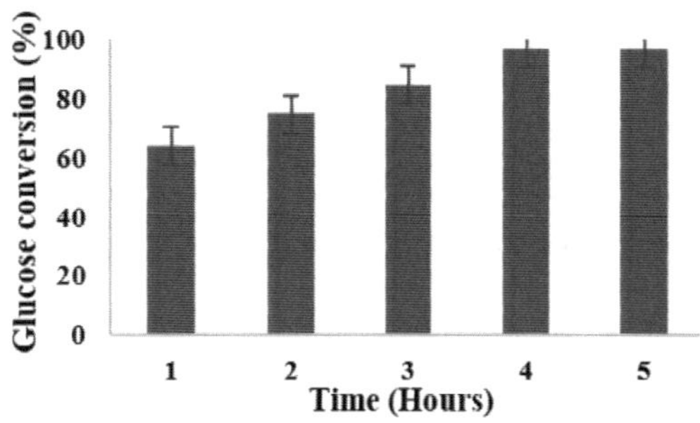

Figure 5. Effect of reaction time to glucose conversion

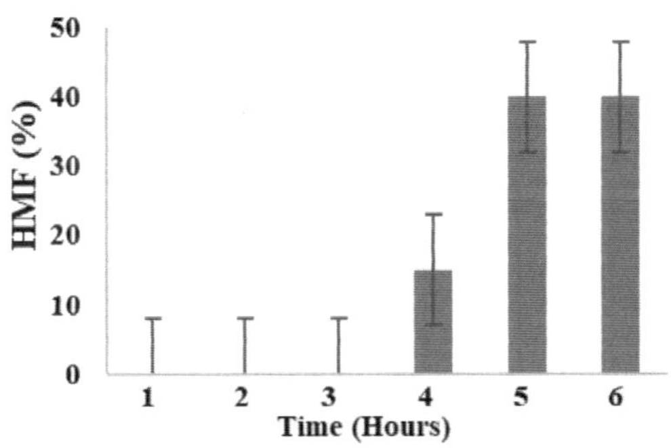

Figure 6. Effect of reaction time to HMF yield

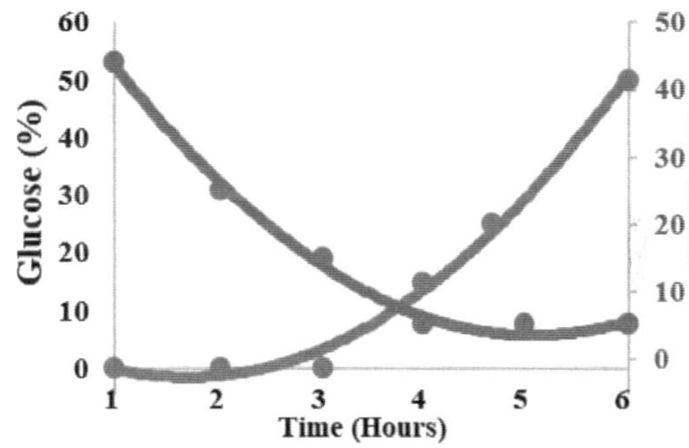

Figure 7. Profile of reaction to glucose conversion (red) and HMF yield (blue)

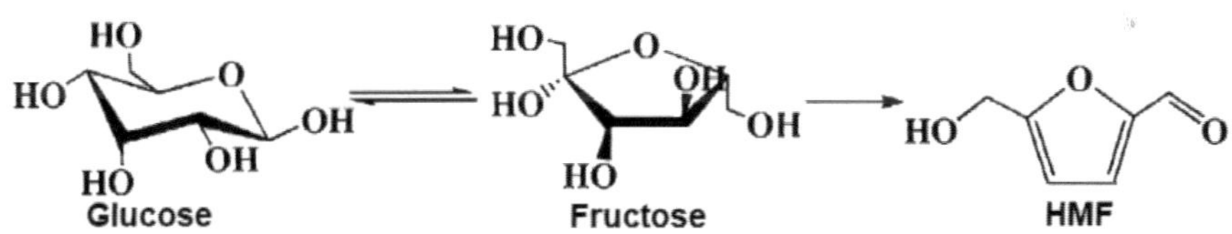

Figure 8. The reaction of HMF formation from glucose

observed that glucose concentration (red) decreases towards time and reaches equilibrium at 4 hours, whereas, HMF concentration (blue) can be observed after 4 hours and rises at 5 hours then the equilibrium is obtained, this is due to the formation of HMF needs a long time, which causes the first fructose to be formed then formed the HMF and fructose in reversible reaction with glucose as shown in figure 8. This supports the fact that the synthesized nanocomposite is capable to be applied as a catalyst for HMF formation from glucose.

4. Conclusions

The nanocomposite of calcium alginate-TiO_2/SiO_2 was successfully synthesized, supported by FTIR, SEM and TEM. The average particle size of TiO_2/SiO_2 composite on the surface of calcium alginate obtained by TEM was about 50 nm. The application of calcium alginate-TiO_2/SiO_2 nanocomposite as catalyst was used for conversion of glucose into 5-hydroxymethylfurfural (HMF) using dimethyl sulfoxide (DMSO) solvents. The best glucose conversion obtained at 140 °C and 4 hours is 97 % and optimum HMF yield obtained at 140 °C and 5 hours is 40 %.

Acknowledgements

This research was supported by PITTA grant University of Indonesia for fiscal year 2018.

References

[1] Venkatesan J, Anil S, Kim S K 2017 *Seaweed Polysaccharides: Isolation, Biological and Biomedical Applications* (Amsterdam: Elsevier)
[2] Paques J P, van der Linden E, van Rijn C J M and Sagis L M C 2017 *Adv. Colloid Interface Sci.* **209** 163–71
[3] He R, Huang X, Zhao P, Han B, Wu T and Wu Y 2018 *Waste Biomass Valorization* **9** 657–68
[4] Bhanja P and Bhaumik A 2016 *Fuel* **185** 432–41
[5] Zhou P and Zhang Z 2016 *Catal. Sci. Technol.* **6** 3694–712
[6] Chatterjee C, Pong F and Sen A 2015 *Green Chem.* **17** 40–71
[7] Xue Z, Ma M G, Li Z, Mu T *RSC Adv.* **6** 98874–92
[8] Zhou C, Zhao J, Yagoub A E A, Ma H, Yu X, Hu J, Bao X and Liu S 2017 *Egypt J. Petroleum* **26** 477–87
[9] Lanziano C S, Rodrigues F, Rabelo S C, Guirardello R, da Silva V T and Rodella C B 2014 *Chemical Engineering Transaction (CET)* **37** 589–94
[10] Kılıç E and Yılmaz S 2015 *Ind. Eng. Chem. Res.* **54** 5220–5
[11] Nait Mohamed FA and Laraba-Djebari F 2016 *Vaccine* **34** 2692–9
[12] Jaypal J J and Dhanaraj S 2017 *Int. J. Biol. Macromol.* **105** 416–21
[13] Helmiyati and Aprilliza M 2017 *IOP Conf. Ser.: Mater. Sci. Eng.* **188** 012019
[14] Helmiyati H and Dini F W 2018 *AIP Conf. Proc.* **2023** 020101
[15] Kibombo H S, Peng R, Rasalingam S and Koodali R T 2012 *Catal. Sci. Technol.* **2** 1737–66
[16] Bhanja P and Bhaumik A 2016 *Fuel* **185** 432–41

Hydroprocessing of kemiri sunan oil (*reutealis trisperma (blanco) airy shaw*) over NiMoCe/γ-Al_2O_3 catalyst to produce green diesel

D Yulia[1,2] and A Zulys[2]

[1] Research and Development Centre for Oil and Gas Technology Lemigas, Jalan Ciledug Raya Kav 109 Cipulir, Jakarta 12230, Indonesia

[2] Department of Chemistry, Faculty of Mathematics and Natural Science (FMIPA), Universitas Indonesia, Kampus UI Depok, Depok 16424, Indonesia

Corresponding author's e-mail: dyulia@lemigas.esdm.go.id

Abstract. Kemiri Sunan is an alternative energy source that has a considerable amount of oil and does not compete for food source. Hydroprocessing of kemiri sunan oil has a big potential for renewable fuels. In this paper, NiMoCe/γ-Al_2O_3 catalyst is used for hydroprocessing of saturated kemiri sunan oil to produce renewable hydrocarbons. The catalyst was prepared by using wet impregnation method with variations of Ce 1%, 5% and 15%. The characterization of catalyst was used by means of Brunauer-Emmett-Teller (BET), Scanning Electron Microscopy (SEM), X-Ray Diffraction (XRD) and X-Ray Fluorescence (XRF). Hydroprocessing was performed using an autoclave batch reactor under operating conditions of 400° C and 35 bar pressure for 5 hours. GC characterization showed that the product of renewable hydrocarbons is dominated by the C_{15}-C_{18} alkane chain.

Keywords: renewable oil, hydroprocessing, catalyst, Kemiri Sunan oil

1. Introduction

Fossil energy is the dominant primary energy supply in Indonesia. Fossil fuels hold responsible for 95.3 % of consumption, such as oil (46.1 %), coal (30.9 %) and natural gas (18.3 %) as the main fuels source [1]. Worldwide energy consumption reaches the highest ever presently, due to the growth of population and social development. One of the main energy consumers is the transportation sector which responsible for around one-fifth from the worldwide energy consumption. Due to the growth of population and more accessible transportation, the fuel requirement, particularly liquid fuels, will become higher in the years to come. Unfortunately, the progressive depletion of petroleum-based fuels created a global crisis of energy. The decreasing availability of petroleum resources takes place simultaneously with an emission increase due to greenhouse gases accountable for global warming. Therefore, methods that can offer greenhouse effect reduction are urgently required, which attracts a great attention to biofuels. Important experiments have been conducted to discover an alternative fuel to substitute gasoline and diesel. The preferable resolution is to discover an alternative fuel equals to the conventional one, namely, suitable with the available infrastructure, as well as sustainable, able to reduce CO_2 emission and to reduce the environmental anthropogenic footprint. Renewable biomass sources can be transformed into an environmentally friendly fuels and a realogical option to generate fuel oil. Fuel oil obtained from biomass is clean as it contains insignificant amount of sulfur, nitrogen and ash content, generating fewer emissions of SO_2, NOx and soot compared to conventional fossil fuels. Zero net emission of CO_2 could be acquired since the CO_2 released from biomass is able for photosynthetically recycling by plants [2].

Biofuels can be derived into four generations according to the nature of feedstock and corresponding production technology. Particularly, the previous biodiesel generation such as fatty acid methyl esters (FAME) is produced via transesterification mainly from vegetable oils and animal fats (complex mixture of triglycerides) by using a varied earth-abundant heterogeneous catalyst in the optimal process conditions. Even though the technology is perfect to be applied commercially, this biodiesel type contains a high amount of oxygenated compounds, suffering from considerable inherent, serious deficiency such as low heating value (or energy density), high viscosity, poor cold flow properties, oxidation instability and incompatibility toward fossil fuel due to its relatively high oxygen content and acidity, which render its practical application as transportation fuels. Therefore, great effort has been concentrated for the upgrading of biomass-derived bio-oils to a more stable and higher quality second-generation hydrocarbon-based biofuels via hydrodeoxygenation (HDO), decarboxylation or decarbonylation over specified catalysts. The produced biofuels obtained from deoxygenation methods comprises a high linear hydrocarbon amount and few oxygen, sulfur and nitrogen. The product carries remarkable physical and chemical characteristics (high cetane number, great cold flow properties, high heating value and excellent storage stability), which are fully compatible with fossil fuels, deemed promising for a large-scale application. Nevertheless, the commercial application of the second-generation biofuels is still under perfection, since some technical drawbacks are required to be resolved [3].

Vegetable oil hydro processing generates hydrocarbon product in the diesel boiling range, hence generally known as green diesel or renewable diesel. The products from hydroprocess method are alkanes in straight chains composed of 15–20 carbons, which shows a high similarity toward the current fossil diesel once it is isomerized. Hydroprocessing has been constructed to create alkanes with straight chain from fatty acid triglycerides of animal fat, and vegetable oils. Metal sulfide catalysts have been commonly applied for vegetable oil hydroprocessing, and other types of catalysts namely noble metal catalysts have also been reported. The transition metal catalysts require the sulfur-containing compounds (such as H_2S) addition to maintain the active form of catalysts. One should note that sulfur might leach out from the catalysts over a long-term condition; therefore the quality of green diesel could be affected should the products be contained of sulfur residuals. On the contrary, the toxic and corrosive properties of H_2S might cause a great danger for the environment, human health and instruments. Although noble metal catalysts were non-sulfided, these substances are more expensive and not applicable for a large-scale process. Generally, the catalysts of transition metal are the sulfide form of NiMo, CoMo and NiW based catalysts containing P, B, and Cu. Nevertheless, lower emphasis has been placed on catalysts containing rare earth metal. Lanthanide or actinide metals incorporated with some metals (namely Ni, Co, Cu or Fe) were reflected as new type of supported catalysts precursors, which can be more active. Reports showed that cerium supports the activity of nickel/alumina catalysts in CO hydrogenation. Metallic catalysts supported by cerium are attractive since the properties of cerium can enhance their thermal stability, metallic dispersion, catalytic activity and decrease the carbon deposition. Thus, the ultimate purpose of this experiment is to promote the non-sulfide NiMo/γ-Al_2O_3 catalyst consists of Cerium (Ce) for the green diesel production. Hydroprocessing of kemiri sunan over NiMoCe/γ-Al_2O_3 catalyst [4] and hydrotreating of Kemiri Sunan oil over NiMo/γ-Al_2O_3 and CoMo/γ-Al_2O_3 catalysts have been reported [5–6]. By taking the biofuels competition problem with food sources into account, the option relies on the non-edible plant oil as the feedstock to generate green diesel for the transportation fuel. This work studies the hydroprocessing of Kemiri Sunan oil over NiMoCe/Al_2O_3 catalyst.

2. Experiments

2.1. Catalysts preparation and characterization

The NiMo catalyst (5wt.%Ni, 10wt.%Mo) was created using wet coimpregnation of aqueous solutions of $(NH_4)_6Mo_7O_{24}$ and $Ni(NO_3)_2$ with the aid of Al_2O_3 grain. Next, the NiMo/Al_2O_3 catalysts were dried overnight at 105 °C and calcined at 400 °C for 4 h. The NiMoCe(x)/Al_2O_3 catalysts (where x indicated Ce content from 1 to 15wt.%) were produced by impregnation of NiMo/Al_2O_3 catalyst with a $Ce(NO_3)_3$ solution. Impregnated samples were dried overnight at 105 °C and calcined at 400 °C for 4 h. BET specific surface areas and pore size distributions were determined by N_2 adsorption-desorption using Quantachrome Surface Area and Pore Distribution Analyzer instrument. The morphology images of samples were taken by Scanning Electron Microscope (SEM) equipped analyzer (OXFORD). X-Ray Diffraction (XRD Empyrian PANanalitycal) was applied to specify the structural properties of the catalysts. Gas Chromatography with Flame Ionisation Detector (GC-FID) Agilent 6890 was used to measure quantitatively the hydrocarbon product. Helium was used as the carrier gas.

Table 1. Specific surface areas and pore size distributions of Catalysts

Catalyst	Specific Surface Area (m^2/g)	Total Pore Volume (Cc/g)	Pore Diameter (Å)	Decrease of Specific Surface Area (%)
γ-Al_2O_3	306.44	0.6846	89.37	100
NiMo/γ-Al_2O_3 sulfided	250.05	0.6577	105.2	18
NiMoCe/γ-Al_2O_3 1 %	209.68	0.5292	100.9	31
NiMoCe/γ-Al_2O_3 5 %	194.10	0.5006	103.2	36
NiMoCe/γ-Al_2O_3 15 %	159.22	0.3588	90.15	48

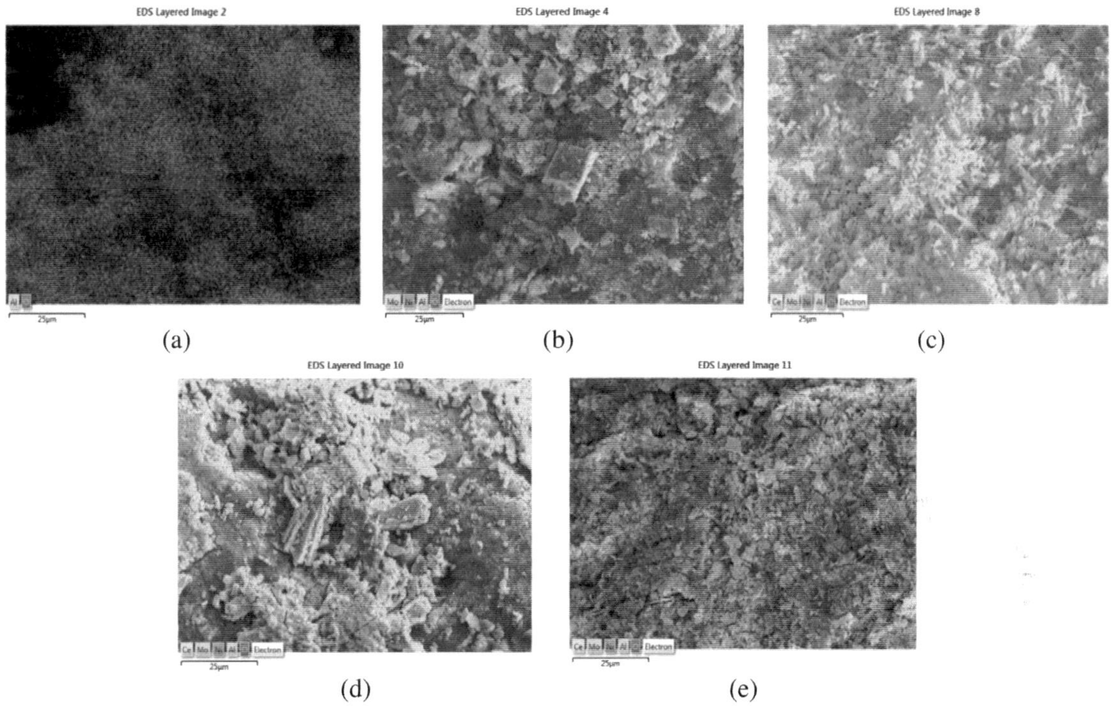

Figure 1. SEM Images of (a) Al_2O_3, (b) NiMo/γ-Al_2O_3 sulfided, (c) NiMo/γ-Al_2O_3 Ce 1%, (d) NiMo/γ-Al_2O_3 Ce 5% and (e) NiMo/γ-Al_2O_3 15%

2.2. Catalytic activity measurements

Kemiri Sunan oil was purchased from Ciawi Bandung. $(NH_4)_6Mo_7O_{24}$, $Ni(NO_3)_2$, $Ce(NO_3)_3$ and Al_2O_3 were purchased from Sigma-Aldrich. The experiments were conducted in an autoclave batch reactor completed with electrically heating system. The sulfided NiMo catalysts were activated in situ prior to the experiments with 5 % (v/v) dimethyl disulfide (DMDS) in gas oil at 400 °C and 3.5 MPa for 5 h. The catalytic hydroprocessing reaction conditions were fixed at 400 °C and 3.5 MPa. The liquid products were analyzed using a GC-FID.

3. Results and discussion

3.1. Catalyst characterization

Specific surface areas and pore size distributions of the support and catalysts are shown in table 1. Al_2O_3 has considerably high surface area (306.44 m^2/g) and large pore diameter (89.37 Å). This characteristic enabled the diffusion of bulky triglyceride molecules. Impregnation of Ni, Mo and Ce on

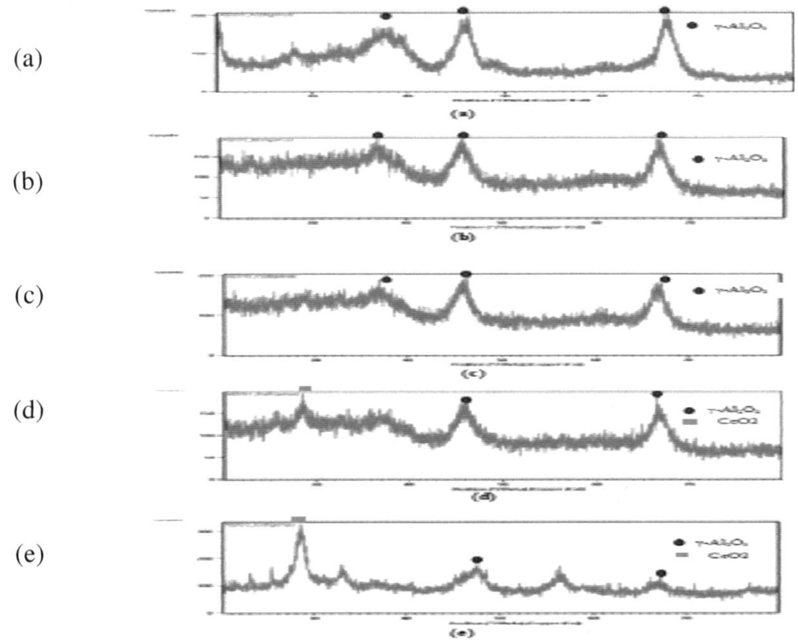

Figure 2. Diffractogram of (a) γ-Al_2O_3, (b) NiMo/γ-Al_2O_3, (c) NiMoCe/γ-Al_2O_3 1% (d) NiMoCe/γ-Al_2O_3 5% and (e) NiMoCe/γ-Al_2O_3 15%

Al_2O_3 generated a decrease in the specific surface area. The impregnation of NiMoCe 15 % on Al_2O_3 can create more decrease in the specific surface area than the impregnation of NiMoCe 1 % and NiMoCe 5 % on Al_2O_3. The decrease of the particular surface area and total pore volume proposes the blocking of some pores in Al_2O_3 by the addition of Ni, Mo and Ce [7]. Figure 1 shows the metal distribution of Al, Ni, Mo, and Ce on the surface of the catalyst.

XRD analysis of catalysts is shown in figure 2. Typically, all catalysts have a uniform peak pattern for γ-Al_2O_3 at 2θ: 37°, 46° and 67° similar to figure 2a [8]. However, these peaks experience a diffraction decline in intensity. This occurrence showed that γ-Al_2O_3 support experienced a decline in specific surface area due to the metal impregnation of Ni, Mo and Ce [9]. In figure 2b, crystalline peak of MoO_3 at around 2θ: 27° [9] was not observed, while in figure 2c, figure 2d and figure 2e the impregnation peak of Ce, which speculated as CeO_2 in γ-Al_2O_3 support after calcination was expected. However, CeO_2 peaks at 2θ: 28.6°, 33.1°, 47.5° and 56.4° [4] can be observed in figure 2d and figure 2e, while it was not obtained in figure 2c, indicating existence of CeO_2 was with the impregnation of Ce 5 % and for Ce 15 % only.

Further measurement of elemental analysis by mean of X-Ray Fluoresence for γ-Al_2O_3 showed the γ-Al_2O_3 content in the support was 99.71 %, while in NiMo/γ-Al_2O_3 catalyst an amount of 4.90 % Ni and 9.94 % of Mo were observed. In the case of NiMoCe/γ-Al_2O_3 (1 %) catalyst, XRF analysis was 0.84% Ce, for NiMoCe/γ-Al_2O_3 (5 %) was of 3.38 % Ce and for NiMoCe/γ-Al_2O_3 (15 %) catalyst was 15.6 % Ce.

3.2. GC analysis

Table 2 shows that the hydroprocessing of Kemiri Sunan oil using Ce 1% produce the highest conversion percentage of diesel fraction (C_{15}-C_{18}) about 57.79%. It indicates that Ce as a promoter can increase a catalytic activity in hydrogenation. This is because cerium oxide metal is well known as lewis base and most probably would donate its electron to NiO therefore enhancing the electron density to metal oxide and forming a complex structure with Ni. However, the low yield of hydroprocessing using Ce 1% is the amount of hydrocarbon C_{18} compared to yield of hydroprocessing using NiMo/γ -Al_2O_3 sulfidedcatalyst.

Table 2. Hydroprocessing of Kemiri Sunan oil (400°C, 3.5 MPa, 5 h)

Catalyst	NiMo/γ-Al_2O_3 sulfided	NiMoCe/γ-		
		Al_2O_3 1%	Al_2O_3 5 %	Al_2O_3 15%
< C_{15}	40.34	20.85	30.02	35.63
C_{15}	24.03	29.72	24.71	22.75
C_{16}	6.10	5.20	4.33	5.04
C_{17}	17.23	20.86	16.97	15.19
C_{18}	2.60	2.01	2.60	1.75
>C_{18}	9.70	21.36	21.37	19.64
Total amount C_{15}-C_{18}	49.96	57.79	48.61	44.73

4. Conclusions

Renewable C_{15}-C_{18} alkanes as the main content of the diesel fraction have been successfully produced from hydroprocessing of Kemiri Sunan oil over NiMoCe/γ-Al_2O_3 1 % catalysts with conversion percentage of 57.79 % at 400 °C, 3.5 MPa and 5h. NiMoCe/γ-Al_2O_3 (1 %) has the highest conversion percentage of Kemiri Sunan oil. The impregnation of Ce to NiMo/γ-Al_2O_3 is proven to increase the conversion percentage of Kemiri Sunan oil. The NiMoCe/γ-Al_2O_3 has the potential to replace the sulfided NiMo/γ-Al_2O_3 in hydroprocessing of Kemiri Sunan oil and is expected to convert it to the renewable hydrocarbons.

References

[1] Dewan Energi Nasional 2016 Indonesia Energy Outlook Energi 2016 (Jakarta: Sekretaris Jenderal Dewan Energi Nasional)
[2] Xu J, Jiang J and Zhao J 2016 *Renew. Sustain. Energy Rev.* **58** 331–40
[3] Li X, Luo X, Jin Y, Li J, Zhang H, Zhang A and Xie J 2018 *Renew. Sustain. Energy Rev.* **82** 3762–97
[4] Liu J, Fan K, Tian W, Liu C and Rong L 2012 *Int. J. Hydrog. Energy* **37** 17731–7
[5] Ilham M 2016, *Hydrotreating of Kemiri Sunan Oil (Aleurites trisperma) by Using CoMo/γ-Al_2O_3 Catalyst to Produce Green Diesel* (Yogyakarta: Universitas Gajah Mada) Undergraduate Thesis.
[6] Setiawan D I 2016 *Hydrotreating of Kemiri Sunan Oil (Reutealis trisperma (blanco) airy shaw) for renewable diesel fuel production* (Bogor: Sekolah Pascasarjana Institut Pertanian Bogor) Master Thesis.
[7] Srifa A, Faungnawakij K, Itthibenchapong V, Viriya-empikul N, Charinpanitkul T and Assabumrungrat S 2014 *Bioresour. Technol.* **158** 81–90.
[8] Okamoto Y 1998 *Appl. Catal. A Gen.* **170** 315–28
[9] Ulfah M and Subagjo 2012 *Reaktor* **14** 151–7

Modification of glassy carbon electrode with NiO nanoparticles prepared using *Ocimum sanctum* leaf extract as urea sensor

P R Arief, Y Yulizar and T A Ivandini

Department of Chemistry, Faculty of Mathematics and Natural Sciences (FMIPA), Universitas Indonesia, Kampus UI Depok, Depok 16424, Indonesia

Corresponding author's e-mail: yokiy@ui.ac.id

Abstract. NiO nanoparticles (NiO NPs) have been prepared with green synthesis method using *Ocimum sanctum* leaf extract (OSLE), which contains secondary metabolite compounds of alkaloids as a weak base source and capping agent. Characterization using Particle Size Analyser (PSA) showed a particle size of NiO about 52.42 nm. NiO NPs was used as modified on glassy carbon (GC) electrode, and then characterized using cyclic voltammetry (CV) and Scanning Electron Microscopy-Energy Dispersive X-Ray (SEM-EDX). GC-NiO NPs was used for preliminary studies of urea sensor. The results showed that GC-NiO NPs can detect urea with limit of detection (LOD) values of 0.12 mM and linearity R^2 of 0.97. In addition, the repeatability of urea sensor studies with GC-NiO NPs electrode shows % RSD of 2.854, while its stability shows % RSD of 7.90 for 3 days.

Keywords : glassy carbon electrode, GC-NiO NPs electrode, *Ocimum sanctum*, urea

1. Introduction

High urea content can cause various adverse effects to the human body. In spite of the low urea content, this can also causes harmful effects such as neurological disorders and hepatitis. Due to the dangerous effects of urea in the environment, it is important to investigate the level of urea in the environment [1]. The organic Ni-metal framework on double-walled carbon nanotubes has been synthesized to detect the presence of non-enzymatic urea showing that the electrode has a high sensitivity of 685 $\mu A.mM^{-1}cm^{-2}$, with a low detection limit of 3 μM and a response time of 10 s. In addition, the sensor also exhibits high stability after 30 days of storage [2]. The non-enzymatic urea sensor prepared using nickel oxide deposited on carbon nanotube (CNT) cellulose showed an excellent sensitivity with good response time. The electrode also exhibited high stability in sensitivity after the storage for 2 months [3].

Nanomaterials have revolutionized in the area of sensor development, due to their materials size and shape, as well as various types to design the sensing platforms for different applications [4]. Metal oxides such as NiO, CuO, and ZnO are known to be low costs substances and well-recognized redox properties, have been utilized for direct electro-catalytic oxidation and to determine numbers of biochemical compounds, namely urea, glucose, and ethanol. Among some transition metals, Ni has excellent anodic electrocatalytic properties [3].

There are various methods for synthesis of metal and metal oxides nanoparticles such as precipitation method, sol-gel [5], hydrothermal [6] and green synthesis. Green synthesis of nanoparticles also provides some beneficials such as convenient, single-stage, low cost, environmentally friendly, comparatively reproducible as well as stable materials [7]. Taken into account the comprehensive potential of plants, this research focuses to utilize a biological green method to synthesize the nanoparticle. In this regard, leaf extract of *Ocimum sanctum* contains secondary metabolites such as alkaloids, flavonoids, saponins, tanins, and terpenoids [8].

In this study, glassy carbon (GC) electrode was modified with NiO NPs (GC-NiO NPs). *Ocimum sanctum* leaf extract (OSLE) was used as a weak base source and capping agent. Although this method is potential for synthesis of nanoparticles, it has never been applied for synthesis of nanoparticle on the surface of carbon electrodes, especially GC. OSLE was used to modify GC surface with NiO NPs. Furthermore, a modified GC electrode with NiO NPs showed promising application as urea sensor.

2. Materials and methods

2.1. Materials

Analytical-grade of nickel nitrate hexahydrate, $Ni(NO_3)_2.6H_2O$ was purchased from Merck. *Ocimum sanctum* leaves as the main raw material were received from conservation unit of biopharmaca (UKBB), center study for tropical biopharmaca, LPPM IPB, Bogor, Indonesia.

2.2. Preparation of OSLE

Ocimum sanctum leaves were washed, dried and mashed to form powder. The extraction was done in two steps. The first, 50 g of leaf powder was macerated in 250 mL of methanol for one week by stirring every day. The filtrate was partitioned using n-hexane. The methanol fraction was isolated and concentrated with a rotary vacuum evaporator. The evaporated methanol fraction was dissolved in water to obtain a stock solution of OSLE.

2.3. Synthesis and characterisation of NiO NPs

The synthesis of NiO NPs followed the method as described in our previous report [9]. 0.05M $Ni(NO_3)_2$ was reacted with OSLE by stirring at 80 °C for 6 h. The formed colloids were heated at 100 °C for 4 h, and then calcined at 450 °C for 2 h. NiO NPs were characterized using Fourier Transform Infrared (FT-IR), UV-Vis Spectrophotometer, X-Ray Diffraction (XRD) and Particle Size Analyser (PSA).

2.4. Modification of GC electrodes with NiO NPs

The modification of glassy carbon electrodes (GCE) with NiO NPs referred to a method as reported by Susanti *et al.* [10] with some modifications. GCE was polished with 0.5μM α-alumina and washed by sonication for 5 min in isopropanol and deionized water, respectively. It was then dried. The cleaned GCE was dipped into 0.05M $Ni(NO_3)_2$ and OSLE by stirring at 80 °C for 6 h. The results were further heated at 100 °C for 4 h and then calcined at 450 °C for 2 h. GC-NiO NPs were characterized using Scanning Electron Microscopy-Energy Dispersive X-Ray (SEM-EDX).

2.5. Preliminary study of sensors urea

GCE-NiO NPs were used for preliminary study of urea sensor. In the electrochemical cell, GC-NiO NPs was applied as the working electrode, while Pt wire and Ag/AgCl as the counter and reference electrodes, respectively and 1M KOH as the supporting electrolyte. Cyclic voltammetry was applied as the determinant for calibration, limit detection, reproducibility and stability.

3. Results and discussion

3.1. Characterisation of NiO NPs

FT-IR characterisation aims to determine the functional groups both NiO and OSLE as shown in figure 1a. OSLE shows the vibration at 3324 cm^{-1}, 2936 cm^{-1}, 1685 cm^{-1} and 1076 cm^{-1}, which corresponds to the vibration of -OH, C-H, C-C, and -CN groups, respectively. The presence of -OH, C-H, C-C and -CN groups assumed that OSLE contains alkaloid compounds, which acts as a weak base source and a capping agent in NiO NPs synthesis. NiO shows no vibration of -OH and C-H groups. However, NiO has strong band at 422 cm^{-1} of which assigned to Ni-O stretching vibration mode [11].

The analysis of both structure and average crystalline size of NiO NPs was characterized by XRD as shown in figure 1b. XRD pattern of NiO NPs shows five diffraction peaks at 37.15°, 43.23°, 62.77°, 75.43° and 79.31° indexed to (111), (200), (220), (311), (222), respectively. These results are in accordance with the COD literature of NiO No. 96-101-0096 [5]. Based on those Miller indices, the crystalline structure of NiO was theoretically known as cubic. The average size of NiO crystals was 28.74 nm calculated using the Debye-Scherrer's formula. According to the PSA characterisation, the

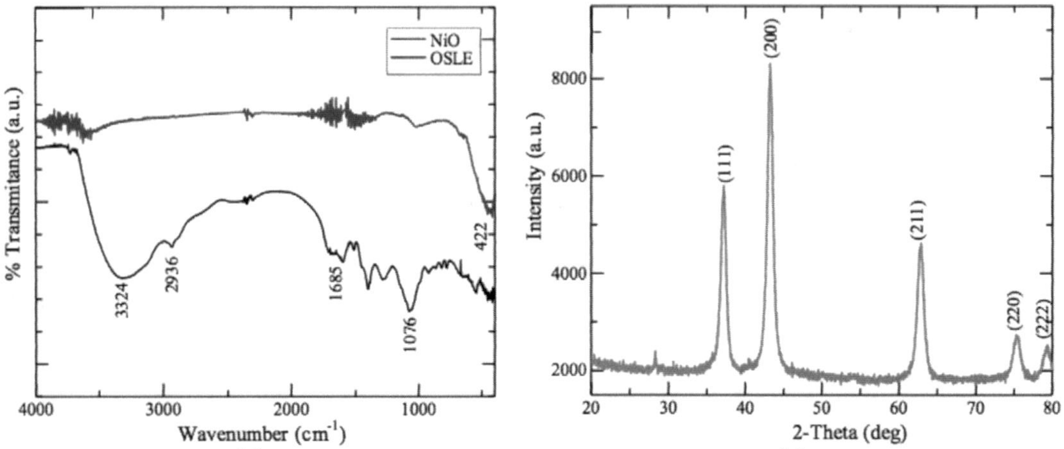

(a) (b)

Figure 1. (a) FTIR spectra of OSLE and NiO (b) XRD pattern of NiO

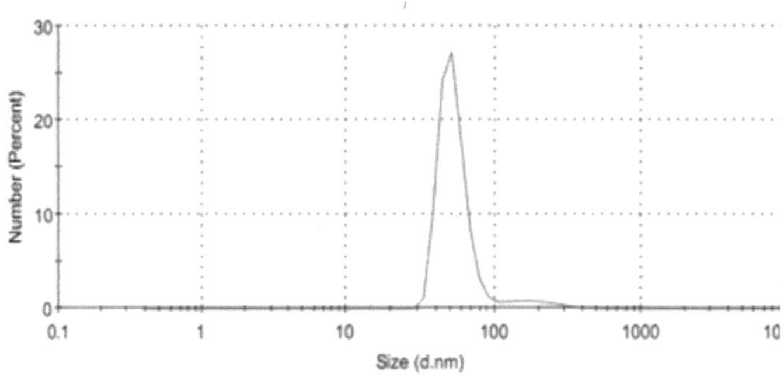

Figure 2. PSA result of NiO NPs

particle size distribution average of NiO NPs was quite homogeneous with sharp peak at 52.42 nm as shown in figure 2.

3.2. Modification of GC electrode with NiO NPs

The morphology and elemental composition of both bare GC and GC-NiO NPs electrodes were respectively characterised by SEM (mapping)-EDS as shown in figure 3. In GC-NiO NPs electrode mapping, NiO NPs exhibited lighter color than bare GC. It demonstrates that NiO has higher conductivity than bare GC. Therefore, NiO has been successfully deposited on the GC surface. This result was also supported by EDS analysis before and after modification with NiO NPs. EDS data indicated that the presence of Ni was around 28.52 %. NiO was only a few deposited on the GC surface due to its tendency to form the non-homogeneous aggregates.

Figure 4 shows the electrochemical properties of Ni, bare GC, and GC-NiO NPs electrodes using CV. When GC-NiO NPs electrode was compared to the bare GCE and Ni electrodes, it has the characteristic of two electrodes combination, which the GC-NiO electrode has a wide potential window but it was not wider than the GC bare, the peaks of oxidation and reduction are also like the Ni electrode. Ni(II)/Ni(III) peaks were observed approximately at 0.5V vs Ag/AgCl. The reaction is represented in alkaline medium through a proton diffusion that formed NiOOH compound.

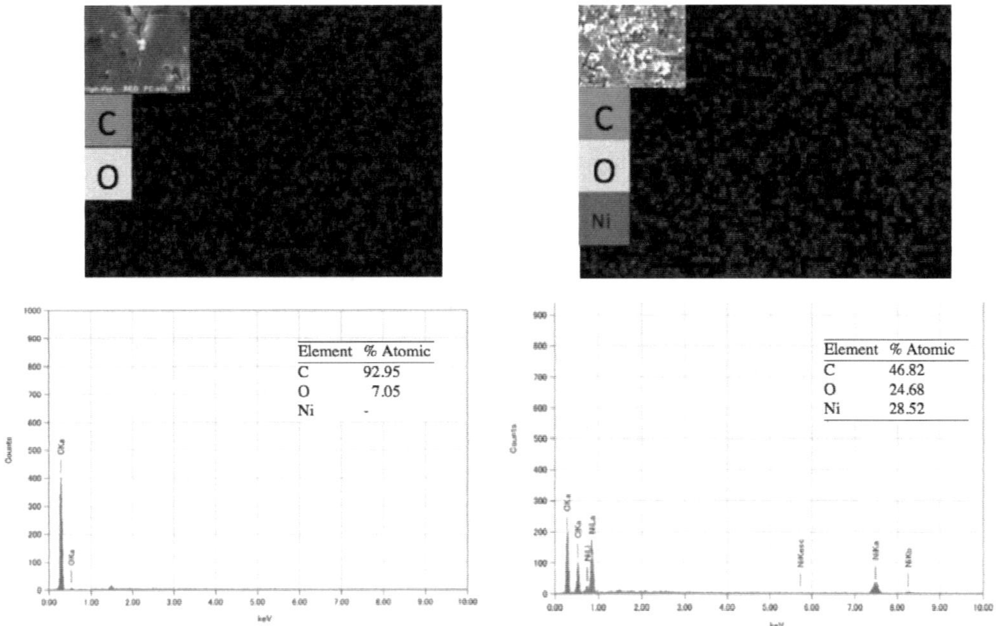

(a) (b)

Figure 3. SEM-EDS characterizations of (a) bare GC and (b) GC-NiO NPs electrodes

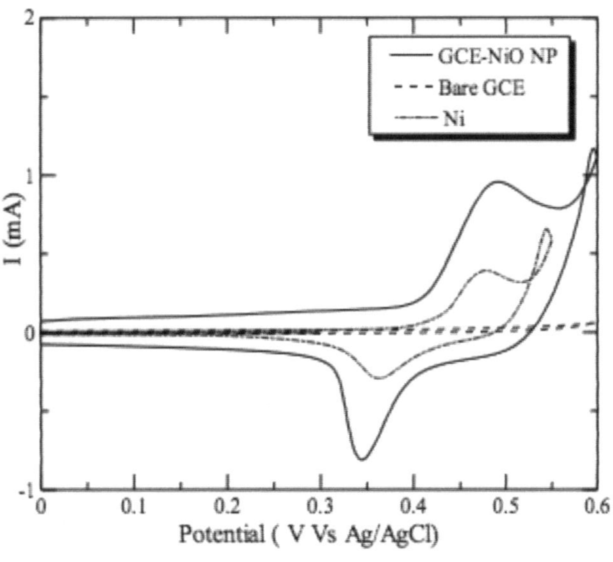

Figure 4. Cyclic voltammogram of Ni, bare GC, and GCE-NiO NPs electrodes

3.3. Preliminary study of urea sensor

Figure 5 shows that the change process of NiO to NiOOH is contributed to the electrocatalytic oxidation of urea in an alkaline medium. The usage of GC-NiO electrode showed the presence of a redox peak of NiO (Ni^{2+}) and NiOOH (Ni^{3+}). The electrocatalytic mechanism occurred when the cathodic peak of Ni^{3+} to Ni^{2+} decreased as long as the urea concentration increased as follows:

$$NiO + OH^- \rightarrow NiOOH + e^- \quad (1)$$

$$6NiOOH + 2CO(NH_2)_{2(aq)} + 2H_2O_{(l)} \rightarrow 6Ni(OH)_2 + 2N_{2(g)} + 3H_{2(aq)} + 2CO_{2(g)} \quad (2)$$

Table 1. The values of reproducibility and repetition of electrode for urea sensor

Electrode	Sensitivity ($mA/mM.cm^2$)	Linearity	LOD (mM)	% RSD Repeatability
Ni	0.0703	0.9884	1.13	4.74
GCE-NiO NPs	0.3001	0.9787	0.12	2.85

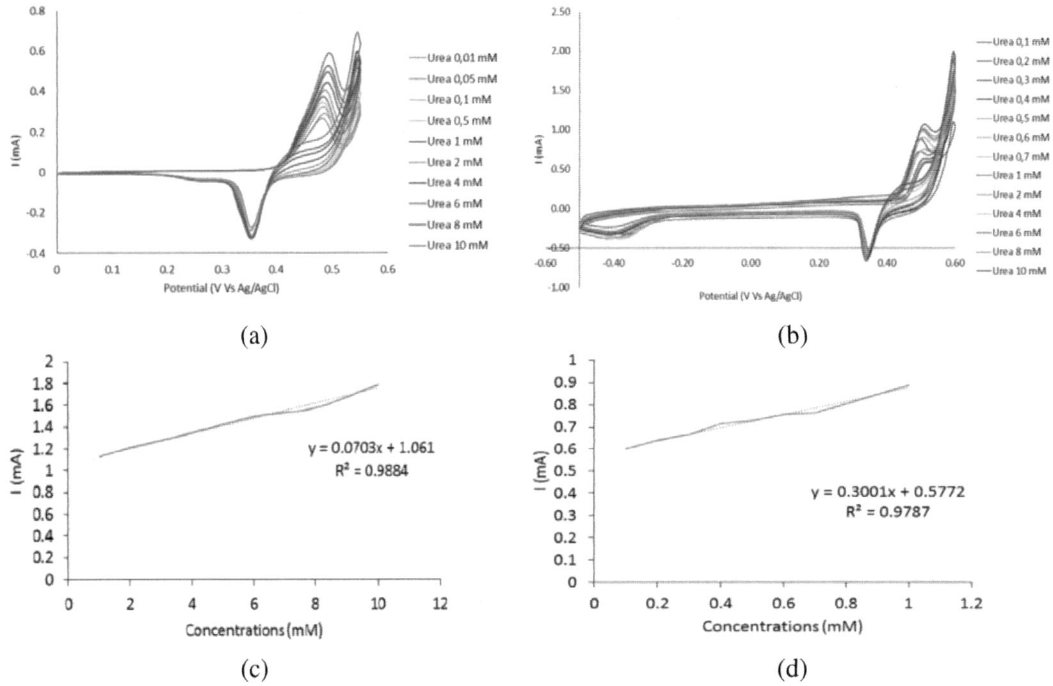

Figure 5. Cyclic voltammogram of (a) Ni and (b) GC-NiO NPs electrodes, (c) plot of anodic peak I vs concentration with Ni and (d) GC-NiO NPs electrodes

Figure 5a and figure 5b show the higher urea concentrations could accelerate the Ni(II)–Ni(III) redox reaction because the Ni(III) species could oxidize the urea. The electron transfer rate could be happened rapidly, which is signed by a higher current at the increase urea concentration. Figure 5c and figure 5d show a liniear correlation between concentration and current from urea oxidation. Ni and GC-NiO NPs electrodes were used to determine the limit of detection (LOD). The LOD of Ni electrode was 1.13 mM with the concentration range of 1–10 mM and R^2 =0.9884, while the LOD of GC-NiO NPs electrode was 0.12 mM with the concentration range of 0.1–1 mM and R^2 = 0.9787. The LOD value of GC-NiO NPs electrode was lower than the Ni electrode. The GC-NiO NPs electrode gave a better result in urea sensor due to its lower background current that could make it more sensitive at a lower concentration.

The reproducibility and repetition of electrode for urea sensor were investigated as shown in figure 6. Ni and GC-NiO NPs electrodes were examined by a relative standard deviation (RSD) using CV in 1 mM urea. The RSD value of Ni and GC-NiO NPs electrodes were 4.74 and 2.85 %, respectively. Table 1 shows the summary of the performance of the prepared sensors. This result concludes that GC-NiO NPs electrode has a better result as an urea sensor than Ni electrode.

IOP Conf. Series: Materials Science and Engineering **763** (2020) 012039 doi:10.1088/1757-899X/763/1/012039

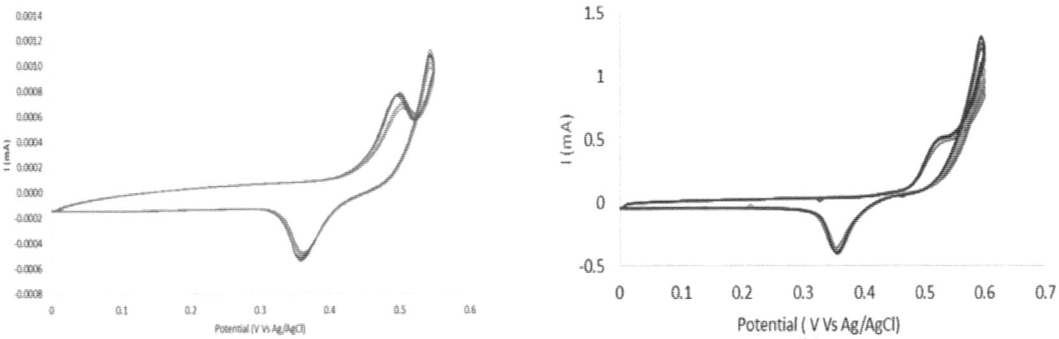

(a) (b)

Figure 6. Cyclic voltammogram using (a) Ni and (b) GC-NiO NPs Electrodes

4. Conclusions

NiO NPs were synthesized using OSLE, which contains secondary metabolites as weak base sources and capping agents. The modification of GC electrode has been successfully performed with NiO NPs. The LOD of both GC-NiO NPs and Ni electrodes were 0.12 and 1.13 mM, respectively. The precision of both Ni and GC-NiO NPs electrodes showed the RSD of 4.74 and 2.85 % (n = 10), respectively. The stability test of GC-NiO NPs electrode with CV measurement showed a fairly good stability with the RSD of 7.90 % for 3 days.

Acknowledgements

The authors would like to thank Universitas Indonesia for funding this research through PIT-9 Grant Universitas Indonesia and PT DKSH (Malvern) for supporting characterization of Particle Size Analyzer.

References

[1] Arain *et al.* 2016 *RSC Adv.* 6 39001–6
[2] Nguyen N S and Yoon H H 2016 *Sens. Actuators* B 236 304–10
[3] Nguyen N S, Das G and Yoon H H 2016 *Biosens. Bioelectron.* 77 372–7
[4] Liu C M, Chen C and Tseng Y C 2012 *J. Electrochem. Soc.* 159 K78–82
[5] Wang X, Yang X, Chen S, Li Q, Wang W, Hou C, Gao X, Wang L and Wang S 2016 *Front. Plant Sci.* 6 1243
[6] Luwang M N, Ningthoujam R S, Jagannath, Srivastava S K and Vatsa R K 2010 *J. Am. Chem. Soc.* 132 2759–68
[7] Mittal J, Batra A, Singh A and Sharma M M 2014 *Adv. Nat. Sci.: Nanosci. Nanotechnol.* 5 043002
[8] Al-Temimi S S M and Al-Mashhedy L A M 2015 *Int. J. Pharm. Bio. Sci.* 6 999–1007
[9] Sulaiman N and Yulizar Y 2018 *Mater. Sci. Forum* 917 167–71
[10] Susanti W, Gunlazuardi J and Ivandini T A 2018 *AIP Conf. Proc.* 2023 020085
[11] Kumary V A, Nancy T E M, Divya J and Sreevalsan K 2013 *Int. J. Electrochem. Sci.* 8 2220–8

IOP Conf. Series: Materials Science and Engineering **763** (2020) 012040 doi:10.1088/1757-899X/763/1/012040

Synthesis of impregnated nickel-mesoporous carbon and its application for reaction of acetylene with carbon dioxide

A D H Kawigraha, Y K Krisnandi and I Abdullah

Department of Chemistry, Faculty of Mathematics and Natural Sciences (FMIPA), Universitas Indonesia, Kampus UI Depok, Depok 16424, Indonesia

Corresponding author's e-mail: iman.abdullah@sci.ui.ac.id

Abstract. Mesoporous carbon has been synthesized using soft-template method with phloroglucinol as carbon precursor and pluronic F127 as a template to form mesosize pores in carbon structure. The Brunauer-Emmett-Teller (BET) results prove the existence of hysteresis loop and adsorption isotherm type IV with total surface area of 432.498 m^2/g and pore diameter of 7.46 nm. The mesoporous carbon was then impregnated with various amount of nickel metal using $Ni(NO_3)_2$ as precursor, followed by four ways of reduction: (1) with ethylene glycol in aqueous solution; (2) with ethylene glycol in 98 % ethanol in water; (3) with $NaBH_4$; and (4) with H_2 gas. Energy Dispersive X-Ray (EDX) analysis shows that nickel content in materials after reduction were (1) 34.48 %, (2) 0.02 %, (3) 0.9 % and (4) 1.73 %, in respective to the reduction methods. X-Ray Diffraction (XRD) patterns of mesoporous carbon reduced with $NaBH_4$ and H_2 show no change in mesoporous carbon structure, especially for peak at 2θ of 24.44° and 43.18°. After impregnation, the materials were further tested for their catalytic activity for acetylene reactions with CO_2. The reactions were conducted in a 100 mL batch reactor with a pressure ratio of acetylene to CO_2 of 0.2:0.8. Some reaction conditions were examined to determine the effect on the reaction: reaction time (3, 6, 12) h and reaction temperature (25, 40 and 80) °C. Product analysis using High Performance Liquid Chromatography (HPLC) shows that new peak appears at retention time of 3.625 minutes. The optimum condition was obtained at 25 °C and 3 hours.

Keywords: nickel catalyst, mesoporous carbon, acetylene, CO_2 utilization

1. Introduction

Carbon dioxide is one of the greenhouse gases that keep the earth's surface to remain warm. However, when its concentration increase excessively, the surface temperature of the earth may increase, causing climate change. Therefore, it is important to reduce CO_2 concentration in the atmosphere to minimize its negative effect to global community. There are two technologies for reducing CO_2 concentration, Carbon Capture and Storage (CCS) and Carbon Capture and Utilization (CCU) [1]. CCU in addition to CO_2 capture also converts CO_2. The suspended CO_2 is more active than the gas phase so conversion is more efficient [2]. CCU technology is able to convert CO_2 into a more valuable chemical [3–4]. Various catalysts have been developed to assist the process associated with the transformation of CO_2 [5]. One of them is the reaction of CO_2 with acetylene to synthesize acrylic acid. There are some reports on Ni-catalyzed carboxylation of ethylene gas with CO_2 to produce acrylic acid [6–7]. Most types of catalysts used were homogeneous catalysts, which, although efficient in chemical reaction, were difficult to be separated from the reaction system. Therefore, solid support is needed to heterogenize those homogeneous catalyst. One of the materials used as catalyst support is mesoporous carbon, which has a large surface, large pore volume and specific physical surface properties. In 2015, Yu *et al.* [8] reported the use of activated carbon as copper support for carboxylation of terminal alkynes [8]. Although the solid catalyst shows high activity but the product was alkyl 2-alkynoate, not alkyl prop-2-enoic acid, the product of Ni-catalyze hydrocarboxylation. Similar result was obtained when using Al_2O_3 as copper support for the same reaction [9]. However, no studies have examined the potential of nickel catalysts impregnated on mesoporous carbon as the catalyst for CO_2 reactions with acetylene.

2. Experimental

2.1. Synthesis of mesoporous carbon (MC)

Synthesis of mesoporous carbon was carried out using previously reported procedure [10–11]. In a typical synthesis, 1.25 g of phloroglucinol and 1.25 g of poly(ethylene oxide)–poly(propylene oxide)–poly(ethylene oxide) triblock copolymer (Pluronic F127) were dissolved into 9.7 g of 10:9 weight ratio of ethanol and water mixture. After the solid dissolved completely, 0.08 ml of 37 % HCl was added to the solution and stirred for 30 min until a light pink color appeared. The solution was then added by 1.25 mL of 37 % formaldehyde. After 30 min, the solution turned cloudy and separated into two layers after 1–2 h. The polymer-rich bottom layer was taken and continued stirring overnight to form elastic, non-sticky monolith. The obtained monolith was then transferred to an autoclave and cured in oven at 100 °C for 24 h. The material was carbonized in tubular furnace under nitrogen flow with heating rate of 1 °C/min from 100 to 400 °C and 5 °C/min from 400 to 850 °C and finally kept at 850 °C for 2 h.

2.2. Impregnation of nickel

The impregnation of nickel was conducted by two methods: dry impregnation (incipient-wetness) and wet impregnation. For wet impregnation, the method was adopted from our previous reported procedure [12]. In a typical procedure, 0.5 g of mesoporous carbon was introduced into a glass cup containing 10 mL of water. A total of 1.23 g $Ni(NO_3)_2{\cdot}6H_2O$ which has been dissolved in 5 mL water was added dropwise. The mixture was then stirred for 6 hours, and the product was filtered. For dry impregnation, 1 g of mesoporous carbon was introduced into a glass cup, then 1.70 mL of $Ni(NO_3)_2$ solution (prepared by dissolving 0.6262 g $Ni(NO_3)_2{\cdot}6H_2O$ in 10 mL water) was added dropwise and stirred for 24 hours.

2.2.1. Reduction with ethylene glycol (water solvent). The mixture from 2.2 (wet impregnation) was placed into a round-bottom flask containing 20 mL of ethylene glycol and heated by reflux at 170 °C for 6 hours. The resulting solid was then filtered and dried.

2.2.2. Reduction with ethylene glycol (water-ethanol solvent). The mixture from 2.2 (wet impregnation) was placed into a round flask containing 20 mL of ethylene glycol and heated by reflux at 170 °C for 6 hours. The resulting solid was then filtered, washed with water-ethanol and dried.

2.2.3. Reduction with $NaBH_4$. A total of 0.2605 g $NaBH_4$ was dissolved in 50 mL of water. Then the Ni-impregnated mesoporous carbon (dry impregnation) was stirred with 5 mL of $NaBH_4$ solution for 30 minutes. Then, the mixture was poured into a quartz tube and dried under H_2 flow with a temperature of 550 °C for 6 hours.

2.2.4. Reduction with H_2. The Ni-impregnated mesoporous carbon (dry impregnation) was poured into a quartz tube and dried under H_2 flow with a temperature of 550 °C for 6 hours.

2.3. Catalysis test

The carboxylation reactions were carried out by using a 100 mL round bottom flask as reactor equipped with thermometer and gas balloon. To the glass reactor, 0.05 g catalyst, 0.032 g promoter ($MgCl_2$), 0.011 g Zn powder, 0.15 mL H_2O, and 10 mL DMF were introduced. CO_2 and acetylene gas with mol ratio of 4:1 were then dialed with a flow rate of 100 mL/min for 1.5 minute. Variation of reaction time (3, 6 and 12) h and temperature (25, 40 and 80) °C were examined to see their effect on reaction.

3. Results and discussion

In figure 1, the material before carbonization (non MC; spectrum number 1) shows an OH stretching peak indicated by a widening peak at about 3500 cm^{-1} originating from phloroglucinol group. There is

Table 1. Comparison of EDX characterization results

Elements (% mass)	Ni@MC reduction methods			
	ethylene glycol (water)	ethylene glycol (water:ethanol)	$NaBH_4$	Ni@MC reduced with H_2
C	35.31	90.15	91.45	89.01
O	30.20	9.83	7.573	9.26
Ni	34.48	0.02	0.9	1.73

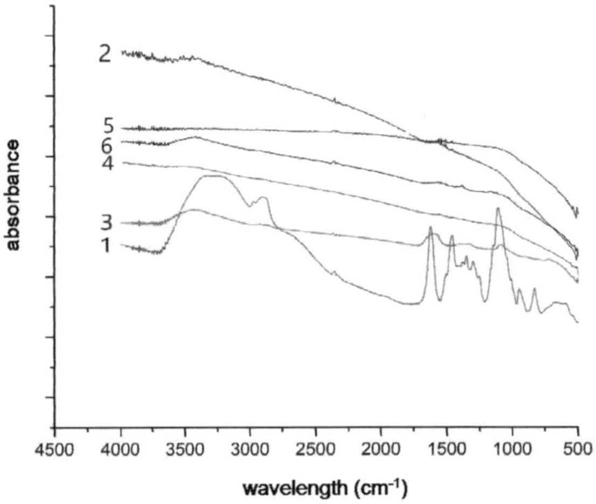

Figure 1. FTIR spectra of (1) Ni non MC, Ni@MC reduced with (2) ethylene glycol (water), (3) ethylene glycol (water:ethanol), (4) H_2, (5) $NaBH_4$, and (6) Ni@MC after carbonization

Figure 2. SEM Images with magnification of 300 times for Ni@MC reduced with (a) ethylene glycol (water solvent), (b) ethylene glycol (water:ethanol solvent), (c) $NaBH_4$ and (d) H_2 gas.

also a methylene group indicated by a peak at a wavelength of 2900 cm^{-1} derived from a bond between formaldehyde and phloroglucinol. The Fourier Transform Infrared (FTIR) spectrum after carbonization (spectrum number 6) shows the loss of the methylene vibration peak, while the peak of the OH stretching still appears but with low absorption intensity. This indicates that the carbonization process has proceeded well. FTIR spectra of the materials after Ni impregnation and reduction are shown in figure 1 spectrum number 2 to 5. Based on the figure, it can be seen that reduction with ethylene glycol still leaves the -OH group in the material as indicated by the appearance of a widening peak at a wavenumber of around 3500 cm^{-1}. Whereas, the reduction with $NaBH_4$ and H_2 reduce or even eliminate the -OH group from the material as indicated by peak loss from the -OH group vibration.

Figure 2 shows the morphological image of the four types of catalysts. Figure 2a shows MC@ Ni reduced with ethylene glycol (water), figure 2b shows MC@Ni reduced with ethylene glycol (ethanol: water), figure 2c shows MC@Ni reduced with $NaBH_4$, and figure 2d shows MC@Ni reduced with H_2. These four images show the surface of the impregnated nickel metal on mesoporous carbon.

Table 1 shows the elemental composition of the materials obtained by four different reduction methods from Energy Dispersive X-Ray (EDX) analysis. Reduction using ethylene glycol with water

IOP Conf. Series: Materials Science and Engineering **763** (2020) 012040 doi:10.1088/1757-899X/763/1/012040

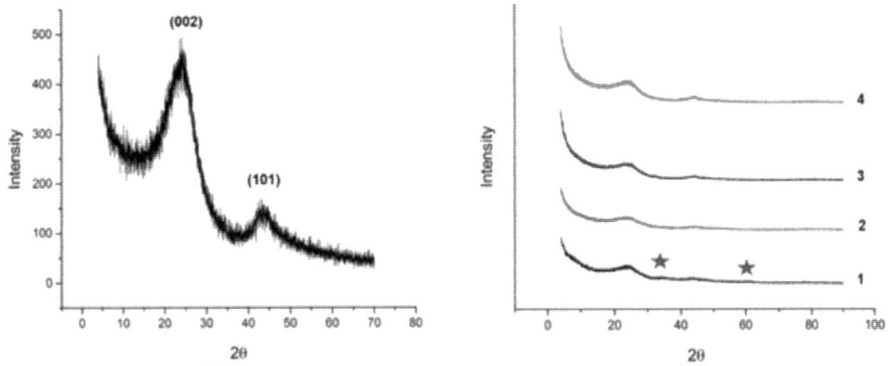

(a) (b)

Figure 3. XRD pattern of (a) mesoporous carbon and (b) Ni@MC that reduced with (1) ethyleneglycol (water), (2) ethyleneglycol (water:ethanol), (3) $NaBH_4$ and (4) H_2.

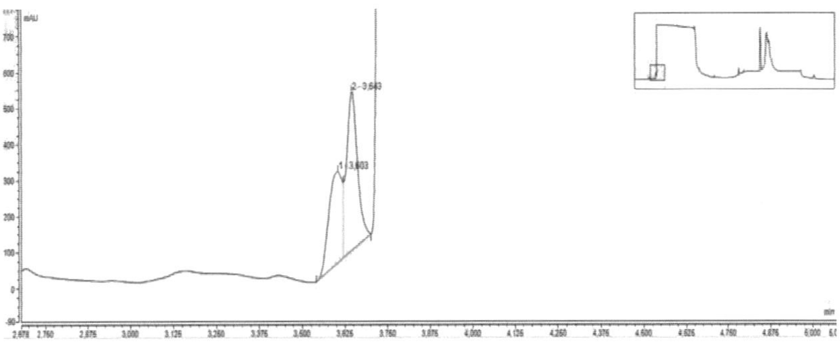

Figure 4. Typical HPLC chromatogram of the reaction mixture (obtained for the reaction conducted at 25 °C for 3 hours)

solvent gave the highest Ni content of 34.48 % while the reduction using ethylene glycol with water:ethanol solvent gave the smallest Ni content of 0.02 %. The big difference in Ni content of the two methods occurs because in the second method, the material that has been produced was then washed again with ethanol-water solvent, which causes the leaching of impregnated Ni species. Similar result is occured for reduction with $NaBH_4$ and with H_2 gas. The reduction result using $NaBH_4$ gave a smaller Ni content than the reduction using H_2 gas. This happens because in reduction with $NaBH_4$, Ni@MC was stirred in $NaBH_4$ solution for 30 minutes, which allows Ni leaching to occur during the process. Whereas in the reduction method with H_2, there was no addition of solvent, and the reduction process was done by flowing H_2 gas at high temperatures.

X-Ray Diffraction (XRD) characterization aims to determine whether the impregnation of Ni metal on mesoporous carbon changes the structure and crystallinity of the support. Figure 3a shows the typical peak of mesoporous carbon at 2θ of 24.44° and 43.18°, which related to (002) and (101) plane [11]. Based on XRD diffraction pattern analysis, there is no peak other than the typical peak of mesoporous carbon. The crystallinity of the mesoporous carbon is indicated by a widened peak, where the crystal is amorphous. This shows that the addition of nickel did not alter the structure of the mesoporous carbon. Based on the XRD diffraction pattern, there is no peak for Ni. This can happen because the % Ni contained in the material is too small to be detected by XRD. In the XRD pattern for the reduction results using ethylene glycol (water), new peaks emerge at 34 ° and 60 ° as indicated by star symbol. Those two emerging peaks indicate the presence of Ni(II)and Ni(0) species [12].

Next, Ni@MC with Ni content of 34.48 % was used as catalyst in the reaction between acetylene and CO_2 at ambient pressure. At first, it was expected that acrylic acid would be the product of the reaction. However, analysis of the reaction mixture with High Performance Liquid Chromatography (HPLC) showed no peak, which indicated the presence of acrylic acid (6.9 minutes of retention time). Instead, new peaks appear at a retention time of around 3.6 minutes as shown in figure 4. The new peak shows the possibility of new compounds formed. Previously these peaks did not appear on the blank solution (obtained from the reaction conducted in the absence of catalyst).

Table 2. Comparison of the peak area of new products in various reaction conditions

Reaction time (h)	Reaction temperature (°C)		
	25	40	80
3	40.6316	34.9610	27.9699
6	38.8019	21.4017	23.1116
12	42.2104	29.8603	27.7182

Complete data from the catalysis reaction between acetylene and CO2 are presented in table 2. The formation of the product seems to be preferred to occur at low temperatures (25 °C). While the time variation does not significantly influence the number of products formed. Generally in the catalytic reaction, the higher the temperature, the more effective the reaction will be. However, at low temperature the gases tend to be easily dissolved in the solvent (DMF), making it possible to come into contact with catalysts and promoter, thus causing more reaction to occur. When the reaction is carried out at high temperatures, there will be a competition between temperatures to lower the activation energy to the temperature that will complicate acetylene to interact with carbon dioxide.

4. Conclusions

Mesoporous carbon impregnated nickel has been sucessfully synthesized without damaging the carbon structure. Mesoporous carbon impregnated nickel has a catalytic ability for the reaction between acetylene with carbon dioxide. Further investigation needs to be conducted to determine the product.

Acknowledgements

The authors gratefully acknowledge Universitas Indonesia for funding this project (PITTA Grant number 2272/UN2.R3.1/HKP.05.00/2018)

References

[1] Songolzadeh M, Ravanchi M T and Soleimani M 2012 *International Scholarly and Scientific Research & Innovation* **6(10)** 900–7
[2] Saha D and Deng S 2010 *J. Colloid Interface Sci.* **345** 402–9
[3] Liu Y, Wang Z U and Zhou H C 2012 *Greenhouse Gas Sci. Technol.* **2** 239–59
[4] Li Y N, He L N, Liu A H, Lang X D, Yang Z Z, Yu B and Luan C R 2013 *Green Chem.* **15** 2825–9
[5] Liu A H, Yu B and He L N 2014 *Greenhouse Gas Sci. Technol.* **5** 17–33
[6] Lejkowski M L *et al.* 2012 *Chem Eur. J.* **18** 14017–25
[7] Vavasori A, Calgaro L, Pietrobon L and Ronchin L 2018 *Pure Appl. Chem.* **90** 315–26
[8] Yu B, Xie J N, Zhong C L, Li W and He L N 2015 *ACS Catal.* **5** 3940–4
[9] Bondarenko G N, Dvurechenskaya E G, Magommedov E S and Beletskaya I P 2017 *Catal. Lett.* **147** 2570–80
[10] Górka J, Zawislak A, Choma J and Jaroniec M 2008 *Carbon* **46** 1159–74
[11] Wang Y, Bai X, Wang F, Qin H, Yin C, Kang S, Li X, Zuo Y and Cui L 2016 *Sci. Rep.* **6** 26673
[12] Yung T Y, Huang L Y, Chan T Y, Wang K S, Liu T Y, Chen P T, Chao C Y and Liu L K 2014 *Nanoscale Res. Lett.* **9** 444

IOP Conf. Series: Materials Science and Engineering **763** (2020) 012041 doi:10.1088/1757-899X/763/1/012041

Study of orthophosphate uptake by diffusive gradient in thin film (DGT) system with Co-loaded chitosan bentonite biocomposite binding gel and the effect of its anion interferences on the phosphate response

P A Sitadevi, R T Yunarti and A Saefumillah

Department of Chemistry, Faculty of Mathematics and Natural Sciences (FMIPA), Universitas Indonesia, Kampus UI Depok, Depok 16424, Indonesia

Corresponding author's e-mail: asep.saefumillah@sci.ui.ac.id

Abstract. The binding gel in DGT has been modified with Co-loaded chitosan-bentonite (Co-CSBent) biocomposite in order to enhance the phosphate binding capacity. The comparison was also performed by Chitosan-bentonite (CSBent). The binding and diffusive gel for this method were made from acrylamide, ammonium persulfate, and N, N-methylenebisacrylamide. The synthesized binding gel and biocomposite were characterized using Fourier Transform Infrared (FTIR), X-Ray Diffraction (XRD) and Scanning Electron Microscopy (SEM). Characterization results showed that biocomposites had been synthesized successfully. For control, DGT Co-CSBent and CSBent were tested with 2 ppm phosphate and without anions, C_{DGT} of orthophosphate value of 1.9127 μg/mL and 1.6643 μg/mL were obtained respectively. Both binding gels in DGT were examined with various anions including Cl^-, SO_4^{2-}, HCO_3^-, and NO_3^- in the concentration ranging from 0.5 mg/L to 2.5 mg/L. At higher SO_4^{2-} inhibitor concentration, C_{DGT} value of 1.0153 μg/mL (CSBent) and 1.2736 μg/mL (Co-CSBent) were obtained. Whereas, at higher Cl^- inhibitor, C_{DGT} value of 1.2934 μg/mL (CSBent) and 1.9584 μg/mL (Co-CSBent) were obtained. At higher HCO_3^- inhibitor concentration, C_{DGT} value of 0.7371 μg/mL (CSBent) and 0.8628 μg/mL (CoCSBent) were obtained. And, at higher NO_3^- inhibitor concentration, C_{DGT} value of 0.459 μg/mL (CSBent) and 0.5889 μg/mL (Co-CSBent) were obtained. Based on the obtained data, Cl^- and SO_4^{2-} do not affect the C_{DGT} of orthophosphate value. However, NO_3^- and HCO_3^- anions gradually reduced the C_{DGT} of orthophosphate value compared with C_{DGT} of the control value. Phosphate binding by biocomposite is controlled by ion exchange, electrostatic force, and Lewis metal ion complexation.

Keywords: orthophosphate, diffusive gradient in thin film, biocomposite, anion interference, chitosan bentonite

1. Introduction

Phosphorus is a macronutrient for the growth of aquatic plants and their metabolism [1]. In the aquatic system, phosphate is divided into colloid, particular, and dissolved. Dissolved phosphate is divided into organic phosphate (dissolved organic phosphate, DOP) and inorganic phosphate (dissolved inorganic phosphate, DIP), which consists of orthophosphate and polyphosphate [2]. Rumhayati [3] subdivided the phosphate ability of filtration process and concentration based on its reactivity to molybdate reagents. The reactive filtrated phosphate of the molybdate reagent is called the filterable reactive phosphate (FRP) containing orthophosphates and polyphosphates and organic phosphates, which are readily hydrolyzed by acids. Meanwhile, the concentration of the filtered organic phosphate (FOP) is determined by the oxidation step before it is reacted with molybdate reagent [3]. Phosphorus in waters comes from a variety of sources, including geological deposits, industrial waste, agricultural activities, household waste, agriculture and mining [4]. The tolerance limit of the phosphate content in water is 0.5 mg/L [5], and many aquatic systems have exceeded the threshold, thereby causing eutrophication of reservoirs, rivers, and lakes.

The diffusive gradient in thin film (DGT) technique has been performed on the measurement of the dissolved phosphate in waters [6]. This DGT apparatus can be used for phosphate analysis in water, soil, or sediment. The principle of DGT apparatus is based on a simple device that accumulates phosphate on a binding agent (binding gel) after passing through a hydrogel that acts as a diffusion layer. It depends on the steady-state concentration gradient of the bonding agent [6]. Accumulation occurs when contacting with phosphate. The accumulated analytes are eluted and determined by using spectrophotometry using the blue phosphomolybdenum method. The capacity of DGT can be determined by using equations derived from Fick's First Law [7–8]

$$C = \frac{M \times \Delta g}{D \times t \times A}$$

The capacity of DGT depends on the adsorbent used on the gel binding. In this study, biocomposite adsorbent was used to be included in gel binding on DGT. The biocomposite was prepared by three main ingredients, including bentonite (in this study is Na-Bentonite), chitosan, with (Co-CSBent) or without Co^{2+}(CSBent). The three compounds have their respective ability to absorb phosphate. But the phosphate is absorbed very little and absorption can only be done under certain conditions. To increase the phosphate sorption capacity and eliminate the deficiencies of each compound, the three compounds are synthesized into biocomposites. It is expected that the compounds formed is stable and created adsorbents that are able to bind phosphates effectively.

2. Experimental

2.1. Materials

Chitosan, Bentonite (PT Madani Alam Lestari), cellulose nitrate membrane (GE Healthcare Life Science), Glacial Acetic Acid (Merck), N,N'-methylenebisacrylamide, Acrylamide (Sigma), Ammonium Persulphate (Merck), N,N,N,N'-tetramethylethylenediamine (TEMED) 99 % (Sigma), Pottasium Antimonyl Tartrate (Sigma), Ammonium Molybdate (Merck), H_2SO_4 p.a (Merck), and Ascorbic Acid.

2.2. Synthesis of Co-CSBent and CSBent biocomposites

2 g of chitosan (CS) was dissolved in 2 % acetic acid solution at 40 °C and stirred. At the same time, 6 g of Na-Bentonite (Bent) was dispersed in 50 mL of distilled water for 2 hours. The bentonite dispersion was added slowly to chitosan solution prepared over 24 hours with strong magnetic stirring at 60 °C. After colloid formation of CSBent biocomposite, metal ions were filled by adding $CoCl_2.6H_2O$ 0.3 % (w/v) solution. Stirred for 1 hour with a strong stirring and obtained biocomposite Co-CSBent in the form of a slurry. The Co-CsBent was stored for 24 hours, then filtered using filter paper and washed with distilled water. Co-CSBent biocomposites were dried in a hot air oven at 45-50 °C until dry to obtain the fine powder. Then, the biocomposite was sieved to obtain a consistent size.

2.3. Preparation of diffusive gel and binding gel (CSBent and Co-CSBent)

The 5 mL gel solution was prepared based on Zhang *et al.* [6] procedure. This gel solution contains 15 % acrylamide solution and 0.3 % cross-linker. Then, added with 35 µL of ammonium persulfate solution, and added 10 µL of TEMED solution into the container. Stirred until homogeneous (approximately 15–20 seconds), immediately the solution was put into a glass mold. The solution which has been inserted into the mold was then heated in an oven at 45 °C for 45 minutes to an hour until the gel formed. The formed gel was washed with distilled water then soaked for 24 hours for hydration. For binding gel preparation, the procedure is almost the same as above, but the gel solution added to biocomposite as much as 2 g.

2.4. DGT component installation

For installation of DGT apparatus, gel sheets (diffusive and binding) were cut in 25 mm diameter, and the filter membrane was soaked in distilled water. DGT devices were washed and rinsed with distilled water. The binding gel pieces were placed first on DGT molding followed by a diffusive gel and membrane filter. The DGT apparatus was perfectly closed.

IOP Conf. Series: Materials Science and Engineering **763** (2020) 012041 doi:10.1088/1757-899X/763/1/012041

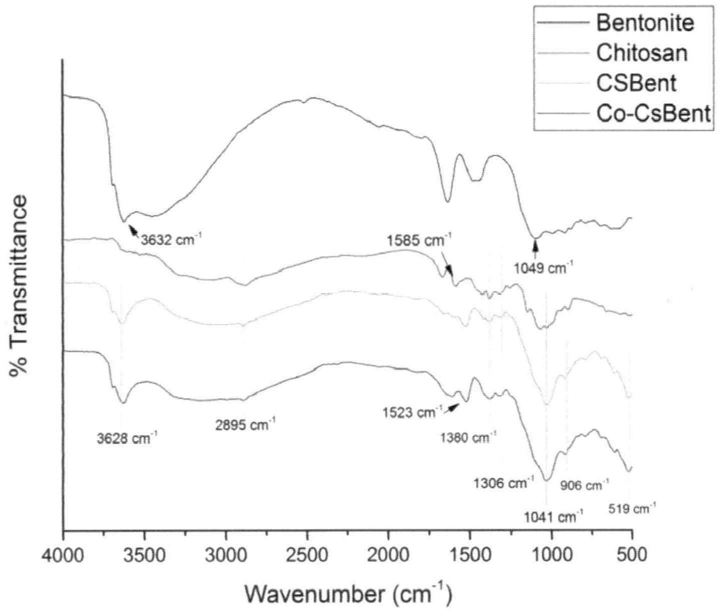

Figure 1. FTIR characterization of bentonite, chitosan, CSBent, and Co-CSBent.

2.5. DGT uptake test
DGT test in this study used 2 ppm phosphate with various circumstances. That included 4 variables, homogeneous binding gel, contact time, pH, and anion with various concentrations.

2.6. Binding gel elution
DGT apparatus that have absorbed the phosphate were dismantled and binding gel that has absorbed the phosphate was taken. The binding gel was immersed in a 0.25M 6 mL H_2SO_4 solution for 16 hours. The elution solution was phosphate, analyzed by blue phosphomolybdenum method.

3. Results and discussion

3.1. Characterization of biocomposite
After biocomposite was made, biocomposites were characterized by Fourier Transform Infrared (FTIR). FTIR results of bentonite, chitosan, CSBent, and Co-CSBent are presented in figure 1. FTIR results from bentonite exhibit a peak at 1049 cm^{-1} which is a stretch of Si-O-Si. The peak at 3632 cm^{-1} is the stretched vibration of OH bound to Al (or Mg) in bentonite [9]. The peak chitosan FTIR results at 1041 cm^{-1} show the presence of C-O-C bonds indicating the presence of ether in the chitosan compound. At peak 1306 cm^{-1} is secondary carbon, that is CH_2-OH, while 1380 cm^{-1} is tertiary carbon, which is CH-OH. Chitosan contains the basic characteristics of the peak at 1585 cm^{-1} and 2895 cm^{-1} because of the amide bond and the presence of the alkyl chain [10]. The small peak at 519 cm^{-1} is the bent and spin vibration of Si-O [11]. The peak at 906 cm^{-1} is a bentonite material that enters the biocomposite, which is Al-O(OH)-Al and Al-Al-OH in the prolonged state [12]. The amide group shifts its peak to 1523 cm^{-1} because the amide is protonated to NH_3^+. In the band 3628 cm^{-1}, there are OH and NH_2 groups derived from chitosan [13]. Based on the FTIR data, it is certain that chitosan and bentonite are integrated with the synthesized biocomposites.

Scanning Electron Microscopy-Energy Dispersive X-Ray Spectroscopy (SEM-EDS) characterization of synthesized biocomposites is given in figure 2. This characterization is to determine the surface of biocomposite. Co-CSBent biocomposites have irregular and large pores characterized by a yellow circle. The pores function is for phosphate uptake in the sample. On the surface morphology, macromolecular compounds is visible, it is caused by impurities during synthesis or when drying in the oven. The pores of the Co-CSBent biocomposite are covered by phosphates

IOP Conf. Series: Materials Science and Engineering **763** (2020) 012041 doi:10.1088/1757-899X/763/1/012041

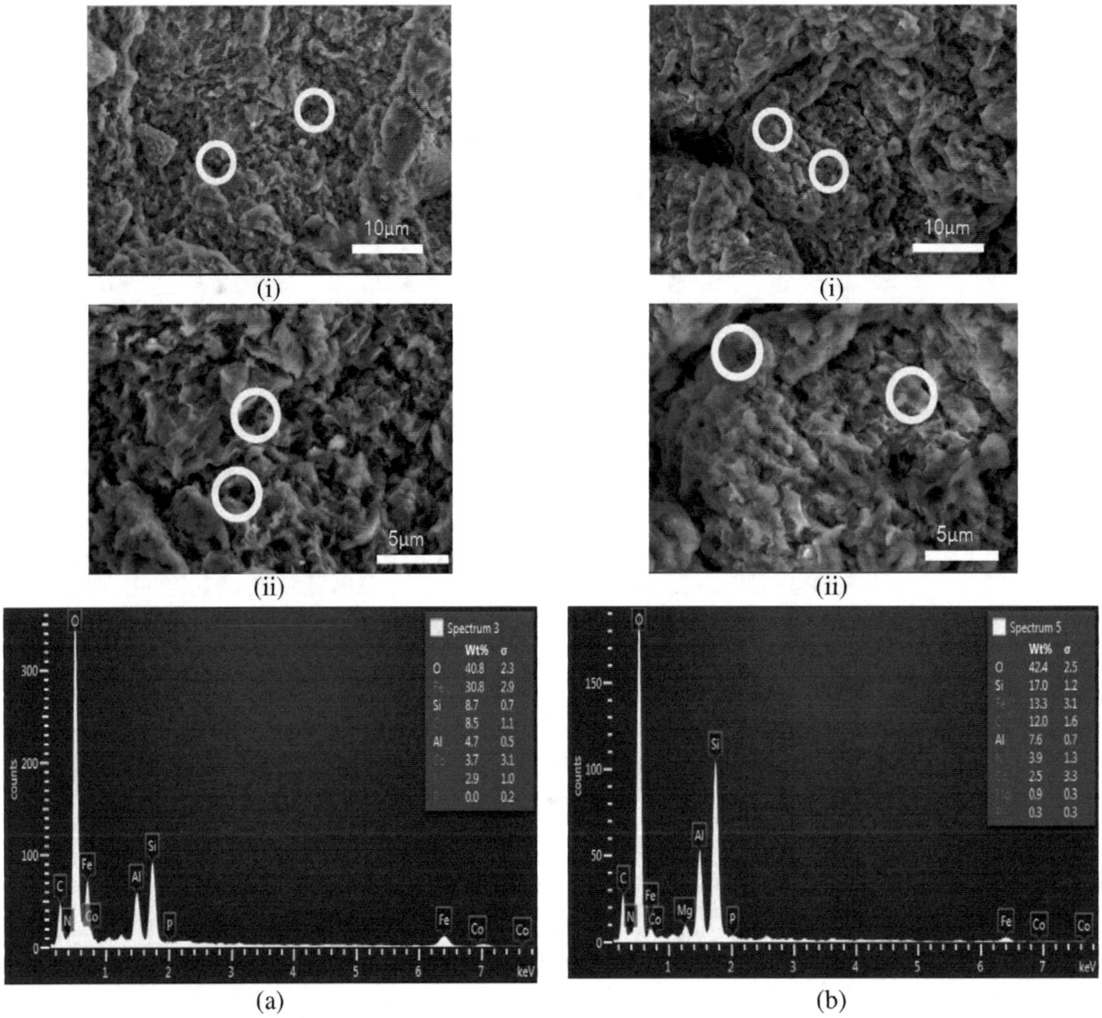

Figure 2. SEM-EDS characterization of biocomposites (a) before phosphate uptake and (b) after phosphate uptake at magnification: (i) 1000 x, (ii) 2500 x and (iii) EDS result.

characterized by a yellow circle. This happens because the biocomposite uptake the phosphate in the sample. EDS characterization show the presence of Co on biocomposite and phosphate element is detected after the uptake reaction. Based on EDS results, Co-CSBent biocomposite had successfully uptake the phosphate.

3.2. Characterization of gels

After gels had been synthesized, gels were characterized by FTIR. FTIR results from acrylamide, diffusive gel, and binding gels are presented in figure 3. On the raw material of acrylamide spectrum, there is a visible peak at 800 cm^{-1}, which is the bending of the NH_2 group out of plane (oop). On the other hand, the 1150 cm^{-1} peak is a deformation vibration of the C-N bond [13]. At 1340-1420 cm^{-1} is a C-C bond presents in the acrylamide compound. Stretching and deformation vibrations corresponded to the C=O appear at 1673 cm^{-1} [14]. The largest peak at 3180-3350 cm^{-1} is the stretching of the NH_2 group having the characteristic of 2 identical peaks. In diffusive peak gel appears almost the same as acrylamide, only different in the transmittance. Significant differences occur at the peak of 1612 cm^{-1}, in diffusive gel peak that wavenumber disappears. It proves that the C=C bond disappears and it can be concluded that the synthesis of acrylamide to polyacrylamide was successful. Diffusive gel, when

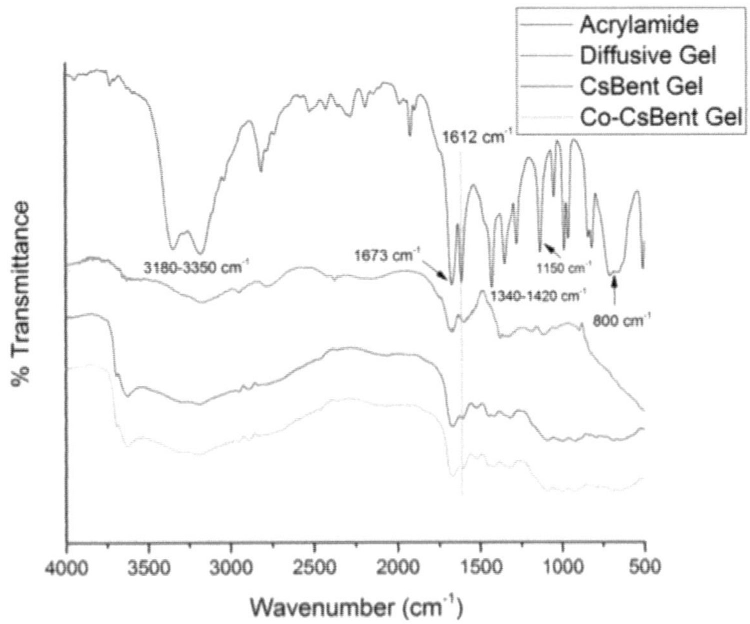

Figure 3. FTIR characterization of acrylamide, diffusive gel, and binding gel (CSBent and Co-CSBent) materials

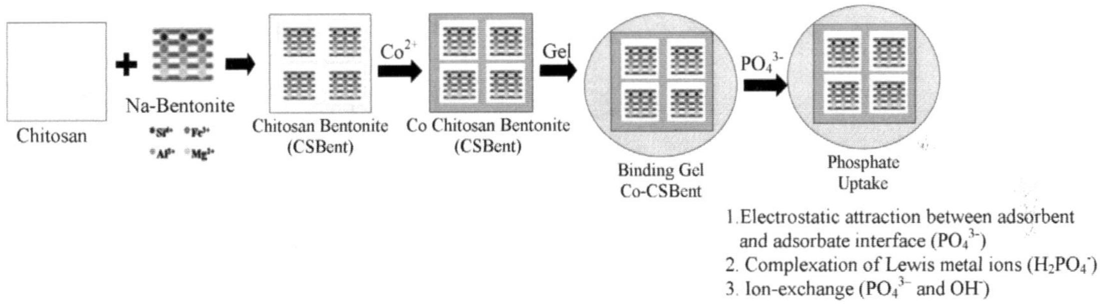

Figure 4. The possible mechanism of phosphate sorption onto biocomposites (adapted from Kumar and Visnawathan [15])

compared with gel binding, did not change significantly. This is because the biocomposites contained in the binding gel is a minor compound, so it is not detected in FTIR.

3.3. Phosphate uptake mechanism

The significance of phosphate sorption onto biocomposites was achieved by electrostatic attraction between adsorbent and adsorbate interface, complexation of Lewis metal ions associated with $H_2PO_4^-$ and ion-exchange [15], which are shown in figure 4.

3.4. DGT uptake test on various experimental variation

This test was performed to determine the ability of DGT-Co-CSBent and DGT-CSBent while ensuring that the binding gel was made to phosphate uptake in various variations. DGT uptake test on various experimental variations is presented in figure 5.

Based on the data on the graph, DGT Co-CSBent and CSBent were tested with 2 ppm phosphate and without anions. The C_{DGT} of orthophosphate value obtained for three repetitions is almost the same with average 1.9127 μg/mL (DGT-Co-CSBent) and 1.6643 μg/mL (DGT-CSBent). This indicates that

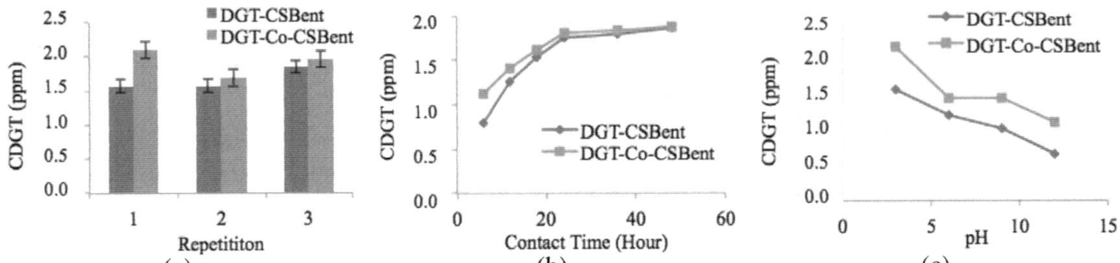

(a) (b) (c)

Figure 5. The result of DGT uptake test on various experimental variations: (a) homogenous binding gel test (control), (b) deployment time and (c) pH of solution

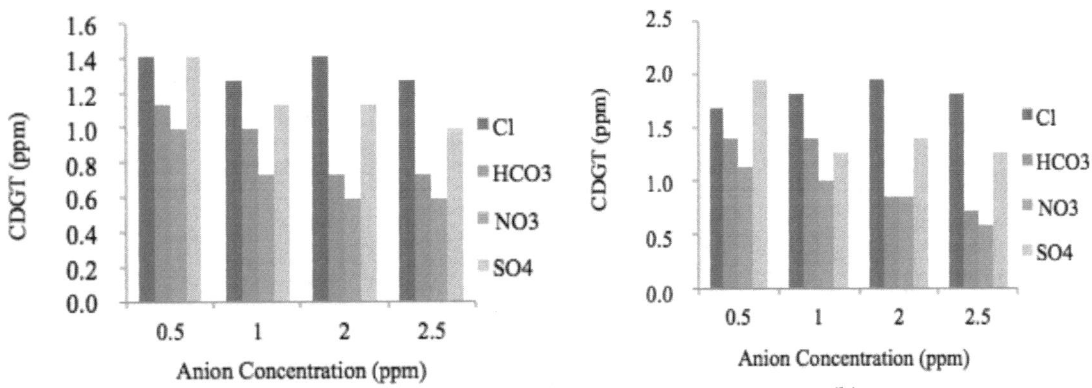

(a) (b)

Figure 6. The result of anion interferences effect on the DGT phosphate uptake of (a) DGT-CSBent and (b) DGT-Co-CSBent

the gel binding made has been quite homogeneous. DGT-Co-CSBent is superior in absorbing phosphate due to the reaction of Lewis metal ion complexation associated with $H_2PO_4^-$ [15]. The optimum deployment time obtained in this experiment was at 24 hours, after that time C_{DGT} was almost stagnant. That is because of the lack of empty sites available on the binding gels that have reached the equilibrium stage. A decrease in C_{DGT} at the elevated pH occurs because an increase in pH value will increase the electrostatic repulsion between phosphate and OH^- ions. Later, phosphates and OH^- ions will compete to be adsorbed by the empty space in the binding gel. While low pH causes electrostatic pull between the opponent of oxyanion of phosphate and the surface of protonated sorbent [16] so as to facilitate phosphate to be bound by the sorbent.

3.5. *DGT uptake test on anion effect*

This test was performed to determine the effect of DGT-CSBent and DGT-Co-CSBent phosphate uptake on anion interferences, which is widely present in water. DGT uptake test on anion effect is presented in figure 6. Based on the obtained data, Cl^- and SO_4^{2-} do not affect the C_{DGT} of orthophosphate value. However, a significant decrease of the C_{DGT} of orthophosphate value from the interferences of HCO_3^- and NO_3^--ions is because the two anions are the hardest bases compared with the other anions. The hardness se quence of an anion according to the theory of HSAB is NO_3^-> HCO_3^-> SO_4^{2-}>PO_4^{3-}>Cl^-. This DGT has a good phosphate uptake in the anion interference of SO_4^{2-} and Cl^-.

4. Conclusions

DGT binding gel Co-CSBent and CSBent biocomposites are able to bind phosphates. The capacity of Co-CSBent binding gel is greater than CSBent binding gel for the phosphate uptake. It is seen from C_{DGT} from DGT-Co-CSBent is higher than DGT-CSBent. The effects of Cl^- and SO_4^{2-} ions are not observed in C_{DGT} of orthophosphate value, whereas NO_3^- and HCO_3^- ions obviously affects the C_{DGT} of orthophosphate value. The results obtained in accordance with HSAB theory which became the basis of this research.

References

[1] Lu J, Liu H, Zhao X, Jefferson W, Cheng F and Qu J 2014 *Colloids Surf.* A *Physicochem. Eng. Asp.* **455** 11–8
[2] McKelvie I D 2000 Phosphates *Handbook of Water Analysis* ed L M L Nollet (New York: Marcel Dekker Inc) Chapter 15 pp 273–95
[3] Rumhayati B 2010 *Jurnal Ilmu Dasar* **11** 160–6
[4] Ruttenberg K C 2003 The global phosphorus cycle *Treatise on Geochemistry* vol 8, ed W H Schlesinger (Amsterdam: Elsevier)pp 585–643
[5] Xu Y, Dai Y, Zhou J, Xu Z P, Qian G and Lu G Q M 2010 *J. Mater. Chem.* **20** 4684–91
[6] Zhang H, Davidson W, Gadi R and Kobayashi T 1998 *Anal. Chim. Acta* **370** 29–38
[7] Zhang H and Davidson W 1995 *Anal. Chem.* **67** 3391–400
[8] Davidson W and Zhang H 1994 *Nature* **367** 546–8
[9] Yoon D H, Jang J W, Seo S S, Jeong S H and Cheong I W 2013 *Adv. Polym. Technol.* **32** 21335
[10] Lee M Y, Hong K J, Kajiuchi T and Yang J W 2005 *Int. J. Biol. Macromol.* **36** 152–8
[11] Darvishi Z and Morsali A 2011 *Ultrason. Sonochem.* **18** 238–42
[12] Zhao D L, Feng S J, Chen C L, Chen S H, Xu D and Wang X K 2008 *Appl. Clay Sci.* **41** 17–23
[13] Jin L and Bai R 2002 *Langmuir* **18** 9765–70
[14] Dumitrescu A M, Lisa G, Iordan A R, Tudorache F, Petrila I, Borhan A I, Palamaru M N, Mihailescu C, Leontie L and Munteanu C 2015 *Mater. Chem. Phys.* **156** 170–9
[15] Kumar I A and Viswanathan N 2017 *Int. J. Biol. Macromol.* **104** 1539–47
[16] Wan Ngah W S, Teong L C and Hanafiah M A K M 2011 *Carbohydr. Polym.* **83** 1446–56

Dissolution profile of partially purified bromelain from pineapple core encapsulated in alginate-guar gum with in vitro study of antiplatelet activity

A A Irfan, S Setiasih and S Hudiyono

Department of Chemistry, Faculty of Mathematics and Natural Sciences (FMIPA), Universitas Indonesia, Kampus UI Depok, Depok 16424, Indonesia

Corresponding author's e-mail: siswati.setiasih@sci.ui.ac.id

Abstract. The aim of this study is to maintain the antiplatelet activity of pineapple core bromelain from degradation in gastric fluid by encapsulating the enzyme in alginate-guar gum crosslinked with glutaraldehyde. Bromelain isolation was followed by purification process such as ammonium sulfate precipitation and dialysis. The specific activity obtained showed an enhancement started for crude enzyme (20.11 U/mg), ammonium sulfate fraction (190.64 U/mg), and dialysis fraction (229.77 U/mg). Dialysis fraction was encapsulated by in situ loading method in alginate-guar gum hydrogel with 1.25% (v/v) glutaraldehyde concentration. The swelling ratio of the hydrogel is 234.37 % in acidity of 1.2; and 1375.91 % in acidity of 7.4; while the encapsulation efficiency amount was 60.42%. I n addition, maximum concentration of bromelain released in dissolution test was higher in artificial intestinal environment (1.79 mg/L) than in artificial gastric fluid (0.13 mg/L), with maximum proteolytic activity of 1.2 U/mL and 0.11 U/mL, respectively. In vitro study of antiplatelet activity showed a good inhibition for dialysis fraction (50.6 %) and dissolution fraction (34.86 %), indicating that the hydrogel is good enough to maintain the activity of bromelain.

Keywords: pineapple core, bromelain, alginate-guar gum hydrogel crosslinked with glutaraldehyde, encapsulation, dissolution, antiplatelet activity

1. Introduction

The frequency of death due to blockage of blood vessels in the brain and heart is one of the issues that became the center of attention in the field of health. World Health Organization (WHO) data in 2012 showed that 17.5 million people worldwide died from cardiovascular disease, equivalent to 31% of 56.5 million of all deaths worldwide. Blockage of blood vessels can be caused by irregular eating patterns, smoking, and complications of certain diseases caused by increasing of platelet aggregation [1]. Aspirin and clopidogrel are often used as agents of antiplatelet aggregation, but their use causes side effects of nausea, headache, stomach cramps, and stomach ulcers [2]. Therefore, it takes another antiplatelet alternatives with fewer side effects, one of which is bromelain contained in pineapple plants.

Bromelain has been studied since 1894 [3] and has been proven to be useful in the therapeutic field as an antiplatelet, fibrinolytic, anticoagulant, and anti-inflammatory agent [4]. However, the application of bromelain as an oral drug tends to be less than optimal. According to the study conducted by Setiasih *et al.* [5], the proteolytic activity of bromelain enzymes decreases with time as they are placed in artificial gastric juices that have a pH of 1.2. The decrease in activity is caused by the interaction of bromelain with fluid and enzymes in the stomach, causing deactivation and degradation of the bromelain [6]. In fact, the intestine can absorb the enzyme bromelain intact without making the enzyme losing the activity or even degraded [7–8].

Encapsulation is the process by which a particular active ingredient is caught in a polymer matrix [9]. This method can be used to protect bromelain from the influence of the stomach environment so that the release of bromelain in the intestine is optimal. In this research, the isolated bromelain from pineapple core is encapsulated into the glutaraldehyde crosslinked alginate-guar gum hydrogel [10–12].

2. Materials and methods

2.1. Materials
The source of bromelain enzyme was pineapple core obtained from a seller of Palembang pineapple in Pasar Induk, Kramat Jati. Synthesis of hydrogel used sodium alginate powder (medium viscosity), guar gum powder, and glutaraldehyde solution 25 % (v/v) (Sigma Aldrich).

2.2. Isolation, ammonium sulfate precipitation, and dialysis of bromelain
The pineapple core were weighed, cut and crushed with a blender in cold condition (± 4 °C) with addition of a phosphate buffer 0.2M pH 7. The filtered solution was centrifuged for 20 minutes (6000 rpm, ± 4° C). The supernatant produced at this stage is called a crude enzyme.

The crude enzyme was then precipitated in multilevel process using ammonium sulfate with a concentration range of 0–20 %, 20–50 %, and 50–80 % of the final crude enzyme volume for one night. The solution was then centrifuged and the resulting precipitate was suspended into phosphate buffer 0.2M with pH 7.

The enzyme fraction was further purified by dialysis using a cellophane bag which had been soaked in demineralized water for 1 hour and then heated for 2 hours in an alkaline EDTA solution (0.001 EDTA and 1 % Na_2CO_3). The sac was immersed in phosphate buffer solution 0.05M pH 7, accompanied by continuous stirring using a magnetic stirrer at 4 °C. The phosphate buffer used was changed regularly every 2 hours and tested using $BaCl_2$ 1 %. If there is no precipitate of $BaSO_4$, then the dialysis is over. The enzyme fraction obtained from each stage was determined for its protein content by Lowry method and its proteolytic activity using Kunitz method

2.3. Synthesis of unloaded alginate-guar gum hydrogel
Alginate-guar gum hydrogel was synthesized with 3 variations of glutaraldehyde concentration, i.e. 1.25 %; 2.5 %; and 5 % (v/v). The synthesis began by dissolving 1.5 g of sodium alginate and 0.5 g of guar gum in 80 mL of water. The mixture was stirred and then added with glutaraldehyde to the desired concentration variation. After homogeneous, the solution was diluted with distilled water up to 100 mL and kept undisturbed for ±12 hours. Then, the hydrogel solution was suspended into a 0.5M $CaCl_2$ solution with syringe (0.1 mm) and let undisturbed for 1 hour. The mixture was filtered and washed with aquadest 3 times. Hydrogels were then dried in an oven at 50 °C until completely dry.

2.4. Synthesis of bromelain loaded alginate-guar gum hydrogel
The bromelain loaded alginate-guar gum hydrogel was also synthesized with the same variation of glutaraldehyde concentrations and a method similar to the unloaded hydrogels. The difference was in the addition of 1 mL of bromelain standard 100 ppm for optimum glutaraldehyde concentration determination process and 1 mL of dialysis fraction of bromelain for dissolution test, after the solution was homogenous with glutaraldehyde. Drying process was done using freeze dry for 4 days. The encapsulation efficiency was determined by dissolving the weighed hydrogel in 10 mL of phosphate buffer 0.2M pH 7.4 and then let undisturbed for 1 day. Subsequently, the solution was sonicated for 1 hour and centrifuged. The obtained filtrate was determined for its protein content. The encapsulation efficiency is determined by the equation [11]:

$$\text{Encapsulation efficiency (\%)} = \frac{[Loaded\ bromelain]\ x\ \frac{Total\ mass\ of\ all\ synthesized\ hydrogel}{Total\ mass\ of\ test\ hydrogel}}{[Added\ bromelain]}\ x\ 100\%$$

2.5. Dissolution test of isolated bromelain loaded alginate-guar gum hydrogel
Dissolution test of bromelain loaded hydrogel was done by dissolution apparatus type 1. The bromelain loaded hydrogel was weighed and then fed into the dissolution apparatus (100 rpm, 37 ± 0.5 °C). For 2 hours, the hydrogel was immersed in KCl-HCl buffer solution pH 1.2 as an artificial gastric fluid medium. Every 30 minutes, 10 mL of filtrate was taken. The hydrogel was then transferred into phosphate buffer 0.2M pH 7.4 as an artificial intestinal environment medium for 8 hours. In the first 2

Table 1. Bromelain activity from each purification process

Samples		Volume (mL)	Total proteolytic activity (U)	Total protein (mg)	Specific activity (U/mg)	Purity (times)	Yield (%)
Crude enzyme		100	351,6	17,48	20,11	1	-
Ammonium sulfat precipitation	F1 (0-20)%	4	29,66	0,33	88,77	4,41	8,43
	F2 (20-50)%	5	50.5	0.26	190,64	9,47	14,36
	F3 (50-80)%	2	2,33	0,02	78,9	3,92	0,66
	Remaining filtrate	78	3,9	1,12	3,48	0,17	1,1
Dialysis	FDialysis	5	256,3	0,22	229,77	11,42	14,78

hours, the filtrate was taken as much as 10 mL every 30 minutes. Then, in the next 6 hours, the filtrate was taken every 1 hour. Each amount of filtrate taken from both mediums was accompanied by the addition of a new medium in the same amount. The dissolution solution was then tested for its protein content by the Bradford method and its proteolytic activity by the Kunitz method.

2.6. In vitro study of antiplatelet activity
Determination of antiplatelet activity was performed on dialysis fraction and dissolution fraction with the highest activity. Determination method refers to Born's method with 1 mg/mL aspirin as positive control and aquadest as negative control. The aggregation percentage was determined by the formula [13].

2.7. Hydrogel characterization
The unloaded and bromelain loaded alginate-guar gum hydrogel was characterized using Fourier Transform Infrared (FTIR) to see any changes in functional groups in the hydrogel and to determine the crosslinking and bromelain interaction with the matrix.

3. Results and discussion

3.1. Isolation, ammonium sulfate precipitation, and dialysis of bromelain
The source of bromelain was a Palembang pineapple core, which was selected because it has a relatively larger amount of bromelain than other pineapple waste. The obtained crude enzyme was then purified by precipitation with ammonium sulfate and dialysis. The precipitation stage was principally referred to salting out process in which the water that solvated the protein will be attracted by ammonium sulfate salt which has a higher solubility than the protein. As a result, the interaction between proteins with their solvents will be weaker than their interactions with other proteins, so that proteins will aggregate to form large molecules with lower solubility. Then, at the dialysis stage there was a concentration gradient between the environment inside the sac and the environment outside the sac, so the molecules inside that are smaller than the pore size of the sac will diffuse out. In contrast, the solvent in the outside environment of the sac will go inside to balance the change in concentration due to the molecular diffusion process. The bromelain activity of each purification stage is tabulated in table 1. From the table, it can be concluded that the specific activity of the enzyme increases with purification. The highest specific activity is the dialysis fraction of 229.77 U/mg with a purity level of 11.42 times the crude enzyme. This fraction was encapsulated into the hydrogel matrix and determined for its antiplatelet activity.

3.2. FTIR spectra and visual appearance of unloaded and bromelain loaded alginate-guar gum hydrogel
Visually (figure 1), the unloaded and bromelain loaded hydrogels have a non-uniform spherical shape. The difference lies in the brown color of unloaded and white color on loaded hydrogels, which are the effect of different drying methods. The Fourier Transform Infrared (FTIR) spectra (figure 2) of the hydrogel matrix exhibit several specific spectrum of sodium alginate and guar gum, among which are broad peaks of 3400-3000 cm^{-1} waves typical for the vibration mode of hydroxyl groups (O-H), peak at

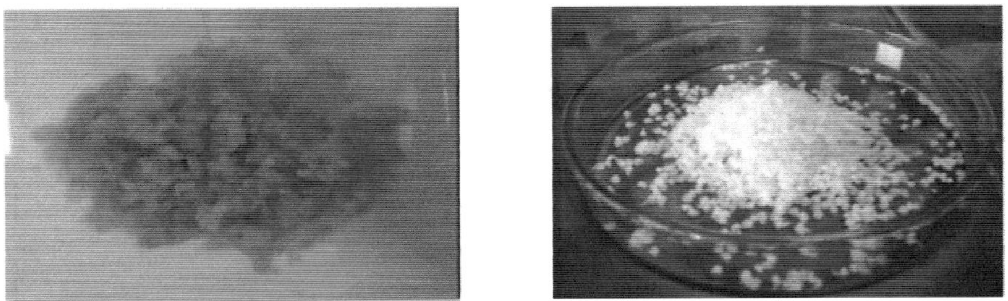

Figure 1. Visual appearance of (a) oven dried unloaded alginate-guar gum hydrogel and (b) freeze dried bromelain loaded alginate-guar gum hydrogel

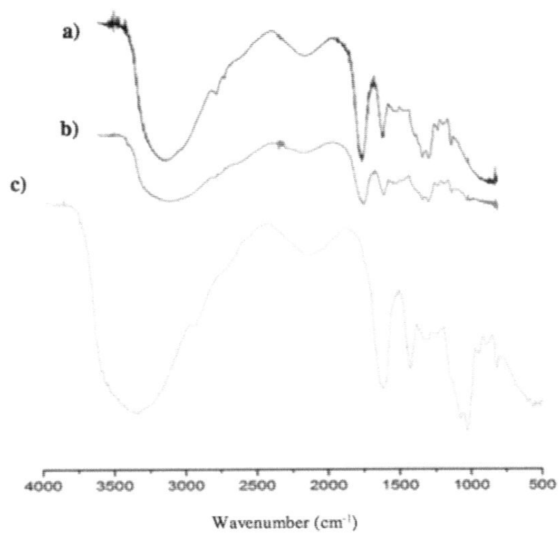

Figure 2. FTIR spectra of (a) alginate-guar gum hydrogel, alginate-guar gum hydrogel (b) crosslinked with glutaraldehyde and (c) loaded with bromelain

1579 cm^{-1} and 1400 cm^{-1} which are the asymmetric and symmetric anionic carboxylic acid groups (COO-) vibration, peak at 1032 cm^{-1} which is a marker of the carbonyl groups (C=O) and the peak at 960 cm^{-1} and 828 cm^{-1}, respectively, denote the C-H bond for mannuronic acid and guluronic acid in the sodium alginate chain [14].

Crosslinked hydrogel spectra showed a decrease in absorption at wave number 3400–3000 cm^{-1}. This is due to the acetyl formation, which converts 2 equivalents of the hydroxy group from sodium alginate and guar gum to ether bonds. In addition, there is an increase in absorption at wave number 1250 cm^{-1}, which also can be associated with the formation of acetal rings and ether bonds, the product of the reaction between the hydroxy groups of sodium alginate and guar gum with aldehyde groups of glutaraldehyde [15].

Crosslinked hydrogel spectra showed a decrease in absorption at wave number 3400–3000 cm^{-1}. This is due to the acetyl formation, which converts 2 equivalents of the hydroxy group from sodium alginate and guar gum to ether bonds. In addition, there is an increase in absorption at wave number 1250 cm^{-1}, which also can be associated with the formation of acetal rings and ether bonds, the product of the reaction between the hydroxy groups of sodium alginate and guar gum with aldehyde groups of glutaraldehyde [15].

FTIR spectra between unloaded and bromelain loaded hydrogels showed the same absorption peaks. This indicates that there is no chemical interaction between bromelain and the hydrogel matrix. However, there is a change of intensity in the wave number 3400–3000 cm^{-1} indicating a physical interaction of hydrogen bond between bromelain and matrix [16].

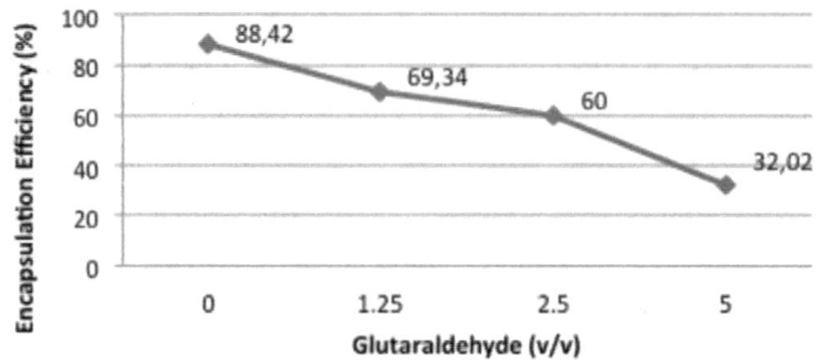

Figure 3. Encapsulation efficiency of alginate-guar gum hydrogel loaded with 1 mL standard bromelain 100 ppm

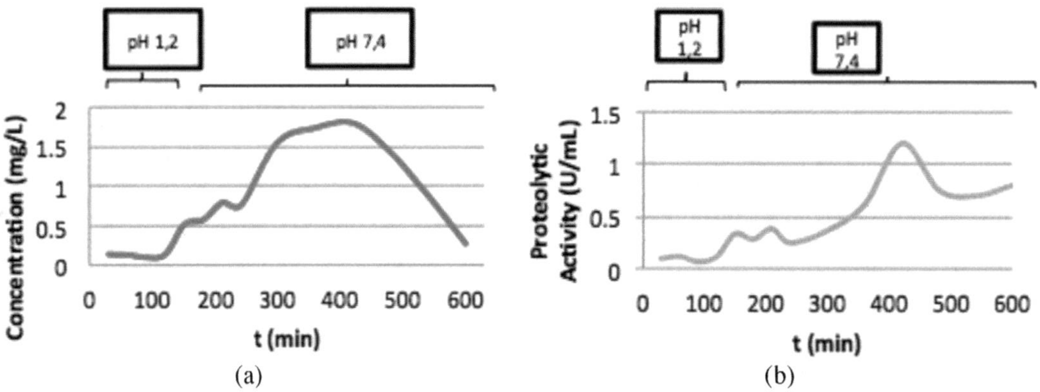

Figure 4. Dissolution test result of bromelain loaded alginate-guar gum hydrogel for (a) protein content and (b) proteolytic activity

3.3. Encapsulation efficiency of alginate-guar gum hydrogel

From figure 3, it can be seen that the greater the concentration of glutaraldehyde added, the smaller the encapsulation efficiency. This is relevant to the structure of the hydrogel matrix, which becomes increasingly dense due to the addition of glutaraldehyde, resulting in the trapped bromelain to become smaller. This rigid structure also causes the trapped bromelain to have no chance of escaping the matrix when immersed in phosphate buffer [17], so its release tends to decrease with the increasing of glutaraldehyde concentrations. The highest efficiency of 88.42 % is owned by hydrogel without the addition of glutaraldehyde (0 %, v/v), but the swelling ratio is still too high in both artificial environments, which is lead to a large release in the gastric fluid and uncontrolled release in the intestinal environment. Therefore, hydrogel with glutaraldehyde concentration of 1.25 % (v/v) is considered to be the best variation and selected to encapsulate the isolated bromelain. Encapsulation of isolated bromelain into the alginate-guar gum hydrogel crosslinked with glutaraldehyde concentration of 1.25 % generate an efficiency of 60.42 %.

3.4. Dissolution test of isolated bromelain loaded alginate-guar gum hydrogel

Based on figure 4, the dissolution fraction in artificial gastric fluid has a relatively lower protein content than the artificial intestinal environment. The maximum measured protein content in artificial gastric fluid was 0.13 mg/L, much smaller than the maximum protein content in the artificial intestinal environment of 1.79 mg/L. The proteolytic activity of dissolution fraction also tends to be higher in the artificial intestinal environment rather than artificial gastric fluid. The maximum proteolytic activity in artificial gastric fluid occurred at 120 minutes at 0.11 U/mL, while in the artificial intestinal environment occurred at 420 mins of 1.2 U/mL. The dissolution fraction with the highest proteolytic activity was then determined for its antiplatelet activity.

Table 2. Antiplatelet activity of bromelain fraction

Samples	Absorbance with ADP Addition		Aggregation (%)	Inhibition (%)
	Before	After		
Aquadest	0.494	0.025	94.93	-
Aspirin	0.493	0.485	1.62	98.29
Dialysis fraction	0.499	0.265	46.89	50.6
Dissolution fraction	0.467	0.187	59.95	36.84

3.5. In vitro study of antiplatelet activity
Based on table 2, the dialysis fraction before the encapsulation shows an inhibition percentage of 50.6 %. After encapsulation, the percent of inhibition was relatively unchanged at 36.84 %. Therefore, it can be concluded that the encapsulation method is good enough in maintaining the antiplatelet activity of bromelain from degradation in gastric fluid.

4. Conclusions
Bromelain can be encapsulated in alginate-guar gum hydrogel crosslinked with glutaraldehyde. The maximum protein content of bromelain in the artificial intestinal environment is greater than in the artificial gastric fluid. Furthermore, the maximum proteolytic activity of dissolved bromelain in the artificial intestinal environment is greater than in the intestinal environment. The inhibition percentage of antiplatelet aggregation is relatively unchanged between bromelain in before and after encapsulation process. This suggests that the activity of bromelain can be maintained up to the intestine.

Acknowledgements
This research was funded by Hibah Publikasi Internasional Terindeks untuk Tugas Akhir Mahasiswa Universitas Indonesia (PITTA) 2018.

References
[1] Musfiroh F F, Setiasih S, Handayani S, Hudiyono S and Ilyas N M 2017 *IOP Conf. Ser.: Mater. Sci. Eng.* **299** 012017
[2] Kurnia S D, Setiasih S, Handayani S and Hudiyono S 2017 *AIP Conf. Proc.* **2023** 020077
[3] Febriani K, Wahyuni I, Setiasih S and Hudiyono S 2017 *AIP Conf. Proc.* **1862** 030095
[4] Manzoor Z, Nawaz A, Mukhtar H and Haq I 2016 *Braz. Arch. Biol. Technol.* **59** e16150010
[5] Setiasih S, Adimas A C D, Dzikria V and Hudiyono S 2017 *IOP Conf. Ser.: Mater. Sci. Eng.* **299** 012016
[6] Shiew P S, Fang Y L and Majid F A A 2010 *Proc. 3rd Int. Conf. Biotechnology for the Wellness Industry* 8-9 October 2010 Kuala Lumpur (Kuala Lumpur: Universiti Teknologi Malaysia)
[7] Chobotova K, Vernallis A B and Majid F A A 2010 *Cancer Lett.* **290** 148–56
[8] Castell J V, Friedrich G, Kuhn C S and Poppe G E 1997 *Am. J. Physiol.* **273** G139–46
[9] Chen Y E and Chi H 2017 *Mater. Sci. Eng.* C **83** 233–46
[10] Bosio V E, Basu S, Abdullha F, Villalba M E C, Güida J A, Mukherjee A and Castro G R 2014 *React. Funct. Polym.* **82** 103–10
[11] George M and Abraham T E 2007 *Int. J. Pharm.* **335** 123–9
[12] Roy I, Sardar M and Gupta M N 2005 *Biochem. Eng. J.* **23** 193–8
[13] Moriyama H, Hosoe T, Wakana D, Itabashi T, Kawai K, Iizuka T W, Hoshi K, Fukushima K and Lau F C 2009 *J. Health Sci.* **55** 103–8
[14] Helmiyati H and Syarifudin A 2018 *AIP Conf. Proc.* **2023** 020080
[15] Yeom C K and Lee K H 1997 J. Appl. Polym. Sci. **67** 209–19
[16] Setiasih S, Prabowo H A, Budianto E and Hudiyono S *Journal of Applied Pharmaceutical Science (JAPS)* **8(10)** 17–24
[17] Eswaramma S and Rao K S V K 2017 *Carbohydr. Polym.* **156** 125–34

IOP Conf. Series: Materials Science and Engineering **763** (2020) 012043 doi:10.1088/1757-899X/763/1/012043

Conversion of cellulose from sweet sorghum (*Sorghum bicolor*) bagasse to levulinic acid over hierarchical Mn/ZSM-5 heterogeneous catalyst

D A Nurani, Y K Krisnandi, M B Simanjuntak, D U C Rahayu and E Saepudin

Department of Chemistry, Faculty of Mathematics and Natural Sciences (FMIPA), Universitas Indonesia, Kampus UI Depok, Depok 16424, Indonesia

Corresponding author's e-mail: endang.saefudin@sci.ui.ac.id

Abstract. The non-food biomass, sweet sorghum (*Sorghum bicolor*) bagasse (SSB), can be a possible source for synthesis of platform chemical such as levulinic acid (LA). It contains high sugar derived from cellulose and hemicellulose. Delignification process of SSB becomes important due to accessibility to those sugars. In this work, the effect of different chemical pretreatments on the delignification process was studied, including alkali pretreatment (NaOH) and oxidative pretreatment ($NaOCl_2$). The products were characterized by Fourier Transform Infrared (FTIR) and X-Ray Diffraction (XRD). The pretreated SSB was then converted into LA using hierarchical Mn/ZSM-5 heterogeneous catalyst at 100 °C with various concentrations of phosphoric acid and fixed concentration of hydrogen peroxide. The conversion reaction was based on Fenton-like reaction. Quantitative analysis of the products was performed using High Performance Liquid Chromatography (HPLC). The result shows that lignin content was significantly reduced from 28.42 % to 16.81 % using 10 % NaOH. Alkali pretreatment on SSB also produces higher yield of LA than oxidative pretreatment. The optimum % yield of LA 14.37 % was achieved using derived cellulose in alkali pretreatment in reaction condition 30 % of H_3PO_4 (v/v), 30 % H_2O_2 (v/v), and 10 % of Mn/ZSM-5 at 100 °C for 10 h reaction time. The catalyst shows good selectivity to LA due to the absence of HMF as intermediate. Beside LA, formic acid and acetic acid were also found in the conversion products.

Keywords: sweet sorghum bagasse (SSB), levulinic acid (LA), hierarchical Mn/ZSM-5, heterogeneous catalyst

1. Introduction

Recently, the conversion of renewable biomass into non-petroleum derived fuels and chemicals becomes attractive due to steep increment of fossil fuels utilization and CO_2 emission. Lignocellulosic biomass is the most abundant renewable resources, which is potential for chemical and fuels production.

It has been reported that Sweet Sorghum Bagasse (SSB) contained a very high sugar content derived from cellulose and hemicellulose [1], which could be converted to levulinic acid, a platform chemical that has many uses in the food industry, the pharmaceutical and medicine with a price range of 8–13 US $ per kg [2].

Since lignocellulosic biomass was very complicated, the pretreatment was not easy either. The crystallinity of cellulose, its accessible surface area, many phenolic compound in lignin content, degree of cellulose polymerization are the main factors considered as affecting the rate of conversion reaction [3]. Biomass pre-treatment such as physical, chemical, and physico-chemical pre-treatments can be used to resolve the obstacles in biomass conversion. Some studies reported that chemical treatment by alkaline solution can increase glucose levels and decrease the levels of lignin [3–4]. In addition, Cao *et al.* [5] also compared the effect of five methods of pre-treatments: (1) dilute NaOH solution autoclaving pretreatment, (2) high concentration NaOH solution immersing pretreatment, (3) dilute NaOH solution autoclaving and H_2O_2 immersing pretreatment, (4) alkaline peroxide pretreatment and (5) autoclaving pretreatment on enhancing enzymatic processes and the production of ethanol from sweet sorghum waste.

Several researchers have conducted research on cellulose conversion to LA via Fenton-like reaction [6]. Sano *et al.* [7] has made soy sauce degradation into levulinic acid using sulfuric acid. Rackemann and Doherty [8] also reported the conversion of lignocellulose from sugarcane bagasse into LA. Chen *et al.* [6] reported that cellulose was degraded to HMF (Hydroxy Methyl Furfural) and LA in system consisting of H_2O_2 and Mn/ZSM-5.

In this study, physical and chemical pre-treatment were conducted in order to decrease lignin content in sweet sorghum bagasse (SSB) and reduce the crystallinity of cellulose. The effect of H_3PO_4 concentration, which has important role in cellulose conversion to LA and time of reaction, was also studied in the catalytic reaction using hierarchical Mn/ZSM-5 as heterogenous catalyst.

2. Experimental

2.1. Materials

Sweet sorghum bagasse (SSB) as biomass source was monohybrid white sweet sorghum (SEAMEO BIOTROP, Tajur-Bogor, West Java). Glucose, levulinic acid (LA), HCOOH, H_2SO_4 96–97%, NaOH pellet, CH_3COOH 100%, H_3PO_4 85%, H_2O_2 30%, NaCl 32% were pro analysis grade from Merck. Technical grade n-hexane, ethanol, and $NaOCl_2$ were used as received. Cation exchange resin Amberlite was from Merckand. Hierarchical Mn/ZSM-5 zeolite (the synthesis of catalyst was described elsewhere [9]).

2.2. Instrumentation

X-Ray diffraction (XRD) with parameter operation at 40 kV and 40mA in a 2θ range between 5° and 60°, using Cu Kα radiation (NanoTech Serpong), Fourier Transform Infrared (FTIR) Shimadzu Prestige 21, High Performance Liquid Chromatography (HPLC) Shimadzu Prominance 20 (Shimadzu Japan) under the following conditions: column with RezexTM ROA-Organic Acid H^+ (8%); flow rate, 0.6 mL/min; mobile phase, H_2SO_4 0.005 N; detector, UV 220 nm; retention time, 45 min.

2.3. Pretreatment of SSB

SSB was cut to 8 cm and soaked in water for 1 h. SSB was then separated from water fraction and dried for 72 h. Dry SSB was ground to pass through 315 μm sieve. The homogenous SSB was then extracted with n-hexane and ethanol using Soxhlet for 6.5 h (dewax process).

SSB was pretreated in order to reduce lignin content. Pretreatment process was carried out in two ways. The first way is alkaline treatment: 5 g of SSB was soaked in 125 mL NaOH solution (5–15% (w/v)) at 328 K for 4 h under vigorous stirring. Second way was alkaline-oxidative chemical pretreatment. SSB was soaked at 328 K for 4 h in the mixed solution (alkaline: oxidative chemicals, 10 % NaOH:2 % H_2O_2 or 10 % NaOCl) with ratio of 1:25. Residue obtained from pretreatment process was then neutralized and dried in the oven at 333 K for 6 h. Both raw and pretreated SSB were characterized by FTIR and XRD. Whereas, α-cellulose, holocellulose, and lignin content were determined by Klason Lignin [10].

2.4. Catalytic conversion of SSB to LA

Pretreated SSB (1.0 g) was added to the mixture of 0.5 mL H_2O_2 30 % and 20 mL H_3PO_4 (30 % or 40 %). The catalyst 10 % Mn/ZSM-5 was then added to the solution. The conversion reaction was carried out in three-necked round bottom flask at 373 K for 10 h with simultaneous sampling every 2 h. As control, the conversion reaction was performed without catalyst under similar conditions. The conversion products were analyzed by HPLC. Calculation of the products yield follows equation (1):

$$\%\, yield\ LA\ (wt\%) = \frac{LA\ weight\ (g)}{feedstock\ (delignfied\ SSB) weight\ (g)} \times 100\% \tag{1}$$

3. Results and discussion

3.1. Pretreatment effect on the chemical composition of SSB

Table 1. Chemical components of untreated SSB.

Chemical Component	% composition
Moisture	11.13
Ash	8.97
Extractive	3.47
Hemicellulose	20.26
α- cellulose	26.94
Lignin	28.42
Total	99.19

Table 2. Effect of pre-treatment to the lignin content of SSB.

Name of Sample	Chemical pretreatment	Lignin content (%)
uSSB	Untreated SSB	28.42
SSB5	NaOH 5%	22.74
SSB10	NaOH 10%	16.81
SSB15	NaOH 15%	19.67
SSB10-p2	NaOH 10% + H_2O_2 2%	24.01
SSB10-p3	NaOH 10% + H_2O_2 3%	21.54
SSB10-o2	NaOH 10% + NaOCl 2%	12.24

Table 1 shows that lignin was the main component in SSB followed by sugar content (α-cellulose and hemicellulose). Lignin has a large group of aromatic polymers resulting from the oxidative combinatorial coupling of 4-hydroxyphenylpropanoids. Structure of lignin is mostly complex. In fact, it presents virtually in all types of cellulosic plant biomass and acts as a protective sheet of the cellulose and hemicellulose components of the biomass material [11]. Pre-treatment role is to break down lignocellulose bonds between lignin and cellulose, remove lignin from the substrate, thus exposing the accessible surface area of cellulose [12]. With proper pre-treatment, optimum yield of desired conversion products could be expected.

In this work, the pretreatment of SSB was done by physical (grinding) followed by chemical pretreatment (alkaline and alkaline-oxidative pretreatment). Physical pretreatment could decrease the crystallinity and degrees of polymerization of cellulose in SSB and improve hydrolysis susceptibility by reducing the particle size. The effect of chemical pretreatment was observed by measuring the lignin content. Table 2 shows that lignin content was reduced after pretreated by both alkaline and alkaline-oxidative chemical solution. Lignin content reduced in alkaline (SSB5, SSB10, and SSB15) while unavoidably, cellulose and holocellulose were also dissolved. The optimum NaOH dosage was 10%. It is suggested that α-O-4 in phenolic and β-O-4 bond in non-phenolic compounds were broken by OH contained in alkaline solution. It eliminates residual lignin fraction as well as the termination of C-C bonds or carbohydrate degradation [13]. The bonds between ether phenyl propane units have been also degraded into monomers.

In order to improve delignification process, the effect of oxidizers addition H_2O_2 and NaOCl to alkaline solution (NaOH 10%) was examined in the range 2–3%. Lignin was significantly removed in alkaline oxidative pretreatment. The lowest lignin content (12.24%) was found in NaOCl 2% pretreatment. The negatively charged hypochlorite or perhydroxyl ions attacks the carbonyl group and a conjugated double bond contained in lignin within the substrate [13]. Alkaline solution (basic condition) assists the dissolution of the remaining lignin that has been oxidized to polar compounds. Therefore, the lignin was readily soluble in polar solvents.

Pretreatment methods with high delignification are preferable feedstock for LA production. Since alkaline solution can also reduce the cellulose content, solid fraction generated after pretreatment was analyzed to identify three primary components (lignin, hemicellulose, and α-cellulose) of SSB. As shown in table 3 pretreated SSB comprises high sugar (α-cellulose and hemicellulose) and low lignin.

Table 3. Compositional analysis of SSB10-o2.

Component	Amount (%)
α-cellulose	18.81
Hemicellulose	18.07
Lignin (Klanson)	12.24

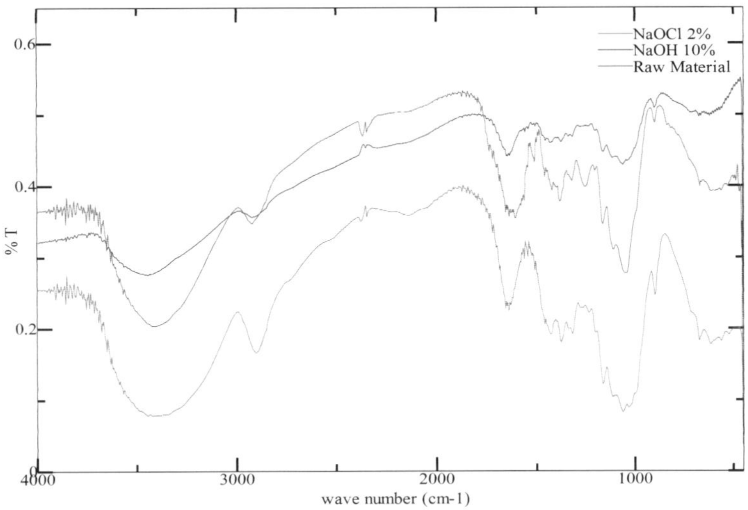

Figure 1. FTIR spectra of untreated (green) and pretreated (red and blue) SSB.

3.2. Characterization of pretreated SSB

Figure 1 shows the FTIR spectra of SSB before (uSSB) and after pretreatment (SSB5 and SSB 10). It can be seen from figure 1 that methylene groups (2900 cm^{-1}) was appeared in SSB after pretreatment followed by the decrease intensity of hydroxyl groups (3400–3500) cm^{-1}. This indicates the phenolic compound was reduced during pretreatment [14,15]. There is also a sharp peak at 895 cm^{-1}, attributed to β 1-4 glycosidic bond which typical bonds of cellulose [16]. The intensity of this peak was increased after the substrate faced alkaline or alkaline-oxidative pretreatment.

FTIR measurement was also performed to evaluate the crystallinity of cellulose. Figure 2 shows the infrared spectra of pretreated SSB and pure α-cellulose as reference. In general, the spectra of alkaline-oxidative treated SBB look similar to that from pure α-cellulose. On the other hand, in the spectra of SSB after alkaline treatment, the intensity of peak at 900 cm^{-1} decreased. Moreover, the broad band related to hydrogen-bonded water in (3700–3000) cm^{-1} region and a peak at 1620 cm^{-1} were lower. It is suggested that the lower intensity decrease could be related to the lower crystallinity in the alkaline pretreated SSB.

Figure 2 shows that pretreated SSB has both crystalline (1430 cm^{-1}) and amorphous structure. There is also aromatics group (1425 cm^{-1}), which indicates intramolecular hydrogen bond in cellulose monomer (C6).

3.3. Cellulose conversion to LA over hierarchical Mn/ZSM-5

In general, cellulose was converted to LA in three main stages. The first stage is termination of intra-molecular and intermolecular hydrogen bond assisted by the presence of acidic molecule (H_3PO_4). Second stage is the termination of β-1-4 glycoside bond. In this stage, hierarchical Mn/ZSM-5 catalyzed the reaction through Fenton-like reaction [6]. The third stage was conversion of cellulose monomers (glucose) to LA via intermediate 5-hydroximethylfurfural (HMF).

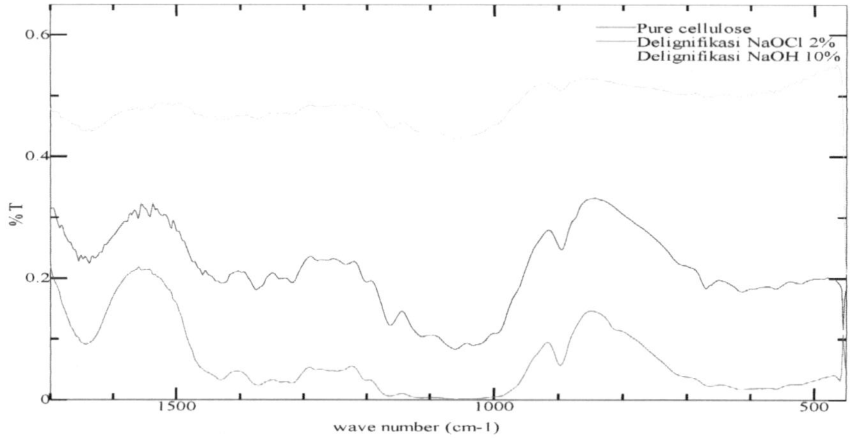

Figure 2. FTIR spectra of SSB after pretreatment at wavenumber 1500-700 cm^{-1}.

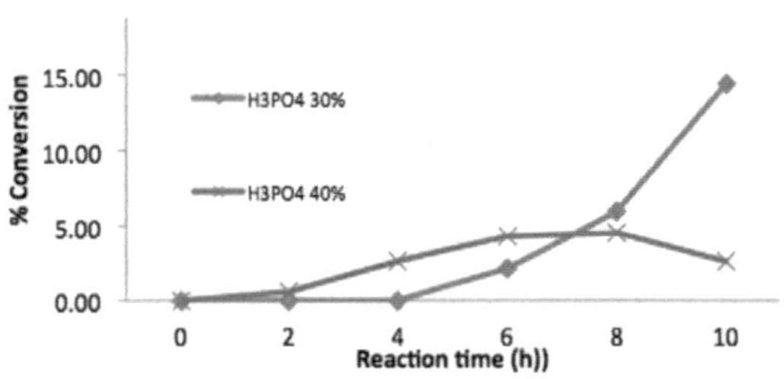

Figure 3. Effect of phosphoric acid concentration to LA production.

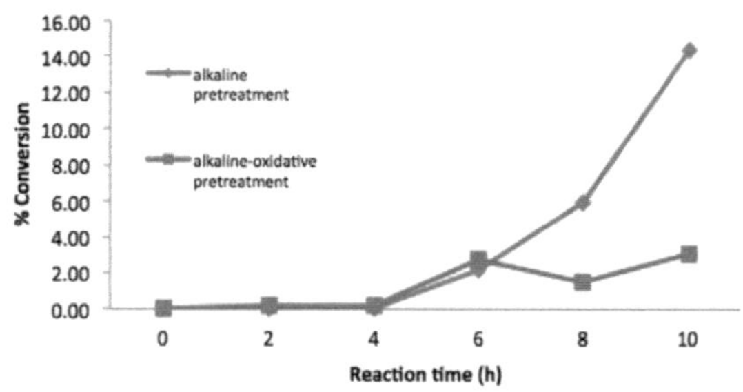

Figure 4. Effect of pretreatment method to LA production.

To study the effect of phosphoric acid concentration, the conversion reaction of cellulose to LA was performed using SSB-10 in the presence of 10% Mn/ZSM-5 catalyst at T 373K. Different concentration of phosphoric acid, 30% and 40%, was tested to find the best condition for LA production.

By examining figure 3, it can be seen that an increase in reaction time has an effect in increasing % conversion of cellulose to LA in both H_3PO_4 30 % and 40 %. The reaction using H_3PO_4 30 % gave a higher % conversion of LA (14.37 %) at 10 h reaction time. The inter- and intra-molecular hydrogen bond of cellulose could be broken by $H_4PO_4^-$ ions resulting from phosphoric acid auto-protolysis [6]. The aggregation state of cellulose was easily destroyed and depolymerized to soluble sugar. The degradation of cellulose was investigated for different pre-treatment methods alkaline SSB10 and alkaline-oxidative pre-treatment SSB10-o2.

The highest yield of LA was achieved by using alkaline pretreatment (figure 4). The amount of cellulose especially amorphous structure may also decrease in alkaline-oxidative chemical pretreatment and affecting the yield. Alkaline pretreatment increases the cellulose digestibility and it is more effective for lignin solubilisation and exhibit minor cellulose and hemicellulose solubilization than alkaline-oxidative pretreatment. It also known that NaOH causes swelling and increases the internal surface of cellulose [13,17].

The reaction without catalyst was also performed under optimum condition of 30 % H_3PO_4 and SSB10 to study the effect of Mn/ZSM-5 to LA production (figure 5). It can be seen that maximum yield of LA (14.37 %) was obtained from the reaction that was catalyzed by Mn/ZSM-5. As described by Chen *et al.* [6], in Fenton-like reaction, hydroxyl radical OH can be formed by the presence of metal oxide that was initiated by H_2O_2 molecule. Mn/ZSM-5 also inhibited charcoal formation, generating more products. Furthermore, heterogeneous catalyst was not only decreasing the reaction temperature but also served as an easy way for catalyst separation.

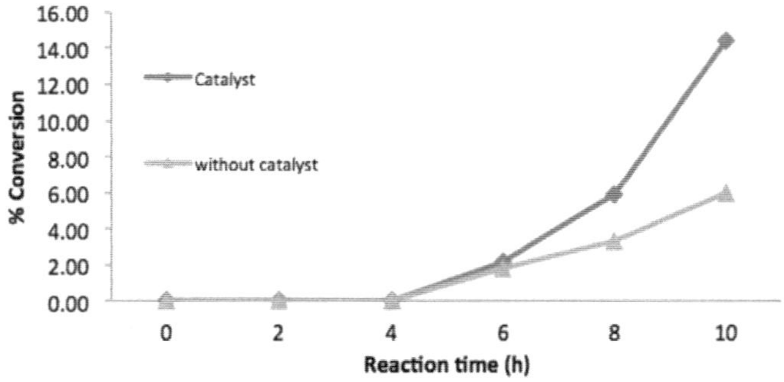

Figure 5. Effect of Mn/ZSM-5 catalyst to LA production.

4. Conclusions

Pretreatment method of biomass affected the cellulose conversion to LA due to degradation of biomass structure by reducing lignin content and decreasing the polymerization degree of cellulose. The use of Mn/ZSM-5 also contributed in increasing LA yield through Fenton-like reaction. The highest LA yield was obtained from the reaction using pretreated SSB under the following condition 1 g 30 % H_3PO_4 in 10 h time of reaction.

Acknowledgements

We are grateful for the financial supports of this research from PUPT BOPTN 2014 No. 1086/H2.R12/HKP.05.00/2014 and PUPT BPPTN 2018 No. 481/UN2.R3.1/HKP.05.00/2018.

References

[1] Patni N, Pillai S G and Dwivedi A H 2013 *Procedia Eng.* 51 355–62
[2] Bevilaqua D B, Rambo M K D, Rizzetti T M, Cardoso A L and Martins A F 2013 *J. Clean. Prod.* 47 96–101
[3] Taherzadeh M J and Karimi K 2008 *Int. J. Mol. Sci.* 9 1621–51
[4] Hendriks A T and Zeeman G 2009 *Bioresour. Technol.* 100 10–8
[5] Cao W, Sun C, Liu R, Yin R and Wu X 2012 *Bioresour. Technol.* 111 215–21
[6] Chen Y, Li G, Yang F and Zhang S M 2011 *Polym. Degrad. Stab.* 96 863–9
[7] Sano A, Satoh T, Oguma T, Nakatoh A, Satoh J I and Ohgawara T 2007 *Food Chem.* 105 1242–7
[8] Rackemann D W and Doherty W O S 2011 *Biofuel. Bioprod. Biorefin.* 5 198–214
[9] Krisnandi Y K, Nurani D A, Agnes A, Pertiwi R, Antra N F, Anggraini A R, Azaria A P and Howe R F 2019 *Indonesian Journal of Chemistry* 19 115–23
[10] Dence C W 1992 The determination of lignin *Methods in Lignin Chemistry (Springer Series in Wood Science)* ed S Y Lin and C W Dence (Berlin: Springer) Chapter 2.2 pp 33–61
[11] Shahzadi T, Mehmood S, Irshad M, Anwar Z, Afroz A, Zeeshan N, Rashid U and Sughra K 2014 *Adv. Biosci. Biotechnol.* 5 246–51
[12] Mosier N, Wyman C, Dale B, Elander R, Lee Y Y, Holtzapple M and Ladisch M 2005 *Bioresour. Technol.* 96 673–86
[13] Alvira P, Tomás-Pejó E, Ballesteros M and Negro M J 2010 *Bioresour. Technol.* 101 4851–61
[14] Jahan M S, Saeed A, He Z and Ni Y 2011 *Cellulose* 18 451–9
[15] Rosa S M L, Rehman N, de Miranda M I G, Nachtigall S M B and Bica C I D 2012 *Carbohydr. Polym.* 87 1131–8
[16] Lanthong P, Nuisin R and Kiatkamjornwong S 2006 *Carbohydr. Polym.* 66 229–45
[17] Chen H, Liu J, Chang X, Chen D, Xue Y, Liu P, Lin H and Han S 2017 *Fuel Process. Technol.* 160 196–206

Synthesis Pb(II)-ion imprinted polymer using tannins from mango leaves (*Mangifera indica L.*)

C Sekarharum, D U C Rahayu and D A Nurani

Department of Chemistry, Faculty of Mathematics and Natural Sciences (FMIPA), Universitas Indonesia, Kampus UI Depok, Depok 16424, Indonesia

Corresponding author's e-mail: d.arifa@sci.ui.ac.id

Abstract. This work reports the synthesis of Pb(II)–Ion Imprinted Polymer (Pb-IIP) using tannin extract from Mango leaf (*Mangifera Indica L.*) as functional ligand (chelating agent and monomer). Pb^{2+} ion found to form a stable compound at 1:1 complex with tannin in aqueous solution. The resulting complex of Pb^{2+} and tannin extract was polymerized using phenol-formaldehyde as cross-linker in acidic medium. The structure and properties of Pb-IIP were compared to Pb-IIP gallic acid and Non Imprinted Polymer (NIP) that were synthesized with similar conditions. These polymers were charaterized by Fourier Transform Infrared (FTIR) and Scanning Electron Microscopy-Energy Dispersive X-Ray (SEM-EDX). The template of Pb(II) ion was completely removed from the polymeric matrix using 1 mol/L HNO_3. The optimum pH of maximum sorption was 5 on IIP-tannin and IIP-gallic acid at 120 mins contact time. Maximum capacity of tannin Pb-IIP, gallic acid Pb-IIP, and NIP were 153.159 mg/g, 147.959 mg/g, and 138.218 mg/g, respectively. It was found that the tannin Pb-IIP sorbent is more selective than gallic acid Pb-IIP and NIP.

Keywords: ion imprinted polymer, lead, tannin, *Mangifera Indica L.*

1. Introduction

Recently, industrial activity is growing rapidly to fulfill the people's needs. But, as we know, the impact of industrial wastes contributes to the environmental pollution. One of the wastes that give major effect to human body is Pb [1]. Low level of lead in blood can cause lifelong health effects. Therefore, the removal of Pb(II) in the wastewater is very important [2].

There are so many metal separation methods that had been developed; however, these conventional technologies show lots of disadvantages, such as secondary pollution [3,4]. Ion imprinted polymer (IIPs) known as selective adsorption that widely used for metal separation [1,5,6]. The important things in IIP synthesis are the template ions, ligand, cross-linking agent, and solvents [7].

Mango (*Mangifera indica L.)* is rich in phenolic compounds [8]. Tannins due to high content of phenolic compounds are inexpensive and can be easily isolated from plants. With phenolic compounds, tannins can have the ability to chelating the metallic compound [8,9].

In this study, a new ion imprinted polymer, adsorption and desorption behavior of Pb-IIP tannin have been compared to Pb-IIP gallic acid as tannin monomers and non-imprinted polymer (NIP). The aim of this research is to prepare a solid-phase adsorbent which has a high capacity and selectivity to Pb(II) ions.

2. Experimental

2.1. Chemicals

Na_2CO_3 (Merck); dichloromethane; H_2SO_4 97 % (Merck); acetic acid anhydride; chloroform; n-hexane;

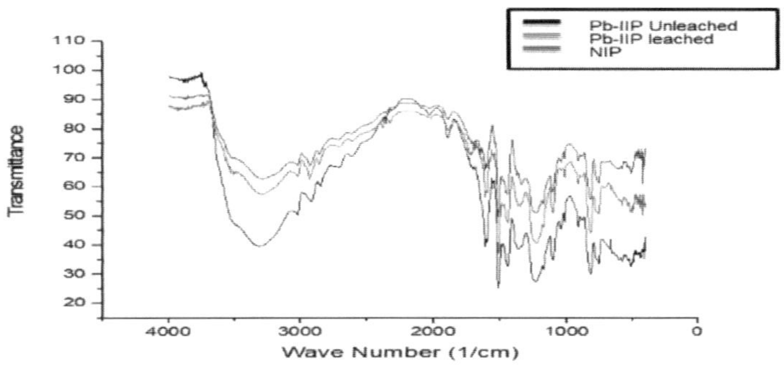

Figure 1. FTIR spectra of Pb-IIP unleached, leached and NIP

ethyl acetate (EtOAc); NaOH; $FeCl_3$; Wagner test; Aquadest; HCl 37 % (Merck) ; HNO_3 69 % (Merck); $Pb(NO_3)_2$; phenol; double distilled water; $Pb(NO_3)_2$ (Merck); $NiSO_4.5H_2O(s)$, $CuSO_4.5H_2O(s)$, $Fe(NO_3)_3.5H_2O(s)$ were from Merck; formaldehyde, HNO_3 and nitrogen gas.

2.2. Apparatus
Freeze dryer (Buchner), magnetic stirrer with hot plate, Scanning Electron Microscopy-Energy Dispersive X-Ray (SEM-EDX), UV-vis Spectrophotometer (Shimadzu 2450), Fourier Transform Infrared Infrared (FTIR) Spectrophotometer (Shimadzu Prestige 21) and Atomic Absorption Spectrophotometer (AAS) (Shimadzu AA 6800).

2.3. Preparation of Pb-IIP
Preparation of Pb-IIP comprises two steps. First, 1 mmol $Pb(NO_3)$ was dissolved in 100 mL of double distilled water. The functional ligand (tannins isolated from mango leaves) was added dropwise to achieve mole ratio of metals/ligand 1:1. The complex solution was stirred for 3 h. 100 mmol phenol, 50 mmol formaldehyde and 50 mL 0.4 M H_2SO_4 were added to the complex solution, respectively. The mixture was stirred until homogeneous and the nitrogen gas was removed from the solution by purging for 10 minutes. The polymerization was carried out in a reflux system with a temperature of 65 °C–70 °C for 12 h. The resulting polymer particles were filtered and washed with double distilled water to remove unreacted materials. In the second step, the imprint metal ions Pb(II) were leached from polymer matrix by 1M HNO_3 and washed by double distilled water until neutral pH. The polymer was dried overnight. The Non Imprinted Polymer (NIP) was prepared by the same manner in the absence of Pb^{2+} template ions. Characterization of IIP and NIP were performed using FTIR and SEM-EDX.

2.4. Extraction procedure
Adsorption test was performed using batch method. 50 mg dried polymer was added to the 10 mL aqueous solution consists of various concentrations of Pb(II) and stirred to reach an optimum contact time with a magnetic stirrer. pH was adjusted to pH 2–12 to obtain optimum condition. The reusability of polymer was tested by desorption step. Elution Pb(II) ions from IIP was performed in 1M HNO_3. The concentration of Pb(II) in the solution was confirmed by FAAS.

3. Results and discussion

3.1. Characterization
IIP was synthesized by polymerization of Pb^{2+}-tannin complex as a template and phenol-formaldehyde as cross-linker monomer using thermal method. The absorbance of ligand (tannin) solution was regarded as a function of metal ion/ ligand mole fraction in order to obtain stoichiometry and stability of complex. The mole fraction plot revealed a sharp inflection point at mole fraction 0.5, indicating the formation of 1:1 ligand to metal complex in aqueous solution.

FTIR spectroscopy was used to study the chemical structure of the synthesized polymer (figure 1).

(a) (b)

Figure 2. SEM images of Pb-IIP (a) unleached IIP and B)leached IIP

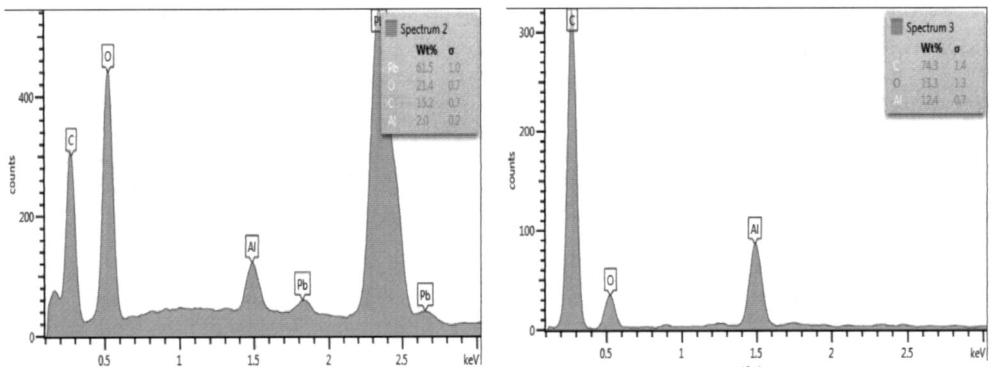

(a) (b)

Figure 3. EDX spectra of unleached Pb-IIP (left) and leached Pb-IIP (right)

The typical and intense tannin backbone could be observed in all FTIR spectra, O-H stretching vibration at 3414 cm^{-1}, C=O and C-O ester stretching band at 1639 cm^{-1} and 1310 cm^{-1} appeared respectively. The sulphonic acid adsorption S=O at 1096 cm^{-1} was observed in the imprinted polymers. The presence of Pb^{2+} ions in Pb-IIP polymer leads in decreasing adsorption intensity and shifts the adsorption to lower wave numbers.

The morphology of Pb^{2+}-tannin was assessed by SEM, and the resulting SEM pattern is shown in figure 2. In figure 2a, the polymerization resulted in irregular shape and surface particle with the size about (300–400) μm. The morphology of the particle depends on type of solvent, metal salt, cross-linker, ligand, polymerization temperature, time and stirring speed on polymerization process [10]. Moreover, the surface morphologies of polymer after leaching process were rougher than that of the unleached Pb-IIP. As shown in figure 2b the circular mold was formed in leached Pb-IIP, indicating Pb^{2+} removal from polymer matrix.

The EDX analysis was used to determine the composition of polymerized particle before and after leaching process. It can be seen in figure 3 that unleached Pb-IIP contains C, O, and Pb whereas in Pb-IIP after leaching, there are C and O. This result confirms that Pb^{2+} ions were completely leached from matrix of polymer using 1 M HNO_3.

3.2. Effect of pH and adsorption time

pH is an important parameter for adsorption of metals in aqueous solution. Batch experiments were carried out to study the effect of pH on adsorption performance. As shown in figure 4, the absorbance of Pb^{2+} increases slowly as the pH increases. In lower pH, the amount of H^+ formed increased, so H^+ competed with Pb(II) to be bonded with cavities in polymer matrix. At the higher pH shows the decrease of Pb^{2+} absorbance. It is because OH^- was formed, had interaction with Pb(II) and precipitated at pH> 8. Based on the result, pH 5 was selected as the optimal pH value for adsorption study in this work.

IOP Conf. Series: Materials Science and Engineering 763 (2020) 012044 doi:10.1088/1757-899X/763/1/012044

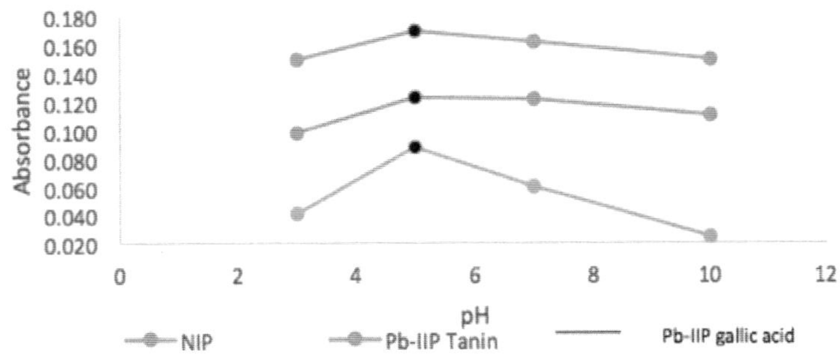

Figure 4. Effect of pH in adsorption performance, 10 mL 5 ppm $Pb(NO_3)_2$, 50 mg IIP

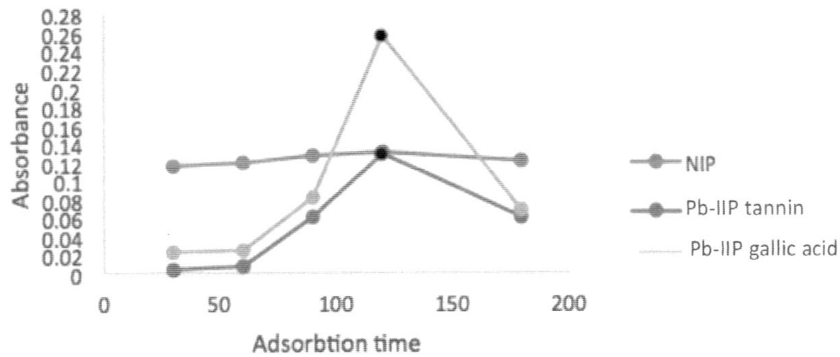

Figure 5. Effect of absorption time, 10 mL 5 ppm $Pb(NO_3)_2$, 50 mg IIP

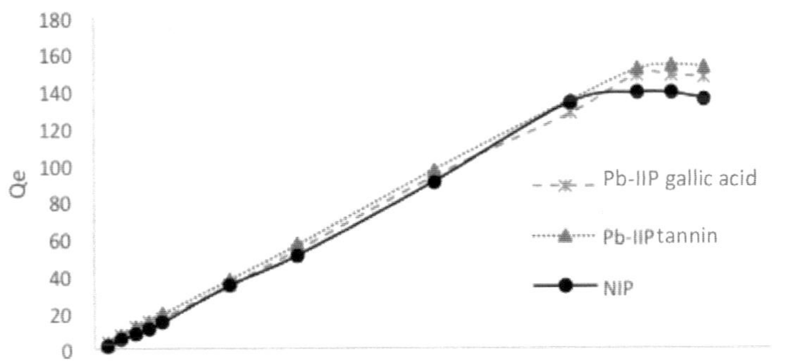

Figure 6. Adsorption capacity of Pb-IIP tannin (red triangle, dotted line) Pb-IIP gallic acid (asterisk, dashed line) and NIP (filled circle, solid line) 10 mL Pb^{2+} 20-1000 mg/L, 50 mg adsorbent

Adsorption rate is an important factor that should be determined in the application. Variation of adsorption time as the function of absorbance was investigated and the result is shown in figure 5. The absorbance slightly increases at the beginning and sharply increases at 60–120 min. The optimum contact time was found at 120 min with the highest adsorption of Pb^{2+} ions. At longer contact time, the adsorption of Pb^{2+} was decreased due to the weak bonding between polymer matrix and Pb(II). From figure 5 and figure 6, it is obtained that NIP, Pb-IIP tannin, and Pb-IIP gallic acid have optimum adsorption at the same pH and contact time.

Table 1. Selectivity coefficient of Pb-IIP tannin, Pb-IIP gallic acid and NIP

Metals	Selectivity Coefficient		
	Pb-IIP Tannin	Pb-IIP Gallic Acid	NIP
Fe	8.767	3.661	0.072
Ni	44.102	8.034	0.161
Cu	57.475	14.550	0.242
Pb	-	-	-

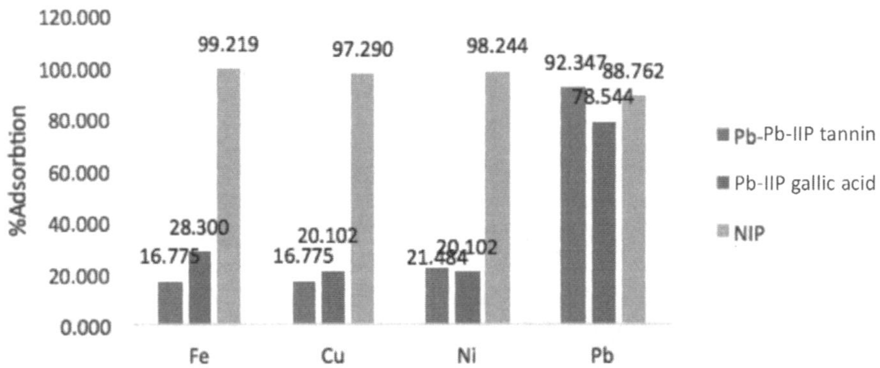

Figure 7. Adsorption of Pb^{2+} and other competing ion from mixed solution, 50 mg adsorbent, 10 mL 5.0 mg/L each metal concentration

3.3. *Adsorption capacity*

The adsorption capacity is defined as the amount of adsorbate in mg, units which can be absorbed in 1 gr of adsorbent. Figure 6 shows that the adsorbent being saturated at different concentrations. In order to reach the saturation point, the Pb(II) concentration was increased until the adsorption capacity values were obtained. The adsorption capacity was calculated using following equation:

$$Qe = \frac{(Co-Ce)V}{W} \tag{1}$$

where Co is the initial concentration of Pb(II), Ce is the concentration of Pb(II) that did not adsorb by the adsorbent, V is the volume of Pb(II) and W is the mass of adsorbent. From figure 6, It can be observed that the adsorption capacity of Pb-IIP tannin, Pb-IIP gallic acid, and NIP were 52.87 mg/g, 50.75 mg/g, and 48.50 mg/g, respectively. Pb-IIP tannin has a higher adsorption capacity than Pb-IIP and NIP. The uniform and accessible imprinted binding sites could enhance the adsorption performance of Pb^{2+} by imprinted polymer [11]. Furthermore, the high molecular weight of tannin might affect the amount of binding sites cavities in the polymer matrix.

3.4. *Selectivity*

This test aims to compare the adsorption performance between lead-ion imprinted polymer (Pb-IIP) and non-imprinted polymer (NIP) in samples containing various types of metal ions i.e. Cu^{2+}, Ni^{2+}, Fe^{3+}. In figure 7 it can be seen that Pb-IIP tannin has a higher adsorption capacity to Pb metal ions than other polymers. Since Pb-IIP absorbs the metal not only on the basis of affinity but also on the shape and size of the metal ions. This is due to the cavity of recognition site possessed by Pb-IIP, whereas the NIP absorbs the metal only based on the affinity of the metal ion.

The selectivity coefficient represents the degree of Pb-IIP and NIP selectivity towards other metal ions. From table 1 it can be seen that Pb-IIP tannin has the highest selectivity coefficient compared to Pb-IIP gallic acid and NIP. High selectivity coefficient value indicates that the metal ions Pb (II) can still be detected even though there are other metal ions, so that the presence of other metal ions against Pb metal ions as an interfering metal is not very significant. Ion imprinted polymer possess an even allocation and cavities which related to the imprint ion size and coordination geometries [7]. The chelating ligand also affects the selective affinities for the imprint ion.

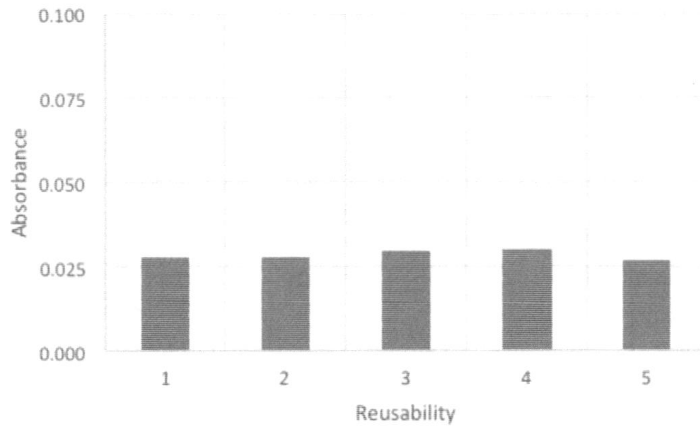

Figure 8. Repetition test of Pb-IIP tannin, 10 mL 5.0 mg/L Pb(II), eluent 0.5 M HNO_3

3.5. Reusability
Reusability is one of the important performances of IIP due to its application. In this research five times repetition of adsorption-desorption test were performed using the same Pb-IIP tannin. Figure 8 shows that adsorbent could be repeatedly used without significant loss of binding cavities. It was also found that %RSD of five replicates were 0.304 % and CV Hortwitz 1.314 %. Pb-IIP tannin has good performance and it has a lower %RSD value than CV Hortwitz.

4. Conclusions
A novel Pb-IIP using tannins as functional ligand from mango (*Mangifera indica L.*) leaves with phenol-formaldehyde crosslinker in acidic medium has been successfully synthesized. Pb-IIP tannin was more selective than Pb-IIP gallic acid and NIP because of its recognition sites and it is also effective as chelating agent in synthesis of IIP due to high selective coefficient value. Pb-IIP tannin has higher adsorption capacity than the others due to the large amount of active group.

Acknowledgements
This research was funded by PITTA Grant from Universitas Indonesia with with contract No. 2253/UN2.R3.1/KKP/05.00/2018

References
[1] Liu H, Kong D, Sun W, Li Q, Zhou Z and Ren Z 2016 *Chem. Eng. J.* 303 348–58
[2] Shukla V, Shukla P and Tiwari A 2018 *Indian J. Med. Spec.* 93 146–9
[3] Juang R S and Shiau R C 2000 *J. Membrane Sci.* 165 159–67
[4] Qin J J, Wai M N, Oo M H and Wong F S 2002 *J. Membrane Sci.* 208 213–21
[5] Behbahani M, Bagheri M, Taghizadeh M, Salarian M, O Sadeghi, Adlnasab L and Jalali K 2013 *Food Chem.* 138 2050–56
[6] Hu S, Xiong X, Huang S and Lai X 2016 *Anal. Sci.* 32 975–80
[7] Shamsipur M, Besharati-Seidani A, Fasihi J and Sharghi H 2010 *Talanta* 83 674–81
[8] Luo F, Fu Y, Xiang Y, Yan S, Hu G, Huang X, Guodi H, Sun C, Li X and Chen K 2014 *J. Funct. Foods* 8 282–91
[9] Fan R, Xie F, Guan X, Zhang Q and Luo Z 2014 *Bioresour. Technol.* 63 167–71
[10] Shamsipur M and Besharati-Seidani A 2011 *React. Funct. Polym.* 71 131–9
[11] Zhou Z, Kong D, Zhu H, Wang N, Wang Z, Wang Q, Liu W, Li Q, Zhang W and Ren Z 2018 *J. Hazard. Mater.* 341 355–64

IOP Conf. Series: Materials Science and Engineering **763** (2020) 012045 doi:10.1088/1757-899X/763/1/012045

Effect of Nonidet SF-5 as surfactant in preparation of microspheres based on polyblend of poly(lactic acid) and polycaprolactone

F Zilvania and E Budianto

Departement of Chemistry, Faculty of Mathematics and Natural Sciences (FMIPA), Universitas Indonesia, Kampus UI Depok, Depok 16424, Indonesia

Corresponding author's e-mail: emilb@ui.ac.id

Abstract. Microspheres of biodegradable polymers have been widely investigated in delivery system for bioactive compounds. In this study, microspheres were prepared from polyblend of poly(lactic acid) and polycaprolactone using water-in-oil (w/o) emulsion solvent evaporation method with Nonidet SF-5 as surfactant. The effect of Nonidet SF-5 on the size of microsphere and its distribution was studied by varying the volume of surfactant (1, 1.5, and 2) mL. Additional variations such as emulsion stirring speed (700, 800, and 900) rpm and dispersion stirring time (0.5, 1, 1.5, and 2) h were also conducted. Microspheres were characterized using Particle Size Analyzer (PSA), Fourier Transform Infrared (FTIR), and optical microscope. The results showed that the addition of surfactant volume decreased the microspheres size from 34.58 μm to 28.70 μm and the most uniform microspheres size was obtained at 1 mL of Nonidet SF-5. While the microspheres that produced through variation of emulsion stirring speed showed the same size, that was 31.50 μm and the most uniform microspheres size was obtained at 800 rpm. Moreover, the increasing in dispersion stirring time also decrease the microspheres size from 31.50 μm to 19.76 μm, but it did not affect the microspheres size distribution.

Keywords: microspheres, poly(lactic acid), polycaprolactone, Nonidet SF-5, solvent evaporation method

1. Introduction

The rapid advancement in modern drug delivery system was begun with the use of biodegradable polymers as carriers to obtain a therapeutic effect by a simultaneous release of the drug within a certain period of time and appropriate dosage. Controlled drugs release has been widely developed which is capable of controlling the rate of release of drugs, maintaining the duration of therapeutic activity or targeting drug delivery to tissues [1]. Poly(lactic acid) (PLA) and poly(ε-caprolactone) (PCL) are biodegradable polymers that are widely used in biomedical applications because they have good biodegradability and biocompatibility. In general, PLA presents good mechanical properties and high modulus strength, but the degree of its fragility is also high. Various ways to resolve the physical properties limitations of PLA have been investigated, one of which is mixing or blending [2]. The toughness properties of PLA can be improved by blending it with PCL [3]. PCL possesses a high crystallinity property, still with a low degradation rate. Poly(lactic acid) in the form of D, L-PLA is degraded more quickly than PCL [4]. Therefore, the development of drug delivery system based on the blend of PCL and PLA is presumed to generate a suitable polymer with a higher speed degradation rate and preferable permeability to improve the controlled release of drugs [5].

Microspheres of biodegradable polymers are one of the biomedical applications that have been widely investigated in delivery systems for bioactive compounds such as drugs. Microsphere is tiny

Table 1. The three variations in preparation of microspheres.

Volume of Nonidet SF-5 (mL)	Emulsion stirring speed (rpm)	Dispersion stirring time (h)
1	700	0.5
1.5	800	1
2	900	1.5
		2

spherical particle in the diameter size ranges from 1–1000 μm [6]. Kemala *et al.* [6] reported that the concentration and volume of PVA (an emulsifier) affected the size of microspheres that were synthesized from the blend of PLA and PCL by a modified (o/w) emulsion solvent evaporation method. In another study by Zhou *et al.* [7] microspheres of PLA and PCL were prepared by a modified double emulsion method using PVA as an emulsifier. The formed microspheres presented spheres contains small holes on the surface and a spongy porous inner structure [7]. In this study, microspheres were prepared from blend of PLA and PCL by a modification in water-in-oil (w/o) emulsion solvent evaporation method to see the effect of Nonidet SF-5 as surfactant on microspheres size and its distribution. Nonidet SF-5 is a nonionic surfactant with low foaming, which enhanced emulsification properties. Low foaming surfactant can be an alternative in microspheres production. In the process of emulsion, stirring with high-speed makes more air will enter into the emulsion and form foam which can disrupt the formation of microspheres and affect the size of the microsphere [6,8]. The size of microspheres and its distribution is one of the important factors that affect the process of drug delivery system. The uniform size of microspheres will provide good dissolution rates and efficiency of drug encapsulation [6].

2. Materials and methods

2.1. Materials
S-lactic acid (90%) was used as monomer which was purchased from Merck. Polycaprolactone was purchased from Perstop CAPA-6800. Distilled water (aquadest) was purchased from PT. Brataco. Nonidet SF-5 was obtained from Evonik Indonesia, and dichloromethane was purchased from Merck.

2.2. Synthesis of poly(lactic acid)
Synthesis of PLA was performed by a modified direct polycondensation without a catalyst, initiator, or solvent [9]. 50 mL of lactic acid (LA) was poured into the three-neck flask and then put it in the reactor. LA was stirred and heated on the hotplate until the temperature of 120 °C, and the temperature was kept constant for 1 h. In this step, the nitrogen gas was flowed to the system. Furthermore, the temperature system was increased to 150 °C and maintained for 22 h. The nitrogen gas was continuously flowed to keep the system free of oxygen until the formed PLA was taken out from the reactor. After 22 h, the heating process was stopped, but the formed PLA was still stirred for 2 h. The formed PLA was cooled for 24 h at room temperature.

2.3. Preparation of 10% (w/v) polyblend solution
The polyblend of PLA and PCL was made from the composition of 60:40 (PLA:PCL) with the blend weight of 0.5 g. The polymers were mixed in dichloromethane (DCM) and stirred with a magnetic stirrer until completely dissolved to obtain the 10% (w/v) polyblend solution [6].

2.4. Variations in preparation of microspheres
Microspheres were made by using a modified w/o emulsion solvent evaporation method [6]. In the preparation of microspheres, three variations were used to see their effect on microspheres size and its distribution, i.e. variations of surfactant volume, emulsion stirring speed, and dispersion stirring time. First, 5 mL of 10% polyblend solution was emulsified in 1 mL of distilled water and 1.5 mL of Nonidet SF-5. The solution was stirred at 700 rpm for 1 h. The emulsion was then dispersed in 350 mL of distilled water and stirred at 900 rpm for 1 h to evaporate DCM. The formed microspheres were filtrated, washed with distilled water, and dried in the oven at 40 °C for 24 h. The variations that were used in this study are listed in table 1.

IOP Conf. Series: Materials Science and Engineering **763** (2020) 012045 doi:10.1088/1757-899X/763/1/012045

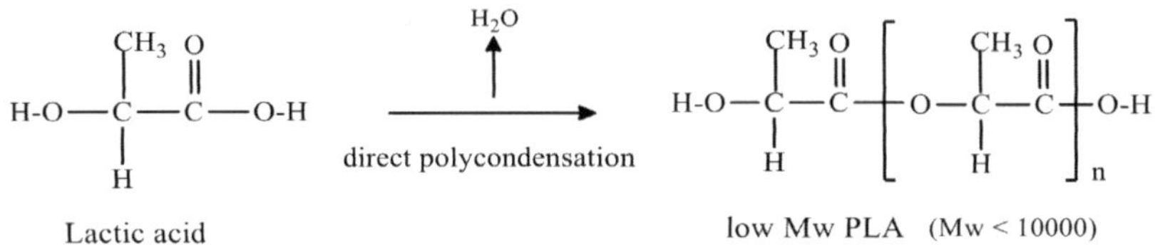

Figure 1. Schematic representation of PLA synthesis.

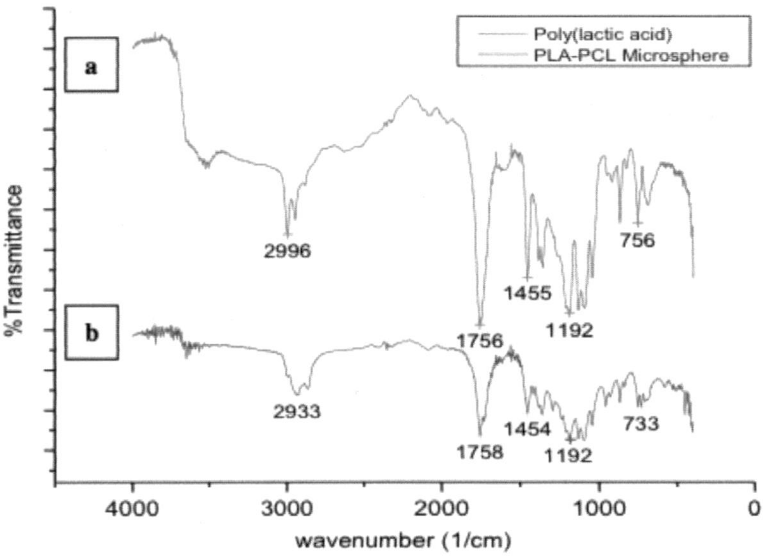

Figure 2. The FTIR spectra of (a) poly(lactic acid) and (b) microsphere.

2.5. Characterization of PLA and microspheres
The functional groups of PLA and microspheres were analyzed using Fourier Transform Infrared (FTIR) Shimadzu Prestige-21 spectrophotometer. The size of microspheres produced from each formula, as well as its distribution, were measured using Particle Size Analyzer (PSA) LS-100 Shimadzu. The microspheres shape and appearance were observed by optical microscope at 13.5x magnification.

3. Results and discussion

3.1. Synthesis of poly(lactic acid)
Synthesis of poly(lactic acid) has been successfully performed using a modified direct polycondensation method without a catalyst, initiator, or solvent. Figure 1 shows the schematic representation of PLA synthesis. The uncatalyzed direct polycondensation method is a simple method, but does not produce high molecular weight PLA (Mw <10000) [9]. The state of the system is made oxygen-free by flowing the nitrogen gas during the polymerization process. The flow of nitrogen gas can eliminate the oxygen gas contained in the system. Oxygen may oxidize the reagent (lactic acid), which causes the color of poly(lactic acid) turns to dark brown. The color change of PLA indicates that the formed PLA has damaged.

The FTIR spectrum of PLA synthesized is shown in figure 2a. IR spectrum of PLA showed characteristic absorption peaks of C-H sp^3 stretching at 2996 cm^{-1}, C=O carbonyl at 1756 cm^{-1}, and C-O ester at 1192 cm^{-1}. The results indicated that the functional groups of formed PLA are in accordance with the results of previous research by Choksi and Desai [10]. They characterized the PLA that was synthesized from lactic acid by polyesterification process, which showed the absorption peaks of CH sp^3, C=O, and C-O at 2988.70, 1716.98, and 1119.50 cm^{-1} in IR spectra, respectively.

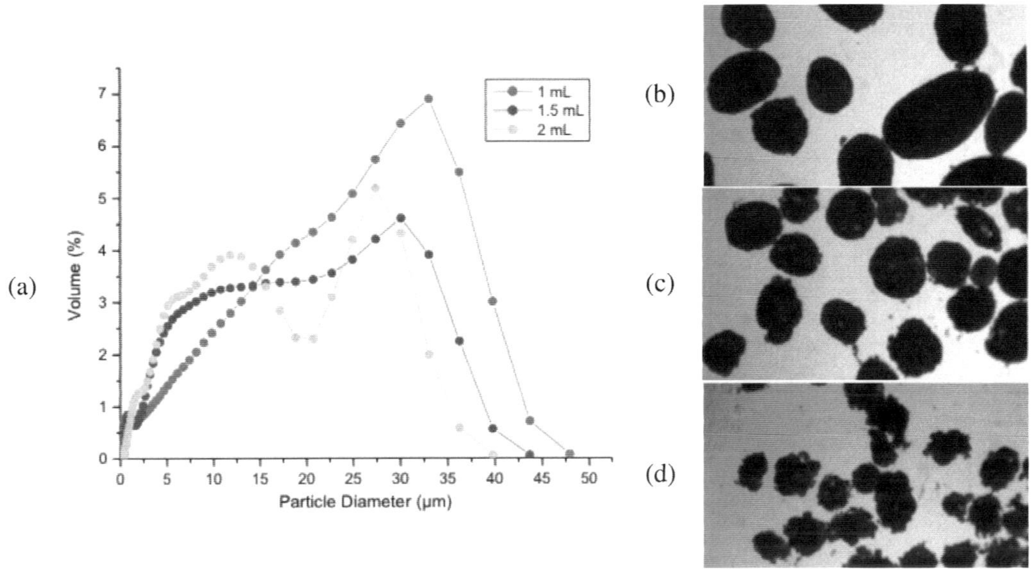

Figure 3. The results of Nonidet SF-5 volume variation (a) The PSA graph of microspheres diameter size and its distribution. Optical microscope images of microspheres (b) 1 mL, (c) 1.5 mL, (d) 2 mL. (13.5x).

3.2. Variations in preparation of microspheres

Microspheres were made by a modified w/o emulsion solvent evaporation method. PLA and PCL were dissolved in a volatile organic solvent (dichloromethane). The solution is then emulsified in other solvent to form droplets and dispersed in continuous phase to precipitate microparticles called microspheres [6]. The emulsion is not a stable system thermodynamically because of the unfavorable interactions between hydrophobic and hydrophilic molecules in the interface [11]. Nonidet SF-5 was applied to stabilize the emulsion between the organic and water phase to prevent coalescence of the droplet, which later can damage the formation of microspheres.

The functional groups of microspheres were characterized by FTIR spectrophotometer to see the interaction between PLA and PCL in microsphere. Poly(lactic acid) and polycaprolactone are aliphatic polyesters with identical structures. The C=O, C-O-C, and C-C peaks are obviously shown at (1754, 1175, and 1200) cm^{-1} in the IR spectra [12]. The IR spectrum of microspheres is shown in figure 2b. Gokalp *et al.* [13] have been successfully synthesized PCL through ring opening polymerization. IR spectra of PCL in their results revealed the same absorption peaks (CH sp^3, C=O, and C-O at (2945.32, 1720.78, and 1238.73) cm^{-1}, respectively) as PLA. The microspheres spectrum was also compared with PCL spectrum from the result of their research. The results showed that the C-H sp^3, C=O, and C-O peaks in PLA and PCL spectra reappeared on the microsphere spectrum at (2933, 1756, and 1192) cm^{-1}, respectively, and there were no new absorption peaks on it. Thus, it indicates that the functional groups in the microspheres are a combination of constituent components of PLA and PCL, so it can be concluded that the interaction of PLA and PCL in microspheres is physical interaction.

3.2.1. Volume of Nonidet SF-5. Microspheres size and its distribution that have been measured from the variations of Nonidet SF-5 volume can be seen in figure 3a. The result indicated that the addition of Nonidet SF-5 volume affected the microspheres size and its distribution. The microspheres size of 34.58 μm, 31.50 μm, and 28.70 μm was obtained at 1 mL, 1.5 mL, and 2 mL of Nonidet SF-5, respectively. The addition of surfactant volume decreased the microspheres size. It may be caused by the formed droplet on the emulsion process becomes smaller along with the addition of surfactant volume. The broader size distribution concomitant the addition of Nonidet SF-5 volume was also observed in the PSA graph. The excessive volume of Nonidet SF-5 can cause the microspheres to become more brittle, so it will decrease the stability of the formed microspheres and affect the size distribution. The most uniform microspheres size was obtained from the formula that used 1 mL of Nonidet SF-5.

(a)

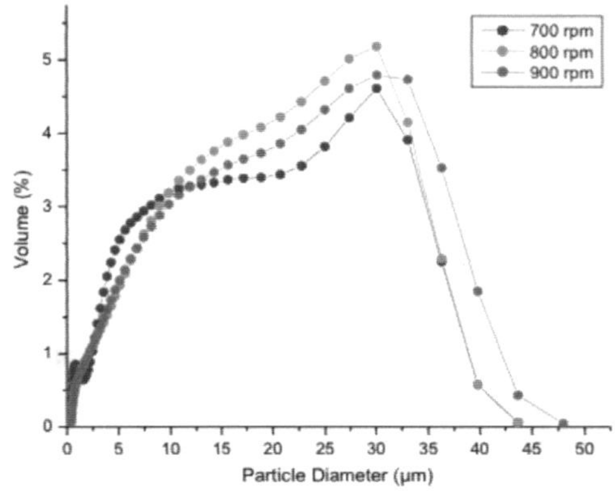

(b)

(c)

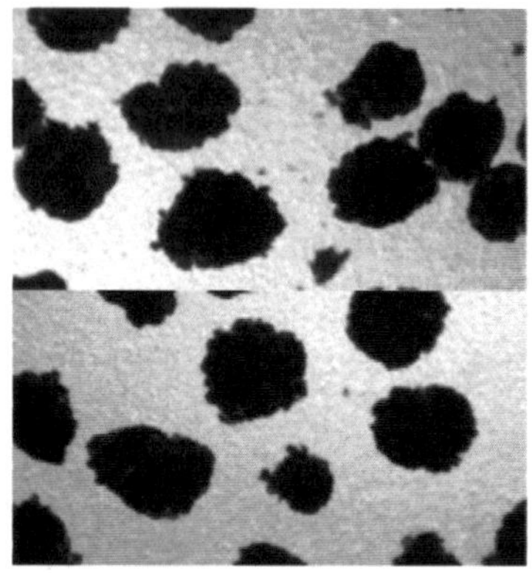

Figure 4. The results of emulsion stirring speed variation. (a) The PSA graph of microspheres diameter size and its distribution. Optical microscope images of microspheres: (b) 800 rpm, (c) 900 rpm (13.5x).

The microspheres shape and appearance were also observed using optical microscope. Optical microscope images for variation of Nonidet SF-5 surfactant volume are shown in figure 3b, figure 3c, and figure 3d. The results showed that the formed microsphere had more damaged and perforated along with the addition of Nonidet SF-5 volume. The use of excessive volume of Nonidet SF-5 affects the solvent diffusion rate resulting in the fragility of the microsphere during the solvent evaporation process thereby damaging the formed microspheres.

3.2.2. Emulsion stirring speed. Microspheres size and its distribution that have been measured from experiments with variations of emulsion stirring speed can be seen in figure 4a. The results indicated that the increased emulsion stirring speed did not affect the size of microspheres but affected the size distribution. It can be caused by the range of stirring speed is too near, so it does not affect the formation of droplet and the size of microspheres. The same microspheres size of 31.50 μm was obtained at 700 rpm, 800 rpm, and 900 rpm. The size distribution of microspheres at 800 rpm and 900 rpm did not differ significantly. But, if it was compared with the size distribution of microspheres at 700 rpm, the increase emulsion stirring speed produced more uniform sized microspheres. The most uniform microspheres size was obtained at 800 rpm.

Optical microscope images for variation of Nonidet SF-5 surfactant volume can be seen in figure 4b and figure 4c. The results showed there were no significant differences in the shape and appearance of the microsphere surface between the speed variations of 800 rpm and 900 rpm. However, when it was compared with the optical microscope results for emulsion stirring speed of 700 rpm in figure 3c, the increase emulsion stirring speed produced microspheres with coarser surface.

3.2.3. Dispersion stirring time. The variation of dispersion stirring time was used to see the effect of solvent evaporation time on the size of microspheres during the solidification process. The size of microspheres and its distribution from the experiment with variations of dispersion stirring time can be seen in figure 5a. The results indicated that the increased dispersion stirring time affected the microspheres size, but it did not affect the size distribution. The microspheres size of 31.50 μm, 31.50 μm, 23.81 μm, and 19.76 μm was obtained at 0.5 h, 1 h, 1.5 h and 2 h, respectively. The decrease in the size of the microspheres by increasing the dispersion time can be due to the droplet contact time with the continuous phase and dichloromethane diffusion rate to evaporate into the atmosphere. The longer the droplet is in the continuous phase; the solvent evaporation is more completed so that the size of the microsphere is getting smaller. Optical microscope images for variation of dispersion stirring time can be seen in figure 5b, figure 5c, and figure 5d. The results showed that the microspheres were not perfectly formed and had a rough surface obtained at 0.5 h. Meanwhile, the surface of the microsphere at the dispersion stirring time of 1.5 h was noticeably smoother than at the dispersion stirring time of 0.5 h. While at dispersion stirring time of 2 h, the surface of the microspheres became rougher and more perforated compared to the dispersion stirring time of 1.5 h. The duration of the diffusion contact with the droplet affects the microsphere solidification process in

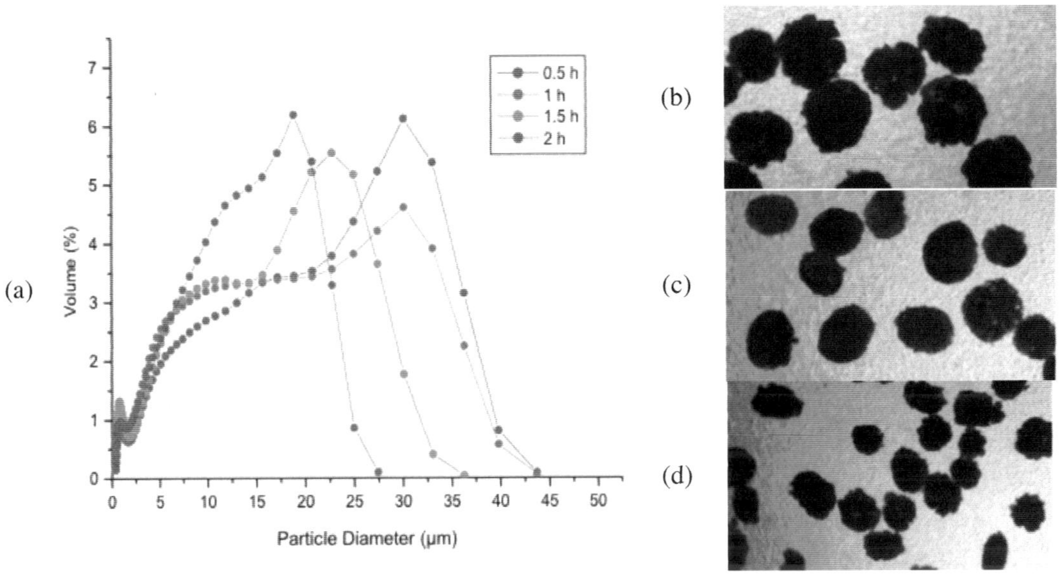

Figure 5. The results of dispersion stirring time variation. (a) The PSA graph of microspheres diameter size and its distribution. Optical microscope images of microspheres: (b) 0.5 h, (c) 1.5 h, (d) 2 h (13.5x)

the continuous phase. When the droplets are in continuous phase in less time, the microsphere solidification process is incomplete and the solvent residue still entrapped in the droplet [14]. Whereas, if the contact time of the continuous phase with microspheres and the stirring time of dispersions are too long, it can cause the microsphere to become more brittle and easily damaged so that holes and rough surfaces formed on the microspheres.

4. Conclusions

Poly(lactic acid) has been successfully synthesized using a modified direct polycondensation method. The microspheres were formed based on polyblend of PLA and PCL with physical interaction therein. The FTIR spectrum of microspheres showed that the functional groups in the microspheres are a combination of constituent components of PLA and PCL. The results of variations in preparation of microspheres showed that the volume of Nonidet SF-5 (a surfactant) affected the microspheres size, size distribution, and structure of the microspheres surface. The most uniform microspheres size was generated from the formula that used 1 mL of Nonidet SF-5. The stirring speed of emulsion affected the size distribution and structure of the microspheres surface. The most uniform microspheres size was obtained at 800 rpm. The stirring time of the dispersion affected the size and structure of the microspheres surface.

Acknowledgements

This research was fully funded by Hibah Publikasi Terindeks Internasional Tugas Akhir Mahasiswa Universitas Indonesia (Hibah PITTA UI) with contract No.233/UN2.R3.1/PPM.00/2018.

References

[1] Freiberg S and Zhu X X 2004 *Int. J. Pharm.* **282** 1–18
[2] Xiao L, Wang B, Yang G and Gauthier M 2012 *Poly(Lactic Acid)-Based Biomaterials: Synthesis, Modification and Applications, Biomedical Science, Engineering and Technology* ed D N Ghista (London: IntechOpen Ltd)
[3] Zhao H and Zhao G 2016 *J. Mech. Behav. Biomed. Mater.* **53** 59–67
[4] Porjazoska A, Karal-Yilmaz O, Kayaman-Apohan N, Cvetkovska M and Baysal B M 2004 *Croat. Chem. Acta* **77** 545–51
[5] Gunatillake P A, Adhikari R and Gadegaard N 2003 *Eur. Cells Mater.* **5** 1–16
[6] Kemala T, Budianto E and Soegiyono B 2012 *Arab. J. Chem.* **5** 103–8

[7] Zhou W, Qian H, Yan L, Luo D, Xu N and Wu J 2015 *Mater. Lett.* **152** 293–7
[8] Li M, Rouaud O and Poncelet D 2008 *Int. J. Pharm.* **363** 26–39
[9] Achmad F, Yamane K, Quan S and Kokugan T 2009 *Chem. Eng. J.* **151** 342–50
[10] Choksi N and Desai H 2017 *Int. J. Appl. Chem.***13** 377–84
[11] Weiss J 2005 Emulsion stability determination *Handbook of Food Analytical Chemistry* vol 1, ed R E Wrolstad *et al.* (Hoboken: John Wiley and Sons) Chapter D3.4 pp 591–607
[12] Chen C C, Chueh J Y, Tseng H, Huang H M and Lee S Y 2003 *Biomaterials* **24** 1167–73
[13] Gokalp N, Ulker C and Guvenilir Y A 2016 *J. Polym. Mater.* **33** 87–100
[14] Tiwari S and Verma P 2011 *International Journal of Pharmacy and Life Sciences* (*IJPLS*) **2** 998–1005

Facile synthesis and growth mechanism of mesoporous anatase titanium dioxide nanoparticles *via* sol-gel method

G Rahmaningsih, J Gunlazuardi and M Khalil

Department of Chemistry, Faculty of Mathematics and Natural Sciences (FMIPA), Universitas Indonesia, Kampus UI Depok, Depok 16424, Indonesia

Corresponding author's e-mail: mkhalil@sci.ui.ac.id

Abstract. An investigation on the synthesis of highly monodispersed mesoporous anatase titanium dioxide (TiO_2) nanoparticles has been reported in this work. The mesoporous nanoparticles were obtained by a facile sol-gel method where titanium isopropoxide (TTIP) was used as the titanium precursor in basic solution. Here, the effect of reaction time was studied to study the growing mechanism of TiO_2 nanocrystal. The as-synthesized nanoparticles were characterized with X-Ray Diffraction (XRD), Fourier Transform Infrared (FTIR), Brunauer-Emmett-Teller (BET) Surface Area Analyzer, and High Resolution Tunneling Electron Microscopy (HR-TEM). Based on the result, it is found that a highly monodispersed mesoporous anatase TiO_2 nanoparticles with large surface area could successfully be synthesized. It is also believed that the morphology of as-synthesized particles could easily be controlled by the reaction time. Results show that uniform mesoporous TiO_2 nanoparticles with perfect spherical morphology could be obtained when the reaction was performed in 3 hours of reaction time.

1. Introduction

Over the last several decades, nanostructured titanium dioxide (TiO_2) has attracted many interests due to its excellent physical and chemical properties [1]. Studies have reported that TiO_2 nanoparticles with high surface area could potentially be useful in many different applications such as photocatalysis, electronics, sensing devices, and renewable energy generation [2–4]. It is believed that the ability to control particle size distribution, morphology, crystallographic structures, and phases, as well as porosity of TiO_2 nanoparticles has been the important key factor in obtaining the desired intrinsic properties of such material [5]. For instance, Jang and co-workers have reported that particle size distribution and phase composition of TiO_2 nanoparticles had a significant influence in its photocatalytic activity against various types of dyes [6]. It is believed that the smaller the particle size distribution of TiO_2 nanoparticles, the better the photocatalytic activity [5–6]. Therefore, many efforts have been carried out to develop a facile method for the synthesis of TiO_2 nanostructures, which allows one to control the aforementioned properties.

Typically, there are three main crystalline phases of TiO_2, i.e. rutile, anatase, and brookite. Among these three crystalline phases, anatase has attracted many attentions due to its high thermodynamic stability and ability to show an excellent photocatalytic activity in comparison to other phases [5]. In industry, TiO_2 is commonly manufactured by treating ilmenite mineral ($FeO•TiO_2$) at acidic condition to produce titanyl sulfate ($TiOSO_4$), which further hydrolyzed to anatase or/and rutile crystals in hydrothermal condition [7]. Nevertheless, fabrication of large-scale TiO_2 using such method still posts several disadvantages such as the inability to control particle distribution and morphology of the crystal, the potential of obtaining mixed crystal phases, relatively high cost, and not environmentally friendly. Hence, an alternative synthetic method for the fabrication of large-scale TiO_2 nanoparticles that enables full control of various parameters such as particle size and crystalline phase is highly desired.

According to literature, the sol-gel method is considered as one of the most frequently used technique to prepare TiO_2 nanoparticles with full control in particle size distribution and crystalline phase [6]. The precipitate from the sol-gel method is typically in the form of amorphous TiO_2. However, it can further be transformed into various crystal phases using calcination process. In this work, a facile sol-gel method at basic condition was developed in order to prepare mesoporous anatase TiO_2 nanoparticles where titanium(IV) isopropoxide (TTIP) was used as the titanium precursor. Here, the effect of reaction time was also studied towards the particle size and crystalline phase of the product in order to provide a better insight into the particle's growth mechanism.

2. Materials and methods

2.1. Materials

TTIP (CAS Number 546-68-9) was used as the starting material and purchased from Sigma-Aldrich (purity: 97 %). NH_4OH solution (28–30 % NH_3 in H_2O) (CAS Number 1336-21-6) and ethanol absolute (CAS Number 64-17-5) were also purchased from Sigma-Aldrich and used in the synthesis of TiO_2 nanoparticles.

2.2. Synthesis of mesoporous anatase TiO_2 nanoparticles

Mesoporous anatase TiO_2 nanoparticles were synthesized according to a method reported by Li and co-workers [8]. However, TTIP was used as titanium precursor instead of tetrabutyl titanate. In this work, the nanoparticles were synthesized by dropwise addition of 0.75 mL TTIP into a mixture of 100 mL absolute ethanol and 0.3 mL NH_4OH. The mixture was then vigorously stirred at 45 °C for (3, 6, 12, and 24) hours. The resulting suspension was then centrifuged at 6000 rpm and the obtained white precipitates were collected and dried in an oven at 60 °C for overnight. Finally, the resulting white powder was then calcinated at 450 °C for 2 hours and used for further investigation.

2.3. Characterization

To investigate the physical and chemical properties of the as-synthesized mesoporous anatase TiO_2 nanoparticles, several characterization techniques were carried out. To study the crystallographic structure of the as-synthesized nanoparticles, X-ray Diffraction (XRD) analysis was done using PANanalytical X'Pert Pro MPD (PANanalytical B.V., Almelo, Netherlands) equipped with fast detector X'Celerator. The measurement was performed using Cu Kα radiation as the X-ray source. Meanwhile, Fourier-transform Infrared spectroscopy (FTIR) analysis was done using Shimadzu IR Prestige 21. Brunauer-Emmett-Teller (BET) surface area and adsorption-desorption isotherm were measured using QUADRASORB evo (Quantachrome Instruments) equipped with FloVac Degasser. Before BET analysis, the samples were degassed at 300 °C for 3 hours. Finally, micrographic images of the nanoparticles were collected using TECNAI G2 Spirit Twin High Resolution-Transmission Electron Microscope (HR-TEM).

3. Results and discussion

3.1. Synthesis of mesoporous anatase TiO_2 nanoparticles

In this study, the effect of reaction time during sol-gel synthetic method to the formation of anatase crystal phase was firstly investigated. Figure 1 shows both XRD pattern and FTIR spectra of the as-synthesized TiO_2 nanoparticles at different reaction time. Based on the result, it is demonstrated that all of the obtained white powder after calcination process. Based on the result, it can be observed that the reaction time has contributed no effect on the type of crystalline phase of the reaction product. All the Bragg peaks in XRD diffractogram suggest that the main crystal phase of the reaction product was anatase and no other peak was obtained for both rutile or brookite phases (JCPDS, No. 21-1272) (figure 1a). Furthermore, peak broadening of (101) anatase phases also suggests an indication that very small grain particles were formed during the reaction. It is believed that, even though rutile is widely known as the most thermodynamically stable crystal phase of TiO_2 in bulk, anatase is the preferred phase when the surface is of TiO_2 becomes very large due to the reduction of grain size into nanoscale [9]. Furthermore, the formation of TiO_2 is also supported by the result from FTIR analysis. Based on the result in figure 1b, it is shown that all reaction products synthesized at different reaction time

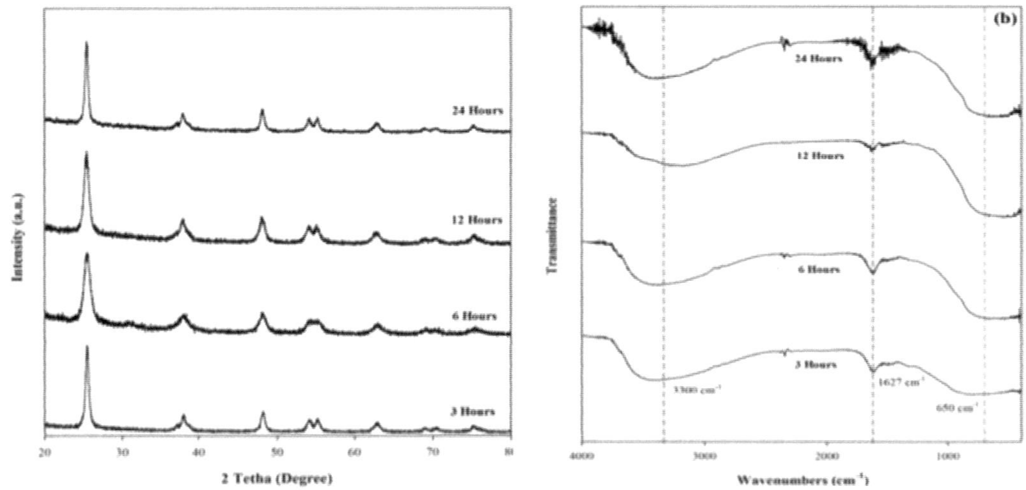

(a) (b)

Figure 1. (a) XRD pattern and (b) FTIR spectra of the as-synthesized TiO_2 nanoparticles at different reaction time.

rendered a very similar IR spectrum containing a strong IR peak at 650 cm^{-1} specifics for the lattice vibration of Ti-O-Ti stretching. In addition, the peaks at 3300 cm^{-1} and 1627 cm^{-1} could also be ascribed for O-H bending and stretching modes, respectively, in surface OH terminal groups or water molecules [8].

Furthermore, the surface physical properties of the as-synthesized TiO_2 nanoparticles were also studied by investigating their adsorption-desorption isotherm using BET surface area analysis. Figure 2 shows the adsorption-desorption isotherm of the as-synthesized TiO_2 obtained different reaction time. Based on the result, it can be observed that all reaction products follow type V of N_2 adsorption-desorption isotherm, which indicates the mesoporous feature of the TiO_2 nanoparticles [5,10]. In addition, the hysteresis loop of the isotherm seems to shift towards higher relative pressure when the sol-gel reaction was carried out more than 3 hours. This suggests that the surface morphology of the obtained mesoporous TiO_2 nanoparticles becomes irregular with the increment of sol-gel reaction time. Moreover, it is also reasonable to believe that hysteresis loop could be attributed to the total contribution of both inter-particles and intra-particles pores due to the mesoporous and nanoparticles co-exist features [5].

In addition, BET surface area and Barret-Joyner-Halenda (BJH) desorption parameters, i.e. pore volume and pore radius were found to be increasing with reaction time (table 1). However, when the reaction was carried out for 24 hours, both BET surface area and BJH parameters were decreased. It is believed that this phenomenon occurs mainly due to the presence of both mesoporous and nanoparticles features [5]. As a result, this coexists two features would result in the formation of bimodal pore distribution consisting of smaller inter-particles and larger intra-particle pores. At low sol-gel reaction time (less than 12 hours), it is believed that the particles are in a regular morphologic feature causing the smaller inter-particles pores becomes predominant. Whereas, longer reaction time would render a more irregular particle assembly causing larger intra-particles pore as the major feature rather than smaller inter-particle pore.

3.2. Growth mechanism

To further study the effect of sol-gel reaction time towards the growth mechanism of the mesoporous anatase TiO_2 nanoparticles, investigation of micrographic images obtained from HR-TEM was also carried out. Figure 3 presents the TEM images of the as-synthesized mesoporous TiO_2 nanoparticles obtained at various reaction time. Based on the result, it is observed that the morphology of the as-prepared mesoporous TiO_2 nanoparticles was significantly affected by the duration of the sol-gel reaction time. Observation under an electron microscope revealed that perfectly spherical particle

Table 1. BET surface area and BJH desorption parameters of the as-synthesized TiO_2 nanoparticles.

Reaction time (h)	BET surface area (m^2/g)	BJH desorption parameters	
		Pore volume (cc/g)	Pore radius (Å)
3	42.197	0.067	19.313
6	99.281	0.229	36.115
12	108.627	0.240	31.195
24	72.427	0.104	19.376

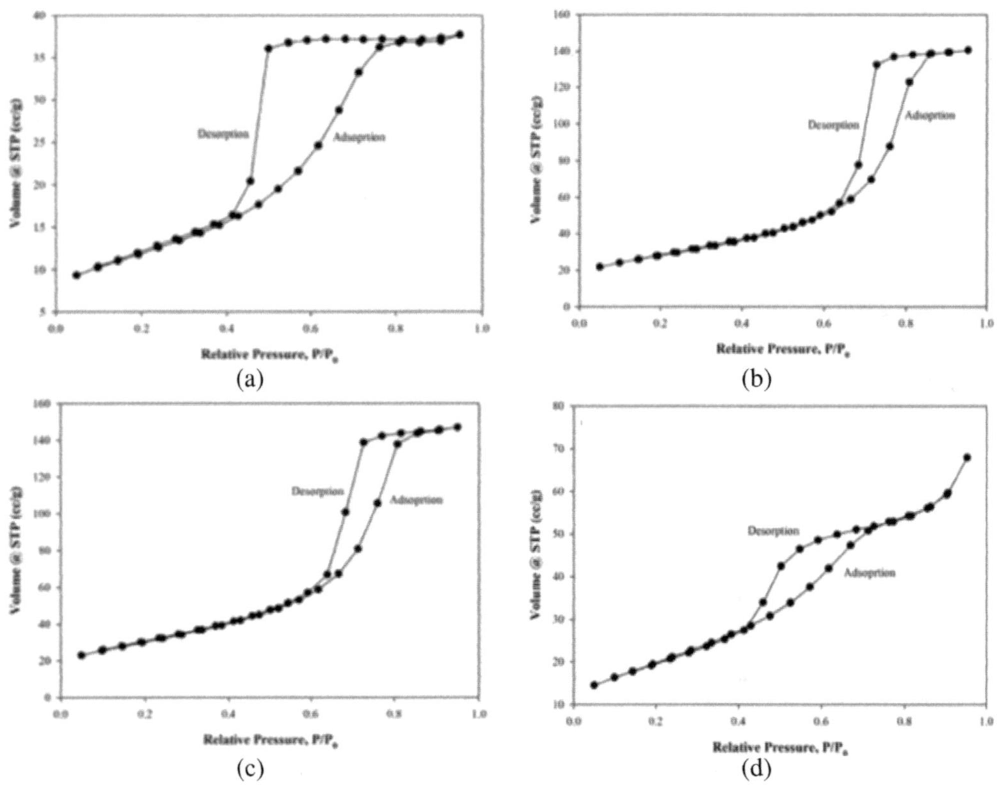

Figure 2. N_2 adsorption-desorption isotherm of the as-synthesis TiO_2 nanoparticles from sol-gel method at (a) 3, (b) 6, (c) 12, and (d) 24 hours of reaction time.

morphology could be obtained when the reaction was carried out less than 6 hours (figure 3a and figure 3b). From the result, the average diameter of the as-synthesized mesoporous TiO_2 nano-sphere particles obtained from 3 h and 6 h sol-gel reaction time were found to be ~457.55 nm and ~649,59 nm, respectively. It is believed that these TiO_2 nano-sphere particles were formed by self-assembly of many small TiO_2 nanoparticles seeds (ø = ~5-10 nm) during sol-gel reaction. As a result, the longer the reactions time the bigger the size of the obtained nanoparticles. Furthermore, this self-assembly is also thought to be responsible for the formation of intra-particle pores, which contributed to the overall mesoporous feature of the as-synthesized TiO_2 nanoparticles.

Nevertheless, TEM images also revealed that a more uniform particle size distribution was obtained when the reaction was carried out for 6 hours instead of 3 hours. This phenomenon could be understood by assuming that, at 6 hours sol-gel reaction time, the LaMer's crystal growth mechanism

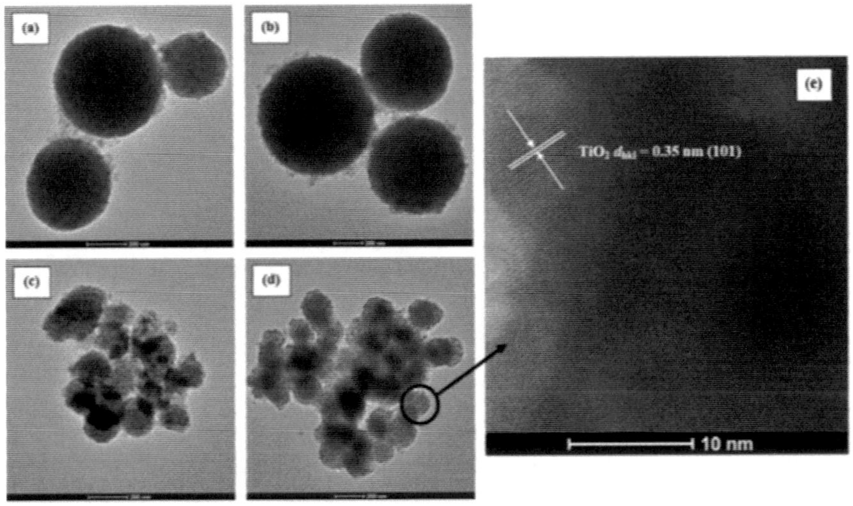

Figure 3. TEM images of the as-synthesized mesoporous TiO_2 nanoparticles obtained at various sol-gel reaction time, i.e. (a) 3, (b) 6, (c) 12, and (d) 24 hours (scale bar: 200 nm); and (e) HR-TEM images of TiO_2 obtained from 24 hours sol-gel reaction.

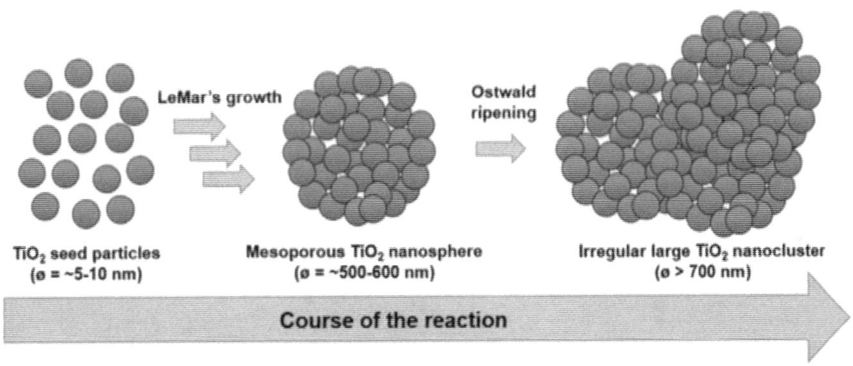

Scheme 1. The proposed schematic growth mechanism of anatase TiO_2 nanoparticles during the course of sol-gel reaction.

has already passed the rapid self-nucleation process and reached the steady-state crystal growth process [11]. As a result, the remaining seed particles would preferably coalesce onto smaller particles clusters due to their higher surface energy creating a very narrow overall particle size distribution. However, when the sol-gel reaction proceeded at longer reaction time and all the remaining seed particles have been used during steady-state crystal growth, the adjacent particle clusters would further coalesce to form more irregular larger cluster particles *via* Ostwald ripening process [12]. This is proven by the fact that no specific particle morphology could be obtained when the reaction was carried out for 12 and 24 hours (figure 3c and figure 3d). Therefore, the overall proposed schematic growth mechanism of mesoporous anatase TiO_2 nanoparticles during the course of the sol-gel reaction could be illustrated according to scheme 1.

4. Conclusions

In summary, a facile method for the synthesis of mesoporous anatase TiO_2 nanoparticles has been investigated in this study. Based on the result, it is observed that the duration of sol-gel reaction had no significant effect on the formation of crystal phase of the TiO_2 nanoparticles. It is demonstrated that all reaction products could unambiguously be indexed as anatase phase regardless of how long the sol-gel reaction proceeded. The result from BET adsorption-desorption isotherm also suggests that all the reaction product follows the type V isotherm which indicates the mesoporous feature of TiO_2.

Furthermore, it is also observed that perfect spherical particles with high surface area could be obtained when the reaction was carried out for no longer than 6 hours. Meanwhile, irregular large particle clusters were obtained when the reaction was done for 12 and 24 hours. It is believed that Ostwald ripening process was responsible for the formation of these irregular large particle clusters due to the coalescence of adjacent two or more TiO_2 nano-spheres.

Acknowledgements

This work is financially supported by Universitas Indonesia through Hibah PITTA (No. 2287/UN2/R3.1/HKP.05.00/2018). We would also like to acknowledge Integrated Laboratory Research Center (ILRC) Universitas Indonesia for helping us to measure the BET analysis.

References

[1] Li W, Wang F, Feng S, Wang J, Sun Z, Li B, Li Y, Yang J, Elzatahry AA, Xia Y and Zhao D 2013 *J. Am. Chem. Soc.* **135** 18300–3
[2] Yang H, Zhang K, Shi R, Li X, Dong X, Yu Y 2006 *J. Alloys Compd.* **413** 302–6
[3] Lee S *et al.* 2010 *Chem. Mater.* **22** 1958–65
[4] Behnajady M A, Eskandarloo H, Modirshahla N and Shokri M 2011 *Desalination* **278** 10–7
[5] Peng T, Zhao D, Dai K, Shi W and Hirao K 2005 *J. Phys. Chem.* B **109** 4947–52
[6] Jang H D, Kim S K and Kim S J 2001 *J. Nanopart. Res.* **3** 141–7
[7] Sugimoto T, Zhou X and Muramatsu A 2003 *J. Colloid Interface Sci.* **259** 43–52
[8] Li W, Yang J, Wu Z, Wang J, Li B, Feng S, Deng Y, Zhang F and Zhao D 2012 *J. Am. Chem. Soc.* **134** 11864–7
[9] Zhang H and Banfield J F 2005 *Chem. Mater.* **17** 3421–5
[10] AlOthman Z A 2012 *Materials* **5** 2874–902
[11] Polte J 2015 *Cryst. Eng. Comm.* **17** 6809–30
[12] Yang H G and Zeng H C 2004 *J. Phys. Chem.* B **108** 3492–5

IOP Conf. Series: Materials Science and Engineering **763** (2020) 012047 doi:10.1088/1757-899X/763/1/012047

In vitro study of DNA-adduct formation (8-OHdG) due to exposure of benzo[a]pyrene and Co(II) through fenton-like reaction

S A Wijanarko, B Budiawan, I C Dani and S Handayani

Department of Chemistry, Faculty of Mathematics and Natural Sciences (FMIPA) Universitas Indonesia, Kampus UI Depok, Depok 16424, Indonesia

Corresponding author's e-mail: dr.budiawan@sci.ui.ac.id

Abstract. Severely increasing environmental pollution with an unhealthy pattern of life in the community can increase the source of exposure of carcinogenic substances that can cause cancer. Benzo[a]pyrene and Cobalt (II) are some examples of carcinogenic substances commonly found in the environment. Both materials can simultaneously enter the body through food or polluted air and damage the DNA. In this study, we studied the effect of B[a]P and Co(II) exposure in-vitro on 2'-deoxyguanosine. An increase in the amount of the 8-hydroxy-2'-deoxyguanosin compound is used as an indicator of the occurrence of DNA damage and a cancer risk biomarker. It was found that the synergistic effect on the combination of B[a]P and Co (II) through Fenton-like reaction is influenced by the concentration of H_2O_2. The amount of 8-OHdG that formed with the combination of B[a]P and Co (II) is 126.23 ppb. It does not vary much with the amount that resulted from B[a]P exposure alone at 136.53 ppb. Exposure of Co(II) through Fenton-like reaction alone gives the highest yield of 8-OHdG formation at 282.67 ppb. Followed by B[a]P and Co(II) at 258.98 ppb, B[a]P and H_2O_2 at 153.16 ppb.

Keywords: Benzo[a]pyrene, Cobalt(II), Fenton-Like reaction, 8-hydroxy-2'-deoxyguanosin.

1. Introduction

The high level of pollution in the environment and an unhealthy lifestyle can leads to an increasing risk of cancer. These two phenomena increase the source of exposure to chemicals that can lead to cancer, referred as carcinogenic substances. Carcinogenic is a trait of a material that can cause the emergence of cancer cells by damaging biomolecules such as DNA.

One example of the carcinogenic substance is benzo[a]pyrene (B[a]P). This compound falls under the class of polycyclic aromatic hydrocarbon formed by an incomplete combustion of organic matter. B[a]P can be easily found in environment due to pollution from burning organic waste, forest fires, and vehicle fumes. B[a]P can also be found in foods, especially from grilled red meat [1]. B[a]P as much as 18.2 +/- 0.5 ng/rod can also be found in cigarette smoke [2].

Cobalt(II) is also another carcinogenic pollutant that can be found in the environment. Since the beginning of the 20th century, mined cobalt has been used to produce metal alloys such as superalloys and magnetic alloys, as well as high-strength steel and hard-metal cemented carbides. In the late 1980s, around one-third of the cobalt employed was in the fabrication of cobalt chemicals, mainly as catalysts and pigments. As metal alloys, cobalt is also widely used in the health equipment such as braces and implants [3].

B[a]P and Cobalt (II) can produce reactive oxygen species (ROS) [4,5]. These radical molecules can induce oxidative damage to DNA and lead to cancer [6]. When DNA is damaged by ROS, it is generally found an increase in the amount of metabolite compound 8-hydroxy-2'-deoxyguanosine

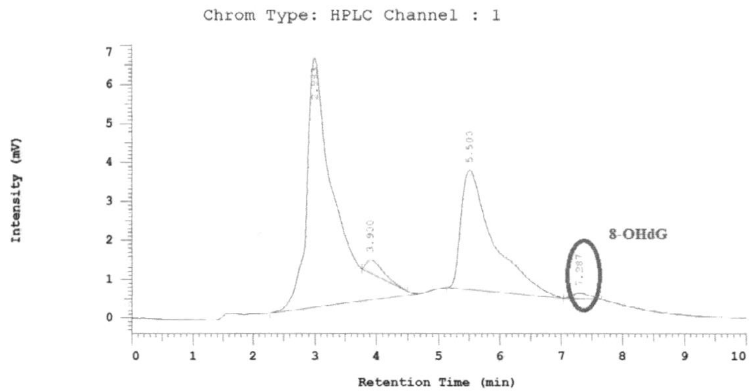

Figure 1. HPLC chromatograph of 8-OHdG that formed from the reaction of B[a]P and 2'-deoxyguanosine.

(8-OHdG). An increase in the number of 8-OHdG compounds can be used as an indicator or biomarker against an increasing risk of cancer due to exposure to carcinogenic compounds [7].

2. Materials and methods

2.1. Chemicals and reagent

2'-deoxyguanosine monophosphate, Benzo[a]pyrene, DMSO, CH_3COONa, K_2HPO_4, KH_2PO_4, NaH_2PO_4, and Na_2HPO_4 were used to make buffer (Sigma-Aldrich). H_2O_2,$Co(NO_3)_2.6H_2O$ (Merck), HCl, NaOH, and aquabidest.

2.2. Instrumentations

Incubator shaker (Julobo SW22), Sonicator and Degasser 8510 (Bronson) were used to remove gas bubble from High Performance Liquid Chromatography (HPLC) solvents, pH meter to measure buffer solution pH (ZORBAX Eclipse Plus C18, 4.6 x150 mm, 5-Micron) and HPLC (Primaide) for chromatography analysis to detect and measure 8-OHdG.

2.3. In vitro studies

The in-vitro studies were done in seven variations of carcinogenic substances exposure, incubated at 37 °C for 18 hours. 200μL of 2'-deoxyguanosine (6 ppm) was dissolved in phosphate buffer 0.01 M (pH 7.4 and 8.4), in variation one, it was reacted with 200μL of benzo[a]pyrene (60 ppm). In variation two, it was reacted with 200 μL of benzo[a]pyrene (60 ppm) and 100μL H_2O_2 (120 ppm). In variation three, reacted with 200μL benzo[a]pyrene (60 ppm) and 100μL Co(II) solution (120 ppm). In variation four, reacted with 100 μL of Co(II) solution (120 ppm) and 100μL of H_2O_2 (120 ppm). In variation five, reacted with 100 μL of Co(II) solution (120 ppm), 100 μL of H_2O_2 (120 ppm), and 200μL of ascorbic acid solution (60 ppm). In variation six, reacted with 200μL of benzo[a]pyrene (60 ppm), 100 μL of Co(II) solution (120 ppm), and 100μL of H_2O_2 (120 ppm). And in variation seven, it was reacted with 200 μL of benzo[a]pyrene (60 ppm), 100 μL of Co(II) solution (120 ppm), 100 μL of H_2O_2 (120 ppm), and 200μL of ascorbic acid solution (60 ppm).

3. Results and discussion

This study aims to provide a scientific information on the effect of combined exposure of carcinogenic substance of both B[a]P and Co(II), to 2'-deoxyguanosine on the formation of 8-OHdG. The reaction of B[a]P and 2'-deoxyguanosine form 8-OHdG can be detected by using HPLC, as seen in figure 1. B[a]P can react with H_2O_2 to form 6-hydroxy-B[a]P. This substance then reacts with O_2 to form 6-oxo-B[a]P and superoxide [8].

$$B[a]P + H_2O \rightarrow 6\text{-OH-}B[a]P \quad (1)$$

$$6\text{-OH-}B[a]P + O_2 \rightarrow O_2 + [6\text{-OH-}B[a]P]\cdot \rightarrow 6\text{-oxy-}B[a]P \quad (2)$$

Moreover, this superoxide can further reacts to form H_2O_2 and hydroxyl radical [8].

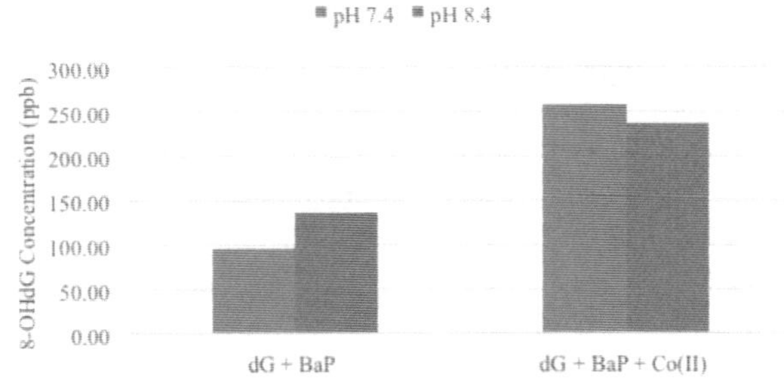

Figure 2. The effect Co(II) addition in 8-OHdG formation

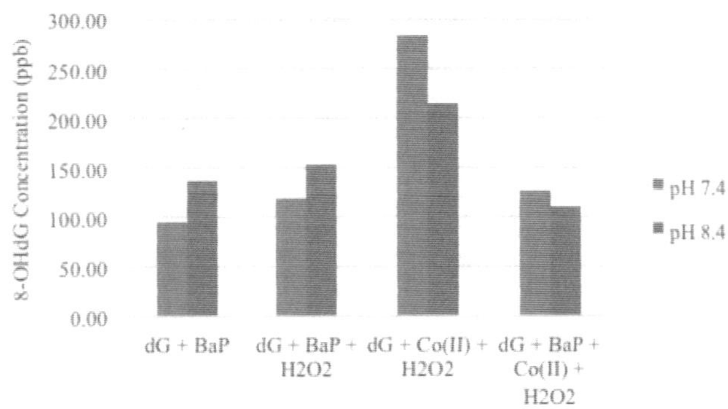

Figure 3. The effect of H_2O_2 amounts in 8-OHdG formation

$$2O_2 + 2H^{\cdot} \rightleftharpoons 2HO_2^{\cdot} \quad (3)$$

$$2HO_2^{\cdot} \rightleftharpoons H_2O_2 + O_2 \quad (4)$$

$$O_2 + H_2O_2 \rightleftharpoons HO^{\cdot} + OH + O_2 \quad (5)$$

The hydroxyl radical that formed then can oxidize 2'-deoxyguanosine and produce 8-OHdG [7]. While Co(II) reacts with H_2O_2 to produce hydroxyl radical through Fenton-like mechanism as follow:

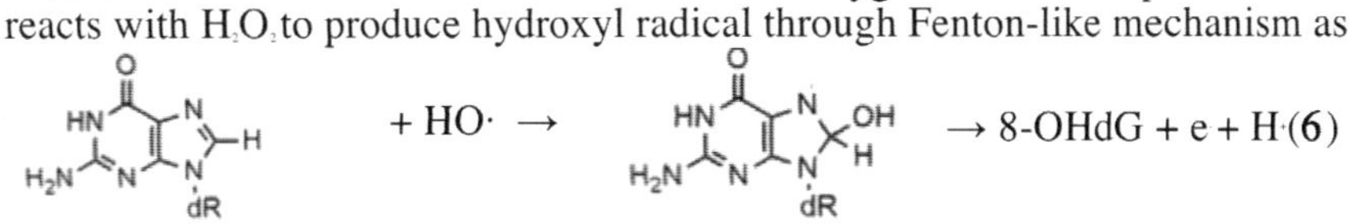

When B[a]P and Co(II) react with 2'-deoxyguanosine, there is an increase in the amount of 8-OHdG, as seen in figure 2. This might be caused by the hydroxyl radical from the reaction of Co(II) and H_2O_2. H_2O_2 is a byproduct from superoxide reaction as mentioned before.

The amount of H_2O_2 gives different results depending on the combination of the carcinogenic substances, as seen in figure 3. The increase of H_2O_2 as the result of 2'-deoxyguanosine and B[a]P reaction can increase the amount of 8-OHdG. While the increase of H_2O_2 in the combination of B[a]P and Co(II) is not followed by the increase of 8-OHdG. It is assumed that there is a synergistic effect of the combination between B[a]P and Co(II). H_2O_2 and Co(II) will react through Fenton-like mechanism effectively. When the amount of H_2O_2 is too high, it is possible that the Fenton-like mechanism also oxidizes the B[a]P[9]. This can decrease the amount of B[a]P in the solution[10]. This might explain why the amount of 8-OHdG that formed were not much different than the one formed in the reaction between 2'-deoxyguanosin and B[a]P alone.

Ascorbic acid is well known due to its antioxidant properties. In fact, many antioxidant substances also contain prooxidant properties. In this study, both antioxidant and prooxidant properties of ascorbic acid can be observed each in different conditions. In figure 4, it shows that when ascorbic acid is added to the mixed solution between 2'-deoxyguanosin and Co(II) + H_2O_2, ascorbic acid

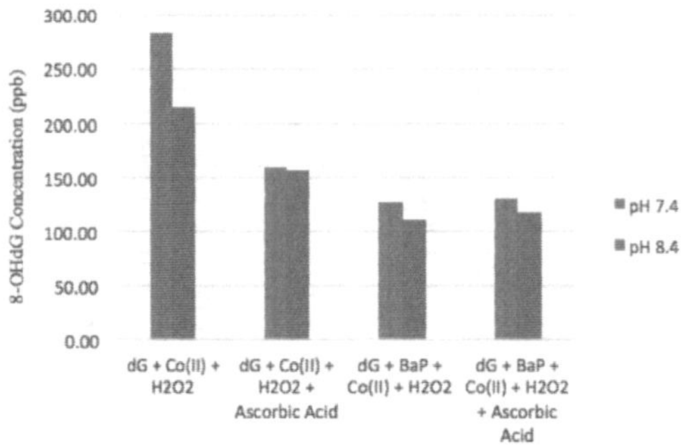

Figure 4. The effect of ascorbic acid in the formation of 8-OHdG

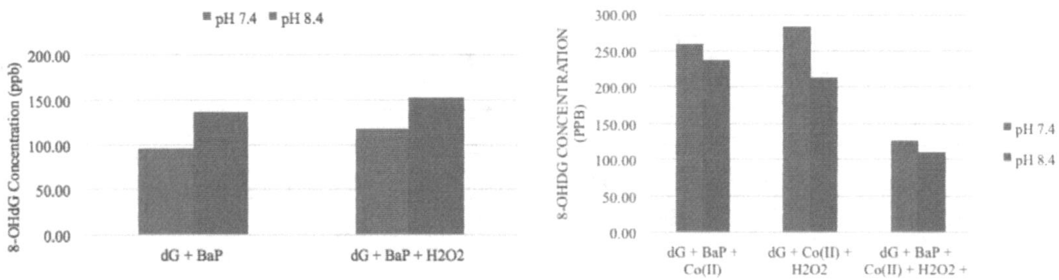

(a) (b)

Figure 5. The effect of pH (a) without and (b) with Co(II) addition in 8-OHdG formation

exhibits its antioxidant properties. While in the mixed solution that contain 2'-deoxyguanosin, Co(II) + H_2O_2, and B[a]P, ascorbic acid exhibits its prooxidant properties. Ascorbic acid can reduce some metal ion thus make them to react easier with H_2O_2 through Fenton or Fenton-like mechanism [11]. In the mixed solution between 2'-deoxyguanosin and Co(II) + H_2O_2, there is high amount of hydroxyl radical, thus there is not enough unoxidized ascorbic acid that can reduce the cobalt ions and make them to react easier with H_2O_2 through Fenton-like mechanism. It indicates that ascorbic acid exhibits its antioxidant properties. While in the mixed solution that contain 2'-deoxyguanosin, Co(II) + H_2O_2, and B[a]P, the amount of hydroxyl radical are decreased by the oxidation of both 2'-deoxyguanosine and B[a]P, thus there are enough unoxidized ascorbic acid that can reduce the cobalt ions and make them to react easier with H_2O_2 through Fenton-like mechanism. It indicates that ascorbic acid shows its prooxidant properties.

pH of the solution also gives different result at different condition. If there is no Co(II) in the solution, a high pH value can increase the amount of 8-OHdG, as seen in figure 5a. On the other hand, if there is Co(II) in the solution, a high pH can decrease the amount of 8-OHdG as seen in figure 5b. The high pH value indicates a low amount of H^+ ions. This will decrease the potential reduction of Co(II) and can turn some of Co(II) into Co(III). The decrease in the amount of Co(II) will produce a small amount of additional hydroxyl radicals through Fenton-like mechanism and it can decrease the amount of 8-OHdG.

4. Conclusions

The exposure of either B[a]P or Co (II) and H_2O_2 at 2'-deoxyguanosine leads to the formation of 8-OHdG compounds. The synergistic effect on the combination of B[a]P and Co (II) through Fenton-like reaction is influenced by the concentration of H_2O_2. At high enough concentration, the synergistic effect is worn off. If there is no Co(II) in the solution, a high pH value can increase the amount of 8-OHdG. On the other hand, if there is Co(II) in the solution, a high pH can decrease the amount of 8-OHdG. Ascorbic acid may act as both antioxidant and prooxidant. In the mixed solution between 2'-deoxyguanosin and Co(II) + H_2O_2, ascorbic acid exhibits its antioxidant properties. On the other hand,

IOP Conf. Series: Materials Science and Engineering **763** (2020) 012047 doi:10.1088/1757-899X/763/1/012047

in the mixed solution that contains 2'-deoxyguanosin, Co(II) + H_2O_2, and B[a]P, ascorbic acid shows its prooxidant properties.

Acknowledgements

We are grateful to Hibah PITTA Universitas Indonesia for funding this research and Balai K3 for providing the HPLC instrument.

References

[1] Aygün S F and Kabadayi F 2005 *Int. J. Food Sci. Nutr.* **56** 581–5
[2] Roemer E, Dempsey R, Lawness-Pyne J, Lukman S, Evans A D, Trelles-Sticken E, Wittke E and Schorp M K 2014 *Regul. Toxicol. and Pharmacol.* **70** S141–53.
[3] International Agency for Research in Cancer (IARC) 2006 *IARC Monographs on the Evaluation of Carcinogenic Risks to Humans (Lyon)* vol 8 (Lyon: International Agency for Research in Cancer) pp 48–55
[4] Kim K B and Lee B M 1997 *Cancer Lett.* **113** 205–12
[5] Simonsen L O, Harbak H and Bennekou P 2012 *Sci. Total Environ.* **432** 210–5
[6] Alberts B, Johnson A, Lewis J, Raff M, Roberts K and Walter P 2014 *Molecular Biology of the Cell* 6th ed (New York: Garland Science) pp 267–8
[7] Valavanidis A, Vlachogianni T and Fiotakis C 2009 *J. Environ. Sci. Health* C **27** 120–39
[8] Sullivan P D 1985 *Environ. Health Perspect.* **64** 283–95
[9] Homem V, Dias Z, Santos L and Alves A 2009 *Journal of Environmental and Public Health* **2009** ID 149034.
[10] Bokare A D and Choi W 2014 *J. Hazard Mater.* **275** 121–35
[11] Yen G C, Duh P D and Tsai H L 2012 *Food Chem.* **79** 307–13

IOP Conf. Series: Materials Science and Engineering **763** (2020) 012048 doi:10.1088/1757-899X/763/1/012048

An efficient one-pot multicomponent synthesis of 1,4-dihydropyridines catalyzed by guanidine hydrochloride

A H Cahyana, B Ardiansah and N R Aisy

Department of Chemistry, Faculty of Mathematics and Natural Sciences (FMIPA), Universitas Indonesia, Kampus UI Depok, Depok 16424, Indonesia

Corresponding author's e-mail: herrykim@ui.ac.id

Abstract. Guanidine hydrochloride has been successfully used in one-pot three component assembly of aromatic aldehyde, ethyl acetoacetate and ammonium acetate to produce 1,4-dihydropyridines. Optimization of reaction conditions revealed that 10 mol% guanidine at room temperature for 4 h in ethanol is the best condition for synthesis of compound **4a** in 85 % yield. Furthermore, compound **4b** was synthesized and obtained in 62 %.

Keywords: 1,4-dihydropyridine, guanidine HCl, multicomponent.

1. Introduction

In the recent decades, nitrogen-containing heterocyclic compounds especially with five- and six-membered ring have displayed powerful pharmacological activities [1]. This fact has influencing the development of new reaction protocols to produce diversity-oriented products through an effective synthetic scheme [2]. Multicomponent reactions (MCRs) are one of the most useful tools in organic synthesis that provide quick access to heterocyclic frameworks [3]. They have attractive features such as high atom economy, time and energy efficiency, practical and environmentally friendly [4,5]. Up to now, many organic reactions have been performed using MCRs strategy, such as Biginelli condensation [6], Kabachnik-Field reaction [7], Knoevenagel-induced domino reaction [8] and 1,3-dipolar cycloaddition based MCRs [9-12]. Therefore, the usefulness of MCRs has enamored a considerable interest from synthetic chemists.

1,4-Dihydropyridine derivatives (DHPs) are widely known to show various biological significances [13]. They have shown antimicrobial [14], antidiabetic [15], anticancer [16] and anti-inflammatory [17] activities. Successful development of 1,4-DHPs has been proven by the availability of some commercial drugs containing 1,4-DHP moieties for treatment of hypertension and cardiovascular diseases, such as nifedipine, felodipine and amlodipine [18]. Because of their widespread of biological activities, numerous strategies have been improved from the classical Hantzsch condensation, such as using microwave irradiation [19], trichloroisocyanuric acid in water media [20], visible light in ethanol [21] and melamine trisulfonic acid [22]. However, some of the reported methods suffer from low product yields, use of excessive or harmful organic solvents, or long reaction time requirement. In continuation of our effort to develop more efficient synthetic protocols for MCRs, herein we reported the use of catalytic amount of guanidine HCl in 1,4-DHPs synthesis.

2. Materials and methods

2.1. General

All chemicals were bought from commercial suppliers, such as Sigma Aldrich and Merck. Thin-Layer Chromatography (TLC) analysis was performed on silica gel Si-Gel 60 F_{254} coated on alumina sheet.

Table 1. Optimization of reaction conditions[a] in 1,4-DHP synthesis

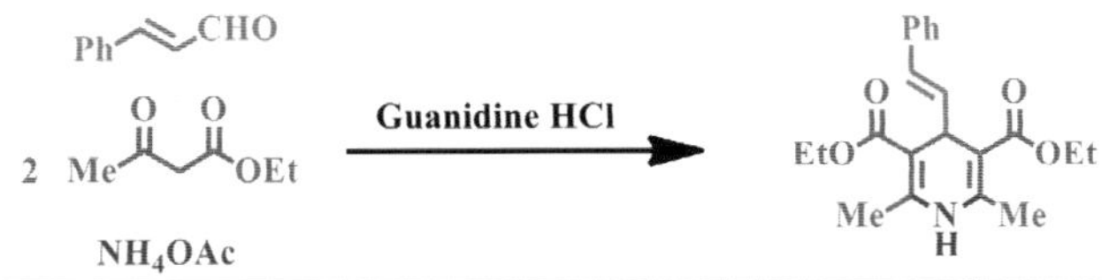

Entry	Catalyst (mol%)	T (°C)	Time (h)	Solvent	Yield[b] (%)
1	-	rt	3	Ethanol	38
2	2.5	rt	3	Ethanol	58
3	5	rt	3	Ethanol	61
4	10	rt	3	Ethanol	77
5	20	rt	3	Ethanol	68
6	10	50	3	Ethanol	79
7	10	60	3	Ethanol	76
8	10	70	3	Ethanol	56
9	10	rt	0.5	Ethanol	45
10	10	rt	1	Ethanol	48
11	10	rt	2	Ethanol	47
12	10	rt	4	Ethanol	85
13	10	rt	3	Water	11
14	10	rt	3	-	51

[a]Cinnamaldehyde (1 mmol), ethyl acetoacetate (2 mmol), ammonium acetate (1 mmol)
[b]Isolated Yield

Maximum absorption of products was measured on Shimadzu 2450 UV-Vis spectrophotometer. Fourier Transform Infrared (FTIR) spectra were recorded on Shimadzu Prestige-21 spectrophotometer. Confirmation of organic product was done by Gas Chromatography and Mass Spectrometry (GC-MS).

2.2. Synthesis of 1,4-dihydropyridines using guanidine HCl

In 20 mL round-bottom flask, cinnamaldehyde (1a, 1 mmol), ethyl acetoacetate (2, 2 mmol), ammonium acetate (3, 1 mmol) in ethanol (2 mL) were stirred along with guanidine hydrochloride (2.5 mol %) at room temperature for 3 h. Solvent was volatilized, then crude result was refined through crystallization from hot ethanol to afford (E)-diethyl-2,6-dimethyl-4-styryl-1,4-dihydropyridine-3,5-dicarboxylate (4a) in 48 % yield (table 1, entry 2). To obtain optimized conditions, temperature, time, solvent, and catalyst amount were varied. Compound 4b was also synthesized using benzaldehyde instead of cinnamaldehyde under optimized protocol.

3. Results and discussion

The results of optimization of reaction conditions are given in table 1. Optimization of reaction conditions was performed using cinnamaldehyde, ethyl acetoacetate and ammonium acetate as substrates and subjected for one-pot three component reaction. For control experiment, we conducted the synthesis in the absence of guanidine hydrochloric acid as catalyst. As expected, low yield (38 %) was obtained (table 1, entry 1). Then, the reaction was attempted with the presence of 2.5 to 20 mol% of catalyst in ethanol at room temperature.

The desired product reached 77 % when 10 mol% of catalyst used. Unfortunately, using 20 mol% of catalyst was ineffective with yield of product of 68 % only. It is noteworthy that increasing temperature did not give better yield. When the reaction was allowed to stir for longer time (4h) at room temperature in ethanol in the presence of 10 mol% catalyst, compound 4a was isolated in 85 %. To our delight, this is the best result among the screened conditions. We have also synthesized

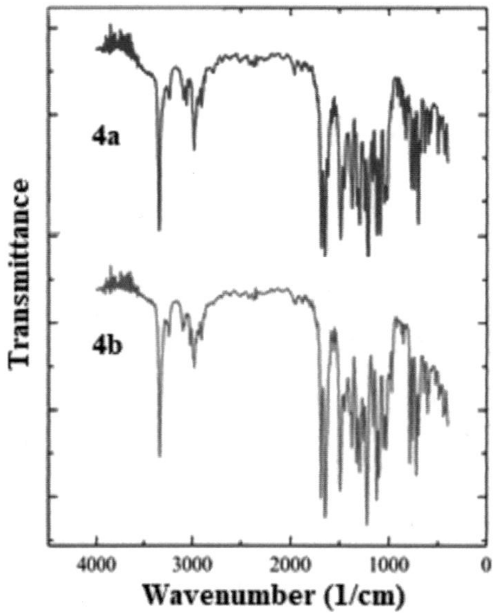

Figure 1. FT-IR spectra of compound 4a and 4b

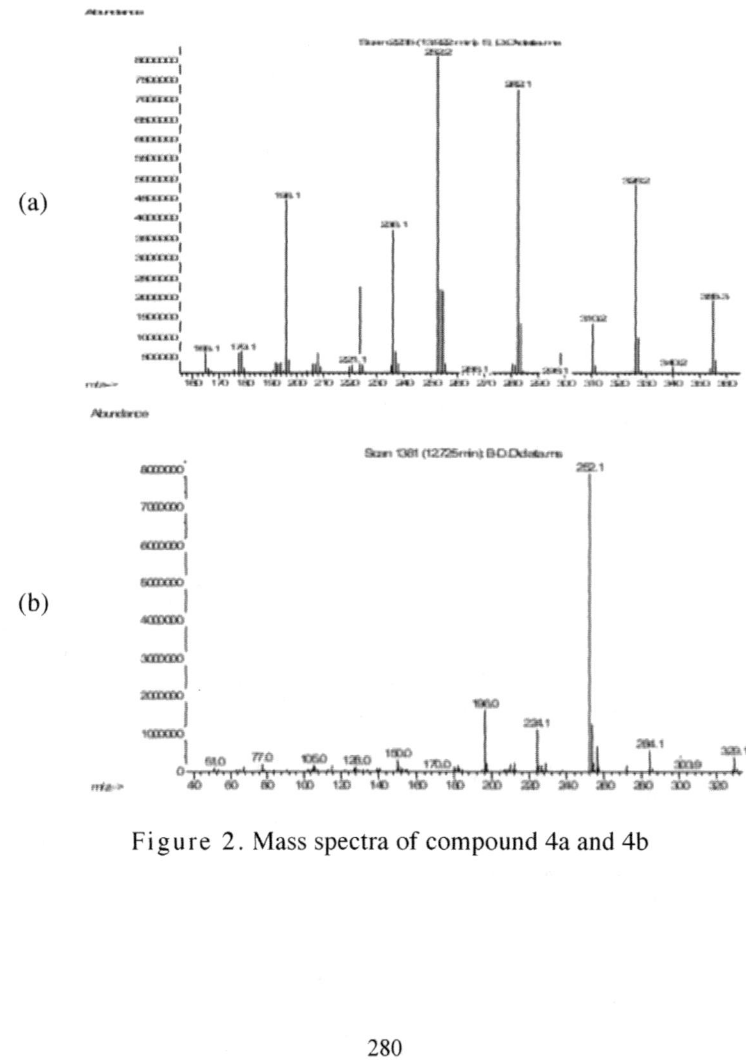

Figure 2. Mass spectra of compound 4a and 4b

compound 4b using benzaldehyde, ethyl acetoacetate and ammonium acetate under the best conditions. The yield of this product was found to be 62 %.

FTIR spectra of compound 4a and 4b are depicted in figure 1. These compounds have similar absorption characteristics. A sharp peak appeared at 3347 cm^{-1} indicates the presence of secondary amine group. Several absorptions around 3017–3030 cm^{-1} are due to olefinic and aromatic C-H having sp^2 hybridization. The band around 2850–2950 cm^{-1} is corresponded to C-H sp^3 stretching vibration. Meanwhile, sharp absorption at around 1639-1687 cm^{-1} is caused by the stretching vibration of C=O as well as C=C bond. Compound 4a and compound 4b have maximum absorption at 353 and 354 nm, respectively.

Confirmation of product structure was done by gas chromatography-mass spectrometry (figure 2). Compound 4a with relative molecular formula of $C_{21}H_{25}NO_4$ is appeared as single peak in GC chromatogram (retention time of 13.94 min); and has m/z value of 355.3 after analysis by mass spectrometer. Compound 4b has m/z value of 329.1, appropriate with molecular formula of $C_{19}H_{23}NO_4$.

4. Conclusions

This work highlighted the use of guanidine as an organic basic catalyst for an efficient synthesis of 1,4-dihydropyridines by the method of multicomponent reaction. Easy work-up and mild condition are the advantages of this method. Compound 4a and 4b were obtained in 85 % and 62% yield, respectively.

Acknowledgements

All authors gratefully acknowledge Universitas Indonesia for supporting this research through PITTA Grant 2018 with contract No. 234/UN2.R3.1/PPM.00/2018.

References

[1] Naik T R R and Shivashankar S A 2016 *Tetrahedron Lett.* **57** 4046–9
[2] Tewari N, Dwivedi N and Tripathi R P 2004 *Tetrahedron Lett.* **45** 9011–4
[3] Davarpanah J, Ghahremani M and Najafi O 2019 *J. Mol. Struct.* **1177** 525–35
[4] Haji M 2016 *Belstein J. Org. Chem.* **12** 1269–301
[5] Wender P A 2014 *Nat. Prod. Rep.* **31** 433–40
[6] Puripat M, Ramozzi R, Hatanaka M, Parasuk W, Parasuk V and Morokuma K 2015 *J. Org. Chem.* **80** 6959–67
[7] Keglevich G and Bálint E 2012 *Molecules* **17** 12821–35
[8] Liu P, Chen D, Zhang S, Zhou R, Ren W, Ouyang L and He G 2015 *Curr. Org. Synth.* **12** 88–94
[9] Rajesh S M, Bala B D and Perumal S 2012 *Tetrahedron Lett.* **53** 5367–71
[10] Mali P R, Shirsat P K, Khomane N, Nayak L, Nanubolu J B and Meshram H M 2017 *ACS Comb. Sci.* **19** 633–9
[11] Mancebo-Aracil J, Nájera C and Sansano J M 2013 *Chem. Commun.* **49** 11218–20
[12] Gayen B, Banerji A and Dhara K 2016 *Synth. Commun.* **46** 293–308
[13] Bade T S, Ebrahimi H P, Alsalim T A, Titinchi S J J, Abbo H S, Bolandnazar Z and Ebrahimi A 2017 *J. Mol. Struct.* **1138** 136–48
[14] Mehta P and Verma P 2013 *Journal of Chemistry* **2013** ID 865128
[15] Briede J, Stivrina M, Vigante B, Stoldere D and Duburs G 2008 *Cell Biochem. Funct.* **26** 238–45
[16] Razzaghi-Asl N, Miri R and Firuzi O 2016 *Iranian Journal of Pharmaceutical Research* **15** 413–20
[17] Tale R H, Rodge A H, Hatnapure G D, Keche A P, Patil K M and Pawar R P 2013 *Medicinal Chemistry Research* **22** 1450–5
[18] Xu L, Li D, Tao L, Yang Y, Li Y and Hou T 2016 *Mol. BioSyst.* **12** 379–90
[19] Ladani N K, Mungra D C, Patel M P and Patel R G 2011 *Chinese Chem. Lett.* **22** 1407–10
[20] Roknaddini M and Sheikhhosseiny E 2016 *Scientia Iranica* C **23** 2756–61
[21] Ghosh S, Saikh F, Das J and Pramanik A K 2013 *Tetrahedron Lett.* **54** 58–62
[22] Aswin K, Logaiya K, Sudhan P N and Mansoor S S 2012 *J. Taibah Univ. Sci.* **6** 1-9.

Gold nanoparticles supported on Al_2O_3 using *Tabebuia aurea* leaf extract and catalytic properties for methylene blue reduction

A Maulidya, Y Yulizar, T Utari and D O B Apriandanu

Department of Chemistry, Faculty of Mathematics and Natural Sciences (FMIPA), Universitas Indonesia, Kampus UI Depok, Depok 16424, Indonesia

Corresponding author's e-mail: yokiy@ui.ac.id

Abstract. In this research, we reported a cost-effective and an environmentally friendly technique for synthesis of gold nanoparticles supported on alumina (Al_2O_3/AuNPs) using *Tabebuia aurea* leaf extract (TAE) as non-toxic reducing agent, efficient stabilizer, and weak base source without adding any surfactants. Al_2O_3/AuNPs was characterized using various techniques including UV-Visible Spectrophotometer, Fourier-transform Infrared (FT-IR) spectroscopy, X-Ray diffraction (XRD), and Particle Size Analyzer (PSA). AuNPs was successfully synthesized using $HAuCl_4$ as precursor. The characterization of UV-Vis Spectrophotometer shows that AuNPs colloidal was formed at 0.15 % TAE. FTIR characterization of Al_2O_3 shows Al-OH band at 1505 cm^{-1} and Al-O band at 680 cm^{-1}. Phase of Al_2O_3 was amorphous confirmed by XRD. The particle size distribution average of Al_2O_3/AuNPs was about 43 nm. The reduction percentage of methylene blue using Al_2O_3/AuNPs catalyst was 70.41 % for 120 min.

Keywords: Al_2O_3/AuNPs, *Tabebuia aurea*, catalytic activity, methylene blue reduction

1. Introduction

Aluminum oxide (Al_2O_3) is a well support catalyst material due to its inert properties, good thermal stability, capable to bind with metals as catalyst, has high surface area and evenly dispersed pores on the surface [1–4]. On the other hand, gold nanoparticle (AuNPs) has important roles in several catalytic processes such as in reduction of nitro-aromatic compounds and azo dyes, low temperature CO oxidation, and organic synthesis [5–8]. However, the use of metal nanoparticles like AuNPs can cause some problems such as homogeneity, reusability, and catalyst separation from the system [9]. Thus, some previous researchers have conducted many studies of metal nanoparticle immobilization on the support material [10–11]. This result is in line with the property of Al_2O_3 as a support catalyst. AuNPs may enhance the catalytic activity of Al_2O_3 due to its active site. The immobilization of AuNPs on Al_2O_3 surface can convert AuNPs phase from colloidal into solid phase [12].

Tabebuia aurea is one of the medicinal plants in tropical countries like Indonesia [13]. This plant has secondary metabolite compounds that can be used to synthesize the nanoparticle. *Tabebuia aurea* leaf extract have been previously reported to contain alkaloid, flavonoid, polyphenols and saponins compounds which act as a weak base source and stabilizing agent in nanoparticles formation [14]. Green synthesis method has been widely used for both metal and metal oxide synthesis. In previous study, we prepared both metal and metal oxide by green synthesis approach using plant [15–19].

To the best of our knowledge, we reported about cost-effective and environmentally friendly synthesis and immobilization of AuNPs on Al_2O_3 surface (Al_2O_3/AuNPs) using TAE. The catalytic activity of Al_2O_3/AuNPs was observed in methylene blue reduction.

2. Materials and methods

2.1. Materials
Tabebuia aurea (TA) was obtained from Jagakarsa, South Jakarta, Indonesia and has been determined at Lembaga Ilmu Pengetahuan Indonesia (LIPI), Bogor, West Java, Indonesia. Methanol and n-hexane were analytical grade from PT. Brataco, Indonesia. $HAuCl_4.3H_2O$ was purchased from Sigma-Aldrich. $Al(NO_3)_3.9H_2O$ and methylene blue were obtained from Merck, Germany.

2.2. Preparation of leaf extract
Dried powder of *Tabebuia aurea* (TA) leave was added into methanol of 300 mL and macerated for 7 days at room temperature. The result was then partitioned by n-hexane. The methanol fraction was evaporated then added by distilled water to obtain TA extract (TAE) for nanoparticles fabrication.

2.3. AuNPs synthesis
0.05, 0.10, and 0.15% of TAE (w/v) were added dropwise into 1.0 x 10^{-4} M $HAuCl_4$ until forming a violet colloidal. The colloidal was further characterized using a UV-Vis spectrophotometer.

2.4. Al_2O_3 synthesis
0.15% of TAE was added dropwise into 0.03M $Al(NO_3)_3$ solution in volume ratio (1:3) and stirred continuously at 80 °C for 5 h to form a white colloid. The colloid was heated at 100 °C for 1 h and annealed at 500 °C for 4 h to obtain white powder of Al_2O_3.

2.5. Al_2O_3/AuNPs synthesis
Al_2O_3 powder was dispersed into 30 mL of 1.0 x 10^{-4} M $HAuCl_4$ with sonication for 20 min. TAE 0.15% (w/v) was added dropwise into the previous mixture and stirred continuously for 100 min at room temperature. The dispersed system was annealed at 500 °C for 4 h to obtain the violet powder of Al_2O_3/AuNPs.

2.6. Characterization
TAE optimum concentration for synthesis of AuNPs and catalytic activity analysis were observed by UV-Vis Spectrophotometer (Shimadzu 2600). The functional group of Al_2O_3/AuNPs was examined by Fourier Transform Infrared (FT-IR) (Shimadzu, Prestige 21). The crystal phase of Al_2O_3 was characterized by X-Ray Diffraction (XRD) Shimadzu 610 at 2θ range of 20°– 60° with Cu Kα (λ = 0.1546 nm) radiation. The particle size distribution of Al_2O_3/AuNPs was characterized by Particle Size Analyzer (PSA) Malvern Zetasizer 1600.

2.7. Catalytic activity analysis
5 mg of Au/Al_2O_3 NPs was added into the mixture of 5 mL of 2.0 × 10^{-5} M methylene blue and 0.1 M $NaBH_4$. The catalytic activity was observed by measuring the absorbance change using UV-Vis spectrophotometer for 120 min.

3. Results and discussion

3.1. Optimization of AuNPs
Synthesis of AuNPs was varied in TAE concentrations of (0.05, 0.10 and 0.15) %. It aims to observe the effect of TAE concentration in AuNPs formation. The formation of AuNPs is signed by the presence of absorption peak at the maximum wavelength of 500–550 nm due to the SPR peak formation [20,21]. A color change of colloid occurs from yellow to violet in the AuNPs formation. AuNPs formation can be described by UV-Vis absorption spectra as shown in figure 1.

0.15 % TAE was the optimum concentration in AuNPs formation. Since this concentration resulted the absorption peak at the smallest maximum wavelength of 540 nm called by blue shift region. It has the conformity with the previous research that the maximum wavelength shift at blue region will produce the small size of nanoparticle [22]. Furthermore, 0.15 % TAE was used as optimum concentration for Al_2O_3/AuNPs fabrication. The particle size of AuNPs was further observed by PSA characterization.

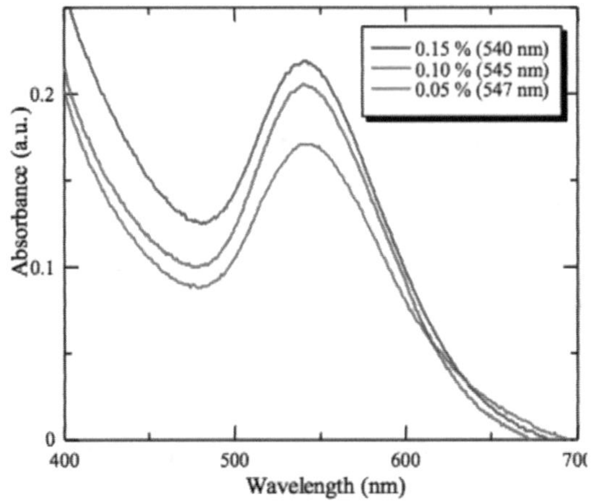

Figure 1. UV-Vis absorption spectra of AuNPs formation.

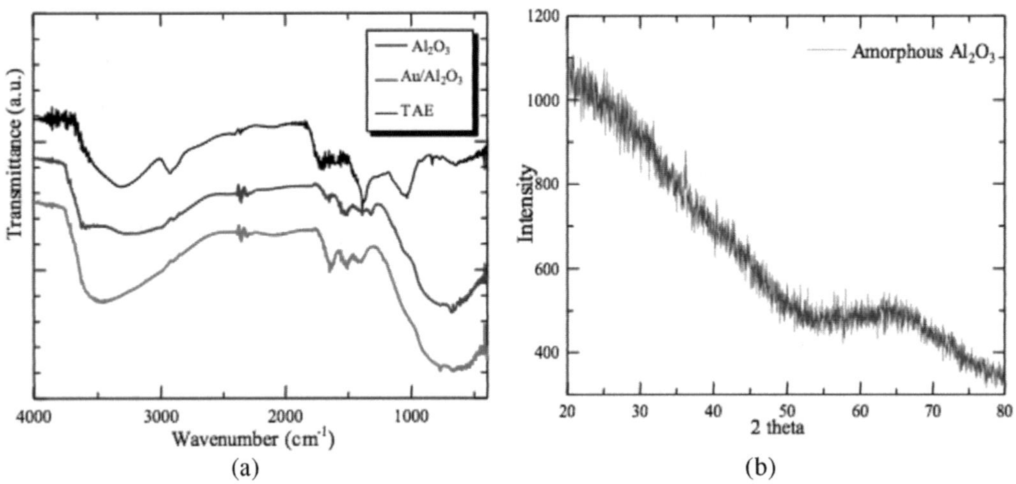

Figure 2. (a) FT-IR spectra of TAE, Al_2O_3 and Al_2O_3/AuNPs, and (b) XRD pattern of Al_2O_3.

3.2. FT-IR spectroscopy and XRD analysis

Functional groups of TAE, Al_2O_3, and Al_2O_3/AuNPs were identified by FT-IR as shown in figure 2a. TAE has broad vibrations band at 3318.7 cm^{-1} corresponds to –OH stretching of polyphenols, flavonoids and saponins that act as capping and reducing agents in AuNPs formation [19,22,23]. The vibrations band at 1384.0 cm^{-1} assigns to C-N stretching from alkaloids compound that plays a role as weak base source for the synthesis of metal oxide nanoparticle [15]. Figure 2 shows that both Al_2O_3 and Al_2O_3/AuNPs reported the typical Al-O stretching vibration at wavenumber of 715.6 and 713 cm^{-1}, respectively. Based on figure 2b, XRD pattern of Al_2O_3 shows no sharp peak, indicating the amorphous phase. The amorphous form of Al_2O_3 has an excellent performance on catalysis due to its better adsorption ability [24].

3.3. Particle size analyzer

PSA characterization was performed to determine the average of particle size distribution of Al_2O_3/AuNPs as presented in figure 3. Figure 3 shows that Al_2O_3 and Al_2O_3/AuNPs were distributed at particle size average of 37.76 and 43.29 nm, respectively. These results demonstrate that both of Al_2O_3 and Al_2O_3/AuNPs are nanoparticles.

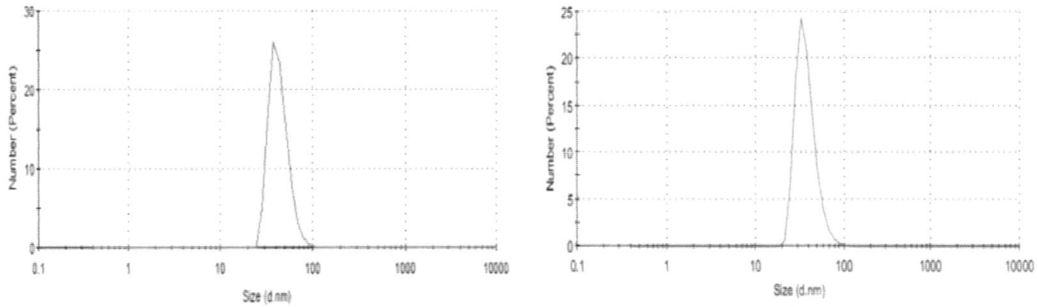

(a) (b)

Figure 3. PSA result of **(a)** Al_2O_3 and **(b)** Al_2O_3/AuNPs.

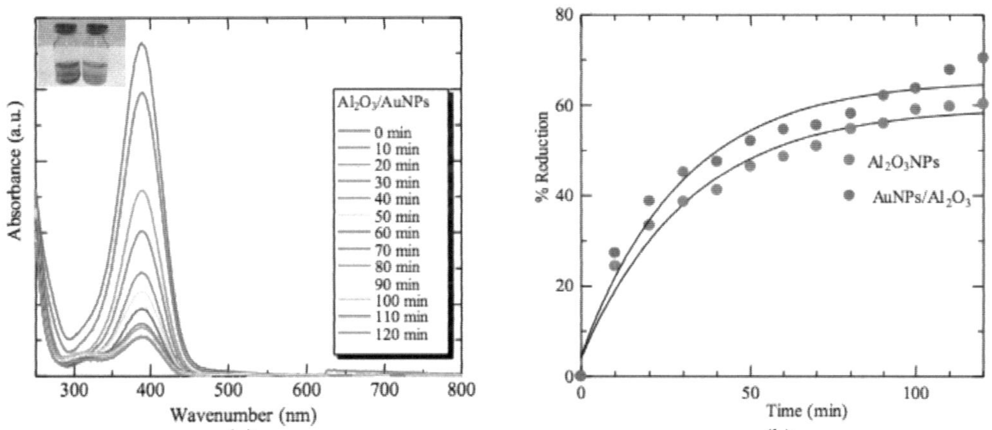

(a) (b)

Figure 4. (a) Decreasing of UV-Vis absorption spectra and (b) reduction percentage of MB with $NaBH_4$ for 120 min using Al_2O_3/AuNPs catalyst.

3.4. Catalytic activity

Figure 4 shows the reduction percentage of MB calculated from the reduction of MB with $NaBH_4$ through the observation of the fading blue color and the decrease of its absorbance. MB was reduced to be a leuco methylene blue (LMB) presented by a new absorption as shown in figure 4a. The formation of LMB occurs at the wavenumber of 300 nm [25]. The reduction percentage of MB using Al_2O_3 and Al_2O_3/AuNPs are 60.26 % and 70.41 %, respectively, as shown in figure 4b. The presence of AuNPs can accelerate the reduction of MB solution. AuNPs plays important role in electron transfer from BH_4^- as a donor to MB as an electron acceptor. BH_4^- are nucleophilic, while MB are electrophilic [26].

4. Conclusions

Al_2O_3/AuNPs has been successfully synthesized using TAE. FTIR characterizations show that there are alkaloid and flavonoid in TAE, which can act as reducing and capping agents as well as weak base source in synthesis of Al_2O_3/AuNPs. AuNPs may enhance the catalytic activity of Al_2O_3 significantly. According to all discussions, we demonstrate that Al_2O_3/AuNPs has a good catalytic activity in the reduction of MB using $NaBH_4$ with reduction percentage of 70.41 %.

Acknowledgements

This work was funded by Hibah Penelitian Kompetensi 2016 from the Directorate General of Higher Education, Ministry of Research, Technology and Higher Education Republic of Indonesia to Universitas Indonesia with contract No.1023/UN2.R12/HKP.05.00/2016. The author would like to thank PT DKSH (Malvern) for supporting characterization of Particle Size Analyzer.

References

[1] Razzaq R, Li C, Usman M, Suzuki K and Zhang S 2015 *Chem. Eng. J.* **262** 1090–8
[2] Chen Z, Kuckling D and Tiemann M 2018 *Nanomaterials (Basel)* **8** E186
[3] Saavedra J, Pursell C J and Chandler B D 2018 *J. Am. Chem. Soc.* **140** 3712–23
[4] Goto T, Itoh T, Akamatsu T, Izu N and Shin W 2016 *Sens Actuators* B *Chem.* **223** 774–783
[5] Shen W, Qu Y, Pei X, Li S, You S, Wang J, Zhang Z and Zhou J 2017 *J. Hazard. Mater.* **321** 299–306
[6] Hinojosa-Reyes M, Camposeco-Solis R, Zanella R, Rodríguez-González V and Ruiz F 2018 *Catal. Lett.* **148** 383–396
[7] Takale B S, Bao M and Yamamoto Y 2014 *Org. Biomol. Chem.* **12** 2005–27
[8] Alegria E C B A, Ribeiro A P C, Mendes M, Ferraria A M, do Rego A M B and Pombeiro A J L 2018 *Nanomaterials (Basel)* **8** E320
[9] Zhang P, Qiao Z A, Jiang X, Veith G M and Dai S 2015 *Nano Lett.* **15** 823–8
[10] Hassanzadeh-Tabrizi S A, Motlagh M M and Salahshour S 2016 *Appl. Surf. Sci.* **384** 237–43
[11] Vedrine J C 2017 *Catalysts* **7** 341
[12] López M, Simakova O A, Murzina E V, Willför S M, Prosvirin I, Simakov A and Murzin D Y 2015 *Appl. Catal.* A **504** 248–55
[13] Agarwal M and Chauhan S 2015 *J. Med. Plants Stud.* **3** 63–8
[14] Prathibha B S and Priya K S 2017 *IOSR Journal of Applied Chemistry (JAC)* **10(6)** 23–9
[15] Yulizar Y, Bakri R, Apriandanu D O B and Hidayat T 2018 *Nano Structure and Nano-Objects* **16** 300–5
[16] Lembang M S, Yulizar Y, Sudirman S and Apriandanu D O B 2018 *AIP Conf. Proc.* **2023** 020093
[17] Yulizar Y, Latifah I, Bakri R and Apriandanu D O B 2018 *AIP Conf. Proc.* **2023** 020097
[18] Sulaiman N, Yulizar Y and Apriandanu D O B 2018 *AIP Conf. Proc.* **2023** 020105
[19] Yulizar Y and Ayun Q 2017 *IOP Conf. Ser: Earth Environ. Sci.* **60** 012006
[20] Apriandanu D O B and Yulizar Y 2017 *IOP Conf. Ser.: Mater. Sci. Eng.* **188** 012013
[21] Gondal M A, Rashid S G, Dastageer M A, Zubair S M, Ali M A, Lienhard J H, McKinley G H and Varanasi K K 2013 *IEEE Photonics J.* **5** 2201908
[22] Tódor I S, Szabó L, Marişca O T, Chiş V and Leopold N 2014 *J. Nanopart. Res.* **16** 2740
[23] Rajakumar G, Gomathi T, Rahuman A A, Thiruvengadam M, Mydhili G, Kim S H, Lee T J, and Chung I M 2016 *Applied Science* **6** 222
[24] Zou J J, Xiong Z, Wang L, Zhang X and Mi Z 2007 *J. Molecular Catal. A: Chem.* **271** 209–15
[25] Pegu R, Majumdar K J, Talukdar D J and Pratihar S 2014 *RSC Adv.* **4** 33446–56
[26] Ganapuram B R, Alle M, Dadigala R, Dasari A, Maragoni V and Guttena V 2015 *Int. Nano Let.* **5** 215–22

Superabsorbent nanocomposite as micronutrient slow-release fertilizer

H Helmiyati and D Nugraha

Department of Chemistry, Faculty of Mathematics and Natural Sciences (FMIPA), Universitas Indonesia, Kampus UI Depok, Depok 16424, Indonesia

Corresponding author's e-mail: helmiyati@sci.ui.ac.id

Abstract. In this research, sodium alginate grafted by poly(acrylate-co-acrylamide) loaded montmorillonite (MMT) and the addition of boric acid as micronutrients into the superabsorbent nanocomposite through in-situ method have been successfully synthesized. We developed superabsorbent nanocomposite with montmorillonite obtained by increasing the amount of montmorolite, which would increase the swelling capacity and by adding boric acid micronutrients will also increase water swelling capacity. The best swelling capacity to water of superabsorbent nanocomposite obtained 426.337 g/g that formulated with MMT contents approximately 20.0 % wt and 0.1 g boric acid. The optimal release capacity of boric acid is 25.561 % for 8000 min or about 6 days. This is supported by the study of kinetics for the water swelling in pseudo second-order while the boric acid release is pseudo first-order obtained. The nanocomposite superabsorbents that are successfully synthesized show that the release of boric acid micronutrients as slow release can be controlled and denoted interesting materials to be applied in agriculture.

Keywords: montmorillonite, micronutrient, nanocomposite, sodium alginate, superabsorbent

1. Introduction

Soil nutrition control for agriculture is extremely vital because when not being fulfilled, plants will show deficiency symptoms, that is slow growth, imperfect fruit development, dry leaves, etc. [1]. Soils' need for nutrient based on the required amount is classified into two classes, macronutrient and micronutrient. Therefore, macronutrient and micronutrient nutrition need to be controlled in order to gain ideal plants [2,3]. Micronutrient is one of the important nutrients used by plants in small amount, because micronutrient in plants usually has a role as a catalyst and thus only needed in a small amount. The amount and quality of agricultural products are highly affected by micronutrient. One of micronutrient elements for plants growth is boron [4].

Although micronutrient is needed only in a small amount, its importance in plants is as significant as macronutrient. Boron is one of micronutrients that is essential for the growth. Boron has a role in cell formation or multiplication, especially in shoots growth point, and also in the growth of pollen powder, flowers, and roots. In boron-deficient plants, sugar translocation and absorption decrease a lot. Boron may also act as inhibitor that controls the activity of enzymes that lead to the formation of toxic phenolic substances. Other symptoms of boron-deficiency are thickening, darkening, and decreasing size of leaves [5]. At present, to distribute nutrients of macro and micro fertilizers is by directly spread to the soil, so that a lot of fertilizer will be wasted. Therefore, to solve those issues, a technology such as superabsorbent nanocomposite that can release nutrients in a controlled behavior is developed, so that the amount of nutrition given to plants will not be excessive [6,7].

Superabsorbent based on alginate grafted with the monomers of acrylic acid and acrylamide have been extensively used in agriculture. The clay addition to superabsorbent nanocomposite has attracted

IOP Conf. Series: Materials Science and Engineering **763** (2020) 012050 doi:10.1088/1757-899X/763/1/012050

Table 1. Superabsorbent nanocomposite variation

Superabsorbent nanocomposite code	MMT (% wt)	Micronutrient boric acid (g)
SN 1	10	-
SN 2	15	-
SN 3	20	-
SN 4	10	0.1
SN 5	15	0.1
SN 6	20	0.1

attention because of its high swelling capacity and controlled release [8–10]. Researches regarding superabsorbent nanocomposite have been done in various fields, such as in industry; diapers [11], drug release [12,13], agriculture [14], and waste water treatment [15].

In this study nanocomposite superabsorbents using boric acid micronutrients referred to Bortolin *et al.* [2], and combined with our previous study [16,17] using sodium alginate biopolymer as a backbone have been synthesized successfully. The swelling-release of superabsorbent nanocomposites of boric acid was studied. The synthesis results were supported by characterization using Scanning Electron Microscopy (SEM) and Fourier Transform Infrared (FTIR).

2. Experiments

2.1. Materials
The materials used were sodium alginate biopolymer as backbone, while served as monomers were acrylic acid and acrylamide, potassium persulfate as initiator, N,N′-Methylenebisacrylamide as crosslinking agent (all from the Merck), montmorillonite as clay (Sigma-Aldrich), and boric acid as micronutrient sources (Sigma-Aldrich).

2.2. Superabsorbent nanocomposite preparation
A superabsorbent nanocomposite preparation was referred to our previous study [17]. By modification, using montmorillonite as inorganic in various concentrations then obtained the optimum conditions and furthermore boric acid micronutrients were added. Whereas, other reagents were constant; sodium alginate is 3 wt.%, 1.2 % KPS initiator is 5 mL, 20% acrylamide is 6 mL, acrylic acid: NaOH (7: 3) is 7 mL and 0.5% MBA 10 mL. The composition of montmorillonite variations can be seen in table 1.

2.3. Swelling-release capacity
In this study the swelling process done in water was determined by gravimetric method and release of boric acid from superabsorbent by visible spectrophotometry. Swelling capacity was calculated using equation (1) [17].

$$Swelling\ capacity\ (g/g) = \frac{w_t - w_o}{w_o} \quad (1)$$

where w_o is the initial weight of the superabsorbent and w_t is the superabsorbent weight after swelling at a certain time. Release capacity was calculated using the equation (2).

$$Release\ capacity\ (\%) = \frac{m_o - m_t}{m_o} x\ 100\ \% \quad (2)$$

where m_o is the superabsorbent weight before release and m_t is the superabsorbent weight after release at a certain time.

3. Results and discussion

3.1. Preparation of superabsorbent nanocomposite

3.1.1. Water swelling capacity of superabsorbent nanocomposite. In this study, preparation of nanocomposite superabsorbents with variations of montmorillonite (SN1, SN2, SN3) which have been synthesized, then determined for the swelling capacity with water, can be seen in figure 1a and

Table 2. Optimal swelling capacity of superabsorbent nanocomposite

Superabsorbent Code	Swelling Capacity (g/g)
SN 1	108.995
SN 2	156.080
SN 3	188.581
SN 4	237.313
SN 5	329.511
SN 6	426.337

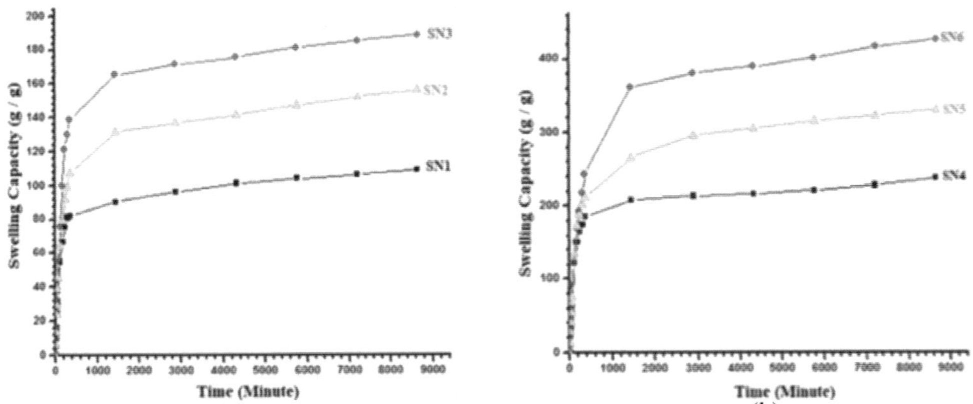

(a) (b)

Figure 1. Swelling capacity of (a) superabsorbent nanocomposite and (b) with micronutrients boric acid addition.

nanocomposite superabsorbents by addition of boric acid micronutrients (SN4, SN5, SN6) can be seen in figure 1b.

As seen in figure 1a, optimal swelling capacity increases by rising montmorillonite concentration. This is due to the addition of montmorillonite (MMT), in which MMT has negative surface charge (OH group), resulting in a repulsion force between the COOH group on the sodium alginate backbone and MBA. The negative charge on the surface of the MMT results in the increasing expansion of the superabsorbent structure, so that the swelling capacity is also high [18]. As seen in figure 1b, similar things also occur with the addition of boric acid. Addition of boric acid will increase swelling capacity, where boric acid or orthoboric acid has three hydroxyl groups with negative charge on the surface, resulting in a repulsion force between the COOH group on the sodium alginate backbone and MBA. The negative charge on the nanocomposite superabsorbents surface comes from boric acid and the high MMT, so the swelling capacity by using boric acid in figure 1b is higher than the swelling capacity in figure 1a. Optimal swelling capacity of superabsorbent nanocomposite can be seen in table 2.

Boric release capacity of superabsorbent nanocomposite. Nanocomposite superabsorbents that have been swollen at the maximum swelling capacity were done with the release capacity test of boric acid determined by the visible spectrophotometer with a wavelength of 545 nm, which can be seen in figure 2. In figure 2a it can be seen swelling capacity to time and obtained an equilibrium time around 8000 min and the highest swelling capacity at the montmorillonite concentration 20 % (SN6) has a release capacity higher than the other montmorillonite concentrations. In figure 2b it can be seen the optimum release capacity of SN4, SN5 and SN6. From the results of release capacity, it can be seen that the obtained nanocomposite superabsorbent is good as a controlled slow release fertilizer with the highest release capacity of 25.561 % for 8000 min or about 6 days.

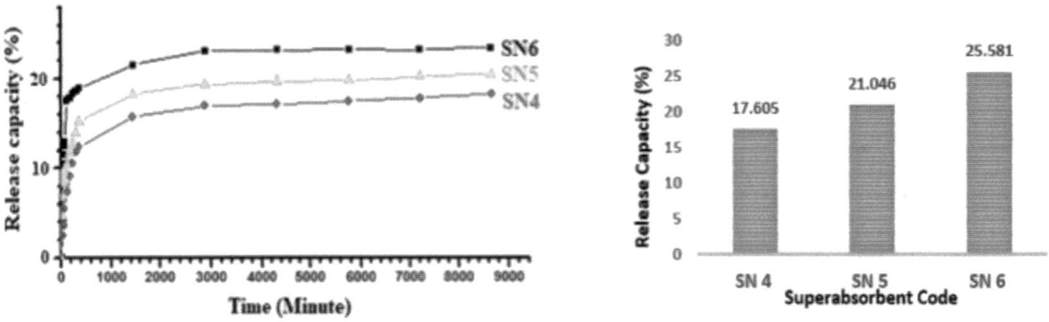

(a) (b)

Figure 2. (a) Release capacity to time and (b) optimum release capacity of boric acid micronutrients

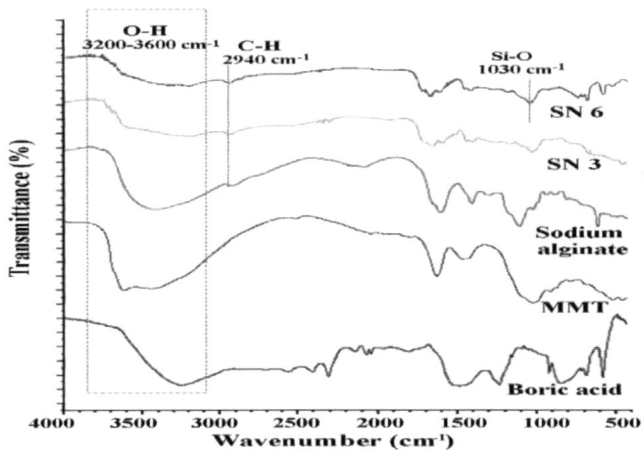

Figure 3. FTIR spectra of (a) pure boric acid, (b) montmorillonite (MMT), (c) sodium alginate, (d) superabsorbent (SN 3) and (e) nanocomposite superabsorbent with in-situ borax acid (SN 6)

3.2. Characterization of superabsorbent nanocomposite

3.2.1. Characterization by FTIR. In figure 3a, it can be seen boric acid has characteristic peaks around 600, 750, 810, 1210, and 1400 cm^{-1}, indicating the presence of BO_3^{3-} ions [2]. In figure 3b, it can be seen montmorillonite (MMT) spectrum, a peak of Si-O stretching appears at 1010 cm^{-1}. Figure 3c can be seen sodium alginate spectrum, the absorption band at 3200–3600 cm^{-1} are OH groups. At 675 cm^{-1} of C-H stretching, the characteristic of sodium alginate at 828 cm^{-1} of C–H vibration from β-mannuronic acid, at 960 cm^{-1} of C-H stretching vibration of guluronic acid at 1603 cm^{-1} of O-C-O asymmetric and at 1410 cm^{-1} of symmetric carboxylate stretching vibration. In figure 3d, nanocomposite superabsorbent without boric acid (SN 3) can be seen the peak of Si-O from montmorillonite shifts to 1100 cm^{-1}, the characteristic of sodium alginate appears and the O-H peak is weakened due to the interaction of this group with the monomer, crosslinking component as forming superabsorbents. Figure 3e is nanocomposite superabsorbents with in-situ boric acid (SN 6) can be seen the peak of Si-O from montmorillonite shifts to 1030 cm^{-1} and the characteristic of sodium alginate appears and peak of the boric acid characteristic can be observed around 600-1400 cm^{-1} with a weaken peak. It indicates the successful synthesis of in-situ borate nanocomposite superabsorbents.

3.2.2. Characterization by SEM. The synthesized superabsorbent characterized using SEM can be seen in figure 4. In figure 4a shows that the surface of superabsorbent nanocomposite has a characteristic of fine fibers, has quite a rough and porous structure from sodium alginates, montmorillonite and it has big porous surface. Figure 4b shows the surface morphology of

Table 3. Kinetics order of superabsorbent nanocomposite

Kinetic	First pseudo order (R^2)	Second pseudo order (R^2)
Kinetics Swelling	0.9325	0.9995
Kinetics Release	0.9994	0.9012

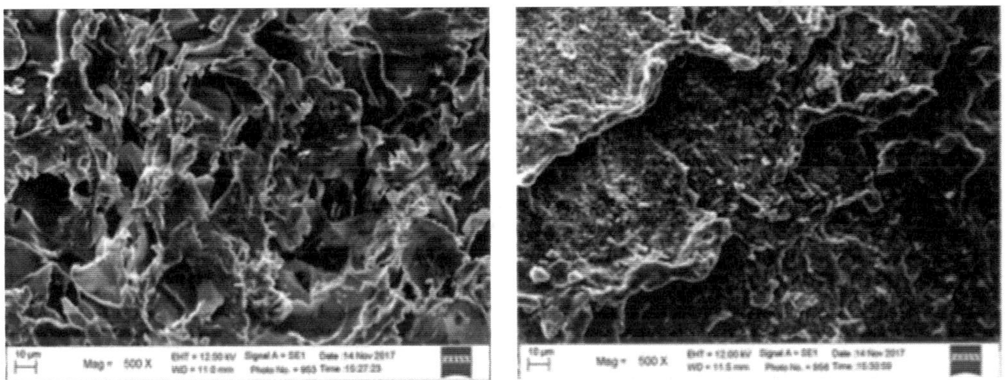

(a) (b)

Figure 4. Morphology of (a) superabsorbent (b) nanocomposite with in-situ boric acid

superabsorbent nanocomposite with in-situ boric acid as slow-release fertilizer, has smaller porous surface than figure 4a, in which its pores were not evenly distributed, due to the interaction between superabsorbents surface with boric acid, since boric acid was acidic and had an interaction with COOH group, so that it has a large swelling capacity.

3.3. Kinetics study of swelling

After obtaining the best nanocomposite superabsorbents, SN6 with a swelling capacity of 426.337 g/g and the release capacity of 25 % for 8000 min or about 6 days, furthermore, the kinetics study was determined. The determination of kinetics was carried out using the first pseudo and second pseudo reaction order. In table 3, it can be seen the R^2 values for first-and second-order kinetics swelling-release. The swelling kinetics of the first-order pseudo reaction rate was obtained at R^2 = 0.9325 and second-order pseudo reaction rate at R^2 = 0. 9995, where R^2 of second pseudo order is higher than R^2 of first pseudo order. The study of kinetics swelling obtained pseudo second order for water. For the release kinetic for micronutrient, the obtained first order kinetics was R^2 = 0.9994 and second order R^2 = 0.9012. From the kinetics study the obtained release kinetics was in first order. This shows that the superabsorbent nanocomposite that was successfully synthesized has the ability to absorb large amount of water and slow release, which controls the boric acid micronutrients. From the results of the swelling and release kinetics studies, it supports the results of the discussed swelling and release capacity, in which the swelling capacity of the nanocomposite in-situ boric acid is greater and the release capacity is small. Nanocomposite superabsorbents that are successfully synthesized show that release boric acid micronutrients can be controlled.

4. Conclusions

Sodium alginate grafted by poly(acrylate-co-acrylamide) loaded montmorillonite (MMT) and the addition of boric acid as micronutrients into the superabsorbent nanocomposite through in-situ method has been successfully synthesized. This result is supported by the characterization with FTIR and SEM. Superabsorbent nanocomposite with the best swelling capacity at 20 % montmorillonite is 426.337 g/g and 0.1 g boric acid. The optimal release capacity of boric acid is 25.561 % for 8000 min or about 6 days. This is supported by the study of kinetics that obtained the swelling order in second-order with R^2 = 0.9995 and the order release is first-order with R^2 = 0.9994. The nanocomposite superabsorbents that are successfully synthesized show that release boric acid micronutrients slow release can be controlled for micronutrient slow release.

Acknowledgements
This study is funded by PITTA Grant Universitas Indonesia with contract No: 2268/UN2.R3.1/HKP.05.00/2018

References
[1] Tilman D, Cassman K G, Matson P A, Naylor R and Polasky S 2002 *Nature* **418** 671–7
[2] Bortolin A, Seafim A R, Aouada F A, Mattoso L H C and Ribeiro C 2016 *J. Agric Food Chem* **64** 3133–40
[3] Li J, Zhuang X, Font O, Moreno N, Vallejo V R, Querol X and Tobias A 2014. *J. Hazard. Mater.* **265** 242–52
[4] Azeem B, KuShaari K , Man Z, Basit A and Trinh T T 2014 *J. Controlled Release* **181** 11–21
[5] Koshiba T, Kobayashi M, Matsuoka K, Fujiwara T and Matoh T 2013 *Soil Sci. Plant Nutr.* **59** 189–94
[6] Helmiyati, Abbas G H and Kurniawan S 2017 *IOP Conf. Ser.: Mater. Sci. Eng.* **188** 012024
[7] Helmiyati, Saefumillah A and Yulianti W 2014 *Asian J. Chem.* **26** 7337–42
[8] Hussien R A, Donia A M, Atia A A, El-Sedfy O F, El-Hamid A R A and Rashad R T 2012 *Catena* **92** 172–8
[9] Lee W F and Chen Y C 2005 *Eur. Polym. J.* **41** 1605–12
[10] Wu J, Wei Y, Lin J and Lin S 2003 *Polymer* **44** 6513–20
[11] Kosemund K, Schlatter H, Ochsenhirt J L, Krause E L, Marsman D S and Erasala G N 2009 *Regul. Toxicol. Pharmcol.* **53** 81–9
[12] Yu S H, Mi F L, Wu Y B, Peng C K, Shyu S S and Huang R N 2005 *J. Appl. Polym. Sci.* **98** 538–49
[13] Samanta H S and Ray S K 2014 *Carbohydr. Polym.* **99** 666-78
[14] Ni B, Liu M, Lü S, Xie L and Wang Y 2010 *J. Agric. Food Chem.* **58** 12373–8
[15] Paulino A T, Guilherme M R, Reis A V, Tambourgi E B, Nozaki J and Muniz E C 2007 *J. Hazard. Mater.* **147** 139–47
[16] Helmiyati and Aprilliza M 2017 *IOP Conf. Ser.: Mater. Sci. Eng.* **188** 012019
[17] Helmiyati H and Syarifudin A 2018 *AIP Conf. Proc.* **2023** 020080
[18] Rashidzadeh A and Olad A 2014 *Carbohydr. Polym.* **114** 269–78

Electroreduction of carbon dioxide using platinum-iridium modified boron-doped diamond (BDD) with various platinum-iridium ratios

M H Saprudin[1], S R Atriardi[1], J Gunlazuardi[1], Y Einaga[2] and T A Ivandini[1]

[1] Department of Chemistry, Faculty of Mathematics and Natural Sciences (FMIPA), Universitas Indonesia, Kampus UI Depok, Depok 16424, Indonesia

[2] Department of Chemistry, Faculty of Science and Technology, Keio University, 3-14-1 Hiyoshi, Yokohama 223-8522, Japan

Corresponding author's e-mail: ivandini.tri@sci.ui.ac.id

Abstract. Modification of boron-doped diamond (BDD) electrodes with bimetallic platinum-iridium was studied. Combination techniques of wet chemical seeding and electrochemical deposition were performed with various Pt-Ir precursor ratios of (1:1), (1:2) and (2:1). These electrodes were then applied for electrochemical reduction of CO_2 dissolved in 0.1 M NaCl solution at the potentials of -1.1 V, -1.2 V, -1.3 V, and -1.5 V (vs Ag/AgCl) for 60 min. Electroreduction of CO_2 applied at PtIr(2:1)-BDD at the potential of -1.1 V (vs Ag/AgCl) produced formic acid, methanol and hydrogen as the main products.

Keywords: boron-doped diamond, bimetal, CO_2 electroreduction, iridium, platinum.

1. Introduction

Total of CO_2 in atmosphere reaches 72 % and increases about 3 % each year [1]. The abundance of CO_2 attracts researcher's interest to produce valuable chemical products. One promising technique to reduce CO_2 is electrochemical reduction because the products can be controlled by changing the electrochemical parameters, such as potentials and electrolytes [2–4]. Moreover, water can be used as both electron and proton sources to produce hydrocarbons [4].

On the other hand, boron-doped diamond (BDD) is established as a superior electrode with many advantages, such as wide working potential and low background current in water solution as well as excellent stability of its surface [5]. However, sometimes it needs to modify the BDD surface with catalysts to improve its electroactive ability. Meanwhile, bimetallic electrocatalysts have attracted great attention as a good candidate to enhance the catalytic activity as well as selectivity over their monometal ones [6-9]. Recently, Kortlever *et al.* [10] reported that modification of carbon with PtPd provide a exceptionally low onset potential for the CO_2 reduction to form formic acid [10].

Platinum, which is known as inert material, provides favorable activation and conversion of CO_2 to CO [11,12]. Electroreduction CO_2 on Pt-BDD electrode at potential of -2.5 V (vs Ag/AgCl) was reported to produce formic acid as the main product with 14.6 % farradaic efficiency [11]. On the other hand, iridium, either in the forms of metal complexes or nanoparticles, has a great electrocatalytic activity [13]. The use of iridium on cobalt electrode in water electrolysis system was reported to increase the product yields [14]. Iridium was also reported to react with water to generate Ir–OH species, that can integrate with Pt–CO species to eliminate surface-adsorbed CO at the lower potential compared to pure Pt [12,15]. Furthermore, our previous research using PtIr(3:4)-BDD for

Figure 1. SEM image of PtIr(1:1)-BDD (a), PtIr(1:2)-BDD (b) and PtIr(2:1)-BDD (c). Inset of (a) shows the image unmodified BDD

electroreduction CO_2 at potential -1.7 V (vs. Ag/AgCl) showed the formation of acetic acid and formic acid [16]. In this research, modification of BDD with bimetallic Pt-Ir with various ratios was studied for electrochemical reduction of CO_2.

2. Experimental

2.1. Materials

BDD electrodes (B:C ratio of the precursor gas was 1:1000) were self-prepared in Einaga Laboratory, Chemistry Departement, Faculty of Science and Technology, Keio University, Japan. Hydrogen hexachloroplatinate (IV), potassium hexachloroiridate (IV) and sodium borohydride ($NaBH_4$) were from Sigma, while NaOH, NaCl, Na_2SO_4 and other chemicals were from Merck. Ultra-pure deionized water was obtained from a Millipore Direct® Q5-UV with a resistivity of 18.2 MΩcm.

2.2. Preparation of PtIr-BDD electrodes

Platinum iridium (Pt-Ir) was deposited onto BDD (PtIr-BDD) by wet chemical seeding followed by electrochemical overgrowth. A volume of 10 μL 1.0M $NaBH_4$ in 0.1M NaOH solution and 40 μL 1mM H_2PtCl_6:K_2IrCl_6 or Pt-Ir solution (mole ratios were 1:1, 1:2, and 2:1) was dropped on BDD surface to provide the wet chemical seeding technique. Then, chronoamperometry in Pt-Ir solution at -0.5 V (vs. Ag/AgCl) for 15 min was conducted to provide the electrochemical overgrowth.

2.3. Electroreduction of CO_2

The prepared PtIr-BDD was employed as the working electrode, while a platinum wire as the counter and an Ag/AgCl system as the reference electrodes. In a two-compartment electrochemical cell, 0.1 M NaCl was placed in the cathode room, while 0.1M Na_2SO_4 in the anode room. Then, CO_2 gas was dissolved in 0.1M NaCl for 30 min with 100 mL/min flow rate. The CO_2 electroreduction were applied using chronoamperometry for 1 h and the products were analyzed using High Performance Liquid Chromatography (HPLC), Gas Chromatography and Mass Spectrometry (GC-MS), and GC with Total Ion Chromatogram (TIC) and Flame Ionization Detector (FID).

3. Results and discussion

3.1. Modification of BDD electrode and its characterization

Preparation of PtIr-BDD electrodes was carried out by electrodeposition method. Prior to the deposition, the BDD surface was treated by Pt and Ir seeds to increase the metal adsorption on the surface. The presence of $NaBH_4$ in the seeding step helps to bind the metal particles on the BDD surface and to reduce Pt and Ir ions to produce the metal forms. The metals then acted as the seeds on the BDD surface, where the metal particles were further grown in the electrodeposition step [17]. The reactions between $NaBH_4$ and Pt-Ir salt solutions are shown below [17,18]:

$$NaBH_4 + 3H_2O + H_2PtCl_6 \rightarrow Pt\downarrow + H_3BO_3 + 5HCl + NaCl + 2H_2 \quad (1)$$

$$NaBH_4 + 3H_2O + K_2IrCl_6 \rightarrow Ir\downarrow + H_3BO_3 + 3HCl + 2KCl + NaCl + 2H_2 \quad (2)$$

The electrodeposition was performed by using chronoamperometry technique at a potential of -0.5 V (vs Ag/AgCl) for 15 min. SEM images in figure 1a and figure 1b show that smaller amount of metal

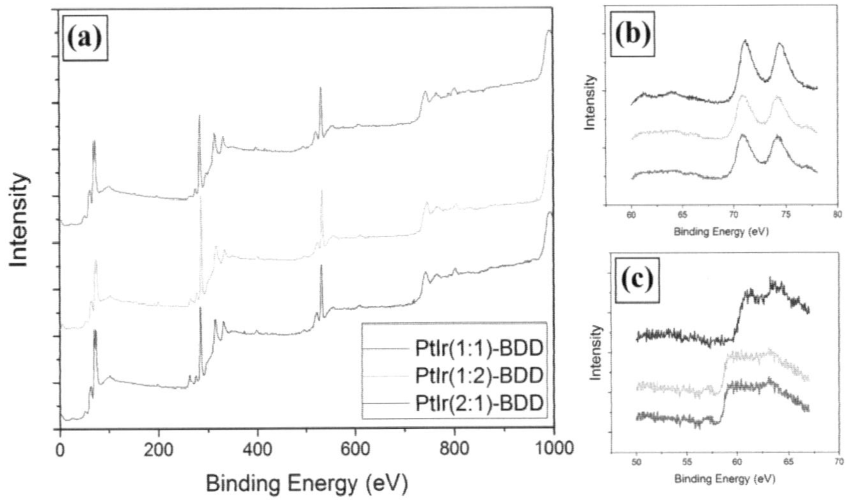

Figure 2. XPS Spectrum of PtIr-BDD Electrodes for (a) the wide scan as well as the narrow scan of (b) Pt and (c) Ir

particles was deposited at PtIr-(1:1)-BDD and PtIr (1:2)-BDD in compare to that at PtIr (2:1)-BDD (figure 1c). However, at PtIr (1:2)-BDD particle aggregation on the surface was observed, indicated the influence of iridium in the deposition of Pt.

Energy Dispersive X-Ray Spectroscopy (EDS) characterization was performed to determine the composition of Pt and Ir at each BDD electrodes. At PtIr(1:1)-BDD surface, 3.67 % of Pt and 1.17 % of Ir could be successfully deposited, while 1.73 % Pt and 0.15 % Ir were deposited at PtIr(1:2)-BDD and 3.84 % Pt and 0.39 % Ir were deposited at PtIr(2:1)-BDD. Roughly, the ratios of Pt-Ir (3:1), (11:1) and (10:1) could be deposited on PtIr(1:1)-, PtIr(1:2)-, and PtIr(2:1)-BDD, respectively. The composition shows that all electrodes exhibit larger mass percentage of Pt than Ir probably because the electrodeposition was generated at potential reduction of Pt [13]. Therefore, Pt particles were easier to be deposited on BDD surface than Ir particles.

The X-Ray Photoelectron Spectroscopy (XPS) spectra for all three electrodes were performed. The wide scans (figure 2a) show typical peaks with high intensity at a binding energy of 284.8 eV and 532.75 eV related to C and O. Beside these peaks, Pt and Ir peaks were observed with smaller intensities. The narrow scans performed to clarify the peak of Pt and Ir at all PtIr-BDD electrodes shows the peak of Pt observed at binding energy around 71 eV (figure 2b) , while the peak of Ir appeared at binding energy of 60 eV (figure 2c). The results indicated that Pt (0) and Ir (0) species were deposited on BDD electrodes [19].

3.2. Electroreduction of CO_2

Prior to CO_2 electroreduction, the reduction potential of CO_2 was determined by using cyclic voltammetry in 0.1M NaCl solution containing of CO_2. Cyclic voltammogram (CV) at PtIr-BDD electrodes with various Pt-Ir ratios show that the reduction peak CO_2 was typically observed at the potential of -1.3 V vs. Ag/AgCl (figure 3). Accordingly, the electrochemical reduction of CO_2 was performed at around -1.1 V, -1.2 V, -1.3 V, and -1.5 V (vs. Ag / AgCl).

3.3. Product analysis of electroreduction CO_2

Product analysis of the electrochemical reduction CO_2 was performed by using HPLC, LC-MS, and GC. The farradaic efficiency of each product obtained was then compared to that of the products resulting from the other electrodes as shown in figure 4. Application of PtIr (2:1)-BDD electrode at the potential -1.1 V resulted the largest total faradaic efficiency (44.88%) composed of 23.71% formic, 13.12% methanol and 7.71% hydrogen.

Furthermore, figure 5 shows that with the decrease of applied potentials from -1.3 V, the faradaic efficiency of formic acid and methanol increased. On the contrary, hydrogen decreased due to the

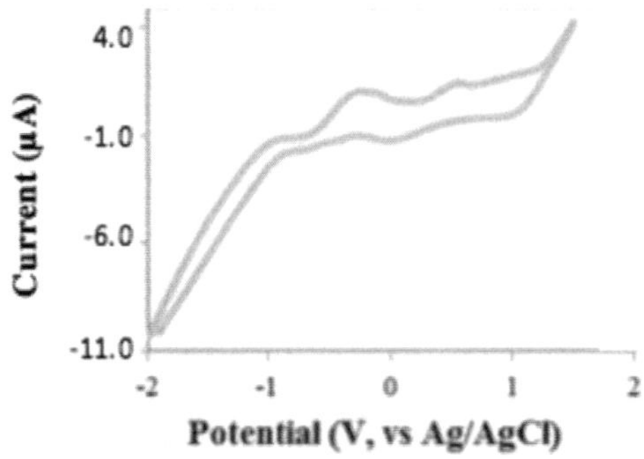

Figure 3. CV of PtIr-BDD in 0.1M NaCl after CO_2 gas dissolved for 15 min.

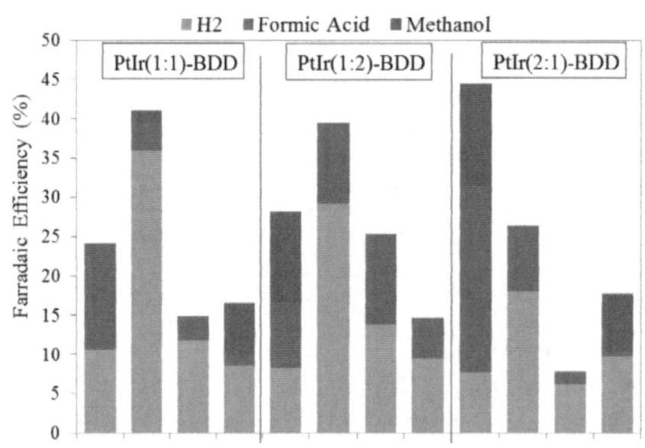

Figure 4. Farradaic efficiency total of the products of electroreduction CO_2 using PtIr-BDD electrodes with various ratios Pt-Ir and various potential applied.

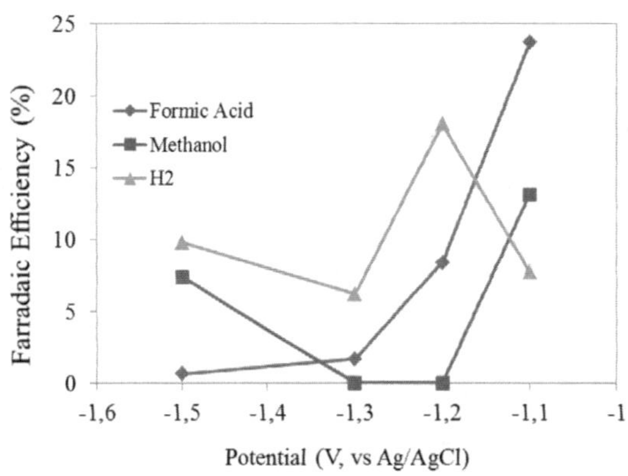

Figure 5. Farradaic efficiency of each products of electroreduction CO_2 using PtIr(2:1)-BDD electrodes with various ratios Pt-Ir and various potential applied.

competition between the direct CO_2 reductions with the hydrogen evolution, which takes place in the same range of potentials [20–22].

3.4. Performance of electrode

Stability of the modified BDD electrodes for electroreduction of CO_2 was checked by before and after used. Table 1 shows a decrease of Pt and Ir mass percentage on BDD surface after used for electroreduction of CO_2. The most stable electrode was PtIr(2:1)-BDD with mass loss percentage of

Table 1. Mass percentage of Pt-Ir particles deposited on BDD before and after used for electroreduction of CO_2

Electrodes	Mass percentage of Pt (%)			Mass percentage of Ir (%)		
	before	after	mass loss	before	after	mass Loss
PtIr (1:1) - BDD	3.67	1.59	56.68	1.17	0	100.00
PtIr (1:2) - BDD	1.73	0.96	44.51	0.15	0.12	20.00
PtIr (2:1) - BDD	4.42	3.84	13.12	0.41	0.39	4.88

13.12% Pt particles and 4.88% Ir. The results showed that although wet chemical seeding has been done before electrochemical deposition, the bimetallic Pt-Ir particles was not deposited very stable on BDD surface. Therefore, it needs to redepositing the particles on BDD surface for applying electroreduction of CO_2 with the same electrode. In other words, performance of Pt-Ir modified BDD electrode by these modification methods are only capable for one time application in electroreduction of CO_2

4. Conclusions

Preparation of PtIr-BDD by using combination techniques of wet chemical seeds and electrodeposition provided homogenous distribution of particles at Pt-Ir ratio of (1:1), whereas the increase of Ir ratio caused aggregation of the deposited particles. Examination of these electrodes for electrochemical reduction of CO_2 indicated the most efficient result was at Pt-Ir ratios of (2:1) using an applied potential of -1.1 V with formic acid, methanol, and hydrogen as the main products.

Acknowledgements

This work was funded by Hibah PITTA Universitas Indonesia, Contract No. 2273/UN2.R3.1/ HKP.05.00/2018

References

[1] Olivier J G J, Schure K M and Peters J A H W 2017 *Trends in Global CO2 and Total Greenhouse Gas Emissions* (The Hague: PBL Netherlands Environtmental Assesment Agency) 2017 Report
[2] Jones J P, Prakash G K S and Olah G A 2013 *Isr. J. Chem.* **54** 1451–66
[3] Zhu D D, Liu J L and Qiao S Z 2016 *Adv. Mater.* **28** 3423–52
[4] Nakata K, Ozaki T, Terashima C, Fujishima A and Einaga Y 2014 *Angew. Chem. Int. Ed. Engl.* **53** 871–4
[5] Asai K, Ivandini T A, Falah M M and Einaga Y 2016 *Electroanalysis* **28** 177–82
[6] Wang J and Swain G M 2003 *J. Electrochem. Soc.* **150** E24–32
[7] Ma M, Hansen H A, Valenti M, Wang Z, Cao A, Dong M and Smith W A 2017. Smith *Nano Energy* **42** 51–7
[8] Sarfaz S, Garcia-Esparza A T, Jedidi A, Cavallo L and Takanabe K *ACS Catal.* **6** 2842–51
[9] Chang Z, Huo S, Zhang W, Fang J and Wang H 2017 *J. Phys. Chem.* C **121** 11368–79
[10] Kortlever R, Peters I, Koper S and Koper M T M 2015 *ACS Catal.* **5** 3916–23
[11] Jasril, Gunlazuardi J and Ivandini T A 2017 *IOP Conf. Ser.: Mater. Sci. Eng.* **188** 012037
[12] Sahin O, Kivrak H, Karaman M and Atbas D 2015 *Am. J. Mater. Sci. Eng.* **3** 15–20
[13] Dewandaru R K P, Gunlazuardi J and Ivandini T A 2017 *AIP Conf. Proc.* **2023** 020098
[14] Tae E L, Song J, Lee A R, Kim C H, Yoon S, Hwang I C, Kim M G and Yoon K B 2015 *ACS Catal.* **5** 5525–9
[15] Papaioannou E I, Siokou A, Comninellis C and Katsaounis A 2013 *Electrocatalysis* **4** 375–81
[16] Atriardi S R, Dewandaru R K P, Gunlazuardi J and Ivandini T A 2019 *IOP Conf. Ser.: Mater. Sci. Eng.* **496** 012040
[17] Gao F, Yang N, Smirnov W, Obloh H and Nebel C E 2013 *Electrochim. Acta* **90** 445–51
[18] La-Torre-Riveros L, Abel-Tatis E, Méndez-Torres A E, Tryk D A, Prelas M and Cabrera C R 2011 *J. Nanopart. Res.* **13** 2997–3009
[19] Ikeo N, Iijima Y, Niimura N, Sigematsu M, Tazawa T, Matsumoto S, Kojima K and Nagasawa Y 1991 *Handbook of X-Ray Photoelectron Spectroscopy* (Tokyo: JEOL)

[20] Kortlever R, Shen J, Schouten K J P, Calle-Vallejo F and Koper M T M 2015 *J. Phys. Chem. Lett.* **6** 4073–82
[21] Whipple D T and Kenis P J A 2010 *J. Phys. Chem. Lett.* **1** 3451–8
[22] Gattrell M, Gupta N and Co A 2007 *Energy Convers. Manag.* **48** 1255–65

IOP Conf. Series: Materials Science and Engineering **763** (2020) 012052 doi:10.1088/1757-899X/763/1/012052

Bangka Kaolin and Bayat Zeolite-based ZSM-5 zeolite as heterogeneous catalyst for hexadecane cracking reaction

E S H Basri[1], Y K Krisnandi[1], R Ekananda[2] and I R Saragi[1]

[1] Department of Chemistry, Faculty of Mathematics and Natural Science (FMIPA), Universitas Indonesia, Kampus UI Depok, Depok 16424, Indonesia

[2] Pertamina Research and Development, Jalan Raya Bekasi Km 20, Jakarta 13920, Indonesia

Corresponding author's e-mail: yuni.krisnandi@sci.ui.ac.id

Abstract. Microporous ZSM-5 material was synthesized using natural aluminosilicate minerals, i.e. Bayat zeolite and Bangka kaolin as silica and alumina sources. Prior to use, Bayat zeolite was treated with several treatments, i.e. purification, fragmentation and dealumination. While Bangka Kaolin was used as the source of additional silica, in which the silica was extracted using aqua regia. After that, ZSM-5 zeolite was synthesized using tetrapropylammonium hydroxide (TPAOH) as template. Then, it was characterized using Fourier Transform Infrared (FTIR), X-Ray Diffraction (XRD) and Scanning Electron Microscopy-Energy Dispersive X-Ray (SEM-EDX). Based on XRD pattern, mordenite inverted framework (MFI) structure was obtained. SEM-EDX images showed that the natural aluminosilicate-based ZSM-5, so called ZSM-5 MA, has a coffin-like shape with average crystal size and Si/Al ratio are 1.6 µm and 14.26, respectively. As reference, another as-synthesized ZSM-5 was prepared from pro-analysis reagents, and labelled as ZSM-5MS. It also has mordenite inverted framework (MFI) structure, Si/Al ratio of 21.45 and average crystal size of 0.8 µm. ZSM-5 was tested with micro activity test (MAT) method for hexadecane cracking reaction at 500 °C. Prior to be used, the ZSM-5 was subjected to NH_4-exchanged followed by calcination at 550 °C to obtain H/ZSM-5. The products that obtained from hexadecane cracking reaction are gas (H_2, C_1–C_4) and gasoline (C_5–C_{12}) fractions. The catalytic test shows that both H/ZSM-5MA and H/ZSM-5MS gave similar results, despite their differences in Si/Al ratio and particle sizes. Therefore, zeolite synthesis using natural minerals as Si and Al sources is promising to be explored further.

1. Introduction

Fluid catalytic cracking (FCC) is one of the main transformation technologies in the oil refinery industry and generates the main of the world's gasoline. The demands of gasoline increase the use of this process. Catalytic cracking is a process that have been applied intensively, because of several benefits, such as faster reaction, better product selectivity, and ideal product combustion [1].

One of the most used zeolite in catalytic cracking is Zeolite Socony Mobil-5 (ZSM-5) that has form of mordenite inverted framework (MFI) structure [2]. ZSM-5 is an active catalyst and/or promotes for several processes such as cracking, aromatization, hydrocarbon isomerization, and alkylation because of the activity, structure selectivity, ion-exchange properties, and certain porous structure [3].

Zeolite production occurs in a big scale and usually synthesized using high cost chemical reagent. So, it needs to find an alternative to reduce the cost. Because of that, the research about synthesis zeolite from natural feedstock, such as natural aluminosilicate minerals, fly ash, rice husk, and others are challenging [4–6]. This research focuses on the preparation of microporous ZSM-5 zeolite with

Bayat zeolite and Bangka kaolin as silica and alumina sources, followed by using it as heterogeneous catalyst for hexadecane cracking.

2. Materials and methods

2.1. Materials.
Bayat zeolite, Bangka kaolin, hydrogen chloride (Merck), nitric acid (Merck), sodium acetate (Merck), acetic acid (Merck), sodium dithionite (Merck), trisodium citrate (Merck), hydrogen peroxide (Merck), sodium hydroxide (Merck), ammonium chloride (Merck), tetrapropylammonium hydroxide (Sigma Aldrich).

2.2. Pre-treatment on Bayat zeolite.
Purification was carried out by mixing Bayat zeolite with buffer acetate pH 5 (1:3) w/v. After that, the mixture was filtered and 200 mL deionized water and 20 mL H_2O_2 were added to the residue. Then, buffer citrate bicarbonate and sodium dithionite were added. After that, the mixture was filtered and dried the residue at 100 °C. Fragmentation was carried out by mixing Bayat zeolite with certain amount of NaOH pellet and deionized water [7], and then heated at 250 °C for 2 h in a furnace. Dealumination was conducted by mixing the Bayat zeolite with 250 mL HCl 2M in the reflux system at 100 °C for 4 h.

2.3. Extracting silica from Bangka kaolin.
Silica extraction was carried out by following the method of Wang *et al.* [8,9]. First, 100 g Bangka kaolin was heated at 800 °C for 6 h. After that, 40 g Bangka kaolin was mixed with 15 mL deionized water, 280 mL HCl 37 % and 94 mL HNO_3 in the reflux system at 100 °C for 4 h. After that, 90 mL HCl and 30 mL HNO_3 were added into the mixture and continued being refluxed at 100 °C for another 4 h. The precipitate was filtered and washed until its pH was neutral, then dried at 100 °C.

2.4. Synthesis ZSM-5.
Synthesis was carried out by following the published method [8,9]. First, 2.19 g Bayat zeolite was mixed with 5 mL deionized water, and stirred until homogeneous. After that, 10.97 mL TPAOH was added, stirred for 1 h. Then, 0.55 g extracted silica and remaining deionized water was added, stirred for 1 h. After that, pH of mixture was adjusted with glacial CH_3COOH until reached pH 11. The mixture was aged at 100 °C for 3 h, and at room temperature for 24 h. The obtained gel was transferred into Teflon-lined autoclave and heated at 170 °C for 6 days. The white precipitate was then washed until pH 8 and calcined at 550 °C for 5 h.

2.5. Modification ZSM-5 become H/ZSM-5.
Modification was conducted by the means of ion-exchange process with 1M of NH_4Cl solution at 60 °C for 12 h. After that, NH_4/ZSM-5 was calcined at 550 °C for 5 h to obtain H/ZSM-5.

2.6. Application.
Application was carried out using micro activity test (MAT) method. Prior to the reaction, catalyst was heated at 300 °C while nitrogen was flown through the reactor. After that, the reactor temperature was set at 500 °C. Approximately 2 mL Hexadecane was flown for 75 sec. The gas and liquid products were characterized by GC-RGA and GC simulated distillation, respectively.

3. Results and discussion

3.1. Pre-treatment of Bayat zeolite
Based on X-Ray Diffraction (XRD) pattern in Treacy and Higgins [2], it is confirmed that Bayat natural zeolites contains mainly of mordenite structure but has some impurities. Therefore, it was subjected to several treatments to remove the impurities [10]. After that, fragmentation was carried out to breakdown zeolite structure to become its fractions, i.e. SiO_4^{4-} and AlO_4^{5-} [7]. Figure 1a shows the results from XRD and figure 1b for infrared measurement. XRD patterns of raw and purified Bayat zeolite are unchanged, while significant change discovered in pattern from defragmented and dealuminated Bayat zeolite. The lost crystallinity and observed amorphous patterns are due to the destruction of Si-O-Al ring during fragmentation, and removal of Al after dealumination, respectively.

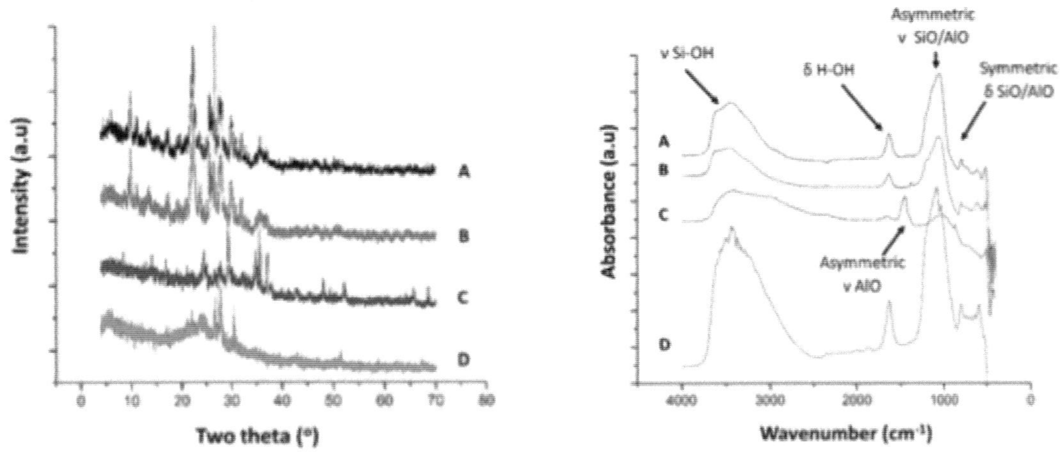

(a) (b)

Figure 1. (a) Powder XRD patterns and (b) FTIR spectra for Bayat Zeolite (A) raw and after (B) purification, (C) fragmentation and (d) dealumination process

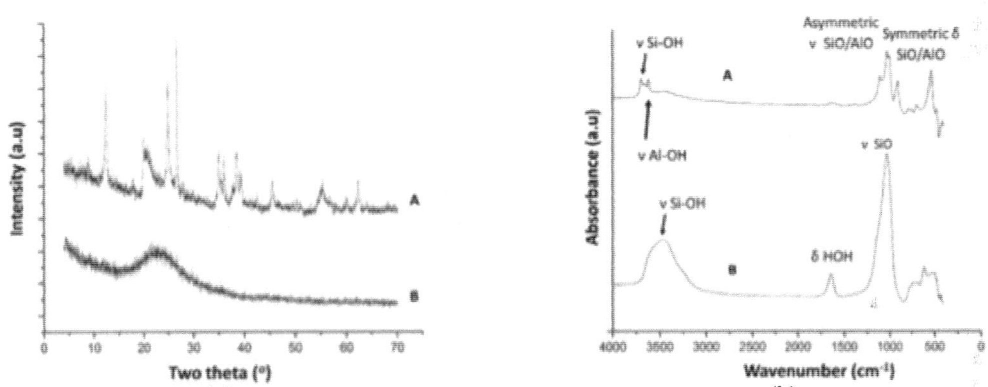

(a) (b)

Figure 2. (a) XRD patterns and (b) FTIR spectra for (A) kaolin and (B) extracted silica (from kaolin)

From Fourier Transform Infrared (FTIR) measurement shown in figure 1a spectrum C, it can be seen that new peak was observed in the fragmented Bayat zeolite spectra at 1300 – 1600 cm^{-1}, assigned for $NaAlO_2$ [5–6]. Energy Dispersive X-Ray (EDX) measurement on final dealumination Bayat zeolite indicates that it contains 34.74% Si, 1.34% Al, 1.39% Na and 63.53% O, to make the Si/Al ratio of 25.9.

3.2. Extraction silica from Bangka kaolin

In order to cover the lack of silica that was required for synthesis, extracted silica from Bangka kaolin was added. Figure 2 shows that both XRD pattern and infrared spectra of extract silica are different from those of original kaolin. In figure 2a pattern A, it can be seen that kaolin has distinct peaks at $2\theta \approx 12.32°$ and 24.86°, while extracted silica (figure 2a pattern B) has amorphous phase and some peaks were disappeared. Furthermore, infrared spectra from kaolin (figure 2b spectrum A) shows the typical peaks of Si-OH and Al-OH at 3600 cm^{-1} and 3500 cm^{-1}, respectively [11], indicative of kaolin as phyllosilicate material. They are not observed in extracted silica (figure 2b spectrum B)), but a broad band appeared in 3600–3000 cm^{-1} instead, assigned for hydrogen-bonded silanol groups [11]. From EDX measurement, the concentrations of Si and Al in extracted silica are 39.71 % and 0.07 %, respectively and the ratio of Si/Al was ∞. This indicates that the extraction process was successfully performed.

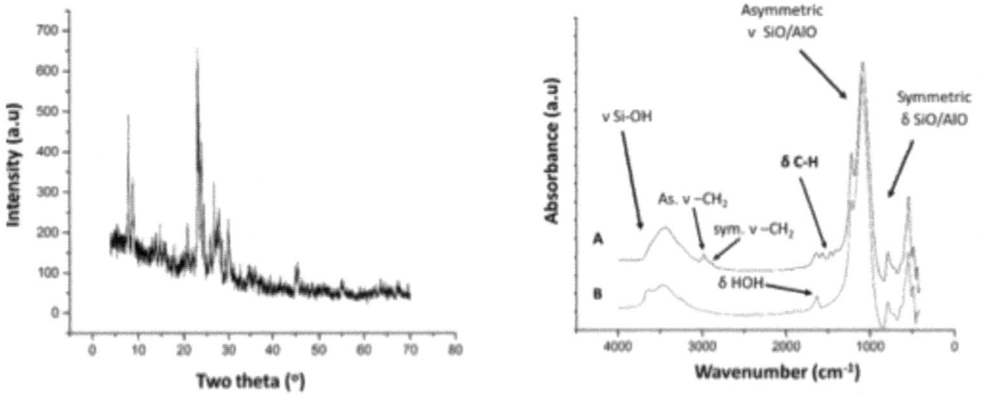

(a) (b)

Figure 3. (a) XRD patterns and (b) FTIR spectra of ZSM-5 MA (A) before and (B) after calcination

(a) (b)

Figure 4. SEM Images of (a) ZSM-5MS (25.000x magnification) and (b) ZSM-5MA (20.000x magnification).

3.3. Synthesis of ZSM-5 zeolites

Two microporous ZSM-5 zeolites were synthesized using single template method. One batch used pre-treated Bayat zeolite and extracted Bangka kaolin (called ZSM-5 MA), and as control, another batch was synthesized using pro analysis precursors (labelled as ZSM-5MS). TPAOH was used as template, to obtain MFI structure, since the crystallinity of ZSM-5 was high when using TPAOH as template [12].

Based on XRD pattern in figure 3a, it is shown that as-synthesized ZSM-5MA has MFI structure with the typical peaks at 2θ = 7°–8° and 2θ = 22°–25° [2]. Furthermore, figure 3b spectrum A shows infrared spectra from ZSM-5MA before calcination with peaks at 2850–2950 cm^{-1} (attributed to ν – CH_2) and at 1450–1350 cm^{-1} (attributed to δ–CH_3) coming from organic template TPAOH.

EDX analyses provide Si/Al ratio of 14.26 and 21.45, for ZSM-5MA and ZSM-5MS, respectively. The lower Si/Al ratio for ZSM-5MA indicates that the framework has more Al, thus its acidity is higher than the ZSM-5MS [13,14]. From SEM characterization in the figure 4, it can be seen that the crystals of ZSM-5MA also has a hexagonal or coffin-like shape, although not as distinct as that from ZSM-5MS. This is in agreement with typical crystal habit for ZSM-5 [9,15]. In addition, ZSM-5MA has average particle size of 1.6μm, while ZSM-5MS has smaller size, about 0.8 μm.

3.4. H/ZSM-5 zeolites

ZSM-5 zeolites were treated with NH_4^+ multiple-exchange and calcination at 500 °C to obtain H/ZSM-5 zeolites. Figure 5 shows the infrared spectra from ZSM-5 MA, while the spectra from ZSM-5 MS,

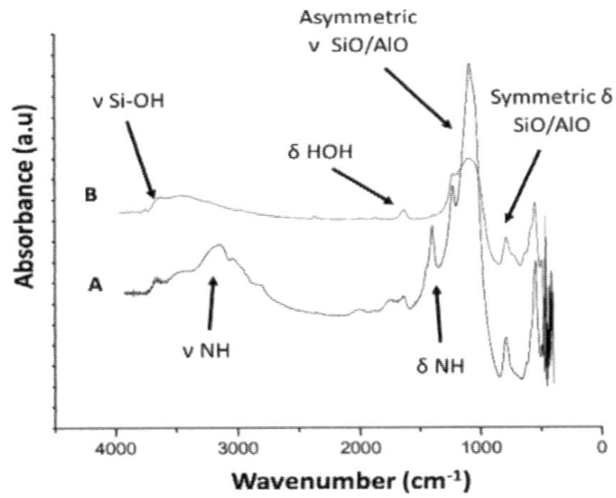

Figure 5. FTIR spectra of NH_4/ZSM-5MA (spectrum A) and H/ZSM-5MA (spectrum B)

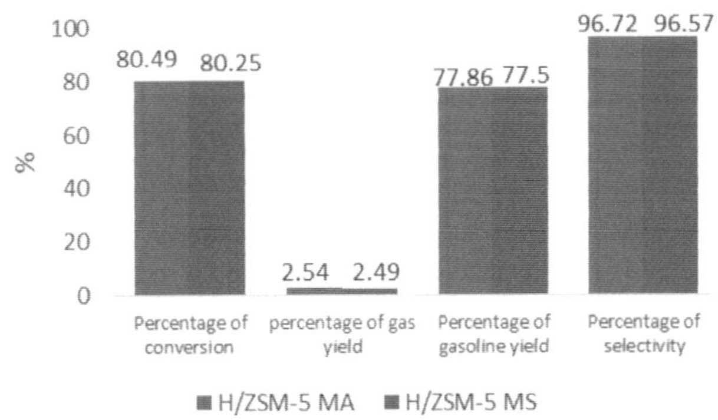

Figure 6. Graphic of % conversion, % yield and % selectivity from hexadecane cracking product

since similar, were not shown for brevity. Figure 5 spectrum A shows that a broad band at 3200–2800 cm^{-1} and a sharp peak at 1600 cm^{-1} were observed, which were attributed to stretching (ν) and bending (δ) vibrations of N-H from NH_4^+ [16], respectively. After calcination, they disappeared due to ammonia removal, and H/ZSM-5 was obtained (figure 5 spectrum B).

3.5. Cracking reaction application

The materials that tested were natural aluminosilicate-based H/ZSM-5 (H/ZSM-5MA) and H/ZSM-5 synthetic (H/ZSM-5MS). The results of the cracking reaction product are summarized in figure 6. From figure 6, it is suggested that the natural aluminosilicate-based H/ZSM-5 has similar catalytic activity with the synthetic H/ZSM-5. It could happen because the Si/Al ratio of H/ZSM-5MA (14.26) is lower than H/ZSM-5MS (21.45) to compensate the larger particle size of ZSM-5MA (1.6 μm) than that of ZSM-5MS (0.8 μm). Lower Si/Al ratio will increase the cation exchange capacity [15], thus more NH_4^+ will replace the Na^+. As consequence, H/ZSM-5MA will be more acidic than H/ZSM-5MS [11].

4. Conclusions

Microporous ZSM-5 was successfully synthesized from Bayat zeolite as silica and alumina sources and Bangka kaolin as additional silica source. Furthermore, it was successfully applied as catalyst in hexadecane cracking reaction, with conversion, yield and selectivity results of 80.49 %; 77.86 % and 96.72 %. The results are comparable to those from synthetic-based ZSM-5 due to benefit of its lower Si/Al ratio and more defective crystal morphology.

Acknowledgements

This work was funded by Hibah PITTA 2018 from Universitas Indonesia No.2341/UN2.R3.1/HKP.05.00/2018. The use of MAT reactor was part of collaboration research project between PT. Pertamina (Persero) and Universitas Indonesia year 2016–2018 number 004/E20400/2016-S0.

References

[1] Wang Y, Xian X, Hou X, Zhang X, Wang L and Liu G 2015 *J. Anal. Appl. Pyrolysis* **113** 133–6
[2] Treacy M M J and Higgins J B 2001 *Collection of Simulated XRD Powder Patterns for Zeolites* (Amsterdam: Elsevier)
[3] Bellussi G and Pollesel P 2005 *Stud. Surf. Sci. Catal.* **158** 1201–12
[4] Liu H, Shen T, Wang W, Li T, Yue Y and Bao X 2015 *Appl. Clay Sci.* **115** 201–11
[5] Liu H, Shen T, Li T, Yuan P, Shi G and Bao X 2014 *Appl. Clay Sci.* **90** 53–60
[6] Mohiuddin E, Isa Y M, Mdleleni M M and Key D 2017 *Micropor. Mesopor. Mater.* **237** 1–11
[7] Rohayati, Krisnandi Y K and Sihombing R 2017 *AIP Conf. Proc.* **1862** 30094
[8] Bakri R, Utari T and Sari I P 2008 *Makara Sains* **12** 37–43
[9] Wang L, Zhang Z, Yin C, Shan Z and Xiao F S 2010 *Micropor. Mesopor. Mater.* **131** 58–67
[10] Ming D W 1987 *Clays Clay Miner.* **35** 469–72
[11] Bräuer P, Situmorang O, Ng P L and D'Agostino C 2018 *Phys. Chem. Chem. Phys.* **20** 4250–62
[12] Gontier S and Tuel A 1996 *Zeolites.* **16** 184–95
[13] Munthali M W, Elsheikh M A, Johan E and Matsue N 2014 *Molecules* **19** 20468–81
[14] Octaviani S, Krisnandi Y K, Abdullah I and Sihombing R 2012 *Makara J. Sci.* **16** 155–62
[15] David C W 1996 *J. Chem. Educ.* **73** 46–50
[16] Zahara Z, Krisnandi Y K, Wibowo W, Nurani D A, Rahayu D U C and Haerudin H 2018 *AIP Conf. Proc.* **2023** 20088

Effect of surfactant mixing in microspheres preparations using polyblend of poly(lactic acid) and polycaprolactone

M A Annisa and E Budianto

Department of Chemistry, Faculty of Mathematics and Natural Sciences (FMIPA), Universitas Indonesia, Kampus UI Depok, Depok 16424, Indonesia

Corresponding author's e-mail: emilb@ui.ac.id

Abstract. Microencapsulation has an important role in the drug delivery system in the medical and pharmaceutical world. As an attempt to improve the therapeutic effect of the drug, microspheres materials are developed by optimizing the size of microspheres and their size distribution to be suitable for drug delivery applications. In this study, poly(lactic acid) and polycaprolactone polyblend with 60:40 (w/w) composition, have been used in the preparation of microspheres through the solvent evaporation (O/W) method using a mixed of common non-ionic surfactants, Tween 80 and Span 80 were used as emulsifier. The microspheres size distribution was observed at various compositions (60:40, 50:50, 40:60) and concentrations (0.5–2.5 % (v/v)) of Tween 80/Span 80. The resulting microspheres were round and have diameter distributed around 30.07 μm. Microspheres were also characterized by Fourier Transform Infrared (FTIR) and Particle Size Analyzer (PSA). The effect of mixing Tween 80 and Span 80 was shown in microspheres size formed, which was similar to the one of Tween 80 (0.375–43.67 μm) and larger than the one of Span 80 (0.375 - 0.791 μm). The distribution size of the resulting microspheres was slightly narrower than the one with the use of Tween 80. The microspheres physical appearances were observed by optical microscope (OM).

1. Introduction

Drug delivery system aims to release a treatment at specific desired anatomical location and to maintain drug concentration within therapeutic window as long as desired [1] and to optimize the therapeutic effects of the drugs, as well as minimizing the adverse effects if possible [2]. Polymer-based drugs delivery devices have been developed to provide a safe path for drugs to pass through the physiological areas that are not friendly inside the body. A controlled molecular structures polymer can be constructed to provide a well-defined response to external conditions [1]. The polymer biomaterials used were usually one single type of polymer or a polyblend, which deliberately blended to obtain the new material with combined superior properties of each polymer constituents, to be more suitable for the application [3].

One of polymer device applications such as microencapsulation, is a process when solids, liquids or gases are coated in the form of thin coating walls surrounding the substance to form microscopic particles called microspheres [4]. Solvent evaporation method is frequently used to encapsulate hydrophobic drugs in an oil-in-water emulsion (o/w) technique [5]. The polymer is dissolved in a soluble, volatile organic solvent. The resulting solution then emulsified into a large volume of water in the presence of a surfactant, followed by removal of solvent by evaporation, resulting in solid microspheres.

To produce small particle size microspheres, it is required the addition of surfactant to decrease the interfacial tension and reduce the coalition on droplets [6]. Mixing two or more different types of surfactants are designed to enhance the properties of the mixed-end products. Positive synergistic

IOP Conf. Series: Materials Science and Engineering **763** (2020) 012053 doi:10.1088/1757-899X/763/1/012053

Table 1. The various concentrations of surfactant used in preparation of microspheres

Mixed surfactants composition (v/v)	Tween 80 concentrations ((v/v) %)	Tween 80 volume (mL)	Span 80 volume (mL)
50:50	0.5	8.5	0.043
	1.5	8.5	0.128
	2.0	8.5	0.170
	2.5	8.5	0.213

nature of mixed-end product is obtained from the interaction between two or more compatible surfactants which have better performance than those achieved by each component when used separately [7]. In most cases, combinations of commonly used surfactants consist of hydrophilic surfactant and lipophilic surfactant [7] are mixed between sorbitol ester (Span) and polyoxyethylene (POE) sorbitol ester (Tween). Due to the greater interaction of Tween with the water phase, the hydrophilic group in Tween extends further into the water than the unsubstituted ester with the polyoxyethylene chain in Span, and this is believed to allow the hydrophobic portion of the two surfactants to interact more closely with each other and with the interface layer than if it was used separately [8].

The aim of this paper is to study microspheres (blank microcapsules) prepared by solvent evaporation method, using polyblend of poly (lactic acid) and polycaprolactone with mixed of two non-ionic surfactants Tween 80 and Span 80 as emulsifier. And to study certain effect from using mentioned mixed surfactant to the microspheres size distribution and its polydispersity. The technical procedure in this study refers to what has been done by Elfrida [9] with modifications and adjustments in some parts.

2. Experimental

2.1. Synthesis of PLA.

PLA were synthesized from its monomer, lactic acid (LA) (Merck) by direct polycondensation method under nitrogen gas flow without adding catalysts, solvents, and initiator. Stirring was started by raising the starting material temperature to 120 °C. Then, continued by kept it constant for 1 h. After 1 h., the system was purged by nitrogen gas about 5–10 min and remained within nitrogen for the rest of the works. The temperature was further raised until reaches 150 °C, then, kept constant for 22 h. Afterwards the temperature was turned off but the system remained the same for approximately 2 h. Once the system temperature dropped, the PLA formed was transferred to the vapor plate and allowed to stand for 24 h until it is hardened [9].

2.2. Preparation of 10 % polyblend solution.

A 10 % polyblend solution was prepared by weighing 0.5 g of polyblend with 60:40 (w/w) composition of PLA and PCL (Perstorp CAPA-6800) [9]. Then, it was dissolved into 5 mL dichloromethane (Merck) [9] followed by additions of Span 80 (TEGO SMO V from PT. Evonic) with volume variation as presented in table 1 and table 2. Continued by stirring for 30 min at 500 rpm to obtain 10 % polyblend solution [10].

2.3. Preparation of Tween 80 solutions.

The Tween 80 solutions were made in various concentrations by dilutions of Tween 80 solution. 10 mL of Tween 80 solution was dissolved in 80 mL aquadest (PT. Demin Indonesia) followed by stirring using magnetic stirrer. The solution was then added into a measuring flask and diluted with the rest of aquadest until the volume reach exactly 100 mL. From the solution, it can be obtained Tween 80 solutions with v/v concentrations 0.5 %, 1.5 %, 2 % and 2.5 % through further dilutions [9].

2.4. Preparation of microspheres.

Microspheres were prepared in sets of various concentrations and various compositions. 5 mL of 10 % polyblend solution was added into Tween 80 solution with each concentration and volume as shown in table 1 and table 2. For the stirring emulsion stage, the new solution was stirred at 700 rpm for 1 h.

Table 2. The various compositions of surfactant used in preparation of microspheres

Mixed surfactants[a] composition (v/v)	Tween 80[b] volume (mL)	Span 80 volume (mL)
60:40	10.2	0.136
50:50	8.5	0.170
40:60	6.8	0.204

[a] Tween 80 : Span 80.
[b] Tween 80 with 2 % concentration (v/v).

Table 3. The average diameter of microspheres at any variation concentration.

Variation concentration (%)	Average diameter (µm)
0.5	1.739
1.5	5.329
2.0	19.51
2.5	19.43

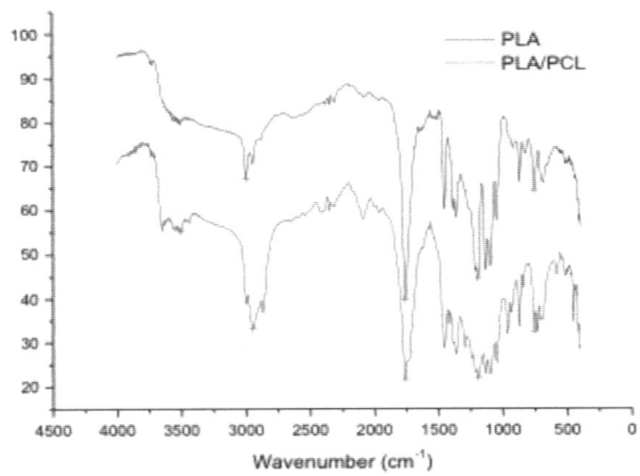

Figure 1. FT-IR spectrum of PLA and PLA/PCL microsphere.

After 1 h, the emulsion formed was inserted into a spray bottle. For the dispersion stage, the solution was then sprayed into 350 mL of aquadest which already rotated at 900 rpm for 1 h. The microcapsules formed were then washed, filtered and dried at room temperature of approximately one night [9].

3. Results and discussion

3.1. Analysis of Functional Groups in Poly(lactic acid) by FT-IR.

Characterization by Fourier Transform Infra Red (FT-IR) is intended to determine the functional groups consisting the chemical structure of synthesized PLA. It could also prove the level of purity and shows whether most of the monomers (LA) have been polymerized into PLA, by comparing FT-IR spectra of the product (PLA) and starting material (LA). From the obtained results, there is a band shift related to vibration of the C = O carbonyl group from monomer spectrum at 1727.06 [11] compared to 1758.19 cm-1 in the polymer spectrum. The band shift between the LA spectrum and the PLA also has a difference in the peak intensity which indicates that there is a structural change in the polymer chain which was carboxylic acid to ester. The spectrum of PLA is shown in figure 1.

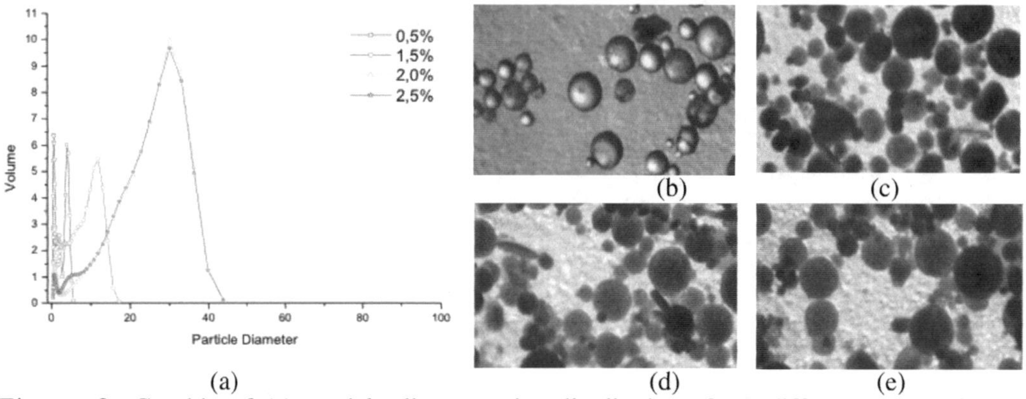

Figure 2. Graphic of (a) particle diameter size distributions from different microspheres formed with varied surfactant concentrations (0.5, 1.5, 2.0 and 2.5) % and photo of each microspheres from different surfactant concentrations (b) 0.5 %, (c) 1.5 %, (d) 2.0 % and (d) 2.5 % observed with optical microscope at 25x magnification

3.2. Analysis of functional groups in microspheres PLA/PCL by FT-IR.

Specific interactions that are often found in polyblends are hydrogen bonding, dipole-dipole interactions, and ionic interactions. Some of the techniques can be used to understand the specific interaction on polyblends one of which is by FT-IR [12]. PLA and PCL are aliphatic polyester with similar structures. The spectrum shows the bands from the C = O, C-O-C, and C-C group vibrations at 1754, 1175 and 1200 cm^{-1}, respectively [13]. As can be seen in figure 1, the bands showed on the FT-IR spectrum of PLA is also found in the spectrum of PLA / PCL microspheres. This indicates that the functional groups on the formed microspheres are combination of its constituent components.

3.3. Effect of surfactant concentration variation on microsphere size distribution.

Particle size diameter of the microspheres formed were measured using the LS 100Q Beckman Coulter Particle Size Analyzer (PSA) with micro volume module. The measurements of microspheres diameter are used to determine the particle size distribution of microsphere formed in each variation of surfactant concentrations (0.5, 1.5, 2.0 and 2.5) %.

As can be seen in figure 2, the addition of surfactant concentration from 0.5 % to 2.0 % has caused the particle diameter size to increase and the polydispersity's levels are relatively decreased, while from 2.0–2.5 % it stays relatively the same. The changes could be seen more clearly in the terms of the average particle diameter (mean) of the microspheres at each concentration variation as shown in table 3.

In this case, the tendency of particle diameter to be increased is likely have been caused by other potential factors that also can affect the particle size. Such as the dispersed phase density. According to literature, it is frequently done to vary the viscosity of the dispersed phase. The particle size will decrease exponentially with viscosity [14]. According to André-Abrant *et al.* [15] and Yang *et al.* [16], as the viscosity of the dispersed phase increased, the size of the resulting microspheres would get bigger. The addition of surfactant Span 80 into the dispersed phase may contribute in the increase of viscosity of the dispersed phase. The more amount of Span 80 added, the thicker dispered phase would be, so that the particle size increases.

3.4. Effect of variation of mixed surfactant composition on microsphere size distribution.

Figure 3a shows the measured data in graphical form of the particle diameter (μm) plotted against volume (%). The three graphs were multimodal, with one narrow peak at the highest intensity (in the area of 30.07 μm) and two wide low-intensity peaks. This indicates that the resulting microspheres have high polydispersity (the size still varies) but the size distribution is dominated by microspheres, which have particle diameter around 30.07 μm. If the graph in figure 3a is enlarged at the highest peak area around 30.07 μm, it can be seen that there are slight intensity differences in terms of amount of volume (%) which was 10.1 %, 9.98 % and 9.67 % for the sample 60:40, 50:50 and 60:40, respectively.

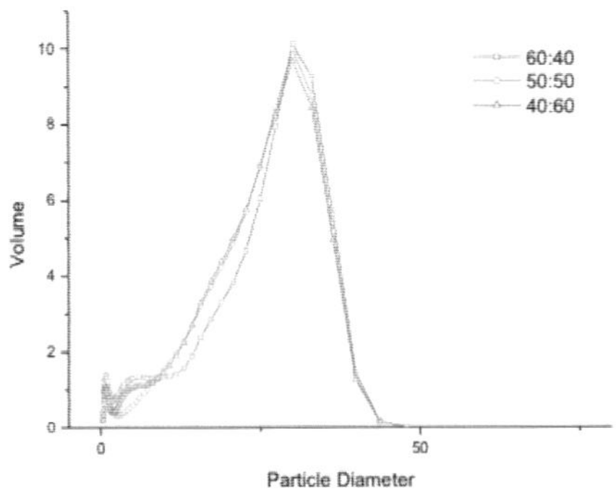

(a)

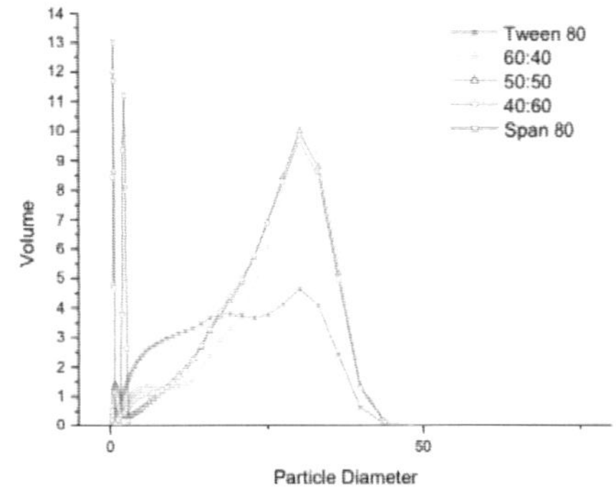

(b)

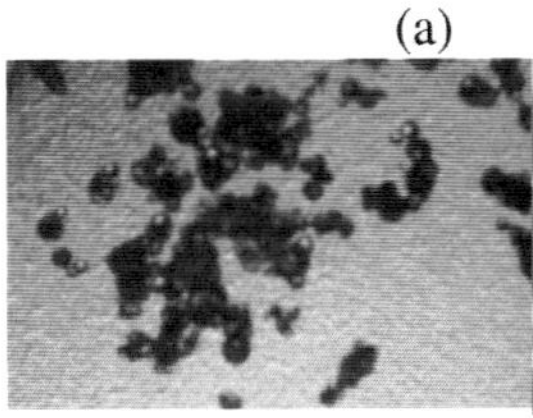

(c)

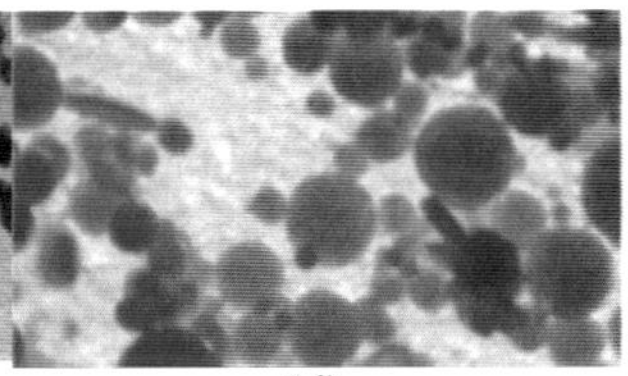

(d)

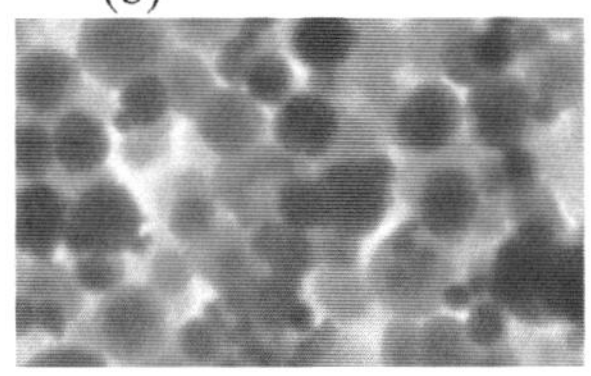

(e)

Figure 3. Graphic of (a) particle diameter size distributions with variation of mixed surfactant compositions, (b) comparison of microspheres size distributions based on different surfactant and microspheres formed with each surfactant concentrations (c) Span 80, (d) Tween 80/Span 80 (50:50), (e) Tween 80 observed with optical microscope at 25x magnification.

Figure 3b shows the comparison of microspheres size distribution based on different surfactant usage and composition between Tween 80 and Span 80. The microspheres formed using Tween 80 have a relatively high polydispersity and wide size distribution area ranging from 0.375 to 43.67 μm.

The size range resembles the ones formed using mixed surfactants. However, microspheres with mixed surfactants have lower polydispersity levels and one of the highest peaks is narrower than those of Tween 80 microspheres have. It is indicated that the microspheres are dominated by a certain size. For microspheres using Span 80 surfactants, only have two groups of particle size distributions based on peaks seen in the graph (0.375-0.791 μm and 1.520-2.920 μm). It is shown a relatively smaller size. However, its polydispersity was lower due to its relatively narrow size range (0.375-0.791 μm), when compared to those microspheres formed with mixed surfactants, which still varied in size. From the different surfactant forms in figure 3c, figure 3d and figure 3e it is obtained that mixing Tween 80 with Span 80 can also affect the different forms of microspheres particle.

If the graph in figure 4a is enlarged around the highest peak area (30.07 μm), it would show difference intensity in the form of volume (%) is 10.1 %, 11.0 % and 9.65 % for sample 700, 800 and 900 rpm, respectively. All three graphs were similar, but when viewed in the form of a group of sizes made based on peaks shown in figure 4a (0.375–3.205, 3.205–5.610 and 5.610–43.67) μm. There are differences in volume (%) between three microsphere sample charts. It was obtained that in 900-rpm sample, the amount of particles with size less than 3.5 μm was more abundant than those found in 700 and 800 rpm samples. According to Heisnaken *et al.* [6] and Gabor *et al.* [17] the increasing of stirring speed will increase the shear force, which would work to break up the droplet to form a smaller one. In this study, it might cause a few portion of bigger droplets in 900 rpm sample to split into smaller droplets.

3.5. Effect of varied stirring time on microsphere size distribution.

If the graph is inspected around the highest peak area (30.07 μm), which can be seen in figure 4b there were intensity differences in the form of volume (%) between every samples. There are 11.6 %, 10.1 %, 10.5 % and 9.95 % for each samples 0.5, 1.0, 1.5 and 2.0 h, respectively. Sample of 0.5 hours has the highest amount of % volume distributed around 30.07 μm area, which is 11.6 %, indicating that the 0.5 sample has more microspheres with diameter around 30.07 μm. This might be related to

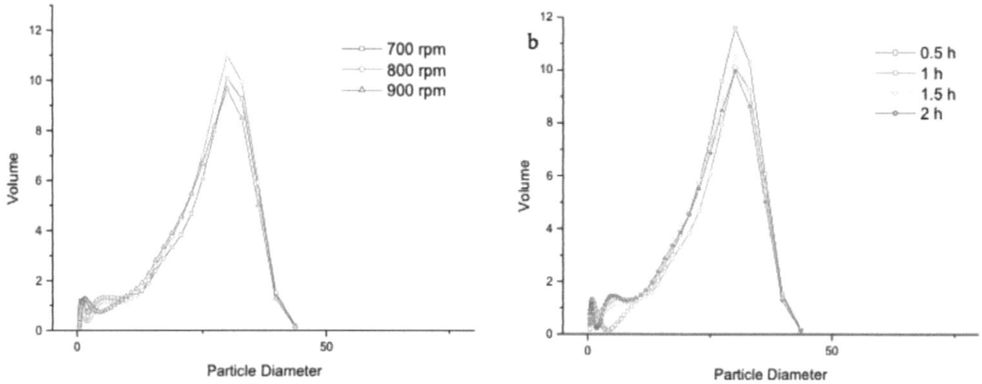

(a) (b)

Figure 4. Graphic of particle diameter size distributions with variation of (a) stirring speed, and (b) stirring time.

the duration of stirring. It is possible that there may be some portion of microspheres in which the size has not reduced yet according to evaporation of the solvent. Thus, means there are still few amount of solvents that have not been evaporated from the surface of the microspheres which eventually affecting the size of the microspheres formed. As for the other three variation samples, there are no significant differences between one another.

4. Conclusions

Microspheres formed using polyblend of poly(lactic acid) and polycaprolactone through the solvent evaporation method (o/w) were spherical, having a small size of < 50 µm and were fine as powder but the particle size distribution was still not uniform. The uniformity of microspheres might be related to the pressure given during spraying or stirring which could affect the distribution of the droplet size formed. The usage of mixed surfactants Tween 80 and Span 80 in microspheres preparation process affected the size of microspheres formed and its polydispersity. The size of the microspheres produced using mixed surfactants with compositions (60:40, 50:50, and 40:60) were similar to the Tween 80 microspheres sizes (0.375–43.67) µm and greater than microspheres produced using Span 80 surfactants (0.375–0.791) µm. The mixed surfactants microspheres polidispersity were lower compared to Tween 80 microspheres but is relatively higher if compared to 80 Span microspheres.

Acknowledgements

Authors would like to thank Universitas Indonesia for funding this research through PITTA Grant with contract number 233/UN2.R3.1/ppm/2018

References

[1] Liechty W B, Kryscio D R, Slaughter B V and Peppas N A 2010 *Annu. Rev. Chem. Biomol. Eng.* 1 149–73
[2] Wen H, Jung H and Li X 2015 *AAPS J.* 17 1327–40
[3] Middleton J C and Tipton A J 2000 *Biomaterials* 21 2335–46
[4] Venkatesan P, Manavalan R and Valliappan K 2009 *Journal of Pharmaceutical Sciences and Research (JPSR)* 1 26–35
[5] Park J H, Ye M and Park K 2005 *Molecules* 10 146–61
[6] Heiskanen H, Denifl P, Hurme M and Pitkänen P 2010 *Chem. Eng. Technol.* 33 1635–44
[7] Rosen M J 2004 *Surfactant and Interfacial Phenomena* 3rd ed (Hoboken: John Wiley & Sons)
[8] Boyd J, Parkinson C and Sherman P 1972 *J. Colloid Interface Sci.* 41 359–70
[9] Elfrida J 2012 *In Vitro Efficiency, Dissolution and Degradation Test of Ibuprofen Microencapsulation Using Polybend of Poly(lactic acid) and Polycaprolactone* (Depok: Universitas Indonesia) Undergraduate Thesis
[10] Kemala T, Budianto E and Soegiyono B 2010 *Arab. J. Chem* 5 103–8
[11] Lasprilla A J R, Martinez A G R, Lunelli B H, Figueroa J E J, Jardini A L and Filho R M 2011 *Chem. Eng. Trans.* 24 985–90

[12] Thomas S, Grohens Y and Jyotishkumar P 2015 *Characterization of Polymer Blends: Miscibility, Morphology and Interfaces* 1st ed (Weinheim: Wiley-VCH Verlag GmbH & Co)
[13] Chen C C, Chueh J Y, Tseng H, Huang H M and Lee S Y 2003 *Biomaterials* **24** 1167–73
[14] Li M, Rouaud O and Poncelet D 2008 *Int. J. Pharm* **363** 26–39
[15] André-Abrant A, Taverdet J L and Jay J 2001 *Eur. Polym. J.* **37** 955–63
[16] Yang C Y, Tsay S Y and Tsiang R C C 2000 *J. Microencapsul.* **17** 269–77
[17] Gabor F, Ertl B, Wirth M and Mallinger R 1999 *J. Microencapsul.* **16** 1–12

Determination of the effect of EDTA and PCMB on purified bromelain activity from pineapple core and in vitro antiplatelet activity

W A Hidayani, S Setiasih and S Hudiyono

Department of Chemistry, Faculty of Mathematics and Natural Sciences (FMIPA), Universitas Indonesia, Kampus UI Depok, Depok 16424, Indonesia

Corresponding author's e-mail: sumi.hudiyono@sci.ui.ac.id

Abstract. Bromelain is an enzyme belongs to the cysteine protease. In this study, bromelain isolated from pineapple core (*Ananascomosus* [L.] Merr) was purified by fractionation using ammonium sulfate followed by dialysis and then ion exchange chromatography. The fraction of bromelain obtained from each purification step showed an increase in specific activity. The highest specific activity of protease was found in 20-50% ammonium sulfate fraction of 104.018 U/mg with a purity level 3.2-fold compared to crude extract. Further purification by ion exchange chromatography using DEAE-Cellulose, the fraction of bromelain showed an increase in specific activity to 278.333 U/mg with a purity level 8.8-fold compared to crude extract. The determination of kinetics parameter of purified bromelain using Lineweaver-Burk plot gives Km value of 0.15 % (w/v) and Vmax of 0.056 U/min. This bromelain can be strongly inhibited by EDTA and PCMB. The addition of EDTA and PCMB at a concentration of 0.5 mM can decrease the activity of the enzyme up to 88.50% by showing the competitive and mix-inhibition types of inhibition, respectively. The antiplatelet activity of the bromelain fraction was tested in-vitro based on the Born method, by using plasma (PRP), acetosal as a positive control and ADP as an aggregator. The purified bromelain showed the ability of an antiplatelet agent with percentage of aggregation 29.51% and percentage of inhibition 68.91%.

Keywords: Pineapple core, bromelain, inhibitor, enzyme kinetics, antiplatelet

1. Introduction
Bromelain is a group of cysteine proteases whose its catalytic function is to hydrolyse the peptide bonds of proteins [1]. This enzyme is commonly found in various tissues of pineapple plants (Ananas comosus [L.] Merr) and the highest content is in the pineapple core. Bromelain which is an endopeptidase enzyme has a sulfhydryl group (-SH) on its active side that can be inhibited by oxidizing compounds, alkylators, and heavy metals. In 2012, Marshall and Golden characterized bromelain from Morindacitrifolia (Noni) [2] and observed the non-competitive inhibition of bromelain by HgCl2 [3]. Protease group is inhibited by PCMB (p-Chloromercuriobenzoate), PMSF (Phenylmethylsulphonyl fluoride), EDTA (Ethylenediaminetetraacetic acid), benzamidine, and pepstatin A [4]. Bromelain has been shown to be antiplateletagent [5]. Bromelain will work on the parts that have a high platelet aggregation activity in stroke and infark miokard and it can reduce platelet sensitivity in ADP (Adenosine 5' diphosphate) induction aggregation on in-vitro test [6].

2. Materials and methods

2.1. Materials

The pineapples were purchased from Pasar Induk Kramat Jati. Ammonium sulfate (Merck), $BaCl_2$ (Merck), NaCl (Merck), EDTA (Merck), PCMB (Sigma), DEAE Cellulose (Pharmacia Inc.), Platelet Rich Plasma (PRP) was obtained from PMI DKI Jakarta, ADP (Sigma), and acetosal (Sigma).

2.2. Fractionation with ammonium sulfate
The crude enzyme was slowly added by ammonium sulfate (0-20%, 20-50%, 50-80%) with a constant stirring. It was stored overnight in the refrigerator [7]. Then, the enzyme was tested for the protein content by Lowry method and proteolytic activity test with Kunitz method [8].

2.3. Dialysis
The fraction enzyme was put into a cellophane bag that immersed in 0.05 M phosphate buffer pH 7.0. The buffer was changed every two hours and tested with $BaCl_2$ 5% and 0.05 M HCl.

2.4. Ion exchange chromatography
The DEAE Cellulose column was eluted with 0.05 M tris-HCl buffer pH 8.0. Then, the subsequent elution with 0.05 M tris-HCl buffer pH 8.0 containing NaCl solution with variation concentration of 0.25 M; 0.50 M; 0.75 M; and 1.0 M [9].

2.5. Kinetics studies
Kinetic parameters such as Km and Vmax were calculated using Lineweaver Burk Plot by plotting 1/V with 1/[S]. Bromelain activity was assayed in the absence and presence of EDTA and PCMB (0.1; 0.3; and 0.5) mM with variation of substrate concentrations (0.25; 0.5; 0.75; 1.0; 1.25; 1.5 %).

2.6. In vitro antiplatelet activity test
The antiplatelet activity test of the enzyme bromelain fraction was carried out by Born method. Calculation of antiplatelet action can be seen from the inhibition percentage of platelet aggregation and percentage of aggregation that occurs [6].

3. Results and discussion

3.1. Bromelain purification from crude extract
Crude enzyme from pineapple core has a spesific activity of 31.523 U/mg proteins. Purification by fractionation with ammonium sulfate is based on the principle of salting out. Ammonium sulfate can separate the protein in solution by precipitation without denaturing the protein. Fraction 2 has the highest spesific activity than fraction 1 and 3, that is 104.018 U/mg protein with purity level of 3.2-fold to the crude enzyme extract. It means the ammonium sulfate concentration 20-50% can precipitate the bromelain protein at the best.

The next stage to purify enzyme fraction was the dialysis. The principle of dialysis is diffusion process that occurs in small molecules through semi-permeable membrane. The spesific activity is 125.365 U/mg protein with purity level of 3.9-fold to the crude enzyme extract.

The dialysis fraction was further purified using ion exchange column chromatography method to separate the bromelain protein with other proteins based on the difference in ion charge on the molecular surface. Resin used in the column is DEAE Cellulose, which is a weak anion exchange resin. 117 fractions were obtained and all fractions were measured for their protein absorbance at 280 nm and proteolytic activity. The highest specific activity of bromelain is fraction 4 (AP 4) that eluted by NaCl 1.0 M. The spesific activity is 278.333 U/mg protein with purity level of 8.8-fold to the crude enzyme extract. Bromelain of this fraction has a negative charge and bonded first with positively charged resin and eluted when the salt concentration is high.

3.2. Kinetics studies

3.2.1. Determination of bromelain kinetics parameters. Kinetic parameters (Km and Vmax) are obtained by transforming the Michaelis-Menten equation to the Lineweaver Burk Plot by mapping the function 1/V to 1/S. Based on the Lineweaver Burk plot, the linear equation is y=2.792x+18.057 with slope value of 0.990 was obtained, so the Km and Vmax values were 0.154 % (w/v) and 0.056 U/min.

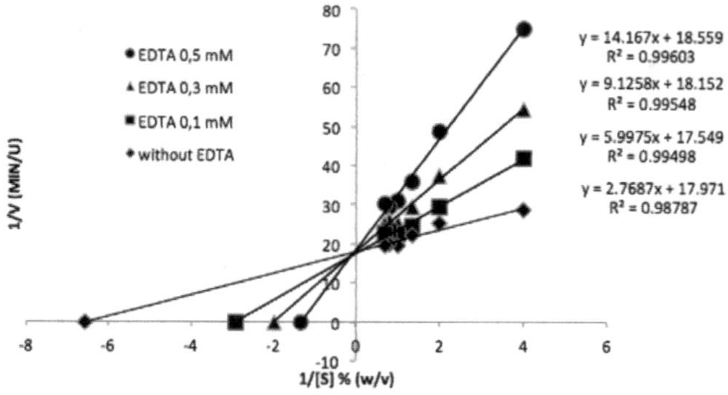

Figure 1. Lineweaver Burk plot for the hydrolysis of Casein in the presence of EDTA (0.1; 0.3; 0.5) mM by purified bromelain (AP4)

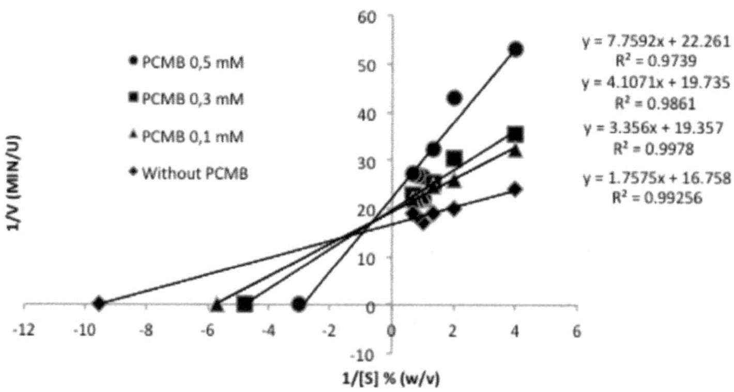

Figure 2. Lineweaver Burk plot for the hydrolysis of Casein in the presence of PCMB (0.1; 0.3; 0.5 mM) by purified bromelain (AP4)

3.2.2. The Effect of EDTA and PCMB on bromelain activity. Proteolytic activity of bromelain in the presence of EDTA in various concentrations obtained the Km value which increased with addition of EDTA (figure 1). Km with the absence and presence of EDTA was increased to 0.763 % (w/v) at 0.154 % (w/v). However, Vmax remained unchanged of 0.055 U/min implying a competitive type of inhibition for bromelain by EDTA. EDTA can decrease the activity of the bromelain up to 82.29 %.

The presence of EDTA on the fraction of bromelain causes the metal ions on the enzyme's active side is chelated by EDTA, so the structure of enzyme is changed and proteolytic activity is decreased. Based on the mechanism of EDTA inhibiting this protease, it can be said that EDTA is a metalloprotease inhibitor.

Proteolytic activity of bromelain in the presence of PCMB in various concentrations obtained the Km value, which increased with addition of PCMB (figure 2). Km with the absence and presence of PCMB was increased from 0.154 % (w/v) to 0.349 % (w/v). However, Vmax was reduced to 0.045 U/min from 0.056 U/min, implying mix-inhibition type. PCMB can decrease the activity of the bromelain up to 88.50 %.

The effects of bromelain inhibitor PCMB were also analyzed. In this case, bromelain contains thiol group (-SH), which reacts with Hg^{2+} ions on PCMB. Hg^{2+} ions will bind to the enzyme-substrate complex to form the enzyme-substrate-Hg complex, so that the bromelain will be inhibited. Thus, the purified bromelain is a cysteine protease group because it can be inhibited by PCMB, which is cysteine inhibitor.

3.3. Antiplatelet activity test in vitro

Antiplatelet activity was assayed by observing percentage of platelet aggregation and inhibition. The percentage of platelet aggregation is the effect of a compound on the process of platelet clot formation in the blood, while the percentage of platelet inhibition is the ability of a compound to inhibit the aggregation process. If the aggregation percentage was low then the inhibition percentage was high, it shows that the compound was more active as well as the ability of these compounds as antiplatelet agents. The more pure fraction of bromelain has a better ability to inhibit platelet aggregation. The highest antiplatelet activity was found in bromelain from DEAE Cellulose column chromatography (AP4) with aggregation percentage of 29.51 % and inhibition percentage of 68.91 %.

4. Conclusions

Bromelain was successfully purified through several of purification stages, such as extraction of crude enzyme, fractionation with ammonium sulfate, dialysis, and ion exchange chromatography. Bromelain purified by ion exchange chromatography DEAE Cellulose gave the highest specific activity of 278.333 U/mg protein with purity level of 8.8-fold to the crude enzyme extract. The determination of kinetics parameter of purified bromelin using Lineweaver-Burk plot gives Km value of 0.154 % (w/v) and Vmax of 0.056 U/min. Bromelain fraction was inhibited by EDTA and PCMB with inhibition type of competitive and mix-inhibition, respectively. The addition of EDTA and PCMB can decrease the activity of bromelain up to 82 %. The antiplatelet activity test of the purified bromelain showed its ability as an antiplatelet agent. The highest antiplatelet activity was found in bromelain fraction DEAE Cellulose (AP4) with percentage of aggregation 29.51% and percentage of inhibition 68.91 %.

Acknowledgements

This research was funded by Hibah Publikasi Internasional Terindeks untuk Tugas Akhir Mahasiswa Universitas Indonesia (PITTA) 2018 on behalf of Dra. Siswati Setiasih, Apt., M.Si.

References

[1] Borrelli F *et al.* 2011 *Neurogastroenterol Motil.* **23** 745–e331
[2] Golden K D and Smith-Marshall J 2012 *Journal of Scientific Research (JSR)* **4** 445–56
[3] Kaur T, Kaur A and Grewal R K 2015 *J. Food Sci. Technol.* **52** 5954–60
[4] Walsh G 2002 *Protein Biochemistry and Biotechnology* (New York: John Wiley and Sons)
[5] Manzoor Z, Nawaz A, Mukhtar H and Haq I 2016 *Brazillian Arch. Biol. Technol.* **59** 1–16
[6] Moriyama H, Hosoe T, Wakana D, Itabashi T, Kawai K, Iizuka T, Hoshi K, Fukushima K and Lau F C 2009 *Journal of Health Science* **55** 103–8
[7] Gautam S S, Mishra S K, Dash V, Goyal A K and Rath G 2010 *Thai J. Pharm. Sci.* (*TJPS*) **34** 67–76
[8] Musfiroh F F, Setiasih S, Handayani S, Hudiyono S and Ilyas N M 2018 *IOP Conf. Ser.: Mater. Sci. Eng.* **299** 012017
[9] Setiasih S, Adimas A C D, Dzikria V and Hudiyono S 2018 *IOP Conf. Ser.: Mater. Sci. Eng.* **299** 012016

Study of microalgae (*Scenedesmus Sp.)* utilization as phosphate bioremediator (PO_4^{3-}) in domestic wastewater medium

M Maisarah[1], A Saefumillah[1] and H Ambarsari[2]

[1] Department of Chemistry, Faculty of Mathematics and Natural Sciences (FMIPA), Universitas Indonesia, Kampus UI Depok, Depok 16424, Indonesia

[2] Laboratory for Water and Waste Water Treatment Technology (BTPAL), The Agency for the Assessment and Implementation of Technology (BPPT), Building 820 Geostech Kawasan Puspitek Serpong, Tangerang Selatan, Banten 15314, Indonesia

Corresponding author's e-mail: asep.saefumillah@sci.ui.ac.id

Abstract. In this research, the use of *Scenedesmus sp.* as a phosphate bioremediator agent is expected to be applied for lowering the phosphate levels in the water as well as improving the quality of the environment. *Scenedesmus sp.* used in this research was isolated from Sungai Cideng. Variation of phosphates were tested to see the effect of various types of phosphate found in the waters in the form of a variation of organic phosphates using the Adenosine 5'-monophosphate (AMP), phytic acid and variation of inorganic phosphates using potassium dihydrogen phosphate (KH_2PO_4) and sodium tripolyphosphate (STPP). It was found that the use of monopotassium phosphate (KH_2PO_4) was better in decreasing phosphate levels and increasing the number of *Scenedesmus sp.* cells with a phosphate decrease percentage of 87.5 %. Phosphate test was performed using UV-Vis spectroscopy instrumentation and the abundance of cell count was performed using a microscope and hemocytometer. The environmental quality test parameters are temperature and pH. Direct application in wastewater from Muara Angke was also found the decrease of phosphate levels and increase the number of *Scenedesmus sp.* cells. *Scenedesmus sp.* has the ability to reduce nutrients in wastewater to below wastewater quality standard of 86.0 % from the initial phosphate concentration of 0.2255 ppm to 0.0314 ppm.

Keywords: *Scenedesmus Sp.*, bioremediator, phosphates, wastewater, Muara angke

1. Introduction

Environmental pollution is the result of the production process, which produces waste. Several types of waste contain phosphate compounds in high concentrations before discharged into the water. It is known, that there are several types of microalgae that can live and thrive in environments with high phosphate concentrations. One of the parameters measured in determining the quality of wastewater treatment results is phosphate content. Phosphate levels in the waters still exceed the predetermined quality standard. Phosphate in waste can be organic phosphate, inorganic orthophosphate and polyphosphate complex. High phosphate content in wastewater can cause eutrophication. Bioremediation technology is a technology that plays an important role in achieving the goal of this research in regard to the environmental side. Starts with a passive metabolic process, bioremediation technology generally involves a process of biosorption or absorption in which the amount of contaminants can be removed and eliminated by sorbent in kinetic equilibrium and compounds contained on the cellular surface of sorbent, then the pollutants can be absorbed into the cellular structure, and finally there is an active metabolic bioaccumulation process from living organisms and requires respiration where absorbing contaminants are transferred to cell [1]. The opportunity to utilize microalgae in the bioremediation process of household waste is very possible. Wastewater is a very good medium and has the potential to be used by microalgae as a growth medium because its effectiveness and also supported by low cost management [2]. So, this research is needed to obtain more detailed scientific facts. Therefore, a research in the form of microalgae simulation is conducted with the aim to know the extent of bioremediation effect using microalgae to increase the quality of household waste especially to decrease the phosphate level. The microalgae used in this study was *Scenedesmus sp.* which is a species of microalgae that has a high growth rate, cosmopolitan characteristics, and not motile. Microalgae *Scenedesmus sp.* is also able to take advantage of various minerals contained in the water to the bottom of the water. This fact became the basis for the use of *Scenedesmus sp.* as an absorbent pollutant contained in the waste. Observation of bioremediation process is conducted on *Scenedesmus sp.* microalgae with the measurement of phosphate for 18 days and environmental quality parameters, once in every 3 days.

2. Materials and methods

2.1. Materials

The materials used were microalgae culture Scenedesmus sp. from Sungai Cideng as the raw material, ethanol (C_2H_5OH) (Sigma) for sterilization, medium bold basal for medium growth of Scenedesmus sp., potassium dihydrogen phosphate (KH_2PO_4) (Merck), sodium tripolyphosphate ($Na_5P_3O_{10}$) (Merck), adenosine 5'-monophosphate (C10H14N5O7P) (Merck), phytic acid ($C_6H_{18}O_{24}P_6$) (Merck) for variation of phosphate, sulfuric acid (H_2SO_4) (Merck), phenolpthalein indicator ($C_{20}H_{14}O_4$) (Sigma), ammonium heptamolybdate [$(NH_4)_6Mo_7O_{24}.4H_2O$] (Merck), potassium antimonyl tartrate [$K(SbO)C_4H_6O_6$. ½ H_2O] (Sigma), ascorbic acid ($C_6H_8O_6$) (Merck) for molibdate reagent, perchloric acid ($HClO_4$) (Merck) for destruction, and wastewater as the main ingredient from household wastewater in Muara Angke estuary.

2.2. Preparation of Scenedesmus sp. inoculum

In the first step, *Scenedesmus sp.* was isolated by referring to previous studies [3]. The cultivation of *Scenedesmus sp.* contained of 490 mL of bold basal medium (BBM) and 210 mL volume of inoculum *Scenedesmus sp.* at 30 % concentration (Vinoculum/Vmedia) in a HDPE bottle incubator for 8 days, under the following cultivating conditions placed the illumination of light and fixed the aeration rate at 1 L/min of air supplied into the bottle by using an air pump. The working volume for the experiment was 700 mL for each bottle and the remaining space was for the headspace to promote gas exchange. Cell calculations were performed using a microscope by the Improved Neubaur count chamber method.

2.3. Preparation of phosphate solution

To examine the effect of phosphate type on bioremediation activity, several phosphate compounds were used as model including sodium Tripolyphosphate (STPP) salt, potassium dihydrogen orthophosphate (KH_2PO_4) which is an inorganic phosphate type and Adenosine 5'-monophosphate (AMP) and Phytic acid (IHP) which is a type of organic phosphate. Preparation of KH_2PO_4, STPP, AMP and IHP from phosphate solution was conducted by dissolving variation of phosphates with distilled water in a measuring flask. Then, 500 mL of phosphate solution was made with concentration of 0; 0.2; 0.4; 0.6; 1 ppm.

2.4. Sampling research

The wastewater was taken from the waters of Muara Angke, Teluk Jakarta. The determination of the waste source was also selected based on the initial survey of the sampling. The location of Muara Angke, Teluk Jakarta is close to the household activities. Surface water was taken from the bridge on the waters of Muara Angke, Teluk Jakarta.

2.5. Bioremediation

Cultivated algae was taken 15 mL then centrifuged at 3000 rpm for 15 minutes to separate the media. The water sample was inserted into a beaker glass 1000 mL, then the water sample was filtered using a Buchner filter to remove the unwanted microalgae as well as suspended material. The same thing was

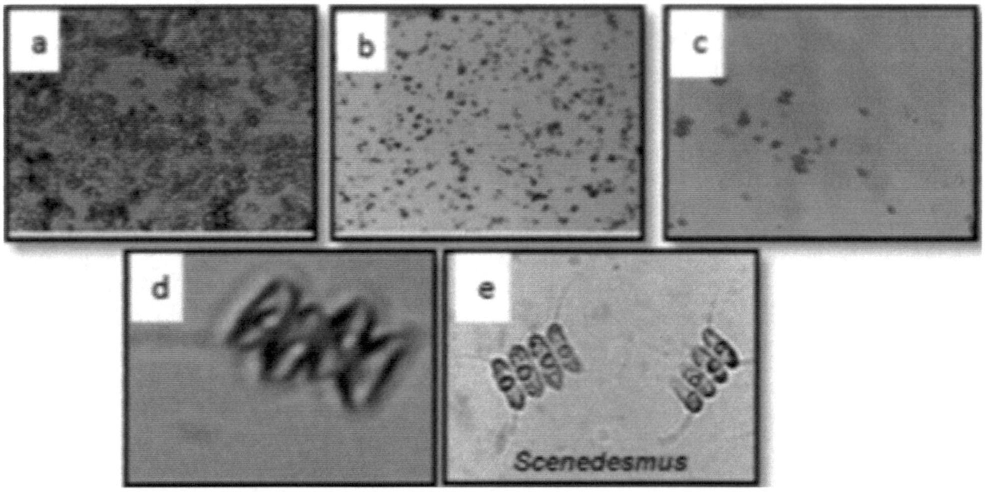

Figure 1. Observations using a microscope with 1000 x magnification of *Scenedesmus sp.* (a) without dilution and with dilution (a) 10^{-4} and (b) 10^{-2}

done on the sample solutions of organic and inorganic phosphate types. Experiments on the study were observed daily until 18th. There was also an addition of solution of 1 ppm Cd metal in wastewater and phosphate compounds to see the effect of 500 mL of metal ions with a mixture of wastewater and 50:50 phosphate compound then the sample was transferred into a plastic gallon.

2.6. Determination of phosphate levels

The data obtained from laboratory test results were analyzed using simple comparative phosphate (PO_4^{3-}) levels before and after treatment to know and see the effectiveness of *Scenedesmus sp.* in lowering phosphate levels (PO_4^{3-}). Phosphate measurements used a spectrophotometric method using the principle of molybdate reagent.

3. Results and discussion

3.1. Results of algae cultivation observation

To know the morphology of microalgae, an observation using microscope with 1000 x magnification was used. Based on the results of green algae, it can be concluded that the green algae is a green algae *Scenedesmus sp.* Based on the structure morphology obtained in figure 1, it shows that the morphological structure of the green algae is *Scenedesmus sp.* with the obtained characteristics similar to the reference literature in (e). The cylindrical algae are obtained, the terminals are round and there is spine in the form of fine hair at the edges of each corner, having colonies of 2-4 cells laterally. The results can be seen in figure 1.

Microalgae *Scenedesmus sp.* was taken from the algae stock in Geostech BPPT Serpong derived from the river Cideng as the mother cultivation microalgae. Cultivation was done among others to obtain a more pure and not polluted algae to multiply the number of microalgae in the sufficient amount for the research. In addition, cultivation was also done to see the growth curve of *Scenedesmus sp.* in its growth media.

The condition of culturing is obtained from measuring the condition of the media after autoclaving on the preparation of media for the nutritional needs of *Scenedesmus sp.* microorganisms. The nutrients used are derived from the bold basal medium or hereafter referred to as BBM media. The pH of the media measured using pH-meter is 6.6 and the temperature is 26.9 °C. This is in accordance with the optimum conditions for the growth of algae culture *Scenedesmus sp.*

3.2. Growth Curve Data

The growth phase of *Scenedesmus sp.* graph shows an increase in the number of cells during the 18 days of observation as seen in figure 2. The measurement of cell count was performed using a

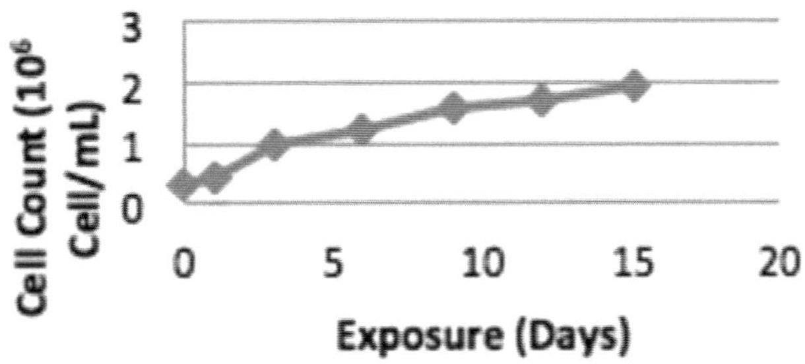

Figure 2. Graph growth *Scenedesmus sp.* in BBM

hemocytometer to determine the growth of microalgae. For 18 days, there was a qualitative observation corresponding with the color that the growth of microalgal cell counts also increased. The original light green color changed on the first day to a darker color in the next day. The color change from bright green to solid green occurs due to the increase population of *Scenedesmus sp.* Microcalgae increased the number of *Scenedesmus sp.* cells indicating the occurrence of utilization of nutrients contained in the fuel medium for microalgae is proliferating. Based on the observation, it was found that the number of cells experiencing a maximum increase is on the 6th day of 1.00 x 10^6 cells/mL indicating that on the 6th day the exponential phase occurs and then undergoes the stationary phase. Thus, the microalgae is taken on the 6th day due to the maximum metabolism of the day and the phase.

3.3. Experimental treatment of variation of inorganic phosphate solution

The resulting cell count, phosphate concentration, temperature and pH can be seen in figure 3. In figure 3b, it can be seen that abundance of *Scenedesmus sp.* in KH_2PO_4 inorganic phosphate solution shows an increase every day. From the curve, it can be seen that the number of cells *Scenedesmus sp.* continues to increase with the highest abundance in the solution in a concentration of 1 ppm of 4.88 x 10^5 cells/mL of the initial abundance of 4.8 x 10^4 cells/mL with a decrease in phosphate levels of 28.2 % in the initial concentration of phosphate 1.0057 ppm decreases to 0.7215 ppm. The increase in growth shows the microalgae is experiencing a logarithmic phase caused by the availability of nutrients especially phosphate. The increase in the number of cells is thought to be caused by inorganic phosphate in the form of KH_2PO_4 which can be well used to support the growth of cells as indicated by curves that tend to rise from the first day until the last day of observation. The increasing number of *Scenedesmus sp.* was also followed by a decrease in phosphate levels in the medium. After the phosphate measurements, it can be seen in figure 3a it was found that the decrease in phosphate content with the highest decrease of 87.5 % in the solution with the initial phosphate concentration of 0.2234 ppm decreased to 0.0278 ppm with the result of the concentration of negative value due to the measured value has exceeded the existing standard curve of exceeding the limit of measurement detection. Furthermore, a decrease of 62.2 % indicated in solution with initial phosphate concentration of 0.4244 ppm decreased to 0.1600 ppm, while the reduction of 50.9 % in the solution with the initial phosphate concentration of 0.6111 ppm decreased to 0.3000 ppm.

Based on figure 4, it shows the ratio of inorganic phosphate absorption also used phosphate source solution in the form of sodium tripolyphosphate (STPP). In figure 4b, the number of cells *Scenedesmus sp.* shows a daily increase with the highest abundance in a solution in a concentration of 1 ppm of 3.68x10^5 cells/mL from an initial abundance of 4.8x10^4 cells/mL which also correlated positively with a decrease in phosphate levels with decreases in the initial concentration phosphate of 1.79 % at a concentration of 1.0057 ppm decreased to 0.9876 ppm. The increasing growth is due to the availability of nutrients especially phosphate but the amount of abundance is found to be lower than that of KH_2PO_4 so it is suspected that *Scenedesmus sp.* can use KH_2PO_4 inorganic phosphate solution.

3.4. Experimental treatment of variation of organic phosphate solution

The organic phosphate solution used in this experiment was Adenosine Monophosphate (AMP) and Myo-Inositol Hexacis Phosphate (IP6). Observation results of the influence of abundance of cell *Scenedesmus sp.* to organic phosphate solution as a cultivation medium is presented in figure 5.

IOP Conf. Series: Materials Science and Engineering **763** (2020) 012055 doi:10.1088/1757-899X/763/1/012055

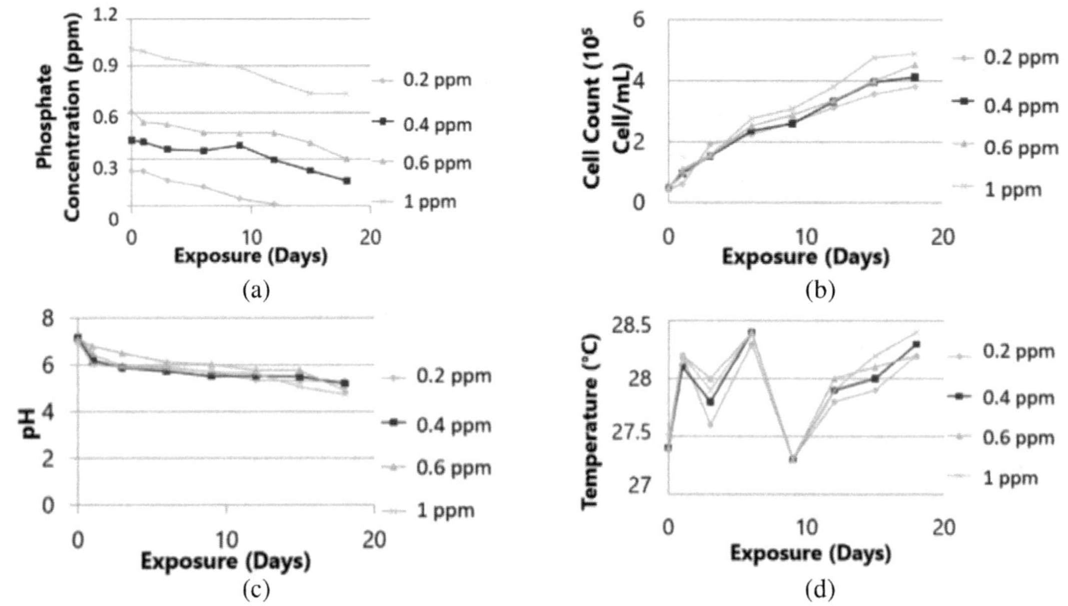

Figure 3. Bioremediation using inorganic phosphate KH_2PO_4 (a) decreased phosphate levels, (b) growth of *Scenedesmus sp.* (c) effect of pH and (d) effect of temperature

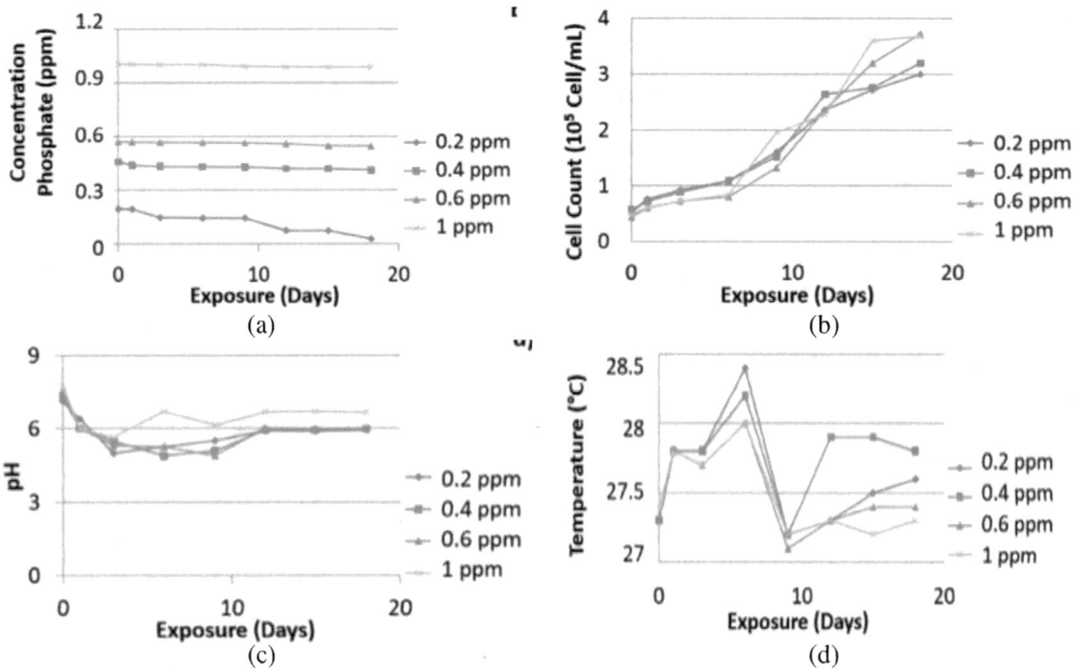

Figure 4. Bioremediation using inorganic phosphate STPP (a) decreased phosphate levels, (b) growth of *Scenedesmus sp.* (c) effect of pH and (d) effect of temperature

On the visible curves, the number of cells *Scenedesmus sp.* continues to rise each day with the highest abundance in a solution in a concentration of 1 ppm of 3.48 x 10^5 cells/mL from the initial abundance of 5.2 x 10^4 cells/mL from the first day to the last day, indicated by a curve that tends to rise. On the curve, it is visible the number of cells *Scenedesmus sp.* continues to increase each day

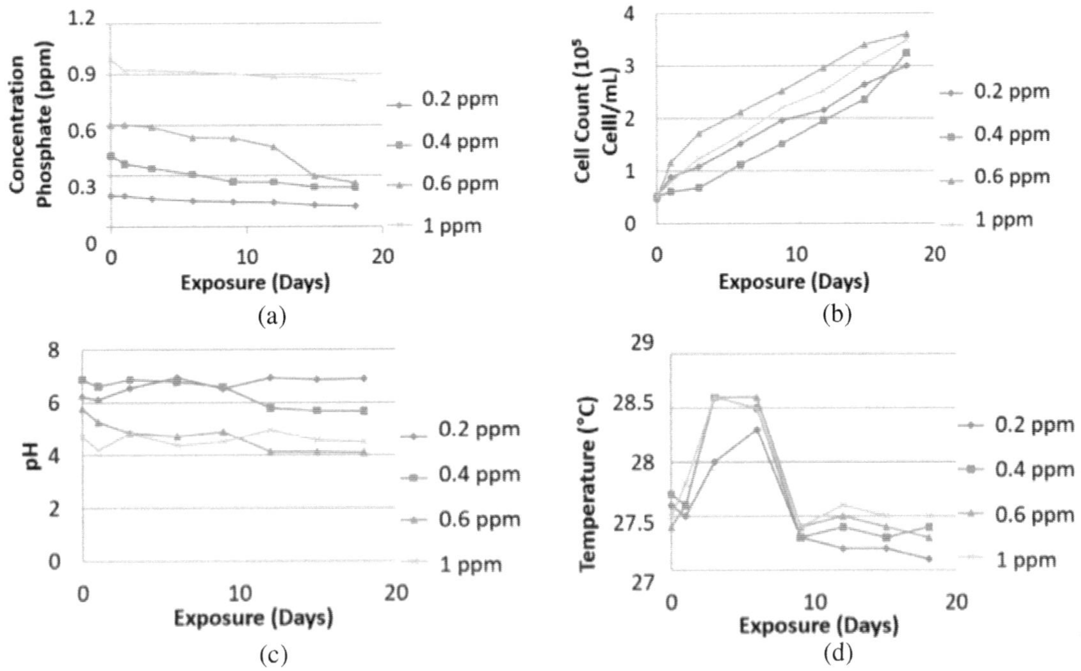

Figure 5. Bioremediation using inorganic phosphate AMP (a) decreased phosphate levels, (b) growth of *Scenedesmus sp.*, (c) effect of pH and (d) effect of temperature

with the highest abundance in a solution with concentration of 1 ppm of 3.48 x 10^5 cells/mL from an initial abundance of 5.2 x 10^4 cells/mL from the first day to the last day, indicated by an upwardly rising curve. It is also demonstrated and correlated with a decrease in phosphate levels that the *Scenedesmus sp.* absorbed the organic phosphate nutrient content. The result of the decrease of phosphate content was obtained with decrease in the solution of 34.3 % with initial concentration of phosphate AMP 0.1815 ppm decreased to 0.1191 ppm. Furthermore, 45.4 % with initial phosphate AMP concentration 0.4216 ppm decreased to 0.2300 ppm. Then, in a solution of 57.4 % with initial concentration of AMP phosphate of 0.6016 ppm decreased to 0.2560 ppm and a decrease of 13.17 % with initial concentration of AMP phosphate of 0.9898 ppm decreased to 0.8594 ppm.

Characterization using organic phosphate phytic acid can be seen in figure 6. Based on figure 6, it shows the suspected *Scenedesmus sp.* microalgae. It may be better to use an organic AMP phosphate solution due to the complex phytic acid (IP6) compound form and the unavailability of the fitase enzyme to metabolize the compound. Figure 6 shows a decrease in phytic acid phosphate levels obtained with a decrease of 4.2 % in a solution with an initial phosphate concentration of 0.9946 ppm decreased to 0.9525 ppm. Furthermore, in solution with initial concentration 0.1893 ppm decreased to 0.1688 ppm with percent decrease of 10.82 %. Continued in a solution of initial concentration of 0.4109 ppm decreased to 0.3833 ppm with a decrease percentage of 6.71 %, and in a solution with an initial concentration of phytic acid of 0.6133 ppm decreased to 0.5560 ppm with a decrease percentage of 9.34 %. Number of cells *Scenedesmus sp.* in IP6 organic phosphate solution showed an increase every day as shown in figure 6b. Number of cells *Scenedesmus sp.* continues to increase daily with the highest abundance at 1 ppm concentration of 3.20 x 10^5 cells/mL of an initial abundance of 4.8 x 10^4 cells/mL. The availability of nutrients, especially phosphates, leads to an increase in the growth of microalgae cells.

3.5. Results of bioremediation of wastewater

Observation results of the influence of abundance cells *Scenedesmus sp.* to household wastewater as a cultivation medium is presented in figure 7 with varying abundance data in each treatment each day. The calculation of abundance is intended to see how far the *Scenedesmus sp.* microalgae can utilize the existing content in the culture medium. The abundance curve shows that in the wastewater the

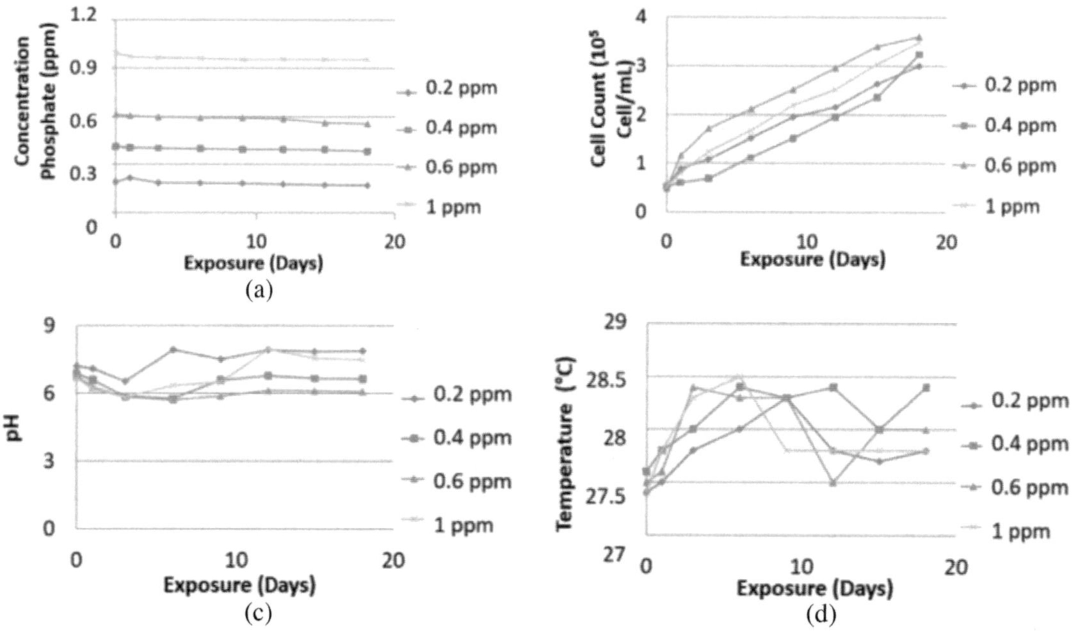

Figure 6. Bioremediation using inorganic phosphate IP6 (a) decreased phosphate levels, (b) growth of *Scenedesmus sp.* (c) effect of pH and (d) effect of temperature

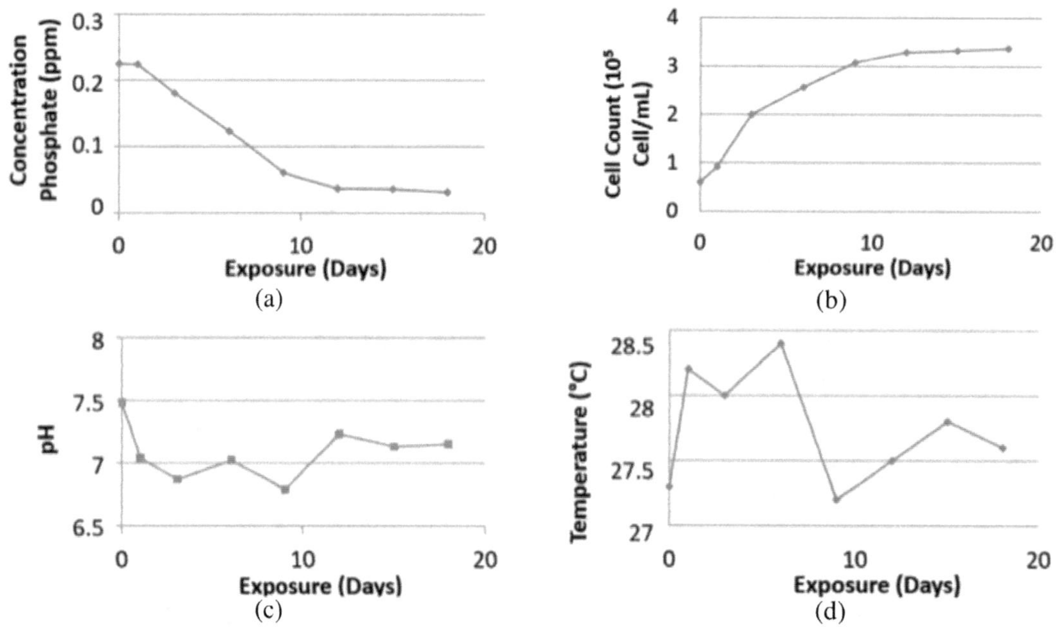

Figure 7. Bioremediation using wastewater (a) decreased phosphate levels, (b) growth of *Scenedesmus sp.*, (c) effect of pH and (d) effect of temperature

number of cells *Scenedesmus sp.* keeps increasing every day at the same time the phosphate decreases. The pattern of growth and reduction of nutrients can be seen in figure 7. The abundance is thought to be due to the amount of nutrient derived from the amount of phosphate used to support cell growth and growth from day one to the last day indicated by a curve that tends to rise, but on the 9th day onwards the cell growth begins to show a stationary phase indicating that the amount of nutrients available in the media began to decline and not enough for its further growth. The cells have utilized the nutrients for their metabolism. There was a growth of cell number from 4.8 x 10^4 cells/mL to 3.2 x 10^5 cells/mL while the decrease of phosphate concentration of 86.0 % from the initial phosphate 10^5 cells/mL while the decrease of phosphate concentration of 86.0 % from the initial phosphate concentration of 0.2255 ppm decreased to 0.0314 ppm. Thus, it is shown that *Scenedesmus sp.* has the ability to reduce nutrients in the form of phosphate in wastewater and use it for growth.

Based on wastewater quality standards according to the Regulation of the Minister of Environment of the Republic of Indonesia No. 5 2014, the phosphate concentration is 2 mg/L and according to Government Regulation No. 82 of 2001 on Water Quality Management and Water Pollution Control phosphate criteria as total P reaches class I & II with maximum limit of 0.2 mg/L and reaches class III with number maximum limit of 1 mg/L. So, the use of *Scenedesmus sp.* as bioremediator agent of wastewater has fulfilled the wastewater quality standard that is 0.0314 ppm. The optimum temperature for *Scenedesmus sp.* is 26–32 °C, at temperature 34–36 °C the *Scenedesmus sp.* will stop growing and inhibit the metabolism process to death. So, the temperature obtained indicates that the wastewater is still in the optimum temperature for the growth of *Scenedesmus sp.* The abundance curves tend to increase along with the pH of the medium, which tends to be alkaline.

4. Conclusions

Based on the results of research, *Scenedesmus sp.* can use variation of phosphate types, both inorganic phosphate and organic phosphate for the necessities of life. The use of inorganic phosphate type in the form of KH_2PO_4 is better than STPP. While on the variation of organic phosphate type, the AMP is better than phytic acid. The abundance of *Scenedesmus sp.*, which is applied directly by using wastewater, is increased daily and at the same time the phosphate levels are decreased. So, it can be concluded that *Scenedesmus sp.* has the ability to reduce nutrients in wastewater by 86.0 %. This phosphate decrease is caused by *Scenedesmus sp.*, which can use organic or inorganic phosphate contained in wastewater for the purpose of life metabolism because phosphate is a macro element for the necessities of life. After bioremediation, water quality in the environment is observed from the temperature and pH. The result after bioremediation showed to be in accordance with the temperature and pH in the threshold of the quality standard of environmental wastewater. So, the bioremediation by using *Scenedesmus sp.* can be said to last well.

Acknowledgements

This research was funded by Insinas 2017 and 2018 Kementerian Ristek Dikti.

References

[1] Kumar K S, Dahms H U, Won E J, Lee J S and Shin K H 2015 *Ecotoxicol. Environ. Saf.* **113** 329–52
[2] Delrue F, Álvarez-Díaz P D, Fon-Sing S, Fleury G and Sassi J F 2016 *Energies* **9** 1–19.
[3] Hariz H B and Takriff M S 2017 *Transactions on Science and Technology* **4** 298–311
[4] Lavens P and Sorgeloos P 1996 *Manual on the Production and Use of Live Food for Aquaculture* (*FAO Fisheries Technical Paper* vol 361) (Rome: FAO UN)
[5] Yewalkar-Kulkarni S, Gera G, Nene S, Pandare K, Kulkarni B and Kamble S 2017 *Indian J. Microbiol.* **57** 241–9
[6] Reynolds C S 2006 *The Ecology of Phytoplankton* (Cambridge: Cambridge University Press)
[7] Larsdotter K 2006 *Microalgae for Phosphorus Removal from Wastewater in a Nordic Climate* (Stockholm: School of Biotechnology) PhD Thesis
[8] Jalal K C A, Alam M Z, Matin W A, Kamaruzzaman B Y, Akbar J and Toffazel H *2011 IIUM Engineering Journal* **12**(4) 125–32
[9] Xin L, Hong-ying H, Ke G and Ying-xue S 2010 *Bioresour. Technol.* **101** 5494–500
[10] Williams J D H, Shear H and Thomas R L 1980 *Limnol. Oceanogr.* **25** 1–11

Chemical vapour deposition (CVD) of molybdenum into medium pore H-zeolites

S Suwardiyanto[1], S Svelle[2] and R F Howe[3]

[1] Department of Chemistry, University of Jember, Jalan Kalimantan 37, Kampus Tegal Boto, Jember 68121, Indonesia

[2] Department of Chemistry, University of Oslo, Chemistry Building, Sem Sælands vei 26, 0371 Oslo, Norway

[3] Department of Chemistry, University of Aberdeen, King's College Aberdeen AB24 3FX, United Kingdom

Corresponding's author: antok.fmipa@unej.ac.id

Abstract. Several medium pore H-zeolites (HZSM-5, HIM-5, HTNU-9 and HZSM-11) were evaluated toward molybdenum deposition *via* chemical vapor deposition using $Mo(CO)_6$ as the molybdenum precursor. The deposition was through a high temperature adsorption of $Mo(CO)_6$ vapor onto dehydrated zeolites. The progress of deposition was monitored gravimetrically. *Ex-situ* infrared (FTIR) spectroscopy was employed to observe the interaction between $Mo(CO)_6$ and the zeolites during the deposition. X-ray photoelectron spectroscopy (XPS) was used to scrutinize the nature of molybdenum deposit within the zeolite. High temperature adsorption of $Mo(CO)_6$ vapor is an irreversible adsorption resulted in a molybdenum deposition onto the zeolites whereas the adsorption conducted at room temperature is a reversible one. Interaction of $Mo(CO)_6$ and the zeolites at high temperature led to the reaction of $Mo(CO)_6$ and hydroxyl group within zeolites i.e. silanol group and Brønsted acid site. The molybdenum dispersion within the zeolites was governed by the particle size of the zeolite. More concentrated molybdenum deposited on zeolite surface occurred on the zeolite with bigger particle size. The highest surface molybdenum deposition was observed on ZSM-11 surface as it has the biggest particle size.

1. Introduction

Molybdenum supported zeolite, especially medium pore zeolites, is an active catalyst toward light alkane activation such as methane conversion into aromatics and dehydrogenation of propane [1, 2]. The most applied method on the preparation of this material is wet impregnation using ammonium heptamolybdate solution [3]. A wide range of synthesis methods has been explored to gain a better catalyst. Some reviews claim that molybdenum introduced by wet impregnation result in catalyst for methane dehydroaromatization with highest activity [3,4].

The most crucial issue on the preparation of molybdenum-supported zeolite is the molybdenum dispersion throughout the zeolite. Conventional method cannot achieve a good molybdenum dispersion [5]. This is important since the reaction is believed to require close proximity of the active molybdenum phase to internal acid sites in the zeolite [6]. An improvement on the molybdenum dispersion was achieved by Sun *et al.* [7] by introducing ammonium heptamolybdate into Zeolite Socony Mobil-5 (ZSM-5) at high pH. Catalyst prepared had a higher performance attributed to a better diffusion of monomeric MoO_4^{2-} species at high pH into zeolite pore. The ability of an anionic molybdenum precursor to diffuse uniformly throughout an anionic zeolite framework is however unclear.

$Mo(CO)_6$ is an alternative precursor to be introduced into microporous framework. This uncharged organometallic precursor with significant vapor pressure at room temperature (~0.14 torr) is readily to be deposited in atomic level since it is a discrete molecule [8]. This method has been successfully applied to synthesize catalyst with high uniform molybdenum loading in large pore zeolite such as faujasite [9] and EMT [10]. Deposition of molybdenum into large pore zeolite conducted by room temperature adsorption of $Mo(CO)_6$ is used to synthesize an active catalyst but this protocol does not work in the zeolite with medium pore such as ZSM-5. $Mo(CO)_6$ with kinetic diameter more than 6 Å exceeds the notional pore diameter of a 10-ring zeolite such as ZSM-5. Therefore this molecule cannot diffuse into the medium pore zeolite by adsorption at room temperature [11]. This paper explained about the work on Molybdenum deposition on various medium pore zeolites, pore diameter of c.a. 5.5 Å, i.e. HZSM-5, HIM-5, HTNU-9 and HZSM-11, *via* chemical vapor deposition at elevated temperature using $Mo(CO)_6$ as the molybdenum precursor.

2. Experimental

2.1. Materials

ZSM-5 (Si/Al ratio: 40, zeolyst), TNU-9 (Si/Al = 13), IM-5 (Si/Al = 13) and ZSM-11 (Si/Al = 22) were synthesized at the University of Oslo. $Al(NO_3)_3.9H_2O$ and Cabosil were used as aluminum and silicon source. The synthesis was conducted in a 52 mL Teflon liner steel autoclave. As synthesized zeolites were calcined at 550 °C for 5 hours in 50:50 oxygen and nitrogen. The calcined zeolites were treated with three-time ion exchange with excess 1M NH_4NO_3 for 2 hours at 70 °C. Calcination was applied at 550 °C for 5 hours to obtain H-form of zeolites $Mo(CO)_6$ was from Koch-Light Labs (Cat. Number 164115-69-9).

2.2. Synthesis

Deposition of molybdenum into the zeolites was performed in a vacuum microbalance. Prior to the deposition, the zeolites were dehydrated at 500 °C under vacuum condition. Freeze vacuum treatment was applied to $Mo(CO)_6$ for purification to be ready for deposition. Adsorption of $Mo(CO)_6$ vapor onto the zeolites was carried out at 120 °C to deposit molybdenum until reaching the required level of deposition. Subsequent evacuation followed the deposition to complete the decarbonylation accompanied by gradual heating at 120 °C, 200 °C and 300 °C. Room temperature deposition of molybdenum was also studied through a room temperature adsorption. Exposure of $Mo(CO)_6$ vapor onto the zeolites at room temperature was conducted until reaching saturation and stepwise desorption was conducted via evacuation and gradual heating treatment (room temperature, 100 °C, 200 °C and 300 °C).

2.3. Characterization

Study on the deposition process was carried out by means Fourier Transform Infrared (FTIR) spectroscopy. Pellet of the zeolites was mounted into a vacuum FTIR cell and dehydrated at 500 °C under vacuum condition. The pellet was exposed to $Mo(CO)_6$ vapor at room temperature or 120 °C followed by subsequent evacuation at room temperature, 100 °C, 200 °C and 300 °C. IR spectra were collected after each treatment with a Nicolet Nexus 870 FTIR spectrometer with a 4 cm^{-1} resolution.

X-ray photoelectron spectroscopy (XPS) was employed to observe the molybdenum dispersion within the zeolites. The dispersion is defined as comparison of surface molybdenum concentration to the bulk molybdenum concentration whereas molybdenum concentration is the molar ratio of molybdenum to the silicon. Surface molybdenum concentration was obtained from XPS studies whereas bulk molybdenum concentration was determined gravimetrically. XPS spectra were collected at EPSRC XPS facilities at the University of Newcastle using Thermo k-Alpha instrument from Thermo Scientific.

3. Results and discussion

3.1. Molybdenum deposition

Gravimetric studies on the adsorption of $Mo(CO)_6$ vapor onto the zeolites (HZSM-5, HIM-5, HTNU-9 and HZSM-11) shows the reversibility of room temperature adsorption, where the uptake of $Mo(CO)_6$ vapour at room temperature with vapour pressure of 0.14 torr was about 1% weight for all zeolites. For brevity, typical graph on $Mo(CO)_6$ adsorption on the zeolite was represented by the adsorption on

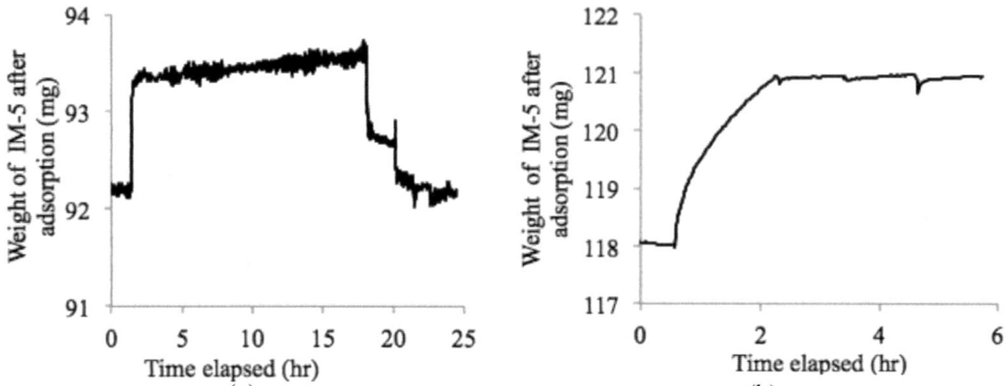

(a) (b)

Figure 1. Weight-time profile of HIM-5 after adsorption of $Mo(CO)_6$ into HIM-5 at (a) room temperature and (b) 120 °C

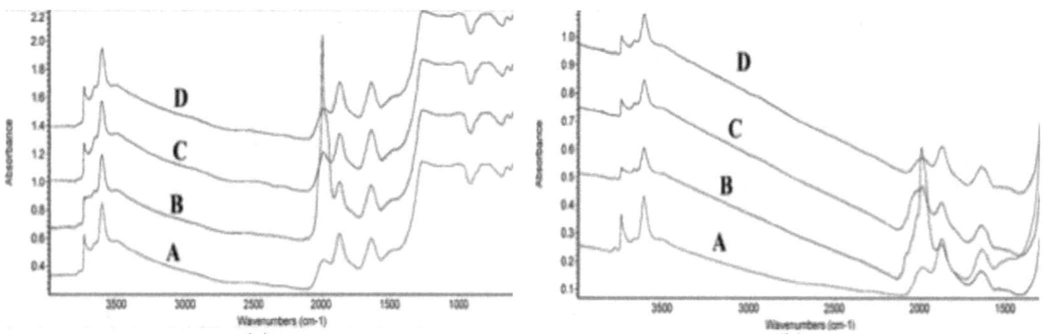

(a) (b)

Figure 2. IR spectra of $Mo(CO)_6$ adsorption into HTNU-9 at (a) room temperature for (A) outgassed zeolite, (B) exposed to $Mo(CO)_6$, (C) evacuated at room temperature and (D) evacuated at 100 °C) and (b) 120 °C for (A) outgassed zeolite, (B) exposed to $Mo(CO)_6$, (C) outgassed at room temperature and (D) outgassed at 100 °C

HIM-5 (figure 1). The adsorbed molybdenum species was almost completely removed by outgassing at 100 °C. Evacuation performed at 300 °C restore the weight of the zeolite into original weight (figure 1a). The reversibility of room temperature adsorption was also confirmed by infrared spectroscopy (FTIR). Figure 2a is a sample of the studies i.e. room temperature adsorption of $Mo(CO)_6$ onto HTNU-9. Exposure of $Mo(CO)_6$ vapour onto HTNU-9 generates some vibration at 2137, 2119 2083 cm^{-1} (shoulder), 2028 cm^{-1} (shoulder) and 1970 cm^{-1} (shoulder), as described in Ohta *et al.* [8] and Abdo and Howe [9]. These peaks are due to the vibration of carbonyl group of adsorbed $Mo(CO)_6$ whereas the band at 2001 cm^{-1} refers to the vibration of free gas phase $Mo(CO)_6$. The bands of adsorbed $Mo(CO)_6$ in the ν(CO) region show similar splitting to that seen for the molecule physically adsorbed on alumina [12] or silica [13] surfaces due to lowering of the octahedral symmetry. It also decreases terminal silanol peak at 3745 cm^{-1} whereas the vibration of Brønsted acid site at 3614 cm^{-1} remains unchanged. Evacuation at 100 °C restores the IR spectra of HTNU-9 into the origin [14].

In contrast to the room temperature gravimetric studies of $Mo(CO)_6$ adsorption, exposure of the zeolites to $Mo(CO)_6$ at its room temperature vapor pressure at 120 °C results in a larger and irreversible uptake. Figure 1b shows the molybdenum uptake on HIM-5 at 120 °C. A 2% uptake was achieved after 2 hours exposure and only a slight weight change on the subsequent outgassing at higher temperature. A longer exposure of zeolite at 120 °C to $Mo(CO)_6$ vapor could be applied to achieve a higher molybdenum uptake and intermittently evacuating CO which was produced during the adsorption.

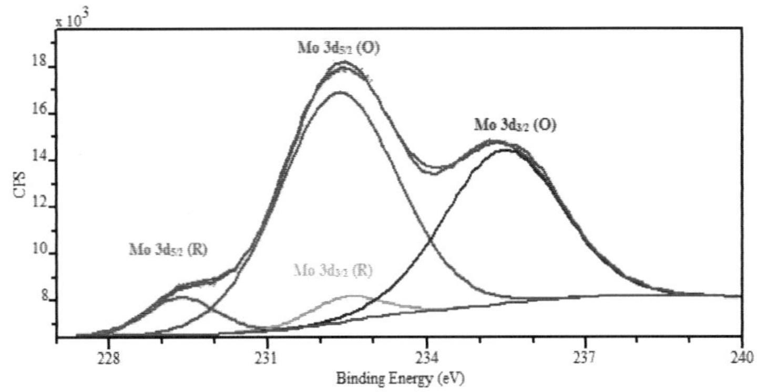

Figure 3. XPS spectra of molybdenum 3d of HZSM-5 with 2% molybdenum deposition

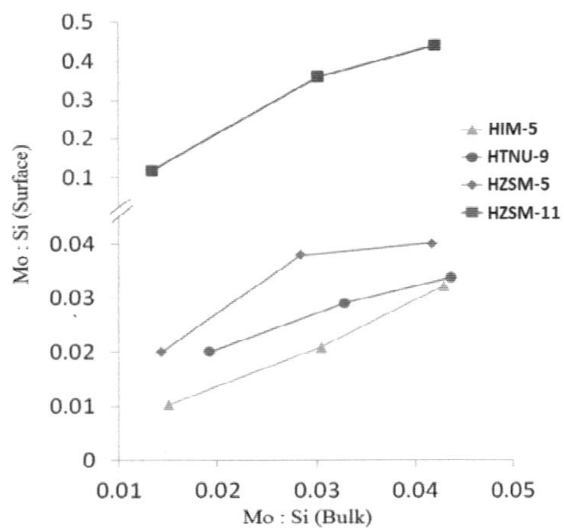

Figure 4. Comparison of surface and bulk molybdenum concentration determined by XPS and gravimetric analysis.

The succeed of molybdenum deposition by adsorption of $Mo(CO)_6$ vapor at 120 °C was also supported by infrared spectroscopy studies (figure 2b). HTNU-9 exposed with $Mo(CO)_6$ vapor at 120 °C gives peaks due to the vibration of carbonyl group of adsorbed $Mo(CO)_6$. The peaks are at 2117 cm^{-1} (shoulder), 2089 cm^{-1} (shoulder), 2054 cm^{-1} (shoulder), 2026 cm^{-1} (shoulder) and 1975 cm^{-1}, whereas the peak at 1998 cm^{-1} shows the presence of gas phase $Mo(CO)_6$ in the IR cell. Those peaks are in accordance with the results reported by Areán *et al.* [14]. Evacuation at 100 °C restores the IR spectra of HTNU-9 at the region of zeolite framework and leaves permanent decrease of Brønsted acid site peak at 3614 cm^{-1} and terminal silanol at 3745 cm^{-1}. This evidence supports the gravimetric studies that the high temperature adsorption leads to the deposition of molybdenum through its reaction with the Brønsted acid site and terminal silanol.

The interaction of molybdenum with Brønsted acid hydroxyl group of zeolites was convincingly shown by the IR spectra of deposited molybdenum at 120 °C as well as the interaction with surface silanol group of zeolite. Further characterization on the molybdenum dispersion was carried out by XPS. Figure 3 is the XPS spectra of molybdenum 3d of HZSM-5 with 2 % of molybdenum deposition. A small amount of molybdenum with lower binding energy of 229.36 and 232 eV for $3d_{5/2}$ and $3d_{3/2}$ respectively is in lower oxidation state, such as molybdenum in MoO_2 or MoOOH, whereas most molybdenum is in higher oxidation state with binding energy of 232.33 and 235.43 eV for $3d_{5/2}$ and $3d_{3/2}$ respectively, this binding energies are close to the molybdenum in MoO_3 [15]. Concentration of molybdenum on zeolite surface obtained by XPS is compared to the bulk molybdenum concentration determined by gravimetric analysis is defined as the molybdenum dispersion over zeolites. The

IOP Conf. Series: Materials Science and Engineering **763** (2020) 012056 doi:10.1088/1757-899X/763/1/012056

(a) (b)

Figure 5. Electron micrograph of (a) HZSM-5 and (b) HZSM-11

molybdenum dispersion over zeolites is presented in figure 4. From XPS and gravimetric characterization on the Mo/H-zeolites, the highly dispersed molybdenum should show a uniform molybdenum concentration throughout the zeolite grain. Figure 4 shows the comparison of molybdenum concentration on the zeolite surface and bulk zeolite. HIM-5, HTNU-9, and HZSM-5 have good molybdenum dispersion into bulk, whereas HZSM-11 has surface molybdenum concentration about 10 fold of the bulk concentration. It is suggested that this difference is caused by the difference of the zeolite particle size. IM-5, TNU-9 and ZSM-5 are in hundreds nanometer where as ZSM-11 has particle size about 10 μm as shown in figure 5. A smaller particle size allows the diffusion of molybdenum easily throughout the zeolite particle whereas a bigger one gives an obstruction to molybdenum to reach the inner part of the zeolite particle lead to the accumulation on the zeolite surface.

4. Conclusions

Adsorption of $Mo(CO)_6$ onto H-zeolites (HZSM-5, HIM-5, HTNU-9 and HZSM-11) conducted at room temperature is reversible whereas adsorption at 120 °C is irreversible and leads to the molybdenum deposition onto the zeolite. Deposition of molybdenum is through decomposition of adsorbed $Mo(CO)_6$ and reaction of molybdenum with Brønsted acid site and terminal silanol of zeolites. Molybdenum dispersion over zeolites is governed by particle size of zeolites i.e. zeolite with bigger particle size has more molybdenum on the surface than smaller particle size.

Acknowledgements

BPLN DIKTI and partial support from Dr. Y. Krisnandi through PUPT BPPTN 2018 grant No. 481/UN2.R3.1/HKP.05.00/2018.

References

[1] Wang L, Tao L, Xie M, Xu G, Huang J and Xu Y 1993 *Catal. Lett.* **21** 35–41
[2] Solymosi F, Németh R, Óvári L and Egri L 2000 *J. Catal.* **195** 316–25
[3] Ismagilov Z R, Matus E V and Tsikoza L T 2008 *Energy Environ. Sci.* **1** 526–41
[4] Atta A Y, Shuwa S M, Naser J T and Jibril B Y 2015 *Asian J. Chem.* **27** 307–12
[5] Hassan A and Sayari A 2006 *Appl. Catal.* A **297** 159–64
[6] Ha V T T, Tiep L V, Meriaudeau P and Naccache C 2002 *J. Mol. Catal.* A-*Chem* **181** 283–90
[7] Sun C, Yao S, Shen W and Lin L 2008 *Catal. Lett.* **122** 84–90
[8] Ohta T, Cicoira F, Doppelt P, Beitone L and Hoffman P 2001 *Chem. Vap. Deposition* **7** 33–7
[9] Abdo S and Howe R F 1983 *J. Phys. Chem* **87** 1713–22
[10] Bécue T, Manoli J M, Potvin C, Djéga-Mariadassou G and Delamar M 1997 *J. Phys. Chem* B **101** 6429–35
[11] Yong Y S and Howe R F 1986 *Stud. Surf. Sci. Catal.* **28** 883–9
[12] Brenner A and Burwell Jr R L 1978 *J. Catal* **52** 353–63
[13] Kurhinen M, Venäläinen T and Pakkanen TA 1994 *J. Phys. Chem* **98** 10237–42

[14] Areán C O, Palomino G T, Platero E E and Mentruit M P 1997 *Dalton Trans.* **19** 873–9
[15] Weckhuysen B M, Rosynek M P and Lunsford J H 1998 *Catal. Lett.* **52** 31–6

Wood plastic composites from chopped and pelleted waste plastic bags: effect of concentrations and reprocessing on flexural modulus and strength

B Ramadhoni[1], L A Prasetyorini[2], O Ujianto[1] and W Wuryanti[2]

[1] Center for Polymer Technology, The Agency for the Assessment and Implementation of Technology (BPPT), Kawasan Puspitek Serpong, Tangerang Selatan, Banten 15314, Indonesia

[2] Research Institute for Human Settlements Agency (RIHS), Ministry of Public Works The Republic of Indonesia, Jalan Panyaungan Cileunyi Wetan, Bandung 40393, Indonesia

Corresponding author's e-mail: benni.ramadhoni.bppt.go.id

Abstract. The utilization of waste plastic bag, especially as a matrix for wood plastic composites, is expected to reduce the environmental problems throughout the world caused by its use. Wood plastic composites were manufactured via melt mixing of chopped or pelleted waste plastic bags, sawdust and polyethylene grafting maleic anhydrate (MAPE) according to Extreme Lattice Mixture - Design of Experiment (DoE). The effects of concentrations and reprocessing of waste plastic bags on flexural modulus of elasticity and strength of composites were analyzed. In general, flexural modulus of composites from chopped and pelleted waste are increased but still below the value of SNI 8154–2015. Moreover, the flexural strength of the resulted composites from chopped waste also increased than that of the pelleted one. Some of the formulas of composites from chopped waste met the value required by SNI 8154 – 2015. On the contrary, the flexural strengths of composites from pelleted waste are decreased than that of the virgin one. This is probably due to the existence of interaction between pelleted waste and sawdust that prevented the waste plastic to perfectly melted, hence acted as stress concentration sites. Furthermore, the resulted composites were optimized for flexural modulus using response optimizer plot in order to achieve modulus 2000 MPa that required by the standard. However, the optimized flexural modulus of composites from chopped and pelleted waste was in the range of 1500 MPa, which is far below the standard. Therefore, reprocessing of chopped waste into pellets is not necessary due to inferior properties of composites from pelleted waste than that of the chopped one.

Keywords: wood plastic composites, waste plastic, chopped, DoE, twin screw

1. Introduction

Wood plastic composites (WPC) have been studied since more than a decade ago. The technology has been applied in many products as construction materials as well as automotive. It can be produced from recycled materials as polyethylene (PE). However, most of recycled PE used for wood plastic composites was from rigid polymer as milk bottle.

Table 1. Formula of composites using chopped and pelleted waste plastic bags

Run	waste plastic (wt.%)	sawdust (wt.%)	MAPE (wt.%)
1	60.0	35.0	5.0
2	70.0	22.5	7.5
3	52.5	45.0	2.5
4	45.0	55.0	0.0
5	35.0	55.0	10.0
6	70.0	27.5	2.5
7	80.0	20.0	0.0
8	80.0	10.0	10.0
9	47.5	45.0	7.5

According to Parker [1], people around the world use billions of tons of plastic products for the last ten years. 20% of the plastic products were plastic bags made from PE [1]. As reported by Halweil [2], around 160,000 plastic bags are produced every second globally. In Indonesia, 11 million plastic bags are produced annually, giving Indonesia as the second largest producer of plastic bag after China. However, the huge number of plastic bag production leaves problems to environment and waste management. At present, only 14 % of waste plastic bags are incinerated while another 5% are recycled [3]. Nevertheless, these are also huge amount of potential recycled PE to be utilized.

Most of these waste plastic bags are recycled into another plastic bags, which has short lifetime. In order to increase the lifetime of waste plastic bags, those wastes can be used as matrix for wood plastic composites. Many groups study the utilization of recycled plastics from different sources such as bottle, water can, etc., as the matrix in wood plastic composites [4–8]. However, there are limited reports on the study of waste plastic bags as the matrix of wood plastic composites. Morris and Cooper [9] processed waste plastic bags into wood plastic composites using compression molding machine and studied the decaying thereof [9]. While El-Haggar *et al.* [10] manufactured wood plastic composites using waste plastic bags as the matrix by adding wood fiber and talc as the filler [10]. Therefore, further studies is needed to the advancement of utilizing of waste plastic bags in wood plastic composites.

In this research, we study the effect of concentration and reprocessing of wood plastic composite manufactured from waste plastic bags and sawdust on flexural strength and modulus elasticity of wood plastic composites in accordance with SNI 8154 – 2015.

2. Materials and methods

2.1. Materials

Random waste plastic bags from laystall were washed by water and chopped into small pieces. Some of the chopped were extruded to form dense and compact pellets. Chopped and pelleted waste plastic bags were used as matrices in wood plastic composites. Admix of hardwoods sawdust (< 80 mesh) was used as filler. High density polyethylene (HDPE) Titanvene 6070EA (MFR 7.5 g/10 min, flexural modulus 1700 MPa) purchased from Lotte Chemical Titan Nusantara, maleic anhydride and benzoyl peroxide both purchased from Merck, were used to prepare polyethylene-graft-maleic anhydride (MAPE) according to the route optimized by Ujianto and Radini [11]. The resulted MAPE was used as compatibilizer for manufacturing wood plastic composites.

2.2. Manufacturing WPCs

Formulas for compounding wood plastic composites (WPC) were generated in Minitab 16, based on Extreme Lattice Mixture - Design of Experiment (DoE). In order to obtain the high flexural properties of composites, the maximum concentrations of chopped or pelleted waste plastics bags; sawdust and compatibilizer were set at 80%, 55% and 10%, respectively [12–15]. Table 1 shows all formulas for this experiment.

The composites were compounded using twin-screw extruder Collins ZK 25 according to conditions shown in table 2. Sawdust was dried at 80 °C in an oven for 24 hours prior to be processed.

Table 2. Process conditions for manufacturing wood plastic composites (WPC)

Temperature (°C)								Screw Speed (rpm)
T_1	T_2	T_3	T_4	T_5	T_6	T_7	T_8	
40	170	175	180	185	190	190	190	60

Table 3. Process conditions for compression moulding

Parameter	Steps				
	1	2	3	4	5
Temperature (°C)	190	190	190	190	40
Pressure (bar)	5	5	5	5	5
Time (minute)	5	5	5	5	15

Table 4. MFR and ΔH_m of chopped and pelleted waste plastic bags

waste plastic bag	MFR (g/10 min)	ΔH_m (J/g)	T_m(°C)
chopped	1.638	60.225	125.90
pelleted	2.722	70.374	128.50

To assure homogeneity, all materials were hand-mixed shortly before compounding. The resulted composites compound was air-cooled and granulized.

The granules were moulded to form sheet using Collins P300P machine according to conditions listed in table 3. The resulted sheet was cut into 150 mm length and 50 mm width specimens. All specimens were pre-conditioned at temperature of 23 ± 2 °C and relative humidity of 50 ± 10 % for more than 24 hours in advance of further testing.

2.3. Measurements and testing

2.3.1. Measurement of melt flow rate, melting temperature and degree of crystallinity. Melt flow rates (MFR) of chopped and pelleted waste plastic bags were measured using CEAST MFI 7062 instrument following ASTM D1238-2013. The melt density of PE at 190 °C (0.761 g/cm^3) was used to calculate MFR. Differential Scanning Calorimetry (DSC) Perkin Elmer was used to obtain heat of fusion (ΔH_f) and melting temperature (T_m) of chopped and pelleted waste plastic bags. Degree of crystallinity (X) was calculated using equation (1) from ΔH_f and theoritical ΔH_m of 100% crystalline HDPE (290 J/g).

$$X = \frac{\Delta H_f}{\Delta H_{f0}} x\ 100\% \tag{1}$$

2.3.2. Measurement of flexural modulus and strength of WPCs. All composites were tested in accordance with ISO 16978 using Universal Testing Machine (UTM) Shimadzu AG-X Plus 50 kN to characterize their flexural modulus and strength. The crosshead speed was 0.3 MPa/sec. The flexural modulus was calculated in the load ranging from 10 % to 40 % of maximum load.

3. Results and discussion

3.1. Melting enthalpy and melt flow rate.

Table 4 shows MFR, ΔH_m and T_m of chopped and pelleted waste plastic bag. The MFR of the pelleted waste is higher than that of the chopped one, giving that the pelleted waste is easier to be processed due to its lower viscosity. It also means that average molecular weight of the pelleted waste is lower than that of the chopped one, attributed to chains scission during reprocessing of the former [16–17].

Table 4 shows that ΔH_m of chopped and pelleted waste plastic bags are 60 J/g and 70 J/g, which corresponds to 23 % and 27 % degree of crystallinity, respectively. The pelleted waste has slightly higher crystal content compared with the chopped one. This is probably due to the lower molecular weight pelleted waste crystallize more facile [18]. During the reprocessing of the pelleted waste, long

Table 5. Flexural modulus of elasticity and strength of wood plastic composites

run \ waste plastic	modulus (E) (MPa)		strength (σ) (MPa)	
	chopped	pelleted	chopped	pelleted
control	554	587	16.7	19.4
1	961	1173	19.3	17.1
2	972	998	23.1	16.2
3	1505	1230	18.1	13.6
4	1528	1437	14.0	11.3
5	1462	1537	16.8	13.4
6	1034	1123	20.9	17.9
7	916	974	20.7	16.7
8	674	745	19.2	14.9
9	1491	1228	20.8	14.5
Average	1171	1161	19.2	15.1

polymer chains were truncated by the screw's shear resulted in shorter and less entangled molecules that promote easier crystallization [17]. This is in agreement with the melting temperature results. The pelleted waste melted in higher temperature compared with the chopped one due to the difference in crystal content [19].

3.2. Flexural modulus and strength of WPCs

Table 5 shows flexural modulus of elasticity and strengths of chopped and pelleted waste plastic bags as well as the composites. Flexural properties of the pelleted waste are lower than that of the chopped one due to the differences in degree of crystallinity of both samples. It is well known that for HDPE, the higher the degree of crystallinity, the higher the modulus and strength [9].

Furthermore, it can be seen in table 5 that the average flexural modulus of the composites from chopped and pelleted waste plastic bags are twofold increase in comparison with the virgin counterpart (1171 MPA and 1160 MPa vs 550 MPa). Generally, the higher the amount of sawdust, the higher the elastic modulus. Unfortunately, no flexural modulus of resulted composites from waste plastic bags meet the specification required by SNI 8154 – 2015 (2000 MPa). This is probably due to the intrinsically low flexural modulus of waste plastic bags. The highest modulus achievable is 1537 MPa of composite from pelleted waste plastic formula 5.

On the other hand, flexural strength of four composites from the chopped waste satisfies the SNI 8154–2015 criterion (20 MPa). The highest flexural strength (23 MPa) is obtained from formula 2 (chopped waste 70.0 %, sawdust 22.5 %, MAPE 7.5 %). On the contrary, no flexural strength of the composites from pelleted waste complies with the standard. This is probably due the interaction between pelleted waste and sawdust during composites processing. Sawdust of hardwood is three times more thermal insulative in comparison with HDPE (thermal conductivity of sawdust is 0.16 W/m °C vs 0.51 W/m °C of HDPE). In the presence of more thermal insulative sawdust, pelleted waste experienced obstruction to melt in the barrel and required higher processing temperature. This resulted in some poorly melted polymers and phase separation between matrix and filler, which acted as stress concentration sites. Moreover, denser and more compact pelleted waste entered the barrel more rapidly than sawdust which resulted in less homogeneous compound. On the other hand, chopped waste melted easier due to its lower T_m. Further more, chopped waste produced more shear that generated extra heat during the processing due to its higher melt viscosity that enhanced the liquefying of the polymer [20].

3.3. Model generation and optimization.

All results were analyzed in Minitab 16 to generate models that describes the characteristics of the composites. Equation (2) to equation (5) are the resulted models for each composite type and properties. As stated previously, no flexural modulus of the composites meet the specification that required by the standard. Therefore, all data were optimized using response optimizer plot in Minitab 16 for flexural modulus of elasticity. The results are depicted in figure 1. Figure 1 shows that in the

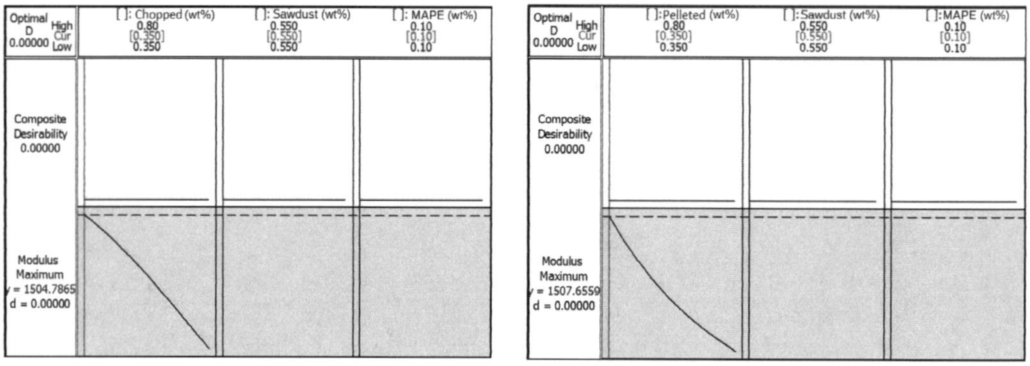

(a) (b)

Figure 1. Response optimizer plot modulus elasticity for composite from (a) chopped waste and (b) pelleted waste

$$E_{chopped} = 503[chopped] + 2444[sawdust] - 172[MAPE] - 2972[sawdust]*[MAPE] + 8580[chopped]*[sawdust]*[MAPE] \quad (2)$$

$$E_{pelleted} = 778[pelleted] + 1925[sawdust] - 412[MAPE] - 7455[sawdust]*[MAPE] + 9979[pelleted]*[sawdust]*[MAPE] \quad (3)$$

$$\sigma_{chopped} = 25.7[chopped] + 3.9[sawdust] - 71.8[MAPE] + 709.6[chopped]*[sawdust]*[MAPE] \quad (4)$$

$$\sigma_{pelleted} = 20.6[pelleted] + 3.1[sawdust] - 823.3[MAPE] + 872.2[pelleted]*[MAPE] + + 1014.6[pelleted]*[sawdust]*[MAPE] \quad (5)$$

range of formulas used in this study. Neither composites from chopped nor pelleted waste can reach the requirement (2000 MPa). Maximum values obtained from the models are in the range of 1500 MPa, which predicted to be achieved with the same formula for both chopped and pelleted waste, i.e. 35 % chopped or pelleted waste, 55 % sawdust, and 10% MAPE.

4. Conclusions

Wood plastic composites from chopped and pelleted waste plastic bags were successfully manufactured using twin-screw extruder. Flexural modulus and strength of the resulted composites were measured. No flexural modulus of the composites satisfied the requirement value of SNI 8154 – 2015 standard due to the low intrinsic property of waste plastic bags. The flexural moduluses of composites from chopped and pelleted waste were about similar. However some of the composites from chopped waste managed to comply with the flexural strength value that required by the standard, probably due to good liquefying of polymer and blending with fillers. Therefore, based on the data resulted from this study, further processing of chopped waste plastic bag into pellet is not required for the manufacturing of wood plastic composites.

References

[1] Parker L 2017 *Here's How Much Plastic Trash Is Littering The Earth* (Washington: National Geography Society) July 19th, 2017 edition.
[2] Halweil B 2004 *Plastic Bags* (Washington: Worldwatch Institute)
[3] Mintarsih T H 2016 *Kebijakan Kantong Belanja Plastik Tidak Gratis* (Jakarta: Direktorat Jenderal Pengelolaan Sampah, Limbah Dan Bahan Beracun Berbahaya).
[4] Najafi S K, Kiaefar A, Hamidina E and Tajvidi M 2007 *Reinforced Plastics and Composites* **26** 341
[5] Beslin V T, Senturk U and Berndt C C 1998 *Resources Conservation and Recycling* **23** 243–58

[6] Carrol D R, Stone R B, Sirignano A M, Saindon R M, Gose S C and Friedman M A 2001 *Resources Conservatioan and Recycling* **31** 241–51
[7] Rajendran S, Scelsi L, Hodzic A, Soutis C and Al-Maadeed M A 2012 *Resources Conservatioan and Recycling* **60** 131–9.
[8] Yan H 2013 *Applied Mechanics and Materials* (AMM) **253–255** 326–9
[9] Morris P and Cooper P A 1998 *Forest Prod. J.* **48** 86–8
[10] El-Haggar S M and Kamel M A 2011 Wood plastic composite *Advanced in Composite Materials* ed P Těšinová London: IntechOpen) Chapter 13 pp 325–44
[11] Ujianto O and Radini F A 2016 *Mater. Sci. Forum* **864** 13–7
[12] Feng D, Caulfield D F and Sanadi A R 2001 *Polym. Compos.* **22** 506–17
[13] Ismail H, Abdullah A H and Abu Bakar A 2010 *J. Reinf. Plast. Compos.* **29** 2489–97
[14] Lin Z, Chen C, Guan Z, Tan S and Zhang X 2011 *J. Appl. Polym. Sci.* **122** 2789–97
[15] Zabihzadeh M, Dastoorian F and Ebrahimi G 2010 *J. Reinf. Plast. Compos.* **29** 123–31
[16] Rideal G R and Padget J C 1976 *J. Polym. Sci.* **57** 1–15
[17] Klyosov A A 2007 *Wood-Plastic Composites* (Hoboken: John Wiley & Sons) pp 493–584
[18] Richards R B 1951 *J. Appl. Chem.* **1** 370–6
[19] Lampman S 2003 *Characterization and Failure Analysis of Plastics* (Ohio: America Society of Metals)
[20] Yacu W A 1985 *J. Food Process Eng.* **8** 1–21

The incident electrical field angle effect in localized surface plasmon resonance (LSPR) of bimetallic Ag-Au nanorod using MNPBEM simulation

M S M Kuddah[1], M H Putra[1,2] and D Djuhana[1]

[1] Department of Physics, Faculty of Mathematics and Natural Sciences (FMIPA), Universitas Indonesia, Kampus UI Depok, Depok 16424, Indonesia

[2] Institute for Computational Physics, University of Stuttgart, Allmandring 3, Stuttgart 70569, Germany

Corresponding author's e-mail: dede.djuhana@sci.ui.ac.id

Abstract. In this study, we have performed the incident electrical field angle effect in localized surface plasmon resonance (LSPR) spectra of bimetallic Ag–Au nanorod using Metallic Nano-Particle Boundary Element Method (MNPBEM) based on the boundary element method. The diameter and length of nanorod are 40 nm and 100 nm, respectively. The fraction of Ag in bimetallic $Ag_xAu_{(100-x)}$ are 0, 20, 40, 60, 80, and 100 within weight percent. The dielectric function of Ag–Au was generated by linear approximation and the incident angle used from 0 to 90 degrees. We found that the intensity of the extinction curve shows decreasing follow logistic type curve as the incident angle increases whereas the resonance peak is constant. For a given silver fraction in bimetallic Ag-Au, the resonance peak of LSPR shifted to higher energy or blue-shift and the intensity increased. Interestingly, the LSPR of bimetallic Ag–Au occurred in the visible range and the transition mode from longitudinal to transverse mode was observed at the angle greater than 60 degrees.

Keywords: MNPBEM, alloy gold-silver, nanoparticle, localized surface plasmon resonance

1. Introduction

In the last two-decades, noble nanoparticles, such as Au and Ag had been attracting both in fundamental and application. The development of technology contributes to the noble nanoparticle applications, such as drug delivery, cancer detection and cancer therapy, and tissue imaging was found in the medical area [1–3], biological imaging [4], hydrogen sensor [5,6] and enhanced Surface-Enhanced Raman Spectroscopy (SERS) signal [7,8]. These applications used an excitation property phenomenon of noble nanoparticles known as localized surface plasmon resonance (LSPR), which depend on size, shape, medium, structure, material selected and polarization [7,9,10]. Besides the pure Au and Ag, alloying them have also attracted many researchers both experimentally [4,8,11] and theoretically, such as Discrete Dipole Approximation (DDA) [9,11,12], Finite Difference Time Domain (FDTD) [5], and Time-Dependent Density Functional Theory (TDDFT) [1]. Silver has higher intensity and resonance energy than gold but easier to oxidize. So, alloying gold with silver is pomising to get alloy with good intensity, energy and stability. However, few studies have discussed the electrical field angle effect in LSPR spectra in bimetallic Ag–Au as the composition variation [12].

IOP Conf. Series: Materials Science and Engineering **763** (2020) 012058 doi:10.1088/1757-899X/763/1/012058

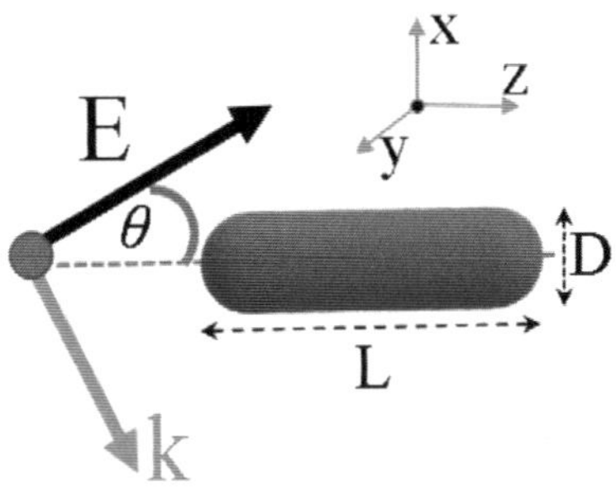

Figure 1. The geometry and dimension of Ag–Au nanorod with the diameter is D and the length is L. The electric field polarization is defined by E and the wavevector by k. The electric field is applied in the angle variation (θ) from 0 to 90°.

In this study, we have observed the electrical field angle effect in LSPR spectra of bimetallic Ag–Au nanorod using optical simulation, Metallic Nano-Particle Boundary Element Method (MNPBEM) with various compositions. The angle was given from 0 to 90° and the dielectric function of Ag-Au was generated by linear approximation. We analysed the LSPR spectra both the peak of resonance and intensity. The intensity decreased as the incident angle increased and the peak of resonance shifted to blue-shift as the fraction of Ag increased.

2. Methods

We have performed the angle effect in LSPR spectra of bimetallic Ag-Au nanorod using MNPBEM [13] based on the boundary element method [14]. A full Maxwell approximation was used to produce the LSPR spectra, such as the extinction cross-section. Single bimetallic nanorod $Au_xAg_{(100-x)}$, which are corresponded to the value of x = 0, 20, 40, 60, 80 and 100 within weight percent [10,15] embedded in air with refractive index, n=1.0 [12]. The diameter and length of nanorod are 40 nm and 100 nm, respectively. The dielectric of bimetallic Ag–Au nanorod was generated by linear approximation [16] and the dielectric function of pure Au and Ag were used from the Johnson and Christy experiment [17]. Then, we applied the electromagnetic wave in the z-direction with the angle variation from 0 to 90° with an increment 10 degree, as illustrated in figure 1.

3. Results and discussion

The extinction curve as the wavelength function of bimetallic Ag-Au nanorod for the fraction and the incident angle variation is presented in figure 2. The result of the Au and Ag curve shows two things. First, the intensity of the longitudinal and transverse mode strongly depends on the incident angle. Second, the resonance peak of wavelength is less sensitive with the incident angle. These results agreed with Tuersun's work by Discrete Dipole Approximation (DDA) simulation [12]. The peak of the extinction curve of bimetallic Ag–Au happened in the visible range and shifted to lower wavelength (blue-shift) as Ag fraction increased [3,16,18]. Shifting to blue-shift of the peak is related to the imaginary part of the dielectric function where the dielectric of Au greater than Ag in visible range [17,18]. Looking detailed in the inset of figure 2, we have observed two peaks happened at a certain incident angle then became a single peak as the fraction of Ag increased. For each composition, two peaks occurred at different angles, for instance, Au pure at an angle 60°, $Au_{80}Ag_{20}$ at 70°, and $Au_{60}Ag_{40}$ at 80°. On the other hand, it was not observed for $Au_{40}Ag_{60}$, $Au_{20}Ag_{80}$, and Ag pure. Changing of two peaks to one peak is related to the direction of the incident light from parallel (0°) to perpendicular (90°) or longitudinal mode to transverse mode.

For understanding, we have also plotted the maximum intensity of extinction as the incident angle (figure 3a) for various compositions. The maximum intensity of extinction decreased following a logistic curve type as the incident angle increased. This result is similar to Fu *et al.* work with DDA simulations [19]. Interestingly, we found that the intensity of extinction was a little increased for a fraction of Ag less than 40 weight percent whereas for larger than 40 weight percent the intensity of extinction significantly increased. This result was consistent with the previous explanation about the imaginary part of dielectric function of Ag. Increasing of Ag fraction contributed to the increasing the intensity of extinction. In figure 3b, we have also presented the peak of Ag–Au as the incident angle for composition variation. The peak showed a constant value as the incident angle increased [12] and

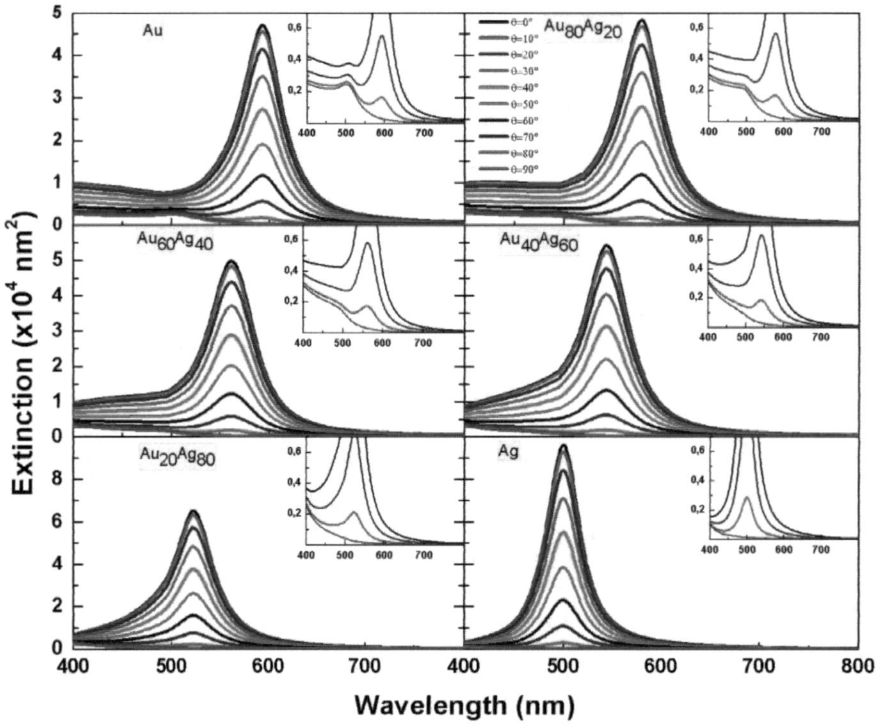

Figure 2. The extinction curve as the function of the wavelength of Ag-Au bimetallic nanorod to the incident angle and composition Ag-Au nanorod variation. The inset figure shows changing the LSPR spectra from two peaks to one peak due to increasing of the Ag fraction, such as $Au_{80}Ag_{20}$ and $Au_{60}Ag_{40}$.

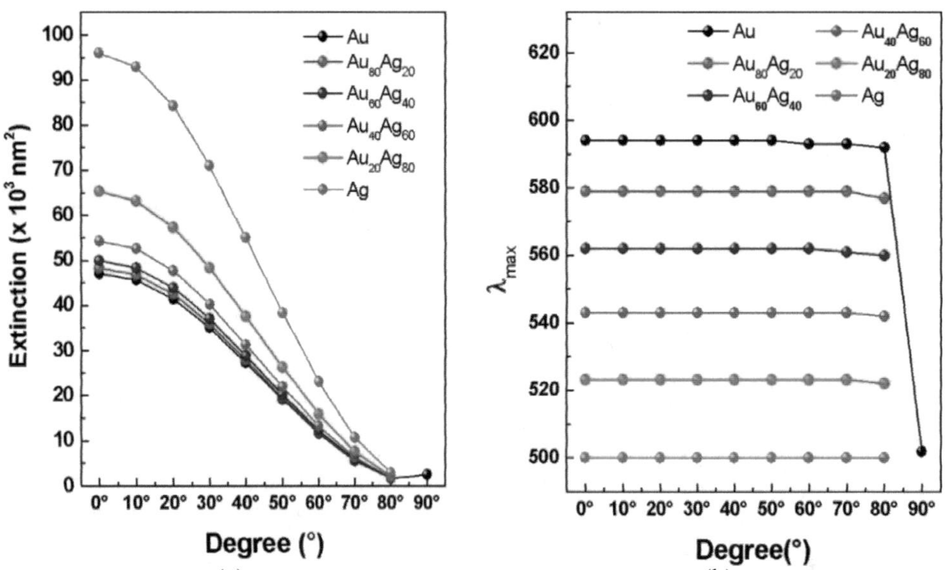

(a) (b)

Figure 3. (a) The peak value of the intensity and (b) the LSPR wavelength as the function of the incident angle in the composition Ag-Au nanorod variation.

shifted to blue-shift as the fraction of Ag increased [16]. More interestingly, we observed the peak of bimetallic Ag-Au slightly shifted for the incident angle around 70° and 80°. This means that we can observe a transition state from longitudinal to transverse mode in bimetallic Ag–Au around 70° and 80°. The peak of transverse mode resonance of bimetallic Ag–Au, however, falls below visible range. Our results showed that the angle effect affected to the LSPR spectra of bimetallic Ag-Au nanorod especially the intensity and a transition from longitudinal to transverse mode.

4. Conclusions

In conclusion, we have observed the effect of angle in LSPR spectra of bimetallic Ag-Au nanorod using MNPBEM simulation. The result showed that the intensity of LSPR spectra decreased as the incident angle increased while the peak of LSPR spectra was relatively constant as shown in figure 3. Increasing of Ag fraction in bimetallic Ag-Au nanorod showed the peak of LSPR shifting to blue-shift and increasing the intensity. Interestingly, the LSPR spectra happened in visible range. We have also observed a transition mode from longitudinal to transverse mode in changing the incident angle direction in the visible light range. The peak resonance for pure Au are 508 nm (transverse mode) and 593 nm (longitudinal mode) happen at 60°, Au_{80} are 497 nm (transverse mode) and 579 nm (longitudinal mode) at 70°, Au_{60} are 496 nm (transverse mode) and 560 nm (longitudinal mode) at 80° while Au's fraction of more than 60, just one peak appears as longitudinal mode.

Acknowledgements

The authors would like to thank Universitas Indonesia for funding this work through PITTA Grant No: 2246/UN2.R3.1/HKP.05.00/2018.

References

[1] Guidez E B and Aikens C M 2013 *J. Phys. Chem.* C **117** 12325–36
[2] Lee K S and El-Sayed M A 2005 *J. Phys. Chem.* B **109** 20331–8
[3] Rioux D, Vallières S, Besner S, Muñoz P, Mazur E and Meunier M 2013 *Adv. Optical Mater.* **2** 176–82
[4] Kim S, Kim S K and Park S 2009 *J. Am. Chem. Soc.* **131** 24 8380–1
[5] Alekseeva S, da Silva Fanta A B, Iandolo B, Antosiewicz T J, Nugroho F A A, Wagner J B, Burrows A, Zhdanov V P and Langhammer C 2017 *Nat. Commun.* **8** 1084
[6] Nugroho F A A 2018 *Nanoplasmonic Alloy Hydrogen Sensors: A Quest for Fast, Sensitive and Poisoning-Resistant Hydrogen Detection* (Gothenburg: Chalmers University of Technology) Doctoral Thesis
[7] Mahmoud M A and El-Sayed M A 2013 *J. Phys. Chem. Lett.* **4** 1541–5
[8] Bansal A and Verma S S 2014 *AIP Adv.* **4** 057104
[9] Link S, Mohamed M B and El-Sayed M A 1999 *J. Phys. Chem.* B **103** 3073–7
[10] Nishijima Y and Akiyama S 2012 *Opt. Matter. Express* **2** 1226–35
[11] Bok H M, Shuford K L, Kim S, Kim S K and Park S 2009 *Langmuir* **25** 5266–70
[12] Tuersun P 2016 *Optik* **127** 3466–70
[13] Trügler A 2016 *Optical Properties of Metalic Nanoparticles: Basic Principle and Simulation* (*Springer Series in Material Science* vol 232) (Cham: Springer International Publishing)
[14] de Abajo F J G and Howie A 2002 *Phys. Rev.* B **65** 115418
[15] Okamoto H, Schlesinger M E and Mueller E M *Alloy Phase Diagram* (*ASM Handbook* vol 3) (Cleveland: ASM International)
[16] Link S, Wang Z L and El-Sayed M A 1999 *J. Phys. Chem.* B **103** 3529–33
[17] Johnson P B and Christy R W 1972 *Phys. Rev.* B **6** 4370–9
[18] Lee K S and El-Sayed M A 2006 *J. Phys. Chem.* B **110** 19220–5
[19] Fu L, Liu Y, Wang W, Wang M, Bai Y, Chronister E L, Zhen L and Yin Y 2015 *Nanoscale* **7** 14483–8

IOP Conf. Series: Materials Science and Engineering **763** (2020) 012059 doi:10.1088/1757-899X/763/1/012059

Effect of titanium substitution on the structural and optical properties of lanthanum orthoferrite nanoparticles

U Hanifah, D Triyono and D Y Jufri

Department of Physics, Faculty of Mathematics and Natural Sciences (FMIPA), Universitas Indonesia, Kampus UI Depok, Depok 16424, Indonesia

Corresponding author's e-mail: djoko.triyono@sci.ui.ac.id

Abstract. A series of $LaFe_{1-x}Ti_xO_3$ (x = 0.4, 0.5 and 0.6) nanoparticles was succesfully synthesized via sol-gel method. All samples were characterized by X-ray diffraction (XRD), Fourier transform infrared spectroscopy (FTIR), and ultraviolet-visible (UV-vis) spectrophotometry. XRD characterization revealed an orthorhombic crystal structure with the *pnma* space group. In the FTIR spectra, the vibrational modes of tilting, bending, and symmetric stretching of the Fe/TiO_6 octahedra were observed. UV-Vis analysis showed the increasing of the energy gap with Ti content.

Keywords : $La(Fe,Ti)O_3$, crystal structure, vibrational mode, energy gap

1. Introduction

Perovskite orthoferrite represents an interesting class of materials and is used for a variety of applications, including electrode materials for high-temperature solid oxide fuel cells, catalysts for the complex oxidation of hydrocarbons, and sensor materials for humidity, alcohol, and oxygen detection [1]. Among the various perovskite orthoferrite compounds, the $LaFeO_3$–based one is an attractive research topic because of its significance contributions to advanced technological applications, such as multilayer capacitors, photocatalyst, gas sensors, and magnetic storage devices [2]. As reported by Tang *et al.* [3], pure perovskite $LaFeO_3$,with *Pnma* space group and orthorhombic structure, has a strong visible absorption and an optical bandgap (E_g) of 2.36 eV.

Several studies reported changes in $LaFeO_3$'s structural and optical properties, which were caused by substitutions on the La and/or Fe sites [1,4–5] such as Ca doping. This resulted a decreased crystallite size in $La_{1-x}Ca_xFeO_3$ with increased Ca content [1]. Janbutrach *et al.* [4] reported that Al substitution on La sites, resulting in $La_{1-x}Al_xFeO_3$ nanopowders, decreases crystallite size and the calculated bandgap in the 2.05–2.09 eV range with increasing Al content. Swatsitang *et al.* [5] observed that Sm doping on La sites leads to decreased lattice parameters, crystallite size, and E_g (from 2.218 to 1.880 eV) with increasing the Sm content. Several researchers focused on the Ti substitution in $LaFeO_3$. Sasikala *et al.* [2] reported that Ti doping (0–80 mol %) on the Fe sites of $LaFeO_3$, synthesized by co-precipitation, increases E_g in the 2.05–2.61 eV range with increasing Ti content. Łańcucki *et al.* [6] observed that, in three-dimensionally ordered (3DOM) materials based on $LaFe_{1-x}Ti_xO_3$ with different Ti amounts (0–20 mol %) using polystyrene nanospheres as a template agent, increasing Ti content leads to higher phase stability followed by a decreased crystallite size.

Based on previous research, Ti substitution on the Fe sites of $LaFeO_3$ induces unexpected properties compared to its parent compounds In addition; the synthesis method also affects the materials' physical properties. Hence, by using the sol–gel method for synthesis, we aim to investigate, in detail, the structural and optical properties of $La(Fe,Ti)O_3$, analyzed by X-ray diffraction (XRD), Fourier

Table 1. Structural parameters of $LaFe_{1-x}Ti_xO_3$ (x = 0.4, 0.5, and 0.6) nanoparticle

Parameter	$LaFe_{1-x}Ti_xO_3$		
	x = 0.4	x = 0.5	x = 0.6
	Crystallography parameters		
Lattice parameters			
a (Å)	5.584	5.565	5.563
b (Å)	7.863	7.836	7.816
c (Å)	5.548	5.572	5.557
Volume (Å^3)	243.5	242.9	241.6
Crystallite size (nm)	18.50	23.82	19.26
Crystallite strain	0.177	0.137	0.077
Density g/cm^3	6.631	6.635	6.667
Atomic Position			
La (x,y,z)	(0.470, 0.250, 0.007)	(0.470, 0.250, 0.007)	(0.470, 0.250, 0.007)
Fe/Ti (x,y,z)	(0.000, 0.000, 0.000)	(0.000, 0.000, 0.000)	(0.000, 0.000, 0.000)
O1 (x,y,z)	(0.198, 0.029, 0.281)	(0.198, 0.029, 0.281)	(0.198, 0.029, 0.281)
O2 (x,y,z)	(0.515, 0.250, 0.580)	(0.515, 0.250, 0.580)	(0.515, 0.250, 0.580)
Wyckoff Factor			
La	4c	4c	4c
Fe/Ti	4a	4a	4a
O(1)	4c	4c	4c
O(2)	8d	8d	8d
	Geometrical parameters		
Bond length (Å)			
Fe/Ti-O1 (s)	1.923	1.929	1.924
Fe/Ti-O1 (l)	2.089	2.089	2.087
Fe/Ti-O2 (m)	2.018	2.010	2.006
<Fe/Ti-O>	2.010	2.009	2.006
Bond angle (°)			
Fe/Ti – O1 – Fe/Ti	157.06	157.11	157.12
Fe/Ti – O2 – Fe/Ti	154.14	153.91	153.93
Tilt angle (°)	14.91	14.95	14.94
Tolerance factor <t>	0.8577	0.8597	0.8611

transform infrared spectroscopy (FTIR), and ultraviolet–visible (UV–Vis) spectrophotometry, and compare the results to those obtained in previous studies.

2. Experimental details

All samples were synthesized by the sol–gel method. The stoichiometric amounts of lanthanum (III) oxide, titanium (IV) oxide, iron (III) nitrate nonahydrate, and citric acid monohydrate were mixed and stirred into solution until the gel phase was obtained. Then, the gel was dried to remove water, citric acid, and other organic compounds and successively calcined until the nanoparticles were formed. The details of the sample preparation have been explained in a previous work [7].

The nanoparticles' XRD patterns were recorded using a PANalytical X'Pert PRO diffractometer with a Cu Kα radiation (λ = 1.5405 Å), 2θ between 20° and 90°, and a step size of 0.02°/s. The Rietveld refinement of the XRD results was performed via GSAS/EXPGUI software. The molecular and metal oxide bonds were analyzed in the 400–4000 cm^{-1} range with a Perkin Elmer Spectrum Two FTIR spectrometer. The E_g values were determined by using an Agilent Cary 100/300 UV–Vis spectrophotometer.

3. Results and discussion

3.1. XRD analysis

XRD patterns of $LaFe_{1-x}Ti_xO_3$(x= 0.4, 0.5, and 0.6) nanoparticles are shown in figure 1. The refinement confirmed that all samples have orthorhombic crystal phase without secondary phase. The structural parameters derived from the XRD patterns are summarized in table 1; there was no change in

Table 2. Fourier transform infrared spectra (FTIR) band assignment for some specific absorption modes of the $LaFe_{1-x}Ti_xO_3$ (x = 0.4, 0.5, and 0.6) nanoparticles

$LaFe_{1-x}Ti_xO_3$	FTIR band positions		
	Fe/Ti−O−Fe/Ti Bending	Fe/Ti−O Stretching	O−Fe/Ti−O Bending
x = 0.4	455	537	758
x = 0.5	434	539	757
x = 0.6	435	542	757

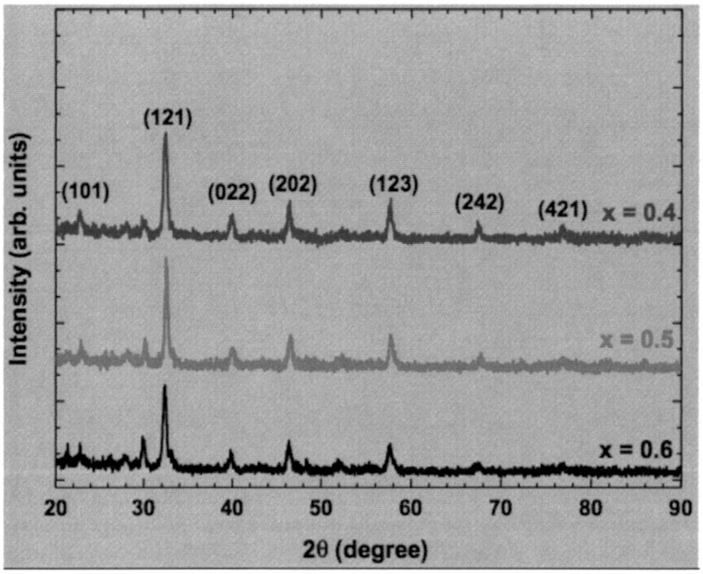

Figure 1. The XRD patterns of $LaFe_{1-x}Ti_xO_3$ (x = 0.4, 0.5, and 0.6) nanoparticles

structural symmetry (orthorhombic) with the *Pnma* space group, similar to the results reported by Sasikala *et al.* [2] and Pokha *et al.* [8] for $LaFe_{1-x}Ti_xO_3$ compounds, and the crystallite size ranged between 18 and 24 nm. The lattice parameters and/or the unit cell volume slightly decreased with increasing Ti content. This might be caused by the Ti^{4+} (ionic radii: 0.610 Å) ions substitution on the Fe^{3+} (ionic radii: 0.645 Å) sites of $La^{3+}Fe^{3+}O_3^{2-}$, introducing Fe^{4+} (ionic radii: 0.580 Å) ions in the lattice for electric charge compensation. Since the ionic radii of Ti^{4+} and Fe^{4+} are smaller than that of Fe^{3+}, the Ti substitution decreased lattice parameters, as reported by Pokha *et al.* [8]. However, the difference among these ionic radii is not significant and, hence, no significant changes were observed in the geometrical parameters (table 1). The Fe/Ti–O bond length slightly decreased with increasing Ti content. The tolerance factor (t) describes the crystal structure stability [2]. Its value was below 0.96, indicating that all samples had stable orthorhombic crystal structures, as confirmed by XRD in a previous study [2].

3.2. FTIR analysis

Figure 2 shows the FTIR spectra of the $LaFe_{1-x}Ti_xO_3$ nanoparticles; they were consistent with those previously reported for $LaFeO_3$, $La(Fe,Ti)O_3$, and $(La, K)FeO_3$ [2,9,10]. The specific FTIR modes are summarized in table 2. The absorption bands around 400–500 cm^{-1}, attributed to the metal–oxygen (M–O) bond vibration, confirmed the typical perovskite structure (ABO_3) had formed [10].

According to previous reports, the absorption bands in 400–500 cm^{-1} were due to the asymmetric stretching and bending vibration of the Fe/Ti–O–Fe/Ti bond, leading to the Fe/TiO6 octahedra

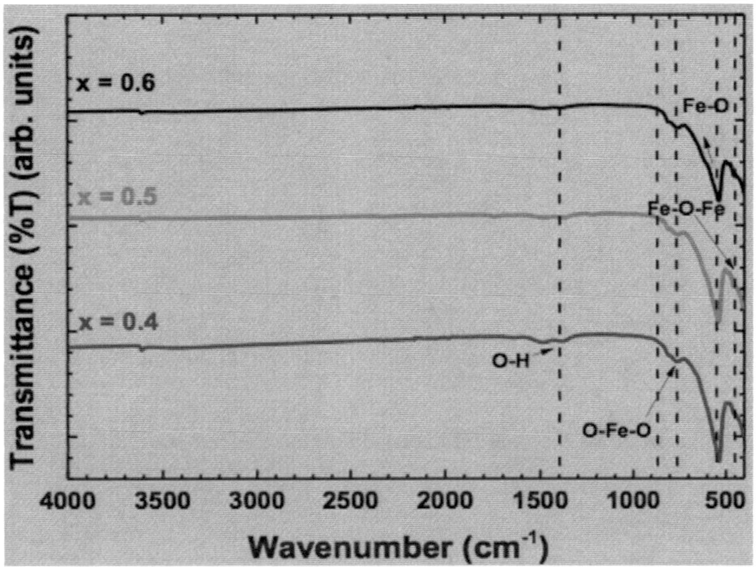

Figure 2. Fourier transform infrared spectra of the $LaFe_{1-x}Ti_xO_3$ (x = 0.4, 0.5, and 0.6) nanoparticles. The dash lines indicate the vibration of symmetry bonding in the lattice.

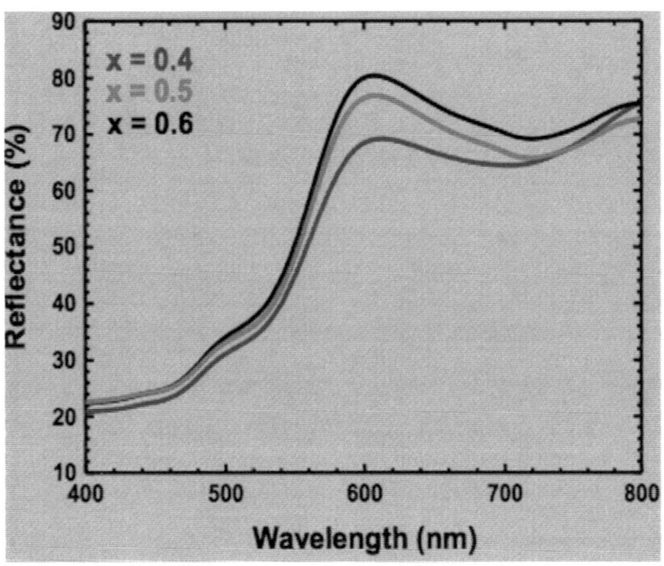

Figure 3. Diffuse reflectance spectra of $LaFe_{1-x}Ti_xO_3$ (x = 0.4, 0.5, and 0.6) nanoparticles

deformation and changes in the corresponding bond angle. The bands in 530–560 cm^{-1} were identified as the stretching vibration modes of the Fe/Ti–O bond, involving the internal motion of a change in thecorresponding bond length. The band around 750 cm^{-1} was assigned to the O–Fe/Ti–O bonds in the lattice, leading to changes in the corresponding bond angle. The bending modes of the O–H bond were observed around 1500 cm^{-1}. Other absorption bands, observed around 3600 cm^{-1}, were assigned to the water molecules' symmetric and asymmetric stretching modes.

As shown in table 2, the Fe/Ti–O band position increased with increasing Ti content, which might be due to the reduced mass of samples at the perovskite structure's B sites [10]. However, the

IOP Conf. Series: Materials Science and Engineering **763** (2020) 012059 doi:10.1088/1757-899X/763/1/012059

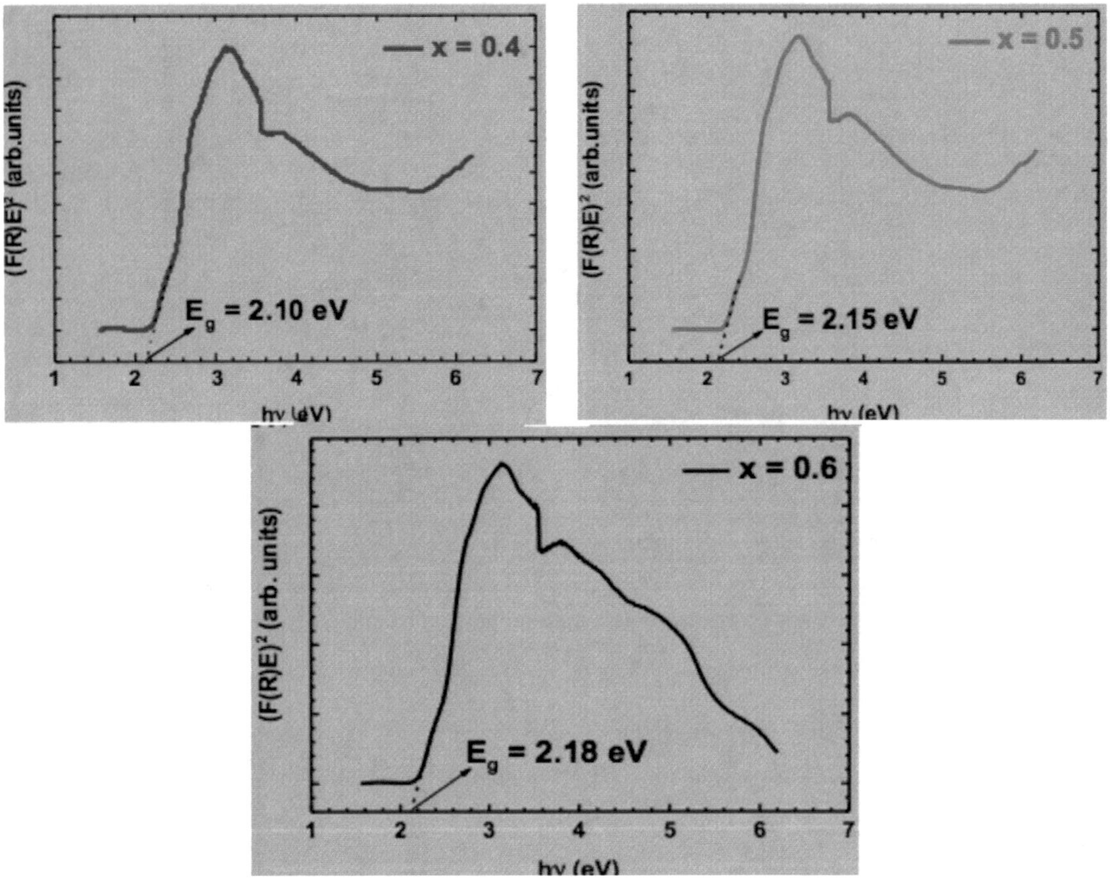

Figure 4. The plot of $(F(R)E)^2$ vs. hν from a series of $LaFe_{1-x}Ti_xO_3$ (x = 0.4, 0.5, and 0.6)

positions of all the bands observed remained almost unchanged, indicating that the bond angle and length did not change (table 1).

3.3. UV–Vis analysis

Figure 3 shows the diffuse reflectance spectra of the $LaFe_{1-x}Ti_xO_3$ (x = 0.4, 0.5, 0.6) nanoparticles in the UV–Vis–near infrared (NIR) range. The Kubelka–Munk function *F(R)* was used to calculate the absorption coefficient (α) based on the diffuse reflectance (*R*) as follows [9,10]:

$$F(R) = \alpha/s = (1 - R)^2/2R$$

where *s* is the scattering coefficient. Then, the energy dependence of the UV–Vis–NIR range was further described.

Figure 4 shows the plots of $(F(R)E)^2$ vs. *hν* for the $LaFe_{1-x}Ti_xO_3$ nanoparticles. Extrapolating the linear parts of the curves toward absorption equal to zero gives E_g for direct transition. The E_g values were in the 2.1–2.2 eV range, which is consistent with those in a previous work for $LaFe_{1-x}Ti_xO_3$ (x = 0.2, 0.4, 0.6, and 0.8) [2]. However, the E_g obtained was smaller than those previously reported [2], possibly because there was no secondary phase in this experiment. Its increasing with the Ti content suggests a decreased conduction response, which is related to the release of space charge carriers for balancing the columbic interactions in the substitution sites.

The effect of Ti substitution on the Fe sites of $LaFeO_3$ could be explained based on the experimental results. The crystal structure (orthorhombic) and symmetry (*Pnma* space group) remained unchanged. The existence of the Fe/Ti–O vibration was confirmed by the FTIR analysis. The UV–Vis analysis confirmed the increasing of the optical bandgap with the Ti content. The decreased lattice parameters and Fe/Ti–O bond length might be due to the strengthened atomic bonding, which makes charge carriers' (electrons, holes, etc.) movement to other sites in the lattice more difficult. Hence, the energy needed to move to another site, i.e., E_g, tends to increase.

4. Conclusions

$LaFe_{1-x}Ti_xO_3$ (x= 0.4, 0.5, and 0.6) nanoparticles were successfully synthesized via the sol–gel method. The structural and optical analysis confirmed the orthorhombic crystal structure. When Ti content increased, the lattice parameters and/or cell volumes decreased and the optical band gap increased.

Acknowledgements

This work was supported by HIBAH PIT 9 from Universitas Indonesia (No.NKB-0024/UN2.R3.1/ HKP.05.00/2019).

References

[1] Andoulsi R, Horchani-Naifer K and Férid M 2012 *Powder Technol.* **230** 183–7
[2] Sasikala C, Durairaj N, Baskaran I, Sathyaseelan B, Henini M and Manikandan E 2017 *J. Alloys Compd.* **712** 870–7
[3] Tang P, Tong Y, Chen H, Cao F and Pan G 2013 *Curr. Appl. Phys.* **13** 340–3
[4] Janbutrach Y, Hunpratub S and Swatsitang E 2014 *Nanoscale Res. Lett.* **9** 498
[5] Swatsitang E, Karaphun A, Phokha S, Hunpratub S and Putjuso T 2017 *J. Sol-Gel Sci. Technol.* **81** 483–92
[6] Łańcucki Ł, Mizera A, Łącz A, Drożdż E, Bućko M M and Pasierb P 2017 *J. Alloys Compd.* **727** 863–70
[7] Jufri D Y, Triyono D and Laysandra H 2017 *AIP Conf. Proc.* **2023** 020021
[8] Phokha S, Hunpratup S, Pinitsoontorn S, Putasaeng B, Rujirawat S and Maensiri S 2015 *Mater. Res. Bull.* **67** 118–25
[9] Bellakki M B and Manivannan V 2010 *Bull. Mater. Sci.* **33** 611–8
[10] Rai A and Thakur A K 2017 *Mat. Sci. Eng.* B **224** 139–49

Synergistic effect of magnetite and carbonous materials incorporated with iron (III)-doped ZnO on the photodegradation of organic dyes

M I Pratiwi[1,2] **and R Saleh**[1,2]

[1] Department of Physics, Faculty of Mathematics and Natural Sciences (FMIPA), Universitas Indonesia, Kampus UI Depok, Depok 16424, Indonesia

[2] Integrated Laboratory of Energy and Environment, Faculty of Mathematics and Natural Sciences (FMIPA), Universitas Indonesia, Kampus UI Depok, Depok 16424, Indonesia

Corresponding author's e-mail: rosari.saleh@ui.ac.id

Abstract. Fe-doped ZnO (Fe:ZnO) was combined with both magnetite (Fe_3O_4) and carbonous materials (graphene) using a co-precipitation method to form Fe:ZnO/G/Fe_3O_4 nanocomposites with various weight ratios (1:0, 1:0.05, 1:0.1, 1:0.3, 1:0.5, and 1:1). These nanocomposites were characterized using different methods of measurements; such as Fourier transform infrared spectroscopy to record the absorption spectra of certain functional groups, X-ray diffraction spectroscopy to identify the structure of the samples, and thermogravimetric analysis to study the loss in samples' weight under increasing temperature. The magnetite and carbonous materials were used to support Fe:ZnO in the photodegradation of methylene blue dye, as a representation of organic dyes. The photocatalytic activity was examined under ultra violet (UV) light irradiation. The effects of Fe:ZnO/G:Fe_3O_4 ratio, scavenger addition, and catalyst dosage on the photocatalytic activity were also studied. Finally, the reusability of the catalyst was examined and proved.

Keywords: component, formatting, style, styling, insert (key words)

1. Introduction

Zinc oxide (ZnO) semiconductor is considered as a heterogeneous photocatalyst that can generate electron-hole pairs (or excitons) when subjected to irradiation at certain wavelengths. However, excitons often recombine, which decreases the photocatalytic efficiency of ZnO [1]. Therefore, research should focus on extending the lifetime of excitons using several methods, such as transition metal loading [1–4] and incorporation of supporting materials [5–7]. Both methods can be simultaneously applied in the same semiconductor.

It was found that the doping of ZnO with Fe ions could increase its photocatalytic activity because it causes a decrease in the bandgap energy and formation of electron trap [1–3], which prevents the recombination of the excitons. Carbon materials can be also used to prevent the recombination of the excitons in photocatalysts by acting as an adsorbent and electron transfer bridge between the catalyst and the pollutant molecules [6,8]. Moreover, research has focused on increasing the stability and reusability of the photocatalyst by incorporating magnetic materials, such as Fe_3O_4, which enables the magnetic separation of the photocatalyst by applying external magnetic field. This is an easier method compared to filtration and centrifugation, especially when used on a large scale [7].

In this study, the characterization and photocatalytic performance of Fe:ZnO/G/Fe_3O_4 nanocomposites were investigated using various Fe_3O_4 loading. The samples were characterized by Fourier transform infrared spectroscopy (FTIR), X-ray diffraction (XRD), and thermogravimetric analysis (TGA) spectroscopies. The photocatalytic activity was studied under ultraviolet (UV) light irradiation with methylene blue (MB) as the organic dye pollutant. The effects of the weight ratio of Fe:ZnO/G to Fe_3O_4, scavenger addition, and catalyst dosage on the photocatalyst performance were observed. The reusability of the photocatalyst was also examined using it in four cycles.

2. Details of experiments

2.1. Sample preparations

Analytical grade materials were used in this study with no further purification. $Fe_2SO_4.7H_2O$, $ZnSO_4.6H_2O$, ethylene glycol, acetate acid, and NaOH were purchased from Merck, and reduced graphene oxide was purchased from Angstrom Materials. In addition, distilled water was used in all solutions during the experiment.

First, Fe:ZnO and Fe_3O_4 nanoparticles were synthesized using co-precipitation and sol–gel methods, respectively, as described in the authors' previous work [2]. Second, the Fe:ZnO/5 wt.% G nanocomposite was prepared using a hydrothermal method. In this experiment, the graphene sheet was placed in ultrasonic bath for 2 h, and then mixed with Fe:ZnO and stirred for another 2 h. Furthermore, the solution was heated for 3 h at 120 °C. It was then centrifuged and left to dry for 2 h at 100 °C. Next, the Fe:ZnO/G/Fe_3O_4 nanocomposites were prepared using a co-precipitation method. The Fe:ZnO/G and Fe_3O_4 nanoparticles were mixed in alcohol and water, respectively, and then placed in ultrasonic bath for 2 h. The following Fe:ZnO/G:Fe_3O_4 weight ratios were prepared: 1:0, 1:0.05, 1:0.1. 1:0.3, 1:0.5, and 1:1. The solutions were then stirred for 15 min and centrifuged. Finally, the nanocomposites were dried in an oven for 2 h at 100 °C.

2.2. Characterization

The prepared nanocomposites were characterized using several techniques. Nicolet iS5 FTIR spectrometer was used to confirm the existence of graphene and magnetite. The structure and phase of the nanocomposites were studied using XRD (Rigaku Miniflex 600) operated under 30 kV and 15 mA with CuKα (λ = 1.54 Å) as a radiation source and monitored at 2θ = 20°–80°. The stability of the nanocomposites under elevated temperatures was examined at a temperature ranging from room temperature to 1000 °C using Rigaku Thermo Plus EVO2 TG-8121 thermogravimetric analysis/differential thermal analysis (TGA/DTA) with aluminum reference.

2.3. Photocatalytic experiment

The photocatalytic experiment was conducted at room temperature with MB as the organic pollutant in an aqueous solution at pH = 7. The catalysts were stirred with the MB solution in a beaker until the solution reaches adsorption–desorption equilibrium. Thirty minutes after finishing the stirring process, the samples were irradiated for 2 h using two Xe lamps. The degradation of MB was measured at 15-min intervals using a UV-Vis Hitachi UH5300 spectrophotometer. The effects of Fe:ZnO/F:Fe_3O_4 ratio, scavenger addition, and catalyst dosage were studied in the photocatalytic experiment. The reusability and adsorption process were also examined.

3. Results and discussion

FTIR spectroscopy was used to detect the vibrations of the molecules in the prepared samples. Figure 1 shows the absorption spectra of five different Fe:ZnO/G:Fe_3O_4 weight ratios within a wavenumber range of 500–4000 cm^{-1}. The figure indicates that there is an overlapping peak of Zn-O and Fe-O vibrations between 500 and 700 cm^{-1} [9]. The stretching vibration of Fe-O at 590 cm−1 [10] confirms the existence of Fe and Fe_3O_4 doping nanoparticles in the nanocomposites. The absorption peaks at 1200 cm^{-1}, 1552 cm^{-1}, and 1718 cm^{-1}, which correspond to the vibrations of C-O-H, C=C, and C=O, confirm the existence of carbon materials in the nanocomposites [11]. However, their effect was reduced by the increase in the Fe_3O_4 amount. The absorption peaks at 1728 cm^{-1} and between 3000 and 3600 cm^{-1} correspond to the O-H bonds in the H_2O molecules [7].

The structure of the samples was studied using XRD spectroscopy. Figure 2 displays the diffraction patterns of five different weight ratios of Fe:ZnO/G:Fe_3O_4. The recorded spectra indicated that there were two different phases associated with each sample. The peaks denoted by # correspond to the

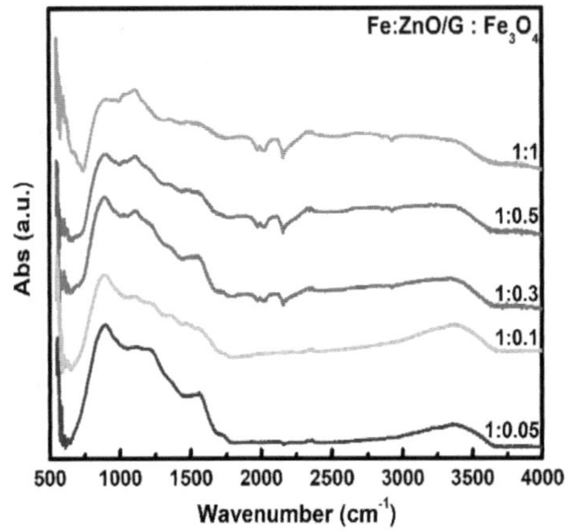

Figure 1. FTIR spectra for various weight ratios of Fe:ZnO/G/Fe_3O_4 nanocomposites

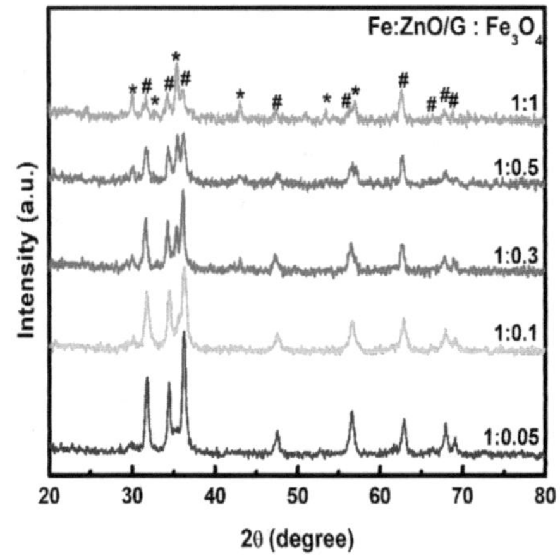

Figure 2. XRD patterns for various weight ratio of Fe:ZnO/G/Fe_3O_4 nanocomposites

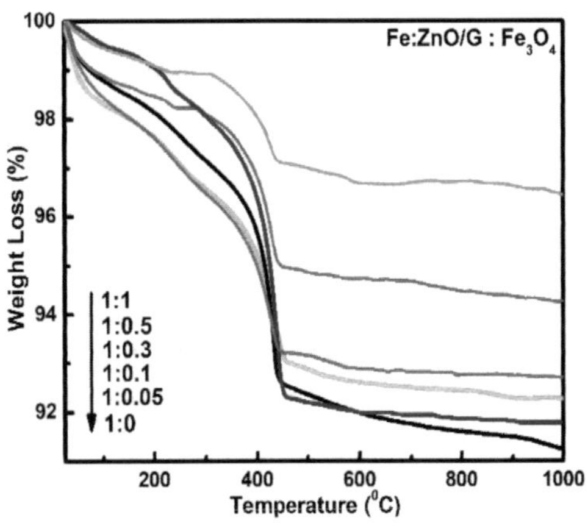

Figure 3. Thermogravimetric analysis curves for various weight ratios of Fe:ZnO/G/Fe_3O_4 nanocomposites

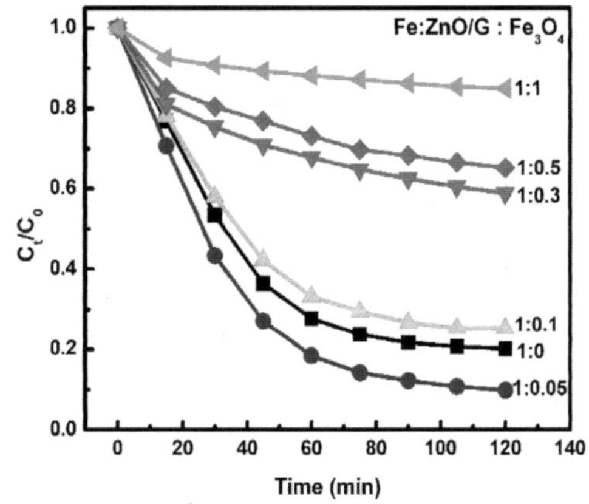

Figure 4. MB photodegradation by different ratios of Fe:ZnO/G/Fe_3O_4 nanocomposites

hexagonal wurtzite structure of ZnO at 2θ = 31.85°, 34.50°, 36.22°, 47.57°, 56.62°, 62.89°, 66.49°, 67.95°, and 69.15° and represent the planes (100), (002), (101), (102), (110), (103), (200), (112), and (201), respectively [3]. The peaks denoted by * correspond to the spinel structure of Fe_3O_4 , which represent the planes (220), (311), (400), (442), (511), and (440) [10]. The amorphous peak of graphene could not be detected in the diffraction spectra because of its low concentration.

Figure 3 shows the relationship between the loss in the sample weight vs temperature of the nanocomposites within the temperature range from room temperature to 1000 °C. The depression in the curves at around 400 °C was caused by the decomposition of carbon materials from the graphene sheet [12], because Fe_3O_4 nanoparticles are known to be sufficiently stable at high temperatures [13–14]. The figure also shows that the percentage weight loss gradually decreased from 8.77 % to 3.56 % with increasing portion of magnetic material in the Fe:ZnO/G sheets. This demonstrates that magnetic materials could increase the stability of the samples at high temperature.

Figure 4 shows the degradation curves of MB. The initial concentration of MB was 30 mg/L, which was reduced by 79.86% before the addition of any magnetic material. The magnetic material, which was added at a ratio of 1:0.05, caused the MB degradation to increase to 90.10%. The combined effect of Fe_3O_4, which is known to enhance the transfer of the photogenerated electrons from the surface of

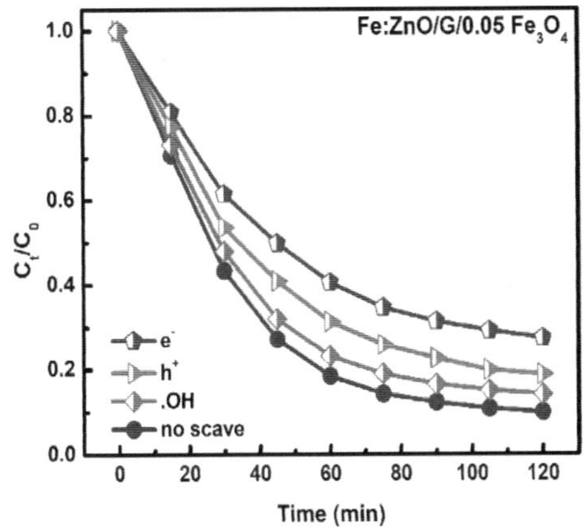

Figure 5. Effect of scavenger addition on the photodegradation of methylene blue

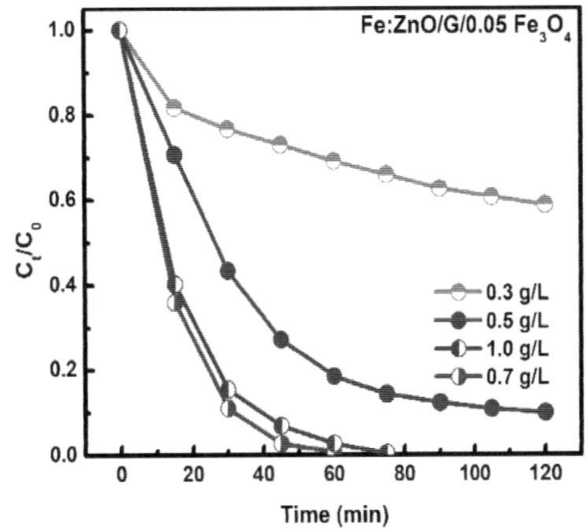

Figure 6. Degradation of MB under different dosage of Fe:ZnO/G/0.05 Fe_3O_4 nanocomposite

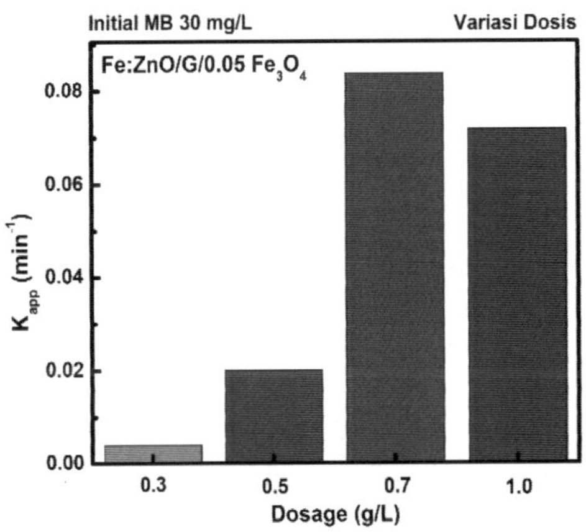

Figure 7. Kapp histogram of MB degradation with different dosages of Fe:ZnO/G/0.05 Fe_3O_4 nanocomposite

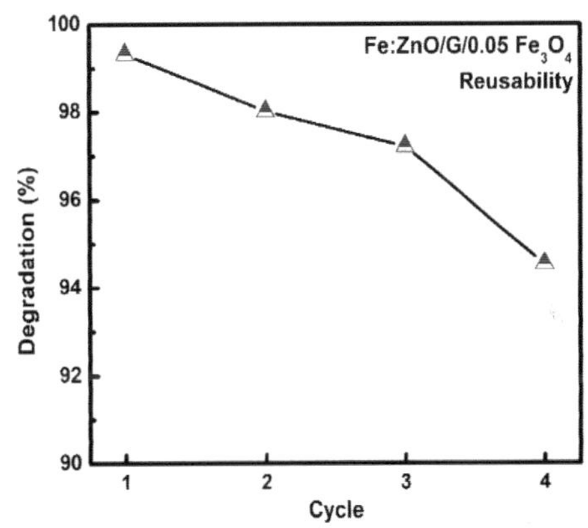

Figure 8. Degradation percentage of MB over four cycles of reuse

the nanocatalyst to the dye solution, and the graphene sheets, which act as electron bridges, improves the photocatalytic performance. However, increasing the Fe_3O_4 material caused the MB degradation to decrease; when Fe:ZnO/G:Fe_3O_4 ratios of 1:0.1, 1:0.3, 1:0.5, and 1:1 were used, they resulted in 74.71%, 41.21%, 34.86%, and 15.05% degradation, respectively. This can be attributed to the hindrance to light penetration into the surface of the catalyst, which is caused by the increased number of Fe_3O_4 nanoparticles. Therefore, the optimum Fe:ZnO/G:Fe_3O_4 ratio for MB degradation was determined to be 1:0.05. These results agree with those obtained in the authors' previous work [15,16]

Several active species, such as electrons, holes, and hydroxyl radicals, play important roles in the photocatalytic activity. Sodium hydroxide, di-ammonium oxalate, and tert-butanol were added to the mixture as scavengers for electrons, holes, and hydroxyl radicals, respectively, to determine the most active species (figure 5). The addition of tert-butanol slightly decreased the MB degradation, which was further decreased by the addition of di-ammonium oxalate and sodium hydroxide. This indicates that electrons are the most active species, followed by holes and then hydroxyl radicals. The result is reasonable enough, because Fe-ion dopants act as an electron trap that prevents the recombination of photogenerated electrons and holes. The graphene sheets and magnetic material had a significant synergistic effect caused by the transferring of the electrons from the catalyst surface or dopant to the solution, which proves that they also prohibit the recombination of the excitons. Finally, the electrons were used in the reduction of water molecules into radicals that cause the decomposition of MB into less toxic products.

The effect of catalyst dosage on dye degradation was studied using various catalyst concentrations (0.3, 0.5, 0.7, and 1.0 g/L). Figure 6 illustrates the results of this test, which indicate that increasing the dosage to 0.7 g/L could enhance the photocatalytic efficiency, because of the increase in active sites that interact with light and enable the adsorption of MB molecules by the graphene sheets. However, at 1.0 g/L catalyst dosage, the MB degradation started to decrease, possibly due to the light scattering effect and high solution opacity [17,18]. Therefore, the optimum dosage of the catalyst, where the maximum reduction took place, was observed at 0.7 g/L. The degradation rate/kapp (min^{-1}) was also calculated using first-kinetic order equation (Figure 7). The values obtained were 0.0039, 0.0201, 0.0839, and 0.0718 min^{-1} with catalyst dosages of 0.3, 0.5, 0.7, 1.0 g/L, respectively.

Figure 8 shows the percentage of MB photodegradation in four cycles of using the Fe:ZnO/G/0.05 Fe_3O_4 catalyst. This test was performed with a catalyst concentration of 0.7 g/L. The figure indicates that the degradation percentage did not significantly decrease after four cycles of reuse, which proves the good stability of the catalyst. However, continuous reuse can reduce magnetic interaction due to weakening of magnetization.

4. Conclusions

Fe:ZnO/G:Fe_3O_4 nanocomposites were successfully synthesized using a co-precipitation method. Different ratios of Fe:ZnO/G:Fe_3O_4 were used. FTIR spectra were used to confirm the existence of Fe_3O_4 and graphene. The structures of ZnO and Fe_3O_4 were examined using XRD, which showed that they had hexagonal wurtzite and cubic spinel structures, respectively. However, XRD spectroscopy could not detect the graphene peaks, due to its low concentration. The TGA results indicated that incorporating Fe_3O_4 into Fe:ZnO/G could enhance its stability at high temperatures. The photocatalytic activity was studied using UV light irradiation with MB as the organic pollutant. It was also proven that incorporating a certain concentration of Fe_3O_4 into the photocatalyst could improve its photocatalytic efficiency. However, exceeding this concentration could decrease the photocatalytic activity of the nanocomposite. Moreover, it was found that electrons are the main active species responsible for the photocatalytic activity. The optimum dosage of the catalyst, which showed maximum degradation of the dye, was found to be 0.7 g/L, and the nanocomposite remained stable after four cycles of reuse.

References

[1] Türkyılmaza Ş Ş, Güya N and Özacar M 2017 *J. Photochem. Photobiol.* A *Chem.* **341** 39–50
[2] Saleh R and Djaja N F 2014 *Superlattices Microstruct.* **74** 217–33
[3] Bousslamaa W, Elhouicheta H and Férid M 2017 *Optik* **134** 88–98
[4] Zhang X J, Chen Y, Zhang S and Qiu C Y 2017 *Sep. Purif. Technol.* **172** 236–41
[5] Chen D, Wang D, Ge Q, Ping G, Fan M, Qin L, Bai L, Lv C and Shu K 2015 *Thin Solid Films* **574** 1–9
[6] Beura R and Thangadurai P 2017 *J. Phys. Chem. Solids* **102** 168–77
[7] Shekofteh-Gohari M and Habibi-Yangjeh A 2015 *Sep. Purif. Technol.* **147** 194–202
[8] Darvishi M, Jamali-Paghaleh F, Jamali-Paghaleh M and Seyed-Yazdi J 2017 *Surf. Interfaces* **9** 167–72
[9] Kanchana S, Chithra M J, Ernest S and Pushpanathan K 2016 *J. Lumin.* **176** 6–14
[10] Song X, Luo X, Zhang Q, Zhu A, Ji L and Yan C 2015 *J. Magn. Magn. Mater.* **388** 116–122
[11] Ossonon B D and Bélanger D 2017 *RSC Adv.* **7** 27224
[12] Loryuenyong V, Totepvimarn K, Eimburanapravat P, Boonchompoo W and Buasri A 2013 *Advances in Material Science and Engineering* **2013 ID** 923403
[13] Fakhri A and Naghipour A 2017 *Comments on Inorganic Chemistry* **37** 201–18
[14] Zhou S, Li Y, Cui F, Jia M, Yang X, Wang Y, Xie L, Zhang Q and Hou Z 2014 *Macromol. Res.* **22** 58–66
[15] Pratiwi M I, Afifah N and Saleh R 2017 *IOP Conf. Ser.: Mater. Sci. Eng.* **188** 012012
[16] Pratiwi M I, Afifah N and Saleh R 2017 *IOP Conf. Ser.: Mater. Sci. Eng.* **196** 012026
[17] He H Y, Huang J F, Cao L Y and Wu J P 2010 *Desalination* **252** 66–70
[18] Bel Hadjltaief H, Ben Ameur S, Da Costa P, Ben Zina M and Galvez M E 2018 *Appl. Clay Sci.* **152** 148–157

Electron transmission probability and density of states of G4 DNA molecule in electric field

E N Sinurat and E Yudiarsah

Department of Physics, Faculty of Mathematics and Natural Sciences (FMIPA), Universitas Indonesia, Kampus UI Depok, Depok 16424, Indonesia.

Corresponding author's e-mail: e.yudiarsah@sci.ui.ac.id

Abstract. Electron density of states and transmission probability of G4 DNA molecule have been calculated. The calculation was carried out on G4 DNA consists of 32 G-quartets. G4 DNA molecule is represented mathematically by using Hamiltonian tight binding model and taking into account the electric field on electron hopping constants. The density of states and transmission probability as a function of energy is calculated using Green's function method. The electric field applied parallel to the length of DNA leads to changes in electron hopping constants. The result shows that the increase in electric field causes electron density of states and transmission probability bands to shift.

1. Introduction

Charge transport on DNA molecules have attracted the attention of researchers [1] because it can be applied for molecular wires in nanotechnology applications [2,3] Charge transport along DNA molecule occurs through the overlap of π-π orbital on adjacent inter-pairs atoms [4,5]. Some variables, the environment and structure of DNA molecule itself, can affect the result of conductivity measurements on this molecule, that is DNA can act as insulator, semiconductor, and even superconductor [6,7].

Beside the canonical Watson-Crick model called the double-stranded DNA (ds-DNA), DNA molecule can be found or synthesized in other forms. One of these structures is G4-DNA called G-quadruplexes, which is formed by a planar arranged tetrad of G-bonded hydrogen base. DNA G4 has a longer size than ds-DNA, so that the charge transport on G4-DNA molecule occurs in a longer distance than on ds-DNA molecule [1]. In addition, the high G-nucleotide, known to have the lowest ionization potential among the nucleotides content makes this structure a highly promising candidate for future bottom-up nanoelectronic applications [8]. To this end, the understanding of charge transport on G4-DNA molecule becomes important.

In this study, we calculated transmission probability and density of states in G4 molecules under the influence of an applied electric field. This study is aimed to have an understanding on the effect of external electric field on DNA molecule, which will affect the charge transport process in it. This effect is rarely considered in the study of charge transport in DNA. The transmission probability and density of states are calculated using Green's function method on G4-DNA molecule consisted of 32 stacks of guanine tetrads. Electric field affects the value of hopping electron constant on G4 DNA molecules based on Miller-Abrahams formula. The results show that both DOS and transmission probability spectra change with the voltage increment. The center of the bands, region with high values of DOS/transmission probability, in the spectra is shifted by the voltage.

2. Theoretical model

The arrangement of hydrogen bonds in G4-DNA molecule is different from the one on the Watson-Crick DNA. Each G base in tetrad G4 forms hydrogen bonds with neighboring two bases through Hoogsteen bonds [9]. The tetrad is stacked together in helix arrangement, with stack spacing and turning angles of 3.25 Å and 30 ° each, smaller than dsDNA. The figure of G4-DNA molecules can be seen in the paper [9]. The Tight binding Hamiltonian model of single electron system on G4-DNA molecule can be stated as:

$$H = \sum_{n=1}^{N}\sum_{\tau=1}^{4}\{(\varepsilon_n^{\tau} + \emptyset_n^{\tau})|n,\tau\rangle\langle n,\tau| + t_{n,n+1}^{\tau,\tau}|n,\tau\rangle\langle n+1,\tau|)\} + \sum_{q=u,d}\sum_{\tau=1}^{4}\{B_n^{\tau,q}|n,\tau\rangle\langle n,q| + t_{n,n}^{\tau,q}|n,\tau\rangle\langle n,q|\} + \sum_{\tau=1}^{4}\{t_{n,n}^{\tau,\tau+1}|n,\tau\rangle\langle n,\tau+1| + t_{n,n+1}^{\tau,\tau+1}|n,\tau\rangle\langle n+1,\tau+1|\} + \text{h.c.} \quad (1)$$

The parameters in equation (1) has been described in paper [10,11] External electric field is applied in direction of G4-DNA stack. Based on Miller-Abrahams [12,13], we wrote the electron hopping constant between two adjacent sites on G4-DNA molecule under the influence of electric field as follows:

$$t_{n,n+1} = t_o \exp\left(-eE\left|\frac{Z(n+1)-Z(n)}{K_B T}\right|\right), \quad (2)$$

$Z(n+1) - Z(n)$ is distance from site $n+1$ to site n, t_0 is the value of electron hopping constant without electric field, $t_{n,n+1}$ is electron hopping constant affected by electric field from initial site $n+1$ to final site n, e is electron charge, k_B is Boltzmann constant, and T is temperature.

The retarded Green's can be calculated from Hamiltonian of the system using

$$G^r(E,\vec{k}) = \left[(E+i\eta)[I] - \{H(\vec{k}) - \Sigma_L^r(\vec{k}) - \Sigma_R^r(\vec{k})\}\right]^{-1}, \quad (3)$$

where E is the electron energy, I is identity matrix, η is very small positive value close to zero, $H(\vec{k})$ is Hamiltonian matrix of DNA molecule, $\Sigma_{L(R)}^r(\vec{k})$ is self-energy due to the contact of DNA molecules with left electrode and right electrode.

After obtaining the retarded Green's function, DOS at each energy can be calculated using

$$DOS(E) = -\frac{i}{\pi} im\left[TrG^r(E,\vec{k})\right] \quad (4)$$

Transmission probability is square of ratio between incoming and transmitted probability amplitudes. From retarded and advanced Green's functions, $G^a(E,\vec{k}) = [G^r(E,\vec{k})]^{\dagger}$, the transmission probability at each energy is calculated using Fisher-Lee relation [14]

$$T(E,\vec{k}) = Tr\left[\Gamma_L G^r(E,\vec{k})\Gamma_R G^a(E,\vec{k})\right], \quad (5)$$

where $\Gamma_{L(R)}$ represents the molecule coupling to electrode $\Gamma_{L(R)} = i\left[\Sigma_{L(R)}^r - \Sigma_{L(R)}^a\right]$, index L for left electrode and R for right electrode.

3. Results and discussion

Density of states as energy function on G4-DNA with 32 G-quartets has been calculated at temperature of 4.2 K and base twisting motion frequency of 4 meV for several voltages. The spectrum of DOS consists of energy region with nonzero DOS (band) and energy region with zero DOS (gaps). The bands represent energy region where the states can be occupied by electron in its way from an electrode to another electrode in the opposite end of DNA molecule. The states in the gap do not participate in the transport process since these states are forbidden for the electron. The density of states calculated in the presence of electric field with voltages of 0, 4.0 mV, 0.01 V, and 0.03 V are presented in figure 1. In the DOS spectrum there are two bands separated by a gap, except at voltage of 0.03 V. At zero voltage, the two bands are in the region of energy around 6.85 eV – 8.9 eV (band 1) and 9.45 eV – 9.8 eV (band 2) and the gap in the energy region around 8.9 eV – 9.45 eV. With voltage increment, the width of the bands becomes narrower and the center of the band is shifted. As a whole, the DOS spectrum becomes narrower and the gap becomes wider with the voltage. This occurs because the model used in this study guarantees that the electron hopping constants get smaller with the voltage. Other change is observed at voltage of 0.03 V, band 1 splits into two.

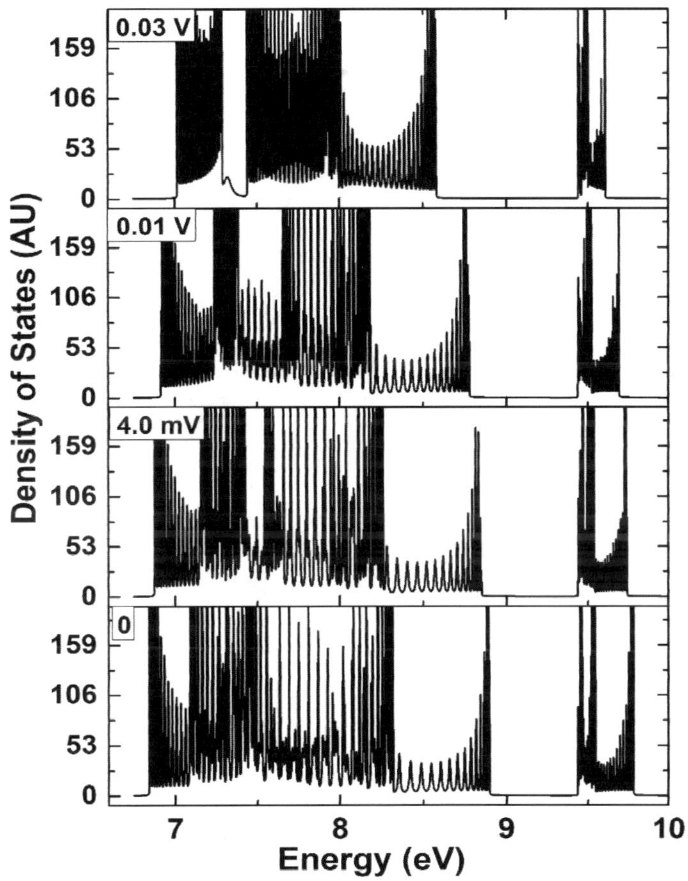

Figure 1. Electron density of states calculated at temperature 4.2 K and twisting motion frequency 4.00 meV for electric field of voltages 0, 4.0 mV, 0.01 V and 0.03 V.

The trend observed in the DOS spectrum is also observed in the transmission probability. The transmission probability is also calculated at temperature of 4.2 K and twisting motion frequency of 4 meV in the applied electric field of voltages 0, 4.0 mV, 0.01 V, and 0.03 V. Those transmission probabilities are presented in figure 2. Similar to the DOS case, at zero voltage transmission probability spectrum shows two regions with nonzero transmission probability (band) and a region with zero transmission probability (gap). Those features are reflection of electron density of states discussed before. The first band is shown in the energy range of around 6.85 eV – 8.9 eV (band 1) and the other around energy range of 9.45 eV – 9.8 eV (band 2). These two bands are separated by a gap in energy range of 8.9 eV – 9.45 eV. With the increase of electric field, the transmission probability spectrum becomes narrower and the width of the regions with nonzero transmission, in which the center band is shifted, also becomes narrower. However, the gap, energy region with zero transmission probability, widens. The gap at zero voltage already discussed in the previous section is the energy range in which the density of states is not affected by electric field. When electric field is increased the energy gap is widened and the energy band becomes narrow. Similar to the DOS spectrum, transmission probability spectrum at 0.03 V volt shows three bands, in energy range of 7.02 eV – 7.25 eV (band 1), 7.45 eV – 7.6 eV (band 2) and 9.45 eV – 9.65 eV (band 3), and two gaps, in energy range of 7.25 eV – 7.45 eV (gap 1) and 7.6 eV – 9.45 eV (gap 2). The increase of electric field causes the transmission probability spectrum to change, some states experience lesser transmission probability. The spectrum changes observed due to voltage increment indicates that lesser states with nonzero transmission probability are involved in the transport process. Nokhrin *et al.* [15] found 1 volt threshold voltage at distance of 5 nm decreases the conductivity of M-DNA by at least 10-fold.

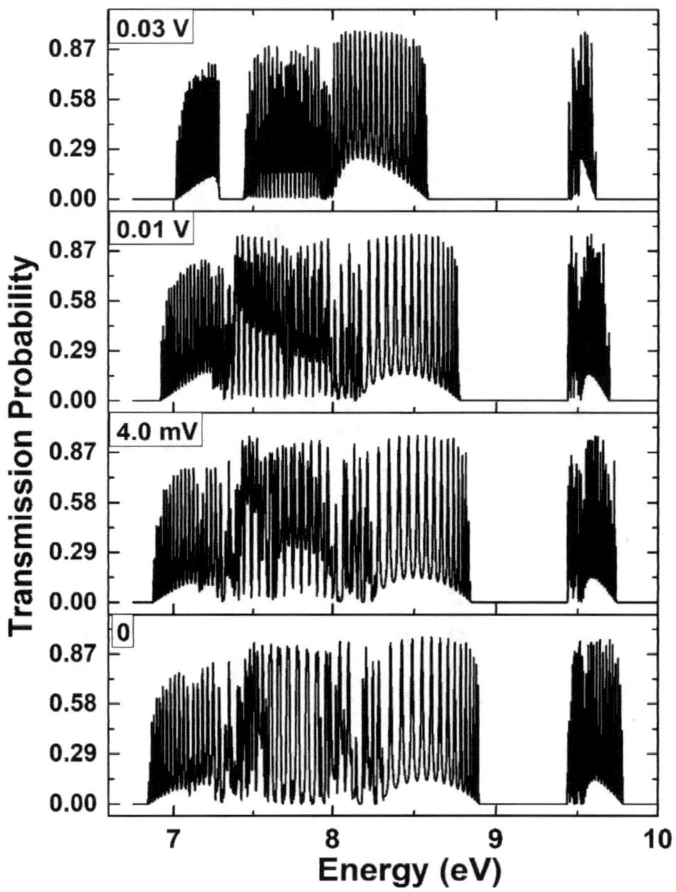

Figure 2. Transmission probabilty of electron calculated at temperature 4.2 K and twisting motion frequency 4.00 meV for electric field of voltages 0, 4.0 mV, 0.01 V, and 0.03 V.

Li *et al.* [12] measures the electrical character of transport hopping in organic semiconductors, in a very high electric field, an effective disruption of electron hopping so that causes a decrease in conductivity.

4. Conclusions

The effect of electric field increment on electron DOS and transmission probability of G4-DNA molecule have been studied. The electron DOS and transmission probability calculated at temperature 4.2 K and twisting motion frequency 4 meV showed the spectrum of DOS and transmission probability changes with the electric fields. The bands of nonzero DOS and transmission probability were shifted as the electric field increases. The effect of electric field on the molecule will affect the electron transport process across the molecule.

Acknowledgements

Authors cannot do this work without help of various parties, either in the form of facilities or materials. Therefore, in this occasion the authors extend their deepest gratitude to Daniel K Suhendro for calculation parameters used in this study and CISCO Computer Laboratory at Department of Physics, FMIPA Universitas Indonesia for facilities used during the research. The authors also thanks to Universitas Indonesia for funding this work through Hibah Pitta 2018 Number: 2259/UN2.R3.1/HKP.05.00/2018.

References

[1] Kang D W, Sun M L, Zuo Z W, Wang H X, Lv S J, Li X Z and Li L B 2016 *Phys. Lett. A* **380** 977–82
[2] Al-Hinai M, Hassanien R, Watson S M D, Wright N G, Houlton A and Horrocks B R 2016 *Nanotechnology* **27** 095704
[3] Dugasani S R, Hwang T, Kim J A, Gnapareddy B, Kim T and Park S H 2016 *J. Phys. D. Appl. Phys.* **49** 105501
[4] Beratan D N, Naaman R and Waldeck D H 2017 *Curr. Opin. Electrochem.* **4** 175–81
[5] de Almeida M L, Ourique G S, Fulco U L, Albuquerque E L, de Moura F A B F and Lyra M L 2016 *Chem. Phys.* **478** 48–54
[6] Qian J, Liu L, Xu B and Tian W 2017 *Org. Electron.* **42** 244–55
[7] Sharipov T I and Bakhtizin R Z *2017 IOP Conf. Ser.: Mater. Sci. Eng.* **256** 012009
[8] Guo A M and Xiong S J 2009 *Phys. Rev. B* **80** 035115
[9] Kang D, Jiang H, Sun Z, Qu Z and Xie S 2011 *J. Phys.: Condens Matter* **23** 055302
[10] Wiliyanti V and Yudiarsah E 2016 *AIP Conf. Proc.* **1729** 020033
[11] Sinurat E N and Yudiarsah E 2017 *AIP Conf. Proc.* **1862** 030038
[12] Li L, Meller G and Kosina H 2007 *Microelectron. J.* **38** 47–51
[13] Jadhav P J , Limketkai B N and Baldo M A 2010 *Adv. Polym. Sci.* **223** 29–44
[14] Qi J, Edirisinghe N, Rabbani M G, and Anantram M P 2013 *Phys. Rev. B* **87** 85404
[15] Nokhrin S, Baru M and Lee J S 2007 *Nanotechnology* **18** 095205

The influence of gap spacing in localized surface plasmon resonance (LSPR) spectra of Ag nanorod-coupled with end-to-end assembly using boundary element method

N Fitria, M Sujak and D Djuhana

Department of Physics, Faculty of Mathematics and Natural Sciences (FMIPA), Universitas Indonesia, Kampus UI Depok, Depok 16424, Indonesia

Corresponding author's e-mail: dede.djuhana@sci.ui.ac.id

Abstract. We have systematically investigated the Localized Surface Plasmon Resonance (LSPR) of Ag nanorod-coupled with end-to-end assembly and influence distance between nanorod using metallic nanoparticle based on the boundary element method (MNPBEM). We focused on the extinction curve of LSPR spectra for investigating the effect of gap spacing for this work. It was found that the extinction curve showed two behaviours, for lower wavelength known as the transversal mode and higher wavelength as the longitudinal mode. For further understanding, we also analyzed the electric field related to LSPR spectra of Ag nanorod-coupled. In end-to-end assembly, the resonance peak of transversal mode shifted to lower wavelength (blue-shift) while for longitudinal mode shifted to higher wavelength (red-shift), both as the spacing gap decreased. Interestingly, for the case in parallel polarization, the peaks of wavelength decay as the gap spacing increased followed a plasmon ruler. We observed the dipole-dipole interaction was the influence in LSPR spectra of Ag nanorod-coupled. According to this result, it is essential to understand the distance dependence in LSPR spectra in coupled noble particle system

Keywords: MNPBEM, Ag nanorod-coupled, end-to- end assembly, LSPR.

1. Introduction

For decades, the investigation on behaviour of Localized Surface Plasmon Resonance (LSPR) spectra of noble nanoparticle (Au and Ag) in array configuration have been intensively studied. The coupling interaction of array configuration has played an essential role in the excitation of the collective electron oscillation on LSPR phenomenon [1] and shows a scaling behavior with an exponential type of LSPR array to the single system [2–5]. However, a few studies have been reported on the influence of gap spacing in LSPR Ag-nanorod-coupled with end-to-end configuration and the threshold value of ratio gap spacing and length of the nanorod

In this paper, we have observed the influence of gap spacing in LSPR spectra of Ag nanorod-coupled with end-to-end assembly using public optical simulation, metallic nanoparticle boundary element method (MNPBEM). The extinction curve showed LSPR spectra consisted of two peaks. The first peak occurred in lower wavelength was the transverse mode and the second peak in higher wavelength known as longitudinal mode, respectively. The plasmon peaks showed blue-shift in the transverse mode whereas red-shift in the longitudinal mode as the gap spacing decreased. Interestingly, ratio LSPR of Ag nanorod-coupled to single showed a scaling behavior following plasmon ruler with exponential function type. Analyzing the ratio of gap spacing and length of nanorod exhibited a threshold value of LSPR Ag nanorod-coupled acts like single nanorod.

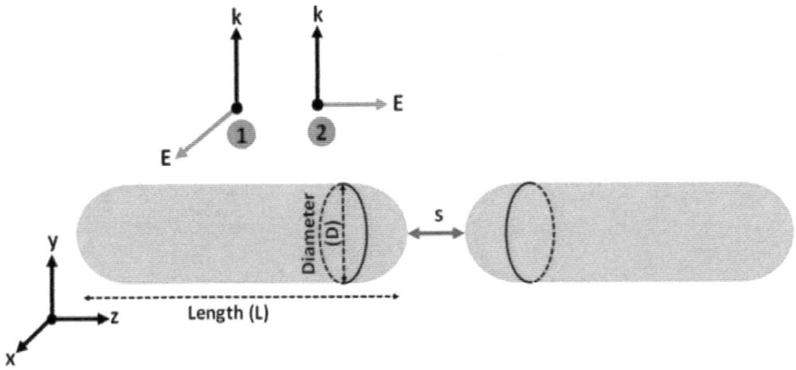

Figure 1. The dimension and configuration of Ag nanorod-coupled with end-to-end assembly with respect to gap spacing *s*. The incident wave is applied in parallel and perpendicular to long pair of Ag nanorod-coupled. The incident wave is presented in the dielectric field **E** and the propagation direction is denoted by wavevector **k**.

2. Methods
We have investigated the influence of gap spacing in Localized Surface Plasmon Resonance (LSPR) spectra of Ag nanorod-coupled with end-to-end assembly using a public optical simulation metallic nanoparticle boundary element method (MNPBEM) [6] based on boundary element method [7]. From this simulation we obtained the absorption C_{abs}, scattering C_{sca}, and C_{ext} extinction cross section as the function of wavelength and was written as

$$C_{sca} = n_b \oint_{\partial\Omega} \Re\{\hat{n}(\mathbf{E}\times\mathbf{B}^*)\}da, \quad C_{ext} = -\frac{1}{n_b}\oint_{\partial\Omega} \Re\{\hat{n}(\mathbf{E}\times\mathbf{B}^*_{inc} + \mathbf{E}^*_{inc}\times\mathbf{B})\}da, \quad C_{abs} = C_{ext} - C_{sca} \quad (1)$$

where $\mathbf{E}$ and $\mathbf{B}$ is the electric and magnetic field vector, $\mathbf{E}^*$ and $\mathbf{B}^*$ is the conjugate of electric and magnetic field vector, and n_b is the refractive index of medium. The diameter was fixed D = 20 nm, and the length was L = 30, 60, and 100 nm or the aspect ratio between the length to the diameter is 1.5, 3, and 5, respectively. The dielectric function of Ag nanorod-coupled used Palik's experiment [8], and the refractive index medium is n = 1.3334. The gap spacing of Ag nanorod-coupled was varied from 5 nm to 500 nm. From this simulation, we produced the extinction curve as the function of wavelength with the polarization of the incident wave in parallel (longitudinal mode) and perpendicular direction (transverse mode) to long pair axis coupled nanorods, as illustrated in figure 1. Besides, we have also produced the extinction curve for a single Ag nanorod for comparing of LSPR shifting.

3. Results and discussion
We have obtained the extinction curve as the function of wavelength. The typical extinction curve of Ag nanorod-coupled is presented in figure 2a for the case D = 20 nm and L = 60 nm or the aspect ratio is 3 and with varying the gap spacing. It was found the extinction curve showed two peaks; firstly, peak occurred in lower wavelength and a second peak in higher wavelength, known as transverse and longitudinal mode, respectively. Further, we observed the peak of LSPR shifted to smaller wavelength as the aspect ratio increased in transverse mode, as shown in figure 2b. It changed around an order of tens. Different in longitudinal, it shifted around the order of hundreds. Its means that varying the aspect ratio will affect the wavelength of LSPR peaks. Increasing the aspect ratio of the nanorod, the plasmon peaks will move to higher wavelength because of the weakening repulsive force between rods. Interestingly in the coupling case, the plasmon peaks tended to shift to higher wavelength in longitudinal while changed to lower wavelength in transverse as the gap spacing decreased. In sense, the transverse and longitudinal mode can be explained by the different polarization that generated the electron oscillation on the surface of nanorod. The electron oscillation occurred along the axis nanorod in longitudinal whereas the short axis of nanorod in transverse.

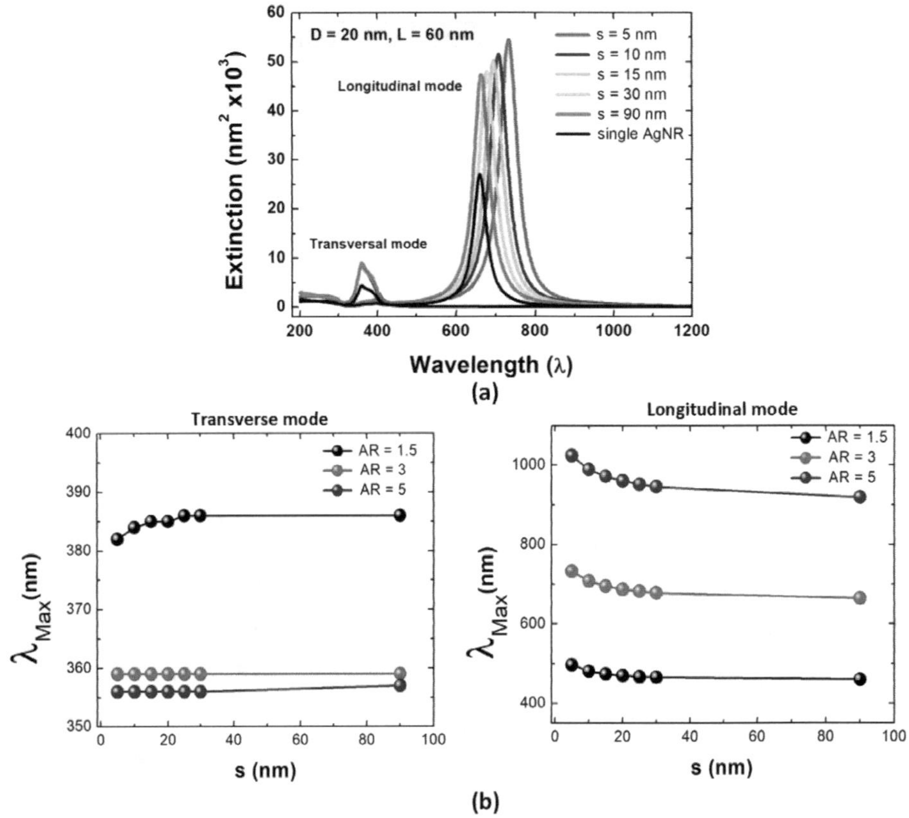

Figure 2. (a) the extinction curve as the function of wavelength for the case D = 20 nm and L = 60 nm or the aspect ratio is AR=3 corresponds to the gap spacing variation (s = 5-90 nm), (b) the peak of the extinction curve relates to the gap spacing in transverse and longitudinal mode, respectively as the aspect ratio variation.

For understanding, we have also produced the imaging of the electric field intensity of Ag nanorod-coupled for the case aspect ratio AR= 3, as presented in figure 3. In the figure, we showed the different LSPR images in transverse and longitudinal modes. The plasmon was formed around the surface of Ag nanorod in transverse while the plasmon occurred at the end of Ag nanorod in longitudinal. Then, these images have also showed the description of LSPR phenomenon in nanorod. For a single Ag nanorod, the radiated electric field caused the charge polarization $\pm q$ and the moment dipole was about **qL**. For the coupling nanorods, the moment dipole was $q\,(2L+s)$ as the gap spacing changed and became $\approx 2qL$ for small the gap spacing.

So, the intensity of Ag nanorod-coupled was higher than single Ag nanorod [9]. The coupling nanorod with end-to-end assembly will excite the opposite surface charges on the end of nanorod in parallel polarization. Then, the attractive forces happened between the nanorod and will reduce the oscillation frequency. Opposite in perpendicular polarization, the nanorod showed the same charge distribution and tended to enhance the repulsive force [10]. This application will increase the resonance frequency and consistent with the extinction curve as explained in figure 2. In simple way, the plasmon resonance of Ag nanorod-coupled can be explained by the dipole-dipole approximation [2].

Furthermore, we have also observed the universal scaling behavior of the extinction peak between Ag nanorod-coupled and single Ag nanorod as a function of the ratio between gap spacing and length of nanorod in longitudinal mode. The peak of plasmon coupling to a single plasmon followed a scaling behavior with exponential decay type to the ratio between the gap spacing and length of nanorod, $\Delta\lambda/\lambda_o = A \exp(-s/L\tau)$ where $\Delta\lambda = \lambda - \lambda_o$ where was the difference peak between plasmon coupling to single plasmon, A was a constant value, s was the gap spacing, L was the length of nanorod, and τ

Table 1. The exponential fitting value from $\Delta\lambda/\lambda_o$ to s/L for the aspect ratio AR=1.5, 3, and 5

Aspect ratio (AR)	Constant (A)	Decay length (τ)
1.5	0.120	0.39
3	0.145	0.26
5	0.150	0.21

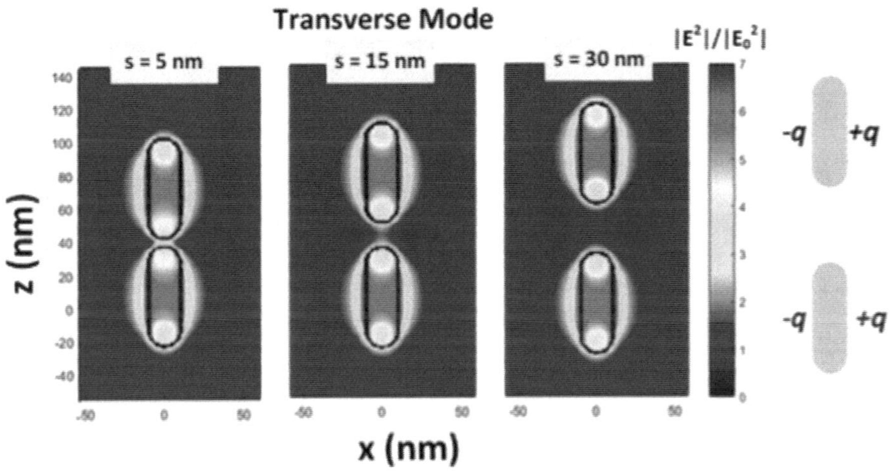

(a)

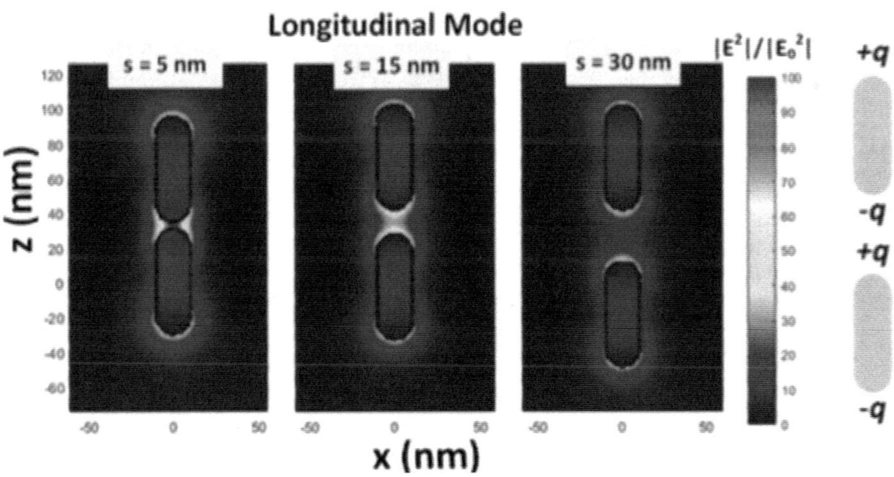

(b)

Figure 3. The electric field image of Ag nanorod-coupled with end-to-end assembly for (a) transvers and (b) longitudinal mode for the case D = 20 nm and L = 60 nm or the aspect ratio is AR = 3 and s = 5, 15 and 30 nm. The colour bar presents the intensity of the electric field.

was the decay length [2]. Figure 4 shows the curve of $\Delta\lambda/\lambda_o$ vs. s/L of Ag nanorod-coupled and the exponential fitting (solid line) for the aspect ratio AR = 1.5, 3, and 5. We used the approximation $A \exp(-x/k)$ for exponential function fitting, where $x = s/L$ and $k = \tau$. The fitting results are given in the table 1.

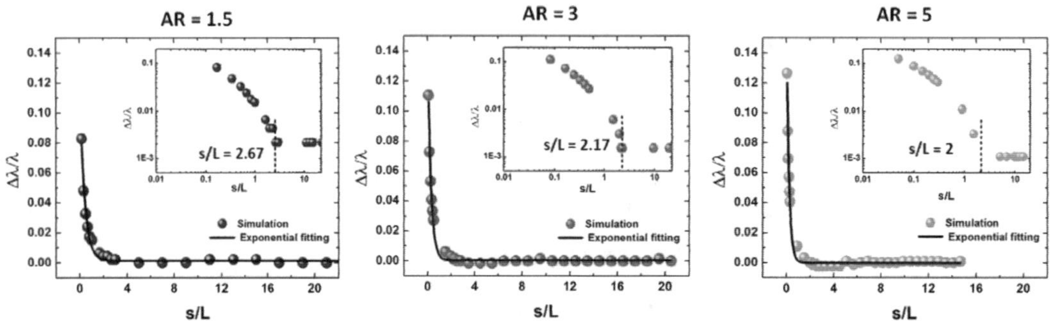

Figure 4. The curve of $\Delta\lambda/\lambda$ vs. s/L and the exponential fitting (solid black line) for the aspect ratio AR=1.5, 3, and 5. The inset figure shows the curve of $\Delta\lambda/\lambda$ vs. s/L in logarithm scale and the value of s/L is denoted by the dot-line.

More interestingly, our data calculation followed a scaling behavior and fitted well with the exponential function, as explained above. From table 1, the aspect ratio increased, the value of A relatively increased, and the decay length tended to decrease. It means, that the parameter of A related to coupling strength that contributed to LSPR peaks shifting and the decay length τ related reducing the electric field intensity as the gap spacing increased. The determination of A was around 0.12 to 0.15, and the decay length τ was around 0.21 to 0.4 showed a good agreement with others results [11–14]. Next, we also analyzed the value of s/L for the aspect ratio variation. With changing in logarithm scale, we can determine a threshold value of s/L, as presented in the inset of figure 4. The value of s/L was 2.67, 2.17, and 2 for AR = 1.5, 3, and 5, respectively (dot-line). The inset figure of figure 4 describes the threshold value. These results show the LSPR peaks of coupling nanorods act like single nanorods at the particular gap spacing.

4. Conclusions

In summary, we have observed the influence of gap spacing of Ag nanorod-coupled with end-to-end assembly using MNPBEM simulation concerning the aspect ratio and gap spacing variation. LSPR spectra showed two peaks that related to transverse and longitudinal modes as the result of polarization directions of the radiated field according to the short and long axis of nanorods. The LSPR peaks of coupling nanorods shifted to lower wavelength (blue-shift) in transverse mode and higher wavelength (red-shift) in longitudinal as the gap spacing decreased. Also, a dipolar coupling model explains this behavior qualitatively as a result of interference of near field each nanorod. The coupling plasmon peaks respect to single as a function of s/L showed a scaling behavior with exponential function and followed with plasmon ruler behavior. The threshold value of s/L in this work showed behavior LSPR spectra of coupling Ag nanorods were same as single nanorods; it is useful to predict the interparticle gap spacing in the biological system

Acknowledgements

The authors would like to thank Universitas Indonesia for funding this work through PITTA Grant No: 2246/UN2.R3.1/HKP.05.00/2018.

References

[1] Jain P K, Eustis S and El-Sayed M A 2006 *J. Phys. Chem.* B **110** 18243–53
[2] Su K H, Wei Q H, Zhang X, Mock J J, Smith D R and Schultz S 2003 *Nano Lett.* **3** 1087–90
[3] Huang C P, Yin X G, Kong L B and Zhu Y Y 2010 *J. Phys. Chem.* C **114** 21123–31
[4] Tabor C, Murali R, Mahmoud M, and El-Sayed M A 2009 *J. Phys. Chem.* A **113** 1946–53
[5] Gunnarsson L, Rindzevicius T, Prikulis J, Kasemo B, Käll M, Zou S and Schatz G C 2005 *J. Phys. Chem.* B **109** 1079–87
[6] Hohenester U and Trügler A 2012 *Comput. Phys. Commun.* **183** 370–81
[7] de Abajo F J G and Howie A 2002 *Phys. Rev.* B **65** 115418
[8] Lynch D W, and Hunter W R 1997 Critiques, metals: an introduction to the data for several metals *Handbook of Optical Constants of Solids vol 2*, ed E D Palik (London: Academic

Press Ltd) pp 341–409
[9] Aizpurua J, Bryant G W, Richter L J, de Abajo F J G, Kelley B K and Mallouk T 2005 *Phys. Rev.* B **71** 235420
[10] Rechberger W, Hohenau A, Leitner A, Krenn J R, Lamprecht B and Aussenegg F R 2003 *Opt. Commun.* **220** 137–41
[11] Jain P K, Huang W and El-Sayed M A *Nano Lett.* **7** 2080–8
[12] Funston A M, Novo C, Davis T J and Mulvaney P 2009 *Nano Lett.* **9** 1651–8
[13] Jain P K and El-Sayed M A 2010 *Chem. Phys. Lett.* **487** 153–64
[14] Ben X and Park H S 2011 *J. Phys. Chem.* C **115** 15915–26

Structural and optical properties of $La_{1-x}Pb_xFe_{0.5}Ti_{0.5}O_3$ (x = 0.1, 0.2, and 0.3) perovskite material prepared by sol-gel method

Y Supriyadi, D Triyono and S K Lukmana

Department of Physics, Faculty of Mathematics and Natural Science (FMIPA), Universitas Indonesia, Kampus UI Depok, Depok 16424, Indonesia

Corresponding author's e-mail: djoko.triyono@sci.ui.ac.id

Abstract. A series of $La_{1-x}Pb_xFe_{0.5}Ti_{0.5}O_3$ (x = 0.1, 0.2, and 0.3) nanoparticles have been successfully synthesized by sol-gel method. All prepared samples were characterized by X-ray diffraction (XRD), Raman scattering, Fourier Transform Infrared spectroscopy (FTIR), and UV-Vis spectrophotometry (UV-Vis). The X-ray diffraction pattern revealed the orthorhombic crystal structure and the increase of crystallite size with Pb content. The Raman scattering exhibited the similar Raman spectrum indicating the crystal structure remain unchanged due to the Pb substitution; indeed, the difference in lattice vibration phonon modes confirmed the variation in structural parameters with Pb-substitution which is also supported by FT-IR analysis. By using UV-Vis characterization, the optical band gap is found in the range of 2.1 – 2.2 eV.

Keywords: $(La,Pb)(Fe,Ti)O_3$, crystal structure, symmetry, energy gap

1. Introduction

Oxide perovskite has general formula ABO_3 [1]. Substitution in A- and B-site of oxide perovskite structure caused the modification in its properties and useful in electrode material, photocatalytic, and chemical sensor application [1–6]. This paper will focus on the modification of perovskite structure with $LaFeO_3$ as a parent compound. $LaFeO_3$ has been considered due to its complex physical behavior, which is strongly related with spin, charge and defects, and orbital state [2]. Several researches have been reported about pure- and doped- on $LaFeO_3$ [2–5]. $LaFeO_3$ has strong visible-light absorption with the optical band gap 2.36 eV [3–4]. Pb-doped on $LaFeO_3$ introduced the colossal dielectric permittivity with multiple relaxations [2]. Ti-doped on $LaFeO_3$ have been reported to cause the decrease of crystallite size, increase the band gap, and introducing the weak ferromagnetic behavior [5].
Based on the previous works [2–5], it is interesting to study the double substitution on La and Fe site of $LaFeO_3$. To the best our knowledge, substitution of Pb- and Ti-ions on the structure and optical properties of $La_{1-x}Pb_xFe_{0.5}Ti_{0.5}O_3$ have not been reported and paid attention on yet. The materials were set-up by sol-gel method and investigated by X-Ray Diffraction (XRD), Raman scattering spectroscopy, Fourier Transform Infrared (FTIR) and UV-Vis.

2. Experimental details

Sol-gel method was used to synthesize the $La_{1-x}Pb_xFe_{0.5}Ti_{0.5}O_3$ with x = 0.1, 0.2, 0.3. The precursors that used were lanthanum oxide (La_2O_3), lead oxide (PbO), iron nitrate ($Fe(NO_3)_3.9H_2O$), titanium oxide (TiO_2), and citric acid as solvent (all analytically grade). The prepared precursors were dissolved in citric acid solution and stirred at 120 °C. To obtain sponge phase, the drying process was applied in certain condition. Then, the sponge phase was calcined to form the powder-phase of $La_{1-x}Pb_xFe_{0.5}Ti_{0.5}O_3$.

Table 1. Structural parameters of $La_{1-x}Pb_xFe_{0.5}Ti_{0.5}O_3$ (x = 0.1, 0.2, and 0.3) as obtained from HighScore Plus software.

	Crystallographic Parameters					
	a (Å)	*b* (Å)	*c* (Å)	volume (Å^3)	crystallite size (nm)	density (gr/cm^3)
x = 0.1	5.566	7.872	5.533	242.5	23.3	6.77
x = 0.2	5.557	7.841	5.680	242.6	24.4	6.91
x = 0.3	5.578	7.860	5.554	243.5	27.2	7.07

	Geometrical Parameters				
	<Fe/Ti-O> (Å)	bond angle (°) B = Fe/Ti: B-O1-B	bond angle (°) B = Fe/Ti: B-O2-B	tilt angle (°)	tolerance factor
x = 0.1	2.007	157.041	154.219	14.883	0.8317
x = 0.2	2.008	157.109	153.966	14.937	0.8304
x = 0.3	2.011	157.075	154.085	14.912	0.8301

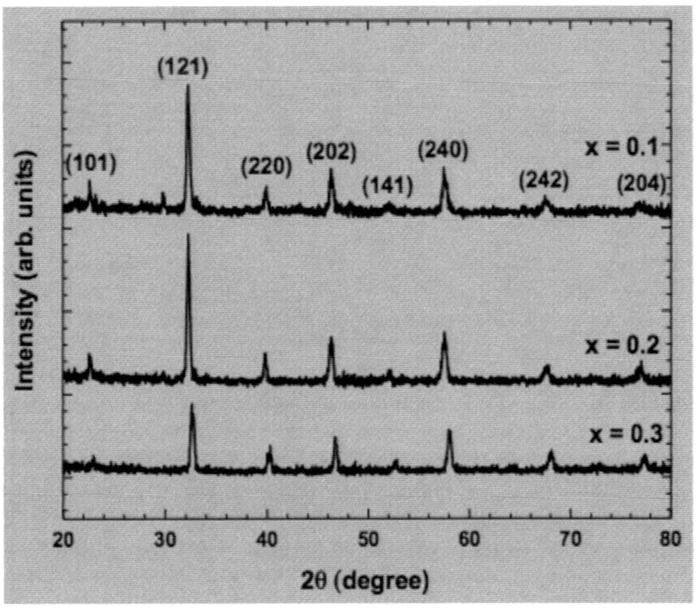

Figure 1. The XRD patterns of $La_{1-x}Pb_xFe_{0.5}Ti_{0.5}O_3$ (x = 0.1, 0.2, and 0.3) calcined at 900 °C.

The structural properties were investigated by using X-Ray Diffractometer with Cu Kα radiation source. Optical properties were studied by using Raman scattering spectra (experimental conditions: laser source of 532 nm, power 2.0 mW), FTIR spectrometer (Perkin Elmer type Spectrum Two), and UV-vis spectrophotometer (Agilent Cary 100/300).

3. Results and discussion

3.1. X-ray diffraction analysis

Figure 1 shows the XRD patterns of $La_{1-x}Pb_xFe_{0.5}Ti_{0.5}O_3$ (x = 0.1, 0.2, and 0.3). It confirms that all compounds are single phase with no measurable secondary phase. The refinement of XRD patterns using HighScore Plus software displays that all samples crystallize in the orthorhombic crystal structure. The details of refinement results are summarized in table 1. As can be seen, there are no significant changes in lattice and geometrical parameters. However, the crystallite size increases with Pb content. The tolerance factor decreases indicating the increase of lattice distortion due to the Pb content.

Table 2. FT-IR mode assignment of $La_{1-x}Pb_xFe_{0.5}Ti_{0.5}O_3$ (x = 0.1, 0.2, 0.3) compounds.

x	FT-IR band position			
	O−Fe/Ti−O Bending	Fe/Ti−O−Fe/Ti Bending	Fe/Ti−O Stretching	O−H Bending
0.1	459	761	535	1477
0.2	459	760	537	1482
0.3	458	754	538	1476

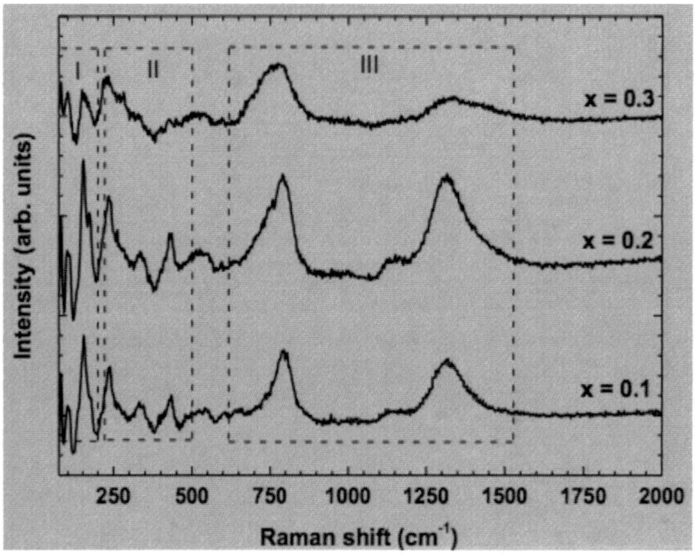

Figure 2. The Raman scattering spectra of $La_{1-x}Pb_xFe_{0.5}Ti_{0.5}O_3$ (x = 0.1, 0.2, and 0.3) compounds. The certain region describing in the dash line is related with vibrational modes occurring in samples.

3.2. Raman scattering spectra

In order to confirm the stability of phase and crystal structure, the Raman scattering was applied [6]. Figure 2 shows the Raman scattering spectra of $La_{1-x}Pb_xFe_{0.5}Ti_{0.5}O_3$ (x = 0.1, 0.2, and 0.3) compounds. All samples show similar Raman spectra indicating the crystal structure remain unchanged due to the variation of Pb content. The spectra are also consistent with previously reported for the undoped and doped of $LaFeO_3$ [6–8]. According to the previous reports [6–8], the modes inside the region I are described as vibrational occurring in La/Pb-site; the modes inside the region II are related with the vibrational in Fe/Ti site; the modes inside the region III should be assigned to multi-phonon processes. The increase of Pb content causes the lowering of frequency phonon modes, broadening of peaks, and reducing of intensity indicating the lattice becomes more distorted as confirmed by XRD analysis.

3.3. FTIR analysis

The vibrational modes occurring in samples are also confirmed by FT-IR analysis. Figure 3 shows the FT-IR analysis of $La_{1-x}Pb_xFe_{0.5}Ti_{0.5}O_3$ (x = 0.1, 0.2, and 0.3) compounds. This result is consistent with previous work for the doped and un-doped $LaFeO_3$ [4, 9]. Some specific vibrational modes are listed in table 2. As can be seen, the vibrational modes of metal oxide appear in the range of 400 – 800 cm^{-1} for all samples. The slightly change in modes wavenumber with Pb substitution confirms the difference in metal oxide bond angle and bond length (table 1).

3.4. UV-Vis analysis

Figure 4 shows the UV-Vis spectra of $La_{1-x}Pb_xFe_{0.5}Ti_{0.5}O_3$ (x = 0.1, 0.2, and 0.3) compounds. As shown in figure 4, the linear parts extrapolation against the absorption which equal to zero describes the direct band gap energy, E_g, for all samples. The E_g is found in the range of 2.1 – 2.2 eV for all samples. These results are consistent with the previous report for direct band gap energy of $LaFeO_3$ [3-4].

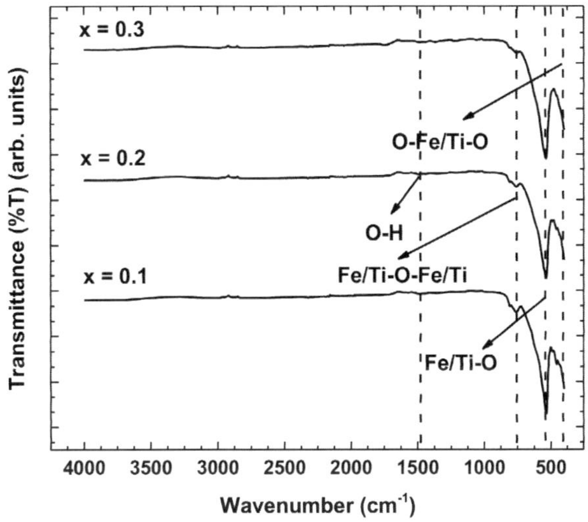

Figure 3. The FT-IR spectra of $La_{1-x}Pb_xFe_{0.5}Ti_{0.5}O_3$ (x = 0.1, 0.2, and 0.3) compounds.

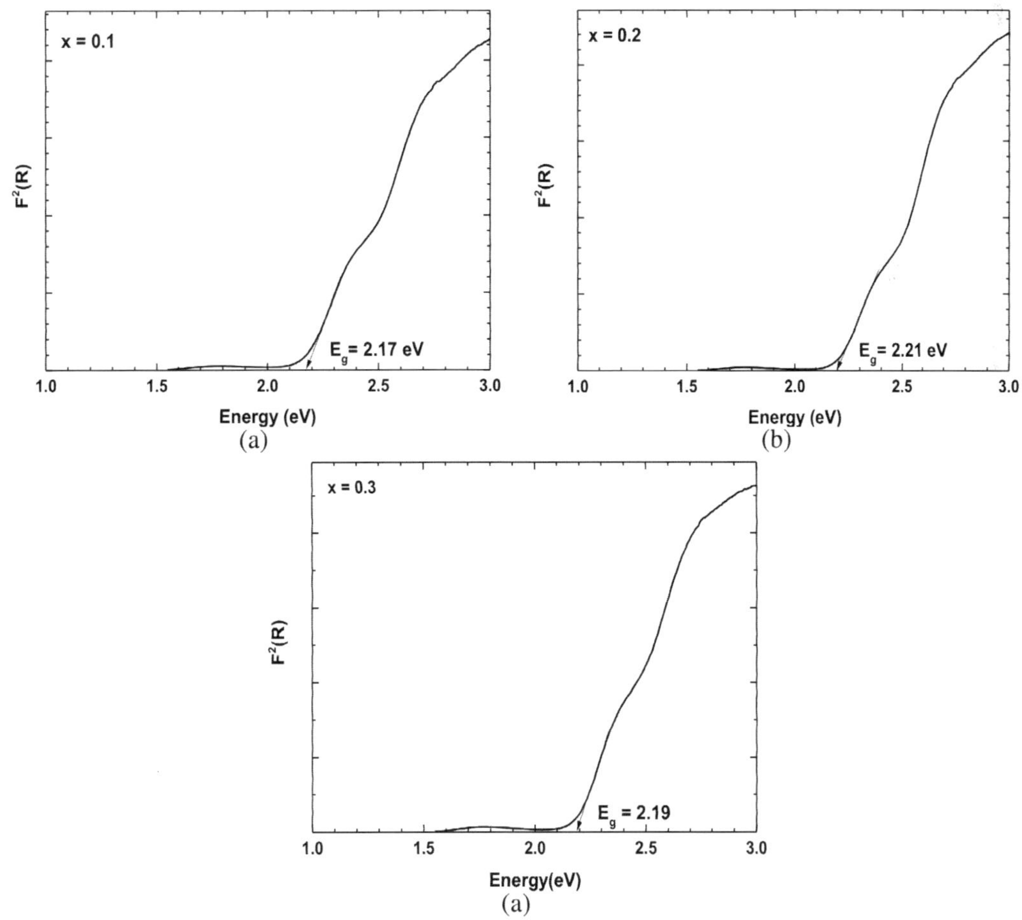

Figure 4. UV-Vis spectra of $La_{1-x}Pb_xFe_{0.5}Ti_{0.5}O_3$ for x = (a) 0.1, (b) 0.2, and (c) 0.3 compounds.

4. Conclusions

The series of $La_{1-x}Pb_xFe_{0.5}Ti_{0.5}O_3$ (x = 0.1, 0.2, and 0.3) have been synthesized by sol-gel method. The structural analysis revealed the orthorhombic crystal structure, which is also confirmed by the identical Raman scattering spectra for all samples. The crystallite size increased with the increase of Pb content. The Raman scattering analysis revealed the presence of metal-oxide vibrational modes in lattice, which also confirmed by FT-IR analysis. The optical band gap for all samples is found in the range of 2.1 – 2.2 eV.

Acknowledgements

The authors would like thanks to Hibah PITTA under contract No. 2256/UN2.R3.1/HKP.05.00/2018 from Universitas Indonesia for financial support.

References

[1] Tien-Thao N, Alamdari H, and Kaliaguine S 2008 *J. Solid State Chem.* **181** 2006–19
[2] Chandrasekhar K D, Mallesh S, Murthy J K, Das A K and Venimadhav A 2014 *Physica* B **448** 304–11
[3] Tang P, Tong Y, Chen H, Cao F, and Pan G 2013 *Curr. Appl. Phys.* **13** 340–3
[4] Ding J, Lü X, Shu H, Xie J and Zhang H 2010 *Mater. Sci. Eng.* B **171** 31–4
[5] Sasikala C, Durairaj N, Baskaran I, Sathyaseelan B, Henini M and Manikandan E 2017 *J. Alloy Compd.* **712** 870–7
[6] Mahapatra A S, Mitra A, Mallick A, Ghosh M and Chakrabarti P K 2016 *Mater. Lett.* **169** 160–3
[7] Runka T and Berkowski M 2012 *J. Mater. Sci.* **47** 5393–401
[8] Martín-Carrón L, de Andrés A, Martínez-Lope M J, Casais M T and Alonso J A 2002 *Phys. Rev.* B **66** 174303
[9] Janbutrach Y, Hunpratub S and Swatsitang E 2014 *Nanoscale Res. Lett.* **9** 498

IOP Conf. Series: Materials Science and Engineering **763** (2020) 012064 doi:10.1088/1757-899X/763/1/012064

Efficient photo-, sono-, and sonophoto-fenton-like degradation of the organic pollutant methylene blue using a $BiFeO_3$/graphene composite

Y D Susanti and R Saleh

[1] Department of Physics, Faculty of Mathematics and Natural Sciences (FMIPA), Universitas Indonesia, Kampus UI Depok, Depok 16424, Indonesia

[2] Integrated Laboratory of Energy and Environment, Faculty of Mathematics and Natural Sciences (FMIPA), Universitas Indonesia, Kampus UI Depok, Depok 16424, Indonesia

Corresponding author's e-mail: rosari.saleh@ui.ac.id

Abstract. In this study, we investigated the degradation of the organic pollutant methylene blue (MB) using photo-, sono-, and sonophoto-fenton processes under visible light, ultrasonic irradiation, and a combination of the two, respectively, in the presence of an iron-containing perovskite catalyst $BiFeO_3$/graphene composite. The sample structure was characterized using X-ray diffraction; thermal analyses were performed using thermogravimetric analysis and differential thermal analysis; and the $BiFeO_3$/graphene composite was prepared using a hydrothermal method. We observed better MB degradation via the $BiFeO_3$/graphene composite using the sonophoto-fenton method relative to either the photo- or sono-fenton methods, with increased MB degradation occurring through the synergistic effects of both the photo- and sono-fenton techniques. Furthermore, addition of a chelating agent, such as ascorbic acid, to the photo- and sono-fenton reactions enhanced MB degradation.

1. Introduction

Advance oxidation processes (AOPs) capable of generating strongly oxidizing hydroxyl free radicals are among the most important processes necessary for decomposing organic pollutants such as methylene blue (MB). AOPs offer several advantages in terms of degrading organic pollutants into a final mineralized form through the production of a hydroxyl free radical (OH). Among the available AOPs, fenton-like degradation is based on the electron transfer between H_2O_2 and a transition metal ion acting as a catalyst and is capable of generating strongly oxidizing OH in the presence of H_2O_2 [1,2].

However, fenton-like degradation has several limitations owing to the limited pH range of the reaction, ion sludge, and difficulty in regenerating the catalyst [3,4]. The classic fenton process requires a low pH (3.0) because of the presence of ion sludge at higher pH levels. The limitations of using fenton-like degradation to decompose organic pollutants can be overcome using heterogeneous catalysts. Furthermore, chelating agents such as ascorbic acid can be used to prevent the formation of ion sludge at higher pH levels [5].

H_2O_2 in the fenton reaction is activated under ultraviolet light and ultrasonic irradiation; therefore, initiation of the fenton reaction in the presence of visible light irradiation requires a new catalyst with a narrow band gap [6]. Recently, increasing attention has focused on the perovskite catalyst $BiFeO_3$ because of its narrow band gap. $BiFeO_3$ is a multiferroic material that exhibits ferroelectric (Tc-1103) ordering and antiferromagnetic (TN-643 K) ordering in the bulk form and has been applied in several catalytic processes owing to its good catalytic performance. However, limitations to its wider application include its rapid electron–hole recombination. Graphene material is two-dimensional and includes a carbon atom with a large surface area, high electron mobility, and good electrical conductivity. There is great interest in graphene material because it can be used as an electron trap, which promotes MB degradation. In this study, we investigated the fenton-like degradation of MB using $BiFeO_3$/graphene under visible light, ultrasonic irradiation, and a combination of the two.

2. Experimental details

2.1. Chemicals
All reagents were analytical grade and commercially available from Merck (Darmstadt, Germany) and Sigma-Aldrich (St. Louis, MO). The reagents used to synthesize the $BiFeO_3$/graphene composite were bismuth(III) nitrate ($Bi(NO_3)_3 \cdot 5H_2O$; 99% purity; Sigma-Aldrich), iron(III) nitrate nonahydrate ($Fe(NO_3)_3 \cdot 9H_2O$; 99% purity; Merck), and potassium hydroxide (KOH; 99% purity; Merck).

2.2. Characterization
The crystallite structure of the sample was investigated using X-ray diffraction (XRD; Rigaku Miniflex 600; Rigaku, Tokyo, Japan), and the thermal stability of the sample was characterized using thermogravimetric analysis (TGA) and differential thermal analysis (DTA) using a Rigaku TG8121 device (Rigaku). The sample was heated from room temperature to 1000 °C at a rate of 2 °C/min, with a flow of argon maintained during analysis.

2.3. Catalyst preparation
The $BiFeO_3$ nanoparticle was synthesized using a hydrothermal method. Equimolar amounts of $Bi(NO_3)_3 \cdot 5H_2O$ and $Fe(NO_3)_3 \cdot 9H_2O$ were stirred using a magnetic stirrer, and KOH solution was added dropwise. The mixed solution was transferred to a 25-mL stainless steel autoclave, and hydrothermal treatment was performed at a reaction temperature of 180 °C for 6 h, after which the product was centrifuged and dried at 120 °C for 3 h. The $BiFeO_3$/graphene composite was also synthesized using a hydrothermal method by adding a specific weight percentage of graphene during the process, with the reaction temperature maintained at 120 °C for 12 h in a 100-mL stainless steel autoclave.

2.4. Photo- and sono-fenton degradation
The photo-fenton reaction was conducted using a 100-mL glass vessel equipped with a magnetic stirrer and a 40-W Xe lamp used as the light source. The sono-fenton reaction was performed in an ultrasonic bath operated at a fixed frequency and power of 40 kHz and 150 W, respectively. The sonophoto-fenton reaction used a combination of the 40-W Xe lamp and the ultrasound source. All three fenton reactions were initiated by the addition of hydrogen peroxide (H_2O_2; 1–4 mL), prior to which the pH of the solution was adjusted from 3 to 11 by adding NaOH and CH_3COOH. The catalyst was added to the dye solution at a range of 0.1–0.4 g/L, with the initial dye concentration at 20 mg/L.

3. Results and discussion
The XRD patterns of the $BiFeO_3$ nanoparticle and the $BiFeO_3$/graphene composite harboring different weight percentages of graphene are shown in figure 1. The $BiFeO_3$/graphene composite displayed a rhombohedral structure representing the $BiFeO_3$ nanoparticle, whereas the peak designating the graphene in the $BiFeO_3$/graphene composite could not be identified by XRD. To identify and characterize the crystal parameters of the $BiFeO_3$ nanoparticle and the $BiFeO_3$/graphene composite, we used a Williamson–Hall plot. The average size of the $BiFeO_3$ crystal was 40 nm. The addition of graphene resulted in a smaller $BiFeO_3$ crystal size (table 1, column 3), which agreed with a previous study reporting that the addition of graphene inhibits the growth of a $BiFeO_3$-nanoparticle crystal [7].
We used TGA to characterize sample stability following heating from room temperature to 1000 °C and to confirm the presence of graphene material in the $BiFeO_3$/graphene composite, given that its presence could not be confirmed by XRD. TGA, as shown in figure 2a, indicated that the weight loss in the $BiFeO_3$/graphene composite occurring from room temperature to 150 °C was associated with the removal of adsorbed water, whereas the weight loss occurring at 400 °C was associated with the combustive effect of the graphene material in the $BiFeO_3$/graphene composite [8]. Additionally, DTA, as shown in figure 2b, indicated that the $BiFeO_3$/graphene composite exhibited exothermic properties at ~380 °C, with the total heat flow of the graphene material during the exothermic process on the the quality of the graphene material on the $BiFeO_3$/graphene composite by calculating the I_D:I_G ratio, as presented in table 1.

Table 1. Lattice parameter, crystallite structure, and heat flow of the $BiFeO_3$ nanoparticle and $BiFeO_3$/graphene composite

Sample	Lattice parameter, $BiFeO_3$		<D> $BiFeO_3$ (nm)	Heat flow (μV/s/mg)
	a (Å); b (Å)	c (Å)		
$BiFeO_3$	5.58	13.88	40	—
$BiFeO_3$/1 wt% graphene	5.58	13.88	39	123
$BiFeO_3$/3 wt% graphene	5.58	13.88	37	520
$BiFeO_3$/5 wt% graphene	5.58	13.88	35	988

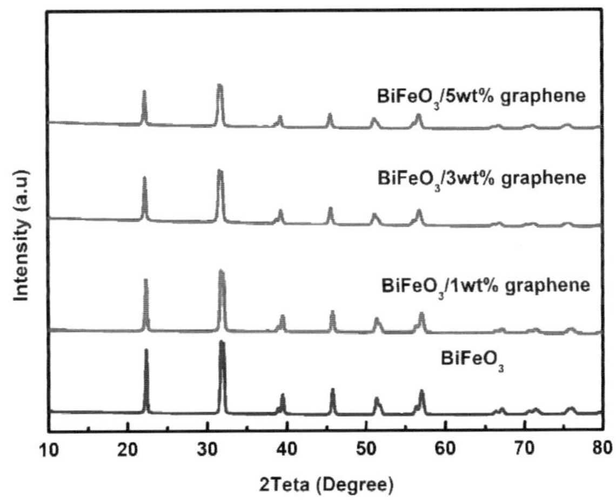

Figure 1. XRD pattern of the $BiFeO_3$/graphene composite.

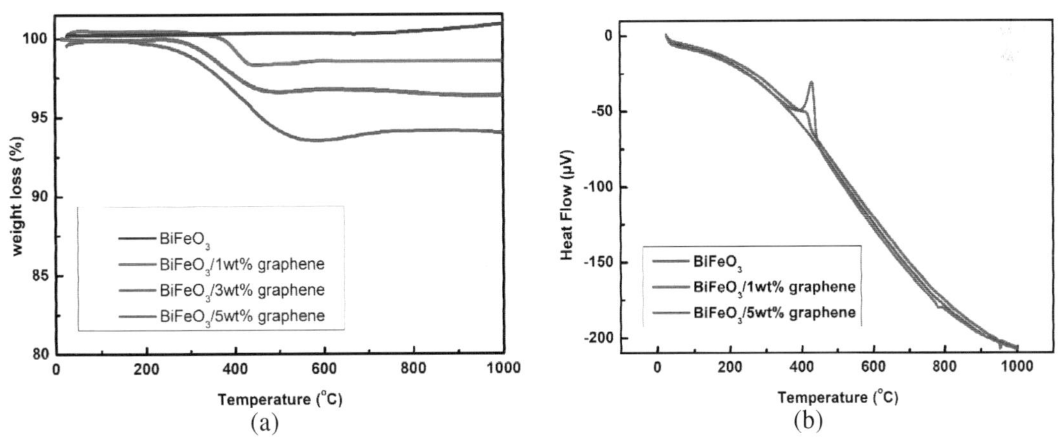

Figure 2. (a) TGA and (b) DTA of the $BiFeO_3$/graphene composite.

We then used Raman spectroscopy to identify graphene material on the $BiFeO_3$/graphene composite according to the presence of a D band (1352 cm^{-1}) and a G band (1595 cm^{-1}). We also used this technique to analyze the quality of the graphene material on the $BiFeO_3$/graphene composite by calculating the I_D:I_G ratio, as presented in table 1.

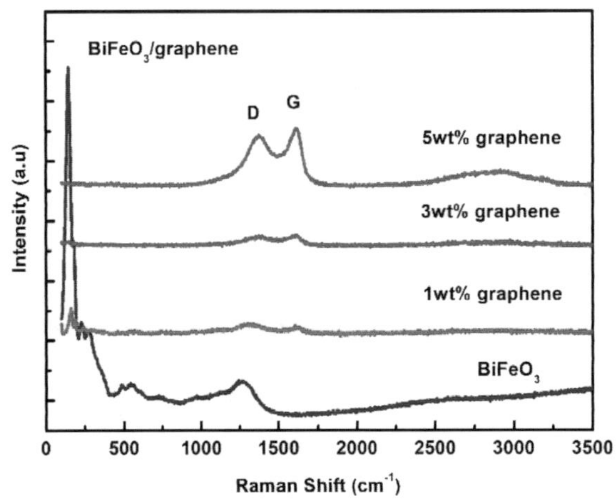

Figure 3. Raman spectra of the BiFeO$_3$/graphene composite.

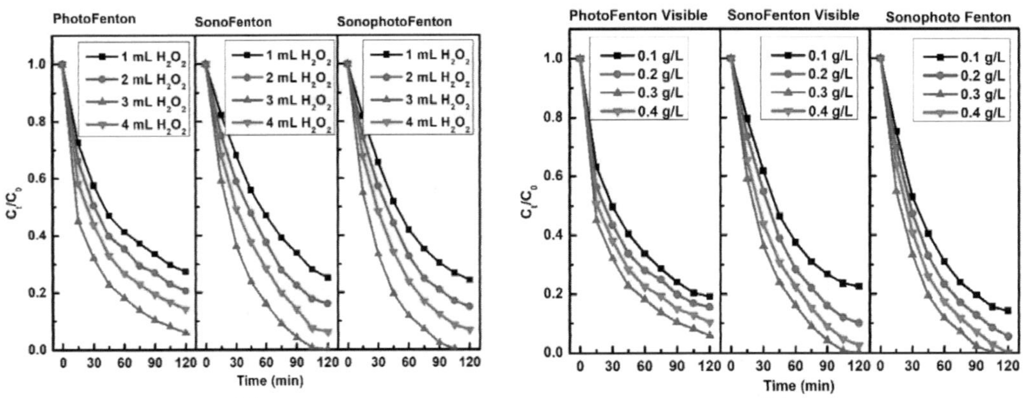

(a) (b)

Figure 4. Effects of (a) different H_2O_2 concentrations and (b) catalyst doses on MB degradation using the BiFeO$_3$/graphene composite.

We then evaluated the effect of pH on each respective fenton reaction (sono-, photo-, and sonophoto-fenton), because this parameter is among the most important influencers of fenton degradation. MB degradation using the BiFeO$_3$/graphene composite was performed at a pH range of 3–11, indicating good results at pH 3 and a decreased degradation efficiency as the pH increased, as shown in figure 3. We attributed this result and our observation of bubbles during the reaction to the elevated O_2 production from the self-decomposition of H_2O_2 [9,10]. Additionally, the higher oxidative capacity might be attributable to the higher concentration of OH produced under acidic conditions. Furthermore, our assessment of the initial dye concentration showed that a higher dye concentration decreased MB degradation.

Fenton degradation using the BiFeO$_3$/graphene composite at a catalyst dose of 0.3 g/L, a pH of 3, and an initial MB concentration of 20 mg/L showed accelerated degradation at higher H_2O_2 concentrations, as shown in figure 4 [10]. Moreover, figure 5 shows that MB degradation using the BiFeO$_3$/graphene composite under both visible and ultrasonic irradiation increased during increases in H_2O_2 concentrations from 1 to 3 mL; however, further increases in H_2O_2 from 3 mL decreased MB degradation. We attributed this to the reaction of excess H_2O_2 with OH to produce HO_2, which has a weaker oxidation capacity than OH [10], thereby explaining the decreased oxidation efficiency in the presence of excess H_2O_2.

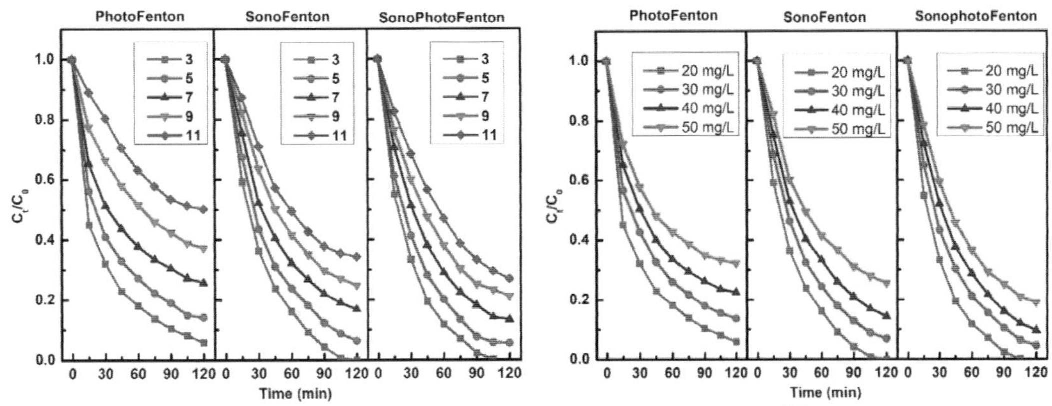

Figure 5. Effects of (a) pH and (b) initial MB concentration on MB degradation using the $BiFeO_3$/graphene composite.

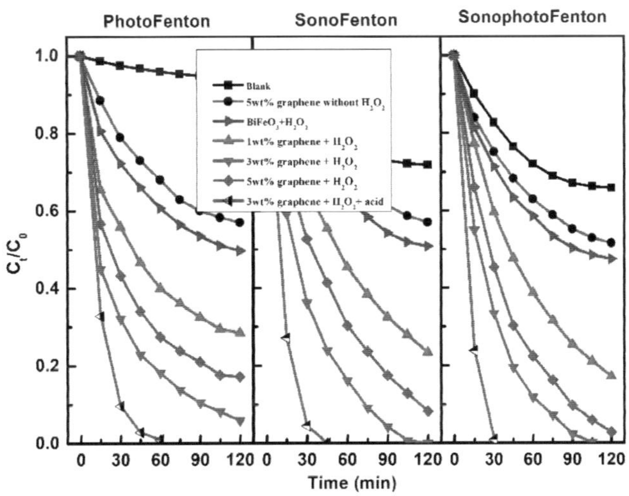

Figure 6. Photo-, sono-, and sonophoto-fenton reactions in the presence of the $BiFeO_3$/graphene composite catalyst.

We then evaluated the effects of catalyst dose ($BiFeO_3$/graphene composite; 0.1–0.4 g/L) on the fenton degradation of MB under visible light and ultrasonic irradiation. As shown in figure 4b, MB-degradation efficiency increased according to increases in the catalyst dose from 0.1 to 0.3 g/L; however, at 0.4 g/L, MB-degradation efficiency decreased, possibly due to increased turbidity at higher levels of $BiFeO_3$/graphene, which can inhibit light penetration [11]. These results suggest an optimal dose of catalyst of 0.3 g/L for subsequent experiments.

Figure 5 shows MB degradation using the $BiFeO_3$/graphene composite via fenton processes in terms of the weight percentage of graphene. In the presence of increasing amounts of graphene, we observed an increased MB-degradation efficiency based on the role of graphene as an electron trapper capable of inhibiting electron–hole recombination. Furthermore, the observed increases in MB degradation occurred up to 3 wt.% graphene, with 5 wt.% resulting in decreased MB degradation. We attributed this to excess graphene inhibiting light/ultrasound penetration [11]. On the basis of these results, we chose 3 wt.% graphene as the optimal amount.

We then investigated the effect of the chelating agent ascorbic acid on MB degradation in the presence of the catalysts, finding that the addition of ascorbic acid increased MB degradation for all the three fenton reactions using the $BiFeO_3$/graphene composite. We attributed this to the prevention of ion-sludge formation that can occur during fenton reactions. The data fitting shown in figure 6

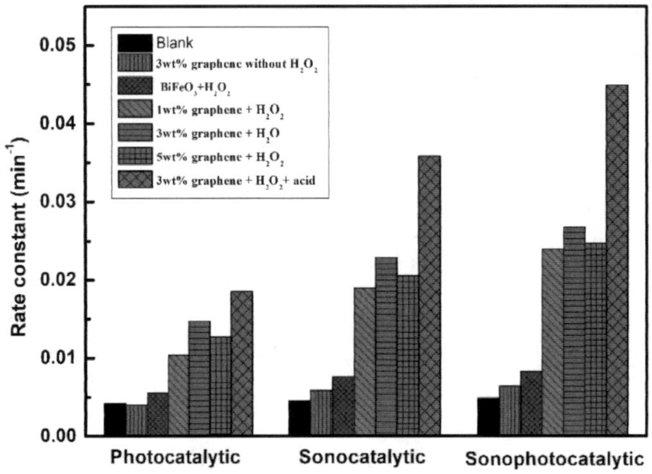

Figure 7. Rate constants for the $BiFeO_3$/graphene composite catalyst for the photo-, sono-, and sonophoto-fenton reactions.

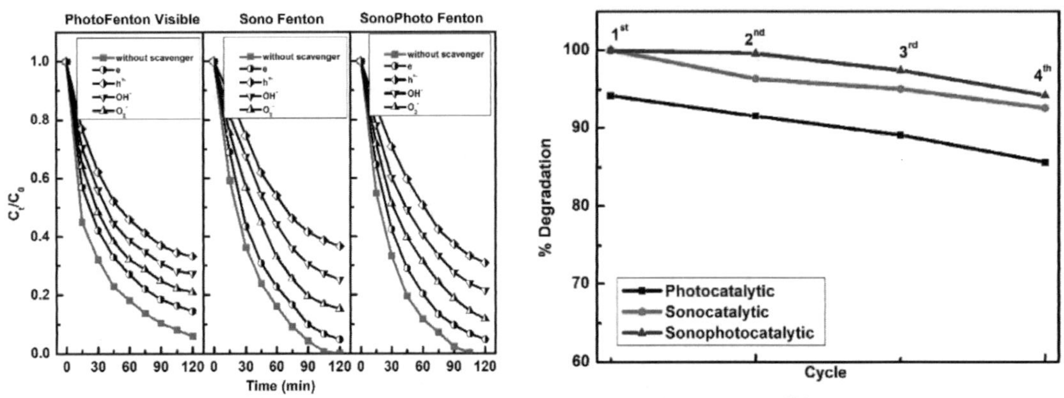

(a) (b)

Figure 8. $BiFeO_3$/graphene (a) scavenging and (b) reusability.

shows that the apparent rate constant (K_{app}) of the different mechanisms increased in the order blank < $BiFeO_3$/graphene without H_2O_2 < $BiFeO_3$+H_2O_2 < $BiFeO_3$/3 wt% graphene < $BiFeO_3$/10 wt% graphene < $BiFeO_3$/5 wt% graphene < $BiFeO_3$/graphene+ascorbic acid. Furthermore, figure 7 shows that MB degradation increased in the order photo-fenton < sono-fenton < sonophoto-fenton, with sonophoto-fenton showing the highest rate of MB degradation due to the synergistic effect of the photo- and sono-fenton reactions.

To determine the contribution of several active species to the fenton reaction, we performed radical trapping by adding an electron scavenger (sodium sulfate), a hole scavenger (diammonium oxalate), a hydroxyl radical scavenger (*tert*-butyl alcohol), and a superoxide radical scavenger (benzoquinone). Figure 8a shows that the addition of various scavengers decreased MB degradation, with the hydroxyl radical scavenger being the most active species at attenuating MB degradation. Additionally, we assessed sample reusability, with figure 8b showing that MB-degradation efficiency decreased during each subsequent cycle.

4. Conclusions

The sonophoto-fenton reaction using a $BiFeO_3$/graphene composite showed the highest rate of MB degradation at pH 3, a catalyst dose of 0.3 g/L, and an initial concentration of 20 mg/L, with the addition of graphene failing to alter the crystal structure of $BiFeO_3$ while also increasing the overall catalytic efficiency because of its role as an electron trapper. Furthermore, the sonophoto-fenton reaction exhibited the highest degradation efficiency because of the synergistic effect between the photo- and sono-fenton reactions.

References

[1] Rusevova K, Köferstein R, Rosell M, Richnow H H, Kopinke F D and Georgi A 2014 *Chem. Eng. J.* **239** 322–31
[2] Andreozzi R, Apuzzo A D and Marotta R 2002 *Water Res.* **36** 4691–8
[3] Tong T, Zhang H, Chen J G, Jin D R and Cheng J R 2016 *Catal. Commun.* **87** 23–6
[4] Georgi A, Schierz A, Trommler U, Horwitz C P, Collins T J and Kopinke F D 2007 *Appl. Catal.* B **72** 26–36
[5] Zheng Y, Yuan Y, Chai Y and Yuan R 2016 *Bioelectron.* **79** 86–91
[6] Jia Y, Wu C J, Kim D H, Lee B W, Rhee S J, Park Y C, Kim C S, Wang Q J and Liu C S 2018 *Chem. Eng. J.* **337** 709–21
[7] Hue J, Men J, Ma J and Huang H 2014 *J. Rare Earth* **32** 1126–34
[8] Shi Y, Chou S L, Wang J Z, Li H J, Liu H K and Wu Y P 2013 *J. Power Sources* **244** 684–9
[9] Subramanian G and Madras G 2016 *Water Res.* **104** 168–77
[10] Xue Z, Wang T, Chen B, Malkoske T, Yu S and Tang Y 2015 *Materials* **8** 6360–78
[11] Afifah N and Saleh R 2017 *IOP Conf. Ser.: Mater. Sci. Eng.* **196** 012048

Primary study of KOH alkali fusion – nitric acid leaching process in extraction of West Kalimantan zircon concentrate

E Yustanti[1], F A F Ode[1] and E Sulistiyono[2]

[1]Department of Metallurgy, Faculty of Engineering, University of Sultan Ageng Tirtayasa, Jalan Jenderal Sudirman Km 3 Cilegon, Banten 42435, Indonesia

[2]Research Center for Metallurgy and Materials, Indonesian Institute of Sciences (LIPI), Gedung 470 Kawasan Puspitek Serpong, Banten 15314, Indonesia

Corresponding author's e-mail: erlina.yustanti@untirta.ac.id

Abstract. This paper aims to prepare the primary study of zirconia concentrate using KOH alkali fusion, nitric acid leaching and coprecipitation method. Alkali fusion with KOH was done with stoichiometric ratio $ZrSiO_4$/KOH of 1:1.5 at 700 °C and 180 min. Variation of water leaching temperature at 25 °C, 95 °C and concentration in nitric acid at 4M and 6M were investigated to determine the optimal parameter to form zirconia. Phase in raw material, residue of water leaching, and product of calcination were investigated by using XRD. High composition of zirconia at 52.12 % and silica 9.64 % in zircon concentrate corresponded to the intensity peak of XRD pattern. Alkali fusion with KOH changed the visual and microstructure into irregular and porous structure in the strong caustic environment. Temperature of water leaching at 25 °C was the optimal condition to produce the highest intensity peak of zirconate and the zirconia formation. Nitric acid at 4M was preferable to form the lower of silica gel formation and produce the highest zirconia.

Keywords: zircon concentrate, KOH alkali fusion, frit, nitric acid leaching, zirconia

1. Introduction

Zircon concentrate is the product of beneficiation process from zircon sand that still contains ilmenite, monazite, and xenotime. Zircon sand is a by-product of gold processing [1]. West Kalimantan is one area in Indonesia that potentially contains of 167 million tons of zircon mineral in various sub-districts [2]. West Kalimantan contains zirconia (ZrO_2) in zircon sand higher than other regions in Indonesia, such as Bangka Belitung and Central Kalimantan [3].

The process of zircon concentrate ($ZrSiO_4$) into zirconia was performed to produce refractory materials that have a high resistant in thermal shock [4]. An aggressive reaction condition is required to decompose the zirconia (ZrO_2) that bonded in stability with silica (SiO_2) [5]. Alkali fusion with NaOH, KOH, and Na_2CO_3 has been developed and resulted in the acquisition of high-grade zirconia [4–10]. Alkali fusion can be done with NaOH or KOH at the temperature of (600–900) °C [7]. Alkali fusion with KOH is more effective than KOH due to the change of Gibbs energy (ΔG). Alkali fusion with KOH produces more negative value of Gibbs energy compared with NaOH at room temperature. Manual calculations of Gibbs energy change in alkali fusion with KOH produces a value at -85.57 kJ/mol and NaOH at -55.68 kJ/mol at room temperature. It shows that the alkali fusion reaction with KOH is more spontaneous than NaOH. Alkali fusion of zircon mineral with KOH is better than Na_2CO_3 due to the required energy. Alkali fusion with Na_2CO_3 is carried out at the temperature up to 1000 °C which requires higher energy than with KOH and generates carbon dioxide (CO_2) as the fusion product which is dangerous to the environment [5].

The stoichiometric ratio between the zircon concentrate and solid alkali in fusion affects the acquisition of the zircon. The higher of solid alkali agent in alkali fusion shows an increase in the acquisition of zirconia [5]. The excess of stoichiometric ratio increases the decomposition of zircon concentrate ($ZrSiO_4$) to form zirconate compound (K_2ZrO_3) [5]. The formation of zirconate has occurred in the theoretical stoichiometric ratio zircon : KOH at 1:1 [8]. The temperature and time of alkali fusion process affect the zirconia formation. The higher temperature and time of alkali fusion show an increase on percent acquisition of zircon that affect the zirconia formation as the final product [9].

Zirconia is able to produce up to 86 % purity through KOH alkali fusion – water leaching – chloride acid leaching [11]. The use of nitric acid (HNO_3) as the leaching agent is one of the alternative parameters that can be applied to optimize the acquisition of zircon over 86 %. The effect of water leaching temperature and nitric acid concentration is a primary study of KOH alkali fusion – water leaching – nitric acid leaching – coprecipitation method which will be examined in this study to produce the highest intensity of zircon formation.

2. Experimental procedures

2.1. Decomposition of zircon concentrate in KOH

Alkali fusion with KOH used to decompose zircon concentrate was from West Kalimantan, Indonesia. Aggregate was put in the iron crucible after zircon concentrate (100 g) evenly mixed with KOH (183.8 g) in beaker glass. Zircon concentrate with size of below 200 mesh was used in alkali fusion process with stoichiometric ratio $ZrSiO_4$/KOH of 1:1.5, fusion temperature was 700 °C, and fusion time was 180 min. in muffle furnace. The microstructure of zircon concentrate with alkali fusion products (frit) were investigated to know the effect of decomposition to the crystalline structure.

2.2. Water leaching process

Frit that contains silica was leached by water with solid percent of 9.09 %. Water leaching process was done in 2000 mL beaker glass, stirred at 200 rpm of magnetic stirrer for 60 min. Variation of water leaching temperatures (25 and 95) °C was investigated to know the effect on zirconate formation in residue. Residue was separated from filtrate using filtration technique and dried in an oven at temperature 110 °C.

2.3. Nitric acid leaching process

The residue of water leaching was leached in 4M and 6M HNO_3 at temperature of 70 °C with a rotational speed of 250 rpm for 1 h. Variation of nitric acid concentrations was investigated to know the effect on zirconia formation as the final extraction process. Nitric acid leaching process was done in three-neck glass with reflux condenser to control the temperature and materials. Filtrate was separated with silica gel as residue using filtration.

2.4. Precipitation and calcination

Zirconyl oxynitrate as filtrate of nitric acid leaching was precipitated with ammonium hydroxide (NH_4OH) until reach the pH 7. Neutral condition (pH 7) was used to produce the highest crystal structure of zirconia formation [11]. Zirconyl oxynitrate was precipitated for 12 hours to produce a high grade of zirconia hydrate as residue. Zirconia hydrate was dried in an oven and calcined in muffle furnace at 800 °C for 3 hours to produce high grade zirconia.

2.5. Instrumentations

Characterization with SEM JEOL JSM - 6390A was conducted to observe the change in microstructure in zircon concentrate and frit. XRF *benchstop* PANalytical Epsilon 1 was used to determine the chemical composition of West Kalimantan zircon concentrate. Compound and phase were identified using X-Ray Diffraction (XRD) Rigaku Miniflex 600 with CuK$_\alpha$ radiation. Sampling preparation of XRD was done and interpreted using software Highscore Plus. Samples that investigated using XRD were raw material of West Kalimantan zircon concentrate, residue of water leaching, and product of calcination.

3. Results and discussion

3.1. Characteristics of West Kalimantan zircon concentrate

Table 1. Chemical composition of West Kalimantan zircon concentrate

Compound	ZrO_2	Na_2O	SiO_2	MgO	CaO	TiO_2	La_2O_3	CeO_2	Y_2O_3	Pr_6O_{11}	Al_2O_3	K_2O
% weight	52.12	24.18	9.64	2.50	1.33	1.17	0.93	0.44	0.38	0.26	1.23	1.3

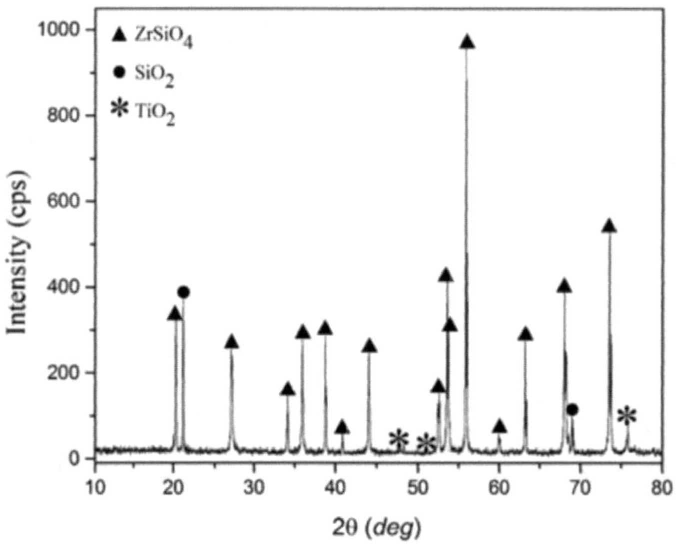

Figure 1 XRD pattern of West Kalimantan zircon concentrate

The chemical composition of West Kalimantan zircon concentrate using XRF identification contains dominant compound of zirconia (ZrO_2) about 52.12 % based on table 1. High impurities (Na_2O, MgO, CaO, Al_2O_3 and K_2O) and rare earth oxides (La_2O_3, CeO_2, Y_2O_3 and Pr_6O_{11}) show that early zircon sand beneficiation process was ineffective. Rare earth oxides in zircon concentrate were monazite and xenotime minerals [3]. High composition of rare earth oxides in zircon concentrate was due to the ineffective separation in the beneficiation process.

XRD identification based on figure 1 shows that West Kalimantan zircon concentrate contains zircon in the form of zirconium silicate ($ZrSiO_4$) with reference code ICDD 01-075-1590, rutile (TiO_2) with reference code ICSD 98-006-9331, and silica (SiO_2) with reference code ICDD 01-083-2471. It was indicated by the highest intensity of zircon ($ZrSiO_4$). The main compound of zircon concentrate was zircon silicate ($ZrSiO_4$) [3].

Silica is a compound that sufficiently contained in West Kalimantan zircon concentrate which in accordance to the high silica peak intensity at 21.2° while rutile is a minor compound. Rutile as minor compound is shown by the lowest intensity of the diffraction pattern and chemical composition, in accordance to XRF data in table 1 which shows that the content of rutile in zircon concentrate is about 1.17 %. The presence of rutile in zircon concentrate is occurred due to titanium (Ti) replaces some atoms of silicon (Si) and forms the TiO_6 octahedral structure [5].

3.2. Effect of KOH alkali fusion in the microstructure of frit

Visual aspect of zircon concentrate before and after alkali fusion presents the color change from brown to dark grey which shown by the strong caustic environment in figure 2. The visual structure changed into more porous and friable. Frit has a porous structure, high friability and caustic environment [4]. The formation of zircon concentrate microstructure in figure 3 presents some prismatic crystals of $ZrSiO_4$ which corresponds to highest intensity of diffraction data in figure 1. Prismatic crystals of zircon concentrate changed into irregular and porous structure as the alkali fusion process as shown by SEM micrograph in figure 3. Zircon particles reacted with molten KOH during alkali fusion as solid – liquid reaction. The solid – liquid reaction by molten KOH destroyed the zircon concentrate crystalline aspect [4].

(a) (b)

Figure 2 Visual aspect of **(a)** zircon concentrate; **(b)** frit after alkali fusion KOH with stoichiometric ratio $ZrSiO_4$/KOH is 1:1.5 at 700 °C and 180 minutes

(a) (b)

Figure 3. Microstructure of (a) zircon concentrate; (b) frit after alkali fusion KOH with stoichiometric ratio $ZrSiO_4$/KOH is 1:1.5 at 700 °C and 180 minutes

3.3. Effect of water leaching temperature in zirconate formation

The residue of water leaching contains potassium zirconium silicate (K_2ZrSiO_4) with reference code ICDD 00-017-0282, potassium zirconium oxide ($K_2Zr_2O_5$) with reference code ICDD 00-020-0963, zirconium titanium oxide ($ZrTiO_4$) with reference code ICDD 00-007-0290, and dipotassium zirconate (K_2ZrO_3) with reference code ICDD 98-001-6264 according to figure 4. The formation of zirconate ($K_2Zr_2O_5$ and K_2ZrO_3) in water leaching residue shows the effectiveness of silicate (K_2SiO_3) dissolution in the alkali fusion product. This is consistent with solubility data which states that the potassium silicate is soluble in water [12].

The formation of potassium zircon silicate (K_2ZrSiO_4) as intermediate compound indicates that the water leaching process is not completely dissolved the silica. Silica can be dissolved as intermediate compound by high concentrate acid leaching [10]. The higher of water leaching temperature reduces the K_2ZrSiO_4 and K_2ZrO_3 compound. That was indicated by the lowering of some K_2ZrSiO_4 and K_2ZrO_3 diffraction peaks. Zirconate compound (K_2ZrO_3) was more soluble in water when the water leaching temperature increased. It was shown by the lowering of intensity peak of zirconate at 54.84°. Potassium zircon silicate (K_2ZrSiO_4) was more easily formed at room temperature of water leaching as indicated by the higher intensity XRD peaks at 19.22°, 31.4°, and 33.18°. The formation of K_2ZrSiO_4 as intermediate compound may be formed as the product of water leaching [13]. Zirconium silicate ($ZrSiO_4$) can be transformed into potassium zircon silicate ($K_2ZrSi_2O_7$, $K_2ZrSi_3O_9$, $K_2ZrSi_6O_{15}$, and $K_2ZrSi_3O_9H_2O$) [13]. The potassium zircon silicate as intermediate compound occured because the

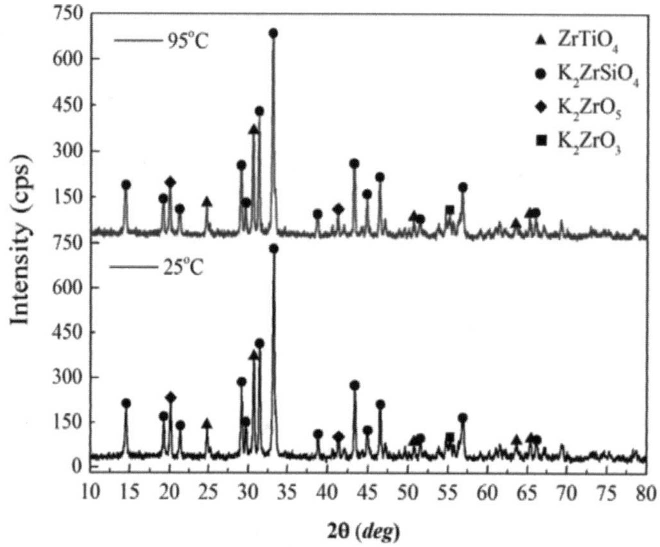

Figure 4. XRD pattern of water leaching residue

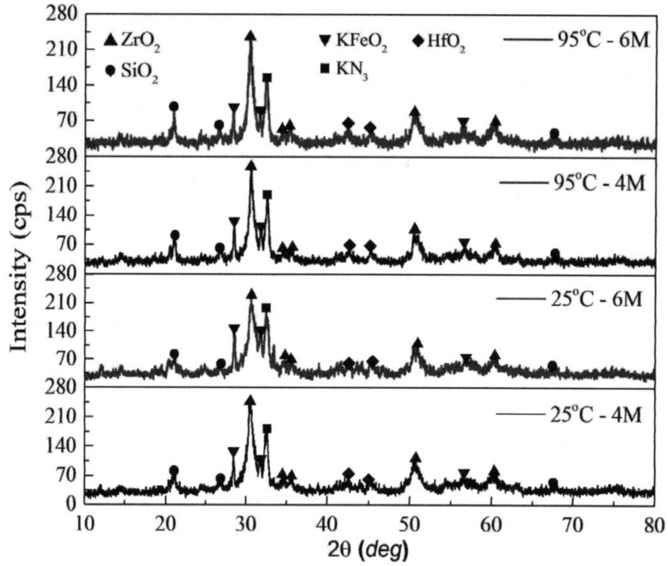

Figure 5 Zirconia formation on different variables

zirconia structure in zirconate increasingly reacts with potassium silicate (K_2SiO_3) by the following equation [14]:

$$ZrO_{2(s)} + K_2SiO_{3(s)} \rightleftharpoons K_2ZrSiO_5$$

3.4. Effect of water leaching temperature and nitric acid leaching concentration in the establishment of zirconia

Water leaching temperature and nitric acid concentration affect the zirconia formation as the final product of the extraction as shown in figure 5. The final product of zirconia extraction contains zirconia (ZrO_2) as the highest compound. The peak intensity of ZrO_2 (ICDD No. 01-088-1007) is a tetragonal phase compound with the dominant intensity of the XRD pattern. The higher of the water leaching temperature and nitric acid concentration in acid leaching reduces the formation of zirconia significantly. That was indicated by the lowering of zirconia's intensity peak at 30.48° and 50.38°.

Room temperature of water leaching process showed the best results to form zirconia. It showed that the increase of water leaching temperature formed a higher silica gel formation. Low acid concentration was preferable in nitric acid leaching to form a lower silica gel formation. It was shown that the intensity of SiO_2 as silica gel (ICDD No. 01-083-2466) at 30.48° increased with the increase of the nitric acid concentration. The increase of the silica gel was formed as the residue of an acid leaching process decreases the recovery of zirconia as final product [4].

4. Conclusions

Zirconia (ZrO_2) was obtained from the decomposition of West Kalimantan zircon concentrate by alkali fusion – nitric acid leaching process. High composition of zirconia (52.12 %) and silica (9.64 %) corresponded to the intensity of XRD pattern which showed the highest peak. Alkali fusion with KOH occurred by the solid – liquid reaction and changed the visual and microstructure into irregular and porous structure in the strong caustic environment. Temperature of 25 °C was the best water leaching temperature to produce the highest zirconate formation. The formation of zirconate ($K_2Zr_2O_5$ and K_2ZrO_3) at 25 °C water leaching temperature in residue was shown to be more effective to solute the silicate (K_2SiO_3) from alkali fusion product. The concentration of nitric acid leaching at 4M was preferable to be used to form the lower silica gel formation and produce the highest zirconia formation.

Acknowledgements

The authors would like to thank the Mineral Extraction Laboratory of Research Center for Metallurgy and Material LIPI and Sultan Ageng Tirtayasa University for facilitating this research.

References

[1] Ayala L, Hurtado J L M, Allwood J A, Tingley D D and Serrenho A C 2015 *Technical Handbook on Zirconium and Zirconium Compounds* (Buckinghamshire: Zircon Industry Association)
[2] Kementerian Energi dan Sumber Daya Mineral 2012 *Kajian Kebijakan Pengembangan Industri Mineral sebagai Kawasan Ekonomi Khusus* (Jakarta: Pusat Data dan Informasi ESDM)
[3] Poernomo H, Biyantoro D and Purwani M V 2016 *Eksplorium: Buletin Pusat Teknologi Bahan Galian Nuklir* **37** 73–88
[4] da Silva R J F, Dutra A and Afonso J C 2012 *Hydrometallurgy* **117–118** 93–100
[5] Kwela Z 2007 *Alkali-Fusion Processes for the Recovery of Zirconia and Zirconium Chemicals of Zircon Sand* (Prectoria: University Pretoria)
[6] Ferreira C A, Formiga T S and Morais C A 2015 *Proc. 24th Int. Mining Congress and Exhibition of Turkey* (IMCET) 14–17 April 2015 Antalya Turkey, ed M G Kilic *et al.* (Ankara: Chambers of Mining Engineers of Turkey) pp 1351–6
[7] Aldila H and Triwikantoro 2015 *Proc. 2nd Int. Conf. on Research Implementation and Education of Mathematics and Sciences* (ICRIEMS) 17–19 May 2015 Yogyakarta (Yogyakarta: FMIPA Universitas Negeri Yogyakarta) pp 37–43
[8] Abdelkader A M, Daher A and El-Kashef E 2008 *J. Alloys Compd* **460** 577–80
[9] Biswas R K, Habib M A, Karmakar A K and Islam M R 2010 *Hydrometallurgy* **103** 124–9
[10] Manhique A, Kwela Z and Focke W W 2003 *Ind. Eng. Chem Res.* **42** 777–83
[11] Abdullah M and Triwikantoro 2015 *Proc. 29th Pertemuan Ilmiah HFI Jateng and DIY* 25 April 2015 Yogyakarta, ed D Rosana *et al.* (Yogyakarta: HFI Cabang Jateng dan DIY) pp 42–5
[12] Budavari S 1989 *The Merck Index: An Encyclopedia of Chemicals, Drugs, and Biologicals* 11th ed, ed M J O'Neil *et al.* (Rahway: Merck)
[13] Ilyushin G D and Dem'yanets L N 1996 *Growth of Crystals* vol 20, ed E I Givargizov and A M Melnikova (Boston: Springer) pp 89–90
[14] Liu R, Qu J, Song J, Qi T and Du A 2013 *Ceram. Int.* **40** 1431–8

IOP Conf. Series: Materials Science and Engineering **763** (2020) 012066 doi:10.1088/1757-899X/763/1/012066

Observation of the domain wall propagation in CoFe and CoFeB nanowires driven by sub-nanosecond magnetic pulse using micromagnetic simulation

S Hawibowo[1], C Kurniawan[1,2] and D Djuhana[1]

[1] Department of Physics, Faculty of Mathematics and Natural Sciences (FMIPA), Universitas Indonesia, Kampus UI Depok, Depok 16424, Indonesia

[2] Research Center for Physics, Indonesian Institute of Sciences (LIPI), Gedung 440–442 Kawasan Puspiptek Serpong, Banten 15314, Indonesia

Corresponding author's e-mail: dede.djuhana@sci.ui.ac.id

Abstract. In this study, we have observed the domain wall (DW) propagation in CoFe and CoFeB nanowires driven by sub-nanosecond magnetic pulse using micromagnetic simulation approach. The length of the nanowire is 2000 nm, the widths are varied of 50, 100, and 150 nm, and the thicknesses are 2.5 and 5.0 nm. The simulation was performed using a rectangular cell size of 5×5×t nm^3 with t is the thickness of the nanowire and the damping factor is 0.05. The sub-nanosecond magnetic pulse length of 0.5 ns was used to move DW through the nanowire. We found that the DW velocity increased as the magnitude of magnetic pulse increased then abruptly decreased which known as the Walker Breakdown (H_{WB}). The transverse type DW structure is observed below the H_{WB} field while vortex/anti-vortex wall structure was formed above H_{WB}. We found that the H_{WB} value of CoFe and CoFeB decreased as the width and thickness of the nanowire increased. The increasing of nanowire thickness also caused the velocity reduction of DW. The energy competition between exchange and demagnetization energy contributed to the DW structure during propagation. The exchange energy is higher than the demagnetization energy as the formation of vortex/anti-vortex wall.

1. Introduction

The domain wall-based memories are exciting topic issue that emerged from the proposal of racetrack memory which introduced by Parkin *et al.* [1] and Parkin [2]. High-speed racetrack memory could be achieved by propagating the domain wall (DW) through the nanowire using the magnetic field, spin-polarized current, or electrical field [3–7]. However, the DW density was relatively lower if we used a standard longitudinal or in-plane magnetization direction. The superparamagnetic effect was also raised in the in-plane magnetization type that can erase the data as the nanowire size decreased [8]. Depending on the external field, DW moves with or without changing its inner structure along on nanowire [3,9]. The perpendicular magnetization type materials were introduced to enhance its performance especially for achieving the smaller, faster, and efficient data storage system [8]. One of the high perpendicular magnetic anisotropic was CoFe-based materials which have been studied by some researchers both in experimental and simulation approaches [10–12]. Hirohata *et al.* [13] studied a multilayer CoFeB/MgO which has a high spin polarization and low damping effect so that potentially can be applied as high-density data storage.

In this study, DW propagation characteristics in CoFe and CoFeB nanowires driven by sub-nanosecond magnetic pulse was observed using micromagnetic simulation. The influence of parameter variation such as wire width and thickness to the DW dynamics was also investigated. The maximum DW propagation speed under a specific magnetic field called a Walker breakdown point (H_{WB}) was examined to improve an understanding in the formation of DW structure and the energy included during propagation.

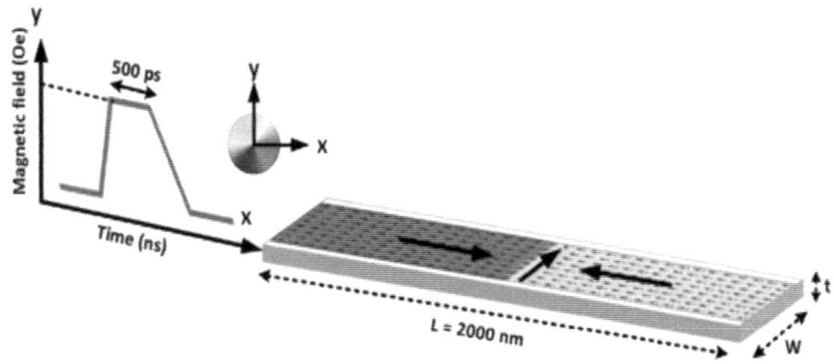

Figure 1. The geometry and dimension of CoFe and CoFeB nanowire with a transverse wall structure at the center of nanowire. A 500 ps time-length magnetic pulse is applied along the nanowire. The color bar is the magnetization direction.

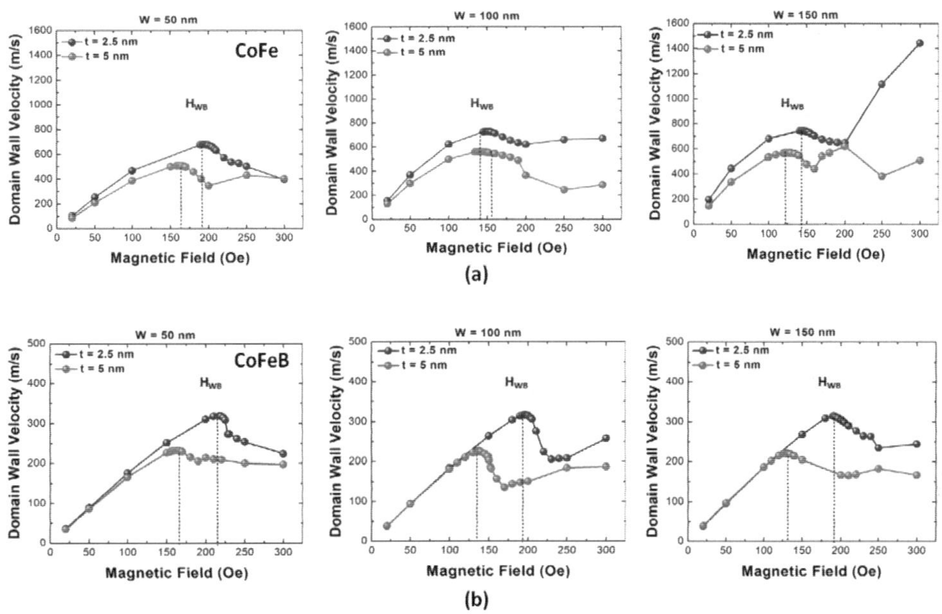

Figure 2. DW velocity as the magnitude of magnetic pulse with the width and thickness variation for (a) CoFe and (b) CoFeB nanowires. The critical magnetic field is known as Walker breakdown field (H_{WB}) denoted by dotted-line.

2. Simulation procedure

The simulation model of nanowires was using a fixed length L of 2000 nm and varied widths W of 50, 100, and 150 nm, and thicknesses t of 2.5 and 5 nm. The material parameters of CoFe using the magnetization saturation M_s, exchange constant A, and anisotropy constant K of 1.4×10^{6} A/m, 1.3×10^{-11} J/m, and 2.7×10^{3} J/m^3, respectively [14]. While, the parameter of CoFeB the magnetization saturation M_s, exchange constant A, and anisotropy constant K of 8.75×10^{5} A/m, 1×10^{-11}, and 5.2×10^{5} J/m^3, respectively [15]. The cell size used in the simulation was $2.5\times2.5\times t$ nm with t is the wire thickness and the damping factor was $\alpha = 0.05$. The DW propagation simulation of CoFe and CoFeB nanowires was performed by public micromagnetic solver OOMMF based on the Landau-Lifshitz-Gilbert (LLG) equation [16,17]. In this simulation, the initial transverse wall (TW) was located at the center of the nanowire, as illustrated in figure 1. Then, a very short magnetic pulse of 500 ps length was applied on the nanowire in x-direction. For the purpose of this study, we have varied the magnitude of magnetic pulse to drive DW along the nanowire.

Table 1. The Walker breakdown field of CoFe and CoFeB nanowires with respect to the width and thickness variation

Thickness (nm)		2.5			5		
Width (nm)		50	100	150	50	100	150
H_{WB} (Oe)	CoFe	194	151	144	162	138	126
	CoFeB	219	197	191	165	138	129

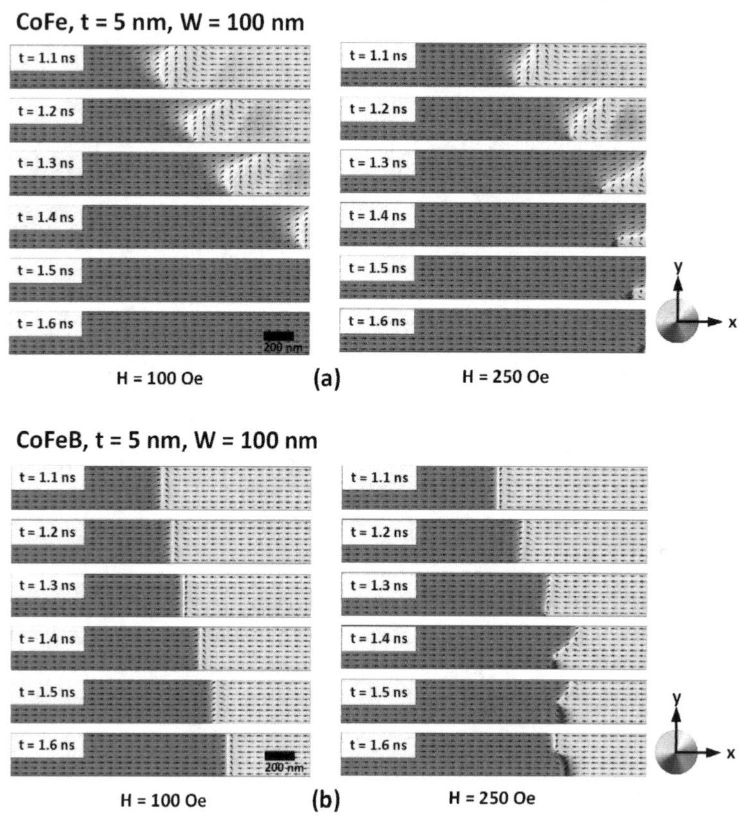

Figure 3. DW structure of CoFe and CoFeB nanowires for thickness t = 5 nm, width W = 100 nm. The structure shows a transverse wall below H_{WB} (100 Oe) and anti-vortex wall after H_{WB} (250 Oe)

3. Results and discussion

With varying the magnitude of magnetic pulse, the magnetic pulse driving the DW along the nanowire is observed. DW velocity for CoFe and CoFeB nanowires to the magnitude of magnetic pulse with the width and thickness variation were presented in figure 2. DW velocity was determined from the change of DW position and time during the magnetic pulse was switched on [18]. In this figure, it is found that DW velocity increased to a certain value and decreased as the magnitude of the magnetic pulse increased. The decrease of the DW velocity was known as Walker breakdown (H_{WB}) [4,19]. It is found that the DW velocity decreased as the thickness increased and relatively constant as the width of nanowire increased. Then, it is observed that the H_{WB} decreased as the thickness and width of nanowire increased. The results exhibited a similar trend to other reported results [3,5,7]. Interestingly, the DW velocity of CoFe and CoFeB nanowire did not abruptly decreased as it was found in Py [3] but slowly decreased as the magnitude of magnetic pulse increased. The DW velocity in CoFe showed higher than CoFeB nanowire. In a sense, the anisotropy properties of material have affected the DW velocity. Table 1 shows H_{WB} values for the width and thickness variation for CoFe and CoFeB nanowires. The table shows that the DW velocity was sensitive to the thickness rather than the width of nanowire.

IOP Conf. Series: Materials Science and Engineering **763** (2020) 012066 doi:10.1088/1757-899X/763/1/012066

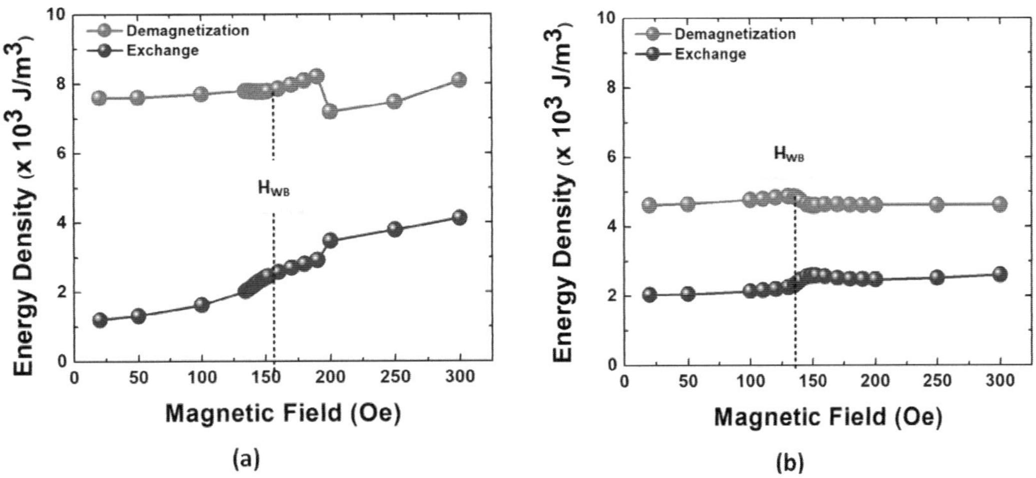

Figure 4. Magnetic energy properties of (a) CoFe and (b) CoFeB nanowires for thickness $t = 5$ nm and width $W = 100$ nm. The energy curve shows increasing exchange energy and decreasing demagnetization energy for magnetic field above H_{WB}.

Furthermore, the DW structure dynamics around H_{WB} was also observed. For example, it is presented the DW structure of CoFe and CoFeB for the case of thickness $t = 5$ nm and width $W = 100$ nm. For sample case below and above H_{WB}, the magnetic field was 100 Oe and 250 Oe, respectively, as shown in figure 3. The DW structure exhibited a transverse structure below H_{WB} while anti-vortex structure is formed above the H_{WB}. This nucleated anti-vortex is seen to move cross the width of the nanowire and producing a turbulent motion of DW leading to lower the average domain wall velocity [20].

The changing of DW structure from transverse to vortex/anti-vortex structure can be explained by the energy density contributed to the system. We have plotted a sample case of the energy density changing from the CoFe and CoFeB nanowire system around H_{BW} value with wire width of 100 nm and thickness of 5 nm as shown in figure 4. We can see that the demagnetization and exchange energy are dominant in the process of DW formation. Generally, the exchange energies are lower below H_{BW} and increases as the magnetic field increases at above H_{BW} value. On the other hand, the demagnetization energies are higher for magnetic field below H_{BW} and decreases at above H_{BW}. Compared to the magnetization evolution image in figure 3, we can see that it is related to the DW structure in the nanowire system both CoFe and CoFeB materials. The increasing of energy exchange around H_{BW} is related to the vortex/anti-vortex DW formation during propagation. Meanwhile, the decreasing of demagnetization energy around H_{BW} is related to the spin interaction of the neighboring magnetization so that the anti-parallel magnetization from vortex type wall produces less demagnetization energy. The geometry of the nanowires were also influenced the energy density. The larger width and thickness in the nanowire geometries produced higher energy density.

4. Conclusions

In this study, we have observed DW propagation characteristics in CoFe and CoFeB nanowires driven by sub-nanosecond magnetic pulse using micromagnetic simulation. The influence of parameter variation such as wire width and thickness to the DW dynamics was also investigated. We found the DW velocity decreased as the thickness increased and relatively constant as the width of nanowire increased. We also observed the H_{WB} decreased as the thickness and width of nanowire increased. Interestingly, the DW velocity of CoFe and CoFeB nanowire did not abruptly decrease, but slowly decreased as the magnitude of magnetic pulse increased. The DW velocity in CoFe showed higher than CoFeB nanowire. In a sense, the anisotropy properties of the material have affected the DW velocity. The exchange energies were lower below H_{BW} and increased as the magnetic field increases at above H_{BW} value. On the other hand, the demagnetization energies were higher at the magnetic field below H_{BW} and decreased at above H_{BW}. The increasing of energy exchange around H_{BW} was related to the vortex/anti-vortex DW formation during propagation.

Acknowledgements
This work is supported by Hibah PITTA No. 2245/UN2.R3.1/HKP.05.00/2018 from Universitas Indonesia program year 2018.

References

[1] Parkin S S, Hayashi M and Thomas L 2008 *Science* **320** 190–4
[2] Parkin S S 2009 *Sci. Am.* **300** 76-81
[3] Nakatani Y, Thiaville A and Miltat J 2005 *J. Magn. Magn. Matter.* **290** 750–3
[4] Djuhana D, Piao H G, Shim J H, Lee S H, Oh S K, Yu S C and Kim D H 2011 *J. Nanosci. Nanotechnol.* **11** 6237–40
[5] Beach G S D, Knutson C, Tsoi M and Erskine J L 2007 *J. Magn. Magn. Mater.* **310** 2038–40
[6] Djuhana D, Piao H G, Yu S C, Oh S K and Kim D H 2009 *J. Appl. Phys.* **106** 103926
[7] Thiaville A, Nakatani Y, Miltat J and Vernier N 2004 *J. Appl. Phys.* **95** 7049
[8] Tudu B and Tiwari A 2017 *Vacuum* **146** 329–41
[9] McMichael R D and Donahue M J 1997 *IEEE Trans. Magn.* **33** 4167–9
[10] Fodor P S, Tsoi G M and Wenger L E 2002 *J. Appl. Phys.* **91** 8186
[11] Sharif R, Shamaila S, Ma M, Yao L, Yu R C, Han X F, Wang Y and Khaleeq-ur-Rahman M 2008 *J. Magn. Magn. Mater.* **320** 1512–6
[12] Zhan Q, Chen Z, Xue D, Li F, Kunkel H, Zhou X, Roshko R and Williams G 2000 *Phys. Rev.* B **66** 134436
[13] Hirohata A, Sukegawa H, Yanagihara H, Žutić I, Seki T, Mizukami S and Swaminathan R 2015 *IEEE Trans. Magn.* **51** 0800511.
[14] Piao H G, Choi H C, Shim J H, Kim D H and You C Y 2011 *Appl. Phys. Lett.* **99** 192512
[15] Wang Y, Wei D, Gao K Z, Cao J and Wei F 2014 *J. Appl. Phys.* **115** 053901
[16] Donahue M J and Porter D G 1999 *OOMMF User's Guide, Version 1.0* (Gaithersburg: National Institute of Standards and Technology)
[17] Gilbert T L 2004 *IEEE Trans. Magn.* **40** 3443–9
[18] Forster H, Schrefl T, Scholz W, Suess D, Tsiantos V and Fidler J 2002 *J. Magn. Magn. Mater.* **249** 181–6
[19] Schryer N L and Walker L R 1974 *J. Appl. Phys.* **45** 5406
[20] Lee J Y, Lee K S, Choi S, Guslienko K Y and Kim S K 2007 *Phys. Rev.* B **76** 184408

IOP Conf. Series: Materials Science and Engineering **763** (2020) 012067 doi:10.1088/1757-899X/763/1/012067

A study of the poly(dA)-poly(dT) DNA electric field-dependence: density of states (DOS) analysis

F Rahmani and E Yudiarsah

Department of Physics, Faculty of Mathematics and Natural Sciences (FMIPA) Universitas Indonesia, Kampus UI Depok, Depok 16424, Indonesia.

Corresponding author's e-mail: e.yudiarsah@sci.ui.ac.id

Abstract. Study of poly(dA)-poly(dT) DNA molecule electronic structure, by analysing its density of states (DOS), had been conducted. This research employed Green's function method as tool to determine electron DOS of the DNA numerically. Electric field that was applied along z-axis is used as perturbation to modify electron hopping constant between sites in the molecule. In order to include electric field in the calculation, Miller-Abraham formula was used. Medium interaction with backbone molecules was considered as well within the calculation in the form of backbone disorder. By increasing electric field, the band in electron DOS spectrum appeared to become narrower and several states experience the decrement of DOS.

1. Introduction

Deoxyribonucleic acid (DNA) is a polymeric bimolecule that can be expected to perform a role in molecular electronics [1]. Some of its quality, which could be beneficial in molecular electronics application, are natural memory storage and self-organizational feature [2]. Moreover, nanoscale fabrication of DNA molecules can be precisely controlled, so it can be very distinctive among other molecular system such as thin-film transistor (TFT) with silver (Ag) nanoparticle inks, terphenyl oligomers, azurin monolayers, etc. [3]. The promising potency of DNA as building block in molecular electronics is one of the main reasons to study the charge transport in DNA [4]. For artificial periodic DNA systems (poly(dA)-poly(dT), poly(dG)-poly(dC), guanin quadruplex, etc.), theoretical approach can be used to determine the energy band structure. It could be a helpful strarter to understand experimental results such as semiconductivity and metal-insulator transition [5].

Theoretical modeling of DNA conductivity is a meticulous and complex task because many factors can alter the charge propagation process, therefore DNA charge transport is defined as highly sensitive to environmental changes. This high sensitivity of charge propagation in DNA is owing to extrinsic (electric and magnetic field, temperature, medium or environment [6]) and also intrinsic (collective base fluctuations caused by phonon [7], sequence variability [8]) influences.

Some observations in electric and magnetic influence in DNA have been done at cellular level, and it showed that the structure of DNA is modified due to physical factors like electric and magnetic field especially in ultraviolet (UV) region [9]. Experimental research by Chunda Zhou et al shows that when double-stranded DNA is placed in AC electrical fields at frequencies of a few hundred Hertz, it exhibits collapses of DNA. The effect is reposed on the limited relaxation times of the counterion cloud, and the next partitioning of the molecule into the common attractive units [10]. Electric field's influence on DNA also has been modeled theoretically by Ceron-Carrasco and Jacquemin [11]. They have computed DNA system affected by fields in the range of 0 to 5 GV/m. The mutagenic mechanism of a single guanine–cytosine DNA base pair is activated by an intense electric field, and might affect its hydrogen bonds in a partially controlled way [11].

In this paper we have analysed the electric field influence on the electron density of states (DOS) of poly(dA)-(dT) DNA molecule model. We also used temperature influence that can lead to fluctuation in twist angle between neighboring base pairs or distance between bases in base pair (bp) [7]. It could be treated as a collective phonon vibration excitation in molecule. We also considered the presence of sugar-phosphate backbone of DNA. The result in our previous work with different type of DNA bases (poly(dG)-poly(dC)) illustrates how electric field influence can decrease conductivity due to voltage increment, it takes place due the application of electric field decreases the base and backbone on-site energy value in the first half of molecule linearly and raises it in the last half [12]. For theoretical model used in this research, we will elaborate it in the next section. Then, the results of our calculation will be expressed in terms of density of states that will be represented in third section. Finally, the paper will be concluded in the final section.

2. Theoretical model

The system observed was a double stranded DNA (ds-DNA) molecule contacted by two leads on each end to give a potential difference in between. As stated in the first section, we used ds-DNA, which consist of adenine (A) and thymine (T) base pair. We assume electron could travel from base to nearest bases within different strand (interstrand hopping), within the same strand (intrastrand hopping) through π orbital overlap [13], and also its pair (watson-crick hopping) through hydrogen bond [14]. Electron also can hop from bases to the nearest backbone or vice versa, as well as between backbone (see ref [15]). Therefore, we picture our model as having four strands (backbone-base pair-backbone), and each base is defined as a site, likewise each backbone.

The tight binding Hamiltonian model is expressed by Rahmi and Yudiarsah [15]:

$$H = \sum_{i=1}^{L}\left[\sum_{\tau,\sigma=1}^{4} (\varepsilon_i^{\tau} + \phi_i^{\tau})|i,\tau\rangle\langle i,\tau| + t_{i.i}^{\tau,\sigma}|i,\tau\rangle\langle i,\sigma| + t_{i,i+1}^{\tau,\sigma}|i,\tau\rangle\langle i+1,\sigma|\right] + h.c \quad (1)$$

Symbol i (sites), τ (base strands), and σ (backbone strands) are for representing electron position in the molecule. The length of the molecule sequence (L) we used is 32 base pairs. Parameter ε_i^{τ} is base onsite energy in site i and strand τ (8.22 eV for A and 9.06 eV for T). $\emptyset_i^{\tau}$ represents the effect of thermal agitation in the form of twisting vibrational motion around bases equilibrium position. Twisting angle of base is randomly distributed by Gaussian distribution. Medium and temperature are also considered randomly distributing backbone on-site energy around 9.36 eV with deviation 0.1 eV. Parameter $t_{i.j}^{\tau,\sigma}$ is electron hopping value between each site. We also included metallic electrodes and molecule-electrode coupling in our tight-binding equation. Onsite energy value for both metallic electrode at each end is 8.45 eV, while hopping constant in metallic electrodes and among molecule-electrode is taken to be 0.75 eV.

External electric field was applied along helix axis of the system as perturbation where, theoretically, it will alter electron hopping constant from site i to j in the molecule. Miller-Abraham formula was used to incorporate electric field influence into our tight binding equation [16]. So, the electron hopping constant from site j to i becomes:

$$t_{ij} = \begin{cases} t_0\, exp\left(\frac{qE}{\kappa k_B T}(z_i - z_j)\right); & if\ (z_i - z_j) < 0 \\ t_0 & ; otherwise \end{cases} \quad (2)$$

where z_i is z coordinate of site i, q is electron charge, E is electric field amplitude, κ is the DNA and medium dielectric constant and $k_B T$ is thermal energy at temperature T.

This research employed Green's function method as tool to calculate electron DOS of the DNA, numerically. Medium interaction with backbone molecules is considered as well in the calculation in the form of backbone disorder. The DOS is calculated using the following formula:

$$DOS(\varepsilon) = -\frac{i}{\pi}Tr\left[G(\varepsilon,\vec{k})\right] \quad (3)$$

Furthermore, the presence of metallic electrodes at both end of DNA molecule is taken into account in the calculation through the self-energy (Σ) of the electrode. Within wideband approximation the self-energy can be formulated as:

$$\Sigma_{i,i}^{\tau,\tau} = -Im\,\pi\, t_{i,p}^{\tau,p} t_{p,i}^{p,\tau} \quad (4)$$

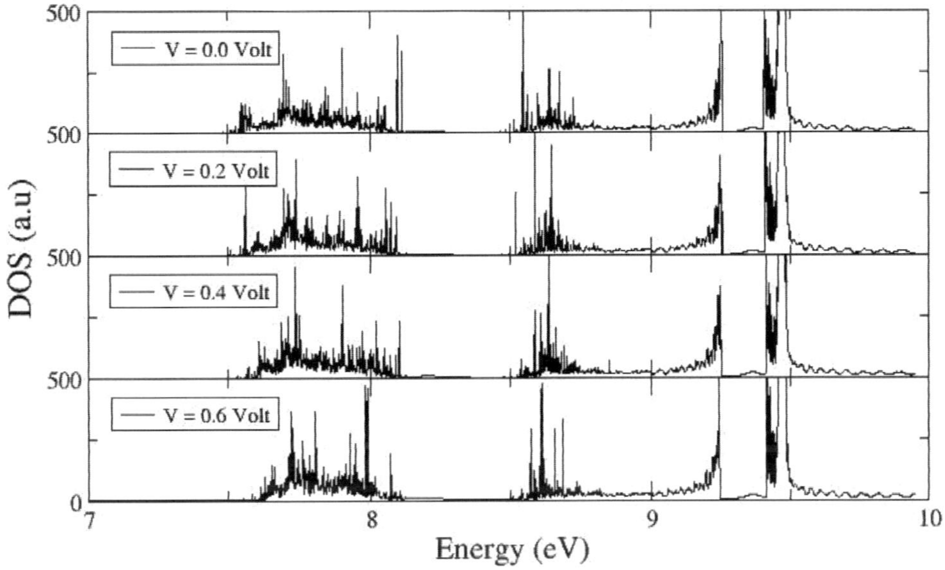

Figure 1. Electron density of states of poly(dA)-poly(dT) DNA in voltage of 0 – 0.6 volt at 4.2 K

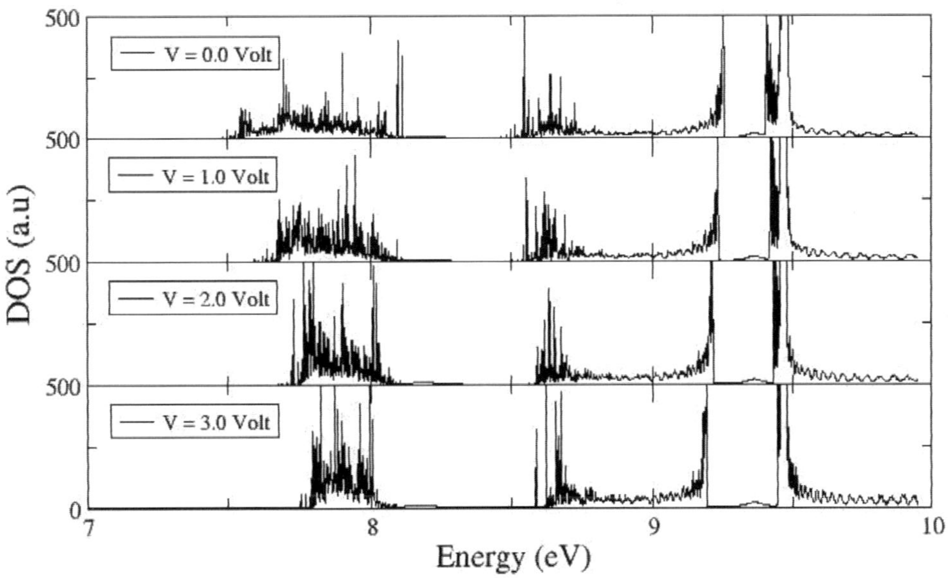

Figure 2. Electron density of states of poly(dA)-poly(dT) DNA in voltage of 0-3 volt at 4.2 K

3. Results and discussion

In this research, the system is treated as wet DNA system, where dielectric constant κ for water medium and DNA is set to 80 in the calculation. The results of the calculations at temperature 4.2 K and twisting motion frequency 0.51 meV for several voltages are shown in figure 1 and figure 2. Both figures show that the electron density of states (DOS) in DNA poly(dA)-(dT) spectrum has three bands, regions of energy in which DOS appears to be not zero or not approaching zero. Those three bands are around 7.6–8.0 eV (band I); 8.75–9.25 eV (band II); and 9.4–10 eV (band III). This means that electron could pass through this region of energy while transporting from one electrode to another. The states in these three bands can participate in the electron transport process. Besides that, in figure 1 and figure 2, it can be seen two regions with very low DOS (almost zero) that are in region 8.0–8.65

eV (gap 1) and 9.25–9.40 eV (gap 2). Because DOS values are very low, the contribution of the state of these two regions are minute on the transport process, since electrons at zero voltage can not occupy DNA states in this energy region.

Electric field applied parallel to the helix axis of the poly(dA)-(dT) DNA could change the electron hopping constant which accordingly change its DOS. The change can be seen from figure 1, as we set the value of electric field higher from 0 to 0.6 V. By increasing the electric field the bands appear to be narrowing where the width of energy gaps seems slightly widen. Figure 1 also exhibits the change of DOS values with electric field increment, some states experience DOS increment while other state experiences DOS decrement. However, most states experience DOS decrement. It shows that within the model used in this study, the electric field disturbs the DNA molecule, which in contradiction with conventional assumption that the electric field only drop at molecule-electrode contact. These results reflect that charge trasport in DNA poly(dA)-(dT) will be affected by the presence of electric field on the molecule. In figure 2, we set the electric field even higher (0–3) V, and it shows more significant constriction in band energy than at the lower voltage. Consequently, the energy gaps look significantly wider, especially in the influence of 3 V electric field. It also shows that several energy in the gap becomes accessible for the electron, since its DOS increases with the voltages.

4. Conclusions

Electron density of states in DNA Poly(dA)-poly(dT) molecule under the influence of electric field along its symmetrical axis at temperature of 4.2 K and twisting motion frequency of 0.512 meV has been studied. Within the model used in this study, the electric field changes the states in the molecule in the sense that the DOS spectrum changes. The width of the bands is narrower accompanied by wider gap with the increase of voltages and many states experience lesser density of states. These phenomenon means that molecule charge transport is compromised by electric field.

Acknowledgements

We would like to thank Daniel K. Suhendro for giving us calculation parameters. We would also give our gratitude to CISCO Laboratory, Physics Department, Universitas Indonesia for letting us using their computational facility. This research is supported by Hibah PITTA Universitas Indonesia 2018 with contract number: 2259/UN2.R3.1/HKP.05.00/2018.

References

[1] Wang H, Marsh R, Lewis J P and Römer R A 2006 *Modern Methods for Theoretical Physical Chemistry of Biopolymers* ed E Starikov *et al.* (Amsterdam: Elsevier) pp 407–28
[2] Church G M, Gao Y and Kosuri S 2012 *Science* **337** 1628
[3] Nokhrin S, Baru M and Lee J S 2007 *Nanotechnol.* **18** 095205
[4] Muren N B, Olmon E D and Barton J K 2012 *Phys. Chem. Chem. Phys.* **14** 13754–71
[5] Iguchi K 2004 *Int. J. Mod. Phys. B* **18** 1845–1910
[6] Guo A M, Xiong S J, Yang Z and Zhu H J 2008 *Phys. Rev. E* **78** 061922
[7] Roche S 2003 *Phys. Rev. Lett.* **91** 108101
[8] Klotsa D, Römer R A and Turner M S 2005 *Biophys. J.* **89** 2187–98
[9] Kuluncsics Z, Perdiz D, Brulay E, Muel B and Sage E 1999 *J. Photochem. and Photobiol.* B **49** 71–80
[10] Zhou C, Reisner W W, Staunton R J, Ashan A, Austin R H and Riehn R 2011 *Phys. Rev. Lett.* **106** 248103
[11] Cerón-Carrasco J P and Jacquemin D 2013 *Chem. Commun.* **49** 7578–80
[12] Rahmani F and Yudiarsah E 2017 *AIP Conf. Proc.* **1862** 030024
[13] Eley D D and Spivey D I 1962 *Trans. Faraday Soc.* **58** 411–5
[14] Nakatani K, Dohno C, Ogawa A and Saito I 2002 *Chem. Biol.* **9** 361–6
[15] Rahmi K A and Yudiarsah E 2016 *AIP Conf. Proc.* **1729** 020037
[16] Li L, Meller G and Kosina H 2007 *Microelectron. J.* **38** 47–51

IOP Conf. Series: Materials Science and Engineering **763** (2020) 012068 doi:10.1088/1757-899X/763/1/012068

The photo-, sono-, and sonophoto-catalytic activity of an $LaMnO_3/TiO_2$/graphene composite for degrading methylene blue in wastewater

Y D Susanti[1,2], D Triyono[1] and R Saleh[1,2]

[1] Department of Physics, Faculty of Mathematics and Natural Sciences (FMIPA), Universitas Indonesia, Kampus UI Depok, Depok 16424, Indonesia

[2] Integrated Laboratory of Energy and Environment, Faculty of Mathematics and Natural Sciences (FMIPA), Universitas Indonesia, Kampus UI Depok, Depok 16424, Indonesia

Corresponding author's e-mail: rosari.saleh@ui.ac.id

Abstract. Here we describe the successful synthesis of an $LaMnO_3/TiO_2$/graphene composite using a co-precipitation method and three different weight percentages (wt%) of graphene. The structure of the sample was characterized using X-ray diffraction (XRD) and Fourier-transform infrared (FTIR) spectroscopy, and the thermal stability of the sample was characterized using thermogravimetric analysis (TGA) and differential thermal analysis. The XRD pattern of the sample confirmed the orthorhombic phase of the $LaMnO_3$ nanoparticle and anatase phase of the TiO_2 nanoparticle; however, the graphene peak unidentifiable by XRD was confirmed using TGA and FTIR spectroscopy. Moreover, we found that the photo-, sono-, and photosono-catalytic activities of $LaMnO_3/TiO_2$ were elevated by the presence of graphene material as an electron trapper, resulting in optimal rates of methylene blue degradation in the presence of 5 wt.% graphene.

1. Introduction

Semiconductor catalysts have attracted increasing attention during the previous decade [1-3], with their subsequent applications in degrading organic pollutants, such as methylene blue (MB), via photo-, sono-, and sonophoto-catalysis [4]. Among catalysts, mixed metal oxides, such as the perovskite oxide $LaMnO_3$, have been used to degrade organic pollutants based on their narrow band gaps, chemical stability, lack of toxicity, and environment safety [5]. However, there remain limitations with $LaMnO_3$ in catalytic processes due to high levels of electron–hole recombination and the presence of low surface areas that can inhibit MB degradation [6,7].

MB degradation can be increased by coupling $LaMnO_3$ and TiO_2 nanoparticles due to the formation of junction levels on the nanocomposite [8]. Moreover, it might be possible to further increase the catalytic activity of the $LaMnO_3/TiO_2$ composite by adding graphene, which is a two-dimensional material comprising a carbon atom with a large surface area, high electron mobility, and good electrical conductivity [9]. Graphene has attracted increased interest due to its ability as an electron trapper capable of promoting MB degradation [10]. The present study reports the photo-, sono-, and sonophoto-catalytic activity of $LaMnO_3/TiO_2$/graphene under visible, ultrasonic, and combined visible and ultrasonic conditions.

2. Experimental detail

2.1. Chemicals

All reagents were of analytical grade and commercially available from Merck Darmstadt, Germany.

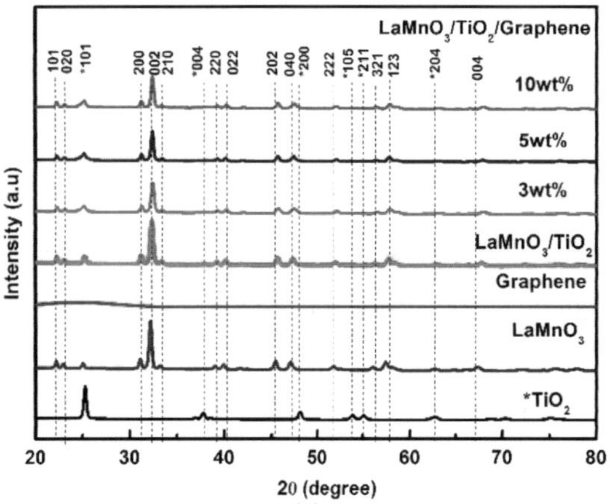

(a)

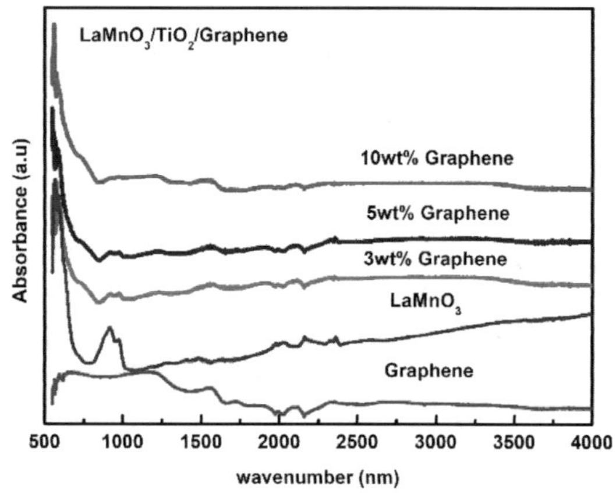

(b)

Figure 1. (a) XRD pattern and (b) FTIR spectra of the $LaMnO_3/TiO_2$/graphene composite at different wt% of graphene.

The reagents used to synthesize the $LaMnO_3/TiO_2$/graphene composite included lanthanum (III) chloride heptahydrate ($LaCl_3$•$7H_2O$; 99% purity; Merck), manganese (II) chloride tetrahydrate ($MnCl_2$•$4H_2O$; 99% purity; Merck), anatase TiO_2 (98% purity; Merck), sodium hydroxide (NaOH; 99% purity; Merck), and ethanol.

2.2. Catalyst preparation
The $LaMnO_3$ nanoparticle was synthesized using a co-precipitation method at 800 °C, as described previously [11]. The $LaMnO_3/TiO_2$ nanocomposite was synthesized using a sol-gel method [12], and the $LaMnO_3/TiO_2$/graphene composite was synthesized using different weight percentages of graphene (3, 5, and 10) wt.% using the co-precipitation method [11].

2.3. Characterization
All prepared samples were characterized using X-ray diffraction (XRD), Fourier-transform infrared (FTIR) spectroscopy, Brunauer–Emmett–Teller (BET) analysis, thermogravimetric analysis (TGA), and differential thermal analysis (DTA). The crystallite structure of the sample was investigated using XRD on a Rigaku Miniflex 600 system (Rigaku, Tokyo, Japan), with the molecular vibration of the sample evaluated using FTIR spectroscopy on a Shimadzu IR Prestige 21 system (Shimadzu, Kyoto, Japan). The thermal stability of the sample was characterized using TGA and DTA using a Rigaku TG8121 system (Rigaku) during sample heating from room temperature to 1000 °C at a rate of 2 °C/min in the presence of a constant flow of argon.

2.4. Catalytic activity
Photo-, sono-, and sonophoto-catalytic activities were analyzed by observing MB degradation in the presence of the $LaMnO_3/TiO_2$/graphene composite under visible light, ultrasonic, and a combination of the two conditions. MB degradation determined under alkaline conditions (pH 13) and within a 2-h window using a Hitachi UH5300 UV-visible spectrophotometer (Hitachi, Tokyo, Japan). Active species present during catalysis were determined using electron (sodium sulfate), hole (diammonium oxalate), and hydroxyl radical (tert-butyl alcohol) scavengers.

3. Results and discussion
Figure 1a shows the XRD results of the samples prepared at different wt.% of graphene. All samples displayed the orthorhombic structure of the $LaMnO_3$ nanoparticle at a 2θ angle of 22.12°, 22.96°, 25.06°, 31.08°, 32.21°, 33.2°, 38.94°, 40.05°, 45.4°, 47.07°, 56.19°, 57.6°, 58.56°, 67.68°, and 75.68° corresponding to reflections of (110), (020), (111), (200), (022), (210), (220), (022), (202), (040), (222), (321), (123), (004), and (412) and anatase phases of the TiO_2 nanoparticle at 2θ of 25.24°, 37.82°, 47.9°, 53.59°, and 62.36° corresponding to reflections of (101), (004), (200), (106), and (215).

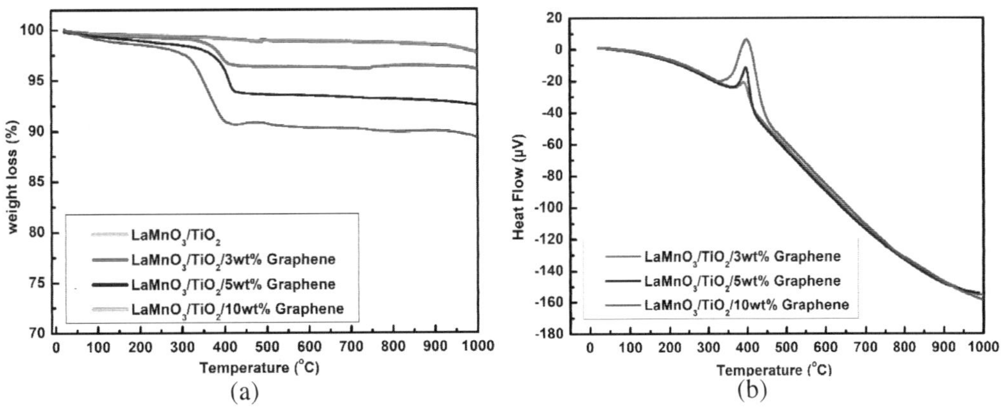

Figure 2. (a) TGA and (b) DTA of the $LaMnO_3/TiO_2$/graphene composite at different wt.% of graphene.

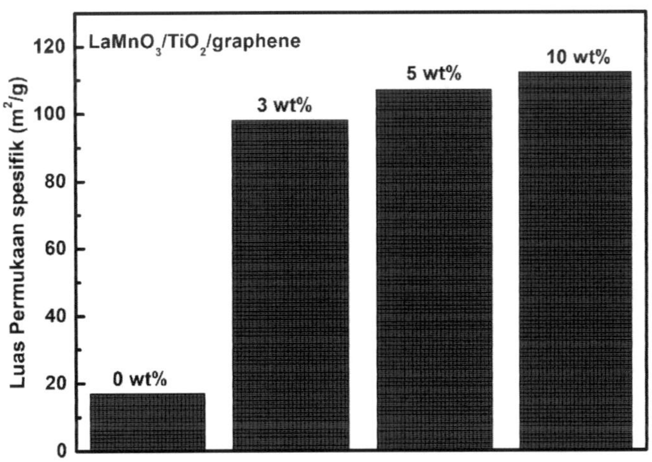

Figure 3. BET analysis of the $LaMnO_3/TiO_2$/graphene composite at different wt.% of graphene.

We were unable to identify a graphene peak using XRD; therefore, its presence was confirmed using FTIR analysis. Figure 1b shows the FTIR spectra of the $LaMnO_3/TiO_2$/graphene composite, indicating strong absorption at 3440 cm^{-1} and corresponding to an O–H-stretching vibration. The presence of graphene was also identified according to the presence of a strong vibration mode at a range of 1460 cm^{-1} to 1750 cm^{-1} and 1260 cm^{-1} corresponding to C=O- and C–OH-stretching vibrations. Furthermore, FTIR analysis indicated that the $LaMnO_3/TiO_2$/graphene composite showed a strong vibration at 555 cm^{-1} corresponding to the Mn–O- and Ti–O-stretching vibrations.

Figure 2a show the TGA results for the $LaMnO_3/TiO_2$ nanocomposite and $LaMnO_3/TiO_2$/graphene composite at different wt.% of graphene composite. We used TGA analysis to investigate sample stability upon heating and total graphene present. We observed a loss of weight in the sample at 150 °C and associated with removal of adsorbed water, as well as at 400 °C in association with the combustion effect of graphene in the $LaMnO_3/TiO_2$/graphene composite, with weight loss at 150 °C suggesting the addition of graphene and indicating the amount of graphene used during synthesis.

Figure 2b shows DTA results for the sample at different wt% of graphene in the $LaMnO_3/TiO_2$/graphene composite. The composite displayed exothermic properties ~380 °C, and increasing amounts of graphene also increased the rate of combustion, with total heat flow in the exothermic process increasing in proportion to the wt% of graphene.

We then used BET analysis to characterize the specific surface area of the sample, as shown in Figure 3. The results showed that the addition of graphene (specific surface area: 400 m^2/g) increased

IOP Conf. Series: Materials Science and Engineering **763** (2020) 012068 doi:10.1088/1757-899X/763/1/012068

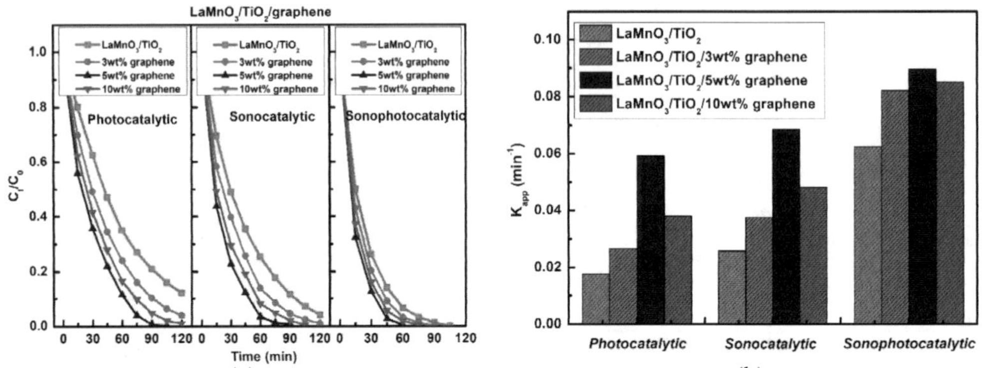

(a) (b)

Figure 4. (a) Catalytic activities and (b) rate constants of the $LaMnO_3/TiO_2$/graphene composite.

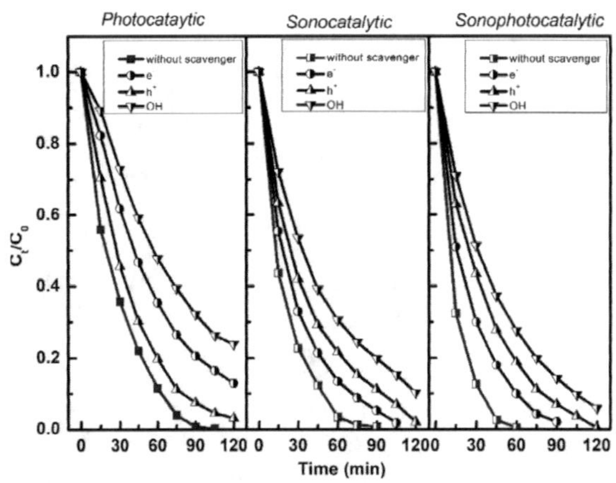

Figure 5. Catalytic activities of the $LaMnO_3/TiO_2$/graphene composite following addition of several scavengers.

the specific surface area of the sample. This parameter represents among the most important aspects of the catalysis process, as a high specific surface area can enhance the contact between the sample surface an organic pollutant, such as MB.

Figure 4 shows the photo-, sono-, and sonophoto-catalytic activities of the $LaMnO_3/TiO_2$/graphene composite, which exhibited higher activity than the $LaMnO_3/TiO_2$ nanoparticle. Previous studies report that the presence of graphene acts an electron trapper during catalysis and promotes higher catalytic rates [13,14]. We found that 5 wt% graphene allowed for optimal catalytic activity; however, higher wt% of graphene (e.g., 10 wt%) resulted in decreased MB degradation rates. This suggested that excess graphene might promote sample aggregation, thereby hindering access to light in the form of irradiation. Furthermore, we found that sonophoto-catalytic activity displayed better MB degradation based on the synergistic effect of visible light and ultrasonic irradiation through the photo- and sono-catalytic processes.

We then evaluated the effects of different scavengers (electron, hole, and hydroxyl radical) on the catalytic activities of the $LaMnO_3/TiO_2$/graphene composite, with the results showing that inclusion of the hydroxyl radical scavenger had the greatest attenuating effect on MB degradation (figure 5).

4. Conclusions

Here, we successfully synthesized an $LaMnO_3/TiO_2$/graphene composite using a co-precipitation method with the addition of graphene enhancing the specific surface area of the sample. Our results showed that 5 wt% of graphene was the optimal, resulting in the highest MB degradation activity by the composite, and that sonophoto-catalysis showed the best results based on the synergistic effects of photo- and sono-catalysis.

References

[1] Hu J, Men J, Ma J and Huang H 2014 *J. Rare Earths* **32** 1126–34
[2] Ahmad M, Ahmed E, Hong Z L, Ahmed W, Elhissi A and Khalid N R 2014 *Ultrason. Sonochem.* **21** 761–73
[3] Soltani R D C, Jorfi S, Ramezani H and Purfadakari S 2016 *Ultrason. Sonochem.* **28** 69–78
[4] Hu J, Ma J, Wang L and Huang H 2014 *J. Alloys Compd.* **583** 539–45
[5] Tang P S, Tong Y, Chen H F, Cao F and Pan G X 2013 *Curr. Appl. Phys.* **13** 340–43
[6] Nguyen-Phan T D, Pham V H, Cuong T V, Hahn S H, Kim E J, Chung J S, Hur S H and Shin E W 2010 *Mater. Lett.* **64** 1387–90
[7] Li F T, Liu Y, Liu R H, Sun Z M, Zhao D S, Zhao D S and Kuo C G 2010 *Mater. Lett.* **64** 223–5
[8] Kumar R D, Thangappan R and Jayavel R 2017 *J. Phys. Chem. Solids* **101** 25–33
[9] Ahn Y, Oh H, Yoon Y, Park W K, Wang W S and Kang J W 2017 *J. Environ. Chem. Eng.* **5** 3882–95
[10] Shaterian M, Enhessari M, Rabbani D, Asghari M and Salavati-Niasari M 2014 *Appl. Surf. Sci.* **318** 213–7
[11] Afifah N and Saleh R. 2016 *Mater. Sci. Forum* **864** 99–105
[12] Afifah N and Saleh R 2017 *Mater. Sci. Eng* **188** 012-060
[13] Afifah N and Saleh R 2017 *IOP Conf. Ser.: Mater. Sci. Eng* **202** 012065
[14] Afifah N and Saleh R 2017 *IOP Conf. Ser.: Mater. Sci. Eng.* **202** 012063

IOP Conf. Series: Materials Science and Engineering **763** (2020) 012069 doi:10.1088/1757-899X/763/1/012069

Study of structure and composition on $La_{0.7}Ba_{0.25}Nd_{0.05}Mn_{1-x}Cu_xO_3$ (x = 0, 0.03, 0.05 and 0.07) prepared by sol-gel method

D R Munazat[1], D S Razaq[1], B Soegijono[1], B Kurniawan[1], S Budiawanti[2] and D Nanto[3]

[1] Department of Physics, Faculty of Mathematics and Natural Sciences (FMIPA), Universitas Indonesia, Kampus UI Depok, Depok 16424, Indonesia

[2] Department of Physics, Faculty of Mathematics and Natural Sciences (FMIPA), Sebelas Maret University, Jebres, Surakarta 57126, Indonesia

[3] Department of Physics Education, Faculty of Tarbiya and Teaching Sciences (FITK), UIN Syarif Hidayatullah, Jalan Ir H Djuanda 95 Ciputat, Banten 15412, Indonesia

Corresponding author's e-mail: bkuru@fisika.ui.ac.id

Abstract. In this work, we have investigated the effect of Cu substitution on structural and chemical composition of $La_{0.7}Ba_{0.25}Nd_{0.05}Mn_{1-x}Cu_xO_3$ (x = 0, 0.03, 0.05 and 0.07) synthesized by sol-gel method. The samples were sintered at 700 °C for 24 hours. Analysis of X-Ray Diffractometer (XRD) patterns revealed that the samples are in cubic structure with space group Pm-3m. The result of Rietveld refinements showed that the increasing Cu doping influences the lattice parameter, unit cell volume and $d_{(Mn,Cu)}$. Goldschmidt's tolerance factor was calculated and confirmed that cubic is the most stable structure. The average crystallite size (D) was determined by applying the Scherrer's equation, indicating the sample in nanocrystalline. Their crystallite sizes are 17.9981, 18.5809, 15.9273 and 16.9486 nm for x = 0, 0.03, 0.05, and 0.07, respectively. The X-Ray Fluorescence (XRF) revealed a spectrum of Cu which means Cu has successfully substituted the Mn site. There is no significant difference between the designated and measured composition.

1. Introduction

Mixed valance perovskite manganites have a general formula of $RE_{1-y}A_yMn_{1-x}TM_xO_3$, where RE represents a trivalent rare earth ion (La, Nd, Pr and others), A stands for a divalent alkaline earth ion (Ca, Sr and Ba) and TM stands for transition metal (Cu, Ni, and Fe). They became the topic of a viable experiment, in the consequence of their attracting physical properties at circa the ferromagnetic (FM)–paramagnetic (PM) transition temperature (Curie temperature, transition metal-insulator temperature), namely the colossal magnetoresistance (CMR), magnetocaloric effects (MCE) (concerning to the magnetic entropy change) as well as the intense relation between structural and magnetic properties [1–8]. Doping of divalent alkaline earth at the La site introduces the mixed valances of Mn^{3+} and Mn^{4+} that was elaborated in the structure of Zener's double exchange (DE) interaction [9]. Furthermore, the Mn-O-Mn bond angle and the Mn-O bond length, which are likewise related to the average ionic radius of the A-site, imply to physical properties of these manganites. The substitution at the Mn-site by other transition metal ions is also crucial because it does not simply modify the Mn^{3+}-O-Mn^{4+} however, generates several new interaction exchange between the Mn ion and the doped transition metal ions as well [1–8].

On the contrary, the preparation method highly influences the physical and chemical properties of

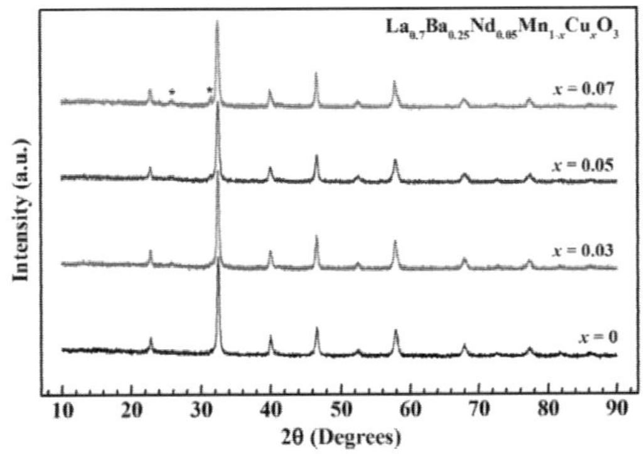

Figure 1. Powder XRD pattern for $La_{0.7}Ba_{0.25}Nd_{0.05}Mn_{1-x}Cu_xO_3$ sintered at 700 °C.

the rare-earth perovskite materials. The sol-gel synthesis has become an interesting method to generate an exceptionally pure and crystalline oxide powders. This method has been used successfully to provide some benefits, namely better homogeneities, lower processing temperatures, short annealing times, high purity of materials and improved material properties [1,11].

In this work, we synthesize the nanocrystalline sample $La_{0.7}Ba_{0.25}Nd_{0.05}Mn_{1-x}Cu_xO_3$ by sol-gel method. Parental compound $La_{0.7}Ba_{0.25}Nd_{0.05}MnO_3$ has been studied by Ünlü *et al.* [8] showed T_C near room temperature which makes the sample a good candidate for CMR and MCE. Because the radius of Cu ions (Cu^{2+} = 0.73 Å, Cu^{3+} = 0.54 Å) is close to that of Mn ions (Mn^{3+} = 0.645 Å, Mn^{4+} = 0.53 Å), the Mn-site can be occupied by Cu ions [1–7]. In this work, we conscientiously examine the influence of the Cu-doping at Mn-site in the nanocrystalline $La_{0.7}Ba_{0.25}Nd_{0.05}MnO_3$ (x = 0, 0.03, 0.05, 0.07) on structural and chemical composition because Cu substituion in mixed valance perovskite manganites have attracted tunable MCE and CMR [2,3]. Meanwhile, further physical properties investigation will be next work of our project.

2. Experimental

Nanocrystalline $La_{0.7}Ba_{0.25}Nd_{0.05}Mn_{1-x}Cu_xO_3$ (LBNMCO) were prepared by sol-gel citrate-nitrate auto combustion method. The precursors used were La_2O_3 (Aldrich, 99.5 %), Nd_2O_3 (Aldrich, 99.5 %), $Ba(NO_3)_2$ (Merck, 99 %), $Mn(NO_3).4H_2O$ (Merck, 98.5 %), $Cu(NO_3)_2.3H_2O$ (Merck, 99.5 %), and citric acid $C_6H_8O_7.H_2O$ (Merck 99.5 %). Stoichiometric amounts of greatly purified metal nitrates were dissolved together in a minimum amount of aquabidest water to obtain a clear solution [11]. We used the precursors in metal nitrate meanwhile oxide metal precursors were diluted by nitrate acid. After each precursor has become a clear solution, the precursor was mixed to synthesize $La_{0.7}Ba_{0.25}Nd_{0.05}Mn_{1-x}Cu_xO_3$ perovskite sample according to the reaction:

$$\begin{aligned} 0.7\mathrm{La(NO_3)_3} + 0.25\mathrm{Ba(NO_3)_2} + 0.05\,\mathrm{Nd(NO_3)_2} + (1-x)\mathrm{Mn(NO_3)_2.4H_2} \\ + x\mathrm{Cu(NO_3)_2.3H_2O} + \beta\,\mathrm{C_6H_8O_7.H_2O} \\ \rightarrow \mathrm{La_{0.7}Ba_{0.25}Nd_{0.05}Mn_{1-x}Cu_xO_3} + \delta\mathrm{N_2} + \gamma\mathrm{H_2O} + \alpha \end{aligned} \quad (1)$$

Citric acid was mixed with the metal nitrates solution. Citric acid carries double functions, serving as the metal ion's complexant as well as the fuel [12]. The solution was stirred continuously and heated up to 80 °C on a magnetic hot plate stirrer. The sample was calcinated at 600 °C for 6 hours to eliminate the organic substance, and then powder LBNMCO was sintered at 700 °C for 24 hours. Crystal structure and phase purity of samples were determined using X-ray diffractometer at room temperature. The chemical composition of samples was characterized using X-ray fluorescence.

3. Results and discussion

The crystal structure of LBNMCO series was characterized by XRD. Figure 1 shows the same XRD patterns for each sample. However, there is an indication of the presence of another peak emerging beside the peak of lanthanum manganite at Cu doped with concentration x = 0.07. The peak is from La_2O_3, which appears as the secondary phase. The peak appears because of sintering process at low temperature. The mixed valance perovskite manganite, the structure for doped lanthanum manganite phase begins to form crystallization above 700 °C [12].

IOP Conf. Series: Materials Science and Engineering **763** (2020) 012069 doi:10.1088/1757-899X/763/1/012069

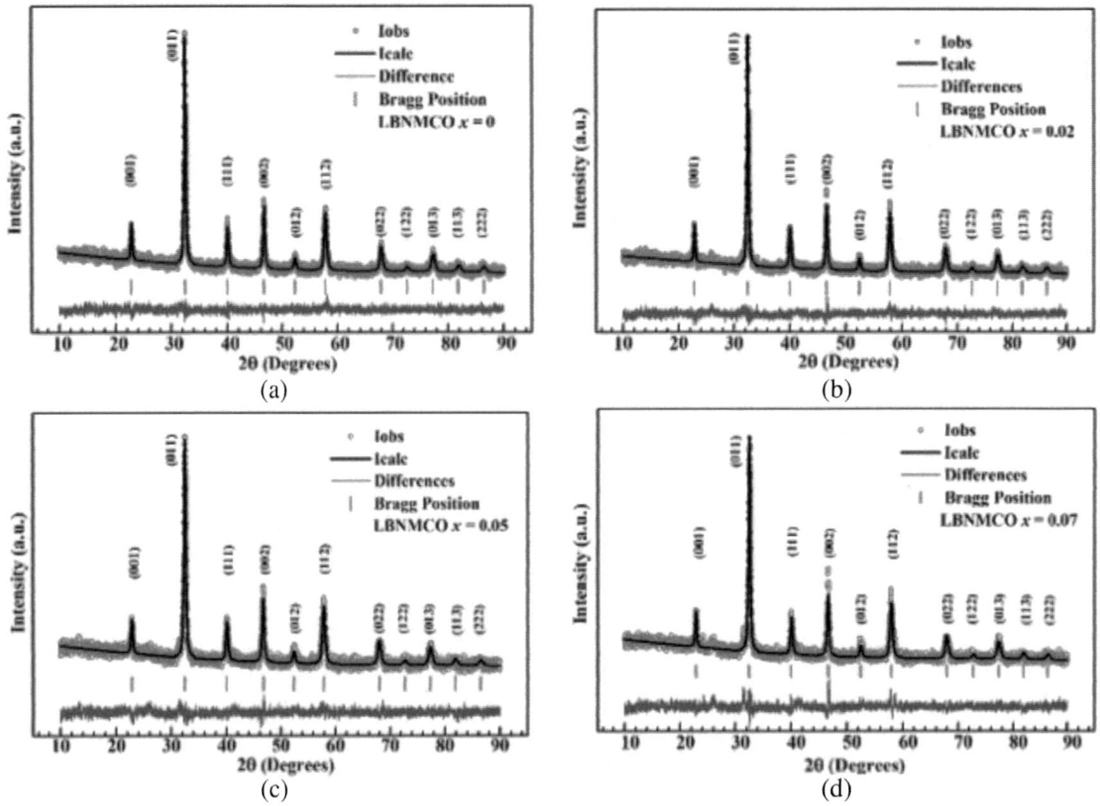

Figure 2. Rietveld analysis result from sample $La_{0.7}Ba_{0.25}Nd_{0.05}Mn_{1-x}Cu_xO_3$ $x = 0$, (b) $x = 0.03$, (c) $x = 0.05$ and (d) $x = 0.07$ sintered at 700 °C.

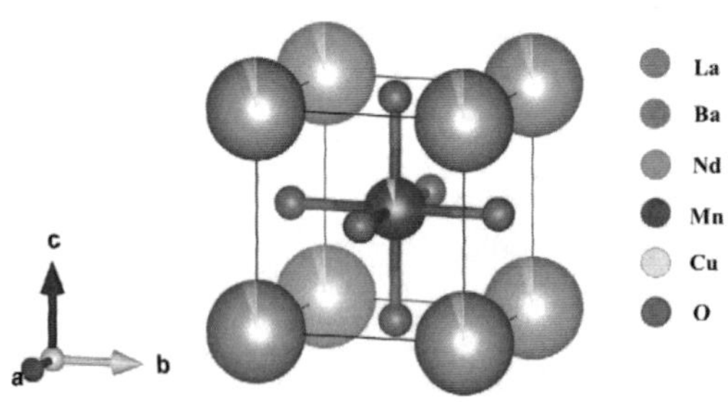

Figure 3. Structure $La_{0.7}Ba_{0.25}Nd_{0.05}Mn_{1-x}Cu_xO_3$.

The diffraction pattern of LBNMCO series has been refined using Rietveld analysis as shown in figure 2. The refinement's product indicated that all samples are cubic structure with space group Pm-3m such as figure 3. The substitution of Cu ions for Mn-site does not change the lattice parameters significantly. The structure stability was confirmed by Goldschmidt's tolerance factor (τ). The value of τ can be determined by using the formulation:

$$\tau = \frac{< r_A > + r_O}{\sqrt{2}(< r_B > + r_O)} \quad (2)$$

with r_A are ionic radii of each element on *A*-site (La^{3+}, Nd^{3+}, and Ba^{2+}), r_B is ionic radii of the *B*-site (Mn^{3+} and Cu^{3+}), r_O is ionic radii of O^{2-}. The stable manganite perovskite structure posesses a τ factor which is

Table 1. Stuctural parameter from Rietveld analysis for $La_{0.7}Ba_{0.25}Nd_{0.05}Mn_{1-x}Cu_xO_3$.

Parameter	$La_{0.7}Ba_{0.25}Nd_{0.05}Mn_{1-x}Cu_xO_3$			
	x			
	0	0.03	0.05	0.07
Structure	cubic	cubic	cubic	cubic
Space Grup	Pm-3m	Pm-3m	Pm-3m	Pm-3m
$a = b = c$ (Å)	3.9020	3.9018	3.8993	3.9003
Volume Unit Cell (Å^3)	59.4096	59.4036	59.2889	59.3349
$d_{<Mn-O>}$ (Å)	1.9510	1.9509	1.9497	1.9502
<Mn-O-Mn> (degrees)	180	180	180	180
Average Crystallites size (nm)	17.9981	18,5809	15.9273	16.9486
Goldschmidt's tolerance factor (τ).	0.97972	0.98126	0.98229	0.98333
discrepancy factor				
GoF	1.1807	1.2393	1.2835	1.4192
RwP	6.6129	6.8466	7.0307	7.3843
RP	5.2261	5.4052	5.5266	5.8040

in the range $0.89 < \tau < 1.02$. The calculation results τ are close to 1 summarized in table 1, which belongs to cubic structure [10]. As reported by Ünlü *et al.* [8] that in the parental compound La0.7Ba0.25Nd0.05MnO3 has a cubic structure.

Table 1 shows the lattice parameter result of Rietveld refinement. The effect of Cu substitution change lattice parameters, unit cell volume, and average Mn-O bond length. Because, it has a cubic structure, the angle of Mn-O-Mn is 180°. For $x = 0.02$ and $x = 0.05$ lattice parameters, unit cell volume, and average Mn-O bond length decreased but increased at $x = 0.07$. It indicates that there is a difference in the percentage of substituted Cu ions. Accordingly, the observed lattice parameter in samples can be defined in terms of the differences in the ionic radii and the variations in the percentages of the Cu^{2+} and Cu^{3+}, Mn^{3+} and Mn^{4+} ion with an increase in Cu doping level [1]. Several XPS study reported that substitution of Cu ions on the Mn-site showed Cu ions existed in the form of mixed Cu^{2+} and Cu^{3+} [1-7].

The *B*-site (six-coordinated) ionic radius of Cu^{2+} (0.73 Å) is greater compared to Mn^{3+} (0.645 Å) and the ionic radius of Cu^{3+} (0.54 Å) is close to that of Mn^{4+}(0.53 Å) and smaller compared to the radius of the high spin state of Mn^{3+} (0.645 Å) [5]. Ghosh *et al.* [6] reported that average radii of Mn^{3+} and Mn^{4+} ion ($R(Mn_{av})$) at lanthanum manganite that is doped on La-site about $x = 0.3$ ($La_{0.7}Sr_{0.3}MnO_3$) are estimated to be 0.619 Å. A smaller ion at the doped site on the subject neighboring Mn–O bonds to centric push reduces the Mn–O bond length. A larger ion at the doped site can subject the neighboring Mn–O bonds to a centric pull and creating an increase in the Mn–O length [4]. The Cu^{3+} (0.54 Å) which is smaller compared to the $R(Mn_{av})$ stretches the to centric push generating a reduction in the Mn–O bond [4,5]. Therefore, one can expect that Cu ions presents in the form of Cu^{3+} dominant at Mn-sites for $x = 0.03$ and $x = 0.05$ whereas Cu^{2+} dominant for $x = 0.07$.

The average crystal structure is calculated using Scherrer formula as follow:

$$D = \frac{0.9\,\lambda}{\beta_{HKL}\cos\theta} \quad (3)$$

where D is crystallite size, λ is Cu-$K\alpha$ radiation which is used during XRD measurement (1.5406 Å), β_{HKL} is full-width at half maximum (FWHM) of each diffraction peak in radian unit, and $\cos\theta$ is value of each diffraction position peak. The calculated result of average crystallite size in the range of 15–19 nm. The low sintering temperature resulted in nanocrsytalline sample [12].

The XRF characterization was used to confirm what elements are contained in the material LBNMCO and confirmed that Cu substitution is successful. XRF measurement results can be seen in figure 4 and the concentration of the contained elements can be seen in table 2. Spectrum Cu confirms that Cu successfully substituted and closed for the calculated concentration.

Table 2. The XRF result sample $La_{0.7}Ba_{0.25}Nd_{0.05}Mn_{1-x}Cu_xO_3$ compared to calculation based on the stoichiometry process.

Element	Concentration							
	Calculated (%)				Measured (%)			
	0	0.03	0.05	0.07	0	0.03	0.05	0.07
La	53.63	53.56	53.51	53.46	54.35	53.79	54.09	54.21
Ba	12.08	12.07	12.05	12.04	14.03	15.09	14.26	13.91
Nd	3.98	3.97	3.97	3.97	3.46	3.63	3.63	3.72
Mn	30.30	29.35	28.72	28.09	28.16	26.34	26.06	25.46
Cu	0	1.05	1.75	2.45	0,00	1.15	1.96	2.70
Total	100	100	100	100	100	100	100	100

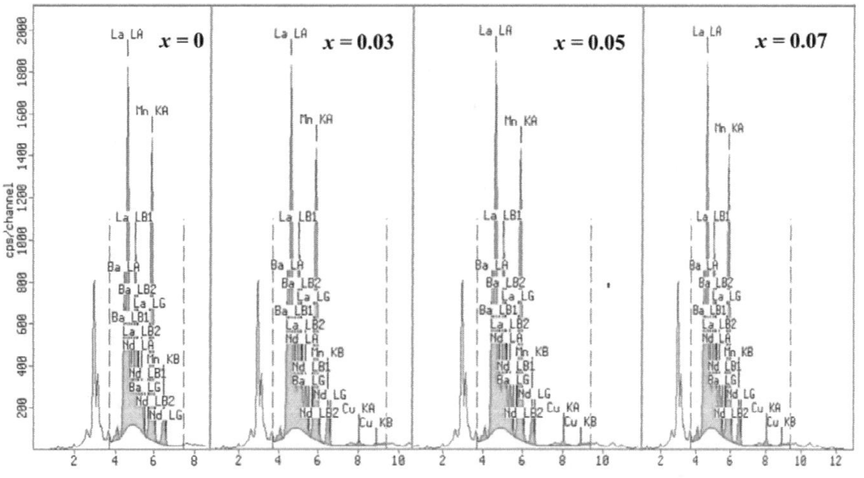

Figure 4. XRF result reveals that three different samples of $La_{0.7}Ba_{0.25}Nd_{0.05}Mn_{1-x}Cu_xO_3$ sintered at different temperature produce the same spectrum without any sign of impurities

4. Conclusions

The $La_{0.7}Ba_{0.25}Nd_{0.05}Mn_{1-x}Cu_xO_3$ (x = 0, 0.03, 0.05, 0.07) material has been successfully prepared by applying the sol-gel method. X-ray diffraction (XRD) characterization results show that samples are cubic structure with space group Pm-3m. Goldschmidt's tolerance factor was calculated and confirmed that cubic is the most stable structure. The effect of Cu substitution changes lattice parameters, unit cell volume, and average Mn-O bond length. The Cu ions exists in the form of Cu^{3+} dominant at Mn sites for Cu ion concentration $x = 0.03$ and $x = 0.05$, whereas Cu^{2+} dominant for $x = 0.07$. The average crystallite size (D) was specified by applying the Scherrer's equation, indicating the sample in nanocrystalline. The X-Ray Fluorescence (XRF) reveal spectrum of Cu that means Cu has successfully substituted the Mn site.

Acknowledgements

This work is funded by Research Grant from UIN Syarif Hidayatullah Jakarta with grant contract Un.01/KPA/234/2018 and support by Department of Physics Universitas Indonesia.

References

[1] Liu Z, Lin W G, Zhou K W and Yan J L 2017 *Ceram. Int.* **44** 2797–802
[2] Chebaane M, Bellouz R, Oumezzine M, Hlil E K and Fouzri A 2018 *RSC Adv.* **8** 7186–95
[3] Wang K Y, Song W H, Dai J M, Ye S L and Wang S G 2001 *J. Appl. Phys.* **90** 6263-7
[4] El-Hagary, Shoker Y A, Mohammad S, Moustafa A M, El-Aal A A, Michor H, Reissner M, Hilscher G and Ramadan A A 2009 *J. Alloy. Compd.* **468** 47–53

[5] Kim M S, Yang J B, Medvedeva J, Yelon W B, Parris P E and James W J 2008 *J. Phys.: Condens. Matter* **20** 255228
[6] Ghosh K, Ogale S B, Ramesh R, Greene R L, Venkatesan T, Gapchup K M, Bathe R and Patil S I 1999 *Phys. Rev.* B **59** 533–7
[7] Cai H L *et al.* 2004 *Powder Diffr.* **19** 329–32
[8] Ünlü C G, Tanış Y E, Kaynar M B, Şimşek T and Özcan Ş 2017 *J. Alloys Compd.* **704** 58–63
[9] Zener C 1951 *Phys. Rev.* **81** 440–4
[10] Manjunatha S O, Rao A, Lin T Y, Chang C M and Kuo Y K 2015 *J. Alloys Compd.* **619** 303–10
[11] Deganello F, Marci G and Deganello G 2009 *J. Eur. Ceram. Soc.* **29** 439–50
[12] Gupta M, Khan W, Yadav P, Kotnala R K, Azam A and Naqvi A H 2012 *J. Appl. Phys.* **111** 093706

IOP Conf. Series: Materials Science and Engineering **763** (2020) 012070 doi:10.1088/1757-899X/763/1/012070

Hysteresis loops observation of FePt and FePd square ferromagnets using micromagnetic simulation

U Fadhilah[1], C Kurniawan[1,2] and D Djuhana[1]

[1]Department of Physics, Faculty of Mathematics and Natural Sciences (FMIPA), Universitas Indonesia, Kampus UI Depok, Depok 16424, Indonesia

[2] Research Center for Physics, Indonesian Institute of Sciences (LIPI), Gedung 440–442 Kawasan Puspiptek Serpong, Banten 15314, Indonesia

Corresponding author's e-mail: dede.djuhana@sci.ui.ac.id

Abstract. In this study, we have investigated magnetic hysteresis loops of squared model FePt and FePd ferromagnets by a public micromagnetic simulation based on Landau-Lifshitz Gilbert (LLG) equation. The hysteresis loops of FePt and FePd ferromagnets are generated by in-plane and out-plane applied field with respect to the size and thickness variation. It is found that the coercivity field for both FePt and FePd squared ferromagnets exhibited different behavior for the size of ferromagnets around 50–100 nm and more than 100 nm. The results showed that the hysteresis loops produced a large coercivity when the in-plane field was applied and zero coercivity for external out-plane field. Interestingly, coercivity field was still observed for materials with size below 100 nm with ranging values between 20–40 mT. From this result, certain values of the coercivity field appeared for out-plane applied field at small thickness indicates a perpendicular magnetic anisotropy (PMA) behavior in FePt and FePd ferromagnets.

1. Introduction

Research on magnetic materials and understanding of magnetization process were such an interesting study in recent years, especially on the development of magnetic recording devices for high-speed data storage. The problem of this topic was raised in order to continue the rapid increasing of both capacity and speed of HDD storage and following the Moore's law on data storage. Both experimental and micromagnetic approaches were utilized to understood the new phenomenon of magnetic based recording system. In addition to the increasing of magnetic recording capacity, the recording technology has also transformed. In the beginning, the longitudinal magnetic recording was applied for magnetic recording and then changed into perpendicular magnetic recording, which the magnetic bits are perpendicular to the film plane which was used on the materials with perpendicular magnetic anisotropy (PMA) [1].

To realize magnetic storage media with heat-stable material characteristic, magnetic materials with large magnetic anisotropy are required so that when it is patterned to the order of nm, the magnetic character is not easy to lose due to the heat effects [2]. One of the materials that fulfill these criteria is FePt and FePd. In recent years, Fe alloy nanoparticle materials such as FePt and FePd attracted huge attention because of their high magnetic anisotropy [3]. These materials have potential for application in magnetic recording media and magnetic random access memory (MRAM) [4–6]. FePt has 50 to 100 times greater magnetocrystalline anisotropy that allowing density area in T/inch^2 order [7]. FePt has anisotropy constants can reach as high as 10^7 J/m^3 [8–10]. The FePd is also one of the feasible Fe-Alloy candidates because of the strong responsiveness of PMA [11]. Based on a research road map of PMA properties on thin films for spintronic applications [12], FePt and FePd thin films exhibit a very large magnetocrystalline anisotropy but have large magnetic damping constants [13].

IOP Conf. Series: Materials Science and Engineering **763** (2020) 012070 doi:10.1088/1757-899X/763/1/012070

Table 1. Material parameters used in micromagnetic simulations [2,3]

Materials	Exchange stiffness A (J/m)	Anisotropy constant K (J/m^3)	Magnetization saturation Ms (A/m)
FePt	10×10^{-12}	6.6×10^{6}	1.14×10^{6}
FePd	6.9×10^{-12}	2.6×10^{5}	1.06×10^{6}

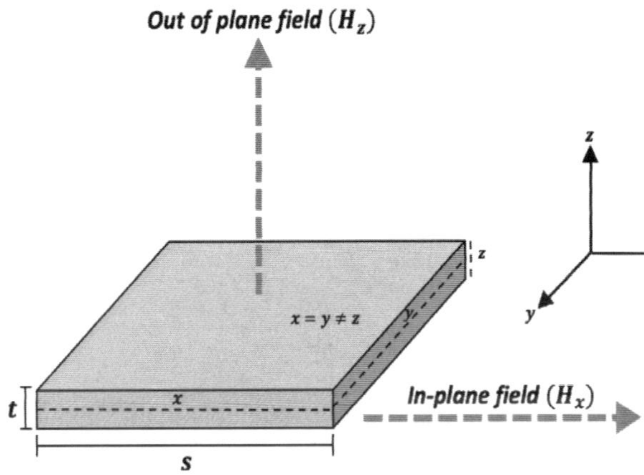

Figure 1. The dimension and geometry of FePt and FePd square ferromagnets. The length size s was varied from 50 nm to 500 nm with external field is applied in-plane and out-of-plane direction.

In this study, we investigated the characteristics of hysteresis loops from FePt and FePd square-shaped model with different sizes and thicknesses using micromagnetic simulation approach. The magnetic field on the in-plane direction and out-of-plane direction are used to analyze the magnetic properties of FePt and FePd thin layers which is connected with the appearance of PMA properties.

2. Simulation procedure

Hysteresis loops observation of FePt and FePd square ferromagnets was carried out using a public micromagnetic simulation software OOMMF (Object Oriented Micromagnetic Framework) [14]. The system works by solving the Landau-Lifshitz-Gilbert (LLG) equation, which is a function of differential equations [15]. The FePt and FePd square shaped were used in the simulation with the parameters given in table 1. The length size, *s* of the square shaped ($x = y$) was varied from 50 nm to 500 nm with two variations of thickness (t) of 5 and 10 nm. The cell size of discretization was fixed to be 2.5 x 2.5 x 2.5 nm^3, and the damping constant α of 0.05. The magnetic field range from -1000 to 1000 mT with for in-plane (H_x) and out-plane (H_z) direction as illustrated in figure 1.

3. Results and discussion

The dependence of magnetization with the angle between the applied field H and the easy magnetization direction caused possible different shapes of hysteresis loop. When the field H is applied parallel to the easy magnetization direction in the material, a straight loops is obtained, whereas a slanted loop is observed if the field H is applied perpendicular to the easy direction [8]. A significant change in the coercivity is observed with respect to different lengths s and thicknesses t, as shown in figure 2. A large H_c value of about 200 mT was obtained for FePt with $s = 50$ nm, which was measured parallel to the plane (in-plane field). It is observed that the coercivity decreases slowly with increasing size of the materials, but still keeps a large value of about 40 to 60 mT and 80 to 100 mT for FePd and FePt, respectively. In the mesoscopic scale, which the length size *s* from 200 nm to 500 nm, coercivity decreases and tends to be constant for FePd with a value of 20 mT ($t = 10$ nm) and 40 mT ($t = 5$ nm). On the other hand, a drastic change of the magnetization loops is observed in the length size *s* between 100 nm and 200 nm for FePt.

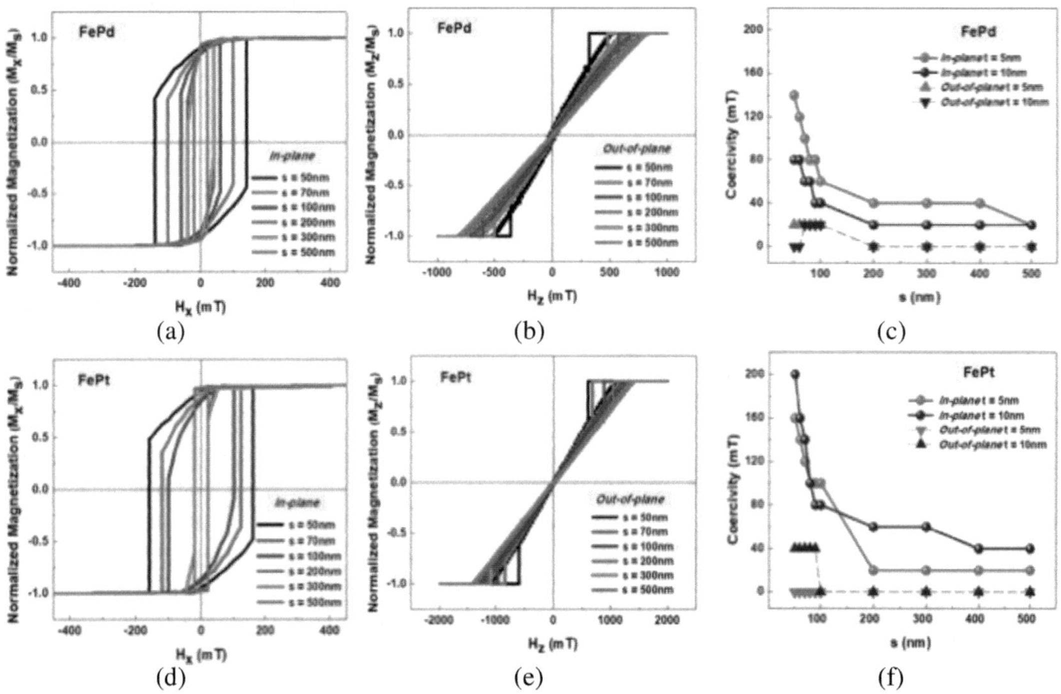

Figure 2. Hysteresis loops for (a, b) FePd and (d, e) FePt square-shaped with different lengths and thicknesses. The magnetic field was utilized in the in-plane direction and out-of-plane (perpendicular) direction. Coercivity H_c for (c) FePd and (f) FePt square-shaped as a function of lengths s and thicknesses t.

When the external field was applied in the out-of-plane direction, the hysteresis loops with different lengths and thicknesses showed nearly zero coercivity. It indicates that such typical of the hysteresis loops is given the external fields toward its hard axis direction. However, it is observed that the coercivity values ranging between 20 to 40 mT as shown in figure 2 (c and f). This small value of the coercivity indicates perpendicular magnetic anisotropy (PMA) behavior in FePt and FePd ferromagnets. According to previous research from Shima *et al.* [16], when magnetic field was employed in the perpendicular direction against the film plane, a huge coercivity of about 40 kOe was obtained for $t = 10$ nm at room temperature.

The domain structure of squared shaped model FePd and FePt with $t = 5$ nm at the coercive state can be seen in figure 3. In the in-plane direction, the multi domain state formation are observed from $s = 50$ nm to 500 nm. Meanwhile, in the out-of-plane direction, vortex structure are observed with reversal mechanism process that occurred was curling mode. This mode avoids creating stray field by passing through a vortex state where the magnetization lies everywhere parallel to the surface [17]. Curling mode requires less energy to make the magnetization minimum. So, generally on the vortex-wall materials, the coercivity field will decrease as the size of material increases. The color represents the magnetization value for the in-plane field direction is shown from the x component and the z component for out-of-plane field direction.

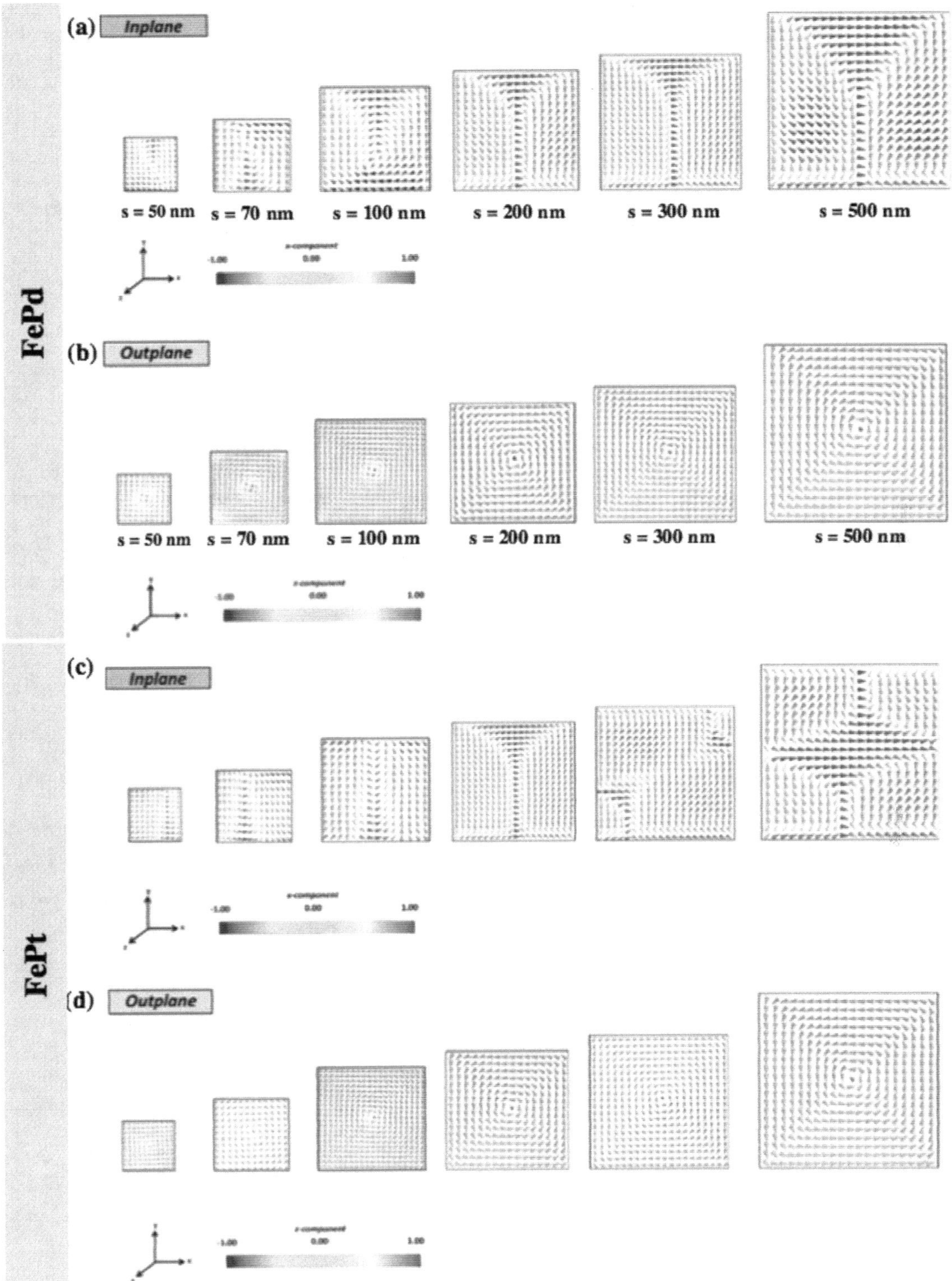

Figure 3. The domain structure in (a, b) FePd and (c, d) FePt of square-shaped t = 5 nm with different length sizes. The magnetic field was employed in the in-plane direction and out-of-plane direction.

4. Conclusions

We have studied the hysteresis loops of FePt and FePd with square-shaped model at different lengths and thicknesses by micromagnetic simulation. The magnetic field was utilized in the in-plane and out-of-plane (perpendicular) directions. It was found that two different shapes of hysteresis loops are influenced by the direction of the applied field. The coercivity decreased as the size of materials increased but tends to be constant in mesoscopic scale. The results showed that the size of materials affected the magnetic properties of FePt and FePd thin layers. The highest coercivity of 200 mT was obtained for FePt, when the external field was applied in the in-plane direction. It was because the anisotropy constant of FePt is higher than FePd. Vortex state was found when the external field was applied in out-of-plane direction with curling reversal mode.

Acknowledgements

The authors would like to thank Universitas Indonesia for funding this research through PITTA Grant with contract number 2245/UN2.R3.1/HKP.05.00/2018

References

[1] Tudu B and Tiwari A 2017 *Vacuum* **146** 329–41
[2] Purnama B, Ismail and Suharyana 2013 *Jurnal Fisika dan Aplikasinya* **9** 30–3
[3] Lukaszew R A, Cebollada A, Clavero C and Garcia-Martín J M 2006 *Physica* B **384** 15–8
[4] Ivanov O A, Solina L V, Demshina V A and Magat L M 1973 *Phys. Metals Metallogr.* **35** 81–5
[5] Mitani S, Tsukamoto K, Seki T, Shima T and Takanashi K 2005 *IEEE Trans. Magn.* **41** 2606–8
[6] Ohtake M, Ouchi S, Kirino F and Futamoto M 2012 *IEEE Trans. Magn.* **48** 3595–8
[7] Scholz W, Fidler J, Schrefl T, Suess D, Forster H, Dittrich R and Tsiantos V 2004 *J. Magn. Magn. Mater.* **272–276** 1524–5
[8] Sung H W F and Rudowicz C 2003 *J. Magn. Magn. Mater.* **260** 250–60
[9] Iwama H, Doi M and Shima T 2016 *J. Magn. Soc. Jpn.* **40** 91–4
[10] Skuza J R, Clavero C, Yang K, Wincheski B and Lukaszew R A 2010 *IEEE Trans. Magn.* **46** 1886–9
[11] Bonell F, Murakami S, Shiota Y, Nozaki T, Shinjo T and Suzuki Y 2011 *Appl. Phys. Lett.* **98** 232510
[12] Hirohata A, Sukegawa H, Yanagihara H, Žutić I, Seki T, Mizukami S and Swaminathan R 2015 *IEEE Trans. Magn.* **51** 0800511
[13] Iihama S, Mizukami S, Naganuma H, Oogane M, Ando Y and Miyazaki T 2014 *Phys. Rev.* B **89** 174416
[14] Donahue M J and Porter D G 1999 *OOMMF User's Guide, Version 1.0* (Gaithersburg: National Institute of Standards and Technology)
[15] Gilbert T L 2004 *IEEE Trans Magn.* **40** 3443–9
[16] Shima T, Takanashi K, Takahashi Y K and Hono K 2002 *Appl. Phys. Lett.* **81** 1050
[17] Coey J M D 2009 *Magnetism and Magnetic Materials* (New York: Cambridge University Press)

Scaling behavior of loop area, magnetic remanence and coercivity in thermal dependence of double perovskite $Ba_{1.7}La_{0.3}FeMoO_6$ compound

F P Sari, L H Suriasugandha, N Franata and D Handoko

Department of Physics, Faculty of Mathematics and Natural Sciences (FMIPA), Universitas Indonesia, Kampus UI Depok, Depok 16424, Indonesia

Corresponding author's e-mail: djati.handoko@ui.ac.id

Abstract. We have detailed investigated the scaling behavior of loop area with respect to temperature for $Ba_{1.7}La_{0.3}FeMoO_6$ double perovskite compound. The hysteresis loop was performed on a vibrating sample magnetometer (VSM-Lakeshore 331) from below to critical temperature with a fixed applied field. Based on field dependence of magnetization, we observed that the loop area remains constant within the range below to near critical temperature then decreases abruptly at around critical temperature, exhibits scaling behavior with respect to temperature. The scaling exponent of magnetic remanence and coercivity field with respect to temperature are analyzed together. The result indicates a coexistence of loop area, magnetic remanence and coercivity field in $Ba_{1.7}La_{0.3}FeMoO_6$ double perovskite compound.

Keywords: scaling, perovskite, magnetic remanence, coercivity

1. Introduction

Double perovskite (DP) materials could be found in many complex phases in $A_2BB'O_6$ formula (A = alkaline-earth or rare-earth elements, B = transition-metal ions, B and B'= reside on both sublattices of a cubic lattice) [1]. Recently, an improved interest in the double perovskite, with an interesting phenomenon such as Curie temperature T_C>300 K in ferromagnetic-paramagnetic (FM-PM) phase transition, high magnetoresistance ratio (MR) in room temperature, strong behavior in half-metallic anti ferromagnetic (AFM) and magnetocaloric effect (MCE) has been increased [2–4]. Base on these properties, DP becomes one promising family material for the fabrication of spin-valve devices such as soft magnetic field detecting [5], the improvement of novel magnetic recording media [6] and strong candidate for the production of enhanced sources of spin-polarized electrons for "spin-tronic" applications [7].

Among the double perovskite family, $Ba_{2-x}La_xFeMoO_6$ becomes the most promising material. A number of comprehensive observations have been done for $Ba_{2-x}La_xFeMoO_6$ compounds [8–10], especially in structure, electrical-magnetic properties, and magnetocaloric effect. By partial substitution of La^{3+} for Ba^{2+} in BFMO, some properties have been changed such as: (1) effective decreasing of ionic radius with stable cubic symmetry unit cell (2) modification of magnetic moment and valence state by electron doping of Fe/Mo ions. Furthermore, this substitution influences the Fe/Mo structural disordering and the magnitude of saturation magnetization (M_s) respectively [11].

However, to our knowledge, no study has been dedicated for the effect of temperature to the hysteresis scaling behavior. In the same time, structure disordering and saturation magnetization plays an important

IOP Conf. Series: Materials Science and Engineering **763** (2020) 012071 doi:10.1088/1757-899X/763/1/012071

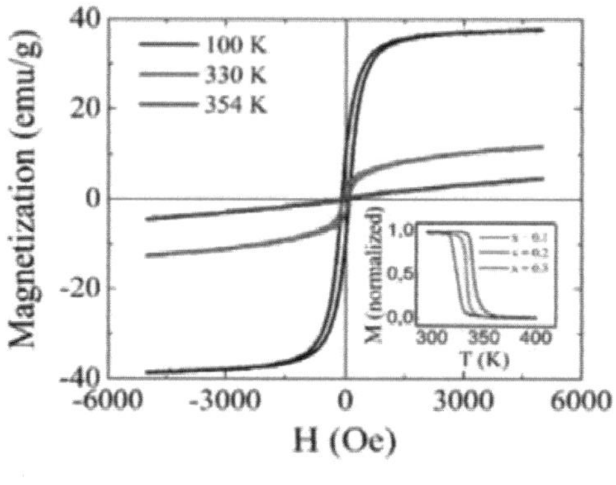

Figure 1. Hysteresis loop that obtained by VSM at 100, 300, and 354 K, the inset shows M-T curve for 3 different $Ba_{1.7}La_{0.3}FeMoO_6$ samples.

role in hysteresis scaling behavior. In this article, we address to discover intensively some properties of partial substitution of La^{3+} for Ba^{2+} on scaling behavior of loop area, magnetic remanence and coercivity in thermal dependence of BFMO materials. Materials with the chemical formula $Ba_{2-x}La_xFeMoO_6$(BLFMO) have been examined completely.

2. Experimental

The standard solid-state reaction method was used to prepare the DP-structured $Ba_{2-x}La_xFeMoO_6$ compound with x = 0.1, 0.2, and 0.3 samples. The starting material, $BaCo_3$, La_2O_3, Fe_2O_3, and MoO_3, were mixed and milled by using small quantity of alcohol. Annealing was performed at 1220 K for 12 hours and at 1270 K for 10 hours. Furthermore, after accomplished to be pellet, the powder was pressed and fired in a stream of 5 % H_2/Ar gas at 1420 K for 21 hours. The temperature-dependent magnetic hysteresis loop measurements have been done by using a vibrating sample magnetometer under 10 kOe maximum applied magnetic field and varied temperature from 100 to 354 K with 2 K increment.

3. Results and discussion

Figure 1 shows field-dependent magnetic hysteresis loops of $Ba_{1.7}La_{0.3}FeMoO_6$ double perovskite compound at 100, 330 and 354 K. As the applied field increases the magnetization rapidly increase and then saturates to a value of Ms. This saturate magnetization has been observed to be decreased linearly as the ambient temperature varied from 100 to 330 K. At further increasing temperature, the magnetization rapidly decreases and tends to be constant. The temperature dependence of magnetization M(T) is depicted in inset of figure 1. The tiny sharp variation in M(T) indicates the occurrence of magnetic ordering in the $Ba_{1.7}La_{0.3}FeMoO_6$ compound.

Below the Curie temperature (T_C), in the range 100 to 300 K, the attributes of the hysteresis loops reveal a very small remanences and coercivities, exhibits a clear soft FM oxide behavior in $Ba_{1.7}La_{0.3}FeMoO_6$ compound. However, the curve corresponding to 354 K reflects practically typical paramagnetic characteristics.

To observe the scaling behavior of $Ba_{1.7}La_{0.3}FeMoO_6$ double perovskite compound, we have determined an area of each hysteresis loop as a function of temperature, as shown in figure 2. In the log- log scale, the loop area tends to be decreased monotonically as the temperature increases. Approaching Curie temperature, the loop area becomes more rapidly decrease. The loop area exhibits a scaling behavior and is fitted by expression, $A{\sim}T^{\gamma}$ where A is a loop area, T is ambient temperature and γ is the scaling exponent. We observed different scaling exponent value for different sample, as shown in inset of figure 2. These differences affected by the different decreasing of coercive field and reducing lattice deformation at elevated temperatures. It exhibits that higher thermal fluctuation promotes the switching process of domain and decreases the activation energy [12].

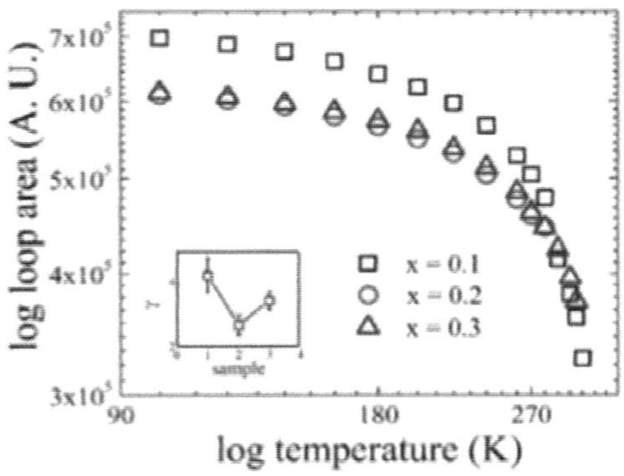

Figure 2. Loop area with respect to temperature for 3 different $Ba_{1.7}La_{0.3}FeMoO_6$ samples.

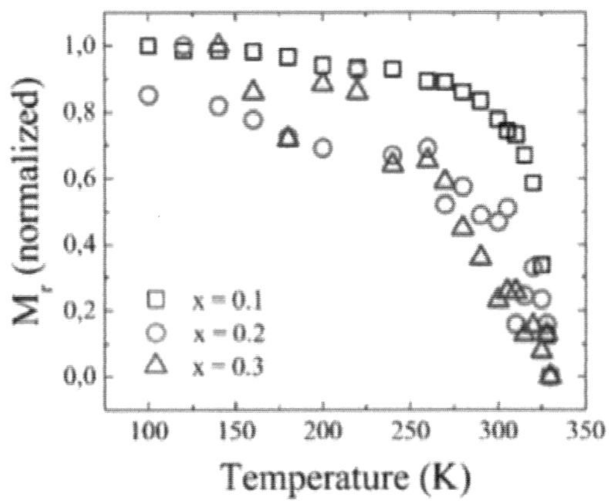

Figure 3. Magnetic remanence as a function of temperature for 3 different $Ba_{1.7}La_{0.3}FeMoO_6$ samples.

For further understanding of scaling behavior, the magnetic remanence of the compound with respect to temperature has been determined. As shown in figure 3, the magnetic remanence shows different behavior for different sample compound. As temperature increases, the magnetic remanence of the sample 1 decreases monotonically then while approaching Curie temperature it rapidly decreases. Different behavior is shown by sample 2 and sample 3. The magnetic remanence decreases almost constant as temperature increases, especially for sample 3, the scaling behavior exhibits a linear relation between magnetic remanence and temperature. The modification of the magnetic remanence scaling behavior should be associated with the phase distribution in $Ba_{1.7}La_{0.3}FeMoO_6$ compound.

For further understanding of scaling behavior, the magnetic remanence of the compound with respect to temperature has been determined. As shown in figure 3, the magnetic remanence shows different behavior for different sample compound. As temperature increases, the magnetic remanence of the sample 1 decreases monotonically then while approaching Curie temperature it rapidly decreases. Different behavior is shown by sample 2 and 3. The magnetic remanence decreases almost constant as temperature increases, especially for sample 3, the scaling behavior exhibits a linear relation between magnetic remanence and temperature. The modification of the magnetic remanence scaling behavior should be associated with the phase distribution in $Ba_{1.7}La_{0.3}FeMoO_6$ compound.

For more detail observation on scaling behavior, the coercive field as a function of temperature has

IOP Conf. Series: Materials Science and Engineering **763** (2020) 012071 doi:10.1088/1757-899X/763/1/012071

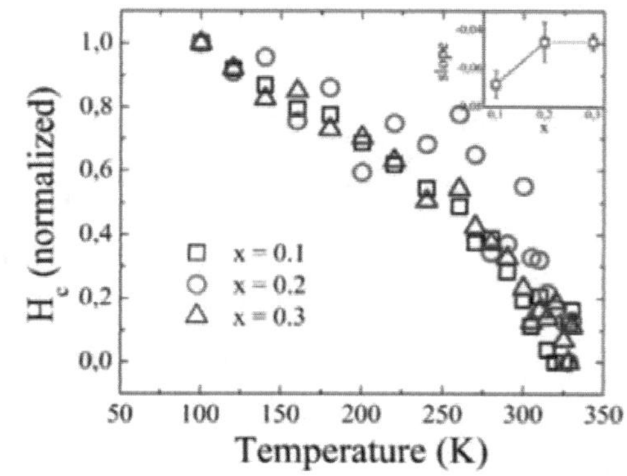

Figure 4. Coercivity as a function of temperature for 3 different $Ba_{1.7}La_{0.3}FeMoO_6$ samples.

been determined. As shown in figure 4, the evolution of coercivity with respect to temperature for 3 different $Ba_{1.7}La_{0.3}FeMoO_6$ compound samples respectively shows similar behavior. The coercivity tends to be linearly decreases as temperature increases with same slope due to the similar temperature-dependent behaviors ascribed to the inherent hysteresis loops for all different $Ba_{1.7}La_{0.3}FeMoO_6$ compound samples within the temperature range. Since the y-intercept could not be meaningfully defined, only the magnitudes of the slopes are shown in inset of figure 4.

4. Conclusions

We have carefully synthesized and observed the $Ba_{2-x}La_xFeMoO_6$ compound of La^{3+} partial substitution that there is scaling behavior of magnetic hysteresis loop parameter with respect to temperature ambient observation. The loop area and coercivity exhibit decreasing value as temperature increases while magnetization has a random behavior.

Acknowledgements

The author would like to thank Universitas Indonesia for PITTA Grant 2018 no 2255/UN2.R3.1/HKP.05.00/2018 which support this work.

References

[1] Serrate D, De Teresa J M and Ibarra M R 2007 *J. Phys.: Condens. Matter* **19** 023201
[2] Han H, Han B J, Park J S, Lee B W, Kim S J and Kim C S 2001 *J. Appl. Phys.* **89** 7687–8
[3] Yuan C L, Zhu Y and Ong P P 2002 *J. Appl. Phys.* **91** 4421–5
[4] Raghunathan A, Melikhov Y, Snyder J E and Jiles D C 2010 *IEEE Trans. Magn.* **46** 1507–10
[5] Xu Y, Memmert U and Hartmann U 2001 *Sens. Actuators* A *Phys.* **91** 26–9
[6] Zavaliche F, Zhao T, Zheng H, Straub F, Cruz M P, Yang P L, Hao D and Ramesh R 2007 *Nano Lett.* **7** 1568–90
[7] Bibes M and Barthelemy A 2007 *IEEE Trans. Electron. Devices* **54** 1003–23
[8] Zhong W, Wu X L, Tang N J, Liu W, Chen W, Au C T and Du Y W 2004 *Eur. Phys. J.* B **41** 213–7
[9] Yang H M, Han H and Lee B W 2004 *J. Magn. Magn. Mater.* **272–276** 1831–3
[10] Kim K S, Yu S C, Kang B S, Zhang P, Kim D H and Oh S K 2012 *IEEE Trans. Magn.* **48** 1352–5
[11] Yang H M, Lee W Y, Han H and Lee B W 2003 *J. Appl. Phys.* **93** 6987
[12] Du G, Liang R, Wang L, Li K, Zhang W, Wang G and Dong X 2013 *Appl. Phys. Lett.* **102** 142903

IOP Conf. Series: Materials Science and Engineering **763** (2020) 012072 doi:10.1088/1757-899X/763/1/012072

Thermal behavior, dielectric and corrosion resistance of polyurethane/carbon/nanoclay hybrid materials

I H Ramadhan, B Soegijono, O Kurniawan, E Virgawati and I Mudzakir

Department of Physics, Faculty of Mathematics and Natural Sciences (FMIPA), Universitas Indonesia, Kampus UI Depok, Depok 16424, Indonesia

Corresponding author's e-mail: bambangs@sci.ui.ac.id

Abstract. Recently, research on hybrid materials of Polyurethane/Carbon/Nanoclay are challenging because of their applications. In this study, these materials were fabricated with various amounts of carbon and organoclay (1, 3, and 5) wt.%). The samples were characterized by thermogravimetry analysis, differential scanning calorimetry, optical microscope, RLC meter, Fourier transform infrared spectroscopy (FTIR) and salt spray test. The results showed that the samples with 5 wt.% carbon/organoclay have higher dielectric constant. Thermal behavior of the samples depends on the content of the filler. Salt spray test showed that the samples with 1 wt.% carbon/organoclay have higher corrosion resistance.

Keywords: polyurethane, organoclay, corrosion resistance

1. Introduction

Polyurethane (PU) has various unique properties such as good elasticity, elongation, tensile strength, highly resistant against abrasion, great resistant towards weather, exceptional luster, color retention and good resistant against corrosion characteristics [1–2]. But in harsh conditions, PU fails to provide satisfactory thermal and mechanical resistance performance, so we need to induce such structural properties in PU through mixing into composites to offer a great performance even in the extreme environmental circumstances. It was noted that the synergistic influence among filler and matrix (PU) increases its ability to stabilize, solubilize, as well as improve the properties of thermal, mechanical, electrical and optical [3–6].

Good thermal stability, environmentally friendly and easily synthesized is regarded as the hallmarks of conducting polymer materials in general. The invention of conducting polymer nanocomposites has revolutionized polymer applications due to their advantages such as high load ratio, high efficiency, light weight and low cost [6–7]. Conducting polymer such as carbon nanotubes (CNT), graphene, and graphite were carried out in a good dispersion of mixtures [8], and also shows good dielectric properties [9]. They provide optimum effect in a sufficiently small amount (0.5–5) % as fillers in a polymer matrix to form composites in the production of various electronic and optical devices such as sensors, actuators, light-emitting diodes, coatings and others [6,10,11].

Other materials that can be used in the development of PU polymer composites are clay. Clay might serve a crucial role in regards to provide a barrier and high impedance properties for coating systems. In other words, clay increases corrosion protection by preventing corrosive media from penetrating through layers. Optimization and new properties of clay/polymer nanocomposites compared with pure polymers, and many new nanocomposites based on polymers/clay have been investigated [12–16]. The integration of

IOP Conf. Series: Materials Science and Engineering **763** (2020) 012072 doi:10.1088/1757-899X/763/1/012072

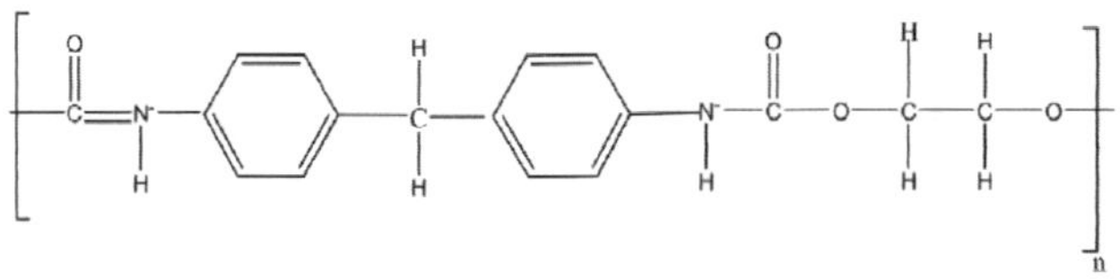

Figure 1. The urethane groups –NH–(C=O)–O–

organoclay into the PU matrix generates the better coating properties namely, adhesion, hardness, barrier properties, thermal resistance, and resistant to corrosion [3,17]. Composite of polyurethane with filler carbon and organoclay were fabricated to obtain a good corrosion resistance and thermal stability hybrid material.

2. Experimental procedure

The precursor was polyurethane methylene diphenylene diisocyanate (MDI) with aluminum addition from the commercial product, MDI based polyisocyanate composition, 4.4'-diphenylmethane, aluminum, aromatic solvent blend and isobutyl acetate. The urethane group –NH–(C=O)–O– link molecular units is shown in figure 1. Carbon was used for graphite filler as one of the allotropes of carbon. The carbon was grounded to the size of 30 mesh to obtain microcarbon. Organoclay was also used as filler from the commercial product. The composite was prepared by mixing PU, carbon and organoclay with various compositions of 1, 3, and 5 wt %, respectively then stirred at 500 rpm for 120 minutes.

Thermogravimetric analysis (TGA) was used to identify the weight loss of the composite as function of temperature and differential scanning calorimetry (DSC) was applied to obtain evidence about heat stability throughout the heating process at 20 °C/min from 30 °C temperature to 950 °C. Fourier transform infrared spectroscopy (FTIR) was also employed to see the functional groups of composite at the wavenumber range of 500–4000 cm^{-1}. Dielectric test was used to see dielectric value of the sample with frequency 100 kHz to 1 MHz. Optical microscopy (OM) was used to analyze morphological changes of samples before and after salt spray tests.

3. Results and discussion

3.1. TGA analysis

TGA pattern offers the evidence of weight loss throughout the heating process as the function of the sample's temperature as shown in figure 2. Figure 2 shows there are two degradation steps, the initial stage is around 340–370 °C and the second is around 680–720 °C. The initial stage is the main stage which has high degradation percentage (45–55) % from 230 °C onset temperature until 420 °C endset temperature. The second stage is minor stage with degradation percentage (5–15) % from 610 °C onset temperature until 790 °C endset temperature. The end of second stage samples displayed higher thermal resistance than pure polyurethane.

Polyurethane has the lowest final weight and the final weight increases with the increase of Carbon/Clay content. It seems the onset temperature is shifting to the higher temperature with the increase of carbon/clay content. The significant difference at the second stage indicates chemical interaction on the surface filler with the matrix. Composite with 3 wt.% carbon/clay shows the highest thermal stability. It probably occurs due to agglomeration of clay or carbon when the content increases. If the agglomeration exists, the chemical interaction between matrix and the surface of the filler decreases.

3.2. DSC analysis

DSC pattern indicate the need or excess heat as temperature function of the samples as shown in figure 3. PU pattern shows a sharp decrease indicating no sufficient resistance against thermal condition, but with the addition of carbon/clay, the pattern becomes relatively stable. It indicates that the sample has thermal stability due to the cross-linking characteristics of the matrix increasing thermal stability by the addition of fillers [18], and as it is seen from figure 3, the best thermal stability is the sample PCClay3.

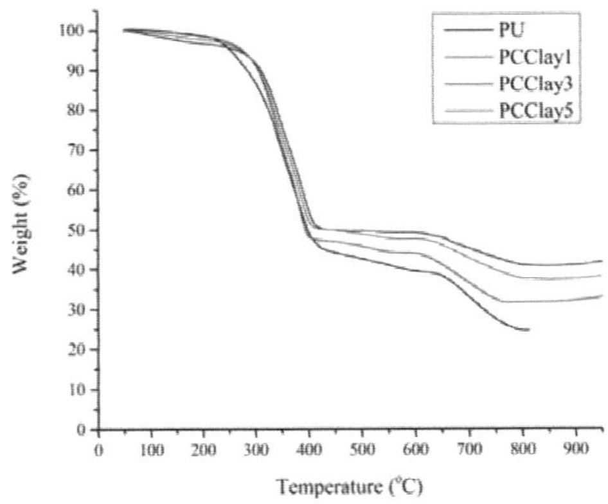

Figure 2. TGA pattern of samples polyurethane/carbon/organoclay with various amount of carbon and organoclay

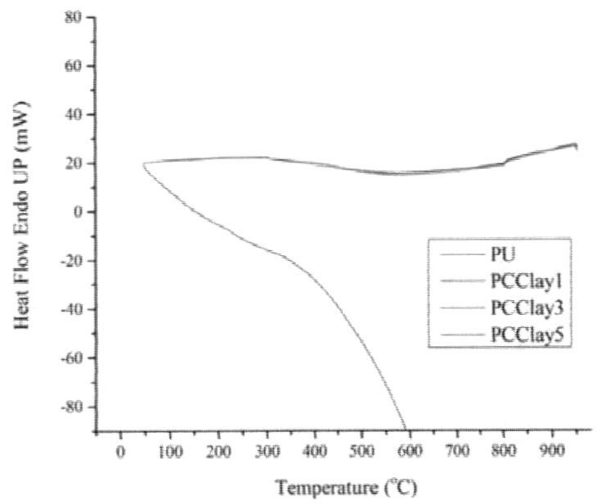

Figure 3. DSC pattern of samples polyurethane/carbon/organoclay (PCClay) with various amount of carbon and organoclay

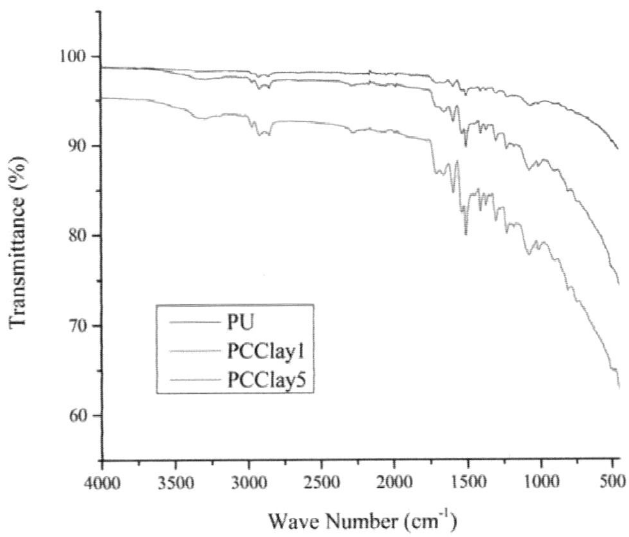

Figure 4. FTIR transmission pattern of the polyurethane/carbon/organoclay with various amount of carbon and organoclay

3.3. FTIR analysis

FTIR transmission pattern as wave number function is shown in figure 4. PU with 1 wt % of carbon/organoclay shows the optimum transmittance for wave number 500–4000 cm^{-1}. Stretching frequency of OH functional group was observed around wavenumber of 3050–3720 cm^{-1}. At wavenumber of 2800–3000 cm^{-1},there was a peak of stretching C-H bond. The C=O (amide I) vibration was identified

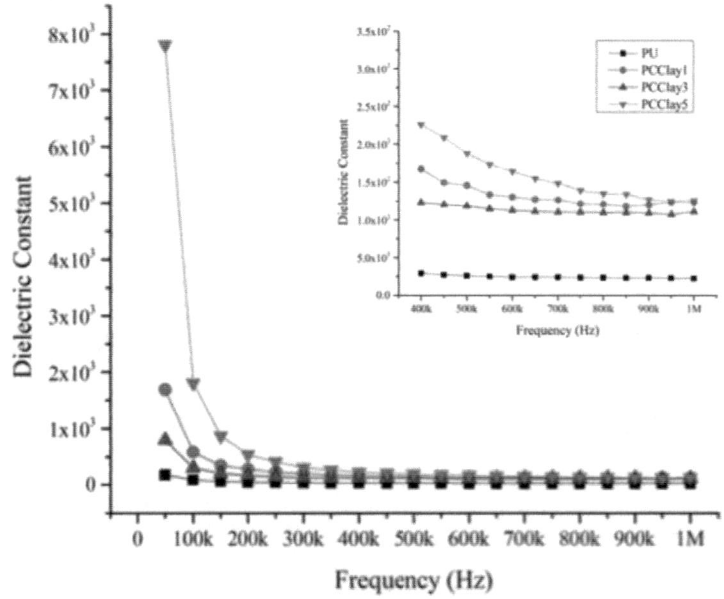

Figure 5. Dielectric curve of the polyurethane/carbon/organoclay with various amounts of carbon and organoclay. Inset in the figure is dielectric constants with different scale for frequency region of 400 kHz to 1 MHz

at the wavenumber of 1665–1780 cm^{-1}. The vibration of C-N, N-H, and C-C (amide II) were identified at the wavenumber of 1500–1590 cm^{-1}. The aliphatic bond was identified at the wavenumber of 767–771 cm^{-1} [4,15]. The transmission pattern showed the vibrations at the wavenumber of 3350 cm^{-1} (OH functional group) as amide I, 1601 cm^{-1} (C-N) and 1510 cm^{-1} (N-H) as amide II, and 760 cm^{-1} as aliphatic bond [4]. The increasing of carbon and clay reduced the transmittance of the samples due to more bond (the peak increase) and due to its opacity properties of the carbon and clay when exposed by UV-vis radiation [19].

3.4. Dielectric analysis

The dielectric value of the sample is found on a linearly inclined curve at frequency 400 kHz–1 MHz in which the values are averaged as shown in figure 5. PCClay5 shows a big difference dielectric value at 100–200 kHz, but the real dielectric value or the dielectric constant is obtained from the average dielectric value of the curve that has been relatively stable. Dielectric constant of the frequency of PU, PC, PClay and PCClay sample comparison which shows a significant difference from the PCClay sample because the highest dielectric constant value reaches 7.8×10^{3} when the frequency is 50 kHz. Inset in figure 5 is a magnification of figure 5 at a frequency of 400 kHz to 1 MHz. The value of the dielectric constant of a sample is taken on a dielectric curve that is relatively linear, then each point is summed and then the average value is determined. After the calculation above is obtained the value of the dielectric constant of the PU sample, PC, PClay, and PCClay respectively are 24.12; 50.25; 42.92; 125.95. Organoclay is dielectric material [20], so it is clear that the increase of the content of clay exhibits the increase of dielectric constant.

3.5. Optical microscopy and salt spray analysis

Figure 6 and figure 7 show surface morphology of the samples of before and after corrosion using 100 x magnification of optical microscopy. Figure 7 shows that PU and PCClay1 have smoother surface

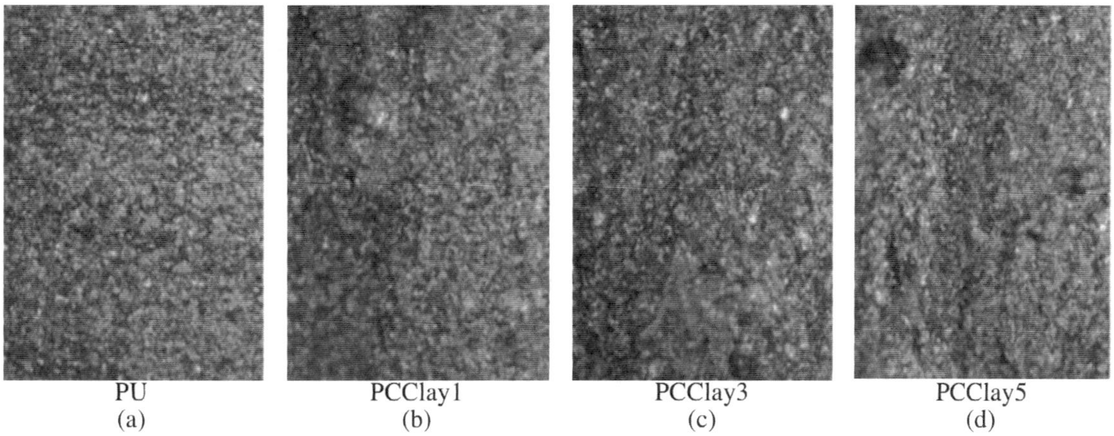

Figure 6. Surface morphology of polyurethane/carbon/organoclay composite with different carbon and organoclay amount before corrosion

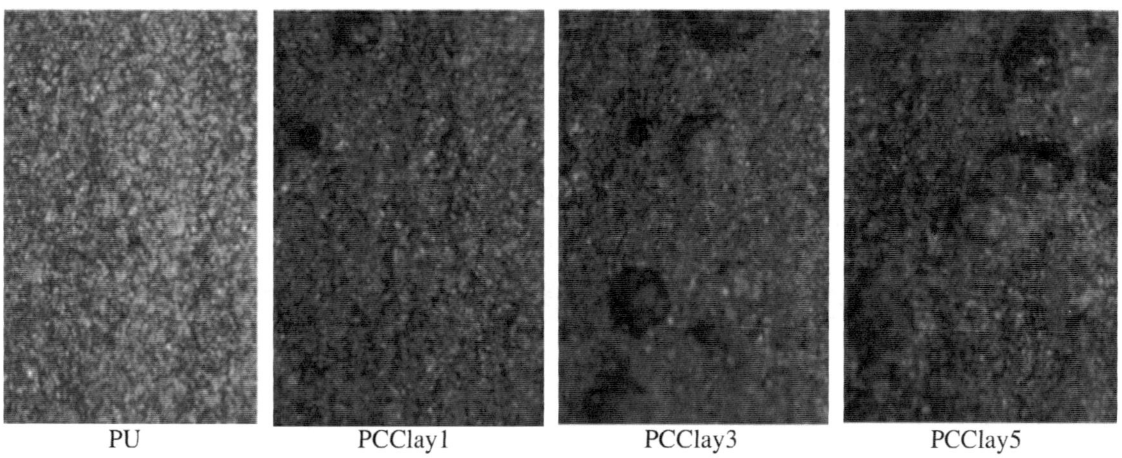

Figure 7. Surface morphology of polyurethane/carbon/organoclay composite with different carbon and organoclay amount after corrosion

morphology than PCClay3 and PCClay5. After corrosion test, the surface morphology was changed. PCClay1 has a better corrosion resistance than samples PCClay3 and PCClay5 (figure 7) because the finer surface of the sample indicates the less amount of agglomeration, which increases the potential for corrosion resistance by PU/M sample [20].

4. Conclusions

Polyurethane/carbon/organoclay composite with various contents of Carbon and Clay have been prepared. PCClay3 provides a higher thermal stability than the other sample. But, PCClay1 has the highest transmittance and good corrosion resistance. On the other hand, PCClay5 has the best dielectric value. Polyurethane/carbon/organoclay composite is possible to be applied as anticorrosive coating for material and also has good thermal and dielectric properties.

Acknowledgements
The authors acknowledge with gratitude the financial supports from the Ministry of Research, Technology and Higher Education of Indonesia (RISTEK-DIKTI), under grant Penelitian Dasar Unggulan Perguruan Tinggi (PDUPT), No. 366/UN2.R3.1/HKP.05.00/2018

References
[1] Cakić S M, Ristić I S and Ristić O Z 2012 Thermal analysis of polyurethane dispersion based on different polyols *Polyurethane* ed F Zafar and E Sharmin (London: IntechOpen Limited) Chapter 5 pp 79–100
[2] Mustapa S R, Aung M M, Ahmad A, Mansor A, TianKhoon L 2016 *Electrochim. Acta* **222** 293–302
[3] El-Fattah M , El Saeed A M, Dardir M M and El-Sockary M A 2015 *Prog. Org. Coat.* **89** 212–9
[4] Gurunathan T, Mohanty S and Nayak S K 2015 *Prog. Org. Coat.* **80** 39–48
[5] Virgawati E and Soegijono B 2018 *J. Phys.: Conf. Ser.* **985** 012032
[6] Khatoon H and Ahmad S 2017 *J. Ind. Eng. Chem.* **53** 1–22
[7] Arianpouya M, Nematollahi M, Arianpouya N and Shishesaz M 2016 *Prog. Org. Coat.* **90** 369–79
[8] Ryszkowska J, Jurczyk-Kowalska M, Szymborski T and Kurzydłowski K J 2007 *Physica E Low Dimens. Syst Nanostruct.* **39** 124–7
[9] Sengupta R, Bhattacharya M, Bandyopadhyay S and Bhowmick A K 2011 *Prog. Polym. Sci.* **36** 638–70
[10] Souri H, Nam I W and Lee H K 2015 *Compos. Sci. Technol.* **121** 41–8
[11] Sun H, Zhang H, Liu S, Ning N, Zhang L, Tian M and Wang Y 2018 *Compos. Sci. Technol.* **154** 145–53
[12] Arianpouya N, Shishesaz M, Arianpouya M and Nematollahi M 2013 *Surf. Coat. Tech.* **216** 199–206
[13] Bayer I S, Steele A, Martorana P J and Loth E 2010 *Appl. Surf. Sci.* **257** 823–6
[14] Heidarian M, Shishesaz M R, Kassiriha S M and Nematollahi M 2010 *Prog. Org. Coat.* **68** 180–8
[15] Kumari S, Mishra A K, Krishna A V R and Raju K V S N 2007 *Prog. Org. Coat.* **60** 54–62
[16] Kurniawan O and Soegijono B 2018 *J. Phys.: Conf. Ser.* **985** 012035
[17] Tcherbi-Narteh A, Hosur M, Triggs E, Owuor P and Jelaani S 2014 *Polym. Degrad. Stab.* **101** 81–91
[18] Xiong J, Zheng Z, Jiang H, Ye S and Wang X 2007 *Compos. Part A* **38** 132–7
[19] Karak N 2012 *Vegetable Oil-Based Polymers* (Cambridge: Woodhead Publishing)
[20] Al-Saleh M H 2017 *Compos. Sci. Technol.* **149** 34–40

IOP Conf. Series: Materials Science and Engineering **763** (2020) 012073 doi:10.1088/1757-899X/763/1/012073

Hysteresis observation of CoFe and CoFeB model disk using micromagnetic simulation

Q Mustaghfiroh[1], D Djuhana[1] and C Kurniawan[2]

[1] Department of Physics, Faculty of Mathematics and Natural Sciences (FMIPA), Universitas Indonesia, Kampus UI Depok, Depok 16424, Indonesia

[2] Research Center for Physics, Indonesian Institute of Sciences (LIPI), Gedung 440–442 Kawasan Puspiptek Serpong, Banten 15314, Indonesia

Corresponding author's e-mail: dede.djuhana@sci.ui.ac.id

Abstract. In this work, hysteresis loop of CoFe and CoFeB disk ferromagnets are observed by using micromagnetic simulation OOMMF based on LLG equation. The diameter varied from 50–500 nm with the thickness 5 nm and 10 nm. For simulation process, the damping factor of 0.05 and the cell size 2.5 x 2.5 x 2.5 nm^3 were fixed. We applied parallel and perpendicular external field to generate hysteresis loop of CoFe and CoFeB disk ferromagnets. Interestingly, we found two behaviours of coercivity, first less than diameter 100 nm and second, greater than 100 nm. For parallel-applied field of CoFe and CoFeB, the coercivity showed fluctuation around 20–160 mT. Greater than diameter 100 nm, the coercivity in a constant value around 40 mT for CoFe and around 20 mT for CoFeB. For perpendicular applied field of CoFe, we still observed the coercivity around 40 mT but greater than 100 nm, the coercivity dropped to zero. For CoFeB with perpendicular applied field, the coercivity decrease as the diameter increase until reach diameter 100 nm. Greater than the diameter 100 nm, the coercivity is constant at 20 mT. According to the results, we had observed the Perpendicular Magnetic Anisotropy (PMA) behaviour in both CoFe and CoFeB disk ferromagnets with certain value of the coercivity when the field applied in perpendicular direction.

1. Introduction

Recently, perpendicular magnetic anisotropy (PMA) has great potential applications as a high density of non-volatile memories [1]. The nonvolatile memory reduces power consumption than volatile that requiring 50 % power [2]. Perpendicular Magnetic Anisotropy (PMA) as a magnetic recording media has a moment magnet arrangement in perpendicular toward each other so it has high density capacity of more than longitudinal recording media [3]. PMA has a special characteristic with no thermal fluctuation which makes no superparamagnetic effect occurs. Superparamagnetic effect makes material to have no coercivity and information will be unstable and lost. To avoid it, anisotropy material increases the thermal stability and out of plane magnetization. Compared to longitudinal recording media, PMA recording media can save the data with 100 times higher density and smaller size than longitudinal recording media [4].

Currently, nanoparticle based magnetic material, which still in investigation are CoFe and CoFeB. CoFe has unique characteristics that can be used in many applications, which need high flux density such as data storage and electronic devices [5]. Unique characteristics are high saturation, high Curie Temperature (T_C), low magnetocrystalline anisotropy, high permeability, and good hardness [6]. Not only CoFe, CoFeB is also interesting to be investigated because of low magnetocrystalline anisotropy,

IOP Conf. Series: Materials Science and Engineering **763** (2020) 012073 doi:10.1088/1757-899X/763/1/012073

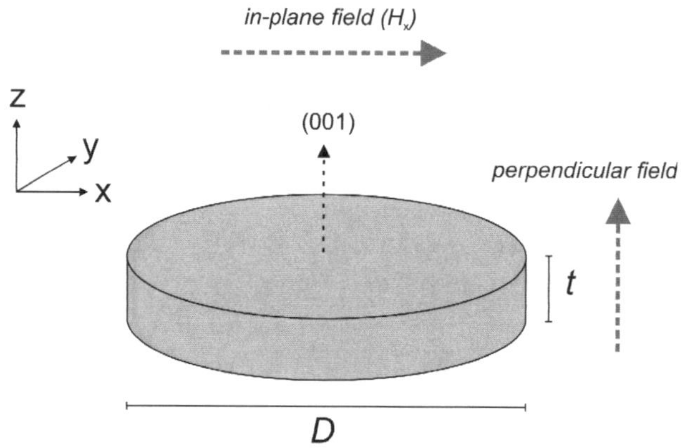

Figure 1. Thin film geometry simulation model disk given external field parallel and perpendicular

high spin polarization (more than 65 %), and moderate saturation [7]. CoFeB can be used to decrease current density to change spin direction for spin transfer torque.

In this work, we have investigated CoFe and CoFeB model disk using micromagnetic simulation. The results from this simulation are hysteresis loop, coercivity field, and domain structure with variation of diameter and thickness. We get the influence of anisotropy material in perpendicular magnetic anisotropy (PMA) and in-plane magnetic anisotropy (IMA). Intrinsic magnetic properties materials were determined by anisotropy material and depend on size.

2. Simulation procedure

PMA phenomena are studied using the public micromagnetic simulation software OOMMF based on Landau Lifshitz Gilbert (LLG) equation [8]. The thicknesses (t) are 5 nm and 10 nm and the diameter of the disk (D) varied from 50 to 500 nm. Damping constant α is set to be 0.05. The cell size used in micromagnetic simulation is 2.5 x 2.5 x 2.5 nm^3. External magnetic field is given in parallel and perpendicular directions (figure 1) with 1000 mT for CoFe and 2000 mT for CoFeB.

Material parameter plays an important role to the results of simulation. Thin film CoFe has magnetic saturation M_s = 1.44 x 10^6 A/m, anisotropy constant 2.7 x 10^3 J/m^3 and exchange constant 13 x 10^{-12} J/m based on Piao *et al.* [9]. CoFe has cubic magnetocrystalline anisotropy with the direction (100) and (010). For CoFeB material parameter based on Chaves-O'Flynn *et al.* [7] has magnetic saturation 9.57 x 10^5 A/m, anisotropy constant 3.86 x 10^5 A/m, and exchange constant exchange 13 x 10^{-12} J/m. CoFeB has uniaxial anisotropy in (001) direction [7].

3. Results and discussion

3.1. Hysteresis loop

To investigate the effects of directions external magnetic field and anisotropy material, we have performed a systematic study of hysteresis loop for diameter 50 nm and 5 nm thickness shown in figure 2. As shown in figure 2, CoFe and CoFeB given parallel and perpendicular external magnetic field represented by hysteresis loop. The hysteresis loop of CoFe has two characteristics when applied in parallel and perpendicular field. Different shape of hysteresis loop in two ways applied field is determined by anisotropy of materials. As given, parallel external field showed square loop hysteresis. Square loop hysteresis indicates that easy axis moment magnet orientation has the same direction with external magnetic field [10]. On the other side, when CoFe given perpendicular field hysteresis loops, it showed the same character as in hard axis. Interestingly, low coercivity field can be seen in hysteresis loop. When CoFe has low coercivity, it needs low external magnetic field to make all of moment magnet parallel with external magnetic field (saturation state). One of PMA material's characteristics is low coercivity that exist when perpendicular is applied field. The same hysteresis loop of CoFe was revealed by Tekgul *et al.* [11].

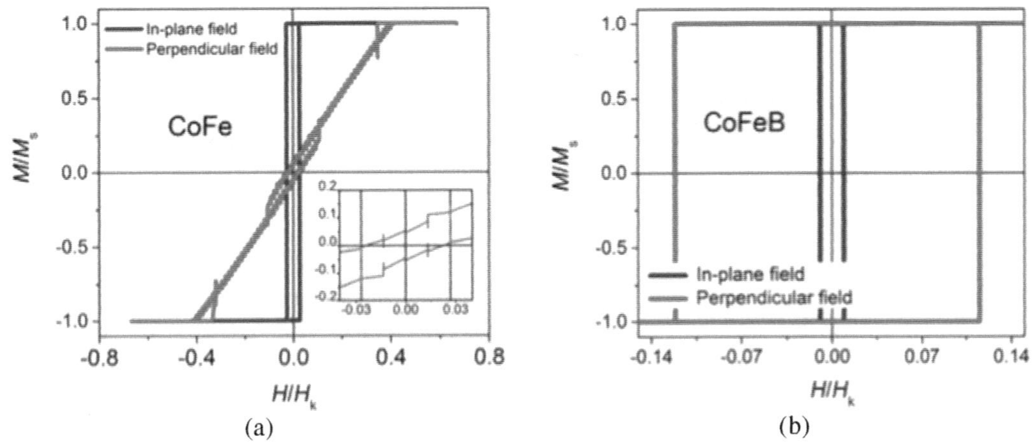

Figure 2. Hysteresis loop of (a) thin film CoFe disk (b) thin film CoFeB disk given external field parallel and perpendicular at $D = 50$ nm and $t = 5$ nm

Boron atom has diamagnetic characteristic, which influences magnetic properties to reduce oxidation and diffusion. Boron has smaller radius than Co and Fe thus stability of material will be increased [12]. With boron's influence, the magnetic mechanism reversal of CoFe showed different shape as shown in figure 2b. CoFeB with parallel external magnetic field showed square loop hysteresis with low coercivity. Coercivity in hysteresis loop showed ability of material to save data in recording media. When applied perpendicular external field CoFeB showed high coercivity (~780 mT). It indicates that CoFeB has strong PMA behavior. Different shape of hysteresis loop when applied parallel and perpendicular indicates anisotropy effect in CoFe and CoFeB. Anisotropy of CoFeB in (001) in z-axis. So easy axis of CoFeB in perpendicular external field and proved with square hysteresis loop in perpendicular. Similar hysteresis loop was observed in the work of Meo *et al.* [2] although the investigated system has different shape as nanodot with thickness 1 nm.

3.2. Coercivity field

Magnetic behaviour can be represented by coercivity field from hysteresis loop. Coercivity is a magnetic external field needed by material to make magnetization normalization comes to zero. Coercivity of CoFe and CoFeB given parallel and perpendicular external field with variation of thickness is shown in figure 3. Coercivity of CoFe with parallel external magnetic field tends to increase under diameter 100 nm until reach maximum diameter of nanoscale 100 nm. At the value greater than 100 nm coercivity fields tends to be constant at 40 mT. Constant coercivity in mesoscopic scale is caused by the effect of size dependent material. Ratio between surfaces to volume ratio in nanoscale is one of special characteristics for nanoscale material. Surface ratio to volume ratio gives an effect to increase magnetic properties of CoFe [13]. In perpendicular magnetic external field thin film CoFe has low coercivity ~40 mT. The coercivity drops to zero until diameter of thin film reach almost 100 nm. Above 100 nm diameters, thin film CoFe has no coercivity field, which indicate that the size of diameter affects magnetic properties. PMA behaviour in CoFe only exists in small diameter less than 100 nm.

From hysteresis loop we can find magnetic behaviour with coercivity. Thin film CoFeB given parallel and perpendicular external magnetic field. Interestingly, with boron's influence in CoFe, the coercivity field decreases as the diameter increase. With thickness varied in 5 nm and 10 nm in perpendicular magnetic external field showed coercivity field decrease as the thickness increase. It follows the relation of reduce anisotropy energy by increasing thickness contribution in thin film [12]. In figure 3b, it shows that as the diameter increase, the coercivity field decreases until upper limit of nanoscale in 100 nm. In greater diameter than 100 nm, the coercivity field tends to be constant at 20 mT and 40 mT. Proportional with theoretic diameter vs coercivity field curve after through critical diameter, the coercivity will decrease and then constant [14]. In perpendicular external field treatment, showed existence of coercivity and proved that CoFeB has Perpendicular Magnetic Anisotropy (PMA) behaviour.

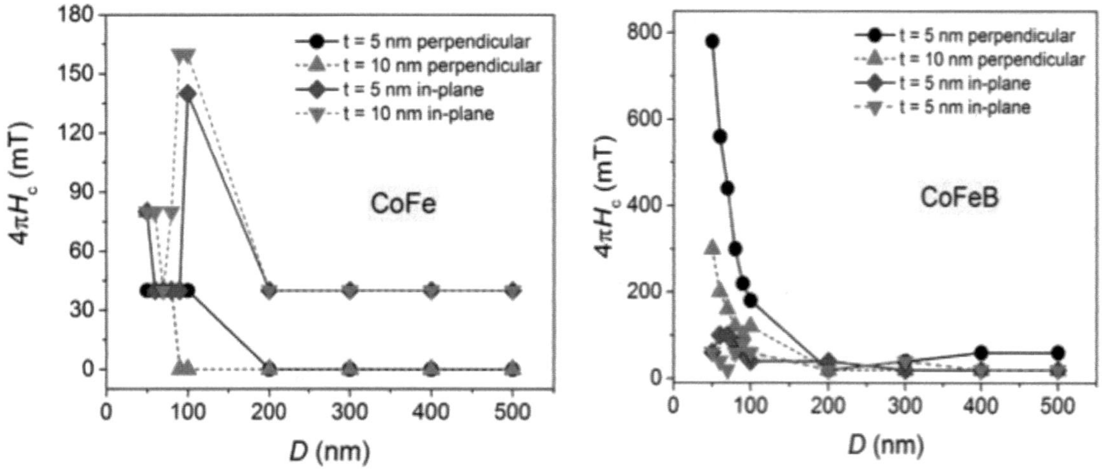

(a) (b)

Figure 3. Coercivity of (a) thin film CoFe disk (b) thin film CoFeB given external field parallel and perpendicular to the disk for two variation in thickness of 5 nm and 10 nm

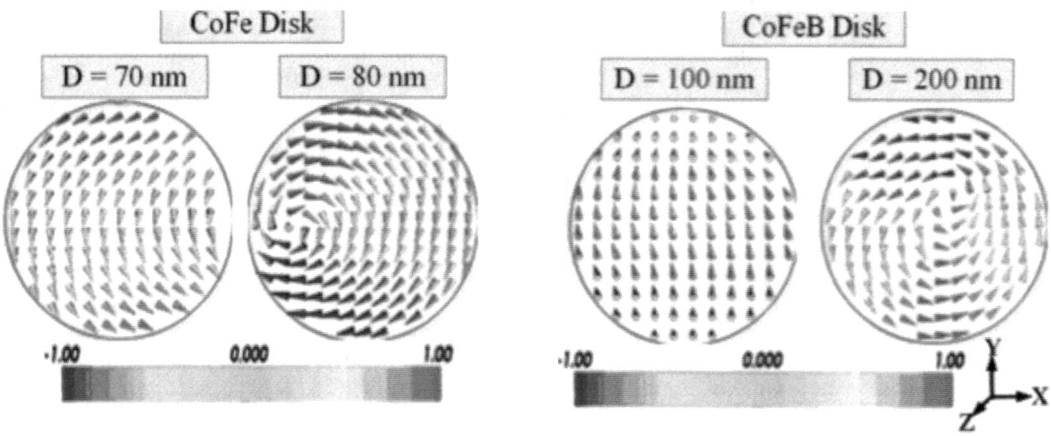

(a) (b)

Figure 4. Domain structure of (a) CoFe disk and (b) CoFeB disk in critical diameter

3.3. Domain structure

One of special from micromagnetic is efficient to show the domain structure transition in every state of hysteresis. Domain structure of coercive state CoFe and CoFeB in critical diameter is shown in figure 4. Thin film CoFe applied parallel external field has multidomain vortex structure in diameter 80 nm and in diameter 70 nm start to change the structure from single to multidomain as coherent rotation. In small size diameter (less than 80 nm) domain structure try to make multidomain state even in small area to minimize demagnetization energy, which tends to make single domain state [14]. CoFe showed different structure as a vortex state in diameter 80 nm. Vortex state exists from diameter 80 nm to mesoscopic scale diameters, which indicate that critical diameter (Dc) in 70 nm. From Kittle equation of critical diameter, it is found Dc of thin film CoFe is in 70 nm. It is proved with domain structure of CoFe made multidomain structure after crossed Dc.

CoFeB thin film in figure 4b showed transition between single to multidomain state with parallel external magnetic field. Diameter 100 nm showed single domain coherent rotation as impact of high demagnetization energy. In diameter 200 nm coercive state of CoFeB showed multidomain structure with vortex rotation. It caused by high exchange energy, which tends to make moment magnet antiparallel [10]. In diameter 100 nm domain structure of CoFeB has coherent rotation and in diameter 200 nm multidomain vortex structure of CoFeB is exist. It can be inferred from domain structure that critical diameter (Dc) of CoFeB is in 100 nm.

4. Conclusions

CoFe and CoFeB thin film have been studied by micromagnetic simulation. CoFe thin film performs PMA behaviour with low coercivity in small diameter less than 100 nm. On the other hand, boron's influence increased the coercivity field in CoFe. CoFeB showed PMA behaviour when applied perpendicular magnetic field. It revealed by hysteresis loop of CoFeB. Anisotropy effect made different shape of hysteresis loop in CoFe and CoFeB. From micromagnetic, the domain structure of CoFe and CoFeB can be seen in detail at transition state in critical diameter. CoFe has multidomain structure in 50 nm dimeter and CoFeB change to multidomain state at 200 nm diameters. In conclusion, by Boron addition in CoFe as CoFeB disk, PMA behaviour will exist at size fewer than 100 nm diameters.

Acknowledgements

The authors would like to thank Universitas Indonesia for funding this research through PITTA Grant Universitas Indonesia with contract number 2245/UN2.R3.1/HKP.05.00/2018

References

[1] Sundar R S and Deevi S C 2005 *Int. Mater. Rev.* **50** 157–92
[2] Meo A, Chureemart P, Wang S, Chepulskyy R, Apalkov D, Chantrell R W and Evans R F L 2017 *Sci. Rep.* **7** 16729
[3] Gubin S P, Koksharov Y A, Khomutov G B and Yurkov G Y 2005 *Russ. Chem. Rev.* **74** 489
[4] Cai G, Wu Z, Guo F, Wu Y, Li H, Liu Q, Fu M, Chen T and Kang J 2015 *Nanoscale Res. Lett.* **10** 126
[5] Negusse E, Lussier A, Dvorak J and Idzerda Y U 2007 *Appl. Phys. Lett.* **90** 092502
[6] Han X, Ma J, Wang Z, Zuo Y and Xi L 2014 *Acta Metall.* 27 1099–104
[7] Chaves-O'Flynn G D, Wolf G, Pinna D and Kent A D 2015 *J. Appl. Phys.* 117 170705
[8] Donahue M J and Porter D G 1999 *OOMMF User's Guide, Version 1.0* (Gaithersburg: National Institute of Standards and Technology)
[9] Piao H G, Choi H C, Shim J H, Kim D H and You C Y 2011 *Appl. Phys. Lett.* 99 192512
[10] Coey J M D 2010 *Magnetism and Magnetic Materials* (New York: Cambridge University Press)
[11] Tekgül A, Alper M, Kockar H and Haciismailoglu M 2015 *J. Mater. Sci. Mater. Electron.* **26** 2411–7
[12] Ngo D T, Quach D T, Tran Q H, Møhave K, Phan T L and Kim D H 2014 *J. Phys. D: Appl. Phys.* **47** 445001
[13] Mukherjee P K 2016 *J. Applicable Chem.* **5** 714–8
[14] Kittel C 2005 *Introduction to Solid State Physics* 8th ed (Hoboken: John Wiley and Sons Inc)

Effect of copper substitution on the structure of $La_{0.7}Sr_{0.25}Nd_{0.05}Mn_{1-x}Cu_xO_3$ (x = 0, 0.03, 0.05, and 0.07) manganites

D S Razaq[1], D R Munazat[1], B Soegijono[1], B Kurniawan[1], S Budiawanti[2] and D Nanto[3]

[1] Department of Physics, Faculty of Mathematics and Natural Sciences (FMIPA), Universitas Indonesia, Kampus UI Depok, Depok 16424, Indonesia

[2] Department of Physics, Faculty of Mathematics and Natural Sciences (FMIPA), Sebelas Maret University, Jebres, Surakarta 57126, Indonesia

[3] Department of Physics Education, Faculty of Tarbiya and Teaching Sciences (FITK), UIN Syarif Hidayatullah, Jalan Ir H Djuanda 95 Ciputat, Banten 15412, Indonesia

Corresponding author's e-mail: bkuru@fisika.ui.ac.id

Abstract. The effect of copper substitution on the structure of $La_{0.7}Sr_{0.25}Nd_{0.05}Mn_{1-x}Cu_xO_3$ (x = 0, 0.03, 0.05, and 0.07) has been studied by using X-ray diffraction (XRD). The samples were successfully synthesized by using citrate-nitrate sol-gel method. Rietveld refinement analysis showed that all samples crystallized in rhombohedral system with $R\bar{3}c$ space group without any detectable trace of impurities. Further analysis on the crystal structural of the samples revealed that copper substitution decreases the average Mn-O-Mn bond angle (<Mn-O-Mn>) and increases the Mn-O bond length ($d_{<Mn-O>}$). Scherrer equation showed the samples are in nanocrystalline. X-ray fluorescence (XRF) results indicate that copper has successfully substituted the manganese ion.

Keywords: LSNMCuO, sol-gel method, crystal structure, bond angle

1. Introduction

Perovskite manganite with general formula $La_{1-x}A_xMn_{1-y}B_yO_3$ (A = Ba, Ca, Sr or Nd; B = Ni, Cu or Ti) has attracted the interest of many researchers for quite a long time. This is due to the fact that such material can exist in many types of structure and possesses diverse physical properties such as magnetoresistance (MR), magnetocaloric effect (MCE), and charge ordering. Several earlier studies have proven that perovskite manganite with different structure will present different electrical and magnetic properties [1–3]. One of most studied of perovskite manganite is $La_{1-x}Sr_xMnO_3$ (LSMO). LSMO is an interesting and promising material to be studied because it has high Curie temperature and relatively low resistivity compared to other kind of lanthanum manganite [4,5]. Therefore, LSMO is attracted to be modified in order to achieve a Curie temperature close to room temperature while maintaining its low resistivity. A close ambient temperature of Curie temperature will be needed in order to be applied in daily life.

A more advanced modification of LSMO can be done by introducing another rare-earth metal into the A-site of LSMO. Cherif *et al.* [3] has shown that when neodymium was substituted into LSMO, it can change the crystal structure of $(La_{1-x}Nd_x)_{0.7}Sr_{0.3}MnO_3$ which affected the magnetization and resistivity. They also have shown that even when there is no significant change in crystal structure, in particular only lattice parameter occurred, the resistivity and maximum magnetization of the material still change according to the substituent concentration [3]. $(La_{1-x}Nd_x)_{0.7}Sr_{0.3}MnO_3$ is an interesting material that need a further study because it still has relatively low temperature while relatively low Curie temperature compared to LSMO.

The $(La_{1-x}Nd_x)_{0.7}Sr_{0.3}MnO_3$ can be a candidate for magnetic refrigerant material based on MCE because the lanthanum rare-earth metal doped manganite have Curie temperature close to room temperature and relatively low resistivity [3]. On the other hand, it has been proven that *B*-site substitution by copper can reduce Curie temperature of the substituted lanthanum manganite [6,7]. Therefore, we suggest that there is a possibility to reduce Curie temperature of $(La_{1-x}Nd_x)_{0.7}Sr_{0.3}MnO_3$ by substituting some manganese ion with copper ion. It will be our next work to report the result. However, the aim of this research is to study the effect of copper substitution on the structure of $La_{0.7}Sr_{0.25}Nd_{0.05}Mn_{1-x}Cu_xO_3$ (LSNMCuO) with $0 \leq x \leq 0.07$. This study is important because it has been proven that even a slight change in the crystal structure can affect the resistivity and magnetization of a substituted lanthanum manganite system [3].

2. Experimental

Nanocrystalline samples of LSNMCuO with x = 0, 0.03, 0.05, and 0.07 (referred as LSNMCuO-0, LSNMCuO-03, LSNMCuO-05, and LSNMCuO-07) were synthesized by citrate-nitrate sol-gel method. Stoichiometric amount of $Sr(NO_3)_2$, $Mn(NO_3)_2{\cdot}4H_2O$, $Cu(NO_3)_2{\cdot}3H_2O$, and citric acid ($C_6H_8O_7$) were diluted in double distilled water until a clear transparent solution was obtained. The citric acid amount can be calculated using molar ratio of metal nitrate to citric acid amount equal to 1:1.2. Stoichiometric amount of La_2O_3 and Nd_2O_3 were converted into metal nitrate ($M(NO_3)_3$) form by reacting it with dilute nitric acid until a clear and transparent solution were obtained. All precursors in solution form were then mixed inside a beaker glass followed by constant stirring on a magnetic hotplate. Afterwards, dried gel was ground into a fine powder and calcined at 550 °C for 5 h. The sample was re-ground and sintered at 700 °C for 24 h. continued by cooling to room temperature inside the furnace. Powder X-ray diffraction (XRD) measurement at room temperature was conducted using Cu $K\alpha$ radiation. Diffraction data were collected from $10° \leq 2\theta \leq 90°$ with step size of 0.02° and scan time over one second for each step. Chemical composition of the samples was examined using X-ray fluorescence (XRF).

3. Results and discussion

Figure 1 shows the result of Rietveld refinement from XRD pattern of LSNMCuO. Based on figure 1, all samples have achieved a single-phase. All main diffraction peaks are matched between experiments compared to the calculated value. Further refinement process reveals that all samples are rhombohedral structure with $R\bar{3}c$ space group.

According to Rietveld refinement, it can be concluded that copper-doped does not change crystal structure up to 7% of all the manganese ion significantly. The structural parameters obtained from Rietveld refinement are displayed in table 1. Based on the values of discrepancy factors yielded from Rietveld refinement, it shows that the refinement process was good. Moreover, GOF factor values are close to unity with small Rwp and Rp values.

From table 1, it can be seen that the lattice *a*-parameter increases; meanwhile *c*-parameter decreases monotonously with increasing copper substitution content. Slight crystal disorder suggests that copper ion has successfully substituted manganese ion at *B*-site of perovskite structure. Furthermore, the change in lattice parameter leads to reduction of unit cell volume. Liu *et al.* [8] reported that the reduction of unit cell volume in copper substitution on manganite occurred in order to maintain the ionic stability. Copper substitution on manganite will induce the existence of Mn^{4+} with smaller ionic radius (0.53 Å) compared to Mn^{3+} ion (0.645 Å). Thus, it is expected that the unit cell volume will decrease as the copper content increases [8].

The Goldschmidt tolerance factor (t_G) has been calculated in this research and the result is presented at table 1. Goldschmidt tolerance factor can be calculated using a mathematical expression:

$$t_G = \frac{0.7\, r_{La} + 0.25 r_{Sr} + 0.05\, r_{Nd} + r_O}{\sqrt{2}\,[(1-x)r_{Mn^{3+}} + x\, r_{Cu^{2+}} + r_O]} \tag{1}$$

Table 1. Rietveld refinement on structural parameter obtained for LSNMCuO.

	LSNMCuO-0	LSNMCuO-03	LSNMCuO-05	LSNMCuO-07
crystal structure	rhombohedral	rhombohedral	rhombohedral	rhombohedral
space group	$R\bar{3}c$	$R\bar{3}c$	$R\bar{3}c$	$R\bar{3}c$
lattice parameter				
a (Å)	5.4928(7)	5.4936(6)	5.4999(6)	5.5006(7)
c (Å)	13.354(3)	13.345(2)	13.351(2)	13.347(2)
unit cell volume (Å^3)	348.9248	348.8023	349.7532	349.7255
discrepancy factor				
GOF	1.165	1.177	1.167	1.151
Rwp (%)	6.138	6.151	6.237	6.204
Rp (%)	4.881	4.908	4.923	4.909
Goldschmidt factor (t_G)	0.966	0.965	0.964	0.963
average crystallite size (nm)	20.114	21.220	21.832	20.609
$d_{<Mn-O>}$ (Å)	1.951	1.953	1.955	1.958
<Mn, Cu – O – Mn, Cu > (o)	166.417	165.453	165.13	163.819

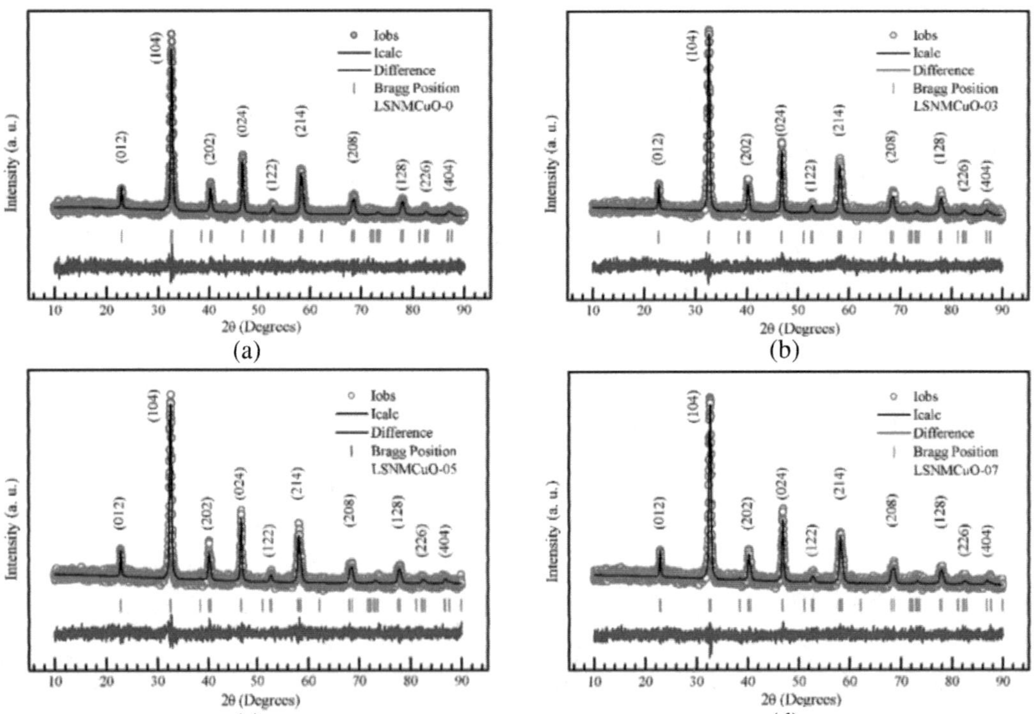

Figure 1. Rietveld refinement of XRD pattern for LSNMCuO samples at room temperature (a) LSMNCuO-0, (b) LSMNCuO-03, (c) LSMNCuO-05, and (d) LSMNCuO-07

After t_G values are known, the structure of perovskite can be deduced theoretically. It has been proven that t_G value above 0.96 will form a rhombohedral structure [9]. We confirm that all the samples are rhombohedral structure obtained from Rietveld refinement analysis. Table 1 shows that Goldschmidt tolerance value decreases as copper content increases. Reduction in t_G value suggests that there is a tendency of a distorted crystal structure compared to cubic perovskite. One may observe an increasing trend in average Mn-O bond length ($d_{<Mn-O>}$) followed by a decreasing trend in average Mn, Cu–O–Mn, Cu bond angle (<Mn, Cu–O–Mn, Cu>). Radaelli *et al.* [10] reported that crystal structural disorder in substituted of lanthanum manganite system was indicated from the changes in average

Table 2. Comparison of metal concentration obtained from XRF measurement compared to theoretical measurement

	Concentration measured (%)				Concentration calculated (%)			
Element	$x = 0$	$x = 0.03$	$x = 0.05$	$x = 0.07$	$x = 0$	$x = 0.03$	$x = 0.05$	$x = 0.07$
Mn	27.54	26.62	25.77	25.21	30.30	29.35	28.72	28.09
Cu	0	1.12	1.89	2.60	0.00	1.05	1.75	2.45
Sr	11.47	11.64	11.83	11.79	12.08	12.07	12.05	12.04
La	55.92	55.65	55.55	55.42	53.63	53.56	53.51	53.46
Nd	5.06	4.97	4.97	4.98	3.98	3.97	3.97	3.97
Total	100.00	100.00	100.00	100.00	100.00	100.00	100.00	100.00

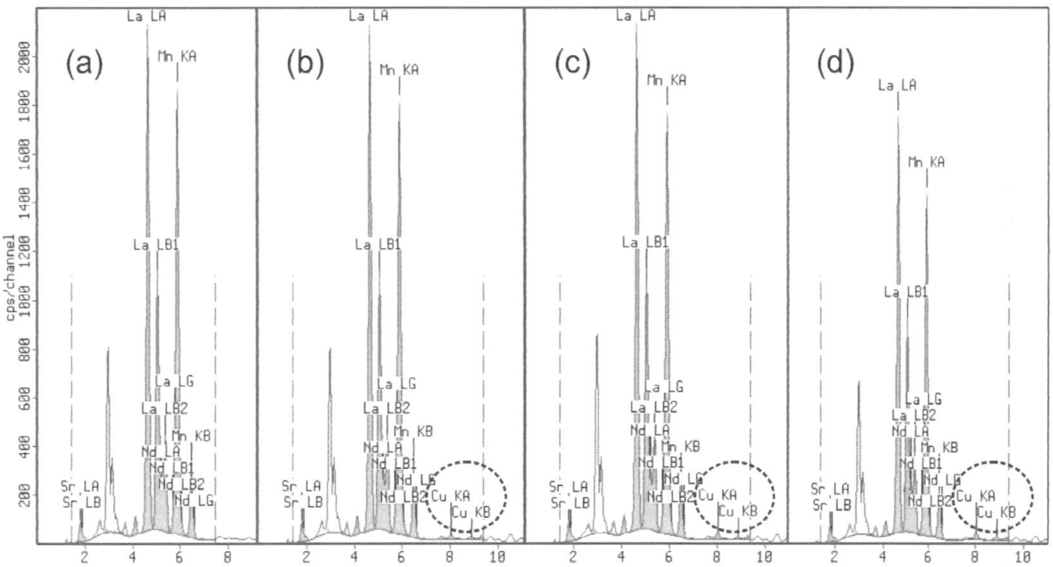

Figure 2. X-ray fluorescence spectra of all metal detected for (a) LSNMCuO-0, (b) LSNMCuO-03, (c) LSNMCuO-05, and (d) LSNMCuO-07. Dashed-circle shows copper spectra.

bond length and average bond angle because of the electrical transport and magnetic properties of the material. Table 2 also shows the average crystallite size of samples, which have been calculated using Scherrer equation as follow:

$$D_{sc} = \frac{0.9\,\lambda}{\beta_{HKL}\cos\theta} \tag{2}$$

where D_{sc} is average crystallite size, λ is Cu-$K\alpha$ radiation (1.5406 Å), β_{HKL} is full-width at half maximum (FWHM) of each diffraction peaks in radian unit, and θ is diffraction peak position for each diffraction peaks. It is interesting to note that the average crystallite size increased as copper concentration increased, and it reaches maximum value at 5% of copper content. The average crystallite size was found to decrease when copper content in the sample were bigger than 5%.

The chemical compositions of all the synthesized samples were examined by using X-ray fluorescence (XRF) measurement. The XRF spectra results of LSNMCuO are presented in figure 2. It can be seen that copper spectrum started to emerge from LBSNMCuO-03. These results support the previous argument of structural analysis that copper ions have successfully substituted manganese ion. Additionally, table 2 also provides information that no other traces of impurities are detected in the sample. This fact supports previous claim that sample has reached single-phase.

4. Conclusions

The effect of copper-doped on structural properties of $La_{0.7}Sr_{0.25}Nd_{0.05}Mn_{1-x}Cu_xO_3$ (x = 0, 0.03, 0.05, and 0.07) synthesized by sol-gel method has been studied in this research article. X-ray diffractometer analysis indicates that copper-doped does not affect the crystal structure, remain rhombohedral with $R\bar{3}c$ space group. Furthermore, as copper-doped increase implies to a decrease unit cell volume of the samples. Copper-doped concentration influences the average bond length and average bond angle. This result indicates that there will be a change in the transport and magnetic properties of the sample.

Acknowledgements

This work is funded by Research Grant from UIN Syarif Hidayatullah Jakarta with grant contract Un.01/KPA/234/2018 and support by Department of Physics Universitas Indonesia.

References

[1] Roy S, Guo Y Q, Venkatesh S and Ali N 2001 *J. Phys.: Condens. Matter* **13** 9547–59
[2] Cherif K, Zemni S, Dhahri J, Dhahri J, Oumezzine M, Ghedira M and Vincent H 2005 *J. Alloys Compd.* **396** 29–33
[3] Cherif K, Dhahri J, Dhahri E, Oumezzine M and Vincent H 2002 *J. Solid State Chem.* **163** 466–71
[4] Im H S, Chon G B, Lee S M, Koo B H, Lee C G and Jung M H 2007 *J. Magn. Magn. Mater.* **310** 2668–70
[5] Mandal P and Ghosh B 2003 *Phys. Rev.* B **68** 014422
[6] Chebaane M, Bellouz R, Oumezzine M, Hlil E K, and Fouzri A 2018 *RSC Adv.* **8** 7186–95
[7] El-Hagary M, Shoker Y A, Mohammad S, Moustafa A M, El-Aal A A, Michor H, Reissner M, Hilscher G and Ramadan A A 2009 *J. Alloys Compd.* **468** 47–53
[8] Liu Z, Lin W G, Zhou K W, and Yan J L 2018 *Ceram. Int.* **44** 2797–802
[9] Manjunatha S O, Rao A, Lin T Y, Chang C M and Kuo Y K 2015 *J. Alloys Compd.* **619** 303–10
[10] Radaelli P G, Iannone G, Marezio M, Hwang H Y, Cheong S W, Jorgensen J D and Argyriou D N 1997 *Phys. Rev.* B **56** 8265–76

The radiochromic indicator using methyl red dye solution as a high-dose gamma-ray dosimeter application

W Chairunisa and C Imawan

Department of Physics, Faculty of Mathematics and Natural Sciences (FMIPA), Universitas Indonesia, Kampus UI Depok, Depok 16424, Indonesia

Corresponding author's e-mail: cuk.imawan@sci.ui.ac.id

Abstract. The radiochromic indicator using the methyl red dye solution has been studied as a high-dose gamma-ray dosimeter. The indicator has been tested by irradiating using gamma rays with a ^{60}Co source. The radiation dose was varied from 5 to 100 kGy with a dose rate of 5.5 kGy/h. The color change of the indicator solution caused by the radiation was characterized by using a UV-Visible spectrophotometer and a camera photograph. In addition, the color stability of the indicator is also studied by observing the effect of time and the storage in light and dark conditions. The effect of time after irradiation on the color of the solution has also been investigated. The methyl red solution indicator showed a radiochromic phenomenon color changes from red to dark red for a dose of 5-20 kGy, faded gradually then became colorless at a dose of 25–100 kGy. These color changes correspond to the peak of the UV-Visible absorbance spectrum at a wavelength of 492 nm. The color of the indicator solution is more stable if kept in dark conditions than in the light conditions. The color of the solution also shows its stability for up to 12 days after irradiation. These results conclude that the indicator solution of methyl red can be used as a dosimeter for gamma rays in the dose range of 5–40 kGy.

1. Introduction

Dosimeters have an important role in the quality control of the radiation process [1]. There are several types of dosimeters, one of which is a radiochromic dosimeter that make use of the chemical structure changes of the dye of radiochromic indicator due to exposure to ionizing radiation. Ionizing radiation on a chemical compound can produce chemical changes to the compound and if the changes can be quantified it can be used as a technique to measure the absorbed dose [2]. Many dye solutions were used for this purpose, where the range of doses absorbed by the indicator can be determined by observing the color changes that occur in the indicator and the change in absorbance values, which can be observed by means of characterization tools such as photographic cameras and UV-Visible spectrophotometers.

Radiochromic dosimeter film has been widely developed as a high-dose gamma-ray dosimeter to observe the change in methyl red color due to gamma radiation in basic and acidic solutions [3,4]. The result in the base medium was exponentially reduced as the absorbed dose increased at the range of 50 Gy to 6 kGy, whereas the solution response in the acidic state decreased linearly as the absorbed dose increased up to 200 Gy. The PVB matrix added to the methyl red dye in various concentrations can be applied as a gamma radiation dosimeter with a range of 5–100 kGy, where dose sensitivity increases with increasing absorbed doses and concentrations of methyl red dye [5]. Kattan and Daher [6] reported that the radiochromic dosimeter of methyl red with the addition of PVC can be applied as a gamma radiation dosimeter at the range of 0–150 kGy.

The purpose of this study was to use a methyl red solution as a high-dose gamma-ray dosimeter, in which the solution would be observed both qualitatively and quantitatively to measure the range of

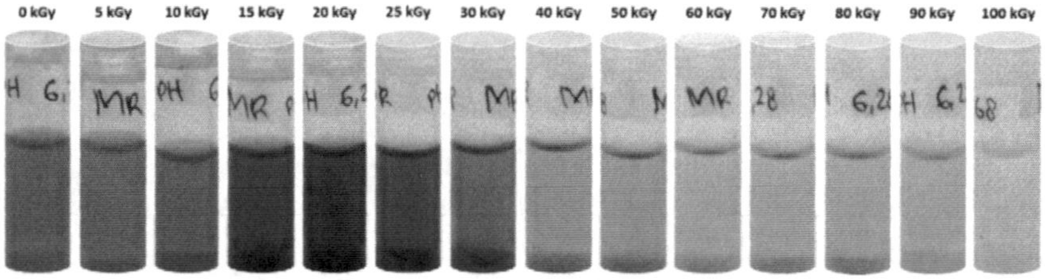

Figure 1. A photograph of color changes of methyl red indicator solutions after irradiated by gamma-ray with a dose range of 0-100 kGy.

absorbed doses. Radiochromic dosimeters used in the routine dosimeter system are usually affected by the storage lighting conditions before and after irradiation. Therefore, an observation was conducted to know the shelf life and post-irradiation stability of the methyl red solution indicator, which will be irradiated at dose ranges from 5–100 kGy.

2. Research procedure

2.1. Preparation of methyl red solution

Preparation of methyl red solution was done by dissolving 0.05 g of methyl red powder (Merck, Germany) in 100 mL of ethanol solution 99.9 % (JT Baker), with the pH value of the solution was 6.28. The solution was poured into a 2 mL ampoule by ¾ parts, which later on used as an indicator of methyl red solution. From the preparation of the solution, there were 50 samples of methyl red solution indicator weighed 1.24 ± 0.01 g.

2.2. Characterization of methyl red solution before irradiation process

Methyl red solution characterization before irradiation process using a UV-Visible Spectrophotometer (GENESYS) was conducted to observe the initial absorbance spectrum of the indicator prior to the exposure of gamma irradiation and the effect of environmental conditions. Characterization was done at wavelengths of 200–800 nm in ultraviolet and visible regions.

2.3. Gamma rays irradiation on methyl red solution

The methyl red solution indicator was irradiated using a Gammacell 220 irradiator with a Cobalt-60 source at the dose ranges of 5–100 kGy with dose rate of 5.5 kGy/h. The accuracy level of the methyl red solution indicator was observed by irradiation of 3 samples for each radiation dose.

2.4. Characterization of methyl red solution after irradiation process

Characterization of methyl red solution after irradiation using UV-Visible Spectrophotometer was performed to see changes in absorbance spectrum occurring on the indicator after exposure to gamma-ray radiation in various absorbent doses. An observation on absorbance spectrum stability was performed for 12 days post irradiation with 4 days time interval.

2.5. *Test on the stability of methyl red solution to lighting condition*

The stability test of methyl red solution was carried out by storing the solution in light condition (the solution was left in the room with lights) and in dark condition (where the solution was wrapped with aluminum foil to isolate the solution from the light). Afterwards, the characterization of methyl red solution was conducted to see the changes of absorbance value of the solution due to the lighting conditions.

3. Results and discussion

3.1. The response of methyl red solution indicator to the dose of radiation

The response of the methyl red solution indicator to the radiation dose has been investigated by irradiating the indicator with gamma rays at various absorbed doses. The color change from the indicator occurs with an increase in absorption dose up to 20 kGy, then undergoes a color fading of up to 100 kGy (figure 1). The color of the methyl red solution changes from red to dark red for a dose of

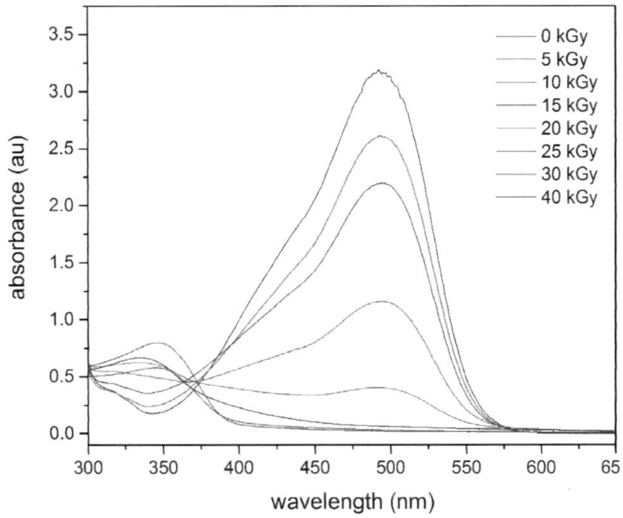

Figure 2. Absorption spectra of methyl red solution dosimeter as a function of wavelength for various absorbed doses.

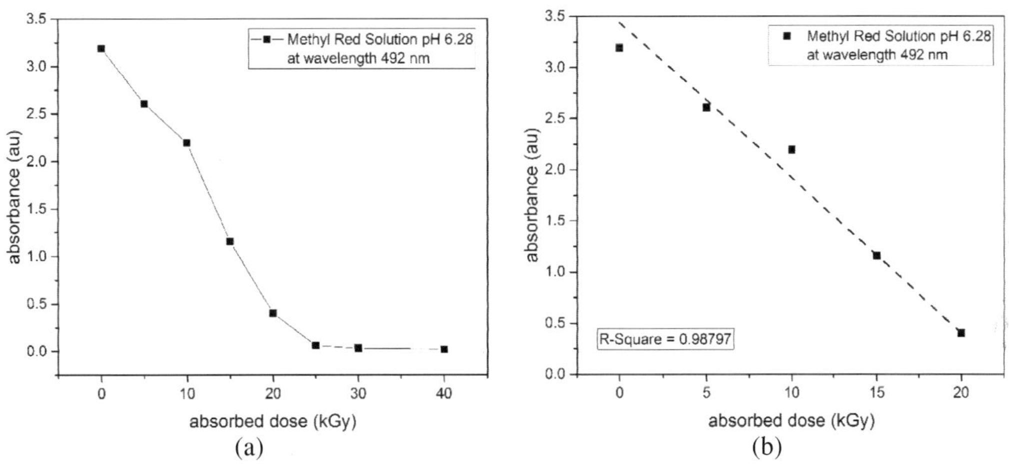

Figure 3. The absorbance of methyl red solution at 492 nm as a function of absorbed dose (a) up to 40 kGy and (b) up to 20 kGy.

5–20 kGy, then fades and turns colorless gradually at a dose of 25–100 kGy. These color changes correspond to changes in absorbance spectra obtained from UV-Visible spectrophotometers.

Measurement of the change in absorbance peak of the methyl red solution was performed using a UV-Visible Spectrophotometer, where 1:8 dilution was performed on the methyl red solution test (1) using ethanol solvent (8). The absorption spectrum of methyl red solution at various absorption doses has been measured in the range 300–650 nm. Figure 2 shows a decrease in the absorption spectrum of the methyl red dye to an increase in absorption dose in the range of 0–40 kGy.

The dose response curve has been observed based on changes in absorption peaks defined at 492 nm against the absorbed dose. The dose response of the methyl red solution is shown in figure 3. The color bleaching of methyl red solution gradually intensifies as the absorbed dose is intensified up to 100 kGy, this can be confirmed by the lower absorbance value of the solution by the increase in absorption dose. as the absorbed increases, more hydrated electrons and free radicals are resulted, creating a breakage of an azo group of methyl red dye, generating the disappearance of a chromophore [7].

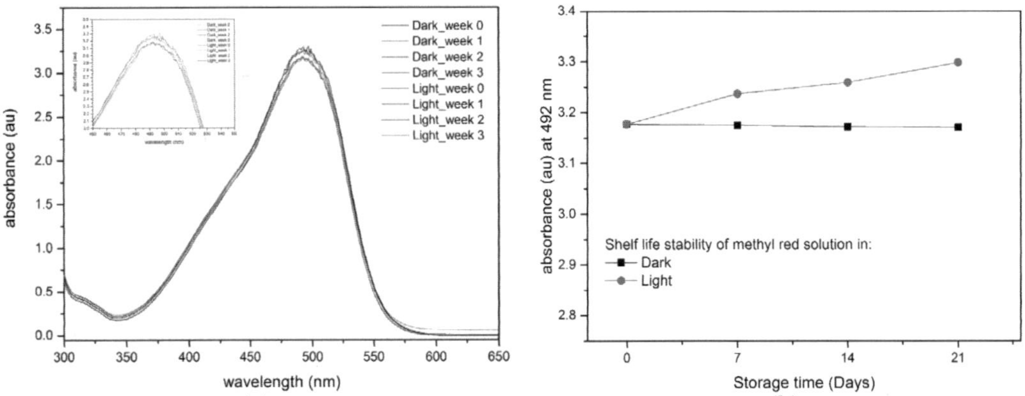

(a) (b)

Figure 4. (a) UV-vis spectra at 300 – 650 nm and (b) absorbance at 492 nm of methyl red solution in dark and light condition at ambient temperature

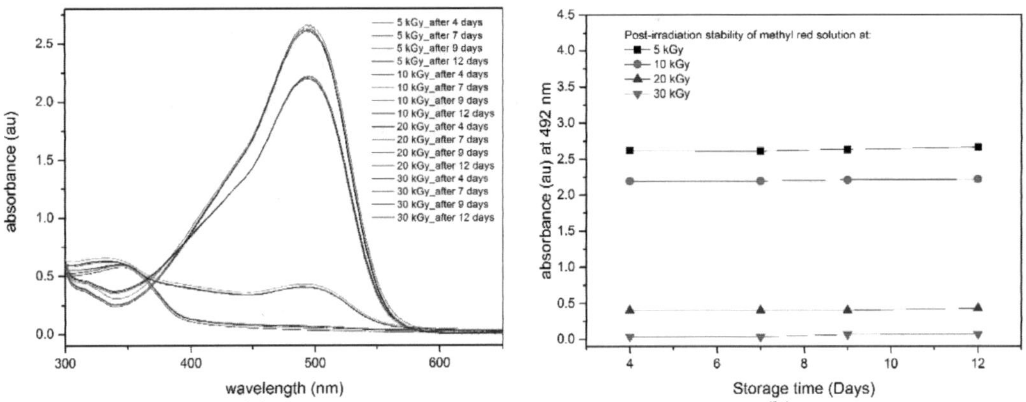

(a) (b)

Figure 5. (a) UV-vis spectra, and (b) absorbance at 492 nm of methyl red solution during 12 days of storage after irradiated at a various absorbed dose.

The accuracy of the methyl red solution as gamma-ray dosimeters has been observed by irradiating as many as 3 samples for each absorbent dose and obtained a precision error of <0.89%. The response curve of methyl red solution obtained showed that there was a sharp decrease in the absorbance value of the solution at a dose range of 5-20 kGy then at a dose range of 25-40 kGy, there was no significant decrease.

3.2. Stability of methyl red solution before irradiation (shelf life)

Measurement of stability of the methyl red solution before irradiation is done by storing the solution in dark and light conditions at room temperature. Figure 4 shows absorbance peak changes of the solution for 3 weeks at 1 week intervals (with 1: 8 dilution using ethanol solvent).

Observations were made at a wavelength of 492 nm. It was observed that the absorbance value of the methyl red solution in dark storage conditions did not change during 3 weeks of storage. Meanwhile, the methyl red solution stored in light conditions has increased the absorbance value gradually. This indicates that the methyl red solution indicator is more stable when kept in dark conditions for 3 weeks of storage before irradiation.

3.3. Post-irradiation stability

The indicator stability of methyl red solution after irradiation was tested by observing the change in absorbance of methyl red solution at a wavelength of 492 nm. Observations were made on methyl red solution which was irradiated at doses of 5, 10, 20 and 30 kGy and measured every 4 days for 12 days after irradiation. The results are shown in figure 5, where there was no significant change in absorbance from the indicator of methyl red solution up to 12 days measurement. This indicates that the methyl red solution has a good stability for up to 12 days after irradiation with a wide range of doses.

4. Conclusions

The methyl red solution indicator is sensitive to the dose of radiation, where the color change and the absorbance value decreased as the radiation dose increased. The methyl red solution indicator can be observed both qualitatively and quantitatively. Qualitative methods was done by observing the color change of the indicators using photographic cameras, while the quantitative method used UV-Visible Spectrophotometer. The methyl red solution has a good linearity at the dose range of 5–20 kGy with a precision error <0.89%. The methyl red solution indicator is more stable when kept in dark conditions for up to 3 weeks of storage before irradiation and stable for up to 12 days post-irradiation.

Acknowledgements

This research was financially supported by Hibah PITTA 2018 No.2244/UN2.R3.1/HKP.05.00/2018 from Universitas Indonesia

References

[1] McLaughlin W L 1970 *Manual on Radiation Dosimetry* ed N W Holm and R J Berry (New York: Marcel Dekker) Chapter 5 p 313
[2] Akram N G, Bhutto W A and Sharif I N 2016 *African Journal of Chemistry* **3** 182–7
[3] Barakat M F, El-Salamawy K, el Banna M, Abdel-Hamid M and Taha A A 2001 *Radiat. Phys. Chem.* **61** 129–36
[4] Ajji Z 2006 *Radiat. Meas.* **41** 438–42
[5] Al Zahrany A A, Rabaeh K A and Basfar A A 2011 *Radiat. Phys. Chem.* **80** 1263–7
[6] Kattan M and Daher Y 2016 *Int. J. Radiat. Res.* **14** 263–7
[7] Bhat N, Nate M M, Bhat R M and Bhatt C 2007 *Indian J. Pure Appl. Phys.* **45** 545–8

The sensitivity calculation of localized surface plasmon resonance (LSPR) Au nanorod by applying a boundary element system simulation

M Sofani[1], M H Putra[2] and D Djuhana[1]

[1] Department of Physics, Faculty of Mathematics and Natural Sciences (FMIPA), Universitas Indonesia, Kampus UI Depok, Depok 16424, Indonesia

[2] Institute for Computational Physics, University of Stuttgart, Allmandring 3, Stuttgart 70569, Germany

Corresponding author's e-mail: dede.djuhana@sci.ui.ac.id

Abstract. In this study, we have calculated the sensitivity of localized surface plasmon resonance (LSPR) Au nanorod by a public metallic particle simulation based on the boundary element method (Metallic Nano-Particle Boundary Element Method, MNPBEM). The diameter of nanorod D is 20 nm, 60 nm, and 80 nm. The variation of aspect ratio is 1.5 to 3.5. The dielectric of Au nanorod based on the Christine-Johnson experiment. To understand sensitivity sense, we have also varied the refractive index medium by Lorentz-Lorentz approximation from a mixture of water and glycerol concentration. The refractive medium index is n = 1.3334 (100 % water pure), n = 1.3605 (80 % water and 20 % glycerol), n = 1.3881 (60 % water and 40 % glycerol), n = 1.4164 (40 % water and 60 % glycerol), and n = 1.4452 (20 % water and 80 % glycerol). From MNPBEM simulation, we have produced LSPR spectra such as absorption, scattering, and extinction curve as the function of wavelength. Then, the sensitivity of LSPR Au nanorod is determined by the gradient of the peak of wavelength to the refractive index medium variation for all aspect ratio. Interestingly, we have found the LSPR Au nanorod consisted of longitudinal and transversal mode in LSPR Au nanorod curve. The longitudinal mode appeared a higher wavelength than the transversal mode in LSPR spectra. In longitudinal mode, the peak of wavelength increased as the aspect ratio increased (red-shift) while in transversal mode, the peak of wavelength was relatively constant. Furthermore, the sensitivity in longitudinal mode increased as the aspect ratio increased, whereas the sensitivity in transversal decreased as the aspect ratio increased. Increasing the sensitivity in longitudinal mode related to red-shift as the nanorod volume increased and the refractive medium index change. According to the results, the sensitivity determination is useful to understand the refractive index medium changes that it is important to design a sensor device.

Keywords: LSPR, au nanorods, boundary element method simulation

1. Introduction

Nowadays, nanoparticles become one of the topics of interest for researchers to develop science and technology, because nanoparticles are quite important topics in both the scientific and application sides. Nano-scale materials can improve the physical, mechanical, and chemical properties of a material without damaging its atomic structure. Nanoparticles can be applied in various aspects of the biomedical field [1], textile industry [2], energy [3], electronics [4] and agriculture [5]. Research in the field of nanotechnology shows the results of new products with better performance than the large size materials.

Utilization of nanoparticles involves a material surface, where nanoparticles have high reactivity,

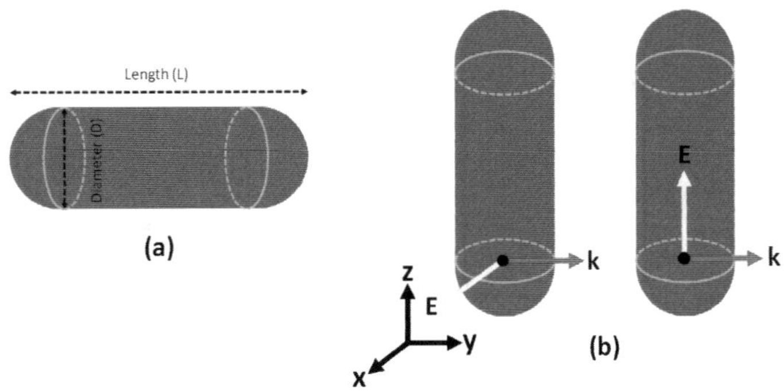

Figure 1. Geometry and dimension of Au nanorod. (a) The aspect ratio is determined by length and diameter of rod and (b) Polarization direction of the electric field

because atoms have a great opportunity to interact. This evident is found in the development of nanoparticles of gold or Au material. The selection of a gold material is due to its stable properties in the air and water without undergoing oxidation unlike iron or other metals [6]. The SPR redshift is related to the enhance of the Au NP extinction cross-section since LSPR increases with a wavelength in the 500–700 nm range [7] and the shape of the nanorod influences of how light is absorbed [8]. Nanorod sensitivity is a value that represents the ability of nanorods to stimulate the environmental refractive index around nanorod [9]. The sensitivity value is very important especially in nanoparticle applications as sensors. Therefore, knowledge of the sensitivity of Au nanorod is very important in its use as a sensor [10].

This study aims to observe the Local Surface Plasmon Resonance (LSPR) phenomenon occurring in Au material from the rod model, analyzing the LSPR shift pattern from pneumatic on the ratio aspect ratio curve to Au material from the rod model, determining the aspect ratio in the LSPR in the visible light region due to gold material which only looks visible, observing the effect of the medium on the material sensitivity of the LSPR phenomenon.

2. Simulation procedure

This study was performed the sensitivity calculation of single gold nanorod using MNPBEM simulation based on boundary element method approximation. Single gold nanorod with various aspect ratios was used and embedded in a different medium. Then, we scaled up the value of each aspect ratio within three different sizes. First, the small diameter was 20 nm for length (30, 35, 40, 45, 50, 55, 60, 65, and 70) nm, respectively. Second, intermediate diameter was 60 nm for the length (90, 105,120, 135, 150, 165, 180, 195, and 210) nm, respectively. Third, the larger diameter was 80 nm for the length (120, 140, 160, 180, 200, 220, 240, 260, and 280) nm, respectively. The geometry of nanorod is shown in figure 1a with length (L) and diameter (D). Then, the electric field is given by two different polarizations, which are parallel and perpendicular to long axis as shown in figure 1b. To calculate the medium refractive index, we have applied the Lorentz-Lorentz approximation by equation (1):

$$\frac{n_{12}^2-1}{n_{12}^2+2} = \frac{n_1^2-1}{n_1^2+2}\phi_1 + \frac{n_2^2-1}{n_2^2+2}\phi_2 \qquad (1)$$

where n_{12} is the refractive index of mixed solution, n_1 and n_2 are the refractive index of water (1.33) and glycerol (1.47), and ϕ_1 and ϕ_2 are volume fractions of water and glycerol, respectively. We used the volume fractions for water (ϕ_1) are 100 % ($\phi_2 = 0$ %), 80 % ($\phi_2 = 20$ %), 60 % ($\phi_2 = 40$ %), 40 % ($\phi_2 = 60$ %), 20 % ($\phi_2 = 80$ %), and 0 % ($\phi_2 = 0$ %), respectively. The dielectric function of gold used the data from Johnson and Christy's experiment. Then, we analysed the extinction spectra to study the sensitivity for different diameter 20 nm, 60 nm, and 80 nm both longitudinal mode and transverse mode, respectively.

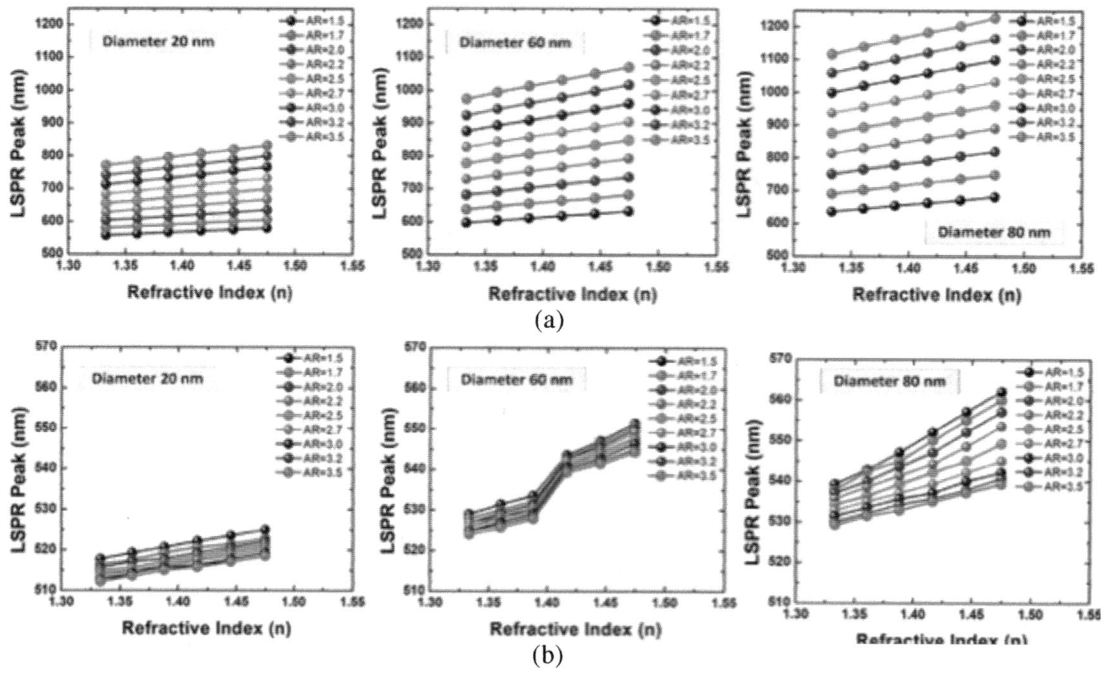

Figure 2. (a) The LSPR peak value for longitudinal mode and (b) transversal mode against the refractive index change in diameter 20 nm, 60 nm, and 80 nm, respectively.

3. Results and discussion

Figure 2 shows the resonance peak position in a various aspect ratio of both longitudinal (figure 2a) and transverse mode (figure 2b) against the refractive index changes in diameter 20 nm, 60 nm, 80 nm, respectively. We found the refractive index change affected the resonance peak position and this result agreed with previous work both theoretically [11] and experimentally [1]. Increasing the refractive index, however, the resonance peak of both transversal mode (TM) and longitudinal mode (LM) shift to longer wavelength [12]. The resonance peak of TM in whole diameter and aspect ratio occurs in visible range as the refractive index increases, but for LM, there is a limit range aspect ratio which is the whole peak resonance occurs in visible range, diameter 20 nm for range 1.5–2.7, diameter 60 nm for range 1.5–2.0, and diameter 80 nm for range 1.5–1.7.

Furthermore, for a given refractive index, we found the resonance peak in each diameter of LM shift to longer wavelength as the aspect ratio increased whereas it was vice versa for the TM. The peak position of TM for a small and intermediate diameter was difficult to be distinguished, while in larger diameter the peak resonance of both TM and LM can be seen. The gradient of linear fit can determine the sensitivity of gold nanorod.

In figure 3, we plotted the sensitivity for different diameter of both longitudinal mode and transverse mode, respectively. The sensitivity was calculated by the wavelength peak displacement per unit changes of medium refractive index ($S=\Delta\lambda_{peak}/\Delta n$) [1], this one is very important for an application like sensor. We have calculated the sensitivity of gold nanorod, in LM for diameter 20 nm was 150.31 RIU/nm, for 60 nm was 218.2 RIU/nm, and for 80 nm was 224.97 RIU/nm. In TM for diameter 20 nm was -52 RIU/nm, for 60 nm was -14 RIU/nm, and for 80 nm was -2 RIU/nm. Interestingly, increasing the diameter affected in the increase of the sensitivity of both longitudinal and transverse mode while aspect ratio just affected in increasing the longitudinal mode sensitivity while the transverse mode tended to decrease [9]. The highest sensitivity of longitudinal and transverse mode was observed in diameter 80 nm. The result show that TM is less sensitive to the aspect ratio changes than LM. The refractive index of LSPRs is wavelength dependent, that is the refractive index will not depend on the mode resonance of the geometry structure once the sensing wavelength is fixed with theoretical results by quasi-static approximation [13].

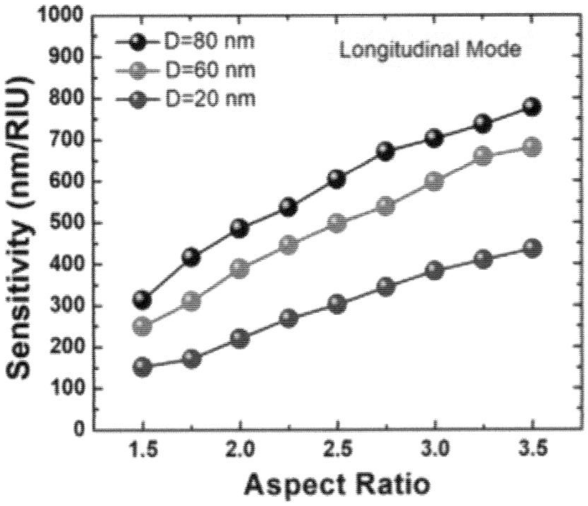

(a)

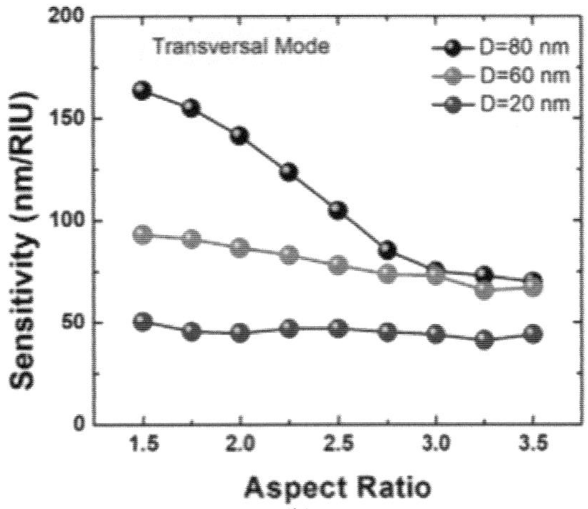

(b)

Figure 3. Sensitivity of Au nanorod for (a) longitudinal and (b) transversal mode.

4. Conclusions

In conclusion, we have successfully calculated the sensitivity of Au nanorod with two different polarizations. The result shows, for longitudinal mode the sensitivity with diameter 20 nm was 150.31 RIU/nm, for 60 nm was 218.2 RIU/nm, and for 80 nm was 224.97 RIU/nm. For transverse mode, the sensitivity with diameter 20 nm was -52 RIU/nm, for 60 nm was -14 RIU/nm, and for 80 nm was -2 RIU/nm. We have also determined the limit range aspect ratio of gold nanorod that occurs in visible range. In transverse mode, for aspect ratio 1.5–3.5 with diameter 20 nm, 60 nm, and 80 nm occurs in visible range. In longitudinal mode, diameter 20 nm for range 1.5–2.7, diameter 60 nm for range 1.5–2.0 and diameter 80 nm for range 1.5–1.7 while the transverse mode occurs in visible range for the whole given aspect ratio. The higher sensitivity was observed for longitudinal mode and transverse mode with diameter 80 nm, respectively.

Acknowledgements

The authors would like to thank Universitas Indonesia for funding this research through PITTA Grant with contract number 2246/UN2.R3.1/HKP.05.00/2018

References

[1] Cao J, Sun T and Grattan K T V 2014 *Sens. Actuators* B *Chem.* 195 332–51
[2] Zhang F, Wu X, Chen Y and Lin H 2009 *Fibers Polym.* 10 496–501
[3] Notarianni M, Vernon K, Chou A, Aljada M, Liu J and Motta N 2014 *Sol. Energy* 106 23-37
[4] Shur M 2014 *J. Phys.: Conf. Ser.* 486 012025
[5] Sabir S, Arshad M and Chaudhari S K 2014 *Scientific World Journal* 2014 ID925494
[6] Mock J J, Barbic M, Smith D R, Schultz D A and Schultz S 2002 *J. Chem. Phys.* 116 6755
[7] Amendola V, Pilot R, Frasconi M, Maragò O M and Latì M A 2017 *J. Phys.: Condens. Matter* 29 203002
[8] Kreibig U and Vollmer M 1995 *Optical Properties of Metal Clusters* (Berlin: Springer Verlag)
[9] Chen H, Kou X, Yang Z, Ni W and Wang J 2008 *Langmuir* 24 5233–7
[10] Yang M, Kostov Y, Bruck H A and Rasooly A 2009 *Int. J. Food Microbiol.* 133 265–71
[11] Trügler A 2011 *Optical Properties of Metallic Nanoparticles: Basic Principles and Simulation* (*Springer Series in Materials Science* vol 232) ed R Hull *et al.* (Cham: Springer International Publishing)
[12] Link S, Mohamed M B and El-Sayed M A 1999 *J. Phys. Chem.* B 103 3073–7
[13] Yong Z, Lei D Y, Lam C H and Wang Y 2014 *Nanoscale Res. Lett.* 9 187

AUTHOR INDEX

Institute of Physics
Dirac House, Temple Back
Bristol BS1 6BE UK

ISSN: 1757-8981
ISBN 978-1-7138-4704-5